T0180524

Advances in Intelligent Systems and Computing

Volume 931

Series Editor

Janusz Kacprzyk, Systems Research Institute, Polish Academy of Sciences,
Warsaw, Poland

Advisory Editors

Nikhil R. Pal, Indian Statistical Institute, Kolkata, India
Rafael Bello Perez, Faculty of Mathematics, Physics and Computing,
Universidad Central de Las Villas, Santa Clara, Cuba
Emilio S. Corchado, University of Salamanca, Salamanca, Spain
Hani Hagras, Electronic Engineering, University of Essex, Colchester, UK
László T. Kóczy, Department of Automation, Széchenyi István University,
Gyor, Hungary
Vladik Kreinovich, Department of Computer Science, University of Texas
at El Paso, El Paso, TX, USA
Chin-Teng Lin, Department of Electrical Engineering, National Chiao
Tung University, Hsinchu, Taiwan
Jie Lu, Faculty of Engineering and Information Technology,
University of Technology Sydney, Sydney, NSW, Australia
Patricia Melin, Graduate Program of Computer Science, Tijuana Institute
of Technology, Tijuana, Mexico
Nadia Nedjah, Department of Electronics Engineering, University of Rio de Janeiro,
Rio de Janeiro, Brazil
Ngoc Thanh Nguyen, Faculty of Computer Science and Management,
Wrocław University of Technology, Wrocław, Poland
Jun Wang, Department of Mechanical and Automation Engineering,
The Chinese University of Hong Kong, Shatin, Hong Kong

The series "Advances in Intelligent Systems and Computing" contains publications on theory, applications, and design methods of Intelligent Systems and Intelligent Computing. Virtually all disciplines such as engineering, natural sciences, computer and information science, ICT, economics, business, e-commerce, environment, healthcare, life science are covered. The list of topics spans all the areas of modern intelligent systems and computing such as: computational intelligence, soft computing including neural networks, fuzzy systems, evolutionary computing and the fusion of these paradigms, social intelligence, ambient intelligence, computational neuroscience, artificial life, virtual worlds and society, cognitive science and systems, Perception and Vision, DNA and immune based systems, self-organizing and adaptive systems, e-Learning and teaching, human-centered and human-centric computing, recommender systems, intelligent control, robotics and mechatronics including human-machine teaming, knowledge-based paradigms, learning paradigms, machine ethics, intelligent data analysis, knowledge management, intelligent agents, intelligent decision making and support, intelligent network security, trust management, interactive entertainment, Web intelligence and multimedia.

The publications within "Advances in Intelligent Systems and Computing" are primarily proceedings of important conferences, symposia and congresses. They cover significant recent developments in the field, both of a foundational and applicable character. An important characteristic feature of the series is the short publication time and world-wide distribution. This permits a rapid and broad dissemination of research results.

** Indexing: The books of this series are submitted to ISI Proceedings, EI-Compendex, DBLP, SCOPUS, Google Scholar and Springerlink **

More information about this series at http://www.springer.com/series/11156

Álvaro Rocha · Hojjat Adeli ·
Luís Paulo Reis · Sandra Costanzo
Editors

New Knowledge
in Information Systems
and Technologies

Volume 2

Editors
Álvaro Rocha
Departamento de Engenharia Informática
Universidade de Coimbra
Coimbra, Portugal

Luís Paulo Reis
Faculdade de Engenharia/LIACC
Universidade do Porto
Porto, Portugal

Hojjat Adeli
The Ohio State University
Columbus, OH, USA

Sandra Costanzo
DIMES
Università della Calabria
Arcavacata di Rende, Italy

ISSN 2194-5357   ISSN 2194-5365 (electronic)
Advances in Intelligent Systems and Computing
ISBN 978-3-030-16183-5   ISBN 978-3-030-16184-2 (eBook)
https://doi.org/10.1007/978-3-030-16184-2

Library of Congress Control Number: 2019934961

© Springer Nature Switzerland AG 2019, corrected publication 2019
This work is subject to copyright. All rights are reserved by the Publisher, whether the whole or part of the material is concerned, specifically the rights of translation, reprinting, reuse of illustrations, recitation, broadcasting, reproduction on microfilms or in any other physical way, and transmission or information storage and retrieval, electronic adaptation, computer software, or by similar or dissimilar methodology now known or hereafter developed.
The use of general descriptive names, registered names, trademarks, service marks, etc. in this publication does not imply, even in the absence of a specific statement, that such names are exempt from the relevant protective laws and regulations and therefore free for general use.
The publisher, the authors and the editors are safe to assume that the advice and information in this book are believed to be true and accurate at the date of publication. Neither the publisher nor the authors or the editors give a warranty, expressed or implied, with respect to the material contained herein or for any errors or omissions that may have been made. The publisher remains neutral with regard to jurisdictional claims in published maps and institutional affiliations.

This Springer imprint is published by the registered company Springer Nature Switzerland AG
The registered company address is: Gewerbestrasse 11, 6330 Cham, Switzerland

Preface

This book contains a selection of papers accepted for presentation and discussion at The 2019 World Conference on Information Systems and Technologies (WorldCIST'19). This Conference had the support of IEEE SMC (IEEE Systems, Man, and Cybernetics Society), AISTI (Iberian Association for Information Systems and Technologies/Associação Ibérica de Sistemas e Tecnologias de Informação), GIIM (Global Institute for IT Management), and University of Vigo. It took place at La Toja, Galicia, Spain, April 16–19, 2019.

The World Conference on Information Systems and Technologies (WorldCIST) is a global forum for researchers and practitioners to present and discuss recent results and innovations, current trends, professional experiences and challenges of modern Information Systems and Technologies research, technological development and applications. One of its main aims is to strengthen the drive toward a holistic symbiosis between academy, society, and industry. WorldCIST'19 built on the successes of WorldCIST'13 held at Olhão, Algarve, Portugal; WorldCIST'14 held at Funchal, Madeira, Portugal; WorldCIST'15 held at São Miguel, Azores, Portugal; WorldCIST'16 held at Recife, Pernambuco, Brazil; WorldCIST'17 held at Porto Santo, Madeira, Portugal; and WorldCIST'18 took place at Naples, Italy.

The Program Committee of WorldCIST'19 was composed of a multidisciplinary group of more than 200 experts and those who are intimately concerned with Information Systems and Technologies. They have had the responsibility for evaluating, in a 'blind review' process, the papers received for each of the main themes proposed for the Conference: (A) Information and Knowledge Management; (B) Organizational Models and Information Systems; (C) Software and Systems Modeling; (D) Software Systems, Architectures, Applications and Tools; (E) Multimedia Systems and Applications; (F) Computer Networks, Mobility and Pervasive Systems; (G) Intelligent and Decision Support Systems; (H) Big Data Analytics and Applications; (I) Human–Computer Interaction; (J) Ethics, Computers and Security; (K) Health Informatics; (L) Information Technologies in Education; (M) Information Technologies in Radiocommunications; and (N) Technologies for Biomedical Applications.

The Conference also included workshop sessions taking place in parallel with the conference ones. Workshop sessions covered themes such as: (i) Air Quality and Open Data: Challenges for Data Science, HCI and AI; (ii) Digital Transformation; (iii) Empirical Studies in the Domain of Social Network Computing; (iv) Health Technology Innovation: Emerging Trends and Future Challenges; (v) Healthcare Information Systems Interoperability, Security and Efficiency; (vi) New Pedagogical Approaches with Technologies; (vii) Pervasive Information Systems.

WorldCIST'19 received about 400 contributions from 61 countries around the world. The papers accepted for presentation and discussion at the Conference are published by Springer (this book) in three volumes and will be submitted for indexing by ISI, Ei Compendex, Scopus, DBLP, and/or Google Scholar, among others. Extended versions of selected best papers will be published in special or regular issues of relevant journals, mainly SCI/SSCI and Scopus/Ei Compendex indexed journals.

We acknowledge all of those that contributed to the staging of WorldCIST'19 (authors, committees, workshop organizers, and sponsors). We deeply appreciate their involvement and support that was crucial for the success of WorldCIST'19.

April 2019

Álvaro Rocha
Hojjat Adeli
Luís Paulo Reis
Sandra Costanzo

Organization

Conference

General Chair

Álvaro Rocha   University of Coimbra, Portugal

Co-chairs

Hojjat Adeli   The Ohio State University, USA
Luis Paulo Reis   University of Porto, Portugal
Sandra Costanzo   University of Calabria, Italy

Local Chair

Manuel Pérez Cota   University of Vigo, Spain

Advisory Committee

Ana Maria Correia (Chair)   University of Sheffield, UK
Andrew W. H. Ip   Hong Kong Polytechnic University, China
Cihan Cobanoglu   University of South Florida, USA
Chris Kimble   KEDGE Business School and MRM, UM2, Montpellier, France
Erik Bohlin   Chalmers University of Technology, Sweden
Eva Onaindia   Universidad Politecnica de Valencia, Spain
Eugene H. Spafford   Purdue University, USA
Gintautas Dzemyda   Vilnius University, Lithuania
Gregory Kersten   Concordia University, Canada
Janusz Kacprzyk   Polish Academy of Sciences, Poland

João Tavares	University of Porto, Portugal
Jon Hall	The Open University, UK
Karl Stroetmann	Empirica Communication & Technology Research, Germany
Kathleen Carley	Carnegie Mellon University, USA
Keng Siau	Missouri University of Science and Technology, USA
Salim Hariri	University of Arizona, USA
Marjan Mernik	University of Maribor, Slovenia
Michael Koenig	Long Island University, USA
Miguel-Angel Sicilia	Alcalá University, Spain
Peter Sloot	University of Amsterdam, the Netherlands
Reza Langari	Texas A&M University, USA
Robert J. Kauffman	Singapore Management University, Singapore
Wim Van Grembergen	University of Antwerp, Belgium

Program Committee

Abdul Rauf	RISE SICS, Sweden
Adnan Mahmood	Waterford Institute of Technology, Ireland
Adriana Peña Pérez Negrón	Universidad de Guadalajara, Mexico
Adriani Besimi	South East European University, Macedonia
Agostinho Sousa Pinto	Polytecnic of Porto, Portugal
Ahmed El Oualkadi	Abdelmalek Essaadi University, Morocco
Alan Ramirez-Noriega	Universidad Autónoma de Sinaloa, Mexico
Alberto Freitas	FMUP, University of Porto, Portugal
Aleksandra Labus	University of Belgrade, Serbia
Alexandru Vulpe	Politehnica University of Bucharest, Romania
Ali Alsoufi	University of Bahrain, Bahrain
Ali Idri	ENSIAS, Mohammed V University, Morocco
Almir Souza Silva Neto	IFMA, Brazil
Amit Shelef	Sapir Academic College, Israel
Ana Isabel Martins	University of Aveiro, Portugal
Ana Luis	University of Coimbra, Portugal
Anabela Tereso	University of Minho, Portugal
Anacleto Correia	CINAV, Portugal
Anca Alexandra Purcarea	Politehnica University of Bucharest, Romania
André Marcos Silva	Centro Universitário Adventista de São Paulo (UNASP), Brazil
Aneta Poniszewska-Maranda	Lodz University of Technology, Poland
Angeles Quezada	Instituto Tecnologico de Tijuana, Mexico
Ankur Singh Bist	KIET, India
Antoni Oliver	University of the Balearic Islands, Spain
Antonio Borgia	University of Calabria, Italy

Antonio Jiménez-Martín	Universidad Politécnica de Madrid, Spain
Antonio Pereira	Polytechnic of Leiria, Portugal
Armando Toda	University of São Paulo, Brazil
Arslan Enikeev	Kazan Federal University, Russia
Benedita Malheiro	Polytechnic of Porto, ISEP, Portugal
Borja Bordel	Universidad Politécnica de Madrid, Spain
Branko Perisic	Faculty of Technical Sciences, Serbia
Carla Pinto	Polytechnic of Porto, ISEP, Portugal
Carla Santos Pereira	Universidade Portucalense, Portugal
Catarina Reis	Polytechnic of Leiria, Portugal
Cédric Gaspoz	University of Applied Sciences Western Switzerland (HES-SO), Switzerland
Cengiz Acarturk	Middle East Technical University, Turkey
Cesar Collazos	Universidad del Cauca, Colombia
Christophe Feltus	LIST, Luxembourg
Christophe Soares	University Fernando Pessoa, Portugal
Christos Bouras	University of Patras, Greece
Ciro Martins	University of Aveiro, Portugal
Claudio Sapateiro	Polytechnic Institute of Setúbal, Portugal
Cristian García Bauza	PLADEMA-UNICEN-CONICET, Argentina
Cristian Mateos	ISISTAN-CONICET, UNICEN, Argentina
Daniel Lübke	Leibniz Universität Hannover, Germany
Dante Carrizo	Universidad de Atacama, Chile
David Cortés-Polo	Fundación COMPUTAEX, Spain
Edita Butrime	Lithuanian University of Health Sciences, Lithuania
Edna Dias Canedo	University of Brasilia, Brazil
Eduardo Albuquerque	Federal University of Goiás, Brazil
Eduardo Santos	Pontifical Catholic University of Paraná, Brazil
Egils Ginters	Riga Technical University, Latvia
Eliana Leite	University of Minho, Portugal
Emiliano Reynares	CONICET-CIDISI UTN FRSF, Argentina
Evandro Costa	Federal University of Alagoas, Brazil
Fatima Azzahra Amazal	Ibn Zohr University, Morocco
Fernando Bobillo	University of Zaragoza, Spain
Fernando Moreira	Portucalense University, Portugal
Fernando Ribeiro	Polytechnic Castelo Branco, Portugal
Filipe Portela	University of Minho, Portugal
Filippo Neri	University of Naples, Italy
Fionn Murtagh	University of Huddersfield, UK
Firat Bestepe	Republic of Turkey, Ministry of Development, Turkey
Fouzia Idrees	Shaheed Benazir Bhutto Women University, Pakistan
Francesca Venneri	University of Calabria, Italy

Francesco Bianconi	Università degli Studi di Perugia, Italy
Francisco García-Peñalvo	University of Salamanca, Spain
Francisco Valverde	Universidad Central del Ecuador, Ecuador
Frederico Branco	University of Trás-os-Montes e Alto Douro, Portugal
Gabriel Pestana	Universidade Europeia, Portugal
Galim Vakhitov	Kazan Federal University, Russia
George Suciu	BEIA, Romania
Ghani Albaali	Princess Sumaya University for Technology, Jordan
Gian Piero Zarri	University Paris-Sorbonne, France
Giuseppe Di Massa	University of Calabria, Italy
Gonçalo Paiva Dias	University of Aveiro, Portugal
Goreti Marreiros	ISEP/GECAD, Portugal
Graciela Lara López	University of Guadalajara, Mexico
Habiba Drias	University of Science and Technology Houari Boumediene, Algeria
Hafed Zarzour	University of Souk Ahras, Algeria
Hamid Alasadi	University of Basra, Iraq
Hatem Ben Sta	University of Tunis at El Manar, Tunisia
Hector Fernando Gomez Alvarado	Universidad Tecnica de Ambato, Ecuador
Hélder Gomes	University of Aveiro, Portugal
Helia Guerra	University of the Azores, Portugal
Henrique da Mota Silveira	University of Campinas (UNICAMP), Brazil
Henrique S. Mamede	University Aberta, Portugal
Hing Kai Chan	University of Nottingham Ningbo China, China
Hugo Paredes	INESC TEC and Universidade de Trás-os-Montes e Alto Douro, Portugal
Ibtissam Abnane	Mohammed V University in Rabat, Morocco
Imen Ben Said	Université de Sfax, Tunisia
Ina Schiering	Ostfalia University of Applied Sciences, Germany
Inês Domingues	University of Coimbra, Portugal
Isabel Lopes	Instituto Politécnico de Bragança, Portugal
Isabel Pedrosa	Coimbra Business School ISCAC, Portugal
Isaías Martins	University of Leon, Spain
Ivan Lukovic	University of Novi Sad, Serbia
Jan Kubicek	Technical University of Ostrava, Czech Republic
Jean Robert Kala Kamdjoug	Catholic University of Central Africa, Cameroon
Jesús Gallardo Casero	University of Zaragoza, Spain
Jezreel Mejia	CIMAT Unidad Zacatecas, Mexico
Jikai Li	The College of New Jersey, USA
Jinzhi Lu	KTH Royal Institute of Technology, Sweden
Joao Carlos Silva	IPCA, Portugal

João Manuel R. S. Tavares	University of Porto, FEUP, Portugal
João Reis	University of Lisbon, Portugal
João Rodrigues	University of Algarve, Portugal
Jorge Barbosa	Polytecnic Institute of Coimbra, Portugal
Jorge Buele	Technical University of Ambato, Ecuador
Jorge Esparteiro Garcia	Polytechnic Institute of Viana do Castelo, Portugal
Jorge Gomes	University of Lisbon, Portugal
Jorge Oliveira e Sá	University of Minho, Portugal
José Álvarez-García	University of Extremadura, Spain
José Braga de Vasconcelos	Universidade New Atlântica, Portugal
Jose Luis Herrero Agustin	University of Extremadura, Spain
José Luís Reis	ISMAI, Portugal
Jose Luis Sierra	Complutense University of Madrid, Spain
Jose M. Parente de Oliveira	Aeronautics Institute of Technology, Brazil
José Machado	University of Minho, Portugal
José Martins	Universidade de Trás-os-Montes e Alto Douro, Portugal
Jose Torres	University Fernando Pessoa, Portugal
José-Luís Pereira	Universidade do Minho, Portugal
Juan Jesus Ojeda-Castelo	University of Almeria, Spain
Juan M. Santos	University of Vigo, Spain
Juan Pablo Damato	UNCPBA-CONICET, Argentina
Juncal Gutiérrez-Artacho	University of Granada, Spain
Justyna Trojanowska	Poznan University of Technology, Poland
Katsuyuki Umezawa	Shonan Institute of Technology, Japan
Khalid Benali	LORIA, University of Lorraine, France
Korhan Gunel	Adnan Menderes University, Turkey
Krzysztof Wolk	Polish-Japanese Academy of Information Technology, Poland
Kuan Yew Wong	Universiti Teknologi Malaysia (UTM), Malaysia
Laila Cheikhi	Mohammed V University, Rabat, Morocco
Laura Varela-Candamio	Universidade da Coruña, Spain
Laurentiu Boicescu	E.T.T.I. U.P.B., Romania
Leonardo Botega	University Centre Eurípides of Marília (UNIVEM), Brazil
Leonid Leonidovich Khoroshko	Moscow Aviation Institute (National Research University), Russia
Letícia Helena Januário	Universidade Federal de São João del-Rei, Brazil
Lila Rao-Graham	University of the West Indies, Jamaica
Luis Alvarez Sabucedo	University of Vigo, Spain
Luis Mendes Gomes	University of the Azores, Portugal
Luiz Rafael Andrade	Tiradentes University, Brazil
Luis Silva Rodrigues	Polytencic of Porto, Portugal

Luz Sussy Bayona Oré	Universidad Nacional Mayor de San Marcos, Peru
Magdalena Diering	Poznan University of Technology, Poland
Manuel Antonio Fernández-Villacañas Marín	Technical University of Madrid, Spain
Manuel Pérez Cota	University of Vigo, Spain
Manuel Silva	Polytechnic of Porto and INESC TEC, Portugal
Manuel Tupia	Pontifical Catholic University of Peru, Peru
Marco Ronchetti	Università di Trento, Italy
Mareca María PIlar	Universidad Politécnica de Madrid, Spain
Marek Kvet	Zilinska Univerzita v Ziline, Slovakia
María de la Cruz del Río-Rama	University of Vigo, Spain
Maria João Ferreira	Universidade Portucalense, Portugal
Maria João Varanda Pereira	Polytechnic Institute of Bragança, Portugal
Maria José Sousa	University of Coimbra, Portugal
María Teresa García-Álvarez	University of A Coruna, Spain
Marijana Despotovic-Zrakic	Faculty Organizational Science, Serbia
Mário Antunes	Polytecnic of Leiria and CRACS INESC TEC, Portugal
Marisa Maximiano	Polytechnic of Leiria, Portugal
Marisol Garcia-Valls	Universidad Carlos III de Madrid, Spain
Maristela Holanda	University of Brasilia, Brazil
Marius Vochin	E.T.T.I. U.P.B., Romania
Marlene Goncalves da Silva	Universidad Simón Bolívar, Venezuela
Maroi Agrebi	University of Polytechnique Hauts-de-France, France
Martin Henkel	Stockholm University, Sweden
Martín López Nores	University of Vigo, Spain
Martin Zelm	INTEROP-VLab, Belgium
Mawloud Mosbah	University 20 Août 1955 of Skikda, Algeria
Michal Adamczak	Poznan School of Logistics, Poland
Michal Kvet	University of Zilina, Slovakia
Miguel António Sovierzoski	Federal University of Technology - Paraná, Brazil
Mihai Lungu	Craiova University, Romania
Milton Miranda	Federal University of Uberlândia, Brazil
Mircea Georgescu	Al. I. Cuza University of Iasi, Romania
Mirna Muñoz	Centro de Investigación en Matemáticas A.C., Mexico
Mohamed Hosni	ENSIAS, Morocco
Mokhtar Amami	Royal Military College of Canada, Canada
Monica Leba	University of Petrosani, Romania

Muhammad Nawaz	Institute of Management Sciences, Peshawar, Pakistan
Mu-Song Chen	Dayeh University, China
Nastaran Hajiheydari	York St John University, UK
Natalia Grafeeva	Saint Petersburg State University, Russia
Natalia Miloslavskaya	National Research Nuclear University MEPhI, Russia
Naveed Ahmed	University of Sharjah, United Arab Emirates
Nelson Rocha	University of Aveiro, Portugal
Nelson Salgado	Pontifical Catholic University of Ecuador, Ecuador
Nikolai Prokopyev	Kazan Federal University, Russia
Niranjan S. K.	JSS Science and Technology University, India
Noemi Emanuela Cazzaniga	Politecnico di Milano, Italy
Noor Ahmed	AFRL/RI, USA
Noureddine Kerzazi	Polytechnique Montréal, Canada
Nuno Melão	Polytechnic of Viseu, Portugal
Nuno Octávio Fernandes	Polytechnic Institute of Castelo Branco, Portugal
Paôla Souza	Aeronautics Institute of Technology, Brazil
Patricia Zachman	Universidad Nacional del Chaco Austral, Argentina
Paula Alexandra Rego	Polytechnic Institute of Viana do Castelo and LIACC, Portugal
Paula Viana	Polytechnic of Porto and INESC TEC, Portugal
Paulo Maio	Polytechnic of Porto, ISEP, Portugal
Paulo Novais	University of Minho, Portugal
Paweł Karczmarek	The John Paul II Catholic University of Lublin, Poland
Pedro Henriques Abreu	University of Coimbra, Portugal
Pedro Rangel Henriques	University of Minho, Portugal
Pedro Sobral	University Fernando Pessoa, Portugal
Pedro Sousa	University of Minho, Portugal
Philipp Brune	University of Applied Sciences Neu-Ulm, Germany
Piotr Kulczycki	Systems Research Institute, Polish Academy of Sciences, Poland
Prabhat Mahanti	University of New Brunswick, Canada
Radu-Emil Precup	Politehnica University of Timisoara, Romania
Rafael M. Luque Baena	University of Malaga, Spain
Rahim Rahmani	Stockholm University, Sweden
Raiani Ali	Bournemouth University, UK
Ramayah T.	Universiti Sains Malaysia, Malaysia
Ramiro Delgado	Universidad de las Fuerzas Armadas ESPE, Ecuador

Ramiro Gonçalves	University of Trás-os-Montes e Alto Douro and INESC TEC, Portugal
Ramon Alcarria	Universidad Politécnica de Madrid, Spain
Ramon Fabregat Gesa	University of Girona, Spain
Reyes Juárez Ramírez	Universidad Autonoma de Baja California, Mexico
Rui Jose	University of Minho, Portugal
Rui Pitarma	Polytechnic Institute of Guarda, Portugal
Rui S. Moreira	UFP & INESC TEC & LIACC, Portugal
Rustam Burnashev	Kazan Federal University, Russia
Saeed Salah	Al-Quds University, Palestine
Said Achchab	Mohammed V University in Rabat, Morocco
Sajid Anwar	Institute of Management Sciences, Peshawar, Pakistan
Salama Mostafa	Universiti Tun Hussein Onn Malaysia, Malaysia
Sami Habib	Kuwait University, Kuwait
Samuel Fosso Wamba	Toulouse Business School, France
Sanaz Kavianpour	University of Technology, Malaysia
Sandra Costanzo	University of Calabria, Italy
Sandra Patricia Cano Mazuera	University of San Buenaventura Cali, Colombia
Sergio Albiol-Pérez	University of Zaragoza, Spain
Shahnawaz Talpur	Mehran University of Engineering and Technology, Jamshoro, Pakistan
Silviu Vert	Politehnica University of Timisoara, Romania
Simona Mirela Riurean	University of Petrosani, Romania
Slawomir Zolkiewski	Silesian University of Technology, Poland
Solange N. Alves-Souza	University of São Paulo, Brazil
Solange Rito Lima	University of Minho, Portugal
Sorin Zoican	Polytechnica University of Bucharest, Romania
Souraya Hamida	University of Batna 2, Algeria
Stefan Pickl	UBw München COMTESSA, Germany
Sümeyya Ilkin	Kocaeli University, Turkey
Syed Asim Ali	University of Karachi, Pakistan
Taoufik Rachad	Mohammed V University, Morocco
Tatiana Antipova	Institute of certified Specialists, Russia
The Thanh Van	HCMC University of Food Industry, Vietnam
Thomas Weber	EPFL, Switzerland
Timothy Asiedu	TIM Technology Services Ltd., Ghana
Tom Sander	New College of Humanities, Germany
Tomaž Klobučar	Jozef Stefan Institute, Slovenia
Toshihiko Kato	University of Electro-Communications, Japan
Tzung-Pei Hong	National University of Kaohsiung, Taiwan

Valentina Colla	Scuola Superiore Sant'Anna, Italy
Veronica Segarra Faggioni	Private Technical University of Loja, Ecuador
Victor Alves	University of Minho, Portugal
Victor Georgiev	Kazan Federal University, Russia
Victor Hugo Medina Garcia	Universidad Distrital Francisco José de Caldas, Colombia
Vincenza Carchiolo	University of Catania, Italy
Vitalyi Igorevich Talanin	Zaporozhye Institute of Economics and Information Technologies, Ukraine
Wolf Zimmermann	Martin Luther University Halle-Wittenberg, Germany
Yadira Quiñonez	Autonomous University of Sinaloa, Mexico
Yair Wiseman	Bar-Ilan University, Israel
Yuhua Li	Cardiff University, UK
Yuwei Lin	University of Roehampton, UK
Yves Rybarczyk	Universidad de Las Américas, Ecuador
Zorica Bogdanovic	University of Belgrade, Serbia

Workshops

First Workshop on Air Quality and Open Data: Challenges for Data Science, HCI and AI

Organizing Committee

Kai v. Luck	Creative Space for Technical Innovation, HAW Hamburg, Germany
Susanne Draheim	Creative Space for Technical Innovation, HAW Hamburg, Germany
Jessica Broscheit	Artist, Hamburg, Germany
Martin Kohler	HafenCity University Hamburg, Germany

Program Committee

Ingo Börsch	Technische Hochschule Brandenburg, Brandenburg University of Applied Sciences, Germany
Susanne Draheim	Hamburg University of Applied Sciences, Germany
Stefan Wölwer	HAWK University of Applied Sciences and Arts Hildesheim/Holzminden/Goettingen, Germany
Kai v. Luck	Creative Space for Technical Innovation, HAW Hamburg, Germany

Tim Tiedemann   Hamburg University of Applied Sciences,
                                    Germany
Marcelo Tramontano   University of São Paulo, Brazil

Second Workshop on Digital Transformation

Organizing Committee

Fernando Moreira   Universidade Portucalense, Portugal
Ramiro Gonçalves   Universidade de Trás-os-Montes e Alto Douro,
                                    Portugal
Manuel Au-Yong Oliveira   Universidade de Aveiro, Portugal
José Martins   Universidade de Trás-os-Montes e Alto Douro,
                                    Portugal
Frederico Branco   Universidade de Trás-os-Montes e Alto Douro,
                                    Portugal

Program Committee

Alex Sandro Gomes   Universidade Federal de Pernambuco, Brazil
Arnaldo Martins   Universidade de Aveiro, Portugal
César Collazos   Universidad del Cauca, Colombia
Jezreel Mejia   Centro de Investigación en Matemáticas A.C.,
                                    Mexico
Jörg Thomaschewski   University of Applied Sciences, Germany
Lorna Uden   Staffordshire University, UK
Manuel Ortega   Universidad de Castilla–La Mancha, Spain
Manuel Peréz Cota   Universidade de Vigo, Spain
Martin Schrepp   SAP SE, Germany
Philippe Palanque   Université Toulouse III, France
Rosa Vicardi   Universidade Federal do Rio Grande do Sul,
                                    Brazil
Vitor Santos   NOVA IMS Information Management School,
                                    Portugal

First Workshop on Empirical Studies in the Domain of Social Network Computing

Organizing Committee

Shahid Hussain   COMSATS Institute of Information Technology,
                                    Islamabad, Pakistan
Arif Ali Khan   Nanjing University of Aeronautics
                                    and Astronautics, China
Nafees Ur Rehman   University of Konstanz, Germany

Program Committee

Abdul Mateen	Federal Urdu University of Arts, Science & Technology, Islamabad, Pakistan
Aoutif Amine	ENSA, Ibn Tofail University, Morocco
Gwanggil Jeon	Incheon National University, Korea
Hanna Hachimi	ENSA of Kenitra, Ibn Tofail University, Morocco
Jacky Keung	City University of Hong Kong, Hong Kong
Kifayat Alizai	National University of Computer and Emerging Sciences (FAST-NUCES), Islamabad, Pakistan
Kwabena Bennin Ebo	City University of Hong Kong, Hong Kong
Mansoor Ahmad	COMSATS University Islamabad, Pakistan
Manzoor Ilahi	COMSATS University Islamabad, Pakistan
Mariam Akbar	COMSATS University Islamabad, Pakistan
Muhammad Khalid Sohail	COMSATS University Islamabad, Pakistan
Muhammad Shahid	Gomal University, DIK, Pakistan
Salima Banqdara	University of Benghazi, Libya
Siti Salwa Salim	University of Malaya, Malaysia
Wiem Khlif	University of Sfax, Tunisia

First Workshop on Health Technology Innovation: Emerging Trends and Future Challenges

Organizing Committee

Eliana Silva	University of Minho & Optimizer, Portugal
Joyce Aguiar	University of Minho & Optimizer, Portugal
Victor Carvalho	Optimizer, Portugal
Joaquim Gonçalves	Instituto Politécnico do Cávado e do Ave & Optimizer, Portugal

Program Committee

Eliana Silva	University of Minho & Optimizer, Portugal
Joyce Aguiar	University of Minho & Optimizer, Portugal
Victor Carvalho	Optimizer, Portugal
Joaquim Gonçalves	Instituto Politécnico do Cávado e do Ave & Optimizer, Portugal

Fifth Workshop on Healthcare Information Systems Interoperability, Security and Efficiency

Organizing Committee

José Machado	University of Minho, Portugal
António Abelha	University of Minho, Portugal

| Luis Mendes Gomes | University of Azores, Portugal |
| Anastasius Mooumtzoglou | European Society for Quality in Healthcare, Greece |

Program Committee

Alberto Freitas	University of Porto, Portugal
Ana Azevedo	ISCAP/IPP, Portugal
Ângelo Costa	University of Minho, Portugal
Armando B. Mendes	University of Azores, Portugal
Cesar Analide	University of Minho, Portugal
Davide Carneiro	University of Minho, Portugal
Filipe Portela	University of Minho, Portugal
Goreti Marreiros	Polytechnic Institute of Porto, Portugal
Helia Guerra	University of Azores, Portugal
Henrique Vicente	University of Évora, Portugal
Hugo Peixoto	University of Minho, Portugal
Jason Jung	Chung-Ang University, Korea
Joao Ramos	University of Minho, Portugal
José Martins	UTAD, Portugal
Jose Neves	University of Minho, Portugal
Júlio Duarte	University of Minho, Portugal
Luis Mendes Gomes	University of Azores, Portugal
Manuel Filipe Santos	University of Minho, Portugal
Paulo Moura Oliveira	UTAD, Portugal
Paulo Novais	University of Minho, Portugal
Teresa Guarda	Universidad Estatal da Península de Santa Elena, Ecuador
Victor Alves	University of Minho, Portugal

Fourth Workshop on New Pedagogical Approaches with Technologies

Organizing Committee

Anabela Mesquita	ISCAP/P.Porto and Algoritmi Centre, Portugal
Paula Peres	ISCAP/P.Porto and Unit for e-Learning and Pedagogical Innovation, Portugal
Fernando Moreira	IJP and REMIT – Univ Portucalense & IEETA – Univ Aveiro, Portugal

Program Committee

| Alex Gomes | Universidade Federal de Pernambuco, Brazil |
| Ana R. Luís | Universidade de Coimbra, Portugal |

Armando Silva	ESE/IPP, Portugal
César Collazos	Universidad del Cauca, Colombia
Chia-Wen Tsai	Ming Chuan University, Taiwan
João Batista	CICE/ISCA, UA, Portugal
Lino Oliveira	ESMAD/IPP, Portugal
Luisa M. Romero Moreno	Universidade de Sevilha, Espanha
Manuel Pérez Cota	Universidade de Vigo, Espanha
Paulino Silva	CICE & CECEJ-ISCAP/IPP, Portugal
Ramiro Gonçalves	UTAD, Vila Real, Portugal
Rosa Vicari	Universidade de Rio Grande do Sul, Porto Alegre, Brazil
Stefania Manca	Instituto per le Tecnologie Didattiche, Italy

Fifth Workshop on Pervasive Information Systems

Organizing Committee

Carlos Filipe Portela	Department of Information Systems, University of Minho, Portugal
Manuel Filipe Santos	Department of Information Systems, University of Minho, Portugal
Kostas Kolomvatsos	Department of Informatics and Telecommunications, National and Kapodistrian University of Athens, Greece

Program Committee

Andre Aquino	Federal University of Alagoas, Brazil
Carlo Giannelli	University of Ferrara, Italy
Cristina Alcaraz	University of Malaga, Spain
Daniele Riboni	University of Milan, Italy
Fabio A. Schreiber	Politecnico Milano, Italy
Filipe Mota Pinto	Polytechnic of Leiria, Portugal
Hugo Peixoto	University of Minho, Portugal
Gabriel Pedraza Ferreira	Universidad Industrial de Santander, Colombia
Jarosław Jankowski	West Pomeranian University of Technology, Szczecin, Poland
José Machado	University of Minho, Portugal
Juan-Carlos Cano	Universitat Politècnica de València, Spain
Karolina Baras	University of Madeira, Portugal
Muhammad Younas	Oxford Brookes University, UK
Nuno Marques	New University of Lisboa, Portugal
Rajeev Kumar Kanth	Turku Centre for Computer Science, University of Turku, Finland

Ricardo Queirós   ESMAD- P.PORTO & CRACS - INESC TEC,
                                   Portugal
Sergio Ilarri   University of Zaragoza, Spain
Spyros Panagiotakis   Technological Educational Institute of Crete,
                                   Greece

Contents

Software Systems, Architectures, Applications and Tools

**Multi-agent Neural Reinforcement-Learning System
with Communication**   3
David Simões, Nuno Lau, and Luís Paulo Reis

**Construction and Integration of a Quadcopter in a Simulation
Platform for Multi-vehicle Missions**   13
Leonardo Ferreira, Álvaro Câmara, and Daniel Castro Silva

**Technological Architecture Based on Internet of Things to Monitor
the Journeys of Artisanal Fishing**   24
Jaime Ambrosio Mallqui, Leysa Preguntegui Martinez,
and Jimmy Armas Aguirre

Gesture Based Alternative to Control Recreational UAV   34
Roberto Ribeiro, David Safadinho, João Ramos, Nuno Rodrigues,
Arsénio Reis, and António Pereira

**Software Modules and Communication to Support Real-Time Remote
Control and Monitoring of Unmanned Vehicles**   45
João Ramos, David Safadinho, Roberto Ribeiro, Patrício Domingues,
João Barroso, and António Pereira

**Communication Modes to Control an Unmanned Vehicle
Using ESP8266**   56
David Safadinho, João Ramos, Roberto Ribeiro, Arsénio Reis,
Carlos Rabadão, and António Pereira

**Automatic Generation of a Sub-optimal Agent Population
with Learning**   65
Simão Reis, Luís Paulo Reis, and Nuno Lau

**Evaluation of Open Source Software for Testing Performance
of Web Applications** 75
Fernando Maila-Maila, Monserrate Intriago-Pazmiño,
and Julio Ibarra-Fiallo

**Blockchain Projects Ecosystem: A Review of Current Technical
and Legal Challenges** 83
Jorge Lopes and José Luís Pereira

App Nutrition Label 93
Sami J. Habib and Paulvanna N. Marimuthu

**Automated and Decentralized Framework for Internet of Things
Systems Using Blockchain and Smart Contracts** 103
Sorin Zoican, Roxana Zoican, and Dan Galaţchi

Improving Ambient Assisted Living Through Artificial Intelligence 110
Alessandro Miguez, Christophe Soares, José M. Torres, Pedro Sobral,
and Rui S. Moreira

**A FIPA-Compliant Framework for Integrating Rule Engines into
Software Agents for Supporting Communication and Collaboration
in a Multiagent Platform** 124
Francisco J. Aguayo-Canela, Héctor Alaiz-Moretón,
Isaías García-Rodríguez, Carmen Benavides-Cuéllar,
José Alberto Benítez-Andrades, and Paulo Novais

**A Chaotic Cryptographic Solution for Low-Range Wireless
Communications in Industry 4.0** 134
Pilar Mareca and Borja Bordel

**Yet Another Virtual Butler to Bridge the Gap Between Users
and Ambient Assisted Living** 145
Marta Carvalho, Ricardo Domingues, Rodrigo Alves, António Pereira,
and Nuno Costa

Usage of HTTPS by Municipal Websites in Portugal 155
Hélder Gomes, André Zúquete, Gonçalo Paiva Dias, and Fábio Marques

**Lego Methodology Approach for Common Criteria Certification
of IoT Telemetry** 165
George Suciu, Cristiana Istrate, Ioana Petre, and Andrei Scheianu

**A Two-Phase Algorithm for Recognizing Human Activities
in the Context of Industry 4.0 and Human-Driven Processes** 175
Borja Bordel, Ramón Alcarria, and Diego Sánchez-de-Rivera

Kitchen Robots: The Importance and Impact of Technology on People's Quality of Life 186
Ema Fonseca, Inês Oliveira, Joana Lobo, Tânia Mota, José Martins, and Manuel Au-Yong-Oliveira

Vehicle-Pedestrian Interaction in SUMO and Unity3D 198
Leyre Artal-Villa and Cristina Olaverri-Monreal

Electronic System for the Detection of Chicken Eggs Suitable for Incubation Through Image Processing 208
Ángel Fernández-S., Franklin Salazar-L., Marco Jurado, Esteban X. Castellanos, Rodrigo Moreno-P., and Jorge Buele

Mobile Application Based on Dijkstra's Algorithm, to Improve the Inclusion of People with Motor Disabilities Within Urban Areas . . . 219
Luis Arellano, Daniel Del Castillo, Graciela Guerrero, and Freddy Tapia

Design and Construction of a Semi-quantitative Estimator for Red Blood Cells in Blood as Support the Health Sector 230
Yadira Quiñonez, Said Almeraya, Selin Almeraya, Jorge Reyna, and Jezreel Mejía

Automatic Forest Fire Detection Based on a Machine Learning and Image Analysis Pipeline 240
João Alves, Christophe Soares, José M. Torres, Pedro Sobral, and Rui S. Moreira

Cybersecurity Threats Analysis for Airports 252
George Suciu, Andrei Scheianu, Ioana Petre, Loredana Chiva, and Cristina Sabina Bosoc

Player Engagement Enhancement with Video Games 263
Simão Reis, Luís Paulo Reis, and Nuno Lau

Multimedia Systems and Applications

VR Macintosh Museum: Case Study of a WebVR Application 275
Antoni Oliver, Javier del Molino, Maria Cañellas, Albert Clar, and Antoni Bibiloni

Web-Based Executive Dashboard Reports for Public Works Clients in Construction Industry 285
Alvansazyazdi Mohammadfarid, Nelson Esteban Salgado Reyes, Amir Hossein Borghei, Alejandro Miguel Camino Solórzano, Maria Susana Guzmán Rodríguez, and Mario Augusto Rivera Valenzuela

Novel Robust Digital Watermarking in Mid-Rank Co-efficient Based on DWT and RT Transform 295
Zahoor Jan, Inayat Ullah, Faryal Tahir, Naveed Islam, and Babar Shah

Multiple Moving Vehicle Speed Estimation Using Blob Analysis 303
Muhammad Khan, Muhammad Nawaz, Qazi Nida-Ur-Rehman,
Ghulam Masood, Awais Adnan, Sajid Anwar, and John Cosmas

Computer Networks, Mobility and Pervasive Systems

New SIEM System for the Internet of Things 317
Natalia Miloslavskaya and Alexander Tolstoy

Standardization Issues for the Internet of Things 328
Natalia Miloslavskaya, Andrey Nikiforov, Kirill Plaksiy,
and Alexander Tolstoy

An Efficient Sybil Attack Detection for Internet of Things 339
Sohail Abbas

Li-Fi Embedded Wireless Integrated Medical Assistance System 350
Simona Riurean, Tatiana Antipova, Alvaro Rocha, Monica Leba,
and Andreea Ionica

Cloud in Mobile Platforms: Managing Authentication/Authorization . . . 361
Vicenza Carchiolo, Alessandro Longheu, Michele Malgeri,
Stefano Iannello, Mario Marroccia, and Angelo Randazzo

**An Ontology-Based Recommendation System for Context-Aware
Network Monitoring** 373
Ricardo F. Silva, Paulo Carvalho, Solange Rito Lima,
Luis Álvarez Sabucedo, Juan M. Santos Gago, and João Marco C. Silva

**Analysis of Performance Degradation of TCP Reno over WLAN
Caused by Access Point Scanning** 385
Toshihiko Kato, Sota Tasaki, Ryo Yamamoto, and Satoshi Ohzahata

**Stratifying Measuring Requirements and Tools for Cloud
Services Monitoring** 396
Paulo Carvalho, Solange Rito Lima, and Kalil Araujo Bispo

**Consensus Problem with the Existence
of an Adversary Nanomachine** 407
Athraa Juhi Jani

Flow Monitoring System for IoT Networks 420
Leonel Santos, Carlos Rabadão, and Ramiro Gonçalves

**The User's Attitude and Security of Personal Information Depending
on the Category of IoT** 431
Basma Taieb and Jean-Éric Pelet

**Development of Self-aware and Self-redesign Framework
for Wireless Sensor Networks** 438
Sami J. Habib, Paulvanna N. Marimuthu, Pravin Renold,
and Balaji Ganesh Athi

Intelligent and Decision Support Systems

**Comparison of Evolutionary Algorithms for Coordination
of Cooperative Bioinspired Multirobots** 451
A. A. Saraiva, F. V. N. Silva, Jose Vigno M. Sousa,
N. M. Fonseca Ferreira, Antonio Valente, and Salviano Soares

**An Actionable Knowledge Discovery System in Regular
Sports Services** 461
Paulo Pinheiro and Luís Cavique

Real Time Fuzzy Based Traffic Flow Estimation and Analysis 472
Muhammad Abbas, Fozia Mehboob, Shoab A. Khan, Abdul Rauf,
and Richard Jiang

**Features Weight Estimation Using a Genetic Algorithm
for Customer Churn Prediction in the Telecom Sector** 483
Adnan Amin, Babar Shah, Ali Abbas, Sajid Anwar, Omar Alfandi,
and Fernando Moreira

**Nonlinear Identification of a Robotic Arm Using Machine
Learning Techniques** 492
Darielson A. Souza, Laurinda L. N. Reis, Josias G. Batista,
Jonatha R. Costa, Antonio B. S. Junior, João P. B. Araújo,
and Arthur P. S. Braga

**Study on the Variation of the A Fund of the Pension System
in Chile Applying Artificial Neural Networks** 502
Alexander Börger and Pedro Vega

**Assessing the Horizontal Positional Accuracy in OpenStreetMap:
A Big Data Approach** 513
Roger Castro, Alfonso Tierra, and Marco Luna

Optimized Community Detection in Social Networks 524
Lamia Berkani, Sara Madani, and Soumeya Mekherbeche

**Requirements for Training and Evaluation Dataset of Network
and Host Intrusion Detection System** 534
Petteri Nevavuori and Tero Kokkonen

A Fuzzy Inference System as a Tool to Measure Levels
of Interaction 547
Ricardo Rosales, Margarita Ramírez-Ramírez, Nora Osuna-Millán,
Manuel Castañón-Puga, Josue Miguel Flores-Parra, and Maria Quezada

Rational, Emotional, and Attentional Choice Models
for Recommender Systems 557
Ameed Almomani, Cristina Monreal, Jorge Sieira, Juan Graña,
and Eduardo Sánchez

Providing Alternative Measures for Addressing Adverse
Drug-Drug Interactions 567
António Silva, Tiago Oliveira, Ken Satoh, and Paulo Novais

Convolutional Neural Network-Based Regression for Quantification
of Brain Characteristics Using MRI 577
João Fernandes, Victor Alves, Nadieh Khalili, Manon J. N. L. Benders,
Ivana Išgum, Josien Pluim, and Pim Moeskops

Information System for Monitoring and Assessing Stress
Among Medical Students 587
Eliana Silva, Joyce Aguiar, Luís Paulo Reis, Jorge Oliveira e Sá,
Joaquim Gonçalves, and Victor Carvalho

Deep Reinforcement Learning for Personalized Recommendation
of Distance Learning 597
Maroi Agrebi, Mondher Sendi, and Mourad Abed

An Approach to Assess Quality of Life Through Biometric Monitoring
in Cancer Patients 607
Eliana Silva, Joyce Aguiar, Alexandra Oliveira, Brígida Mónica Faria,
Luís Paulo Reis, Victor Carvalho, Joaquim Gonçalves,
and Jorge Oliveira e Sá

Big Data Analytics and Applications

Big Data Analysis in Supply Chain Management in Portuguese SMEs
"Leader Excellence" 621
Fábio Azevedo and José Luís Reis

Big Data Analytics and Strategic Marketing Capabilities: Impact
on Firm Performance 633
Omar Anfer and Samuel Fasso Wamba

Study of the Mobile Money Diffusion Mechanism
in the MTN-Cameroon Mobile Network: A Model Validation 641
Rhode Ghislaine Nguewo Ngassam, Jean Robert Kala Kamdjoug,
Sylvain Defo Wafo, and Samuel Fosso Wamba

Geospatial Modeling Using LiDAR Technology 654
Leyre Torre-Tojal, Jose Manuel Lopez-Guede, and Manuel Graña

ND-GiST: A Novel Method for Disk-Resident *k*-mer Indexing 663
János Márk Szalai-Gindl, Attila Kiss, Gábor Halász, László Dobos,
and István Csabai

Human-Computer Interaction

**Sustainable Physical Structure Design of AR Solution
for Golf Applications** 675
Egils Ginters

**The Effect of Multisensory Stimuli on Path Selection in Virtual
Reality Environments** 686
Guilherme Gonçalves, Miguel Melo, José Martins,
José Vasconcelos-Raposo, and Maximino Bessa

**The Impact of Gender, Avatar and Height in Distance Perception
in Virtual Environments** 696
Hugo Coelho, Miguel Melo, Frederico Branco, José Vasconcelos-Raposo,
and Maximino Bessa

Usability Evaluation of a Virtual Assistive Companion 706
Ana Luísa Jegundo, Carina Dantas, João Quintas, João Dutra,
Ana Leonor Almeida, Hilma Caravau, Ana Filipa Rosa,
Ana Isabel Martins, Alexandra Queirós, and Nelson Pacheco Rocha

**Internet Access in Brazilian Households: Evaluating the Effect
of an Economic Recession** 716
Florângela Cunha Coelho, Thiago Christiano Silva, and Philipp Ehrl

**The Impact of the Digital Economy on the Skill Set
of High Potentials** 726
Mariana Pinho Leite, Tamara Mihajlovski, Lars Heppner,
Frederico Branco, and Manuel Au-Yong-Oliveira

**Designing a Dual-Layer 3D User Interface for a 3D Desktop
in Virtual Reality** 737
Hind Kharoub, Mohammed Lataifeh, and Naveed Ahmed

**The Control of a Vehicular Automata Through Brain Waves.
A Case Study** 748
Christian Ubilluz, Ramiro Delgado, Priscila Rodríguez, and Roberto Lopez

**Usability Study of a Kinect-Based Rehabilitation Tool
for the Upper Limbs** 755
Gabriel Fuertes Muñoz, Jesús Gallardo Casero,
and Ramón A. Mollineda Cárdenas

Leadership and Technology: Concepts and Questions 764
Ana Marisa Machado and Catarina Brandão

**Keystroke and Pointing Time Estimation for Touchscreen-Based
Mobile Devices: Case Study Children with ASD** 774
Angeles Quezada, Margarita Ramirez Ramírez, Sergio Octavio Vázquez,
Ricardo Rosales, Samantha Jiménez, Maricela Sevilla, and Roberto Muñoz

**The Use of Virtual Cues in Acquired Brain Injury Rehabilitation.
Meaningful Evidence** 785
Sergio Albiol-Pérez, Alvaro-Felipe Bacca-Maya,
Erika-Jissel Gutierrez-Beltran, Sonsoles Valdivia-Salas,
Ricardo Jariod-Gaudes, Sandra Cano, and Nancy Jacho-Guanoluisa

**Problematic Attachment to Social Media: Lived Experience
and Emotions** 795
Majid Altuwairiqi, Theodoros Kostoulas, Georgina Powell, and Raian Ali

**Gender Differences in Attitudes Towards Prevention and Intervention
Messages for Digital Addiction** 806
John McAlaney, Emily Arden Close, and Raian Ali

**Evaluating of Mobile Applications and the Mental Activation
of the Older Adult** 819
Maricela Sevilla, Ángeles Quezada, Consuelo Salgado, Ricardo Rosales,
Nora Osuna, and Arnulfo Alanis

**SmartWalk Mobile – A Context-Aware m-Health App for Promoting
Physical Activity Among the Elderly** 829
David Bastos, José Ribeiro, Fernando Silva, Mário Rodrigues, Rita Santos,
Ciro Martins, Nelson Rocha, and António Pereira

**A Usability Analysis of a Serious Game for Teaching Stock Market
Concepts in Secondary Schools** 839
B. Amaro, E. Mira, L. Dominguez, and J. P. D'Amato

Ethics, Computers and Security

Macro and Micro Level Classification of Social Media Private Data . . . 853
Paul Manuel

**Implementation of Web Browser Extension for Mitigating
CSRF Attack** 867
Saoudi Lalia and Kaddour Moustafa

What Does the GDPR Mean for IoToys Data Security and Privacy? . . . 881
Esperança Amengual, Antoni Bibiloni, Miquel Mascaró,
and Pere Palmer-Rodríguez

**Factors Influencing Adoption of Information Security in Information
Systems Projects**   890
Landry Tafokeng Talla and Jean Robert Kala Kamdjoug

Privacy Preserving *k*NN Spatial Query with Voronoi Neighbors   900
Eva Habeeb, Ibrahim Kamel, and Zaher Al Aghbari

Analysis of Authentication Failures in the Enterprise   911
Richard Posso and Santiago Criollo-C

**Digital Addiction: Negative Life Experiences and Potential
for Technology-Assisted Solutions**   921
Sainabou Cham, Abdullah Algashami, Manal Aldhayan, John McAlaney,
Keith Phalp, Mohamed Basel Almourad, and Raian Ali

Improving Cross-Border Educational Services with eIDAS   932
Tomaž Klobučar

**Correction to: Keystroke and Pointing Time Estimation for
Touchscreen-Based Mobile Devices: Case Study Children with ASD** . . .   C1
Angeles Quezada, Margarita Ramirez Ramírez, Sergio Octavio Vázquez,
Ricardo Rosales, Samantha Jiménez, Maricela Sevilla, and Roberto Muñoz

Author Index   939

Software Systems, Architectures, Applications and Tools

Multi-agent Neural Reinforcement-Learning System with Communication

David Simões[1(✉)], Nuno Lau[1], and Luís Paulo Reis[2]

[1] DETI/IEETA, Institute of Electronics and Informatics Engineering of Aveiro, University of Aveiro, Aveiro, Portugal
{david.simoes,nunolau}@ua.pt
[2] LIACC/FEUP, Artificial Intelligence and Computer Science Laboratory, Faculty of Engineering, University of Porto, Porto, Portugal
lpreis@fe.up.pt

Abstract. Deep learning models have as of late risen as popular function approximators for single-agent reinforcement learning challenges, by accurately estimating the value function of complex environments and being able to generalize to new unseen states. For multi-agent fields, agents must cope with the non-stationarity of the environment, due to the presence of other agents, and can take advantage of information sharing techniques for improved coordination. We propose an neural-based actor-critic algorithm, which learns communication protocols between agents and implicitly shares information during the learning phase. Large numbers of agents communicate with a self-learned protocol during distributed execution, and reliably learn complex strategies and protocols for partially observable multi-agent environments.

Keywords: Multi-agent systems · Neural networks ·
Emergent communication

1 Introduction

One of the most complex challenges in multi-agent systems (MAS) fields is achieving coordination between agents. A recent increase in hardware capabilities and the re-emergence of neural networks as policy approximators for highly complex environments has led to great results in reinforcement learning, albeit mostly in the single-agent paradigm [2]. One of the most popular technique for MAS is to apply single-agent algorithms and showing that successful policies can be learning with implicit coordination, despite losing theoretical guarantees in non-stationary environments.

Despite this, such techniques are infeasible in situations where information must be explicitly shared to achieve coordination, or where communication is simply useful to complete the scenario's goals. Recent proposals have tackled this issue, focusing on information-sharing implicitly during the learning phase

© Springer Nature Switzerland AG 2019
Á. Rocha et al. (Eds.): WorldCIST'19 2019, AISC 931, pp. 3–12, 2019.
https://doi.org/10.1007/978-3-030-16184-2_1

or explicitly communicating during execution. Communication protocols are not simply hard-coded or derived from existing symbols, but are instead learned *tabula rasa*.

We propose a neural-based reinforcement-learning system, which we call Asynchronous Advantage Actor-Centralized-Critic with Communication (A3C3). We explicitly share information with learned communication protocols during execution, and implicitly share information during the learning phase through the use of a centralized critic. We describe the algorithm and its basis, and compare the advantages and differences of our proposal against the original algorithm in complex environments that require information sharing.

The remainder of this paper is structured as follows. Section 2 lists related work, such as reinforcement or supervised learning techniques for communication, and information sharing techniques through centralized critics. Section 3 introduces and formally describes A3C3, as well as its architecture, modules and limitations. Section 4 shows the results of our proposal obtained in complex environments. Finally, Sect. 5 draws conclusions and lists future work directions.

2   Related Work

Many recent approaches to MAS learning have shown promising results, and focus on learning policies with communication protocols, or policies with centralized learning for implicit coordination. Some techniques generalize better in varied environments, and some have unrealistic assumptions or impractical drawbacks.

The Counterfactual Multi-Agent Policy Gradients (COMA) [4] is an actor-critic extension that supports distributed execution, but requires centralized training. This centralized-learning, distributed-execution framework follows the intuition that algorithms (the value network, in this case) can be augmented with extra information regarding the other agents during the learning phase, while during execution only local information is required, thus allowing agents to run in a decentralized manner. COMA uses the same centralized value network for all agents, with the shared agent observations and their actions as input. The use of agent actions as inputs for the value networks means the environment is now stationary for the critic, even as policies change. COMA addresses the credit-assignment problem by comparing how each agent's action effectively affects the expected value (using the critic network to estimate this). The critic's complexity depends on the amount of agents being trained, and it remains unclear how the network handles varying numbers of agents with this shared observation, and how it scales to large numbers of agents. This approach does not use communication, but instead is based in implicit coordination. Due to the lack of communication, agents may not be able to account for a partially observable environment. Using the same centralized critic for all agents also means the algorithm does not support different reward functions for different agents (like in competitive games).

Multi-Agent Actor-Critic for Mixed Cooperative-Competitive Environments (MADDPG) [8] is another actor-critic extension that supports distributed execution, but requires centralized training, since it also augments the critic with additional information. MADDPG uses a value network for each agent, with the shared agent observations and their actions as input, which allows for agents with different reward functions to learn together (any non fully cooperative environment, for example). The use of agent actions as inputs for the value networks means the environment is now stationary for the critic, even as policies change. Similarly to COMA, agents may not be able to account for a partially observable environment.

CommNet [10] is an algorithm that passes messages between layers of its neural network, where each message is the average of all the messages sent by all other agents. The algorithm learns the agent's policy and continuous communication protocol simultaneously, and requires a differentiable communication channel. It also contains several communication steps between each environment state, which is an uncommon assumption, but it allows the execution of a varying amount of agents. Actions for all agents are output simultaneously, similarly to the Joint-Action Learners (JAL) method [1], which prevents the distributed execution of agents, and may also lead to scalability issues.

Differentiable Inter-Agent Learning (DIAL) [3] handles partial observability with a neural network for message passing through a differentiable communication channel. Messages are discretized and coupled with state observations in a Q-network, and gradients are pushed through the communication channels to optimize the message network. While this algorithm requires centralized learning, an extension based in a memory buffer [5] has been proposed to address the issue.

3   Asynchronous Advantage Actor Centralized-Critic with Communication

Asynchronous algorithms running multiple threads simultaneously in multi-core CPU have outperformed single-threaded GPU-based algorithms [9]. They keep global networks which are updated by multiple workers asynchronously. The Asynchronous Advantage Actor-Critic (A3C), for example, keeps a policy network $\pi(a_t|s_t; \theta)$ and a value network $V(s_t; \theta_v)$, updating them with mini-batches of samples from each worker.

Based on A3C, we propose Asynchronous Advantage Actor-Centralized-Critic with Communication (A3C3), a distributed asynchronous actor-critic algorithm in a multi-agent setting with differentiable communication and a centralized critic, as can be seen in Fig. 1. It combines an implicit information sharing mechanism (the centralized critic) with explicit communication protocols (through differentiable communication), in order to ensure reliability and robustness during the learning phase, and allow agents to handle partially-observable domains, thus improving the coordination level of the team. Each agent has a global actor, critic and communication network, and each worker keeps local copies of the global networks. Workers update these global networks with mini-batches and periodically copy them into their local networks, as can be seen in Fig. 2. A larger amount of workers leads to an increase in the convergence speed of our algorithm.

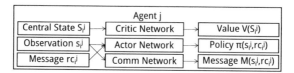

Fig. 1. The architecture of an agent j at time-step i, using three separate networks: a policy (or actor) network, which outputs an action probability $\pi(s_i^j; rc_i^j)$ (from which a_i^j is sampled) based on a given local observation s_i^j and received messages rc_i^j from other agents; a communication network, which outputs an outgoing message sc_i^j based on a given local observation s_i^j; and a value (or critic) network, which outputs a value estimation based on a given centralized observation S_i^j.

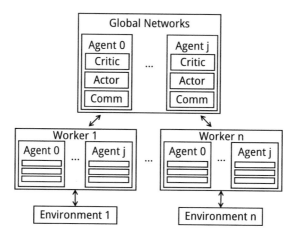

Fig. 2. A3C3 architecture, using n separate workers. Each worker interacts with its own environment and its separate set of j agents. As samples are collected in mini-batches, workers asynchronously update the global networks, and copy those weights into their local networks.

A3C3 is formally described in Algorithm 1. Each worker copies the global networks into local memory and samples the current environment observation. For each worker, an action and outgoing message are computed. Actions are synchronously performed, and a new observation is sampled, as well as an action-state reward. When a terminal state is reached, or a sufficient amount of observations have been sampled, the loss of the local networks is calculated. These gradients are then applied to the global networks, whose weights are then copied into local memory once more. These steps are repeated until convergence is achieved. If the number of agents $J = 1$, there is no communication, and the centralized critic uses the agent's local observation, the algorithm becomes A3C [9]. In other words, A3C3 is a more general version of A3C.

Within each worker, each agent j learns its own policy network θ_a^j, by sampling the environment at time-step t for their own observation s_t^j, and using it as input along with received messages rc_t^j. The environment may or may not be partially observable, and each agent observation may be a local partial observation of the environment's current state. After all agents have sampled the environment and determined an action a_t^j to follow, based on the policy $\pi(a_t^j|s_t^j, rc_t^j, \theta_a^j)$, they execute it on the environment, and a new cycle starts. After $t_{\max}$ time-steps, or if a terminal state is reached, the policy network is optimized with the error between the actual returns and the value network's output $(R - V(S_i^j, \theta_v^j))$, for all time-steps i.

A centralized value network θ_v^j for each agent j is used, whose input is a centralized observation S_t^j of the environment. This centralized observation is environment dependent, and can be the environment state, or a concatenated observation of the environment from each agent's local perspective. This centralized observation S_t^j can also contain the actions of all other agents $k, \forall k \neq j$, the messages received by the current agent j, or both. After each mini-batch, if a terminal state has not been reached, the value network is used to bootstrap the expected value $V(S_t^j, \theta_v^j)$ for the next state S_t^j. The error of the value networks is calculated using the squared error between the actual returns and the value network's output $(R - V(S_i^j, \theta_v^j))^2$, for all time-steps i.

Each agent also learns its own communication network θ_c^j. The communication protocol is problem-dependent, and determines the size of messages, their range, to what agents they are delivered, their reliability, among others. The properties of the communication protocol are captured within a communication map, which lists which messages has each agent received from other agents. A standard value rc_{initial} for a non-received message is necessary for the agent's policy network. At time-step t, after all agents j have sampled an observation s_t^j from the environment and determined a message sc_t^j to send, the message is sent, and a new cycle with time-step $t + 1$ starts. Here, the messages sc_t^j sent by agent j on the previous cycle are received as rc_{t+1}^j by other agents. After $t_{\max}$ time-steps, or if a terminal state is reached, the communication network is optimized with the policy network's loss L_a^j. Gradients are calculated with respect to the received messages, and mapped as loss of sent messages for the communication network.

If agents are homogeneous, agents can use parameter sharing between their networks, to speed up the learning process. This allows the same actor, critic, or communication network to be learned by all agents simultaneously.

4   Results

We test our proposal in two complex games where communication is crucial to achieve good results, a Traffic Intersection simulator, and a Pursuit challenge. We test multiple combinations of communication channels (from none to

Input: Global shared learning rate η, discount factor γ, entropy weight β, number of agents J, network weights θ_a^j, network weights θ_v^j, network weights θ_c^j, batch size $t_{\max}$, maximum iterations $T_{\max}$, and default message value rc_{initial}

Input: Local network weights ϑ_a^j, network weights ϑ_v^j, network weights ϑ_c^j, and step counter t

1:   $t \leftarrow 0$
2:   $rc_0^j \leftarrow rc_{\text{initial}}$ for all agents j
3: **for** iteration T $= 0, T_{max}$ **do**
4:      Reset gradients $d\theta_a^j \leftarrow 0$, $d\theta_v^j \leftarrow 0$, and $d\theta_c^j \leftarrow 0$ for all agents j
5:      Synchronize $\vartheta_a^j \leftarrow \theta_a^j$, $\vartheta_v^j \leftarrow \theta_v^j$, $\vartheta_c^j \leftarrow \theta_c^j$ for all agents j
6:      $t_{start} \leftarrow t$
7:      Sample observation s_t^j for all agents j
8:      Sample or derive centralized observation S_t^j for all agents j
9:      **repeat**
10:         **for** agent $j = 1, J$ **do**
11:             Calculate message sc_t^j to send with $sc_t^j \leftarrow M(s_t^j, rc_t^j, \vartheta_c^j)$
12:             Sample action a_t^j according to policy $\pi(a_t^j | s_t^j, rc_t^j, \vartheta_a^j)$
13:             Map sent communication sc_t^j into received communication rc_{t+1}, and build communication map m_t
14:         **end for**
15:         Take action a_t^j for all agents j
16:         Sample reward r_t^j and new observation s_{t+1}^j for all agents j
17:         Sample or derive centralized observation S_{t+1}^j for all agents j
18:         $t \leftarrow t + 1$
19:      **until** terminal S_t^j for all agents j or $t - t_{start} = t_{max}$
20:      **for** agent $j = 1, J$ **do**
21:         $R^j = \begin{cases} 0 \text{ for terminal state} S_t^j \\ V(S_t^j, \vartheta_v^j) \text{ otherwise} \end{cases}$
22:         $L_c \leftarrow 0$
23:         **for** step $i = t - 1, t_{start}$ **do**
24:             $R \leftarrow r_i^j + \gamma R$
25:             Value loss $L_{v_i}^j \leftarrow (R - V(S_i^j, \vartheta_v^j))^2$
26:             Actor loss $L_{a_i}^j \leftarrow \log \pi(a_i^j | s_i^j, rc_i^j, \vartheta_a^j)(R - V(S_i^j, \vartheta_v^j)) - \beta * H(\pi(a_i^j | s_i^j, rc_i^j, \vartheta_a^j))$
27:         **end for**
28:         **for** step $i = t, t_{start} + 1$ **do**
29:             Received communication loss $L_{rc_i}^j \leftarrow \dfrac{\partial L_{a_i}^j}{\partial rc_i^j}$
30:         **end for**
31:      **end for**
32:      Map received communication loss $L_{rc_{i+1}}$ into sent communication loss L_{sc_i} using communication map m_i for all agents
33:      **for** agent $j = 1, J$ **do**
34:         **for** step $i = t - 1, t_{start}$ **do**
35:             Accumulate gradients $d\theta_c^j \leftarrow d\theta_c^j + \dfrac{\partial L_{sc_i}^j}{\partial \vartheta_c^j}$
36:             Accumulate gradients $d\theta_a^j \leftarrow d\theta_a^j + \dfrac{\partial L_{a_i}^j}{\partial \vartheta_a^j}$
37:             Accumulate gradients $d\theta_v^j \leftarrow d\theta_v^j + \dfrac{\partial L_{v_i}^j}{\partial \vartheta_v^j}$
38:         **end for**
39:      **end for**
40:      Update network weights $\theta_a^j \leftarrow \theta_a^j + \eta d\theta_a^j$, $\theta_v^j \leftarrow \theta_v^j + \eta d\theta_v^j$, and $\theta_c^j \leftarrow \theta_c^j + \eta d\theta_c^j$ for all agents j
41: **end for**

Algorithm 1. Pseudo-code for a worker thread running A3C3.

twenty), and critics (using only agents' local partial observations, or centralized). A demonstration video is available at https://youtu.be/0eJFEUlMwKE.

For our tests, the actor, critic, and central critic networks had two fully connected hidden layers of $40x$ and $20x$ nodes activated with a ReLU function, where x represents a problem-dependent layer size modifier, and the communication network had a single fully connected hidden layer of $20x$ nodes activated with a hyperbolic tangent function, with a non-received message rc_{initial} default value being all zeros. This was the simplest architecture we found that consistently converged to proper policies, and allowed us increase the speed of our tests. All networks' initial random weights were computed with the Xavier initializer [6] with default parameters, were optimized with the Adam optimizer [7] with default parameters, and we used an entropy weight $\beta = 0.01$. The learning rate η chosen was also the highest and fastest we found that consistently allowed the networks to converge. The future reward discount factor γ depends on the importance of future rewards, with 0 leading to policies that focus on short-term rewards, and 1 focusing on long-term expected rewards.

4.1 Traffic Intersection

The Traffic Intersection simulator consists on several four-way road intersections (agent flow from north or west to south or east), which must be crossed by multiple cars. Each agent has a local partial observation of the environment, sensing the cars immediately around themselves, and knowing their intended direction. Agents get small penalties if they do not move, get heavily penalized if they crash, and cars that are turning have priority (heavier penalty for non-turners). Agents can only communicate with the cars within their vision range. For a central critic, the state space consists on each agent's observation, and the desired direction of other visible cars. The optimal strategy is for a car to inform others at intersections whether it needs to turn or not, and if both vehicles would follow the same direction, the car without priority awaits until the intersection is clear.

For our learning phase, we used a team of 20 agents, and a total of 6 by 6 intersecting roads. Agents receive individual rewards, -1 points for stopping or ramming the front car, -10 points if they collide at an intersection without having priority, and -5 points if they collide at an intersection while having priority. We used a learning rate $\eta = 10^{-4}$, a future reward discount factor $\gamma = 0.9$, a network layer size multiplier $x = 1$, and $N = 3$ concurrent worker threads. Despite learning with 20 agents, the policy can be applied to different scenarios with hundreds of agents or roads.

We can see in Fig. 3(a) that, without communication, no suitable policy is found, since agents cannot communicate whether they have to turn or not. This leads to agents always trying to cross the intersection, and randomly colliding, thus averaging at around 30 points. Using a centralized critic increases the complexity of the policy and does not improve it. However, once we include communication channels (a single one is enough), we achieve the optimal policy. Increasing the amount of communication channels makes the optimal policy slower to reach with a decentralized critic, but has the opposite effect with centralized one, where using five communication channels gives the best results. Interestingly enough, agent populations can learn opposite protocols, either indicating that they have to turn, or that they have to move forward, depending on the randomly initialized weights.

4.2   Pursuit

The Pursuit game, also known as Predator/Prey, consists on two teams of agents, where one team must capture the other. The prey team is hard-coded, and each prey runs from the closest predator. Each predator has a local partial observation of the environment, and is able to see a small area around itself as well as its global coordinates. Agents get small penalties as time passes, and get penalized and randomly placed if they collide. At each time-cycle, agents can move in four directions or remain in the same position. Agents can broadcast messages to all other agents. For a central critic, the state space consists on the global observation of both predator and prey's positions. A high-level strategy is for predators to explore the map until a prey is found, and then broadcast the prey's position so that all predators can converge and capture it.

For our tests, we used a team of four predators and four prey in a toroidal map with a time-limit. Prey have global vision, and predators have a local perspective around themselves, equivalent to less than 10% of the total map. Predators receive joint rewards, 1 point with each captured prey, and -1 points for each collision, which randomly places them on the map. Because episodes are longer, and policies harder to learn, we used a learning rate $\eta = 5 * 10^{-5}$, a future reward discount factor $\gamma = 0.95$, a network layer size multiplier $x = 3$, and $N = 12$ concurrent worker threads.

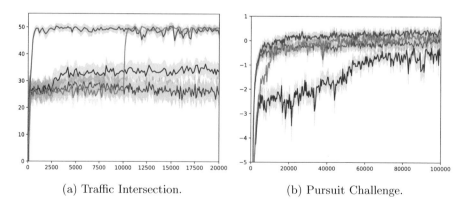

(a) Traffic Intersection.   (b) Pursuit Challenge.

Fig. 3. Results of A3C3 for the tested environments. The plots represent the average reward obtained by independent agents with no communication and no centralized critic (purple), no communication and a centralized critic (red), communication and no centralized critic (green), and communication and a centralized critic (blue), over training episodes.

We can see in Fig. 3(b) that, without communication and a centralized critic, a long time is taken for a suitable policy to be found, since finding prey and coordinating with other predators happens mostly by chance. However, with communication enabled, a good policy is quickly found, since searching for prey is no longer random, and there is also now a communication protocol for coordinating the capture. We can also see how a centralized critic helps accelerating the learning process and achieving a better policy.

5   Conclusion

We propose a reinforcement-learning neural-based system, the Asynchronous Advantage Actor-Centralized-Critic with Communication (A3C3) algorithm. It is a distributed algorithm, with asynchronous updates by worker threads, based in actor-critic methods, for multi-agent settings, capable of handling partially-observable environments. Information is implicitly shared during the learning phase through the use of a centralized critic, and explicitly shared with learned communication protocols during execution. We require differentiable communication and a centralized learning phase, but we support distributed independent execution.

We formally described A3C3 and demonstrated how it achieves successful policies in complex multi-agent partially-observable domains. We tried environments where agents only observe a small fraction of the environment, and where communication is necessary for high-level policies. We show how we can scale our algorithm to large numbers of agents, while still allowing their distributed execution. We showed how a centralized critic helps increase the robustness and speed of the learning process, achieving better policies and communication protocols. We also test different inputs on our critic networks, but do not find any combination to meaningfully improve our solution. Our end-to-end differentiable distributed algorithm is a more general version of previous works, and we have published our source-code at https://github.com/bluemoon93/A3C3.

Future work directions include a coordination protocol such that agents and worker threads can be run on different machines (to further increase the scalability of our system), exploring how memory channels can be exploited by having agents sending messages to themselves in future time-steps, and utilizing Long Short-Term Memory units for the communication network architecture (such that agents not only learn what to communicate, but also what to ignore from received messages). The exploration of additional environments, different parameters, network architectures, team sizes, and messaging protocols would also provide valuable information regarding the versatility of our proposal.

Acknowledgements. The first author is supported by FCT (Portuguese Foundation for Science and Technology) under grant PD/BD/113963/2015. This research was partially supported by IEETA (UID/CEC/00127/2013) and LIACC (PEst-UID/CEC/00027/2013).

References

1. Claus, C., Boutilier, C.: The dynamics of reinforcement learning in cooperative multiagent systems. In: Proceedings of the Fifteenth National/Tenth Conference on Artificial Intelligence/Innovative Applications of Artificial Intelligence, AAAI 1998/IAAI 1998, pp. 746–752. American Association for Artificial Intelligence, Menlo Park (1998). http://dl.acm.org/citation.cfm?id=295240.295800
2. Firoiu, V., Whitney, W.F., Tenenbaum, J.B.: Beating the world's best at super smash bros. with deep reinforcement learning. CoRR abs/1702.06230 (2017). http://arxiv.org/abs/1702.06230
3. Foerster, J.N., Assael, Y.M., de Freitas, N., Whiteson, S.: Learning to communicate with deep multi-agent reinforcement learning. CoRR abs/1605.06676 (2016). http://arxiv.org/abs/1605.06676
4. Foerster, J.N., Farquhar, G., Afouras, T., Nardelli, N., Whiteson, S.: Counterfactual multi-agent policy gradients. CoRR abs/1705.08926 (2017). http://arxiv.org/abs/1705.08926
5. Foerster, J.N., Nardelli, N., Farquhar, G., Torr, P.H.S., Kohli, P., Whiteson, S.: Stabilising experience replay for deep multi-agent reinforcement learning. CoRR abs/1702.08887 (2017). http://arxiv.org/abs/1702.08887
6. Glorot, X., Bengio, Y.: Understanding the difficulty of training deep feedforward neural networks. In: Teh, Y.W., Titterington, M. (eds.) Proceedings of the Thirteenth International Conference on Artificial Intelligence and Statistics. Proceedings of Machine Learning Research, vol. 9, PMLR, Chia Laguna Resort, Sardinia, Italy, 13–15 May 2010, pp. 249–256 (2010). http://proceedings.mlr.press/v9/glorot10a.html
7. Kingma, D.P., Ba, J.: Adam: a method for stochastic optimization. CoRR abs/1412.6980 (2014). http://arxiv.org/abs/1412.6980
8. Lowe, R., Wu, Y., Tamar, A., Harb, J., Abbeel, P., Mordatch, I.: Multi-agent actor-critic for mixed cooperative-competitive environments. CoRR abs/1706.02275 (2017). http://arxiv.org/abs/1706.02275
9. Mnih, V., Badia, A.P., Mirza, M., Graves, A., Lillicrap, T.P., Harley, T., Silver, D., Kavukcuoglu, K.: Asynchronous methods for deep reinforcement learning. CoRR abs/1602.01783 (2016). http://arxiv.org/abs/1602.01783
10. Sukhbaatar, S., Szlam, A., Fergus, R.: Learning multiagent communication with backpropagation. CoRR abs/1605.07736 (2016). http://arxiv.org/abs/1605.07736

Construction and Integration of a Quadcopter in a Simulation Platform for Multi-vehicle Missions

Leonardo Ferreira[(✉)], Álvaro Câmara, and Daniel Castro Silva

LIACC - Artificial Intelligence and Computer Science Laboratory,
FEUP - Faculty of Engineering, University of Porto, Porto, Portugal
{up201305980,up201300764,dcs}@fe.up.pt

Abstract. Unmanned Aerial Vehicles or UAVs have been growing at a rapid pace due to the huge range of operations they can perform on the field. These vehicles are currently present in military applications, agricultural, rescue missions or telecommunications, among many other fields. However, some of the existing UAV solutions in the market are quite costly, not only monetarily but also in the complexity they entail.

The objective of this work is to build a low-cost quadcopter (a type of UAV) and integrate it in a multi-vehicle mission simulation platform. The quadcopter is built with the help of a mounting guide and must be able to perform all basic flight operations. The simulation platform provides a flexible configuration of missions and test scenarios that allows to study the performance of the UAVs. After the integration, we will be able to analyze the interactions between real and simulated UAVs. Several tests (individual and collective) on the components used and interaction with the simulation platform are taken into account to evaluate the performance of the flight and integration with the simulation platform. With the tuning and testing done, it was possible to perform a brief flight and the virtual quadcopter managed to follow the real quadcopter's movement. With these experiments, we can conclude that it is possible to mount a low-cost quadcopter without sacrificing flight or integration in other systems.

Keywords: Unmanned Aerial Vehicles (UAVs) ·
Unmanned Aerial System (UAS) · Simulation · Quadcopter ·
Autonomous vehicles

1 Introduction

Since the advent of Unmanned Aerial Vehicles (UAVs), new methods and technologies have been developed that allowed the expansion of UAVs to several areas [9]. Thanks to recent developments in Aeronautics, several companies and other entities have sought to invest and explore the potential of UAVs [6]. These vehicles are excellent because have unique characteristics (size and characteristics)

© Springer Nature Switzerland AG 2019
Á. Rocha et al. (Eds.): WorldCIST'19 2019, AISC 931, pp. 13–23, 2019.
https://doi.org/10.1007/978-3-030-16184-2_2

that allow them to perform a variety of tasks [1]. Despite some legal problems and the high number of crashes that occurred in the past decade (more than 40 registered accidents), UAVs are very useful in society and play an important role in many countries. They have the ability to reach remote areas with little or no required workforce, requiring little effort, time and energy. However, some of the existing solutions capable of performing these demanding tasks are expensive and difficult to handle due to several features they bring with them [11].

The goal of the work reported herein is to build a low-cost quadcopter and integrate it in a simulation platform for multi-vehicle missions. This work is part of a larger work [12] whose purpose is to use autonomous vehicles to carry out missions of surveillance, search and rescue, airspace management, restoration of communications in disaster situations, among others. This work is divided in two phases: build and configure a quadcopter for flight; and integrate the quadcopter into the multi-vehicle mission simulation platform.

The first phase includes the assembly of the quadcopter and the configuration of its flight controller. The second phase includes setting up the simulation platform to support the use of UAVs, and to integrate the previously built quadcopter in it, establishing a communication protocol between the real and virtual quadcopters. The interaction between real and virtual UAVs is also studied at this phase.

The remainder of this paper is organized as follows. In Sect. 2, we address the state of the art on UAVs. Section 3 introduces the approach followed, and Sects. 4 and 6.2 are dedicated to each of the two phases of this work, namely the assembly of the vehicle and integration in a simulation platform. Section 6 presents the experimental setup and test results, and finally some conclusions and lines of future work are presented in Sect. 7.

2   State of the Art

The state of the art of this work is presented from two different perspectives: low-cost quadcopters and available software; and studies that combine real UAVs with simulated ones.

2.1   Low-Cost Quadcopters

There are several low-cost options available in the market; however, we focused on the handmade ones since our intention is also to build a low-cost quadcopter from scratch. We also refer to some software important for flight operations. The four quadcopters presented in Table 1 were the most simple quadcopter found. Each option is analyzed based on controller board, software and price. Each one has features that makes it ideal depending on the application and the performance scenario. Project YMFC-AL and YMFC-32 are two very similar projects developed to fly the drone for entertainment. The Comelicottero option was also built for flight but while the previous are commanded by a radio controller this is managed through WIFI from any PC. The RZtronics Project distinguishes itself by using any smartphone via Bluetooth to control the flight. Project YMFC-AL was used to manage our flight (see Sect. 3.2) because it is very well documented and possess guides (very important, specially for someone who doesn't have experience with hardware).

Table 1. Low-cost quadcopters

Quadcopter	Controller board	Software	Price (dollars)
Project YMFC-AL[a]	UNO	YMFC-AL (handmade)	150
Comelicottero[b]	Yun	Comelicottero (handmade)	No information
Project YMFC-32[c]	NANO	STM32	135
Quadcopter RZtronics[d]	UNO	MultiWii	120–150

[a]Source: http://www.brokking.net/ymfc-al_main.html
[b]Source: https://blog.arduino.cc/2015/06/08/building-a-quadcopter-running-on-arduino-yun/
[c]Source: http://www.brokking.net/ymfc-32_main.html
[d]Source: http://rztronics.com/arduino-flight-controller-for-your-quadcopter/

2.2   Similar Studies

There are some works that use properties similar to the ones we implemented. Veloso proposed a simulation platform to assist in the design and development of UAVs [13]. The simulation platform supports decision making simulated by multi-agent systems so that tasks such as avoiding collisions can be effectively performed. He explores the concept of symbiotic simulation, the connection between a real quadcopter and a virtual one. Through the implementation and integration of several technologies, like symbiotic simulation, co-simulation and multi-agent systems, it was possible to create a platform that optimizes results in the decision making of the UAV agents in the virtual environment, as in the real one. Despite some data packages losses and failures when updating the virtual quadcopter position, the results were considered a success.

Another approach is presented by Leite [7]. He proposed the improvement of some features presented by the control platform DroidPlanner, by managing multiple UAVs simultaneously. The DroidPlanner control station is an application that controls the UAV flight and operates on the basis of a message exchange system between a mobile device (preferably Android) and a UAV. The messages sent by DroidPlanner to the UAV contain commands to execute. After being received and interpreted by the UAV, it executes the commands and sends feedback and information about its current state to the control platform, so that it can update the graphical interface. Although the results were successful, the system crashes when there is more than one UAV communicating with the platform (due to the high volume of messages).

Cai et al. studied hardware-in-the-loop (HIL) simulation systems for small-scale UAV helicopters [3]. HIL is a technique used to test, verify and validate the real-time performance of complex systems. The framework explored is structured in four modules: onboard hardware (activates helicopter servo actuators and output sensors), flight control (executes automatic control algorithms), ground station module (monitors the helicopter through data view and 3D view interfaces) and software (integrates the previous modules to perform real-time hardware-in-the-loop simulation). They were tested on small-scale UAV helicopters (HeLion and SheLion) and the results proved that the framework is able to effectively predict several situations of potential danger, despite some failures that occur when towards extreme conditions.

Meyer et al. proposed a system for comprehensive simulation of quadrotor UAVs through ROS and Gazebo simulator [8]. This approach allows simultaneous simulation of diverse aspects such as flight dynamics, onboard sensors like IMUs, external imaging sensors and complex environments. The dynamic model explored was tested with wind tunnels and validated by comparing simulated and real data.

Carvalho [5] tries to solve some problems associated with low autonomy and adaptation to different cargoes on quadrotors. She proposed a realistic simulation, where different solutions (like hybrid approaches) can be tested and validated. It takes in consideration a sensing part, which simulates sensors and through the readings estimates the condition of the vehicle, and also a control part, which is in charge of moving the vehicle to a given position and orientation reference. The results were the expected and like the previous approach, were validated by comparing simulated and real data.

3 Problem Approach

The main objective of this work is to assemble and integrate a quadcopter into a simulation platform for multi-vehicle missions.

3.1 System Architecture

The fact that this work follows other proposals [4] made it ideal to continue with several technologies still in use. The simulation platform, underpinned by the principles outlined in [12], has been developed to be a more complete and low-cost solution when compared to others available in the market. It is composed of several modules (about 10), but this work focuses only on a few of them, as illustrated in Fig. 1.

All these modules communicate and are dependent on each other. The main component in Fig. 1 is the Simulator (FSX) which acts as a simulation engine capable of simulating and monitoring the various events that may occur in a simulated environment. It is interconnected with the other modules through Simconnect, providing information to various types of agents and thus enabling them to make decisions.

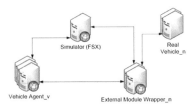

Fig. 1. Architecture of some modules of the simulation platform [12]

The Vehicle Agent module is one of the most important modules for the integration of the quadcopter in the simulation platform. It is responsible for creating and handling UAV agents, receive and send messages to the real UAV, update the virtual UAV position and many other features. The Real Vehicle module explores the assembly and configuration of the quadcopter and the control station (responsible for the connection between the real quadcopter and the simulation platform). The External Module Wrapper is a Vehicle Agent submodule which transforms real data into virtual data so the virtual UAV can move in the simulated environment.

3.2   Main Material and Software

The main material used to build the quadcopter was: F450 Quadcopter frame, 2 pairs of propellers, Battery ZOP Power 11.1V 2200MAH 3S 30C, 4 ESC 30A SimonK, 4 920kv Brushless engines, MPU6050 sensor, HC-SR04 sensor, GPS NEO 6M, 2 NRF24L01 Radio + PA + LNA, Arduino Uno and Arduino Mega2560 and FlySky FS-i6 Kit. For the flight controller software, the following were considered: ArduPilot[1], LibrePilot[2], MultiWii[3], DroneCode[4] and YMFC-ALL[5]. Since there was no need for a very elaborate flight, but rather an efficient one to show a sustainable flight, Project YMFC-All [2] was chosen as the flight controller. This software package is responsible for the controller board setup (Arduino Uno) and gyroscope calibration. It has a lot of documentation and examples of how to mount the quadcopter components. All the steps required to perform the setup and configuration are detailed on the software author's website [2].

4   Quadcopter and Ground Station Control Assembly

All the previously mentioned components were gathered and assembled, in order to build the quadcopter and the ground control station (GCS).

---

[1] Source: http://ardupilot.org/.
[2] Source: https://www.librepilot.org/site/index.html.
[3] Source: http://www.multiwii.com/.
[4] Source: http://autoquad.org/.
[5] Source: http://www.brokking.net/.

The quadcopter, shown in Fig. 2(a), uses an Arduino Mega2560, HC SR04 sensor, MPU6050 sensor, GPS NEO 6-M, NRF24 radio, ESCs, engines and propellers (not included in the image for safety measures). Once the building is finished, it is configured with a flight controller software (provided by YMFC-ALL software) so the quadcopter can fly. The installed sensors support the integration of the quadcopter in the simulation platform by collecting the information needed for the virtual UAV movement (altitude, heading, latitude and longitude).

As for the GCS, shown in Fig. 2(b), it requires only an Arduino UNO and NRF24 radio. It is a crucial component since it enables communication between the quadcopter and the simulation platform. It deals with real data provided by the quadcopter and commands sent by the simulation platform.

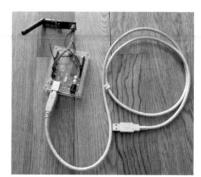

(a) Quadcopter   (b) Ground Station Control

Fig. 2. Assembled Hardware (Quadcopter and GCS)

5   Integration

This section describes the integration of the vehicle in the simulation platform, UAV agents, how the information flows from the real quadcopter to the simulation platform, and other aspects related to integration.

UAV agents differ from existing agents (aircraft, cars, boats and submarines) within the simulation platform because they can receive and interpret external data and send commands to the real UAV. The UAV agent representation in the simulation platform was done with an FSX model for quadcopter. However, models compatible with this type of vehicle weren't found on FSX model banks. Due to this barrier, we used a helicopter model since it is the vehicle with morphology and mechanics most similar to those of a quadcopter. Since the simulation is performed externally, and it is the external data that causes the virtual quadcopter to move, the fact that the internal flight model is very different from the real one has no influence on system tests. Another important aspect analyzed was the virtual UAV movement. Since FSX and SimConnect are not ready to handle UAV objects we had to put some delays in the UAV agent movement in order to control the movement smoothness. This will be explored below, in Sect. 6.

The real UAV sends to the simulation platform information regarding its latitude and longitude (captured by the GPS 6M), altitude (HC-SR04 sensor) and heading (MPU6050 sensor). The simulation platform sends the real quadcopter commands to be executed by quadcopter. The data is transmitted to the control ground station by serial port and then to the quadcopter with NRF24 radio.

The system suffered some changes to be able to handle real data (provided by the sensors), use STANAG as communication protocol and benefit from features such as the control of UAV Engines.

How the simulation platform handles the data from the ground control station and then transposes it into the virtual quadcopter is also an important question. The platform is equipped with two threads, one that extracts the information from the ground control station (quadcopter latitude, longitude, altitude and heading) and another that allows the virtual quadcopter to assume a new position/attitude based on the data collected.

The messages used in the communication follow the 4th edition of STANAG 4586 [10]. This document defines communication protocols and messages format to be exchanged between the vehicles and the ground control station. A STANAG message has several properties (source ID, Destination ID, Data, Message Type, Message Properties and others). In this project, three main STANAG commands were considered to notify the real quadcopter to execute certain events: Vehicle Steering (changes the heading), AV Position Waypoint (changes the quadcopter position) and Engine (turn on the engines). Some customization to these messages was done since some parameters are not relevant to the project and because it is not possible to discern which type of maneuver to perform. The possible maneuvers involve Move UP, Move Down, Move Forward, Move Backwards, Rotate, and Engine. For example, the Move Up has the structure: 2002|7|CMD_UP-altitude, where the altitude is given by the simulation platform.

The features developed involve testing the different movements that the virtual quadcopter is able to perform (Vertical Take Off, Vertical Land, Move Forward, Move Backwards and Rotate); use the real quadcopter sensors (GPS, MPU6050 and HC-SR04) on the simulation platform (when one of the attributes of the data changes, the virtual quadcopter moves with the movements mentioned before); send commands from the platform to the real quadcopter, so as to perform a certain maneuver.

6   Testing and Results Discussion

In this chapter, we discuss all the tests conducted and results obtained. The tests were performed on two levels: quadcopter flight and integration in the simulation platform.

6.1 Quadcopter Flight

Tests were carried out on the quadcopter flight in several domains, such as FlySky FS-i6 Transmitter Radio Inputs (test radio signals like thrust, yaw, roll and pitch); ESCs and engine's rotation (check if all ESCs work correctly and in tune); Gyroscope (verify if it is properly calibrated). It is imperative to do the best calibration possible since these components help the quadcopter achieve a better and sustainable flight. No new tests were conducted in this field because all these components are already being evaluated during the flight controller setup. Plus, these are explained and detailed in the Project-YMFC website [2].

6.2 Integration

Individual and collective tests were performed during the experiments. Individual tests study each component while collective ones test them all together.

The NRF24 radios, GPS and HC-SR04 were tested separately while the gyroscope wasn't, as it was already tested on the quadcopter flight. Regarding the NRF24 radios, the influence that the antenna has on the communication was studied. The tests without antenna are not so important because they limit the communication to a maximum distance of 100 m. After analyzing the results with antenna we concluded that communication time is proportional to distance, with an average communication time of 1700 ms. We also learned that the presence of WIFI Networks can disrupt the communication between the NRF24 radios.

Concerning the GPS, we tested in external and internal environments to see how scenarios affect the calibration time. Figure 3(a) shows that the GPS calibration time is larger in internal environment. We also learned that after the GPS calibrates it stays calibrated, that is, it can receive information from the satellite while powered by the battery.

The HC-SR04 sonar, used to retrieve the altitude, can retrieve information up to a maximum distance of 5m. It was tested on three scenarios that are distinguished by the distance that they are from an object. In both, the objective was to observe what distances the sonar receives when it encounters surfaces that show an inclination to the right or inclination to the left. As can be seen in Fig. 3(b) the first and third scenario present similar results while the second has worse results. This phenomenon is caused by the sensitivity and limitations that sonar presents when faced with high inclinations. The sonar obtains strange readings when the inclination is greater that 45°. This data is relevant because it indicates that the sonar used is not suitable to extract the altitude with uneven and sloping terrains.

As Collective tests we checked vertical movements of the UAV (NRF24 radio with HC-SR04 sensor); horizontal movements (NRF24 radio with GPS) and all the components above (NRF24 radio, GPS, HC-SR04 sensor and MPU6050). This last test was the most important since it evaluates the system with all needed components.

Three tests were performed with the mentioned components and with different types of movements. The movements are applied sequentially.

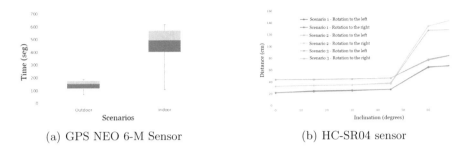

(a) GPS NEO 6-M Sensor   (b) HC-SR04 sensor

Fig. 3. Conducted tests

Table 2. System performance after all movements applied

Movement (cm)	Msg. rate (msg/sec)	Movement diff. (ms)
Up-20; Forward-300; Rotate-20°	13	2104
Down-20; Backwards-300; Rotate 60°	14	2452
Up-10; Backwards-300; Rotate-120°	13	2651

Other metrics such as data losses were studied but not included here because they depend on the NRF24 radio and serial port performance (external components).

Table 2 shows that when there is an increase on the angle rotation the time difference between the real and virtual movement increases as well. Despite the time obtained between the beginning of the actual movement and the end of the virtual movement was too high, it was somewhat expected. The introduction of the movement fluidity is the cause to this behaviour. As we mentioned in Sect. 6.2 to control the movement fluidity we had to put some delays to increase their smoothness. The results obtained are due to a fluid and progressive movement. We could decrease these times but this would result in a poorly perceptive movement, something we wouldn't want.

Another important tested feature that showed good results was sending commands to the real quadcopter. The quadcopter was able to receive all kind of commands from the simulation platform with the appropriate data.

7   Conclusions and Further Work

A low-cost quadcopter was built and several functionalities were developed to allow the integration of this UAV into an existing simulation platform. This vehicle was set up to test the flight and use data from sensors. A number of measures have been taken to comply with the rules imposed by the NATO STANAG 4586 and achieve the best results possible. In terms of UAV flight, we tried to calibrate as much as possible the components integrated in the quadcopter to obtain a more sustainable flight and prevent possible failures. We learn that factors like poorly calibrated gyroscope, poorly calibrated or defective ESCs, wrong engine rotation and bad weight distribution can affect the flight performance negatively. Regarding the integration, we developed a new kind of agent to represent UAVs in the simulated platform. This system is capable of simulating virtual agents that move in the simulated environment with the information provided by the real quadcopter sensors. Besides that, it provides a way to send commands to the real UAV. However, none of this would be possible without the ground station control, a new component build to support the communication between real UAV and simulation platform.

In the course of developing this project, new objectives were established, including improving the combination between the quadcopter flight with the simulation platform by introducing high level commands (go to point and other maneuvers). The use of multiple digital twins in the system as well as the improvement of virtual agent movement (pitch, roll and yaw) in the simulation platform are also important factors that should be explored in the future.

References

1. Austin, R.: Unmanned Aircraft Systems - Human Factors in Aviation. Aerospace Series, pp. 245–280. Wiley, May 2010
2. Brooking, J.: Project YMFC-AL - the arduino auto-level quadcopter (2015). http://www.brokking.net/ymfc-al_main.html. Accessed 20 May 2018
3. Cai, G., Chen, B.M., Lee, T.H., Dong, M.: Design and implementation of a hardware-in-the-loop simulation system for small-scale UAV helicopters. Mechatronics **19**, 1057–1066 (2009). https://doi.org/10.1016/j.mechatronics.2009.06.001
4. Câmara, A.: Air Traffic Control using Microsoft Flight Simulator X. Master's thesis, Department of Informatics Engineering, University of Coimbra, Coimbra, Portugal, July 2013
5. Carvalho, J.C.P.V.: Realistic Simulation of an Aerial Platform. Master's thesis, Faculdade de Engenharia da Universidade do Porto, July 2014. (in Portuguese)
6. Joshi, D.: Exploring the latest drone technology for commercial, industrial and military drone uses, July 2017. http://www.businessinsider.com/drone-technology-uses-2017-7. Accessed 3 Apr 2018
7. Leite, C.E.T.: Control Platform for Multiple Unmanned Aerial Vehicles (UAVs) (in Portuguese). Master's thesis, Federal University of Rio Grande do Sul (2015)
8. Meyer, J., Sendobry, A., Kohlbrecher, S., Klingauf, U., von Stryk, O.: Comprehensive simulation of quadrotor UAVs using ROS and Gazebo. In: Simulation. Modeling, and Programming for Autonomous Robots, pp. 400–411. Springer, Heidelberg (2012)
9. Newcome, L.R.: Unmanned Aviation: A Brief History of Unmanned Aerial Vehicles. American Institute of Aeronautics and Astronautics, February 2005. https://doi.org/10.2514/4.868894
10. North Atlantic Treaty Organization: Standardisation Agreement (STANAG) 4586: Standard Interfaces of UAV Control System (UCS) for NATO UAV Interoperability. Standardization Agreement NSO/0471(2017)JCGUAS/4586, NATO Standardization Agency (NSA), April 2017
11. Safe, A.: Common drone types and capabilities (2016). http://www.airsafe.com/issues/drones/drone-types.htm. Accessed 20 May 2018

12. Silva, D.C.: Cooperative Multi-Robot Missions: Development of a Platform and a Specification Language. Ph.D. thesis, Faculdade de Engenharia da Universidade do Porto, June 2011
13. Veloso, R.F.D.: Platform for the Design, Simulation and Development of Quadcopter MultiAgent Systems. Master's thesis, Faculty of Engineering, University of Porto, February 2014. (in Portuguese)

Technological Architecture Based on Internet of Things to Monitor the Journeys of Artisanal Fishing

Jaime Ambrosio Mallqui, Leysa Preguntegui Martinez[✉],
and Jimmy Armas Aguirre

Peruvian University of Applied Sciences, Lima, Peru
leysajimena@gmail.com

Abstract. In this paper, it was proposed a technological architecture based on Internet of Things to monitor the journeys of artisanal fishing. This proposal is designed based on the needs of artisanal fishermen and engineers of an institution dedicated to the study of the resources of the sea. This architecture employs components in the Cloud to have greater capabilities of storage, processing and analysis of data through sensors, controllers and mobile platforms in the Cloud. The proposed architecture consists of five layers: 1. Internet of Things layer; 2. Presentation layer; 3. Synchronization layer; 4. Cloud Computing layer; 5. Analytical layer. The sample was taken from the shore of Callao, Peru and validated by the Institute of the Sea of Peru (IMARPE). The information is displayed in a mobile application to the fishermen and in a web dashboard to IMARPE engineers, where the architecture was validated. The benefits are the possibility for the fisherman to improve their orientation at sea and know the sea temperature in real time. For IMARPE, the benefit is to have updated data on artisanal fisheries.

Keywords: Artisanal fishing · Technological architecture · Internet of Things · Cloud Computing · Sensor · Controller · Arduino · Bluetooth · Platform as a service · Disruptive innovation

1 Introduction

The fishing sector is divided into inland fisheries, aquaculture and marine fisheries This last is subdivided into industrial fishing and artisanal fishing [1].

In Peru, the artisanal fishing activity is carried out by individuals, groups, relatives or small businesses, its capacity of boat is small and during the operation of fishing, it predominates the manual work. In addition, it is characterized by not having technology or having precarious technology because of the reduced amount of capital [2].

Artisanal fishing has become an important sector because it contributes to the economy, especially in coastal areas and at the national level for its contribution to the indexes of employment [2], it allows many people living in the coast to have the opportunity to move their families forward and develop in this activity [3]. In the last census carried out by PRODUCE and INEI, there were about 44,161 artisanal fishermen [2].

© Springer Nature Switzerland AG 2019
Á. Rocha et al. (Eds.): WorldCIST'19 2019, AISC 931, pp. 24–33, 2019.
https://doi.org/10.1007/978-3-030-16184-2_3

This paper presents as a final product a technological platform based on Internet of Things to monitor the journeys of artisanal fishing in Peru. The motivation of this work lies in two important points: First, to provide a technological infrastructure to artisanal fishermen. Second, to monitor the activities of artisanal fishermen through the sensors and controllers that our architecture provides.

This document is organized as follows: in the first part, Internet of Things solutions related to sea monitoring are analyzed, to identify the most important variable within artisanal fisheries, then the components of Internet of Things, mobile platforms and cloud platforms. The second part describes the case study that was used to evaluate the proposal, detailing the measurement variables considered for its implementation and analyzing the results obtained.

2   Literature Review on Internet of Things Technology Platform

Recent advances in Internet of Things are driving a technological revolution, there could be created prototypes of intelligent systems that improve the quality of life and improve the awareness about resources and the environment [4]. Also, Gartner has predicted that there will be about 26 billion devices in the Internet of Things in 2020 [5].

2.1   Internet of Things Solutions Applied to Fish Monitoring

In the literature, multiple proposals, parameters and controllers with Internet of Things technology related to fish monitoring have been identified.

Table 1 shows seven proposals found in the literature, which are grouped by their similarity and their respective benefit. However, they are not aimed directly at fishing, but at the monitoring of water quality for fish habitat which is important for fishermen.

Table 1. Related proposals.

No	Related proposals	
	Proposals	Benefits
1	Monitoring of water quality [6–9]	This proposal allows for faster decision-making [6, 7], optimizing processes in aquaculture farms, ensuring sustainability in water quality and fish survival [8]. It also allows to pursue changes within water quality [9]
2	Remotely monitor aquaculture farms [10]	This proposal allows to reduce equipment, management costs and work [10]
3	Determine optimal water conditions [11]	This proposal allows the correct development of different fish species [11]
4	Low cost efficient measurement system [12]	This proposal allows to optimize resources, improve profitability and sustainability of fish, know the parameters of water quality, tank environments and fish behavior for decision making [12]

2.2   Evaluation of Sea Monitoring Parameters

The proposals mentioned before took into account four parameters of seawater, which are mentioned in Table 2 and the most significant is temperature. This parameter is a key variable because it influences water quality parameters such as the concentration of dissolved oxygen, the physiochemical characteristics of water and ecological processes such as the nutrient cycle [6, 8, 10]. For these reasons the temperature was chosen.

Table 2. Parameters for water monitoring.

No	Parameters for water monitoring	
	Parameters	Reference
1	Temperature	[6–12]
2	Dissolved Oxygen (DO)	[6, 7, 9, 11]
3	pH	[6, 7, 9, 11]
4	Water level	[8–10, 12]

2.3   Evaluation of Internet of Things Components

Controllers. Table 3 shows the controllers used for the mentioned proposals, of which the most significant is the Arduino controller. This controller has a hardware and software that is easy to use and can be programmed with a simplified version of C. Unlike other boards, it does not require additional hardware to be programmed [9].

Table 3. Controllers used in the proposals.

No	Controllers used in the proposals	
	Controllers	Reference
1	ARM 7	[6, 10]
2	Arduino	[7, 9, 12]
3	Raspberry pi 3 model B	[8]

Data Transmission Method. It was decided to use Bluetooth technology for the following reasons: the majority of modern cell phones have it, works independently of the mobile phone signal, is easier to integrate between IoT-Mobile and consumes less power with respect to wireless communication [13].

2.4   Evaluation of Mobile Platforms

Mobile Platforms. Based on the characteristics of the mobile platforms, a native platform was selected, due to its ability to integrate with the hardware, its great visual impact and its performance [14] as shown in Table 4.

Operating Systems. The best operating systems according to the market presence and the satisfaction of the users are: 1. Android, 2. Windows; 3. iOS [15]. It was decided to

Table 4. Comparison of types of mobile platforms.

		Mobile platforms	
		Native platform	Hybrid platform
Characteristics	Interaction with hardware	Bluetooth, GPS, accelerometer and magnetometer	–
	UX	Great visual impact	It does not have great visual impact
	Performance	Excellent	Medium

use Android because it has the highest market share in the world (74.69%) [16] and the knowledge of programming that is required is more common, which is Java [14].

2.5   Evaluation of Cloud Platforms

According to Forbes, the three main providers of Cloud are: 1. Microsoft; 2. Amazon; 3. IBM [17], in which Microsoft and IBM have a higher rating with respect to PaaS [18]. IBM Cloud was chosen because it has the following benefits: flexibility, efficiency and value strategy [19]. In addition, its platform is more intuitive, has more support and the cost is lower compared to Microsoft [18]. Also, because the University where the architecture is development has an agreement with IBM called "IBM Academic Initiative", which allows to use enterprise cloud for a free period for academic and research purposes [20].

3   Proposed Technology Architecture

3.1   Description of the Proposed Architecture

Based on the scientific documents analyzed, It is proposed a technological architecture based on the Internet of Things and focused on artisanal fishing, sector that has little or no technological support. The proposal focuses on 2 types of users: artisanal fishermen and engineers from research institutions of the resources of the sea. The variables considered are surface temperature, distance from the shore and location, which were chosen based on the literature reviewed and the recommendations of IMARPE. A functional prototype was developed to evaluate the proposed architecture, which has 5 general layers: Internet of Things layer, Presentation layer, Synchronization layer, Cloud Computing layer and Analytical layer. For the validation, specific IoT, Movil and Cloud components were used, however; it can be implemented using other technologies that meet the characteristics mentioned in the architectural layers.

Figure 1 shows the proposed technological architecture that consists of the grouping of components, grouped by layers and that will be detailed below.

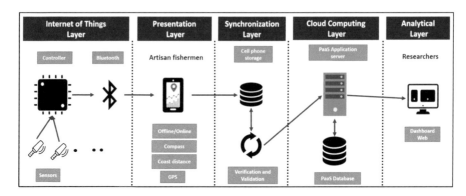

Fig. 1. Technological architecture based on Internet of Things as support for artisanal fisheries.

3.2 Layers of Architecture

Internet of Things Layer. In this layer, the seawater parameters are obtained through the sensors that transmit the information to the Internet of Things controller. This controller propagates the data to a mobile application using a Bluetooth module.

For the prototype, It was used the water temperature sensor DS18B20 [21], an Arduino UNO card that acts as a controller and the Bluetooth module HC-06 [22].

Presentation Layer. This layer is responsible for receiving the data sent from the Internet of Things layer via Bluetooth. Also, in this section the GPS of the mobile device is used to support the proposed architecture and determine the positioning of the boat in degrees, minutes and seconds; as well as calculating the distance from the shore in nautical miles. In addition, this layer will use the magnetometer and accelerometer sensors of the cell phone to show the orientation of the boat. On the other hand, it should be able to also register fishing data obtained. Because GPS and other sensors are already incorporated in the cell phone, the data capture does not require internet connection. When the fisherman is already in a zone with internet access, he can synchronize the data with the server in the cloud and download the required maps in the application.

Synchronization Layer. This layer ensures that the data is validated in the process of synchronization with the cloud. For this, the fisherman has to create a fishing period that relate to their data obtained by selecting the data by dates and times obtained. This process has its execution on the mobile device. If the information is recorded successfully in the cloud, it is removed from the cell phone.

Cloud Computing Layer. This layer is responsible for hosting the application that will provide web services for the treatment, processing and storage of the information sent by the mobile device. The prototype used a server of NodeJS and a PostgreSQL database, both hosted in the IBM Cloud.

Analytical Layer. This layer shows the information of the days of the artisanal fishermen in graphics and maps to the research engineers. This information is shown in a web dashboard developed in NodeJS.

4 Case Study

4.1 Organization

The Institute of the Sea of Peru, also called IMARPE, is a technical organization where the proposed Internet of Things architecture was validated. This institution is dedicated to the investigation, study and knowledge of the Peruvian sea and its natural resources. IMARPE has systems with which it monitors large industrial fisheries, but it is not possible to regularly monitor artisanal fisheries because they do not have technological tools. For this reason, the institution was receptive to the proposal. For the tests of the prototype, boats operated by fishermen were used and 3 navigations were made in the area of Punta, Callao within the first mile because there were no licenses and life insurance to embark within several nautical miles. These boats were made with an average duration of one hour each.

4.2 Implementation

Obtaining Sea Parameters Through the Internet of Things. The developed prototype shown in Fig. 2, is placed in the boat and the temperature sensor is placed on the sea surface. For this work, own rechargeable batteries with a duration of approximately 6 h were used; however, it could be adapted to the boat's batteries for greater autonomy.

Fig. 2. Internet of Things prototype for sensing the surface temperature of the sea.

Receiving Data in the Mobile Application. The Internet of Things prototype sends the values every 0.7 s. The mobile application receives and displays this data and acts as intermediate storage. In addition, the application also stores GPS data and distance to the shore, with the possibility of viewing the orientation of the boat with the compass.

In Fig. 3, the application is shown in operation and connected to the IoT prototype.

Cloud Storage. When the fisherman is in an area with an internet connection, he can synchronize the data collected with the cloud. In addition, he can visualize his previous days of fishing to have a better administration of these. The advantage of using a database in the cloud is the ability to scale in capabilities and pay per use.

Data Collection and Analysis. A statistical and geospatial analysis was performed with the data obtained in the navigations carried out. The IMARPE engineers were

Fig. 3. Screenshots of the mobile application developed and running.

shown how the data could potentially be analyzed and displayed for their studies and monitoring. For example, they were shown that they could see the trips made by artisanal fishermen along with the sea temperature, as they see industrial fishermen, so that they have control of the fishing areas destined for the reproduction and sustainability of the fishing resources [23]. In addition, they will be able to know how many miles away from the shore these fishermen usually are.

4.3   Results

After testing the prototype at sea, more than 10,000 records were obtained with GPS coordinates, temperature and distance to the shore. Figure 4 shows the first 5 records of the database.

	idDataSensor	idFishingTime	temperature	lat	long	shore distance	date time
0	1383	1526225583657	16.12	-12.069417	-77.166030	0.043238	5/13/2018, 11:41:40 AM
1	1728	1526225583657	16.06	-12.069317	-77.166049	0.049077	5/13/2018, 11:41:41 AM
2	1998	1526225583657	16.06	-12.069317	-77.166049	0.049077	5/13/2018, 11:41:42 AM
3	2052	1526225583657	16.00	-12.069317	-77.166049	0.049077	5/13/2018, 11:41:44 AM
4	1618	1526225583657	16.00	-12.069317	-77.166049	0.049077	5/13/2018, 11:41:45 AM

Fig. 4. Preview of the records in the database

The Python programming language was used for data analytics. Of these records the average temperature was 16.5 °C, with a minimum of 14.8 °C and a maximum of 17.8 °C. With respect to the distance from the shore, it was navigated up to 0.46 nautical miles.

To analyze the behavior of the temperature with respect to the distance to the shore, tables and analyzes were made. To see this behavior, a linear regression was applied that is shown in Fig. 5. It can be seen that the line tends to be horizontal and the

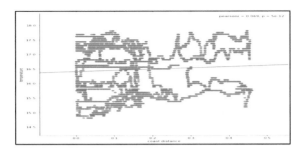

Fig. 5. Chart of the correlation between temperature and distance to the shore.

pearson correlation is almost 0, so there is no linear relationship between the values obtained from the distance of the shore and the temperature.

Based on the data compiled by the application and the information provided by IMARPE, it has been observed that both artisan fishermen and the engineers of this institution are interested in learning about the environment in which fishing takes place. For example, knowing the temperature of the sea water is an important factor during this activity, since it indicates if there could be certain fish. According to IMARPE, the common habitat of the anchoveta is between 14.5 °C and 20 °C [23], while the perico is a highly migratory fish and prefers warm water masses with temperatures greater than 23 °C [24].

On the other hand, they also find it useful to know what distances from the shore they usually carry out their fishing activities. For example, the anchoveta is usually found within 60 nautical miles [25] and if it were close to the seashore, it could mean that an event has occurred at sea, such as the phenomenon of the child, displacement of water, pockets of cold water, upwelling of nutrients, among others [26]. That is why they were also shown a chart where one could visualize in what nautical miles the fishermen are usually.

Also, they were interested to see in a map the fishing journeys made to have a realistic view of the data as shown in the Fig. 6.

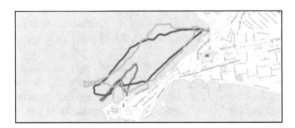

Fig. 6. Map of the routes made during navigation

With respect to the fishermen, they benefit from a mobile application connected to IoT components, which allows them to have a compass, GPS and sensors for their proper navigation and capture of shoals. In addition, with the fishing record they will be able to know their catch rates and thus better plan their trips and resources. IMARPE could also provide fishermen with up-to-date information through the application map, such as restricted areas due to the presence of young or fish in reproduction.

5 Conclusions

This paper presented a technological architecture for the monitoring of artisanal fisheries. The sampling was then carried out and both the functionality and the data obtained with IMARPE were validated. After showing through charts the records obtained, they expressed their interest in the proposal, because they could know more about the artisanal fishermen which would also benefit from the application. That is why the organization verified the great impact that such architecture could have on their research and told us that they had plans to implement it the following year in some artisanal fisheries.

According to the final recommendations of IMARPE, in future scenarios the dependence of the cell sensors must be reduced and the IoT layer must use the batteries of the boat. In addition, this institution recommended us to take samples every minute to reduce energy use. It's recommended to use an expert company in IoT such as Particle.io for the production of the proposal. Particle.io has experts and develops products in large quantities and customized [27].

References

1. Ministerio de la producción: Anuario estadístico pesquero y acuícola (2015). https://www. produce.gob.pe/documentos/estadisticas/anuarios/anuario-estadistico-pesca-2015.pdf
2. IMARPE: Atlas de la pesca artesanal del mar del Perú (2017). http://biblioimarpe.imarpe. gob.pe:8080/handle/123456789/3167
3. Organización de las Naciones Unidas para la Alimentación y la Agricultura: La pesca artesanal (2016). http://www.fao.org/3/a-i5951s.pdf
4. Xu, Y., Helal, S.: Scalable cloud-sensor architecture for the Internet of Things. IEEE Internet Things J. 3(3), 285–298 (2016). https://doi.org/10.1109/JIOT.2015.2455555
5. Razzaque, M.A., Milojevic-Jevric, M., Palade, A., Clarke, S.: Middleware for Internet of Things: a survey. IEEE Internet Things J. 3(1), 70–95 (2016). https://doi.org/10.1109/JIOT. 2015.2498900
6. Viswanadh Boyina, E., Kumar, S.: ARM based aqua monitoring system using Iot environment, vol. 9, no 2, pp. 1347–1353 (2016). International Science Press
7. Encina, C., Ruiz, E., Cortez, J., Espinoza, A.: Design and implementation of a distributed Internet of Things system for the monitoring of water quality in aquaculture. In: Wireless Telecommunications Symposium, pp. 1–7 (2017). https://doi.org/10.1109/wts.2017. 7943540
8. Bokingkito, P.B., Llantos, O.E.: Design and implementation of real-time mobile-based water temperature monitoring system. Procedia Comput. Sci. 124, 698–705 (2017). https://doi.org/ 10.1016/j.procs.2017.12.207
9. Durga Sri, B., Nirosha, K., Priyanka, P., Dhanalaxmi, B.: GSM based fish monitoring system using Internet of Things. Int. J. Mech. Eng. Technol. 8(7), 1094–1101 (2017)

10. Kiruthika, S.U., Suba Raja, S.K., Jaichandran, R.: Internet of things based automation of fish farming. J. Adv. Res. Dyn. Control Syst. **9**(1), 50–57 (2017)
11. Stehfest, K.M., Carter, C.G., McAllister, J.D., Ross, J.D., Semmens, J.M.: Response of Atlantic salmon Salmo salar to temperature and dissolved oxygen extremes established using animal-borne environmental sensors. Sci. Rep., 1–10. https://doi.org/10.1038/s41598-017-04806-2
12. Parra, L., Sendra, S., García, L., Lloret, J.: Design and deployment of low-cost sensors for monitoring the water quality and fish behavior in aquaculture tanks during the feeding process. Sensors **18**(3), 1–23 (2018). https://doi.org/10.3390/s18030750
13. Universiad Oberta de Catalunya: IoT: Dispositivos, tecnologías de transporte y aplicaciones (2017). http://openaccess.uoc.edu/webapps/o2/bitstream/10609/64286/3/agonzalezgarcia 0TFM0617memoria.pdf
14. El-Kassas, W.S., Abdullah, B.A., Yousef, A.H., Wahba, A.M.: Taxonomy of cross-platform mobile applications development approaches. Ain Shams Eng. J. **8**(2), 163–190 (2017). https://doi.org/10.1016/j.asej.2015.08.004
15. CROWD: Best Operating System (2018). https://www.g2crowd.com/categories/operating-system?segment=small-business
16. Statcounter GlobalStats: Mobile Operating System Market Share Worldwide (2018). http://gs.statcounter.com/os-market-share/mobile/worldwide
17. Forbes: The Top 5 Cloud-Computing Vendors (2017). https://www.forbes.com/sites/bobevans1/2017/11/07/the-top-5-cloud-computing-vendors-1-microsoft-2-amazon-3-ibm-4-salesforce-5-sap/#7e74cadb6f2e
18. TrustRadius: IBM Cloud PaaS vs Microsoft Azure vs Amazon Web Services (2018). https://www.trustradius.com/compare-products/ibm-cloud-paas-vs-microsoft-azure-vs-amazon-web-services#platform-as-a-service
19. IBM: Benefits of cloud computing (2018). https://www.ibm.com/cloud/learn/benefits-of-cloud-computing
20. IBM: Iniciativa Académica Perú (2016). https://www.ibm.com/developerworks/community/wikis/home?lang=en#!/wiki/W04da5e7f0a49_4581_8335_3b29447b58dd
21. Terraelektronika - solución de problemas reales de desarrolladores de equipos electrónicos en Moscú: DS18B20 Waterproof Temperature Sensor Cable (2018). https://www.terraelectronica.ru/pdf/show?pdf_file=%2Fz%2FDatasheet%2F1%2F1420644897.pdf
22. Future electronic corporation: HC Serial Bluetooth Products (2018). http://www.fecegypt.com/uploads/dataSheet/1480849570_hc06.pdf
23. Sociedad Nacional de Pesquería: Sostenibilidad de los recursos pesqueros (2018). https://www.snp.org.pe/sostenibilidad-de-los-recursos-pesqueros/
24. IMARPE: Biología y pesquería del perico o dorado Coryphaena hippurus (2018). http://biblioimarpe.imarpe.gob.pe:8080/bitstream/handle/123456789/2949/INFORME%20%2042%281%29-2.pdf
25. IMARPE: Anchoveta (2018). http://www.imarpe.pe/imarpe/archivos/articulos/imarpe/recursos_pesquerias/adj_pelagi_adj_pelagi_anch_mar07.pdf
26. OCEANA: La anchoveta y el niño (2018). https://peru.oceana.org/es/la-anchoveta-y-el-niño
27. PARTICLE: Particle Studios (2019). https://www.particle.io/particle-studios#production

Gesture Based Alternative to Control Recreational UAV

Roberto Ribeiro[1], David Safadinho[1], João Ramos[1], Nuno Rodrigues[1], Arsénio Reis[3(✉)], and António Pereira[1,2]

[1] School of Technology and Management, Computer Science and Communication Research Centre, Polytechnic Institute of Leiria, Campus 2, Morro do Lena – Alto do Vieiro, Apartado 4163, 2411-901 Leiria, Portugal
eng.rob.ribeiro@gmail.com,
davidsafadinho.12@gmail.com,
jr.joaoramos@outlook.com,
{nunorod,apereira}@ipleiria.pt
[2] INOV INESC INOVAÇÃO, Institute of New Technologies – Leiria Office, Campus 2, Morro do Lena – Alto do Vieiro, Apartado 4163, 2411-901 Leiria, Portugal
[3] INESC TEC, University of Trás-os-Montes and Alto Douro, Vila Real, Portugal
ars@utad.pt

Abstract. The arrival of Unmanned Aerial Vehicles (UAV) to the consumer market has been changing the way we interact with our surroundings. Fields like cinema and photography are improving with the possibility of reaching unsafe areas, as much as industry related companies that can now supervise their infrastructures safely. Besides many other professional areas, this fever also got into sports and UAV racing became a new way to compete. There are limited alternatives to control a UAV, each one with its pros and cons. The most reliable is the conventional handheld controller, a bulky equipment that requires practice and dexterity to be a good pilot. Consequently, technologically illiterate individuals, users with low dexterity or hand malformations and elders are departed from this game breaking technology, either for professional or recreational purposes. To expand the control of UAV to more member of our society, a different control equipment should be developed. In this paper we propose a solution based in two lightweight hand worn devices sensitive to motion changes. These changes are used to detect modifications in the orientation of the device, which are then transmitted through Bluetooth Low Energy (BLE) to a mobile app that is responsible for their interpretation as simple input or input patterns. The result of this interpretation should be used as flight commands to control a UAV. Through simple and intuitive hand movements, users can accurately pilot a quadcopter. This alternative presents a new and easier approach to control UAV that decreases the time that is required to learn how to use it, with the outcomes of an accentuated learning curve. The results obtained with the usability tests performed with users with different capacities of interaction, confirmed the viability of this solution, as much as its simplicity and intuition in the control.

© Springer Nature Switzerland AG 2019
Á. Rocha et al. (Eds.): WorldCIST'19 2019, AISC 931, pp. 34–44, 2019.
https://doi.org/10.1007/978-3-030-16184-2_4

Keywords: Gesture control · Internet of Things · Motion processing ·
UAV control · Unmanned Aerial Vehicles · Real-time systems · Wearables ·
Wireless communication technologies · Wireless Personal Area Networks

1 Introduction

Businesses, industry and recreational activities are changing to welcome UAV as a tool
to speed up and ease as many processes as possible. Delivery companies like Amazon
are testing cargo dropping with the Prime Air service [1], to avoid problems like traffic
jams, and the Aerial Sports League is organizing drone competitions for entertainment
purposes. There are other opportunities to improve the processes of our society with the
help of UAV in fields like civil protection (e.g.: natural disasters control) and agriculture (e.g.: field fertilization) [2]. Since 2013, when the UAV became popular in the
consumer market with the Phantom series from DJI, the quality of the aircrafts (e.g.:
speed, endurance, obstacle detection) has been increasing, as much as their affordability
[3]. However, not every civilian is able or feels comfortable to use this new technology
because the existent control alternatives are limited, rely on dexterity and many sessions of practice, or simply have many possibilities of failure. In this group of users, we
can include technologically illiterate individuals, elders and patients with hand defor-
mations or low dexterity, either disease related or caused by accidents. The growth of
aging population also justifies the relevance of a control alternative capable of adapting
elders to new intelligent and unmanned technology like UAV [4, 5].

In this article we propose an alternative control solution for UAV with the objective
of expanding to as many individuals as possible the limited target audience, bearing in
mind that the current market niche is indirectly created by the disadvantages in the
actual control alternatives. With this work, individuals with interaction difficulties
related to physical or mental issues get an easier and more intuitive way to control
UAV, which directly promotes elders to use new technologies to improve their lives,
through adapted ways to interact with informatic systems. We intend to seize the
opportunity created by the evolution of Internet of Things (IoT) and the consequent
improvements in wireless communication and sensors, the ubiquity of mobile devices
and the availability of drones in our quotidian. The cornerstone of this alternative is the
development of a small, lightweight, efficient and low-cost device, capable of generating data related to changes in its orientation and communicate it to a mobile device
through a standard communication technology. The user should wear one of these
devices in each hand and connect them to a mobile app. The input generated by each of
the wearables is transmitted to the app and transformed into flight commands that are
sent to the real UAV. This research is focused in the generation of input, its communication and interpretation. The success of this study is determinant to the next step
that consists in communicating the commands from the mobile app to the real UAV.
Also, for safety purposes, a video game was developed to replace the real UAV.

Moreover, this article follows a previous research about a cloud based platform to
control UAV in real time through the Internet and intends to complement it [5].

The structure of this paper is as follows. In Sect. 2 we present the state of the art
regarding the present control alternatives in the market. Then, in Sect. 3 we analyze the

architecture that supports the proposed control solution. In Sect. 4 the implementation of the prototype is described. Next, in Sect. 5 the test scenario is depicted. Finally, Sect. 6 includes the conclusions and intended future work.

2 State of the Art

The recreational UAV are taking over the market with their many utilities. Depending on the task to perform, there are some control alternatives that provide different characteristics of flight and control (e.g.: accuracy, camera control). The most usual and reliable option to pilot an Unmanned Vehicle (UV), not only drones but also other vehicles (e.g.: cars, boats), is the conventional handheld remote controller, that is very similar to the gamepads used to play video games. These remotes are made of a set of buttons, each one with a specific function, and a pair of analogic joysticks to move the vehicle and/or control an attached camera if there is one. In the case of drones, these joysticks are used to perform movements along the horizontal plane (i.e.: forward, backwards, right and left), positive or negative rotations around the vertical axis and changes in the altitude of the aircraft (i.e.: up and down). Many of these controllers can establish a bidirectional communication through two different radiofrequencies (RF), 2.4 GHz and 5.8 GHz, that depending on the quality of the controller can cover an area of as far as 7 km radius in open field. Big companies like DJI, the World Leader in Camera Drones, send controllers with different technologies (Fig. 1), complexity and, consequently, prices. For instance, Cendence is a top-notch 1000$ controller for high-end quadcopters that provides a more professional approach for tasks related with multimedia (e.g.: cinema, photography), through a functionality that defines a master-slave connection with another controller, to command the flight and the camera separately [7].

Fig. 1. The remote controllers Spark (left) and Cendence (right), both from DJI.

There are two ways of using a mobile device locally to mitigate the need of a remote controller. The first and more usual way corresponds to a mobile app that presents a set of virtual buttons and joysticks through a Graphical User Interface (GUI) the user can interact with to pilot the UAV through touch inputs. However, this alternative is sensitive to the surroundings, since humidity and dust can influence the touch detection on the display, making it a naturally inaccurate solution. FreeFlight Mini from Parrot [8] and Cheerson's app (Fig. 2) are examples used to control aircrafts like Parrot Swing and Cheerson's CX-33 Scorpio [9]. The second alternative depends on the gyroscope module, integrated in the mobile device, to interpret the respective orientation as input to move the vehicle. For instance, considering that the control smartphone is in landscape mode, when the user tilts it forward (i.e.: clockwise roll rotation), the UAV moves forward, like in the same Cheerson's control app described above (Fig. 2). Relatively to the disadvantages, the gyroscope only reads three axes, while the control of a drone depends of four. In this case, one set of commands is discarded (e.g.: altitude changes).

The local hand gestures control in recreative drones was popularized by DJI with their aircraft Spark [10]. Besides interpreting gestures as a control alternative and trigger to execute predefined functions, Spark also identifies the owner through facial recognition, for safety purposes. Through the palm control, the UAV follows the hand of the user and if a frame gesture is performed it takes a picture. Other functionalities like the palm landing and the gestures to start and stop a video recording complete the list of possibilities [11, 12]. The great advantage of hand gestures as a control alternative is the fact that no device is needed to control the drone. However, the lack of precision of the solution restricts it to very simple recreational tasks, focused in the pilot, which makes this solution impracticable even for simple field recognition or multimedia activities.

Fig. 2. App FreeFlight Mini from Parrot (left) and Cheerson's control app (right).

3 Architecture of the System

The architecture conceived to make the proposed solution possible is separated in three modules (Fig. 3). The first is responsible for generating the control input and includes two hand-worn devices. Every time there is a change in their orientation, these devices communicate through BLE with module 2. This module is represented by an Android mobile device running the control app, responsible for the transformation of the received input into flight commands. The app includes a video game to simulate and test the control. If the user wants to control a real UAV the interpreted commands are sent to the UAV that is identified by module 3. As the focus of this document are the wearables and the respective communication with the app, a generic wireless communication is considered to send commands to the UAV.

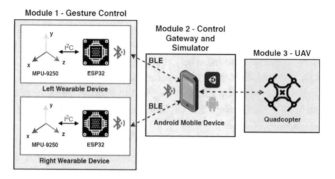

Fig. 3. Architecture of the proposed UAV control alternative.

4 Implementation of the Prototype

To achieve the described architecture, we developed a prototype of the idealized control alternative, composed by an Android mobile app and two wearable devices. Besides the developed wearables, to run the app we used the mid-range smartphone Xiaomi Mi 5, that works as a bridge between the wearables and the UAV. The functional sequence of the system corresponds to the next. After connecting the wearables, the user should start the scan for BLE peripherals that will list the two devices. The app (i.e.: central device) should subscribe the pitch and roll characteristics of both wearables and calibrate them. The calibration consists in stabilizing the wearables parallel to the ground and press a button in the app that writes to the calibration characteristic. In that moment, the notification is activated and each time the pitch or roll values change, the wearables send the respective values to the app. These values are processed into flight commands. This is where the app diverges. The flight commands can be consumed by the simulation video game or transmitted to a real UAV.

4.1   Wearable Devices

The hand-worn devices are responsible for generating the values that describe their orientation relatively to the longitudinal and transversal axes, that is the roll and pitch (Fig. 4).

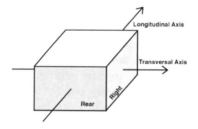

Fig. 4. Representation of the reference axes for the orientation.

We assembled a hand-worn device as defined in the system's architecture. The MPU-9250 motion sensor, a 9-axis module equipped with a gyroscope, an accelerometer and a magnetometer, is generates raw data related to motion. Then, an ESP WROOM 32 dual core microcontroller, developed by Expressif Systems, reads the raw data from the MPU-9250 and calculates the pitch and roll values in Euler angles. The communication between the MPU-9250 and the ESP32 is established through I2C (Fig. 5). This wired connection is also responsible for powering the MPU-9250. To power the wearable, it is used the USB port available in the version of the ESP32 used to build the current prototype. The wearables are represented in Fig. 5, where the letters A, B and C identify the motion sensor, the microcontroller and the USB cable working as power supply, respectively.

Since the accelerometer is susceptible to any kind of external force (e.g.: gravity), there are disturbs in the readings. Regarding the gyroscope, as the time passes, the values start to derivate, making it viable only for short time intervals [13]. To obliterate these problems, a complementary filter is used to apply a factor to the values of the sensors, giving higher priority to the accelerometer as the duration of the interval between readings increases [14]. This filter corresponds to the equation identified by (1), used to calculate the pitch value, where Δt and F correspond to the time between readings and to the factor applied over the raw data values.

$$pitch = F \times (pitch_{actual} + gyroscope_x \times \Delta t) + (1 - F) \times accelerometer_x \qquad (1)$$

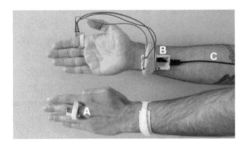 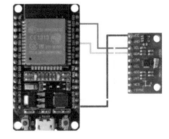

Fig. 5. User equipped with the prototyped wearable devices and scheme representing the connections between the microcontroller (left) and the motion sensor (right).

4.2 BLE Communication

The BLE technology was chosen to establish a bidirectional communication between the wearables and the smartphone running the control app. This is a standard technology, available since 2010 with the introduction of Bluetooth 4.0, which makes it compatible with the majority of the mobile devices [15]. It allows to create WPAN, a close-range communication network that follows a star-bus topology where a central device can connect to many peripherals to consume their data. In this solution, the wearables assume the role of peripherals, while the smartphone corresponds to the central device. Besides, this technology is power efficient and has low transmission latency [16, 17].

Fig. 6. Representation of the GATT profile created by the wearables.

For this communication to work, each wearable creates a GATT profile composed of a service with three characteristics, as represented in Fig. 6. Two of them correspond to the orientation changes in the defined axis (i.e.: pitch and roll) whose notification is activated when the central device writes a value in the calibration characteristic. When the notification is on, every change in the respective value is automatically notified to the central device.

4.3   Control App and Simulation Video Game

The app responsible for transforming the motion data in flight commands was developed for Android devices in the game engine Unity to ease the inclusion of a virtual simulation. The access to the communication functionalities of the device is granted through an Android library, developed in Java, that is included in the Unity project.

The flow of the data until it is sent as flight commands is represented in Fig. 7. The MPU9250 is responsible for the generating motion data based on the gyroscope and accelerometer modules (A). This data is read by the ESP32 that calculates the pitch and roll values of the wearable (B) to transmit through BLE (C and D). Already in the app,

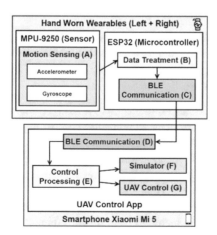

Fig. 7. Motion input flow in the developed solution.

the pitch and roll values are interpreted and transformed in flight commands (E) to be consumed by integrated the video game (F) or transmitted to the real UAV (G).

The interpretation of the commands depends on the pitch and roll values of each wearable. The use of only two axes simplifies the interaction with the developed solution. To control a recreational UAV like Phantom 4 from DJI, there are linear actions and one-time orders like the take off and the landing. The two previous orders are interpreted through a pattern made with both wearables in a manner of confirmation.

To test the solution, we created a virtual simulator (Fig. 8). The purpose of this simulation is to replace the real scenario. It consists in a full-scale environment around which we can fly a quadcopter through the control wearables. The metrics used in the simulation are real, which means that the UAV moves and rotates with a ratio of meters and Euler angles per second, respectively. For testing purposes, we also created a build of the simulation for Windows platforms that allows to control the UAV through a gamepad (e.g.: Dualshock 4) and compare the control alternatives.

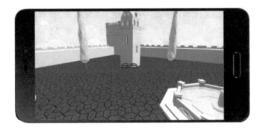

Fig. 8. UAV control simulator developed in Unity.

5   Test Scenario and Results

To test the solution, ten users with different ages, technological capacities and skill, were submitted to usability tests. These tests were carried out in a safe environment where the real scenario and the UAV were replaced by a video game (Fig. 9). The objective of this evaluation is to verify the functionality of the wearables as a UAV control alternative and understand if it viable for user with low interaction capacity, as elders. The tests were divided in two phases. The first consisted in following a pre-defined trajectory without colliding with any obstacle, which allowed to adapt the control sensibility to each user. The second phase was similar to the first one, but instead of using the developed wearables and the mobile app, we used a Dualshock 4, identical to the conventional UAV remote controllers, to control the desktop version of the simulator.

Fig. 9. Elder user testing the developed control alternative with the simulator.

The results confirmed the success of the developed solution through the simulation video game. Users were able to fully control a UAV and felt more comfortable to control the vehicle with the orientation of their hands than with the Dualshock 4. The solution was also considered simpler and more intuitive to the users with no previous contact with both the tested alternatives. Users also demonstrated ease in the learning of the control mechanics through the developed solution. After getting used to this solution, users felt comfortable controlling a UAV around the virtual environment.

6 Conclusions and Future Work

With this research we created a local control alternative for UAV based on wearable devices, capable of interpreting changes in the orientation of the users' hands in two different axes (i.e.: pitch and roll). The data relative to these changes is then transmitted to a mobile app responsible for its transformation in flight controls that can be tested in the video game integrated in the same app. The wearables were built with low-cost equipment and no additional devices were needed, since the solution takes advantage of a ubiquitous device (i.e.: smartphone) to interpret, test and, eventually, send the flight commands to a real UAV. The users testing the resultant flight experience prove higher intuition and ease of control. The flight mechanics become simple to understand.

As future work we intend to improve the control algorithm and customize the control attributes based on user capacities (i.e.: amateur, experienced user, elder) through the creation of profiles. The alteration of these attributes (e.g.: UAV speed, control precision) should also be allowed in real-time to help the user achieve maximum control confidence easily. Also, the communication from the mobile app to a real UAV should be developed and tested with different aircrafts to achieve results in a real scenario where other factors like the psychological state of the pilot or meteorological events (e.g.: wind) can affect the control.

Acknowledgements. This work was supported by the project SAICT-POL/24048/2016 – NIE - Natural Interfaces with the Elderly, with reference NORTE-01-0145-FEDER-024048, financed by the Foundation for Science and Technology (FCT) and co-financed by the European Regional Development Fund (FEDER) through the Northern Regional Operational Program (NORTE2020).

References

1. Shavarani, S.M., Nejad, M.G., Rismanchian, F., Izbirak, G.: Application of hierarchical facility location problem for optimization of a drone delivery system: a case study of Amazon prime air in the city of San Francisco. Int. J. Adv. Manuf. Technol. **95**(9–12), 3141–3153 (2018)
2. Mohammed, F., Idries, A., Mohamed, N., Al-Jaroodi, J., Jawhar, I.: UAVs for smart cities: opportunities and challenges. In: 2014 International Conference on Unmanned Aircraft System, ICUAS 2014 - Conference Proceedings, pp. 267–273 (2014)
3. Horsman, G.: Unmanned aerial vehicles: a preliminary analysis of forensic challenges. Digit. Investig. **16**, 1–11 (2016)
4. Reis, A., Paredes, H., Barroso, I., Monteiro, M.J., Rodrigues, V., Khanal, S.R., Barroso, J.: Autonomous systems to support social activity of elderly people: a prospective approach to a system design. In: 1st International Conference on Technology and Innovation in Sports, Health and Wellbeing, TISHW 2016, Proceedings (2016). https://doi.org/10.1109/TISHW.2016.7847773
5. Reis, A., Barroso, I., Monteiro, M.J., Khanal, S., Rodrigues, V., Filipe, V., Paredes, H., Barroso, J.: Designing autonomous systems interactions with elderly people. Lecture Notes in Computer Science (including subseries Lecture Notes in Artificial Intelligence and Lecture Notes in Bioinformatics), vol. 10279 (2017). https://doi.org/10.1007/978-3-319-58700-4_49
6. Safadinho, D., Ramos, J., Ribeiro, R., Caetano, R., Pereira, A.: UAV multiplayer platform for real-time online gaming. In: Recent Advances in Information Systems and Technologies, WorldCIST 2017. Advances in Intelligent Systems and Computing, vol. 571 (2017)
7. DJI - The World Leader in Camera Drones/Quadcopters for Aerial Photography. http://www.dji.com/cendence. Accessed 30 Oct 2017
8. Swing Quadcopter Minidrone. Parrot Official. https://www.parrot.com/global/minidrones/parrot-swing#perform-the-greatest-aerobatic-manoeuvres. Accessed 31 Oct 2017
9. CX-33 Scorpio. http://www.cheersonhobby.com/en-US/Home/ProductDetail/97. Accessed 31 Oct 2017
10. Spark - Seize the Moment. https://www.dji.com/spark. Accessed 30 Oct 2017
11. DJI's Spark Is a Hand-Gesture Controlled Drone that Flies Off Your Hand - The Drive. http://www.thedrive.com/aerial/10674/djis-spark-is-a-hand-gesture-controlled-drone-that-flies-off-your-hand. Accessed 30 Oct 2017
12. DJI Spark Camera Drone Becomes More Fun and Intelligent with New Video and photo Features. https://www.dji.com/newsroom/news/dji-spark-camera-drone-becomes-more-fun-and-intelligent-with-new-video-and-photo-features. Accessed 10 Jan 2018
13. Coopmans, C., Jensen, A.M., Chen, Y.: Fractional-order complementary filters for small unmanned aerial system navigation. J. Intell. Robot. Syst. Theory Appl. **73**(1–4), 429–453 (2014)
14. Ngo, T., Nguyen, P., Huynh, S., Le, S., Nguyen, T.: Filter design for low-cost sensor in quadcopter, no. 3, pp. 488–493 (2017)

15. Amaro, J.P., Patrao, S., Moita, F., Roseiro, L.: Bluetooth low energy profile for MPU9150 IMU data transfers. In: 5th Portuguese Meeting on Bioengineering, ENBENG 2017. Proceedings, pp. 1–4 (2017)
16. Putra, G.D., Pratama, A.R., Lazovik, A., Aiello, M.: Comparison of energy consumption in Wi-Fi and bluetooth communication in a Smart Building. In: 2017 IEEE 7th Annual Computing and Communication Workshop Conference, CCWC 2017, no. 2014 (2017)
17. Jeon, K.E., She, J., Soonsawad, P., Ng, P.C.: BLE beacons for Internet of Things applications: survey, challenges and opportunities. IEEE Internet Things J. 5(2), 811–828 (2017)

Software Modules and Communication to Support Real-Time Remote Control and Monitoring of Unmanned Vehicles

João Ramos[1], David Safadinho[1], Roberto Ribeiro[1],
Patrício Domingues[1], João Barroso[3(✉)], and António Pereira[1,2]

[1] School of Technology and Management,
Computer Science and Communication Research Centre,
Polytechnic Institute of Leiria, Campus 2, Morro do Lena – Alto do Vieiro,
Apartado 4163, 2411-901 Leiria, Portugal
jr.joaoramos@outlook.com,
davidsafadinho.12@gmail.com,
eng.rob.ribeiro@gmail.com,
{patricio.domingues,apereira}@ipleiria.pt
[2] INOV INESC INOVAÇÃO, Institute of New Technologies – Leiria Office,
Campus 2, Morro do Lena – Alto do Vieiro,
Apartado 4163, 2411-901 Leiria, Portugal
[3] INESC TEC, University of Trás-os-Montes and Alto Douro,
Vila Real, Portugal
jbarroso@utad.pt

Abstract. The usage of unmanned vehicles for professional, recreational and healthy purposes has increased and is a huge signal of their advantages. Among other benefits, they reduce or even cancel the need of having human lives aboard, which means that there is no risk of injuries in dangerous tasks. Although, most of the time the users are near the vehicles, which cannot be possible nor proper for personal or security reasons. Therefore, it is proposed a software solution to allow users to control and monitor unmanned vehicles remotely in real-time just as if they were in the vehicles' place. Then it follows an implementation to control and monitor remotely-piloted cars of different types. This solution has been applied to a real-case scenario for testing purposes and it has been concluded that the software architecture proposed can be generically applied to different kinds of vehicles with transparency to the users that are able to control, from everywhere and with their own personal devices, whatever vehicles they want.

Keywords: "Anytime, anywhere" · Ground-stations · Platform based ·
Real-time · Remote control · Remotely-controlled cars ·
Remotely-piloted vehicles · Software architecture · Unmanned vehicles ·
Unmanned aerial vehicles · *WebRTC* · *WebSockets*

© Springer Nature Switzerland AG 2019
Á. Rocha et al. (Eds.): WorldCIST'19 2019, AISC 931, pp. 45–55, 2019.
https://doi.org/10.1007/978-3-030-16184-2_5

1 Introduction

Unmanned vehicles are being used in several different areas for many tasks. Either aerials, aquatics or terrestrials vehicles, all of them represent huge benefits for professional [1–3], recreational [4] and healthy purposes [5, 6]. Considering that there is no driver inside them, despite the risks for the hardware, no involved human lifes are put in danger [7]. For that reason, there is a need to study and develop new control methods considering that the traditional ones cannot be applied.

The work being developed in the area of control methods can be roughly divided in two main groups: (i) the methods that require the user to be near the vehicle and the methods that allow the user to remotely control the vehicle [7]. The first ones represent, at least, one problem, bearing in mind that sometimes the users cannot be near the vehicle, either for security or logistic reasons. As such, there have been created alternatives that rely in remote control. These, in other way, allow users to control the vehicles from any point of the planet, requiring only a few things such as an Internet connection. The advantages of this kind of control methods are the main motivation for this work, which consists in creating an alternative remote solution to control unmanned vehicles. This work is the continuation of a previous work developed by the same authors [8], in which it has been proposed a generic platform to control and monitor unmanned vehicles. In that proposal, there was introduced and presented an overview of the system, without analyzing and detailing further the solutions' technical and lower-level aspects. As so, this work focuses on the specification and on the implementation of the software modules that must be developed and integrated in the platform in order to allow users to manually control multiple vehicle types as if they were right near them.

The next section reviews research regarding remote control solutions. Analyzed research works are focused on specific vehicle types and do not consider the genericity that this work intends to introduce in order to integrate a virtually infinite group of different types of equipment. Considering the crescent demand on the vehicles of the type quadcopter and the research that have been created around them, there are different solutions of control and monitoring for these vehicles. However, currently these applications limit the maneuverability of the vehicles. Specifically, they do not allow users to manually control vehicles like it is done with regular radio remote-controller: move in the longitudinal and transversal axis, go up and down and rotate to the left and to the right.

Keeping in mind the existing solutions and the disadvantages that they have, in the next section proposes a new remote-control communication method between users and vehicles that allows the manual control of them. This is achieved through two distinct software modules: (i) the user module and (ii) the vehicle module and the protocol that bridge the two modules together.

After the proposal of the modules and the specification of the communication protocol, the paper focuses on the implementation of those modules applied to different remote-controlled cars equipped with four wheels and a video camera. In this implementation, users can control the vehicles remotely and visualize, in real-time, the environment.

The implemented modules for cars control and monitoring have been applied in a real scenario for testing purposes. Then, they were tested during a scholar activity where it was possible to test the solution's functionalities with different vehicles having different characteristics.

The paper ends with a summary of its main conclusions as well as the contributions that it gives to the area of unmanned vehicles and the correspondent control alternatives. The final section also discusses future work needed to complement the current system and to apply to other type of vehicles.

2   Related Work

This section reviews related work. Specifically, different applications for controlling unmanned quadcopter vehicles are reviewed. As a continuation of the study previously done and published, the applications are split in two groups and are analyzed from the point of view of the software architecture. The main flaws found in these solutions bear with their limitations on the vehicle type and model and with the limitation on the control method that does not allow users to manually control them as they do with regular remote-controllers. For example, ordinary remote-controllers allow vehicles to go forward/backwards, rotate, and go up/down [8].

The first group of applications is widely known as ground-stations (Fig. 1). Their main goal is to control one or more vehicles directly through a computer or a mobile device [10]. The user can define a point in the map, to whom the vehicle should move to, waiting for new instructions. Besides that, there is an option in which the user can define a set of points, designated by mission that the vehicle follows, point by point, in a straight line. In both options the user has real-time access to flight variables such as altitude, speed and location. The communication between the user and the vehicle for commands and statistics transmission follows an open-source protocol named MAV-Link [11, 12]. In this way, the control methods available in this kind of applications do not allow a manual operation of the vehicles like it is done with a regular controller.

Beyond ground-stations, the other set of applications does not focus in the control of the aircrafts but, instead, takes benefit of them to accomplish a specific purpose. Two examples of that group are DroneDeploy [13] and SkyCatch [14] that use drones from the brand DJI to take aerial footages and do territorial measurements and mapping [15]. The process is resumed in the following steps (Fig. 2): mission planning, data acquisition and information extraction from the collected data. Initially the user registers, in a map shown in the application, the area to be analyzed. After that, the application itself defines the best path that the vehicle should follow to capture photos of the entire zone. Then, the user gets the mission plan in order to send the commands to the vehicle to which s/he must be connected. During the mission the drone takes photos at strategic locations and the user has metrics shown in the interface of the application such as altitude, speed, duration of the flight and an estimation of the remaining time. At the end, the user retrieves the photos from the vehicle and uploads them to the application that will handle their processing. This way, the application provides analysis tools like 2D and 3D maps of the terrain and measurement tools to calculate approximated distances, areas and volumes.

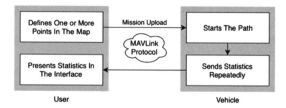

Fig. 1. Ground stations' communication diagram representing how users and vehicles interact.

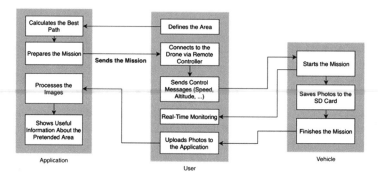

Fig. 2. Software architecture of terrain analysis applications like *DroneDeploy* and *SkyCatch*.

3   Architecture Proposal

The architecture proposed is this paper is comprised of two main modules: user module and vehicle module. Accordingly, there should exist a server application to manage the creation and maintenance of vehicles and users' groups. This proposal does not focus on the platform but on the logical architecture of the main modules. It terminates with a proposal regarding the communication protocol that both user and vehicle modules must adhere to communicate efficiently between themselves, independently of the user input and vehicle types.

3.1   User

The users have at their disposal, in real-time, access to metrics and images that the vehicles capture with the on-board cameras. As such, users must have four different blocks with dedicated purposes (Fig. 3).

There are three communication channels that link the user and the vehicle: control commands transmission, from the user to the vehicle, and metrics and images transmission, from the vehicle to the user. The first block (Fig. 3), on the left, is responsible for transmitting and receiving the data between the vehicle and the user. This block is directly connected to the next block, which is responsible for data processing. The data processing block interprets the data received as metrics and converts them to the correct format to be presented to the user. For instance, it performs unit conversion and/or data filtering. The other main task of the data block is the translation of commands issued by the input method into generic instructions understood by the vehicle. Considering a car and a controlling keyboard as example, this process could translate the key up event into a movement to the front and the key down event into a backwards movement. Finally, the last block corresponds to the graphical interface that shows the camera images, provided directly from the data transmission module, and the properly formatted metrics.

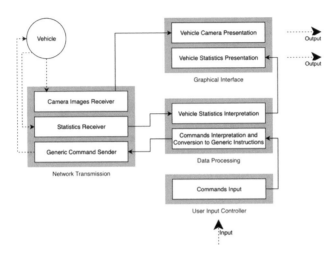

Fig. 3. Users' software architecture with four main blocks for data input, output, processing and transmission over the network.

3.2   Vehicle

The vehicles' software, in a similar fashion to the users' software, must be designed and developed having in consideration the principles of control and monitoring through the network. Furthermore, the entire system must be generic so it can be applied to any kind of vehicle, either aerial, aquatic or terrestrial.

All vehicles must have a data transmission module that operates over the network (Fig. 4). Since users send generic instructions, each vehicle type should know how to translate them into specific actions. This way, it becomes transparent for the user to control vehicles with different characteristics, such as, for example, bikes and cars. Additionally, the vehicle must have at least one camera to capture the images of the front of the vehicle and a set of sensors in order to register and provide metric variables to the user. In parallel, the metrics obtained must be also used to validate the instructions issued by the user to prevent any accidents and inappropriate control. The metrics need to be filtered and sanitized to prevent the access of information that could be sensitive like the exact GPS position. After the processing and the validation of the control instructions, these are sent to the last block where they are transformed in low level actions like, for example, the increase of the engine speed.

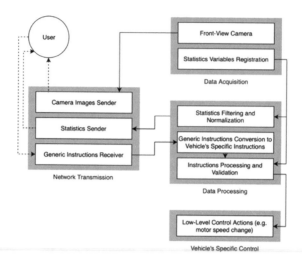

Fig. 4. Vehicles' software architecture with three main blocks for data acquisition, processing and transmission over the network and one block for vehicles' specific control methods.

3.3   Communication Protocol

The communication module regulates how data and commands flow between the user and the vehicle modules.

As stated before, the system relies on three communication channels for (Fig. 5): (i) control instructions'; (ii) metrics and (iii) transmission of the onboard camera images. To keep the vehicles safe from risks, commands must be transmitted periodically, that is, within a specific time interval depending on the characteristics of the vehicles (e.g., speed). This way, if the connection between a vehicle and its user is somehow interrupted, the vehicle will stop right after it detects the non-arrival of the periodic command. Likewise, metrics are sent periodically to the user. However, while commands are critical for the safeness of the whole operation, metrics are not. Thus, if metrics are not transmitted for a while, the system continues to operate, although emitting a warning to the user. On the contrary, the camera feed is highly critical, so it should be delivered in real-time. In fact, if the user stops seeing the on-board feed of the vehicle, there is a high probability of crashing the vehicle, hence the control of the vehicle should be aborted in order to stop the vehicle securely on the place where it started.

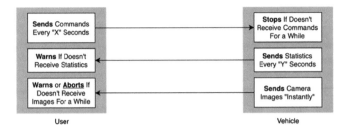

Fig. 5. Communication protocol between user and vehicle, distinguishing the three different channels.

4   Implementation

The software proposal was implemented to control cars using computers, with the focus of creating a generic software to control several different types of vehicles. Among the remotely-piloted cars, usually via radio controllers, there exist two different ones (Fig. 6): cars without steer and with four-wheel-drive, that turn by rotating independently the wheels clockwise or counter-clockwise, and cars with steer on the front wheels and rear-wheel-drive.

Fig. 6. Two remote-controlled cars, with different characteristics: on the left, the car has integral traction with independent wheel control and no steering; on the right, the car has steering in the front wheels and traction in the rear axle.

4.1   Vehicle Module

The vehicles used in this implementation are based on pre-existing structures already equipped with wheels and chassis and fitted with small-sized computers Raspberry Pi and a compatible camera named Raspberry Pi Camera. These devices allowed the installation of custom software developed accordingly the solution proposed before. The camera real-time feed relies on the WebRTC [8] technology that enables a direct connection between the vehicle and the user. The software developed for the transmission of metrics and for receiving commands is based on NodeJS, and on the JavaScript programming language. The software creates a complete service from the commands/metrics transmission through the Internet until the low-level control of the motors. Except for the camera service, all transmitted data is carried over the WebSockets protocol. For this purpose, the Javascript SocketIO is used. This protocol enables the creation of a direct communication channel between the two modules to exchange information in real-time.

The software is adapted to tackle the characteristics differences between the vehicles. In common, both vehicles are ready to receive the following instructions: FRONT, REAR, LEFT, RIGHT. Then, each vehicle type has to decide how to act depending on its own physical features.

One of the vehicles has an extra ultrasonic sensor in the front (Fig. 7) to detect nearby obstacles. Then, it has been improved the block of software responsible for the data processing in order to use the sensor information acquired to validate the instructions sent by the user. In this case, if the car is closer less than thirty centimeters of an object, it ignores the input and stops immediately, issuing a warning message to the user. Since the experimental vehicles do not have other sensors, the prototype implementation does not carry other data in the metrics channel.

Fig. 7. Special car equipped with an ultrasonic sensor in the front to detect obstacles.

4.2   User Module

For the user, the control instructions should be transparent and intuitive for any car: go forward, go backwards, turn left and turn right. It is also possible to combine one movement on each axis (for example: forward and left). The interface for the user is a webpage, using HTML 5 and JavaScript, to make it accessible without the need to install any software [16, 17]. The feed from the on-board camera is shown in real-time. Simultaneously, periodically the instructions read from the keyboard are converted to generic commands: up arrow to move forward, down arrow to move backwards, left arrow to turn to the left and right arrow to turn to the right. This conversion to generic commands could be applied also for, as an example, to joysticks of a USB keypad.

5   Tests and Results

The implementation has been applied in a scholar demonstration event with emphasis for informatics enthusiasts. The main objectives with these tests were to verify all the functionalities of the remote-control and evaluate the user experience of the control based on the cameras' images transmission.

For this purpose, the experimental setup comprised six vehicles and six computers (Fig. 8) connected to a common wireless network. The vehicles used were from the three types mentioned in the previous section. Based on this local network, the connection via WebSockets and WebRTC has been possible and established between the users and the vehicles. Users controlled each vehicle through the web-based interface at the desktop computers. The goal was to be the first to complete a circuit, racing against other opponents (Fig. 8). This test scenario confirmed the feasibility of the project, regarding the control and monitoring of unmanned vehicles outside the common radio-based solutions. Indeed, users could control the vehicles solely by watching the real-time video feed, proving that delay and latency inside the network were not critical. Moreover, users were abstracted from the differences in the controls of the different types of vehicles as these were completely transparent to them. They just had to know how to go forward, backwards, left and right, which were completely intuitive to them considering that many car simulation games also use the keyboard arrows for the same control behavior.

Fig. 8. The three different types of remote-controlled cars used in the test scenario (on the left), a user controlling a car via the keyboard (on the center) and a subset of the computers connected to the vehicles with camera images transmission (on the right).

6   Conclusions and Future Work

This work designed and proposed a complete software architecture that allows the control of any kind of vehicle, either terrestrial, aquatic or aerial. As a proof of concept, the herein presented architecture was applied for controlling several vehicles, all of them with specific and different types of steering controls types. The successful application of proposed architecture in a real-case scenario, with multiple vehicles and users, confirms that this proposal is feasible and relevant. Moreover, it can be easily adaptable to the control of new vehicle types that can surge up. The users can, from their personal devices, control any vehicle they want from where they want, without knowing any technical aspects about how the vehicles work and how they differ between themselves.

To turn this solution more accessible to any user, as future work, this software must be applied in remote scenarios using the Internet, where users are in distant places remotely controlling vehicles placed anywhere. More than distributing the service, there must also be created new modules to control and monitor other types of vehicles such as boats, tanks and drones.

Acknowledgments. This work was supported by: national funds through the Portuguese Foundation for Science and Technology (FCT) under the project UID/CEC/04524/2016; and by the project SAICT-POL/24048/2016 – NIE - Natural Interfaces with the Elderly, with reference NORTE-01-0145-FEDER-024048, financed by the Foundation for Science and Technology (FCT) and co-financed by the European Regional Development Fund (FEDER) through the Northern Regional Operational Program (NORTE2020).

References

1. Mandrekar, A: Autonomous flight control system for an unmanned aerial vehicle (2008)
2. Li, H., Wang, B., Liu, L., Tian, G., Zheng, T., Zhang, J.: The design and application of SmartCopter: an unmanned helicopter based robot for transmission line inspection. In: Proceedings of the 2013 Chinese Automation Congress, CAC 2013, pp. 697–702 (2013)
3. Silva, M.T., Lemos, O.L.: Análise de Falhas no Plantio de Café por Meio de Ortomosaico Produzido com Aeronave Remotamente Pilotada, pp. 1–8 (2011)
4. U. A. S. Special, "Archeology meets UAS Technology", GEO Informatics - Magazine for Surveying, Mapping & GIS Professionals, vol. 18 (2015)
5. Reis, A., Paredes, H., Barroso, I., Monteiro, M.J., Rodrigues, V., Khanal, S.R., Barroso, J.: Autonomous systems to support social activity of elderly people: a prospective approach to a system design. In: Proceedings of the 1st International Conference on Technology and Innovation in Sports, Health and Wellbeing, TISHW 2016 (2016). https://doi.org/10.1109/TISHW.2016.7847773
6. Reis, A., Barroso, I., Monteiro, M.J., Khanal, S., Rodrigues, V., Filipe, V., Paredes, H., Barroso, J.: Designing autonomous systems interactions with elderly people. Lecture Notes in Computer Science (including subseries Lecture Notes in Artificial Intelligence and Lecture Notes in Bioinformatics), vol. 10279 (2017). https://doi.org/10.1007/978-3-319-58700-4_49
7. Koubaa, A., Qureshi, B.: DroneTrack: cloud-based real-time object tracking using unmanned aerial vehicles. IEEE Access **6**, 13810–13824 (2018)
8. Safadinho, D., Ramos, J., Ribeiro, R., Caetano, R., Pereira, A.: UAV multiplayer platform for real-time online gaming. Adv. Intell. Syst. Comput. **571**, 577–585 (2017)
9. Koubaa, A., Qureshi, B., Sriti, M.F., Javed, Y., Tovar, E.: A service-oriented Cloud-based management system for the Internet-of-Drones, Section 3, pp. 26–28 (2017)
10. Koubaa, A., Qureshi, B., Sriti, M.F., Javed, Y., Tovar, E.: A service-oriented Cloud-based management system for the Internet-of-Drones. In: 2017 IEEE International Conference on Autonomous Robot Systems and Competitions, ICARSC 2017, pp. 329–335, September 2017
11. Fuller, B., Kok, J., Kelson, N., Gonzalez, F.: Hardware design and implementation of a MAVLink interface for an FPGA-based autonomous UAV flight control system. In: Australasian Conference on Robotics and Automation, ACRA, 2–4 December 2014

12. Atoev, S., Kwon, K.R., Lee, S.H., Moon, K.S.: Data analysis of the MAVLink communication protocol. In: 2017 International Conference on Information Science and Communications Technologies, ICISCT 2017, pp. 1–3, December 2017
13. DroneDeploy, DroneDeploy (2017). https://www.dronedeploy.com. Accessed 19 Dec 2017
14. Skycatch, Skycatch: Drone Image Processing Platform (2017). https://www.skycatch.com. Accessed 19 Dec 2017
15. Borges, R.O., et al.: Utilização de Drones de Pequeno Porte como Alternativa de Baixo Custo para Caracterização Topográfica da Infraestrutura de Transportes no Brasil, pp. 1–5 (2017)
16. Di Lucca, G.A.: Testing web-based applications: the state of the art and future trends. In: Proceedings of the International Computer Software and Applications Conference, vol. 2, p. 65, August 2005
17. Nabil, D., Mosad, A., Hefny, H.A.: Web-based applications quality factors: a survey and a proposed conceptual model. Egypt. Inform. J. **12**(3), 211–217 (2011)

Communication Modes to Control an Unmanned Vehicle Using ESP8266

David Safadinho[1,2], João Ramos[1,2], Roberto Ribeiro[1,2],
Arsénio Reis[3(✉)], Carlos Rabadão[1,2], and António Pereira[1,2]

[1] School of Technology and Management,
Computer Science and Communication Research Centre,
Polytechnic Institute of Leiria, Campus 2, Morro do Lena – Alto do Vieiro,
Apartado, 4163, 2411-901 Leiria, Portugal
`davidsafadinho.12@gmail.com`,
`jr.joaoramos@outlook.com`, `eng.rob.ribeiro@gmail.com`,
`{carlos.rabadao,apereira}@ipleiria.pt`
[2] INOV INESC INOVAÇÃO, Institute of New Technologies –
Leiria Office, Campus 2, Morro do Lena – Alto do Vieiro,
Apartado 4163, 2411-901 Leiria, Portugal
[3] INESC TEC, University of Trás-os-Montes and Alto Douro,
Vila Real, Portugal
`ars@utad.pt`

Abstract. The Wi-Fi networks are more and more omnipresent in our quotidian. It is a cheap technology that is available in many places, public or private (e.g.: schools, hospitals, public transports). The vulgarization of this technology is related to the impact that the concept of *Internet of Things* (IoT) has been having during the last years, revolutionizing the way to interact with "things". The use of Unmanned Vehicles (UV) tends to increase, for ludic or professional ends. These vehicles allow to assist and minimize the human intervention in many situations, but one of the associated problems settles in their communication architecture that needs the use of an ad-hoc remote controller that restricts the control area. This work intends to explore the ESP8266 microcontroller to control a UV over a Wi-Fi connection. Three architectures that support the interaction with the vehicle are presented and specified. The real test scenario validated all the described architectures to control the vehicles.

Keywords: Anytime, anywhere · Service-oriented architecture ·
Unmanned Vehicles · Internet of Things · NodeMCU - ESP8266 ·
Wi-Fi connection · REST · GET

1 Introduction

The UV are vehicles that can be controlled remotely by electronic and computational means, under the monitorization of humans. Their presence in the market is not recent and they have been a useful tool, since they minimize and help in the human intervention in many situations, for ludic or professional purposes. Despite not being available for everyone, these vehicles are already used by the military for long time to

© Springer Nature Switzerland AG 2019
Á. Rocha et al. (Eds.): WorldCIST'19 2019, AISC 931, pp. 56–64, 2019.
https://doi.org/10.1007/978-3-030-16184-2_6

perform many kinds of missions like area recognition, explosive devices defusing, espionage or aerial space control [1]. Nowadays, with the evolution and reduction of costs of the electronic components, the UV have been increasingly sought after to sectors like agriculture, where vehicles with enough precision to ease the processes are being developed [2]. Also, in areas of health and sports there is an ongoing trend and effort to develop unmanned and intelligent vehicle-based devices to improve communication and assist people [3, 4].

One of the limitations of these vehicles settles in the communication architecture that in most cases are point-to-point where the communication distance is limited to a determined action radius and need a physical ad-hoc remote controller to command the vehicle. There are many architectural patterns that are drawn to ease the interaction of multiple vehicles in collaboration [5], but, to operate, the vehicles need specific communication modules to interpret the commands.

Technological progress, constant size reduction of electronic components and sensor improvements, are aspects that allow to build smaller and reliable board controllers. The IoT concept is increasingly present in ours every day and it is undoubtedly a big influence issue in the construction of new vehicles [6].

According to previsions of the Cisco company, the number of devices connected to the Internet until 2030 will be of 500 billion [7] and the traffic of mobile devices and Wi-Fi technology will represent more than 63% of the IP traffic in 2021, where the remaining 37% will be occupied by wired devices [8]. These numbers encourage the development of other areas that use wireless connectivity [9].

So, considering the previous concepts and following the work developed in [10], the objective of this work is to use a communication alternative to control UV based in Wi-Fi, applying the architectures defined that support the communication between the user and the UV, either locally or through the Internet, using low-cost equipment.

This article has the following structure. In Sect. 2 it is presented the related work. In Sect. 3 it is presented the System's Architecture that allows us to achieve the defined goal. In Sect. 4 it is presented the implementation of the platform to control the vehicle. Then, in Sect. 5, we expose how the test scenario for our prototype was built. Lastly, in Sect. 6 are presented the conclusions of this article and a small discussion related to possible improvements in future.

2   Related Work

This section analyzes research works proposing different architectures to control UVs. There are some architectural models thought for the control of UV, being the most common the point-to-point communication between the signal sender and the vehicle. This solution is the most used architecture in many types of vehicles, being them terrestrial, aerial or aquatic. Their biggest problem is to limit the communication with the vehicle to a determined area and being required an ad-hoc controller for each vehicle.

However, to solve this problem and one of the biggest challenges in the simultaneous control of multiple devices, a new paradigm of wireless communication was studied and called Flying Ad hoc Networks (FANET) [11, 12]. This concept that is more focused in the unmanned aerial vehicles and its objective is to administrate their autonomous movement. The functional architecture of this network is composed by a central point that communicates with a vehicle that works as a leader of a determined group of vehicles. The responsibility of the leader is to receive the information from the central network infrastructure and share it with the other members. Figure 1 illustrates the functioning of this network.

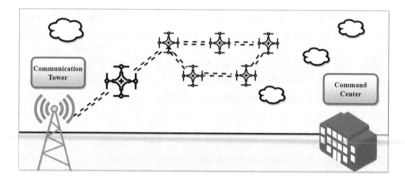

Fig. 1. Examples of a Flying Ad-Hoc Network (FANET)

Directed to the availability of services the work designated by "DroneMap Planner" [13] was analyzed, where the authors proposed the creation of a platform that allows to provide services related to the control of UAV through the Internet. Its functional architecture (Fig. 2) uses the concepts of Internet of Drones (IoD), that enables the control of its vehicles over the Internet. These vehicles, on the contrary of the approach followed by commercial drones, need to be able to connect to a wireless network infrastructure through a Wi-Fi connection. Besides containing a communication module different from the common vehicles, these need the capacity to interpret the MAVLink protocol. This protocol is used to communicate with the Ardupilot that manages the actions of the vehicle. The used architecture also follows a distributed approach, containing a cloud-based platform that is responsible for the management of the entire communication with the vehicles. It is a solution mainly focused in delivering mission-oriented services, being the flight path previously defined in the platform through GPS reference points.

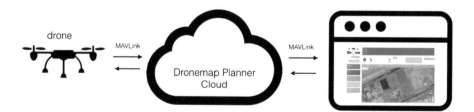

Fig. 2. Functional architecture of the "DroneMap Planner" system.

3 System's Architecture

The system's architecture can be used through three different approaches (Fig. 3). They are point-to-point communication without infrastructure (1), point-to-point communication with a local network infrastructure, without Internet access (2) and lastly, the point-to-point communication through a network infrastructure that allows the access to the Internet (3). Depending on the situations it can be possible to execute the vehicle control using one of the specified architectures. Each of them allows to control a UV and its control service is located in the communication controller, available through a web interface.

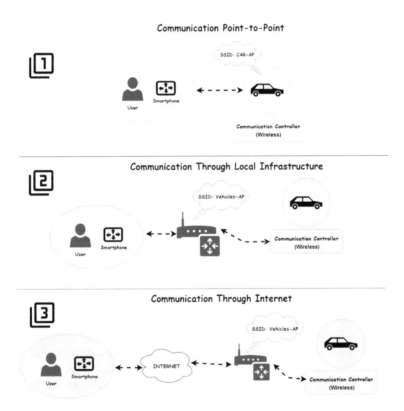

Fig. 3. Architecture of the systems that make possible the control of a UV through a communication controller.

The first architecture, identified by 1 in Fig. 3, is similar to the communication approach of vehicles usually commercialized for entertainment ends. The difference corresponds to the communication mean that is established by radio frequencies and requires a proprietary controller to emit the control commands. In this architecture, the vehicles create a network hotspot, through its communication controller, where it is possible to establish a wireless communication with a controller device (e.g.: computer, smartphones). The user must connect to its access point (AP) and use the Internet browser to reach the IP address of the vehicle. This architecture works well in scenarios where the user is relatively near the device. However, it presents problems when the objective is to control the vehicle with a wider distance. The higher the distance, then the lower will be the communication signal. If we want to apply this solution in multiple vehicles in the same place, it will create conflicts in hotspot networks. The solution for the previously mentioned problems passes by adding the network infras-tructure that eases the communication between the devices, allowing it to obtain wider reach, being in a central point.

In the architecture identified by 2 in Fig. 3, the architectural paradigm changes. In this case, both the vehicle and the user are connected to a network infrastructure supported by a router. The responsibility of this device is to create an AP that works as communication central point between the devices available in the network. In this situation, the user needs to access the IP address of the vehicle and the router directs all the traffic to the destiny. This architecture limits the communication locally, not being possible to control a vehicle outside the defined network.

The chosen solution is the one identified by 3 in Fig. 3. This architecture allows to control the vehicles *anytime and anywhere*. It becomes possible if the vehicle and the user are connected to the Internet. Since it is not a local network, where the IP addresses are accessible locally, it is necessary to open communication ports to the exterior, which lifts the need for security measures. This process needs to be done in the side of the network infrastructure of the vehicle, through the network service known as port forwarding that is responsible for redirecting the traffic coming from a deter-mined port to a device internal to the network, in this case the vehicle.

4 Implementation

The implementation of this system is dependent of many components, these being hardware or software. The required hardware to the functioning of the vehicle was essentially its chassis and the four independent motors connected to a Motor Control Shield [14]. This shield is responsible for all the control of the vehicle's motors. To take advantage of a wireless communication, the NodeMCU development board was used. This device integrates an ESP8266 microcontroller that allows to manipulate digital ports. Its programming is done through the Arduino platform that includes the libraries and drivers required to its functioning. In the next sections the assembling scheme of the vehicle and the description of the implemented software are presented.

4.1 Assembling

The components the vehicle is made of require a specific assembling process. The Fig. 4 illustrates the connection scheme of the vehicle's motors. The connection of them has been done in a way to have each side of the vehicle independent, which means that when it is issued the instruction to go forward, all the motors turn in the same direction. In order to allow the vehicle to rotate, considering that the vehicle has no steer, the wheels on the left of the vehicle should turn in the same direction and the wheels on the right side should turn in the opposite direction.

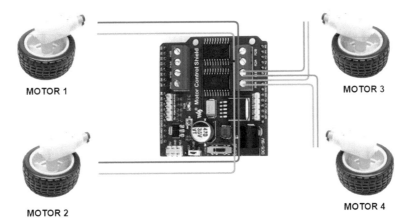

Fig. 4. Connections between the motors and the Motor Control Shield.

The Fig. 5 illustrates the connection scheme between the NodeMCU and the motors' controller. With these connections, the motors receive instructions to turn on or off. The power supply voltage of the motors' controller and the NodeMCU is 5 volts and for this work the power supply unit used was a powerbank.

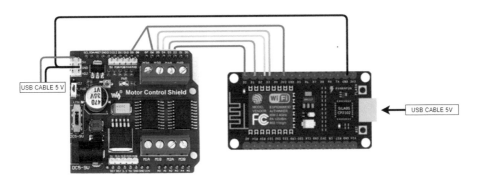

Fig. 5. Connections between the Motor Control Shield and the ESP8266.

4.2 Control Software

The developed software to control the vehicle is installed in the communication controller ESP8266. The control system was created using Web oriented Technologies namely HTML and JavaScript. The Fig. 6 illustrates the Webpage that is presented to the user after connection to the vehicle. In that webpage, there are shown five buttons that correspond to the possible operations for vehicles' control: front, back, left, right and stop. When the user presses the pretended button, it is identified and depending on the direction that he or she intends and enabled or disabled the digital ports according to the vehicle's motors in order to go in the desired direction. The interaction fashion with the system can be done either by a smartphone/tablet or a computer. If it is with a computer, the control can be done by a keyboard. The ESP8266 is programmed to connect to a network infrastructure that already exists, being then a network client. When the buttons are pressed there is created a communication to the device using the communication technology REST, more specifically by the method GET.

Fig. 6. Vehicle control through the ESP8266 microcontroller.

5   Test Scenario and Results

Using the implementation done in the previous section, this solution was used by a group of people in an event that took place in the Polytechnic Institute of Leiria. This group of people used their smartphones to interact with the system. As the Fig. 7 illustrates, the test scenario was composed by 7 vehicles. Due to the lack of power-banks at the moment, the power supply of the vehicles, during this test scenario, was done with a regular power supply connected to an outlet for each vehicle.

Initially it has been implemented the architecture point-to-point without the network infrastructure, described by the number 1 of the Fig. 3, but rapidly it has been concluded the existence of problems related to network collisions. To solve the problem, it has been implemented the architecture defined by the number 2 of the Fig. 3. With this test scenario it has been proved that it has been possible to control the various vehicles through a Wi-Fi communication using low-cost resources.

Fig. 7. Vehicles used in the test scenario.

6 Conclusions and Future Work

With this article it was developed a control system of an unmanned vehicle that will allow the remote control over three different scenarios (e.g.: point-to-point communication without infrastructure, point-to-point communication using a local network infrastructure without Internet access and lastly, point-to-point communication using a network infrastructure that allows the *anytime, anywhere* control). Building the test scenery, it was noticed that in both architectures the vehicle control worked, but when the control of multiple vehicles in the same place is needed, the point-to-point architecture became unviable.

To facilitate interaction with the vehicle, it was developed a simple control interface to be accessible to any type of user. It is hosted in the NodeMCU–ESP8266 microcontroller. This device is cheap and demonstrated that it is capable to maintain a stable Wi-Fi connection, being able to manipulate it digital ports.

In the future it is pretended to implement new services using that type of vehicles, specifically in applications related with elderly people.

Acknowledgments. This work was supported by: national funds through the Portuguese Foundation for Science and Technology (FCT) under the project UID/CEC/04524/2016; and by the project SAICT-POL/24048/2016 – NIE - Natural Interfaces with the Elderly, with reference NORTE-01-0145-FEDER-024048, financed by the Foundation for Science and Technology (FCT) and co-financed by the European Regional Development Fund (FEDER) through the Northern Regional Operational Program (NORTE2020).

References

1. Silva, M.R., et al.: Communication network architecture specification for multi-UAV system applied to scanning rocket impact area first results. In: 2017 Latin American Robotics Symposium 2017, Brazilian Symposium on Robotics, pp. 1–6 (2017)

2. Bascetta, L., Baur, M., Gruosso, G.: Electrical unmanned vehicle architecture for precision farming applications. In: 2017 IEEE Vehicle Power and Propulsion Conference, VPPC 2017, Proceedings, pp. 1–5 (2018)

3. Reis, A., Paredes, H., Barroso, I., Monteiro, M.J., Rodrigues, V., Khanal, S.R., Barroso, J.: Autonomous systems to support social activity of elderly people: a prospective approach to a system design. In: 1st International Conference on Technology and Innovation in Sports, Health and Wellbeing, TISHW 2016, Proceedings (2016). https://doi.org/10.1109/TISHW.2016.7847773

4. Reis, A., Barroso, I., Monteiro, M.J., Khanal, S., Rodrigues, V., Filipe, V., Paredes, H., Barroso, J.: Designing autonomous systems interactions with elderly people. Lecture Notes in Computer Science (including subseries Lecture Notes in Artificial Intelligence and Lecture Notes in Bioinformatics), vol. 10279 (2017). https://doi.org/10.1007/978-3-319-58700-4_49

5. Zafar, W., Khan, B.M.: A reliable, delay bounded and less complex communication protocol for multicluster FANETs. Digit. Commun. Netw. 3(1), 30–38 (2017)

6. Kamal, Z., Mohammed, A., Sayed, E., Ahmed, A.: Internet of Things applications, challenges and related future technologies. WSN World Sci. News 67(672), 126–148 (2017)

7. Cisco: Internet of Things, ISACA J. (2017)

8. Cisco Visual Networking Index: Forecast and Methodology, 2016–2021 - Cisco. https://www.cisco.com/c/en/us/solutions/collateral/service-provider/visual-networking-index-vni/complete-white-paper-c11-481360.html. Accessed 19 June 2018

9. The Internet of Things: 5 Predictions for 2018. https://blogs.cisco.com/innovation/the-internet-of-things-5-predictions-for-2018. Accessed 03 July 2018

10. Safadinho, D., Ramos, J., Ribeiro, R., Caetano, R., Pereira, A.: UAV multiplayer platform for real-time online gaming. In: Recent Advances in Information Systems and Technologies, WorldCIST 2017. Advances in Intelligent Systems and Computing, vol. 571 (2017)

11. Zafar, W., Muhammad Khan, B.: Flying ad-hoc networks: technological and social implications. IEEE Technol. Soc. Mag. 35(2), 67–74 (2016)

12. Gupta, L., Jain, R., Vaszkun, G.: Survey of important issues in UAV communication networks. IEEE Commun. Surv. Tutorials 18(2), 1123–1152 (2016)

13. Qureshi, B., Koubaa, A., Sriti, M.-F., Javed, Y., Alajlan, M.: Poster: Dronemap - a cloud-based architecture for the Internet-of-Drones. In: International Conference on Embedded Wireless Systems Networks, pp. 255–256 (2016)

14. Motor Control Shield - Waveshare Wiki. https://www.waveshare.com/wiki/Motor_Control_Shield. Accessed 30 Nov 2018

Automatic Generation of a Sub-optimal Agent Population with Learning

Simão Reis[1(✉)], Luís Paulo Reis[1], and Nuno Lau[2]

[1] LIACC/FEUP, Artificial Intelligence and Computer Science Laboratory,
Faculty of Engineering, University of Porto, Porto, Portugal
simao.reis@outlook.pt, lpreis@fe.up.pt
[2] DETI/IEETA, Institute of Electronics and Informatics Engineering of Aveiro,
University of Aveiro, Aveiro, Portugal
nunolau@ua.pt

Abstract. Most modern solutions for video game balancing are directed towards specific games. We are currently researching general methods for automatic multiplayer game balancing. The problem is modeled as a meta-game, where game-play change the rules from another game. This way, a Machine Learning agent that learns to play a meta-game, learns how to change a base game following some balancing metric. But an issue resides in the generation of high volume of game-play training data, was agents of different skill compete against each other. For this end we propose the automatic generation of a population of surrogate agents by learning sampling. In Reinforcement Learning an agent learns in a trial error fashion where it improves gradually its policy, the mapping from world state to action to perform. This means that in each successful evolutionary step an agent follows a sub-optimal strategy, or eventually the optimal strategy. We store the agent policy at the end of each training episode. The process is evaluated in simple environments with distinct properties. Quality of the generated population is evaluated by the diversity of the difficulty the agents have in solving their tasks.

Keywords: Computer games · Reinforcement Learning ·
Learning in games · Multi-agent systems

1 Introduction

Modern video games have more difficult than ever to offer ideal experience to players, giving more incentive to find ways to mitigate video games development costs [13]. The entertainment industry has the most interest in refining and tuning their game balancing strategies in order to maintain player retention and increase revenue. One of the main factors to offer an engaging experience is the balance between challenge and skill [3]. In the case of multiplayer variants is game fairness. Game fairness can be defined as a game where players have similar chances of winning [19].

© Springer Nature Switzerland AG 2019
Á. Rocha et al. (Eds.): WorldCIST'19 2019, AISC 931, pp. 65–74, 2019.
https://doi.org/10.1007/978-3-030-16184-2_7

Research on this subject is not yet very wide in literature, perhaps for the lack of incentive. A potential research application lies on Serious Games. The motivation and engagement of patients in training sessions using typical therapy tasks is difficult [6]. One of the directions that has been emerging for rehabilitation is the incorporation of a social dimension. Social interaction brings engagement in terms of player experience [18], increasing the patients' motivation in rehabilitation to keep practicing the exercises they need. Social play can be achieved with the incorporation of competitive and collaborative tasks [1]. The balance between challenge and skill of the player is considered one of the most important variables that affects players enjoyment [3]. We're currently working on a framework where the problem of game balancing is modeled as a meta-game, a game where the action of the player affects the rules of another game. Giving appropriate training a Machine Learning (ML) agent by learning to play a meta-game could learn the most efficient way to adapt the game rules of another game. However, ML solutions require a lot of data, in this case a lot of playing data. Current Massive Multiplayer Online games can obtain this kind of data as active players reaches hundreds of thousands or even millions, and that allows for developers to collect and analyze statistics and make changes to the game. But for development and testing phases, it's sometimes this task is not feasible as the game is suffering constant changes.

During training Reinforcement Learning (RL) agents learn in a trial and error fashion. Agents will gradually become better to perform their task. As such we propose a methodology for the automatic generation of a population of surrogate agents where each member has a distinct skill in beating the game. We preserve the evolutionary phases of an agent training. Each agent will have a distinct skill level in solving their task. This way the population could be used to reproduce game-plays as much as needed, were agents of different skill competes against each other in order to a meta-agent figure out how to best balance the game, in order to become more fair. Our contribution lies in the evaluation of RL as a mean to automatically generate a population of sub-optimal stationary agents in environments with distinct properties.

The rest of this work is organized as follows. In Sect. 2 we briefly discuss some related work in RL in games and video games. In Sect. 3 we introduce some basic background. In Sect. 4 we discuss our test environments and metric to evaluate agents performance. In Sect. 5 we discuss our experiment and results. Finally, in Sect. 6 we discuss our main conclusions and future work.

2   Related Work

Games are a important test bed for RL [9]. Recent advancements in RL have contributed to solve numerous complex single and multiplayer video games.

On the video games domain, Deep Q-Learning (DQN) was developed as the first and tested Atari Games [12]. It was the first deep learning model to successfully learn control policies over high-dimensional input data. OpenAI developed

an algorithm to train RL agents by self play and were able to beat professional Dota 2^1 players [2]. Recently in [8], agents achieve human-level capacity
in Quake III Arena Capture the Flag, with only pixels and game points as
input. They use a novel two-tier optimization process in which a population of
RL agents are trained concurrently. They play thousands of parallel matches in
teams against each other on randomly generated environments. Agents display
high level behaviours such as navigating, following, and defending.

In the domain of strategic sequential games, AlphaGO [14] learned to play the
game of Go using a mix of RL and supervised learning, i.e., with human knowledge. AlphaGO Zero [17] mastered the game of GO, without human knowledge.
AlphaZero [15] mastered the games of Chess, Shogi and Go, with a more general
purpose solution.

State of the art RL is moving towards policy based learning. Deterministic
Policy Gradient (DPG) [16] assumes that the policy is deterministic decision
function. Deep Deterministic Policy Gradient (DDPG) [10] extends DQN by
combining it with DPG. DDPG allows DQN to learn in continuous spaces with an
Actor-Critic framework. Distributed Distributional DDPG (D4PG) [4] improves
DDPG to run in an distributed environment. Multi-Agent DDPG (MADDPG)
[11] extends DDPG to multi-agent coordination environments, where agents
have partial observation or local information. Twin Delayed Deep Deterministic (TD3) [7] like DQN tries to prevent overestimation of the value function by
improving DDPG. It adds three new steps: Clipped Double Q-learning; Delayed
update of Target and Policy Networks; and Target Policy Smoothing.

3   Background

In RL the mapping history of observation the next optimal action is called the
policy of the agent π. Storing all observations and actions may be computationally unfeasible, and large memory may be required. In practice it is usually opted
to use the reactive model of an agent where only the current observation θ is used
for decision-making $\pi(\theta)$. A property that is very desirable for our meta-game
training model is that of minimal computational complexity for decision making
on agents that are playing the to be adapted game. The reflective model fulfills
this property.

Q-Learning is one of the main RL algorithms and serves as base of construction for more advanced algorithms. Temporal-Difference (TD) learning is
model-free (doesn't depend of knowing the world model, or distribution of state
transition) and learns from episodes of experience. TD algorithms can learn
from incomplete episodes. Q-Learning has two main implementations: Table Q-
Learning; and DQN. While the first has a more stable learning and convergence
guarantees, it cannot be applied to continuous environments and doesn't scale
well with very complex environments with large world states. Q-Learning aims
to maximize future expected value, in accordance with the update rule:

$$Q(s_t, a_t) := (1 - \lambda)Q(s_t, a_t) + \lambda[R(s_t, a_t) + \delta \max_{a_{t+1}} Q(s_{t+1}, a_{t+1})] \qquad (1)$$

---

[1] www.dota2.com/.

where Q is the action-state quality value and R is the reward function, given by a world interpreter that evaluates the agent actions. λ is the learning rate and δ is the future discount factor.

4   Methodology

In this section we analyze the main characteristics of both the agent and environment that affects the automatic generation of a population of agents using RL. We also present the main advantages of sampling by learning.

Parallel game is one were each player is competing without interaction, each agent its playing its own single player game. Their task is comparable, for example, with how many points each one has earned. For simplification we will assume that agents playing parallel games are each playing the same game or under the same ruler set. The opposite to parallel games are non parallel games, know as strategic or interactive games. The payoff of taking a decision is affected not only by their action, but by their opponents actions as well.

Deterministic agent is one that always takes the same action in a given state. A non-deterministic (or stochastic) agent is one that has a probability distribution over actions to take, one distribution for each world state. In a deterministic environment an agent always reaches the same state when taking the same action. In a non deterministic environment sometimes the agent may not reach the expected state. Both are very common in video games, so we tested our generation procedure on every combination of deterministic and stochastic agent and environment.

We also define games with pitfall states as games with terminal states which are not goal states. Games with pitfalls can heavily affect population generation, mainly when the environment is non deterministic. Even if TD learning may learn from incomplete episodes, if the agent never reaches the goal state, it will never learn how to clear the game. If the maximum number of training steps per episode is shorter than the distance of the starting state to the closet goal state, the agent will also never learn how to play the game. As these agents don't actually learn how to play the full game, we don't find them interesting and don't include them into the population. To avoid wasting computational iterations, we applied technique we called Mask Blinding. When an agent concludes unsuccessfully an episode by falling to a pitfall state, we blind the last action performed in the last non terminal state the agent reached, by setting $Q(s_{pitfall-1}, a) = -1$ in Table Q-Learning (considering that Q-values are non negative values). This way a deterministic agent will give less priority, and a stochastic agent (using SoftMax function over Q-values as exploration policy for example) will have a very reduced probability of choosing the action that takes the agent to the pitfall state.

To describe and compare the generated population we make use of a population trace. A population trace is an ordered set of agents by their skill level. Its a curve indexed by the agents' id numbers and the co-domain is the skill level of each agent. It is a monotonic growing succession. The most important characteristics of the population trace are the Δ (difference between max and min skill) and Γ (skill density or variety). Large Δ means there is a great gap from the best performing agent to the worst performing agent. Large Γ represents a rich variety of agents.

5   Experimentation

In this section we present the simulations performed for agent learning over multiple simple environments, where the previous discussed characteristics impact over population generation.

OpenAI Gym [5] is a framework that become one of the main standard to develop RL solutions. It provides an interface for generic RL agents to interact with an environment. Their website contains a library of pre-built environment and leader-board of agents' performance.

The first tested scenario was FrozenLake[2], a simple maze environment. The agent has a starting state on one of the lake extremes and walks through a frozen lake towards the other side, the goal state. The lake is filled with holes, pitfall states, so it must be careful to avoid dropping in one of those holes. The action set of the agent is to move up, down, left and right. FrozenLake allows for the re-dimension of the maze. We experimented with a 4×4 board (16 states) and 8×8 board (64 states), with normal floor or slippery floor, were the agent may unintentionally slip and reach and undesired state.

As FrozenLake is a relatively small discrete environments, we use Table Q-Learning for automatic generation of the population. In our tests we alter between agents being deterministic and stochastic. In Fig. 1 we can observe the population trace with every combination scenario of deterministic and stochastic agent and environment in the 4×4 version of FrozenLake.

First result is trivial, with both deterministic agent and environment the agent rapidly finds the optimal path and has no incentive to deviate from its strategy. The resulting population trace is a constant line with skill equal to the number of steps needed to reach the goal state. This results in a population with $\Delta = 0$, and $\Gamma = 1$. When the environment becomes stochastic (slippery floor) the population trace becomes a curve. The most interesting observation in our opinion is the skill distribution. There are little agents with minimal and maximal performance, observable by the rapid growth of the curve close to index zero and in the largest index, while it stays more linear in the middle. The difference with the case of stochastic agent with deterministic environment is that the curve is more smooth. This indicates that the variance of the environment is greater than of the agent. Our stochastic agent uses SoftMax function over Q-values as exploration policy. In the last scenario we were unable to obtain a population, as the stochastic agents constantly slipped to holes. But applying Blind Masking we were able to generate a population, only with a very less smooth skill curve. Δ and Γ values are very similar in the last three.

In the FrozenLake 4×4 experiment we had $\lambda = 0.1$, $\gamma = 0.95$, exploration rate of 0.1, 2000 episodes, max steps equal to 99.

---

[2] https://gym.openai.com/envs/FrozenLake-v0/.

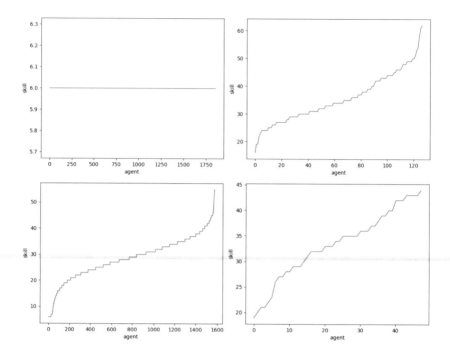

Fig. 1. Population trace in FrozenLake-v0 environment of OpenAI Gym, with 16 states. Skill is the number of the steps the agent took on average to reach the goal state. Deterministic Agent and Environment (Left-Top); Deterministic agent and stochastic environment (Right-Top); Stochastic agent and deterministic environment (Left-Bottom); Stochastic agent and environment with blind masking (Right-Bottom).

We repeated the same test scenarios in FrozenLake with a 8×8 dimension board, or 64 states. The population traces is illustrated in Fig. 2. The deterministic agent obtained similar results to the previous scenario (not illustrated). However using a stochastic agent the generation of population was inconsistent or null. As such we applied Blind Masking to every scenario for comparison. Its observable that the population trace for every case except deterministic agent and environment is very similar. This means that population can be successfully generated as the agent avoids pitfall states, and has population has the same quality regardless the conditions of the generation. In addition, Blind Masking does not make use of environment changes, and still produces reactive agents as intentioned. Also shows that the methodology scales (as long as it is feasible with the method in use, in this case Q-Table).

In the FrozenLake 8×8 experiment we had $\lambda = 0.1$, $\gamma = 0.95$, exploration rate of 0.1, 5000 episodes, max steps equal to 99.

Next we test with two simple continuous environments, with physical dynamics. First is Cart Pole[3], where a cart with and inverted pendulum must be moved

---
[3] https://gym.openai.com/envs/CartPole-v0/.

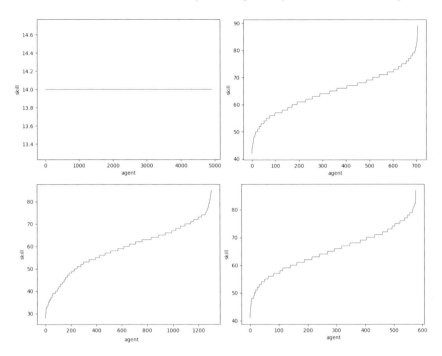

Fig. 2. Population trace in FrozenLake environment of OpenAI Gym, with 64 states. Blinding mask was used in every scenario. Deterministic agent and environment (Left-Top); Deterministic agent and stochastic environment (Right-Top); Stochastic agent and deterministic environment (Left-Bottom); Stochastic agent and environment (Right-Bottom).

(left or right) in order to not let drop the pendulum. In Mountain Car[4], where a car must be balanced until it reaches the goal on the top of the cliff. For automatic generation of population we used DQN, as the environment are continuous. The population trace for both scenarios are illustrated in Fig. 3.

The most relevant characteristic between the two is the type of challenge. While in Cart Pole we can evaluate agent performance by the number of steps before dropping the pendulum, in Mountain Car we can only evaluate the agent when reaches the goal. The Γ value of Cart Pole trace is much greater than Γ value of Mountain Car (both run over a similar number of episodes). Although Mountain Car doesn't have pitfall states, it does suffer from the other discussed problem, of the number of steps needed to reach the goal, which very affected the quality of the population.

In the Cart Pole experiment we had $\lambda = 0.001$, $\gamma = 0.95$, replay memory batch size of 20, exploration rate decay of 0.995, starting on 1.0 until 0.01 minimum. In the Mountain Car experiment we had $\lambda = 0.01$, $\gamma = 0.99$, replay memory batch size of 20, exploration rate decay of 0.9, starting on 1.0 until 0.01 minimum. One hidden layer of 24 Rlu units.

---

[4] https://gym.openai.com/envs/MountainCar-v0/.

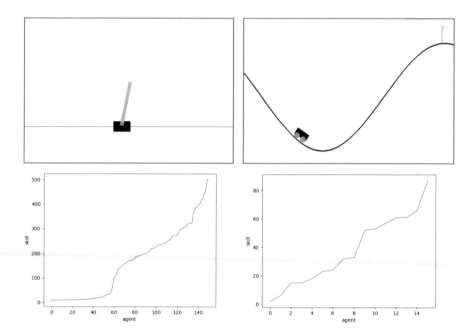

Fig. 3. Population trace in CartPole-v1 environment of OpenAI Gym (left), skill is the number of steps the agent stayed without dropping the inverted pendulum; Population trace in MountainCar-v0 environment of OpenAI Gym (right).

6   Conclusion

With this work, we concluded that RL can be used to successfully generate automatically a population of diverse agents with distinct task proficiency on parallel games. The methodology is seamingless to multiple game characteristic such as an agent or game being deterministic or stochastic, or the environment being continuous or discrete. As such has the potential contribute for minimal development effort in game balancing for even more very complex games.

Blind Masking showed promising results, it only works with table Q-Learning. The technique cannot be directly applied in deep networks which are essential for learning of complex environments. In addition, state of the art are now policy gradient based, most following the Actor-Critic architecture. These advanced algorithms can potentially solve the Mountain Car problem by learning faster a policy. It will also be interesting avenue to explore population generation with more state of the art learning algorithms with the use of Blind Masking.

Also, an open problem remains with evaluation of agent performance in strategic games. In a simple simultaneous game like Rock Paper Scissors, an agent that plays always rock, beats an agent that always plays scissors, one that always plays scissors defeats on that always player paper, etc. Nash Equilibrium (NE) solution is well known, both players play Rock, Paper or Scissor with equal probability. But a stationary player which always plays rock will have a

50% win rate against an stationary opponent playing NE. There is not a strictly full ordering of stationary agents by skill in strategic games.

Future work in population generation will involve two lines of work. First, to improve our methodology with state of the art policy gradient algorithms, integrating Blind Masking. Second, to research the design and generation of non stationary agents and how to evaluate their skill. Both avenues are to be applied in our current work with meta-game learning for game balancing.

Acknowledgments. This work is supported by: Portuguese Foundation for Science and Technology (FCT) under grant SFRH/BD/129445/2017; LIACC (PEst-UID/CEC/00027/2013); IEETA (UID/CEC/00127/2013).

References

1. Alankus, G., Lazar, A., May, M., Kelleher, C.: Towards customizable games for stroke rehabilitation. In: Proceedings of the SIGCHI Conference on Human Factors in Computing Systems, CHI 2010, New York, NY, USA, pp. 2113–2122. ACM (2010)
2. Andersen, P., Goodwin, M., Granmo, O.: Deep RTS: a game environment for deep reinforcement learning in real-time strategy games. In: 2018 IEEE Conference on Computational Intelligence and Games (CIG), pp. 1–8, August 2018
3. Baldwin, A., Johnson, D., Wyeth, P., Sweetser, P.: A framework of dynamic difficulty adjustment in competitive multiplayer video games. In: 2013 IEEE International Games Innovation Conference (IGIC), pp. 16–19, September 2013
4. Barth-Maron, G., Hoffman, M.W., Budden, D., Dabney, W., Horgan, D., Dhruva, T.B., Muldal, A., Heess, N., Lillicrap, T.: Distributional policy gradients. In: International Conference on Learning Representations (2018)
5. Brockman, G., Cheung, V., Pettersson, L., Schneider, J., Schulman, J., Tang, J., Zaremba, W.: OpenAI gym. CoRR, abs/1606.01540 (2016)
6. Burke, J.W., McNeill, M.D.J., Charles, D.K., Morrow, P.J., Crosbie, J.H., McDonough, S.M.: Augmented reality games for upper-limb stroke rehabilitation. In: 2010 Second International Conference on Games and Virtual Worlds for Serious Applications, pp. 75–78, March 2010
7. Haarnoja, T., Zhou, A., Abbeel, P., Levine, S.: Soft actor-critic: off-policy maximum entropy deep reinforcement learning with a stochastic actor. CoRR, abs/1801.01290 (2018)
8. Jaderberg, M., Czarnecki, W.M., Dunning, I., Marris, L., Lever, G., Castañeda, A.G., Beattie, C., Rabinowitz, N.C., Morcos, A.S., Ruderman, A., Sonnerat, N., Green, T., Deason, L., Leibo, J.Z., Silver, D., Hassabis, D., Kavukcuoglu, K., Graepel, T.: Human-level performance in first-person multiplayer games with population-based deep reinforcement learning. CoRR, abs/1807.01281 (2018)
9. Li, Y.: Deep reinforcement learning: an overview. CoRR, abs/1701.07274 (2017)
10. Lillicrap, T.P., Hunt, J.J., Pritzel, A., Heess, N., Erez, T., Tassa, Y., Silver, D., Wierstra, D.: Continuous control with deep reinforcement learning. CoRR, abs/1509.02971 (2015)
11. Lowe, R., Wu, Y., Tamar, A., Harb, J., Abbeel, P., Mordatch, I.: Multi-agent actor-critic for mixed cooperative-competitive environments. In: Neural Information Processing Systems (NIPS) (2017)

12. Mnih, V., Kavukcuoglu, K., Silver, D., Graves, A., Antonoglou, I., Wierstra, D., Riedmiller, M.A.: Playing atari with deep reinforcement learning. CoRR, abs/1312.5602 (2013)
13. Oliveira, S., Magalhães, L.: Adaptive content generation for games. In: Encontro Português de Computação Gráfica e Interação (EPCGI), pp. 1–8, October 2017
14. Silver, D., Huang, A., Maddison, C.J., Guez, A., Sifre, L., van den Driessche, G., Schrittwieser, J., Antonoglou, I., Panneershelvam, V., Lanctot, M., Dieleman, S., Grewe, D., Nham, J., Kalchbrenner, N., Sutskever, I., Lillicrap, T.P., Leach, M., Kavukcuoglu, K., Graepel, T., Hassabis, D.: Mastering the game of go with deep neural networks and tree search. Nature **529**, 484–489 (2016)
15. Silver, D., Hubert, T., Schrittwieser, J., Antonoglou, I., Lai, M., Guez, A., Lanctot, M., Sifre, L., Kumaran, D., Graepel, T., et al.: Mastering chess and shogi by self-play with a general reinforcement learning algorithm. arXiv preprint arXiv:1712.01815 (2017)
16. Silver, D., Lever, G., Heess, N., Degris, T., Wierstra, D., Riedmiller, M.: Deterministic policy gradient algorithms. In: Xing, E.P., Jebara, T. (eds.) Proceedings of the 31st International Conference on Machine Learning, vol. 32 of Proceedings of Machine Learning Research, pp. 387–395, Bejing, China, 22–24 June 2014. PMLR (2014)
17. Silver, D., Schrittwieser, J., Simonyan, K., Antonoglou, I., Huang, A., Guez, A., Hubert, T., Baker, L.R., Lai, M., Bolton, A., Chen, Y., Lillicrap, T.P., Hui, F., Sifre, L., van den Driessche, G., Graepel, T., Hassabis, D.: Mastering the game of go without human knowledge. Nature **550**, 354–359 (2017)
18. Vanacken, L., Notelaers, S., Raymaekers, C., Coninx, K., van den Hoogen, W.M., Ijsselsteijn, W.A., Feys, P.: Game-based collaborative training for arm rehabilitation of MS patients: a proof-of-concept game, pp. 1–10 (2010). conference; GameDays2010; 2010-03-25; 2010-03-26; Conference date: 25-03-2010 Through 26-03-2010
19. Wu, M., Xiong, S., Iida, H.: Fairness mechanism in multiplayer online battle arena games. In: 2016 3rd International Conference on Systems and Informatics (ICSAI), pp. 387–392, November 2016

Evaluation of Open Source Software for Testing Performance of Web Applications

Fernando Maila-Maila[1], Monserrate Intriago-Pazmiño[1(✉)],
and Julio Ibarra-Fiallo[2]

[1] Departamento de Informática y Ciencias de la Computación,
Escuela Politécnica Nacional, Quito, Ecuador
fernando_efml@hotmail.com,
monserrate.intriago@epn.edu.ec
[2] Colegio de Ciencias e Ingenierías, Universidad San Francisco de Quito,
Cumbayá, Ecuador
jibarra@usfq.edu.ec

Abstract. The evaluation of Open Source software for testing performance in Web applications is accomplished. In order to achieve the objective, the research model was followed: literature review, strategy, methods to generate data and the analysis of the data. This study was conducted based on measures of the ISO/IEC 25023 standard, and the software test process proposed by ISTQB. Regarding the software evaluation, it was done considering the characteristics described by their official developers and experimentation. The tool that obtained the best results was JMeter, covering 80% of the selection criteria. It is applied to a case study. This evaluation allowed to know the performance measures of a real web application.

Keywords: Software quality · Open Source · ISO/IEC 25010 ·
ISO/IEC 25023 · ISTQB

1 Introduction

Although, the importance of software today is indisputable, there is a wide and varied market in terms of software products. In the precise case of software that allows testing software, and given the massive use of online services, we will study a broad set of software to achieve web application performance tests. In a recent work [1], a study of various tools to perform tests of web application performance were analyzed, the authors comment that most are expensive and complex to use. Therefore, they propose a general-purpose framework to test the performance of web applications, separating the logic of the application from the performance functionality to be tested. We were not able to acquire and test their proposal. In the publication [2], the consequences when performing web application performance tests in a virtual environment compared to performing it in a physical environment are discussed. The authors conclude that special considerations must be taken when working in a virtual environment to report reliable data. In the article [3], the researchers note that unit tests are not as common as functional unit tests. Therefore, a greater effort is required in this fundamental task of web applications.

© Springer Nature Switzerland AG 2019
Á. Rocha et al. (Eds.): WorldCIST'19 2019, AISC 931, pp. 75–82, 2019.
https://doi.org/10.1007/978-3-030-16184-2_8

Our work differs from the previous ones in the following aspects: we study Open Source software, we analyze the capabilities of each software in relation to quality measures in performance regulated by international standards, and we consider it essential to identify a process to execute the tests. The quality of a software product will determine success in terms of user acceptance. The importance of software quality has led to continuous approaches in international standards to ensure that quality in each software product. Among the most adopted, the ISO/ICE 25000 family of standards has the purpose of establishing a system for quality evaluation of software products [4]. A quality software product is the result of the quality of its elements. Therefore, the ISO/IEC 25010 standard, for this quality characteristics are determined as: functional adequacy, reliability, performance, operability, safety, maintainability compatibility and transferability. Each feature has sub-characteristics that are evaluated and determine the quality of a software product, which can be measured internally and externally. This work has the scope to study the measures of the performance characteristic, which are measured externally. Another important contribution of this work has been to identify an adequate process of execution of software quality tests. Based on the ease of adoption, the testing process proposed by International Software Testing Qualifications Board (ISTQB) [5].

The rest of this work is presented as follows: Sect. 2, describes the formal research process that has been followed; Sect. 3 presents a case study, in which performance tests are developed to a web application; finally, in Sect. 4, some conclusions are pronounced.

2   Methodology

2.1   Research Method

To develop this evaluation of open source tools the research model proposed in [6] has been followed. The adopted stages have been: literature review, strategy, data generation and data analysis. The highlights of each stage are summarized in Table 1.

Table 1. Findings at each stage of the research process.

Stage	Description
Literature review	The literature review was focused on the search of works related to search words: evaluation of software quality in web applications, software quality standards, software quality testing procedure
	It has been possible to identify related works and solutions: quality sub characteristics from ISO/IEC 25010 (Quality Model Division), quality measures from ISO/IEC 25023 (Quality Measurement Division) [4], relevant features of software products, and a testing process from the International Software Testing Qualifications Board (ISTQB) [5]
Strategy	Strategy based on experimentation, that is using software
Method to generate data	Weighted matrix of features of software products
Analysis of the data	Quantitative analysis of the matrix defined in the previous subsection
	These will be described in the following subsections

It is important to distinguish the research process followed by the tool evaluation process. And, on the other hand, the process to perform software quality tests, identified in mentioned research process. The testing process is applied in the case of study, to generate and analyze data obtained by using the most appropriate software tool to measure quality.

2.2 Quality Measures

In order to obtain a quantitative quality level, ISO/IEC 25010 and ISO/IEC 25023 standards define sub-characteristics and quality measures for performance characteristics (see Table 2). Each measure is defined, and it is represented by a mathematical expression for its calculation. For example, the response time is defined elapsed time between giving a command to start a batch of tasks until the first response is received. It includes processing time and transmission time. In the case of the Internet system or other real-time system, sometimes the transmission time is much longer. The expression for measuring response time is given in Eq. (1), where A is the request delivery time, and B is the time to receive the first response.

$$RT = A - B \tag{1}$$

Table 2. Quality measures for performance characteristics.

Characteristic	Sub characteristic	Measures
Performance	Time behavior	Response time, wait time, performance
	Resource utilization	CPU utilization, memory utilization, I/O devices
	Capacity	Number of online requests, number of simultaneous accesses, bandwidth transmission system

2.3 Selection of Software for Testing

The criteria that will be considered for the pre-selection of tools are the following:

a. Be Open Source, score between 0 and 1.
b. Date of the tool version, score between 0 and 1.
c. Documentation, score between 0 and 1.
d. Available installer, score between 0 and 1.
e. Allow to evaluate performance measures, score between 0 and 5. Each performance measure has a weighted score.

The three best scores allowed to pre-select three software products on which to experience performance tests. The selected tools are installed without problem except CLIF, which despite following the installation manuals and installing different versions does not run properly. The execution of the two tools is done on the Gatling test page [12]. We performed the needed scripts for testing. The results obtained by JMeter and Gatlin are presented in Table 4, which summarizes what was obtained by the two tools with a load of 10 users.

After installing and experimenting with these software's, it became clear that JMeter is the most appropriate tool to obtain performance measurements in web applications. Additionally, as reported in Table 3, JMeter is the software that allows you to obtain a greater number of measurements. Their results are like those obtained with another tool. Which gives us confidence in the data. We also can comment that it has a friendly interface.

Table 3. Comparative analysis of several open source software. The three best ones are in bold.

Software	a	b	c	d	e	Total
Cano WebTest [7]	1	1	1	1	0	4.00
CLIF [8]	**1**	**1**	**1**	**1**	**3.12**	**7.12**
D-ITG [9]	1	1	1	1	0	4.00
Fast Web Performance Test Tool [10]	1	1	1	1	1.56	5.56
Funkload [11]	1	0	1	1	0	3.00
Gatling [12]	**1**	**1**	**1**	**1**	**3.58**	**7.58**
Grinder – Java Load Testing Framework [13]	1	0.5	1	1	1.56	5.06
JMeter – Load and Performance tester [14]	**1**	**0**	**1**	**1**	**4.2**	**7.20**
MStone [15]	1	0	0	1	0	2.00
Multi-Mechanize – web performance and load testing framework [16]	1	1	1	1	0	4.00
OpenSTA – Open Systems Testing Architecture [17]	1	1	1	1	0	4.00
Performance Co-Pilot [18]	1	1	1	1	1.56	5.56
Pylot – Performance & Scalability Testing of Web Services [19]	1	0	0	1	0	2.00
SoapUI [20]	1	0.5	1	1	0	3.50
TestMaker [21]	1	0.5	1	1	0	3.50
Tsung [22]	1	0.5	1	1	1.88	5.38
Xceptance LoadTest [23]	1	1	1	1	0	4.00

2.4   Testing Process

In order to execute an evaluation of the software product quality, it is necessary to determine a process. A software testing process involves a set of activities. As previously mentioned, to define everything necessary in this research work, a literature review was carried out. After which, we opted for the software testing process proposed by the International Software Testing Qualification Group (ISTQB) [5]. ISTQB establishes the next activities:

- Test planning and control
- Test analysis and design

- Test implementation and execution
- Evaluating exit criteria and reporting
- Test closure activities

Table 4. Measurements obtained when testing on the gatling test page.

Software	Response time (ms)	Wait time (ms)	Performance (requests/min)	Number of online requests (n/s)	Number of simultaneous accesses (n/s)	Bandwidth transmission system (Kb/s)
JMeter	1313	29	30.1	2.06	0.34	3.44
Gatling	597	17	-	3.33	0.52	-

3 Study Case

In this section, we report, very briefly, the use of JMeter software and the ISTB testing process, to measure the performance of a real web application. Consider like prerequisites of the testing process, have installed JMeter, Blaze-Meter web browser extension, and the web application to test.

3.1 Web Application Description

The performance tests will be made to the web application "IL METALMOTORS". This application supports the company to develop, produce, market and distribute metal-mechanical and automotive products and services of excellent quality.

3.2 Test Planning and Control

The web application "IL METAL MOTORS" has six modules for business administration. The tests are limited to the modules: Authentication, Products and Reports. A work schedule and responsible are defined. It is also established that, the level of risk on the web application to be tested is low.

3.3 Test Analysis and Design

For each functionality of the web application, are established: a test scenario, test cases, creation of scripts and implementation of the tests. The results obtained are recorded in a table in which graphs will be generated and finally, after an analysis, obtain the conclusions.

3.4 Test Implementation and Execution

The modules and their corresponding functionalities were tested with different number of concurrent connections (see Tables 5, 6 and 7).

Table 5. Performance testing with 50 concurrent connections.

Functionality	Response time (ms)	Wait time (ms)	Performance (requests/min)	Number of online requests (n/s)	Number of simultaneous accesses (n/s)	Bandwidth transmission system (Kb/s)	Error (%)
Login	3237	49	2	2.04	1.02	757.89	50.0
Product registration	664	138	2.2	2.17	0.36	283.08	16.7
Inventory report	678	99	2	2.02	0.50	287.97	25.0

Table 6. Performance testing with 100 concurrent connections.

Functionality	Response time (ms)	Wait time (ms)	Performance (requests/min)	Number of online requests (n/s)	Number of simultaneous accesses (n/s)	Bandwidth transmission system (Kb/s)	Error (%)
Login	3724	91	2	2.19	1.09	819.86	60.5
Product registration	664	273	1.47	3.29	0.38	194.38	29.6
Inventory report	761	189	2.1	2.11	0.52	409.08	42.0

3.5 Evaluating Exit Criteria and Reporting

The web application "METALMOTORS S.A" with loads less than 50 did not present significant errors. However, by increasing the concurrency load, the application responds with much less time, and presents more errors. We could infer that; the application responds up to a load of fifty concurrences. In the tests that are carried out with a load of one hundred and two hundred the errors increase considerably.

3.6 Test Closure Activities

There were no more test cycles, and the testing work in the web application was terminated.

Table 7. Performance testing with 200 concurrent connections.

Functionality	Response time (ms)	Wait time (ms)	Performance (requests/min)	Number of online requests (n/s)	Number of simultaneous accesses (n/s)	Bandwidth transmission system (Kb/s)	Error (%)
Login	908	81	4.9	4.93	2.46	1170.79	93.8
Product registration	888	236	5.1	5.08	0.84	640.3	87.4
Inventory report	1078	121	4.9	6.6	1.65	809.99	89.6

4 Conclusions

The formal research process allows us to ensure a high level of confidence in this work and recommend the highest scoring tool as the most convenient to achieve performance tests of web applications. Open Source tools to evaluate web application performance are a good alternative to proprietary software. In this work, it was possible to obtain a list that can be a reference for other researchers or technicians. The results obtained in a first search are seventeen tools, then filtered under certain parameters that reduce the list of tools to eight. Finally, after submitting the tools to tests you can verify a satisfactory operation with one of them.

The performance of a web application is fundamental for its users to obtain a real benefit when using it. However, it would be convenient to study tools that automatically allow testing of other features of web applications.

References

1. Alex, N., Chen, S.: A study of contemporary system performance testing framework. In: Advanced Methodologies and Technologies in Network Architecture, Mobile Computing, and Data Analytics, pp. 1546–1561. IGI Global (2019). https://doi.org/10.4018/978-1-5225-7598-6.ch114
2. Arif, M.M., Shang, W., Shihab, E.: Empirical study on the discrepancy between performance testing results from virtual and physical environments. Empir. Softw. Eng. **23**(3), 1490–1518 (2018). https://doi.org/10.1007/s10664-017-9553-x
3. Stefan, P., Horky, V., Bulej, L., Tuma, P.: Unit testing performance in java projects: are we there yet? In: Proceedings of the 8th ACM/SPEC on International Conference on Performance Engineering, ICPE 2017, pp. 401–412. ACM, New York (2017). https://doi.org/10.1145/3030207.3030226
4. International Organization for Standardization and International Electrotechnical Commission, Systems and Software Quality Requirements and Evaluation (SQuaRE 2014) (2014)
5. International Software Testing Qualifications Board, Certified tester foundation level syllabus. https://www.istqb.org/downloads/send/52-ctfl2011/3-foundation-level-syllabus-2011.html. Accessed 29 Nov 2018
6. Oates, B.: Researching Information Systems and Computing. SAGE Publications, London (2006)
7. Canoo WebTest. http://webtest.canoo.com/webtest/manual/WebTestHome.html. Accessed 29 Nov 2018
8. OW2 Consortium. https://clif.ow2.io/. Accessed 29 Nov 2018
9. Botta, A., Dainotti, A., Pescapè, A.: A tool for the generation of realistic network workload for emerging networking scenarios. Comput. Netw. **56**(15), 3531–3547 (2012). https://doi.org/10.1016/j.comnet.2012.02.019
10. Bogdan, D.: http://fwptt.sourceforge.net. Accessed 29 Nov 2018
11. Nuxeo SAS. https://github.com/nuxeo/FunkLoad. Accessed 29 Nov 2018
12. Gatling Corp Load testing for web applications. https://gatling.io/. Accessed 29 Nov 2018
13. The grinder, a java load testing framework. http://grinder.sourceforge.net/. Accessed 29 Nov 2018
14. Apache jmeter. https://jmeter.apache.org/. Accessed 29 Nov 2018
15. Mstone. http://www.qatestingtools.com/testing-tool/mstone. Accessed 29 Nov 2018

16. Multi-mechanize. https://multi-mechanize.readthedocs.io/en/latest/. Accessed 29 Nov 2018
17. Opensta. http://opensta.org/. Accessed 29 Nov 2018
18. Performance co-pilot. https://pcp.io/index.html. Accessed 29 Nov 2018
19. Pylot (python load tester) - web performance tool. https://code.google.com/archive/p/pylt/. Accessed 29 Nov 2018
20. The most advanced rest & soap testing tool in the world. https://www.soapui.org/. Accessed 29 Nov 2018
21. Testmaker 6 surfaces performance bottlenecks and functional issues. http://www.pushtotest.com/testmaker-open-source-testing. Accessed 29 Nov 2018
22. Tsung. http://tsung.erlang-projects.org/. Accessed 29 Nov 2018
23. The test automation and load test tool for software development. https://www.xceptance.com/en/xlt/. Accessed 29 Nov 2018

Blockchain Projects Ecosystem: A Review of Current Technical and Legal Challenges

Jorge Lopes[1] and José Luís Pereira[2(✉)]

[1] Mestrado Integrado em Engenharia e Gestão de Sistemas de Informação,
UMinho, Guimarães, Portugal
a73263@alunos.uminho.pt
[2] Departamento de Sistemas de Informação, UMinho & Algoritmi,
Guimarães, Portugal
jlmp@dsi.uminho.pt

Abstract. Blockchain is a decentralized transaction and data management technology developed first for Bitcoin cryptocurrency. The interest in Blockchain technology has been increasing since the idea was presented in 2008. The reasons for all the interest around Blockchain comes from its decentralized nature, while providing security, anonymity and data integrity. All these features without any third-party organization in control of the transactions, therefore creating interesting research areas. The Blockchain is a versatile technology, which can be implemented in a vast list of industries and maybe change completely the way we approach some of their use cases. In this paper, we will present some of the most relevant and recent Blockchain projects, classifying them by their main purpose or value proposition. Additionally, we describe the main technical and legal challenges, which may prevent this technology from a more rapid and wider use.

Keywords: Blockchain · Smart contracts · Projects ecosystem ·
Technical and legal issues

1 Introduction

Blockchain technology has many applications that can influence significantly the way we interact in financial markets, among others. The purpose of this technology is the use of a distributed and decentralized ledger for verifying and recording transactions, while allowing parties to send, receive, and record value or information through a peer-to-peer network of computers.

Blockchain has a wide-range of applications, including as a platform for smart contracts. These are transactions or contracts converted into computer code that facilitate, execute and enforce agreements between two or more parties, having the potential to improve financial transactions and operational and counterparty risk associated with monitoring or enforcing contractual obligations.

While the potential of the Blockchain technology is widely recognized, some technical and legal issues have yet to be tackled, in order for this technology to have even more impact on society. This paper aims to identify and describe those challenges.

© Springer Nature Switzerland AG 2019
Á. Rocha et al. (Eds.): WorldCIST'19 2019, AISC 931, pp. 83–92, 2019.
https://doi.org/10.1007/978-3-030-16184-2_9

Regarding the structure of the paper, we first review the basic concepts related to this technology. Further, we describe some of the most promising applications of Blockchain in different use cases. Finally, we will address the major technical and legal difficulties and obstacles that may hinder a broader use of this technology.

2 Blockchain Technology

Indeed, Blockchain technology has many applications in a varied list of business sectors. The purpose of this technology is the use of a distributed and decentralized ledger for verifying and recording transactions, while allowing parties to send, receive, and record value or information through a peer-to-peer network of computers. In the next sections, we briefly present some of the most relevant concepts about this technology.

2.1   Blockchain

In simple terms, a Blockchain is a kind of distributed database that stores time ordered data in a continuously growing list of blocks. The blockchain is maintained using a network of computers with no central "master". Each *block* in the chain contains *transactions* which represent a change of state in the database; for example, the transfer of funds from one account to another. Transactions are verified by multiple nodes in the network and are eventually stored in blocks in the blockchain. Each block contains a signed hash of the contents of the preceding block, making it impossible for a block's contents to be altered [1]. It is important to highlight that a transaction does not need to be coin-based. It can be an asset, which the owner can prove that he/she has it in his/her possession [2].

2.2   Decentralized Consensus

Blockchain aims to produce decentralized consensus, a specific state or set of information to be agreed upon by all agents via rules and protocols, without the need to trust or rely upon a centralized authority. This supposedly makes the consensus more secure and tamper-proof. Moreover, it rewards a community for properly maintaining the consensus, allowing greater recording and processing power in an incentive and typically competitive manner [3].

There are two main approaches to achieve consensus on a blockchain:

- **Proof-of-Work** (PoW) - rewards users who solve complicated cryptographical puzzles in order to validate transactions and create new blocks (i.e., mining). This system ensures that once a block is validated, it cannot be denied because doing so requires the malicious entity to have computing power that can compete with the entire existing network;
- **Proof-of-Stake** (PoS) - the entity in charge of creating the next block is chosen in a deterministic manner, and the chance that an account is chosen depends on its wealth (i.e., the stake).

In both cases and many other consensus generation designs, the goal is to incentivize responsible and accurate recordkeeping, while reducing tampering.

2.3   Hash Pointer

If someone tries to change the contents of a block in the middle of the chain the hash generated for this block, which is stored in the previous node, also would need to change. So, even if the hash of the previous block and the data block were changed, the previous block next pointer is based on a hash of the previous block. Because the previous block is composed of both the data and the previous hash (which has a change) these would also indicate tampering. Thus, the only way to change a block is by changing the entire blockchain, but at that point the head of the blockchain would be incorrect and since this is the value that the users store, the users would be able to detect the tampering [4].

A hash pointer provides the blockchain with a tamper-evident system in a simple way. For example, if someone changes the contents of one block the hash of the next block will not mash up and we will notice the inconsistency.

Even if an attacker has enough computer power to change the block and all the hash pointers of the following blocks in the chain, he/she will arrive at a dead-end because the last hash pointer is the value we remember as being the head of the list and the inconsistency will be noticed inevitably.

2.4   Smart Contracts

A smart contract is like a program, which runs on the blockchain, and has its correct execution enforced by the consensus protocol. A smart contract can encode any set of rules represented in its programming language. For instance, it can execute transfers when certain events occur. Accordingly, smart contracts can implement a wide range of applications, including financial instruments and self-enforcing or autonomous governance [5].

A smart contract is linked to an account, is identified by an address and its code resides on the blockchain. After uploading a contract to the blockchain, it will react accordingly to the code implemented previously and is able to do transactions like a normal member of the blockchain. The main objectives are to satisfy common contractual conditions, minimize both malicious and accidental events, and minimize the need for trusted intermediaries [6]. Related economic goals include lowering fraud loss, arbitrations and enforcement costs.

Regarding Smart Contracts development, *Solidity* stands out as the programming language used by one of the most notorious Blockchain platforms - *Ethereum*.

3   Blockchain Projects Ecosystem

The number of Blockchain projects has increased exponentially since it was presented in 2008 by someone under the pseudonym of Satoshi Nakamoto. According to [7], these projects might be classified into seven categories (Fig. 1). Each category has a different purpose, as we are going to briefly explain in the next sections.

Fig. 1. The blockchain projects ecosystem

Currencies

Currencies are divided into Base Layer Protocols, Payments and Privacy. Mostly, these projects were focused on building a better currency for various use cases and represent either a store of value, medium of exchange, or a unit of account. The Privacy sub-category differs from the other two because of its special effort on maintaining the anonymity of the clients and their transactions. To this category belong companies such as Bitcoin, Decred, Ripple, Interledger, CoinJoin and many more.

Developer Tools

This category is divided into Smart Contracts, Scaling, Oracles, Security, Legal, Interoperability, Privacy and DAGs. This category is primarily used by developers to build blocks for decentralized applications. In order to allow users to directly interact with protocols through the application, most of the current designs need to be proven out at scale. To this category belong companies such as Ethereum, Lisk, Raiden, Oraclize, Zeppelin, BTC Relay and IOTA, among others.

Fintech

This category is divided into Trading/Decentralized Exchanges, Insurance, Lending and Funds/Investment/Management. When interacting with a number of different protocols and applications many may have their own native cryptocurrency. In order to be able to have an economy with multiple currencies certain tools are needed for exchanging one unit of currency for another, facilitating lending, accepting investment, etc. To this category belong companies such as Omise, bitShares, Insurex, Salt, ETHLend, Conomi, Blockchain Capital and many more.

Sovereignty

This category is divided into the User-Controlled Internet, Governance, VPN, Communication, Identity, Security and Stable Coins. As we know, Blockchains still suffer from scalability and performance issues, but the value provided by their trustless architecture can discard performance issues when dealing with sensitive data that requires a third party. The projects in this category eliminate the need to trust in any individual or organization but rather in the incentives implemented through cryptography and economics. To this category belong companies such as Fabric, Aragon, Mysterium Network, Toshi, TrustStamp, Rivetz and many more.

Value Exchange

This category is divided in Non-Fungible and Fungible and these subcategories are also divided into smaller divisions such as Content Monetization, Data, Market Places, Social, File Storage, Computation, Mesh Networking, Energy and Video. Markets that allow users to exchange goods and services that are fungible, assets whose individual units are essentially interchangeable, will commoditize things like storage, computation, internet connectivity, bandwidth, energy, etc. Non-fungible markets don't have the same benefits but still allow providers to earn what their good or service is actually worth rather than what the middlemen think it's worth after they take their cut. To this category belong companies such as Streamium, Enigma, Ethlance, Filecoin, Golem, Power Ledger and many more.

Shared Data

This category is divided into the Internet-of-Things, Supply Chain/Logistics, Attribution, Reputation and Content Curation. Through financial incentives provided by Blockchain-based projects, we create the opportunity to open up numerous markets, except the value no longer will accrue to the aggregator but rather to the individuals and companies that are providing the data. Ideally, the result would be more contributors and higher quality datasets as the market sets the going rate for information and compensates participants accordingly relative to their contribution. Usually, the main client of these companies is a larger organization but using Blockchain technology it will create new opportunities to distribute datasets in decentralized projects that weren't previously possible or profitable. To this category belong companies such as IOTA, TMining, Mycelia and Bloom, among others.

Authenticity

This category is divided into Data and Ticketing. Basically, cryptocurrencies are just digital assets native to a specific Blockchain and projects, these projects are using digital assets to represent either real-world goods or data. In cases where is used sensitive data or markets for goods, it is wise to use a Blockchain to assure the user of their integrity discarding the possibility of fraud or other malicious intentions. To this category belong companies such as Factom, Tierion, Guts, Aventus and many more.

4   Problems and Issues

Unfortunately, the Blockchain technology has several issues and challenges that should be considered. These limitations can be technical or juridical. We will start to explain some of the technical issues and then move on to the legal ones.

4.1   Technical Issues

According to [8] there are five major technical issues that need to be addressed.

Limited Scalability

Blockchains are fundamentally decentralized, meaning that no central party is responsible for securing and maintaining the system. Instead, every single node on the network is responsible for securing the system by processing every transaction and maintaining a copy of the entire state.

Although a decentralization consensus mechanism offers us the core benefits of the Blockchain, it comes at the cost of scalability, since decentralization by definition limits the number of transactions the Blockchain can process to the limitations of a single fully participating node in the network. This means two things:

- **Low throughput** - Blockchains can only process a limited number of transactions in a time period;
- **Slow transaction times** - The time required to process a block of transactions is slow in comparison to the nearly instantaneous confirmations you get when using services like Square or Visa.

Therefore, public Blockchains are forced to make a trade-off between low transaction throughput and a high degree of centralization, that is, as the size of the Blockchain grows, the requirements for storage, bandwidth, and computer power required also increase. At some point, it becomes only possible for the few nodes that can afford the resources to process blocks.

Storage Constraints

Most applications that are built on a public Blockchain require some sort of storage solution, nevertheless, in order to store data on a public Blockchain database, the data has to be:

- Stored by every node in the network;
- Stored indefinitely since the Blockchain is appended only and immutable.

In a decentralized network where every node has to store more and more data, eventually, it will impose a huge cost on each node. As a result, storage remains a huge obstacle for any realistic application that might be built on the Blockchain.

Unsustainable Consensus Mechanisms

On a Blockchain, users don't have to trust anyone else with their transactions. This provides users properties such as autonomy, censorship resistance, authenticity, and permissionless innovation.

The mechanism used over time to enable a trustless Blockchain, not easily subverted by attackers, is called a "consensus protocol." Proof-of-Work is one of the methods to achieve consensus, but it comes with some disadvantages such as:

- **Specialized hardware has an advantage** - One downside of Proof-of-Work is the use of specialized hardware. Devices called "Application-Specific Integrated Circuits (ASICs) were designed solely for the purpose of mining Bitcoin. Ever since mining with a regular computer has become completely unprofitable. This is far from the decentralized nature of the Blockchain, where everyone has the opportunity to contribute to the security of the network and not only the ones who can afford better hardware;

- **Mining pool centralization** - The concept behind a mining pool is that instead of each user mining on their own and having a tiny chance of earning the block reward, they mine for a pool. The pool then sends them a proportionate, consistent pay-out. The problem with mining pools is that since they have more "weight" in the network, large mining pools have less variance in their returns than a single user;
- **Energy waste** - Miners spend massive amounts of computing power to run the computations that solve the Proof-of-Work algorithm. According to Digiconomist's Bitcoin Energy Consumption Index, Bitcoin's current estimated annual electricity consumption stands at 29.05TWh, which represents 0.13% of total global electricity consumption.

Inadequate Tooling

In order to be able to do their work effectively and efficiently developers need adequate tools, and the ones available still need to be improved. As specified by [8], a list what is missing from the tooling ecosystem:

- An IDE that has good linters and all the necessary plug-ins for effective smart contract development and Blockchain analysis;
- A build tool and compiler that is well-documented and easy to use;
- A better deployment tools;
- Technical documentation that actually exists or is not completely out of date for various APIs and frameworks;
- Testing frameworks that aren't lackluster. There are a few tools for Ethereum like Truffle which is okay, but more options and experimentation around testing frameworks are badly needed;
- Debugging tools;
- Logging tools;
- Security auditing;
- Block explorers and analytics.

Quantum Computing Threat

One of the future threats to cryptocurrency and cryptography is the issue of quantum computers. Although quantum computers today are still somewhat limited in what types of problems they can solve, it won't always be that way. The truth is that most popular public-key algorithms can be efficiently broken by a sufficiently large quantum computer. It's important that the Blockchain and the cryptography that underlies it, be prepared to become quantum-proof.

4.2   Legal Issues

According to [9] there are eight major legal issues that need to be addressed.

Jurisdiction

Blockchain can cross jurisdictional boundaries as the nodes on the network can be located anywhere in the world. This can pose many complex jurisdictional issues, which require careful consideration in relation to the relevant contractual relationships. The principles of contract and title differ across jurisdictions and therefore identifying the appropriate governing law is essential. However, in a decentralized environment, it may be difficult to identify the appropriate set of rules to apply.

Liability

The risk to customers of a systemic issue with trading related infrastructure such as Blockchain could be serious if trades are not settled or are settled incorrectly. Likewise, the risk relating to security and confidentiality will be towards the top of the risk issues of any prospective customer. Blockchain poses different risks as a consequence of the technology and manner of operations: one of the main issues affecting public Blockchain is the inability to control and stop its functioning. So, the allocation and attribution of risk and liability in relation to a malfunctioning Blockchain service must be thought through carefully, not just at the vendor-customer level, but also as between all relevant participants.

Data Privacy

One of the main aspects of the Blockchain is that data is tamper-proof, once data is stored it cannot be altered, this clearly has implications for data privacy, particularly where the relevant data is personal data or metadata sufficient to reveal someone's personal details. Equally, the unique transparency of transactions on the Blockchain is not easily compatible with the privacy needs of the banking sector: the use of crypto-addresses for identity is problematic as no bank likes providing its competitors with precise information about its transactions and the banking secrecy must be kept by law.

Decentralized Autonomous Organizations (DAOs)

DAOs are essentially online, digital entities that operate through the implementation of pre-coded rules. These entities often need minimal to zero input into their operation and they are used to execute smart contracts, recording activity on the Blockchain. Modern legal systems are designed to allow organizations, as well as actual people, to participate. Most legal systems do this by giving organizations some of the legal powers that real people have. But what legal status will attach to a DAO? Are they simple corporations, partnerships, legal entities, legal contracts or something else? Since the DAOs management is conducted automatically, legal systems would have to decide who is responsible if laws are broken.

The Enforceability of Smart Contracts

Since Smart Contracts are prewritten computer codes, their use may present enforceability questions if attempting to analyze them within the traditional 'contract' definition. This is particularly true where Smart Contracts are built on permissionless Blockchains, which do not allow for a central controlling authority. Since the point of such Blockchains is to decentralize authority, they might not provide for an arbitrator to resolve any disputes that arise over a contract that is executed automatically. Customers should ensure that Smart Contracts include a dispute resolution provision to reduce uncertainty and provide for a mechanism in the event of a dispute.

Compliance with Financial Services Regulation

Many sourcing arrangements, including the use of certain technology solutions, require regulated entities to include in the relevant contracts a series of provisions enabling them to exert control and seek to achieve operational continuity in relation to the services to which the contracts relate. Using Blockchain, this may well be more of a challenge. The contracts and overall arrangement will need to be carefully reviewed to ensure compliance, as required.

Is Data on a Blockchain "Property"?

In common law, as a general principle, there is no property right in the information itself. While individual items of information do not attract property rights, compilations of data may be protected by intellectual property rights. Where a database of personal information is transacted, if the receiver wants to use the personal information for a new purpose, he/she has to get consent for this from the individuals concerned.

Due Diligence on Blockchain

Public companies and private investors have already begun to make significant capital investments in Blockchain technology startups. This trend is likely to accelerate, as commercial deployments of Blockchain technology become a reality. Transactional lawyers who are tasked with performing due diligence on the buy and/or sell side in connection with these investments need to understand Blockchain technology and the emerging business models based on the technology. Traditional due diligence approaches may need to be adapted. For example, there will be unique issues concerning ownership of data residing on decentralized ledgers and intellectual property ownership of Blockchain-as-a-Service offerings operating on open source Blockchain technology platforms.

5   Conclusion

Blockchain is an emergent technology that is evolving in a daily basis. As this paper has emphasized, it holds a huge transformation potential in a vast number of fields and might radically transform our daily lives. Similar to other new technologies, to realize its full potential, Blockchain will develop through numerous iterations and will inevitably go through trials, evolution, failures and ultimately widespread adoption. The challenge will be to strike the right balance between ensuring the governance, safety and resilience of the system while not infringing on the innovation and development of this fast evolving technology.

This paper aims to contribute to clarify the advantages of Blockchain, such as providing anonymity, security, privacy, and transparency to all its users, while making possible for two entities to trade any asset without the need to trust each other or a third entity to regulate the transaction. However, as this paper has exposed, these advantages come with a lot of challenges and difficulties that still need to be addressed. Some of these are of a technological nature but many others, probably the more difficult ones to solve, are legal.

Acknowledgements. This work has been supported by FCT - Fundação para a Ciência e Tecnologia, within the Project Scope: UID/CEC/00319/2019.

References

1. Bird, G.: Block chain technology, smart contracts and Ethereum. IBM Developer, 6 September 2016
2. How does Bitcoin Blockchain work and what are the rules behind it? https://www.quora.com/ How-does-Bitcoin-Blockchain-work-and-what-are-the-rules-behind-it. Accessed 21 Mar 2018
3. Cong, L.W., He, Z.: Blockchain disruption and smart contracts. SSRN Electron. J., 10 September 2018
4. Hash Pointers – deltadeltaandmoredeltas. https://www.deltadeltaandmoredeltas.com/hash-pointers/. Accessed 21 Mar 2018
5. Luu, L., Chu, D.H., Olickel, H., Saxena, P., Hobor, A.: Making smart contracts smarter. In: Proceedings of the 2016 ACM SIGSAC Conference Computer and Communications Security – CCS 2016, pp. 254–269 (2016)
6. What is a Smart Contract_ – Pactum – Medium. https://medium.com/pactum/what-is-a-smart-contract-10312f4aa7de. Accessed 21 Mar 2018
7. Mapping the blockchain project ecosystem. https://techcrunch.com/2017/10/16/mapping-the-blockchain-project-ecosystem/. Accessed 21 Mar 2018
8. Kasireddy, P.: Fundamental challenges with public blockchains (2017). https://medium.com/ @preethikasireddy/fundamental-challenges-with-public-blockchains-253c800e9428/. Accessed 21 Mar 2018
9. Jessica, B., Duncan, T., John, P.: Blockchain: Background, challenges and legal issues (2017). https://www.dlapiper.com/ ~ /media/Files/Insights/Publications/2017/06/Blockchain_ background_challenges_legal_issues_V6.pdf

App Nutrition Label

Sami J. Habib$^{(\boxtimes)}$ and Paulvanna N. Marimuthu

Computer Engineering Department, Kuwait University,
P.O. Box 5969, 13060 Safat, Kuwait
sami.habib@ku.edu.kw

Abstract. This paper aims to introduce the concept of nutrition label for Apps, and encourages App developers to start assessing the effect of apps before their executions. Presently, the food products, automobiles, equipment, and appliances are all labelled with nutritional/energy consumption facts to provide a sustained lifestyle and environment. Many Apps are developed to compare the nutrition and energy facts of food products. The rapid development of internet technology and mobile devices made App usage ubiquitous and the business and lifestyle are at the fingertip now. However, these Apps have to be defined with standards so as to assist the users in selecting a secured and efficient one suiting to their needs. We have carried out a preliminary analysis to explore the impact of Apps installed on an enterprise network, and we envision that the App nutrition label should comprises of functional connectivity, popularity, energy and security information, as the basic constituents, since these four parameters selected for now have direct effects on the network. We have selected the following quantitative and qualitative metrics, namely: degree of functional connectivity, range of connectivity, power consumed and closeness centrality to derive the four parameters respectively. The empirical results on testing Apps activity installed on an enterprise demonstrate the feasibility of framing common App standards, facilitating the customers to be aware of App usage and the security prior to installation.

Keywords: Energy consumption · Web apps · Mobile apps · Nutrition label · Testing · Standards · Quantitative and qualitative metrics

1 Introduction

Nutrition labelling of packaged food products has been started effectively from 1973 and it is evolving until now. Nutrition label describes the contents and nutrients of a food product [1, 2] and it guides the consumer in food selection. In the same way, since mid-1970s, fuel economy labeling have been started and updated many times over the years. Recent updating includes CO_2 emission, details of electric vehicles and a quick response (QR) code to read vehicle information through smart phones [2]. Energy labeling has been extended to electric appliances from 2008 to inform the consumers about the operating cost and energy usage of the product [3].

Nearly four decades ago, the personal computers have become popular and during that period, the number of software packages and networking of computers are limited. With rapid increase in information and communication technology, the software and

© Springer Nature Switzerland AG 2019
Á. Rocha et al. (Eds.): WorldCIST'19 2019, AISC 931, pp. 93–102, 2019.
https://doi.org/10.1007/978-3-030-16184-2_10

hardware architectures and ease of networking within an enterprise are extended. However, the network across the globe is possible after the introduction of Internet and today, Internet and the growth of mobile devices have played a significant role in peoples' daily routine.

The mobile devices initially released much before with few applications (apps), such as calculator, calendar and few games [4]. More apps are developed now and these Apps cover almost everything needed for our daily routine, such as travel, health, banking, share market, education, lifestyle, news, entertainment, gaming, utilities, social networks and much more. Apps are ubiquitous, as the instant messaging and live streaming are popular among the common person. The report from Google and Apple, the two major players of application market stated that from July 2008 to January 2017, there were 2.7 and 2.5 million applications added to their respective stores, which confirmed the exponential growth of apps [5]. The present mobile apps contain few information, namely: platform, version, type of use, memory to be consumed and resources needed at the time installation and reviews about the product, which are beneficial to a group of tech-savvy users. Apps did not have metrics comprehensible to common individual and so, it is time to add metrics to enlighten the product quality [6].

In this paper, we have explored the impact of few Apps executed on an enterprise network, envisioned to define common labels so as to facilitate the user with some prior knowledge about the selected App prior to installation. The labels are: functional connectivity, popularity, energy and security. We have selected few quantitative and qualitative metrics of the Apps, namely: degree of functional connectivity, range of connectivity, power consumed and closeness centrality to analyze the traffic pattern so as to label the Apps. Even though the effect is collective, we have attempted to define it for a single user, which could be extended to many in future. The simulation results on testing the activity of Apps installed on an enterprise demonstrate the feasibility of framing common App standards, facilitating the customers to be aware of its popularity and security prior to installation.

2   Related Work

Our literature survey on labeling on App reveals that there are no common standards developed yet. Besides, there is not much work discussing the need for defining the standards and adding labels to the App. Akhawe and Finifter [7] briefed about the future of adding labels to mobile Apps, and they discussed about adding security information of the Apps. Jebari and Wani [8] proposed a centroid based approach to classify web pages by genre, where they utilized information sources such as URL, title and anchors to extract the details. Sun et al. [9] discussed to add labels to know the type of user's psychological needs the App satisfy. The authors conducted in-depth interviews and utilized a classifier algorithm to filter the reviews to categorize the Apps. Chen et al. [10] proposed a framework to tag the mobile App with few keywords for improving the search, categories and advertising.

In addition, the survey on quantitative and qualitative analyses of Apps reveal that a very few works discussed the quantitative and qualitative study on analyzing App performance. Hayes [11] carried out a quantitative study of educational mobile Apps to analyze few parameters, such as the grade level, subject, learning ratings and skill category. Jiang et al. [12] utilized a new mobile application crawler and gathered information. Further, they conducted a quantitative analysis on the collected data to study the average latency of mobile applications. Alderson et al. [13] proposed a qualitative approach to model a real-time internet traffic, where they analyzed the degree of connectivity against the bandwidth at a router level.

Furthermore, few studies were conducted to analyze the evolution of social networks. Sarkar et al. [14] utilized locality sensitive hashing technique to predict the neighborhood and derived a non-parametric model for link prediction in dynamic networks. They utilized another approach to observe some measures, such as the centrality and proximity of nodes in the network over a specific period. Leskovec et al. [15] studied a wide range of real-time graphs based on proximity measures and observed two phenomena, such as the densification of graph over time and shrinking of distance between the nodes. Leskovec et al. [16] carried out an experimental study on evolution of social networks, where they analyzed the historical information about the node and edge arrivals. Tong et al. [17] developed a software to track the addition of new links and new nodes continuously in a time-evolving social networks. In this paper, we have made a preliminary attempt to label the Apps so as to reveal App constituents.

3   Labeling of Web-Applications

We have anticipated few labels, such as functional connectivity, popularity, energy and security to an App, as these labels have direct effort on the network and provide awareness in selecting an App from a group of similar kind. We considered a volume of V Kbytes traffic generated from a mobile/web app over a period from $t_0 \rightarrow t_i$, connecting a maximum of N customers. The internet connected topology is a scale-free, but we restricted it to a finite number for modeling purpose. We have defined few metrics both quantitatively and qualitatively to derive statistical characteristics of Apps namely: degree of functional connectivity, range of connectivity, energy consumption, and closeness centrality.

3.1   Functional Connectivity

We have defined the *functional connectivity* of a node as the connectivity or relationship established between the App (target node) and other spatially separated customers over a timespan t, as in Eq. (1). Let d_i be the degree of a node n_i (n_i is the customers' device where the app is going to be installed), which is the sum of incoming and outgoing connections. This label would give an idea about the number of maximum ($n_h \in N$) and minimum ($n_l \in N$) customers. Here the total connectivity is divided into various domains $k = \{d_1, d_2, \ldots, d_m\}$ geographically and the average degree of customers $(A_{\deg})$ using the App from various domains is shown as in Eq. (2). In Eq. (2), the numerator specifies the degree of functional connectivity of the App with customers distributed in m domains.

$$fc(n_i)_t = \sum_{j=1}^{N} (n_{i,j} + n_{j,i}) \tag{1}$$

$$A_{\deg} = \sum_{k=1}^{m} \frac{d_i(fc(n_i), k)}{N_k} \text{ where, } d_i(fc(n_i)) = \left| \sum_{j=1}^{N} (n_{i,j} + n_{j,i}) \right| \tag{2}$$

3.2 Popularity

The term popularity is defined to determine how popular the App among its users, whether restricted to local domain or extended globally. The App with higher average functional connectivity, estimated from Eq. (2) may have more influence or popularity and it is defined in Eq. (3). Hereby the term *th* specifies the threshold value, which is at least greater than 50% of the once connected customers.

$$P(App) = \begin{cases} A_{\deg} < th & \text{limited} \\ A_{\deg} \approx th & \text{normal} \\ A_{\deg} > th & \text{popular} \end{cases} \tag{3}$$

3.3 Energy

We have considered a wireless transmission and the normalized power consumed by the transmitted packet of a defined size is obtained by integrating the area under the generated power spectral density curve $P_w(f)$ as illustrated in Eq. (4). Hereby, $P_w(f)$ has a unit of watts per hertz.

$$P = \int_{-\infty}^{\infty} P_w(f) df \tag{4}$$

We have observed a maximum transmission bit rate of $\frac{1}{T_b}$ from the observed data traffic pattern. We have selected the Manchester NRZ signal [18] coding to encode and the data is sent using user datagram protocol (UDP) protocol. The power consumed is given by Eq. (5) and the energy consumed is estimated utilizing Eq. (6).

$$P_{Manchester-NRZ}(f) = T_b \left(\frac{\sin(\pi f T_b/2)}{\pi f T_b/2} \right)^2 \sin^2(\pi f T_b/2) \tag{5}$$

$$\text{Energy consumed per day } = E = \int_f P_{Manchester-NRZ}(f) df * T \tag{6}$$

3.4 Security

We have selected closeness centrality as an index of node-security. We have redefined the closeness centrality (CC) in [19] for an App as in Eq. (7), which is the ratio of degree of functional connectivity (undirected graph) of the App from customers in m domains to the sum of shortest paths between the App and connected customers ($l(App, c_j^k)$). Here, the term l refers to the shortest distance, the term App refers to the source node, where the App is installed and the term c_j^k refers to the total connected customers in m domains. Then, we have defined the range of closeness centrality from zero to one. An App with a larger closeness centrality may be more secure to targeted attacks.

$$CC = \frac{d_i(fc(App), m)}{\sum_{k=1}^{m} l(App, c_j^k)} \tag{7}$$

4 Framing App Label

We frame the structure of App label as in Fig. 1, where we define the labels, such as connectivity, popularity, energy and security over a time span. The connectivity is categorized into three: light, medium and high to know how the App is communicative among the customers. Popularity is also classified into three, based on the range of connected domains: (i) local, if the customers are from the local hub (domain), (ii) extended, if the customers are connected widely and (iii) global, if connectivity range is high. Then, the third label, energy, details the average energy consumed over the timespan and the fourth label, security, is defined in a scale of zero to one. The App is unsecure, when the value is close to zero and its moderate if it is around the midpoint and highly secure when closes to one.

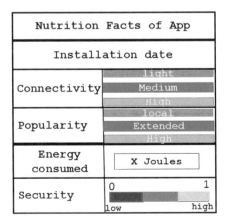

Fig. 1. App label frame.

5 Results and Discussion

We consider a medium enterprise network, extending its service to a finite number of customers through App facility. We have carried out two sets of experiments to estimate the selected quantitative and qualitative metrics in Sect. 3, to observe the App behavior and to label it accordingly. We generated two sets of data over a period of 120 days after the installation of an App to observe its behavior, where we coded our quantifying simulator in Java platform. From the outcome, we deduced a pattern for degree of functional connectivity over the measurement period, as in Fig. 2, and the degree of connectivity is found to follow a monotonically increasing connectivity pattern.

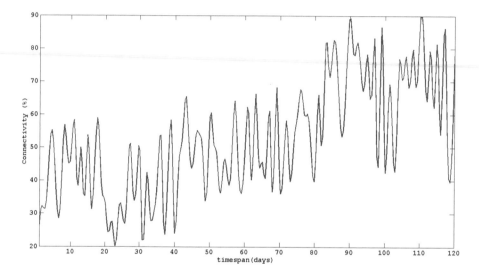

Fig. 2. Connectivity against timespan for App-I.

The connectivity increases since the installation and stays at high. We estimated the degree of connectivity quantitatively as in Fig. 3, which measures the number of customers over the observation period. This label rate the activity of the App as highly connected. Out of the total traffic, 20% of the traffic were light and 33% were in medium range.

We have analyzed the range of functional connectivity (R) of the App, where the domains are partitioned into three regions: local (R $\leq$ 150 m radius), extended (150 $\geq$ R $\leq$ 300 m radius) and global (R $\geq$ 300 m radius). Consequently, the popularity of the App is defined as local popularity, extended popularity and highly popular. The range of connectivity of the App is demonstrated in Fig. 4, which has more extended customers and few global customers. Thus, the popularity of the App is categorized as extended popularity.

The security is measured using closeness centrality factor, where we used the distance vector from the range of functional connectivity, as illustrated in Fig. 5.

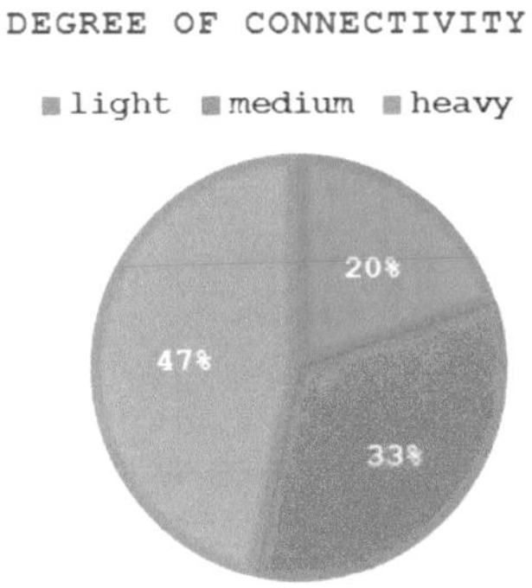

DEGREE OF CONNECTIVITY

■ light ■ medium ■ heavy

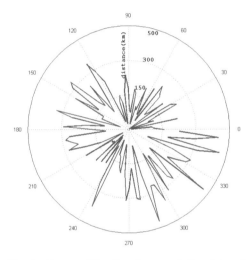

Fig. 3. Degree of functional connectivity in percentage against the maximum users.

Fig. 4. Range of functional connectivity of App-I.

The normalized closeness centrality is found to be 0.4889, which shows that the customers are widespread and the security is moderate.

The experiment is repeated on the second App to quantify the degree of functional connectivity, which is illustrated in Fig. 5. The degree of connectivity is comprehended with a decreasing function. Thus, App-II is rated with less active.

We have analyzed the range of functional connectivity (R) of the App-II, which is shown in Fig. 6. Here, the domain range is found to be local (R ≤ 150 m radius) and this shows that the popularity of App-II is within its neighborhood. We have estimated the normalized closeness centrality from Fig. 6 and it is found to be 0.27599. However increase in connectivity would increase the security. The value of closeness centrality shows that App-II is secured at low level.

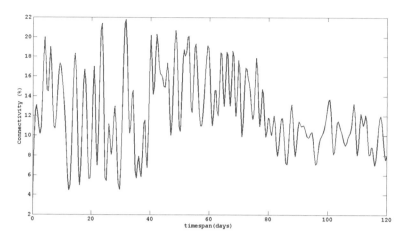

Fig. 5. Connectivity against timespan in App-II.

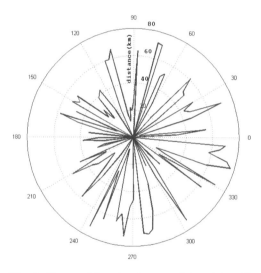

Fig. 6. Range of functional connectivity of App-II.

The average energy consumed by the data transmission generated for App-I and App-II are computed using Eqs. 4–6. The energy consumption is found to be 9866 J for App-I and 697 J for App-II over the timespan. Finally, we draw the labels for App-I and App-II as in Fig. 7.

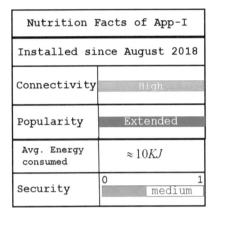

a) App-I   b) App-II

Fig. 7. Nutrition label of apps.

6 Conclusion

We have proposed and developed a labeling scheme for web or mobile applications utilizing quantitative and qualitative metrics derived from the traffic pattern collected after their installations. We have selected the metrics namely: degree of functional connectivity, range of connectivity, power consumption and closeness centrality to demonstrate the quality of the Apps, as these metrics have direct effect on network. Our preliminary analysis is to explore the impact of few Apps resulted in framing the Apps with four labels, namely: connectivity, popularity, energy and security.

We are continuing our research to add more constituents so as to define common standards to Apps community.

Acknowledgement. This work was supported by Kuwait University under a research grant no. QE01/17.

References

1. Nayga, R.M., Lipinski, D., Savur, N.: Consumers' use of nutritional labels while food shopping and at home. J. Consum. Aff. **32**(1), 106–120 (1998)
2. Van den Wijngaart, A.W.: Nutrition labelling: purpose, scientific issues and challenges. Asia Pac. Channel Clin. Nutr. **11**(2), 68–71 (2002)
3. Silitonga, A.S., Atabani, A.E., Mahlia, T.M.I.: Review on fuel economy standard and label for vehicle in selected ASEAN countries. Renew. Sustain. Energy Rev. **16**(3), 1683–1695 (2012)
4. Russo, A.C., Rossi, M., Germani, M., Favi, C.: Energy label directive: current limitations and guidelines for the improvement. Procedia CIRP **69**, 674–679 (2018)
5. Sarwar, M., Soomro, T.R.: Impact of smart phone's on society. Eur. J. Sci. Res. **98**(2), 216–226 (2013)

6. Omar, M.W., Ali, M.N.M., Zakaria, A., AlHady, S.A.: The correlation between label messages and labeling effectiveness. In: Proceedings of International Conference on Science and Social Research, Kuala Lumpur, Malaysia, 5–7 December (2010)
7. Akhawe, D., Finifter, M.: Product labels for mobile application markets. In: Proceedings of Mobile Security Technologies Workshop, 24 May, San Francisco, USA (2012)
8. Jebari, C., Wani, M.A.: A multi-label and adaptive genre classification of web pages. In: Proceedings of 11th International Conference on Machine Learning and Applications, Boca Raton, FL, USA, 12–15 December, pp. 578–581 (2012)
9. Sun, Z., Ji, Z., Zhang, P., Chen, C., Qian, X., Du, X., Wan, Q.: Automatic labeling of mobile apps by the type of psychological needs they satisfy. Telematics Inform. **34**(5), 767–778 (2017)
10. Chen, N., Hoi, S.C.H., Li, S., Xiao, X.: Mobile app tagging. In: Proceedings of 9th ACM International Conference on Web Search and Data Mining, San Francisco, California, USA, 22–25 February, pp. 63–72 (2016)
11. Hayes, T.: Mobile apps for 21st century skills: a quantitative analysis of educational mobile apps on graphite.org. In: Proceedings of 2016 World Conference on Educational Media and Technology, Vancouver, BC, Canada, 28–30 June, pp. 1630–1637 (2016)
12. Jiang, Z., Kuang, R., Gong, J., Yin, H., Lyu, Y., Zhang, X.: What makes a great mobile app? A quantitative study using a new mobile crawler. In: Proceedings of IEEE Symposium on Service-Oriented System Engineering, Bamberg, Germany, 26–29 March, pp. 222–227 (2018)
13. Alderson, D., Li, L., Willinger, W., Doyle, J.: Understanding Internet topology: principles, models, and validation. IEEE/ACM Trans. Networking **13**(6), 1205–1218 (2005)
14. Sarkar, P., Chakrabarti, D., Jordan, M.: Nonparametric Link prediction in large-scale dynamic networks. Electron. J. Stat. **8**(2), 2022–2065 (2012)
15. Leskovec, J., Kleinberg, J., Faloutsos, C.: Graphs over time: densification laws, shrinking diameters and possible explanations. In: Proceedings of International Conference on Knowledge Discovery in Data Mining, Chicago, IL, USA, 21–24 August (2005)
16. Leskovec, J., Backstrom, L., Kumar, R., Tomkins, A.: Microscopic evolution of social networks. In: Proceedings of Knowledge Discovery and Data Mining Conference, Los Vegas, Nevada, USA, 24–27 August, pp. 462–470 (2008)
17. Tong, H., Papadimitriou, S., Yu, P., Faloutsos, C.: Fast monitoring proximity and centrality on time-evolving bipartite graphs. Statistical Analysis and Data Mining **1**(3), 142–156 (2008)
18. Habib, S.J., Marimuthu, P.N., Zaeri, N.: Carbon-aware enterprise network through redesign. Comput. J. **58**(2), 234–245 (2015)
19. Wu, T., Chang, C-S., Liao, W.: Tracking network evolution and their applications in structural network analysis. IEEE Trans. Network Sci. Eng. (Early Access) (2018). https://doi.org/10.1109/TNSE.2018.2815686

Automated and Decentralized Framework for Internet of Things Systems Using Blockchain and Smart Contracts

Sorin Zoican$^{(\boxtimes)}$, Roxana Zoican, and Dan Galaţchi

University Politehnica of Bucharest, Bucharest, Romania
{sorin,roxana,dg}@elcom.pub.ro

Abstract. This paper describes a framework for IoT systems that use blockchain technology and smart contracts. Such systems ensure the storage of information in a decentralized and secure database and can perform transactions between nodes automatically. The architecture of the system is like a LoRaWAN architecture in which the gateway nodes form a blockchain subsystem that runs command scripts for data filtering. The paper focuses on implementing gateways using a powerful microcontroller ADuCRF101 (with integrated ARM Cortex M3 and RF Transceiver) and evaluating performance as computing time and energy consumption.

Keywords: Internet of Things · Smart contract · Energy saving

1 Introduction

In Internet of Things (IoT) smart devices can be connected one to other using technologies that should permit: secured information storage, communication between nodes and low power consumptions. Such systems are more and more used in smart water and agriculture, smart factory, smart cities, smart parking and traffic, e-health, smart environment, and education [1]. Relevant IoT real life examples are: AWS Marketplace IoT solutions for smart home and city, connected healthcare and industrial processes, IBM Connected Vehicle Insights, etc. Given the increasing importance of IoT, motivation of this paper is to see if the blockchain and smart contract technologies (which ensure security and automation of transactions) can be implemented at the gateway level to improve the network performance. The transactions between nodes should be done automatically and at specific moments of time to assure proper relationships between nodes and to minimize the amount of data storage. A general architecture of such system is depicted in Fig. 1. Each end node represents a specific sensor that will send information to gateways. The gateways form the blockchain system which plays two important roles: (a) to maintain a secure decentralized database and (b) to implement smart contracts reducing the amount data stored in blockchain and to automate the transactions between gateways. The information stored in blockchain is then transmitted (upon request) to network servers which ensure the Internet connectivity and communicate with application servers [2].

The architecture is similar to the LoRaWAN [2] architecture. As a novelty element, unlike other architectures [4], in the proposed IoT system, the role of gateway nodes

© Springer Nature Switzerland AG 2019
Á. Rocha et al. (Eds.): WorldCIST'19 2019, AISC 931, pp. 103–109, 2019.
https://doi.org/10.1007/978-3-030-16184-2_11

will be extended to implement blockchain and smart contract that will no longer be deployed in network servers.

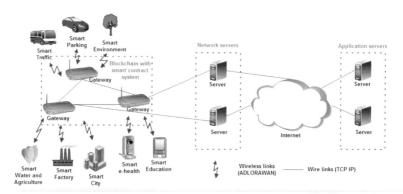

Fig. 1. The general architecture of IoT system

The challenges in this architecture are: (1) how to save the energy consumed by each end node, and (2) how to implement the blockchain system in nodes with relative limited resources.

2 Blockchain Technology and Smart Contracts

The blockchain technology represents a solution to implement a secured and distributed database. An immutable record of the history of transactions done by devices is maintained in successive linked blocks using hash function of previous block content as it is shown in Fig. 2. The modification in previous blocks is not possible, because changing its hash and this breaks references in a blockchain [3].

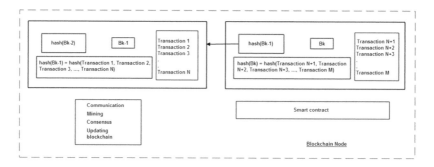

Fig. 2. Blockchain and Smart contracts

The blockchain network is based on ADLoRaWAN wireless protocol stack [8, 10] which has the following features: self-healing, multi-hop wireless network, stateless routing, and instantaneous recovery power. A simplified proof of work (PoW) should be implemented to achieve small compute time and therefore the possibility to integrate it in the gateways [5, 6]. Each node participating in blockchain (miner) has a unique identifier (*MIN* – Miner Identification Number) and a public key, *K*. A hash value of the concatenation of *MIN* and *K* is computed as $H = hash(MIN, K)$. The miners are forced to wait *N* block after their last proposed block. A score is calculated for each proposed block as $score = 2^{256} - |H - MIN|$. The first miner that complete the computation of hash function sends the block for validation, otherwise the block with highest score is selected to validation. The consensus is achieved using Practical Byzantine Fault Tolerance (PBFT) algorithm [7]. A smart contract is a program (script) running on the top of a blockchain. This script verifies a set of rules agreed by the nodes. If these set of rules are met, the script will produce an output. The smart contract allows automation by verifying and enforcing the agreed rules.

The paper will investigate in the following section the computation time and the energy consumption of this implementation.

3   Wireless Sensor Network Node Architecture

The end node architecture has two components of two components: a processing and wireless module (based on ADuCRF101 analog microcontroller) [8] and some sensors (temperature, humidity, ambient light, and infrared light) [9] as are shown in Fig. 3. The microcontroller ADuCRF101 (depicted in Fig. 4) represents an integrated data acquisition and wireless transceiver solution for low power wireless applications. The data acquisition section consists of a 12–bit analog digital converter (ADC) with six input channels powered from an internal low dropout regulator (LDO). The wireless transceiver is dual band (862 MHz to 928 MHz and 431 MHz to 464 MHz) and can use multiple configurations (two levels Frequency Shift Keying) or Gaussian frequency shift keying). This analog microcontroller integrates an ARM Cortex-M3 processor (RISC architecture, 32 bits data bus, and 16 MHz clock), configurable peripheral IO ports with DMA channels (SPI, UART, GPIO, general purpose timers, PWM and RF), 128 kB flash memory and 16 kB SRAM. The RF block includes 128-bit AES encryption/decryption with hardware acceleration with very low power consumption. This circuit is designed to operate in battery powered applications (2.2 V–3.6 V).

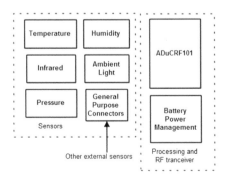

Fig. 3. The WSN node architecture

The device can be configured to operate on different operating modes: normal mode, flexi mode (any peripheral can wake up the device), hibernate mode (an internal wake-up timer remains active) and shutdown mode (only an external interrupt can wake up the device).

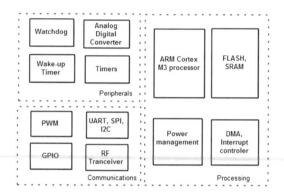

Fig. 4. The ADuCRF101 analog microcontroller architecture

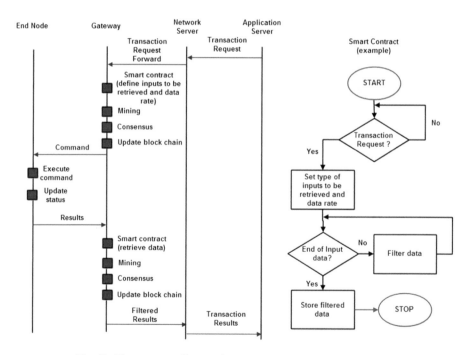

Fig. 5. The message diagram between actors in WSN architecture

The proposed mode of operation of the WSN network in Fig. 1 is depicted in Fig. 5. The application server initiates a transaction that will be transmitted through the network server to the gateway. The gateway takes over the request, records the transaction in blockchain (seen as the transaction database) and transmits the necessary commands to end-nodes (determines which end-node actuators are active and what their operating parameters are). The end-node executes the command, updates its own state, and transmits the results (sensor readings) back to the gateway. The smart contract ensures transaction automation and input data filtering before the transaction. Data filtering means comparing data from input to retrieve only data is within certain limits. This reduces data flow and traffic to gateways and servers. Gateways run the smart contract script, process the sensors data (filters them), update a blockchain database and transmit the filtered data through the network server to the application server. The challenge is to see if the gateway (that has limited resources) is able to perform all the necessary operations to update blockchain (mining, consensus achieving). The advantage of such an approach, compared to the solution in which all blockchain processing is carried out in the network server, is reducing the complexity of the architecture by eliminating a relatively large number of network servers required to form the blockchain system (NB: gateways are already part of the system, to connect end nodes in the Internet and their number is large enough to form blockchain and achieve consensus) and decrease the traffic between gateways and network servers.

4   Main Results and Performance Evaluation

The paper evaluates the possibility to implement the blockchain and smart contracts in the gateways. Two criteria were analyzed: the computation time and the energy consumption. The gateways are realized using the ADuCRF101 microcomputer (Analog Devices). Three profiles (smart contracts) associated with different levels of filtering data from sensors have been considered. For each profile, the processing times for routing, mining, and consensus - with variable number of nodes used for the PBFT (or group view) algorithm have been estimated, as is shown in Fig. 6.

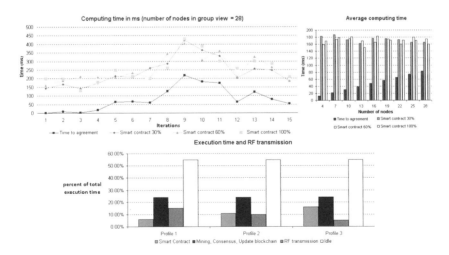

Fig. 6. The computing time (instantaneous and average) and RF transmission

The average total execution time (mining, consensus, routing) is about 180 ms. One can observe that the consensus algorithm takes most of the computation time. The message length may vary from 64 bytes to 128 bytes depend on the type of smart contract. The system activation period is set to T = 1 s and the transfer rate up to 300 kbits/s. Based on estimated computation time and the power consumption in normal or idle state for each component in system (MCU ADuCRF101 – CPU ARM with 3 mA in normal mode, RF module with 20 mA and 2 μA in idle mode, and the sensors: ADXL362 - accelerometer, ADT75 - temperature, SHT21 - temperature and humidity, APSD-9005 - ambient light sensor, PIR - proximity infrared – all with average current 0.5 mA in normal mode and 0.2 μA in idle mode) and assuming that the these components enter idle state after they finish their jobs, the battery life was evaluated accordingly with three profiles as in Table 1. The obtained results are illustrated in Fig. 7.

Table 1. The functioning profiles

Device	Profile 1 (Smart contract – no data filtering)	Profile 2 Smart contract – 30% data filtering)	Profile 3 Smart contract – 60% data filtering)
ADuCRF101 (MCU and RF)	30% T active 55% T idle 15% T idle with RF active	35% T active 55% T idle 10% T idle with RF active	40% T active 55% T idle 5% T idle with RF active
All sensors	20% T active 80% T idle	20% T active 80% T idle	20% T active 80% T idle

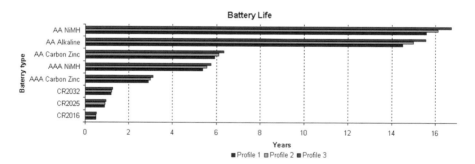

Fig. 7. The average battery life

These results prove the possibility to implement the blockchain in the gateways (the computational effort is not very high for the simplified PoW implementation). The implementation of smart contract will slightly increase the computation time, but the radio transmission will be decreased due the data filtering performed in the smart contract (as one can observe in Fig. 6), so the energy consumption will decrease. In this paper, it was assumed that the battery operates under normal conditions; otherwise the lifetime estimated may drop by up to 20% [11].

5   Conclusion

The paper presents a WSN architecture based on blockchain and smart contracts. The experiments have shown that the gateway nodes can implement in hundreds of milliseconds (real time for an IoT system), all the necessary computations for blockchain and for routing. Several functioning profiles have been considered and the average battery life has been evaluated for different battery types. The obtained results are very good (battery life is about years or more if a smart contract for filtering data is implemented). The proposed framework reduces the complexity of the network architecture by eliminating a relatively large number of network expensive servers required to form the blockchain system and using better the less expensive gateways already existing in the network. The smart contacts improve the overall system performance: automate the transactions, reduce the data traffic between gateways and network servers and reduce the energy consumption. Future work will explore new approaches to PoW and consensus to achieve lower computing times to compensate for the decrease in battery capacity under unfavorable conditions of use.

References

1. Lin, J., Shen, Z., Miao, C., Liu, S.: Using blockchain to build trusted LoRaWAN sharing server. Int. J. Crowd Sci. **1**(3), 270–280 (2017). https://doi.org/10.1108/ijcs-08-2017-0010aaa
2. Al-Kashoash, H.A.A., Kemp, A.H.: Comparison of 6LoWPAN and LPWAN for the Internet of Things. Aust. J. Electr. Electron. Eng. (2017). https://doi.org/10.1080/1448837X.2017.1409920
3. Khan, M.A., Salah, K.: IoT security: Review, blockchain solutions, and open challenges. Future Gener. Comput. Syst. **82**, 395–411 (2018)
4. Niya, S.R., Jha, S.S., Bocek, T., Stiller, B.: Design and implementation of an automated and decentralized pollution monitoring system with blockchains, Smart Contracts, and LoRaWAN. In: Network Operations and Management Symposium (2018). https://doi.org/10.1109/noms.2018.8406329
5. Zoican, S., Vochin, M., Zoican, R., Galatchi, D.: Blockchain and consensus algorithms in internet of things. In: International Symposium on Electronics and Telecommunications 2018 (ISETC 2018) (2018)
6. http://www.chainfrog.com/wp-content/uploads/2017/08/consensus.pdf. Accessed 11 Jan 2018
7. Castro, M., Liskov, B.: Practical byzantine fault tolerance and proactive recovery. ACM Trans. Comput. Syst. Assoc. Comput. Mach. **20**(4), 398–461 (2002). https://doi.org/10.1145/571637.571640
8. https://www.analog.com/media/en/technical-documentation/data-sheets/ADuCRF101.pdf. Accessed 11 Jan 2018
9. https://www.analog.com/en/design-center/evaluation-hardware-and-software/evaluation-boards-kits/eval-wsn.html#eb-overview. Accessed 11 Jan 2018
10. https://www.analog.com/en/design-center/landing-pages/002/apm/wsn-solution-2014.html. Accessed 11 Jan 2018
11. Lu, R., Yang, A., Xue, Y., Xu, L., Zhu, C.: Analysis of the key factors affecting the energy efficiency of batteries in electric vehicle. World Electric Veh. J. **4**, 5–9 November 2010. ISSN 2032–6653, Shenzhen, China

Improving Ambient Assisted Living Through Artificial Intelligence

Alessandro Miguez[1], Christophe Soares[1,2], José M. Torres[1,2],
Pedro Sobral[1,2(✉)], and Rui S. Moreira[1,2,3]

[1] ISUS Unit, University Fernando Pessoa, Porto, Portugal
{afmoreira,csoares,jtorres,pmsobral,rmoreira}@ufp.edu.pt
[2] LIACC, University of Porto, Porto, Portugal
[3] INESC-TEC, FEUP - University of Porto, Porto, Portugal
http://isus.ufp.pt

Abstract. The longevity of the population is the result of important scientific breakthroughs in recent years. However, living longer with quality, also brings new challenges to governments, and to the society as a whole. One of the most significant consequences will be the increasing pressure on the healthcare services. *Ambient Assisted Living* (AAL) systems can greatly improve healthcare scalability and reach while keeping the user in their home environment. The work presented in this paper specifies, implements, and validates a smart environment system that aggregates Automation and *Artificial Intelligence* (AI). The specification includes a reference architecture, composed by three modules, whose tasks are to automate and standardize the collection of data, to relate and give meaning to that data and to learn from it. The system is able to identify daily living activities with different levels of complexity using a temporal logic. It enables a real time response to emergency situations and also a long term analysis of the user daily routine useful to induce healthier lifestyles. The implementation addresses the applications and techniques used in the development of a functional prototype. To demonstrate the system operation three use cases with increasing levels of complexity are proposed and validated. A discussion on related projects is also included, specifically on automation applications, *Knowledge Representation* (KR) and *Machine Learning* (ML).

Keywords: Ambient Assisted Living · Artificial Intelligence ·
Knowledge Representation · Machine Learning · Home automation ·
Daily living activities

1 Introduction

According to the *United Nations* (UN) [1] it is estimated that in the year 2050, for the first time in history of humanity, the elderly population, will surpass the younger population. Many factors explain this phenomenon, especially the technological breakthroughs in the health area, which have increased life expectancy

© Springer Nature Switzerland AG 2019
Á. Rocha et al. (Eds.): WorldCIST'19 2019, AISC 931, pp. 110–123, 2019.
https://doi.org/10.1007/978-3-030-16184-2_12

around the globe. This scenario brings new social, economic, cultural challenges and a greater demand for solutions, including those related with information technologies. One of the challenges that AAL seeks to solve is how to improve the quality of life of people with special needs, such as the elderly, in their homes, so that they can carry out their daily activities with some degree of independence.

The present work proposes a reference architecture for an AAL smart system and a prototype of its implementation. The system is based on a home automation framework to interact with sensors and actuators, takes advantage from an ontology to provide semantic context to sensor data and uses ML to classify user activities.

The paper is organized as follows: in Sect. 2 the state of the art on home automation systems and related work on activity context representation is presented. Section 3 describes a reference architecture for the AAL smart system. Section 4 come up with an evaluation of the system performance in a number of defined use cases. Finally, Sect. 5 presents the paper conclusion.

2 State of the Art on Smart Home Systems

Work in the area of AAL has two approaches, a technological one, which addresses issues related to the devices used and how they should be integrated into a system and an intelligent data processing approach, which tries to model and implement a system architecture that is able to gather context and high level information from sensor data. In the next sections, we present some projects on the state of the art for home automation systems and activity context representation.

2.1 Background on Home Automation Systems

A great deal of effort in the field of solutions for the automation of environments comes from the open source communities, which join efforts around projects that aim to spread the automation with low implementation cost. It should be mentioned that the initial proposal of these community projects is to disseminate the use of *Internet of Things* (IoT) technologies and make them useful in the context of home automation. Each project tries to implement on its own way in order to remove the barrier of protocol diversity. Thus, these applications transcend the original purpose and can be considered useful to overcome the obstacle of the protocols in scientific projects that aim the context of intelligent environments, as is the case of the present work. One of the applications found was Domoticz [2], which is tailored for running on low power computing devices and various operating systems. It is intended to integrate different sensors and actuators, and has a closed list of supported devices in its documentation. The Jeedom [3] application was designed to work only in the Unix environment, having a market place for installing plugins, some of which have to be paid for. The OpenHAB [4] framework, instead of supporting specific devices, operates as a middleware layer between different protocols and an abstraction created by the framework. In this way the support of a new protocol only requires the development of a suitable plugin. The support for a large number of protocols, together with the ease of configuration of different devices, as well as the existence of an active community, were decisive factors in the selection of this framework for this work. A summary of the frameworks review is presented in Table 1. This table uses a scale of plus marks in which the more the marks, the better the rank of the framework in a specific category of evaluation and a scale of check marks to indicate if the application address that constraint.

Table 1. Review of the home automation frameworks

Application	Open Source	Community	Documentation	Protocols	Multi-platform
OpenHAB	✓	+++	+++	+++	✓
Domoticz	✓	++	++	+	✓
Jeedom	✓	+	+	++	✗

2.2   Related Work on Activity Context Representation

The survey [5] analyses 20 AAL papers and creates a table, in which classifies these works according to approaches and subjects discussed, belonging to 4 classes, research fields/technologies: AAL system properties, paper structure and types of ML algorithms. In [6] is proposed an approach to model human activities, that is defined as a hierarchical model of complex activities as a mixture of simple activities. They use sensor fusion, classification, clustering and topic distribution from acceleration and physiological signals. The paper [7] propose an algorithm that pre-process data from wearables in order to identify some events, like walking and falling. They analyze other works and divide them into sliding window approaches and primitive activities. In [8], the authors propose a knowledge-based representation of *Activities of Daily Living* (ADL), for real time monitoring of smart home contexts. The paper presents three challenges that the representation of ADL should address: the degrees of freedom of the individuals with respect to the order of accomplishment of the tasks; the heterogeneity of the sensors used as well as the data formats provided; ADL are composed of sequences of temporally related actions. The paper mentions the different approaches to solve the problem: those using statistical and probabilistic analysis techniques; the discriminative ones, that use techniques of ML; those that use logical formalisms, that uses techniques of KR. Then it describes the ontology modeling process based on the domain of ADL, the entity taxonomy and the modeling of temporal, spatial and environmental contexts, the information of which will be provided by the sensors. The paper [9] presents a framework to support people with mental illnesses that uses sensors of different modalities to recognize daily activities. On temporal ontologies, [10] does an analysis of the Time-OWL ontology, describing its three layers and exemplifying its use with the representation of a leveraged buyout. In [11] an approach is presented for representation and inference in the context of *Web Ontology Language* (OWL) 2.0. It distinguishes the temporal information in quantitative aspect, comprising values for dates, instants and intervals, and qualitative aspect, considered information about temporal events whose durations are unknown, like Allen's temporal relations. The qualitative aspect is guaranteed using rules *Semantic Web Rules Language* (SWRL). The paper then discusses details of temporal expression and inference using examples and algorithms.

3   Smart Home Reference Architecture

Figure 1 presents the smart home system architecture based on three independent modules. The Automation module handles the communication with different sensors, actuators and existing applications over the network, as well as the interaction with different personas and the remaining modules. The KR module performs inference rules based on data input from the Automation module (cf. sensor values, persona activities, etc.). The ML module is able to classify user activity patterns by receiving input features received also from the Automation module.

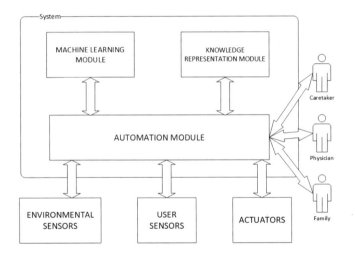

Fig. 1. Smart home system architecture

The implementation of the Automation module was based on the environment automation framework OpenHAB [4]. OpenHAB is an Open Source software that centralizes communications with the most heterogeneous types of technologies and protocols, commonly used for automation and IoT. This broker allows the integration and management of different devices and their data. The OpenHAB facilitates considerably the effort to deal with a vast variety of protocols and home automation devices.

The main goal of the Automation module is to allow OpenHAB to work as a middleware between the hardware layer (sensors and actuators) and the two upper layers, i.e., the ML and KR modules, responsible for the reasoning, either by inductive or deductive approaches. Figure 2 presents the data flow of the Automation Module.

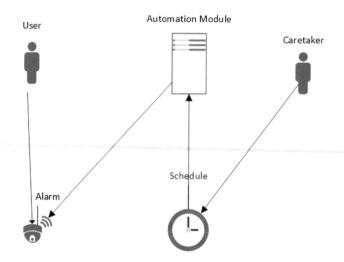

Fig. 2. Data flow of the Automation module

The primary goal of KR module is to provide a semantic representation of data received from the Automation module. This representation is achieved using an Ontology Language called OWL, and implemented by Protégé and Jena *frameworks*. The OWL allowed the creation of SWRL rules to infer complex activities based on a temporal sequence of simple activities. Figure 3 presents the communication between the automation and KR modules, on identifying the *WatchTV* complex activity. more specifically, the Automation module recognizes the following expected sequence of events: (i) enter the room, (ii) sit on the couch and (iii) turn on the television. This information flow is sent to the KR module, which infers the specific activity and reports it to the Automation module, which may afterwards appropriately act on the environment (e.g. dim lights and adjust sound surround system according user profile).

The KR taxonomy used in this project is represented in the ontology of Fig. 4. The model has a root class called *Event* to represent any event. An *Event* has a beginning and an end (respectively *hasBeginning* and *hasEnd* properties). Both properties belong to the OWL-Time [12] ontology. The *Event* has four subclasses: *Localization*, *Device*, *Activity* and *Emergency*.

The *Localization* class represents the presence of a human in a specific room division. This approach considers the temporal and spatial dimension. It is possible to represent the presence of a person in the living room, in a certain time

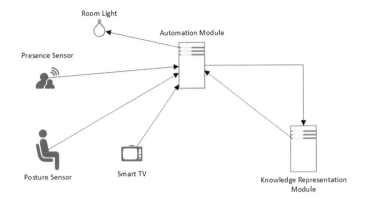

Fig. 3. Data flow of the KR module

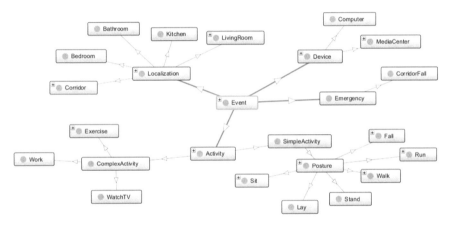

Fig. 4. Taxonomy for the activities representation (Protégé screenshot)

interval, between two time instants (between entering and living the room). The *Device* class represents a device usage within the environment. Examples of meaningful devices are represented by the subclasses *TV* and *Computer*.

A complex activity can be defined as the occurrence of a number of events within the same time frame [13]. For example, *WatchTV* is a complex activity involving three events within the same time frame: the user enters the living room, sits on the couch and turns on the television. This sequence of events is translated as the beginning of *LivingRoom* being earlier than the *Sit* and this happening before the *TV* is turned on.

The ML module is responsible for performing the inductive reasoning algorithms. This task is achieved by receiving events from the Automation module, process those events and send back results to the Automation module. Figure 5 presents the data flow of the ML module.

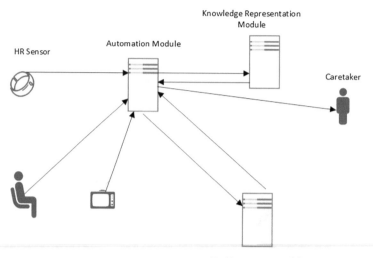

Fig. 5. Data flow of the ML module

Currently, the ML module is implemented in Java and uses the WEKA [14] software library. This module is organized in several classes. For example, the *ClassifierModel* class uses a classifier (*Classifier*), a filter (*Filter*), the training instances (*Instance*) and an estimator (*Evaluation*) that performs the prediction of new data after the training stage. The *AbstractModel* class is a composition of the previous class and a set of classified instances. The goal of this class is to represent the ranking model for various sets of data. A classifier model was created to perform the classification of the heart rate pattern during user activity. Thus, this classifier needs to access data from the user cardiac activity (cf. average, maximum, minimum, and standard deviation heart rate values within the corresponding time period). The class that represents this DataSet extends the *AbstractModel* class, and is created with the *ClassifierModel* using information from the training data.

In this inductive module a *Classifier* can be chosen from a list of existing implementations in the Weka library. The current implementation uses a *Naive Bayes* classifier together with a cost-based classifier to disambiguate between false positives and negatives. The distinction is particularly important, in health contexts, since a false negative classification may lead to consider an emergency situation as normal. The *CostSensitiveClassifier* applies a *Cost Matrix* (CM) to a specific classifier. This matrix has two dimensions where the main diagonal considers the true positive (TP) and true negative (TN), while the secondary diagonal has the false negative (FN) and false positive (FP), as seen in Fig. 6a. The identity CM does not alter the results of the classifier (see Fig. 6b). The CM is used to promote the rank of a false negative, statistical type II errors, and this is achieved when its cost is increased as see Fig. 6c.

$$\begin{bmatrix} TP & FP \\ FN & TN \end{bmatrix} \qquad\qquad \begin{bmatrix} 0 & 1 \\ 1 & 0 \end{bmatrix} \qquad\qquad \begin{bmatrix} 0 & 1 \\ 10 & 0 \end{bmatrix}$$

(a) CM template   (b) Identity CM   (c) Applied CM

Fig. 6. Cost Matrices (CM) used to improve false negative classifications

4 Evaluation of the Smart Home System

All tests were performed on the same computer executing the three modules, as well as a test program that performed all the *Hiper-Text Transfer Protocol* (HTTP) requests and checked the system states using the JUnit test library [15]. The specifications of the computer hardware are listed in Table 2.

Three use cases were defined and evaluated. Each of these use cases has different levels of complexity requiring the interaction with one or more of the system modules.

Table 2. Specifications of the computer used in tests

Parameter	Value
RAM memory	12 GB
Hard drive	500 GB SSD
Processor	Intel i7-7500U
Operating System	Windows 10 64-bit

4.1 Use Case 1: Monitoring Medicine Intake

The first use case considers an user which must intake a dose of medication prescribed by a doctor at a specific time. The medication is stored in a smart dispenser that intercommunicates with the system and it also has an alarm which is triggered at the intake hours. When the user takes its pills, the alarm is dismissed and turned off. Otherwise, when the user misses its medication, the caretaker is notified and can act accordingly to the protocol. This use case only requires the Automation module.

For use case 1, the system checks if the user has failed to take medication. The verification is done by consulting the user's calendar (supported by the CalDav protocol), an intelligent dispenser and an alarm. The goal is to make the system alert the user to the time of taking medication by means of an alarm that will be turned off when the use of the drug dispenser is detected. If within a certain period, for instance 10 min, it does not occur, the alarm will be turned off and an email alert sent to the caregiver or doctor. Figure 7 illustrates the behavior of the system for the events mentioned.

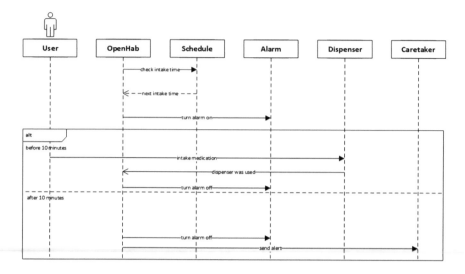

Fig. 7. Sequence diagram for use case 1

Two tests were implemented for this use case 1. The first test aimed to determine that the alarm was triggered to alert the user to the time of drug taking, and that after a certain period of time, in this case 70 s, the latter used the dispenser. The goal is to determine if the order of events is consistent with what is expected, that is, determining a new time to take the medication. If it is already on time, the alarm is triggered, the user takes some time to go to the dispenser, uses it, the alarm goes off and the system waits for the new planned time. Because the rule has a polling condition every minute, a wait of 70 s has been chosen to be able to demonstrate that the rule remains each time polling is performed. Figure 8 demonstrates that the test was successful, with the first two lines referring to the determination of a new take, with the consequent activation of the alarm. Following a waiting period, the user activates the dispenser to determine the shutdown of the alarm and the dispenser itself.

```
2017-09-13 11:31:53.322 [INFO ] [runtime.busevents    ] - NextIntake state updated to 2017-09-13T11:31:53
2017-09-13 11:31:53.386 [INFO ] [runtime.busevents    ] - Alarm received command ON
2017-09-13 11:33:03.320 [INFO ] [runtime.busevents    ] - Dispenser state updated to ON
2017-09-13 11:33:03.336 [INFO ] [runtime.busevents    ] - Alarm received command OFF
2017-09-13 11:33:03.350 [INFO ] [runtime.busevents    ] - Dispenser received command OFF
2017-09-13 11:33:03.360 [INFO ] [runtime.busevents    ] - NextIntake state updated to Undefined
```

Fig. 8. Test 1 results for use case 1

The second test is similar to the previous scenario except that the dispenser is not used. The alarm is activated for one minute and then deactivated and an email is sent to the caregiver or doctor of the user, alerting to the failure to take the medication, as can be verified in Fig. 9.

```
2017-09-13 11:34:38.454 [INFO ] [runtime.busevents           ] - NextIntake state updated to 2017-09-13T11:34:38
2017-09-13 11:34:38.507 [INFO ] [runtime.busevents           ] - Alarm received command ON
2017-09-13 11:36:00.005 [INFO ] [org.openhab.model.script.   ] - Email sent to caretaker
2017-09-13 11:36:00.013 [INFO ] [runtime.busevents           ] - Alarm received command OFF
2017-09-13 11:36:00.021 [INFO ] [runtime.busevents           ] - NextIntake state updated to Undefined
```

Fig. 9. Test 2 results for use case 1

4.2   Use Case 2: Inferring Complex Activities

The second use case demonstrates how the system is able of recognize complex activities. When an user is watching television, for example, the system is able to recognize a sequence of simple activities, such as someone entering a room, sitting on a couch and turning the television on. This sequence of activities in this specific order causes the system to infer the activity: watching television, and may result in reducing the surrounding environment lighting, so the Automation and KR module have to interact. The system data flow for this use case is shown in the Fig. 10.

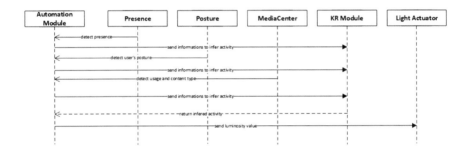

Fig. 10. Sequence diagram for use case 2

The first test carried out was aimed at confirming that the Automation and KR modules can interact in a consistent way, so that a complex activity is properly inferred. Thus, the test client sends several HTTP requests to the Automation module, which contains information about the user entering in the living room, then sitting on the couch, and finally turning the *media center* on. The Automation module will have these *Items* state modified and notifies the KR module about that change. When a complex activity is inferred, this information will be sent from the KR module to the Automation module. Figure 11 present the requests arriving at the Automation module for this second use case.

```
2017-09-19 11:43:34.799 [INFO ] [runtime.busevents        ] - LivingRoomSensor state updated to ON
2017-09-19 11:43:34.813 [INFO ] [runtime.busevents        ] - BathroomSensor state updated to OFF
2017-09-19 11:43:34.831 [INFO ] [runtime.busevents        ] - ActualLocalization state updated to Living Room
2017-09-19 11:43:39.804 [INFO ] [runtime.busevents        ] - ActivityDetector state updated to 5
2017-09-19 11:43:39.810 [INFO ] [runtime.busevents        ] - ActualPosture state updated to Sitting
2017-09-19 11:43:44.809 [INFO ] [runtime.busevents        ] - MediaCenter state updated to ON
2017-09-19 11:43:44.820 [INFO ] [runtime.busevents        ] - ActualDevice state updated to MediaCenter
2017-09-19 11:43:46.061 [INFO ] [runtime.busevents        ] - ActualActivity state updated to WatchTV
```

Fig. 11. Test results for use case 2

4.3   Use Case 3: Classification of Abnormal Behaviours

The last use case illustrates how the system recognizes complex activities by considering observed data and using *ML algorithms* to classify. In this scenario, the system monitors activities such as watching television or doing exercise and collects data from a heart rate sensor. This information is used to assess whether, for these activities, the heart rate values are normal or not. If they are abnormal the system triggers an alert to the caregiver. This use case requires that all system modules are communicating and cooperating. In this use case, the KR module needs to detect a complex activity based on sensor data sent by the Automation module. This activity is then combined with the heart rate sensor values, by the ML module, to verify if the sensor readings are consistent with the detected activity. If they are not, the Automation module will send an alert to a caregiver. Figure 12 presents the interactions between the different modules of the system for use case 3.

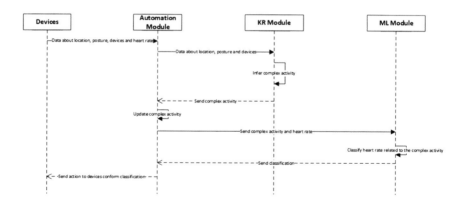

Fig. 12. Sequence diagram for use case 3

User values were obtained using an ECG Band, an HRMI board and an Arduino board. The synthetic values were programmatically created having as parameters the mean value expected for a tachycardia situation (Mean), a standard deviation value (Std) and a random value obtained from a Gaussian distribution (*Gaussian*), according to the formula 1 below:

$$Value = Gaussian * Std + Mean \qquad (1)$$

```
Correctly Classified Instances        263              98.1343 %
Incorrectly Classified Instances        5               1.8657 %
Kappa statistic                         0.939
Mean absolute error                     0.0191
Root mean squared error                 0.1104
Relative absolute error                 6.2672 %
Root relative squared error            28.3336 %
Total Number of Instances             268

=== Confusion Matrix ===

   a   b   <-- classified as
 215   3 |   a = normal
   2  48 |   b = abnormal

=== Detailed Accuracy By Class ===
```

	TP Rate	FP Rate	Precision	Recall	F-Measure	MCC	ROC Area	PRC Area	Class
	0.986	0.040	0.991	0.986	0.989	0.939	0.999	1.000	normal
	0.960	0.014	0.941	0.960	0.950	0.939	0.999	0.996	abnormal
Weighted Avg.	0.981	0.035	0.982	0.981	0.981	0.939	0.999	0.999	

(a) Results of the classification using the unit CM

```
Correctly Classified Instances        261              97.3881 %
Incorrectly Classified Instances        7               2.6119 %
Kappa statistic                         0.9183
Mean absolute error                     0.0322
Root mean squared error                 0.143
Relative absolute error                10.5633 %
Root relative squared error            36.7136 %
Total Number of Instances             268

=== Confusion Matrix ===

   a   b   <-- classified as
 211   7 |   a = normal
   0  50 |   b = abnormal

=== Detailed Accuracy By Class ===
```

	TP Rate	FP Rate	Precision	Recall	F-Measure	MCC	ROC Area	PRC Area	Class
	0.968	0.000	1.000	0.968	0.984	0.921	0.999	1.000	normal
	1.000	0.032	0.877	1.000	0.935	0.921	0.999	0.996	abnormal
Weighted Avg.	0.974	0.006	0.977	0.974	0.975	0.921	0.999	0.999	

(b) Results of classification with a 10x higher CM on FP (false positives)

Fig. 13. Test for use case 3

The results of the classification can be seen in the Figs. 13a and b.

The rationale behind the chosen use cases was related with the various degrees of interaction between the modules of the system and also according to the complexity of the involved tasks.

Therefore, the first use case focus on demonstrating that the automation module alone is sufficient to perform tasks that typically may be compared to a reflexive agent (i.e. an entity that has an internal model of the world, even if rudimentary, and is able to act upon it in a deterministic way).

The second use case demonstrates that the Automation and KR modules can interact in a consistent way. This cooperation corresponds to the traditional knowledge-based agent pattern, i.e., the Automation module adds new sentences to the vocabulary of the KR module, which then could infer new sentences from these additions and influence the decisions made by the Automation module. This KR module proved to be particularly effective in inference of complex activities.

The third and last use case demonstrates the integration of the three modules. Some data provided by the Automation module is processed by the KR module, which is then returned back to the same module along with additional data to the ML module. This interaction between the three modules exhibits a knowledge-based agent pattern, but also evidences the behavior of a learning-based agent, since the ML module provides the ability to classify new forthcoming activities (based on previous pattern behaviours). With proper orchestration, this use case demonstrates the flexibility of the solution created to deal with the diversity of ADL situations.

5 Conclusion

This paper describes a prototype implementation for the proposed smart home reference architecture. This prototype provides a realist and feasible proof of concept by combining the use of ML and KR modules, to create a supporting smart home environment. Such system enables the monitoring and assistance to people in accomplishing their daily home life tasks. The current implementation uses monitoring rules with a *polling-based* strategy, though it can be changed to a more efficient *push-based* mechanism. Much more relevant is the dynamic nature of the architecture and its modules. This is an important contribution and one of its strongest points. For example, the Ontology itself may be extended and improved both with entities and inference rules to tackle a lot more assisting scenarios. Moreover, the ML module can also be enhanced, for example, with additional use of different classifiers and big data collected from different home environments. Such use of information and data sets from multiple home settings and usage scenarios will undoubtedly contribute to train and compare different classifiers and choose the better suited for supporting smart home assisted living.

Future plans include reusing the prototype of the proposed smart home reference architecture for building a complete and comprehensive set of patterns of complex activities. Moreover, it will be enhanced with state of the art ML classifiers, specifically trained to help monitoring and assisting well known daily living scenarios. For example, capturing medication intake patterns and, based on that, help improving treatment adherence; verifying whether or not a user is having more sedentary or less sociable behaviors, all these based on their patterns of past activities. Such information is highly relevant and can also be re-used to infer whether or not a user may be experience depressed behaviours, hence triggering alerts to himself or associated caregivers or relatives.

Acknowledgements. This work was partially funded by: project QVida+, Estimação Contínua de Qualidade de Vida para Auxílio Eficaz á Decisão Clínica, NORTE-01-0247-FEDER-003446, supported by Norte Portugal Regional Operational Programe (NORTE 2020), under the PORTUGAL 2020 Partnership Agreement, through the European Regional Development Fund (ERDF); and by FCT-Fundação para a Ciência e Tecnologia in the scope of the strategic project LIACC-Artificial Intelligence and Computer Science Laboratory (PEst-UID/CEC/00027/2013); and by Fundação Ensino e Cultura Fernando Pessoa.

References

1. UN. Ageing. http://www.un.org/en/sections/issues-depth/ageing/. Accessed 2 Nov 2017
2. Domoticz. Domoticz. https://domoticz.com/. Accessed 1 Nov 2017
3. Jeedom. Jeedom. http://www.jeedom.com/. Accessed 1 Nov 2017
4. OpenHAB. OpenHAB. http://www.openhab.org/. Accessed 1 Nov 2017
5. Dimitrievski, A., Zdravevski, E., Lameski, P., Trajkovik, V.: A survey of ambient assisted living systems: challenges and opportunities. In: Proceedings of the 2016 IEEE 12th International Conference on Intelligent Computer Communication and Processing, ICCP 2016, pp. 49–53 (2016)
6. Peng, L., Chen, L., Wu, X., Guo, H., Chen, G.: Topic model based hierarchical complex activity representation and recognition with classifier level fusion. In: Proceedings of the International Conference on Ubiquitous Computing (UbiComp), vol. 9294(c), pp. 1–11 (2016)
7. Munoz-Organero, M., Lotfi, A.: Human movement recognition based on the stochastic characterisation of acceleration data. Sensors (Switzerland) **16**(9), 1–16 (2016)
8. Chen, L., Nugent, C.D., Wang, H.: A knowledge-driven approach to activity recognition in smart homes. IEEE Trans. Knowl. Data Eng. **24**(6), 961–974 (2012)
9. Stavropoulos, T.G., Meditskos, G., Kompatsiaris, I.: DemaWare2: integrating sensors, multimedia and semantic analysis for the ambient care of dementia. Pervasive Mob. Comput. **34**, 126–145 (2016)
10. Frasincar, F., Milea, V., Kaymak, U.: TOWL: integrating time in OWL. In: Semantic Web Information Management: A Model-Based Perspective, pp. 225–246 (2010)
11. Batsakis, S., Petrakis, E.G.M., Tachmazidis, I., Antoniou, G.: Temporal representation and reasoning in OWL 2. Semantic Web **8**(6), 981–1000 (2017)
12. W3. Time Ontology in OWL. www.w3.org/TR/owl-time/. Accessed 4 May 2016
13. Allen, J.F.: Maintaining knowledge about temporal intervals. Commun. ACM **26**(11), 832–843 (1983)
14. Frank, E., Hall, M.A., Witten, I.H.: The WEKA Workbench, 4th edn., pp. 553–571. Morgan Kaufmann, San Francisco (2016)
15. JUnit. JUnit 5. https://junit.org/junit5/. Accessed 5 Apr 2017

A FIPA-Compliant Framework for Integrating Rule Engines into Software Agents for Supporting Communication and Collaboration in a Multiagent Platform

Francisco J. Aguayo-Canela[1], Héctor Alaiz-Moretón[2],
Isaías García-Rodríguez[2(✉)], Carmen Benavides-Cuéllar[2],
José Alberto Benítez-Andrades[2], and Paulo Novais[3]

[1] IEEE Power & Energy Society, Seville, Spain
francisco.aguayo@ieee.org
[2] Department of Electrical and Systems Engineering and Automation,
Universidad de León, León, Spain
{hector.moreton, isaias.garcia,
carmen.benavides, jbena}@unileon.es
[3] Algoritmi Centre/Department of Informatics,
University of Minho, Braga, Portugal
pjon@di.uminho.pt

Abstract. Production rules have been traditionally considered a good knowledge representation formalism for creating expert systems, and also as a good mechanism for building intelligence within software agents due to the flexibility of their declarative knowledge representation. But the use of rules and rule engines for implementing behaviors inside agents in the context of a multiagent environment has a number of challenges if one wants to adhere and preserve the defining characteristics of agency and support the kind of communication and collaboration mechanisms needed by multiagent systems. This paper describes a framework, and its implementation, for the integration of production rules reasoning mechanisms inside software agents in the context of a multiagent platform, preserving the defining characteristics of agency and accomplishing the well-established FIPA standards for agent interaction.

The implementation has been accomplished by using either CLIPS or Jess as the rule engine, but it can be extended to other formalisms. JADE was chosen as the multiagent platform for developing and testing the solution. The proposed framework includes the uncoupled integration of the rule inference engine into the agent and the adherence to FIPA specifications about using protocols and communication processes. The resulting framework can be used to adapt legacy expert systems whose functionality can be divided to be performed by separate agents or to build new, distributed intelligent systems exploiting the capacities of a multiagent platform at different levels.

To show the validity of the approach, a functional test environment was built and is available for downloading.

Keywords: Rule-based agent · Multiagent systems · JADE · CLIPS · Jess · FIPA · Interaction protocol

© Springer Nature Switzerland AG 2019
Á. Rocha et al. (Eds.): WorldCIST'19 2019, AISC 931, pp. 124–133, 2019.
https://doi.org/10.1007/978-3-030-16184-2_13

1   Introduction

Production rules is a well-known knowledge representation formalism that has been extensively used for building intelligent systems. The spread of microcomputers and their increasing computation power made it possible to study and build parallel and distributed environments for running intelligent applications much faster and with greater flexibility and modularity than with traditional, monolithic rule-based expert systems [1, 2].

With the research on the agent abstraction and the multiagent distributed system paradigm, the rule-based systems found another field of application. Rules, as the building blocks for performing the agent behavior, as well as coordination paradigms for letting the rule-based agents collaborate towards a common objective, resulted in the so-called multiagent production systems, which are conceptually different from parallel and distributed production systems. While in the case of distributed production systems the aim is to distribute a monolithic expert system in order to run it faster, in the case of the multiagent paradigm the objective is to make different, independent production systems, work together with a common goal by coordinating their activities [3].

The first step towards building a multiagent production system is the integration of a rule engine into a software agent. Then, a coordination mechanism and a collaboration architecture are also needed in order to obtain the collaborative nature that multiagent systems require. The integration of the rule engine into the agent is crucial to achieve the coordination of the agents and also the flexibility needed to build a multiagent system of this nature, this integration must be accomplished by having into account the future coordination and communication mechanisms that may be needed.

The objectives of the present work regarding the integration of a rule engine into an agent are:

- The integration must be neutral with respect to the particular technology of the rule engine (no assumption or knowledge is needed for one agent to communicate with another one regarding the particular rule engine that is running in the receiving agent).
- Agents with a rule-based behavior must be able to collaborate in a number of ways with other rule-based agents, both by direct communication between them and also by collaboration using a shared knowledge repository.
- Actions performed in the rule-based system of a given agent must be fully and exclusively controlled by the agent itself, separately from other activities or behaviors.
- The rule engine must not block the agent basal behavior (for example, the ability to continue receiving ACL messages) while performing the reasoning.
- The design must help and ease the development of rule-based multiagent applications.

The platform chosen for building the practical solution is JADE [4]. This election is based on its maturity, the size of the user community, and the use of behavior-based agents, which are susceptible to allow the integration of rule engines as their behaviors.

The rest of the paper is organized as follows. Section 2 introduces previous research aimed at the integration of rule engines and rule-based reasoning services into the agents. Section 3 shows the proposed solution, giving a functional view of the different solutions designed for the framework. In Sect. 4, a test environment built for testing and demonstration purposes is presented. This environment is fully functional can be downloaded from the specified URL. Finally, Sect. 5 gives some discussion and points to future research work.

2   Building Agents that Use Rule-Based Reasoning

Rule-based agents allow the use of declarative statements for defining the agent reasoning capabilities. This is a desirable characteristic when the problem-solving technique is not well defined and/or may need a number of adjustments, even at run time. With rule-based agents, a number of reasoning models can be built, from reactive to deliberative ones, as well as hybrid ones [3].

A distinction is usually made between rule-based agents and rule-enhanced agents. A rule-based agent is an agent whose behavior and/or knowledge is totally expressed using rules, while a rule-enhanced agent is the one that use a rule engine as an additional component for specific reasoning capabilities. In any case, the integration strategy can be the same when using a behavior-based agent as the ones used in the JADE platform [5, 6]. In the rest of the document, the agents with an integrated rule engine will be referred as "rule-based agent".

2.1   Previous Solutions

The approach designed by Cardoso [7] consists in the instantiation of a Jess (Java Expert System Shell) rule engine [8] object within a JADE (one shot or cyclic) behavior. It is one of the first attempts to build the agent reasoning by means of a rule engine. As pointed in [9] this approach has a number of limits:

- The rule engine is completely hidden to other agents of the system, and there is not any support for the cooperation among different rule-based agents.
- The solution is bound to a particular technology: The Java Expert System Shell (Jess) in this case.
- Implementing some of the agent capabilities into the rule engine, as the solution does, is cumbersome (for example parsing ACL messages in Jess).
- The agent may get blocked when the rule engine is running, and then a limit in the number of run cycles of the rule engine must be imposed.

JAMOCHAAGENT [10] introduces a transformation component between the agent and the rule engine that is in charge of transforming FIPA ACL messages (transporting a description of an action to be performed by the rule engine of the receiving agent) into convenient actions for the rule engine. In this case, the CLIPS system [11] is the one used for acting as the rule engine. This functionality achieves a certain degree of flexibility into the system because agents can be re-programmed when the system is running without the necessity of restarting the whole platform. The transformation between ACL messages and CLIPS statements are based upon direct correspondences between ACL speech acts (Agree, Failure, Inform, etc.) and CLIPS engine actions that are evaluated by production rules inside the rule engine.

Having in mind the limitations found for the Jess-JADE integration [7], a research was conducted that proposed the integration of an open source Java-based rule engine, Drools, into JADE (Drools4JADE, or D4J) [5, 6]. This work also performs the integration of the rule engine as an agent behavior. The work stresses the need of using the agent communication acts as a way of interchanging problem-solving knowledge, that is, rules that may have been developed by an agent but may be useful for other/s. This "rule mobility" also raises security concerns that should be faced in a real scenario. The use of Drools, based on the JAVA language, allows referring to ACL messages both in the precondition and in the consequent parts of the rules. This resulted in the design decision of allowing manipulation of facts and rules on Drools agents through ACL messages by exposing an API for invoking these actions. The integration of a Bean-Shell [12] component into the agent allows manipulating the Drools elements by using JAVA as a scripting language.

More recent approaches use the service-oriented architecture to build a reasoning service for the agents in a multiagent platform. These approaches try to solve the interoperability problem when using different reasoning paradigms and representation languages. EMERALD [13] is a framework for interoperation of intelligent agents in the Semantic Web. The framework integrates different reasoning mechanism into a multiagent platform using a Reasoner agent that, actually, acts more like a web service. EMERALD is based on FIPA specifications, achieving a fully FIPA-compliant model and deals with trust issues [13]. The approach in Rule Responder [14] also focuses on the interoperability of different reasoning paradigms, integrating them in the multiagent platform through a service-oriented architecture; but the solution is not fully compliant with FIPA specifications.

3   The Proposed Solution

Based on the requirements stated in Sect. 1 and the previous research presented in Sect. 2, the final integration solution was designed and implemented. The framework proposed in this work uses an integration of the agent with its rule engine that is both uncoupled, in the sense of being independent of the engine technology, but also having a strong link in another sense because a given engine will be exclusively devoted to the agent where it is instantiated. The building blocks of the solution are presented next, with some details of their implementation.

3.1   Building an Interface for Managing the Rule Engine from the Agent

An agent integrates a rule engine by means of a programming interface (called *RBEngine*) that allows this agent to transparently make use of any rule engine technology by building an implementation for one of the possible technologies to be used.

RBEngine objects are created inside the initialization method for the agent. This way, the agent has control over the functions and responses from the inference engine and, also, it is possible to delay, up to the instant of agent creation in the platform, the decision about which kind of inference engine technology is to be used for a given agent.

The object representing the rule engine is unique and is subordinated to the object representing the agent and to its execution thread. The rule engine object is created in the moment of the incorporation of the agent to the platform and not as part of the lifecycle of a behavior of the agent.

3.2   Communication Mechanism Between Agents

The mechanism for communication between rule-based agents adheres to FIPA specifications, using ACL messages and a domain ontology for representing the specific meanings of the message contents.

The set of actions that a rule-based agent can ask other to perform on its rule engine is limited to predefined list that includes the typical activities of these types of systems (loading new facts and rules, execute a number of cycles, query some fact values, perform a clear or reset operation, etc.). These actions are designed in accordance to the interface defined for interacting with the rule engine and constitutes both an efficient and secure way of conceptualizing the set of possible operations to be performed in any rule engine that may be used.

3.3   Communication of the Agent with Its Rule Engine

As was previously stated, the actions to be performed by the rule engine are requested by implementing and invoking the *RBEngine* interface from the agent but, before this, the agent has performed another important task.

Figure 1 shows a schema of the communication mechanism described here. When an agent with a rule engine receives an ACL message with actions to be performed in its engine, the agent first captures the message (first loop in Fig. 1), communicates to the sender the acceptance of the action (in the case it is accepted) and, then, builds and sends itself a new ACL message with the same action than the sender proposed (if accepted) and the identification of the conversation and the sender, see second loop in Fig. 1. The sender has, then, a notification of the acceptance of the required actions and, as the rule engine may take some time to obtain the results, the actions to be achieved are placed again in the receiving agent's message queue in order to be finally processed by the rule engine when it is ready to do so.

This way, the agent is free to continue interacting with other agents in the platform. Once the rule engine process and executes the actions, the results of the execution are obtained, and the sender is informed of the results. This mechanism, based on a double implementation of a FIPA interaction protocol (sender agent to receiving agent and receiving agent to itself – to its rule engine –), allows an ordered, controlled, and privately managed way of using the rule engine of the agent. Using a FIPA interaction protocol is a natural way of implementing communication, both between agents and between an agent and its rule engine.

This behavior of the agent is functionally similar to a proxy: the agent analyzes the request, an acceptance message is sent to the sender, and the request, along with an

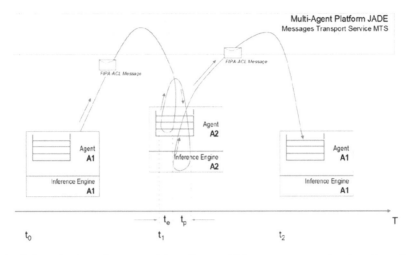

Fig. 1. Schematic view of an agent A1 sending an ACL message targeted at the rule engine in agent A2.

identification of the conversation that originated it, is stored for later processing when the rule engine is ready.

3.4   Execution of the Rule Engine Activities in a Threaded Behavior

When the rule engine is ready to process a new message, a Finite-State-Machine (FSM) behavior is responsible for implementing this processing. The first state of the FSM is a listening behavior, which captures the messages that the agent had previously built and put in its message queue. The second state is an execution behavior that obtains the contents of the message and inserts them into the rule engine. A threaded-behavior wraps this behavior in order to ensure that the execution thread of the agent is not interrupted by the activity of the rule engine. When this activity ends, and the thread of the execution behavior is dead, the third state, the response behavior starts. This behavior captures the output of the engine shell, builds the response message and sends it to the original requester agent by using the original conversation IDs. Finally, the response behavior returns control of the FSM to the listening behavior and the next ACL message for the rule engine in the agent's message queue, if any, is processed.

4   Test Environment

To test the validity of the approach, the integration solution was implemented in the JADE multiagent platform. CLIPS (CLanguage Integrated Production System) and Jess (Java Expert System Shell) were chosen as the rule engines to be integrated due to their popularity, stability and different developing languages (C and Java respectively). Agents with rule engines and the rest of the components were programmed as previously described. These agents were augmented with a graphical user interface where different windows allowing interactions with the agents were programmed:

- A BeanShell component was included in the agent, including a command window, to manage the internals of the agent by using Java as a scripting language, even at runtime. Figure 2 shows a screenshot.
- A synchronous shell was developed to let a human user interact with the rule engine of any rule-based agent in the platform. The user can interact directly with the rule engine by introducing the specific rule language commands. The window gets blocked until a response is received from the remote engine. This functionality is used for programming, testing and debugging. Figure 3a shows a screenshot of this window.
- An asynchronous shell used for interacting with any rule-based agent using the limited set of commands mentioned in Sect. 3.2 and the non-blocking functionality introduced in Sects. 3.3 and 3.4. This graphical interface allows the human to build the same kind of messages that the agents would build when they are in production. The window is not blocked, and the user can continue sending commands. The responses arrive when the remote engine finishes processing. Figure 3b shows a screenshot of this window.

Fig. 2. BeanShell window in a rule-based agent.

A fully functional, self-contained test environment, along with some extra functionalities, is available for download and use at the following URL: http://cekb.unileon.es/MAS/WORLDCIST19/, instructions for deployment and use are included. This environment may be used for educational purposes, or as a multiagent-based development environment, as presented in [15].

(a)   (b)

Fig. 3. Synchronous (a) and asynchronous (b) GUI windows in a rule-based agent.

5   Discussion and Conclusion

This work describes a complete framework for the integration of a rule engine into an agent with the objective of building rule-based (or rule enhanced) agents able to communicate and cooperate in a multiagent platform.

The integration is made by using an interface that is used for interacting with an object representing the rule engine. The communication between rule-based agents is built by using FIPA compliant ACL messages transporting a set of constructs described in an ontology that represent a limited set of actions that can be performed in the rule engines of the agents.

The approach was inspired and is similar, to some extent, to the ones in [6, 7, 10]. But with several important differences that are highlighted next.

Communication mechanisms between the agent and the inference engine have been proposed in [7] using the *store* function from Jess to pass information between the agent and the engine by means of a HashMap structure, or by using data structures (*deftemplate* construct from Jess) for accommodating the fields of the received ACL (Agent Communication Language) messages. These solutions, while being viable, only consider the integration of a given inference engine (Jess in this case), and use native functions of the rule-engine for performing part of the processing. Also, in the opposite direction, when the rule engine has performed its execution cycle and the results are ready to be sent to the original requester, some approaches use the rule engine for building this response by using ad-hoc functions programmed for this purpose. This breaks the decoupling of the agent from its rule engine and makes it dependable on the technology used.

The approach presented in this paper uses a careful separation between the tasks of the agent and those of the rule engine. The listening behavior (se Sect. 3.4) is responsible for processing the ACL message targeted at the rule engine and generate the corresponding calls to the RBEngine API. On the other hand, the response behavior is responsible for accessing the standard I/O of the rule engine to obtain whatever value is needed for building the response message.

The EMERALD framework [13] uses the "reasoning as a service" paradigm where the so-called Reasoner agent may ask any external inference engine that can perform the reasoning needed. This supposes a very loose coupling between the agent and the rule engine and, to some extent, breaks the encapsulation of the agent's functionalities, rising an important concern about trust that must be addressed. The vision used in the present research is quite different in this sense. Though our approach also uses an uncoupled schema because of the use of an interface for accessing the rule engine, this rule engine is created in the moment of the incorporation of the agent into the platform, and not as part of the lifecycle of a behavior of the agent, as is the case in other approaches, and it is exclusively devoted to perform the reasoning activities of that agent in particular.

Apart from the careful decoupling between the actions that the rule engine must perform and those pertaining to the agent, which were discussed in the previous paragraphs, the solution proposed also introduces an original way of interacting between the agent and its rule engine. This interaction consists of the use of a FIPA compliant interaction protocol for this purpose. This interaction protocol-based implementation is used for communications not only with the own internal engine of a given agent but also with other engines in other remote agents in the platform. These interactions, then, adhere to the recommended structure for the dialog sequence among agents having *Rational Effect* (RE) as defined by FIPA under the denomination *FIPA-Request protocol*.

When an agent receives an ACL message, it first analyzes and decides if the requested actions are accepted. In the case they are accepted, the receiving agent auto-assigns again the received requests as a new ACL message aimed at itself (and notifies the sender about this acceptation). This allows to manage, internally and privately, the execution and the responses issued by the inference engine, an also makes room for implementing extra security, privacy or trust verifications in the agent before delivering the message to the rule engine. Though these issues have not been addressed in depth in this research, a simple verification of the sender agent was built in order to see if it is listed in an access list of trusted agents of the platform.

The combination of the integration and communication solutions between the rule engine and the agent allows a non-blocking execution of the agent activities during the message interchange or the inference tasks. Moreover, if new messages arrive at the receiving agent with new actions for the rule engine, these can be stored in the message queue once accepted by the agent, preserving the order in which they may arrive. This may be important for some interaction or coordination mechanisms among agents in a multiagent environment.

Some limitations of the approach, in its current implementation, include the lack of testing of the system by using the most usual agent coordination and collaboration schemas, only some basic ones have been tested. Another limitation is the need of further study for making the approach compatible with knowledge representation and reasoning technologies from the Semantic Web. In this sense, future work includes implementing and studying these limitations, as well as building a proof-of-concept application in a practical and useful scenario. Also, security and trust issues should also be addressed and incorporated into the framework.

Acknowledgments. The work presented in this paper was supported by Junta de Castilla y León [grant number LE078G18].

References

1. Coulouris, G., Dollimore, J., Kindberg, T.: Distributed Systems: Concepts and Design, 5th edn. Cambridge University Press, New York (2012)
2. Durfee, E.H., Lesser, V.R., Corkill, D.D.: Trends in cooperative distributed problem solving. IEEE Trans. Knowl. Data Eng. **KDE-1**, 63–83 (1989)
3. Costin, B., Braubach, L., Paschke, A.: Rule-Based Distributed and Agent Systems (2011)
4. Bellifemine, F., Poggi, A., Rimassa, G.: JADE: a FIPA2000 compliant agent development environment. In: International Conference on Autonomous Agents and Multiagent Systems (2001)
5. Beneventi, A., Poggi, A., Tomaiuolo, M., Turci, P.: Integrating Rule and Agent-Based Programming to Realize Complex Systems. Learning (2004)
6. Poggi, A., Tomaiuolo, M.: Rule engines and agent-based systems. In: Encyclopedia of Artificial Intelligence, pp. 1404–1410. IGI Global (2009)
7. Cardoso, H.L.: Integrating JADE and Jess. http://jade.tilab.com/doc/tutorials/jade-jess/jade_jess.html
8. Friedman-Hill, E.J., et al.: Jess the Java Expert System Shell. Distrib. Comput. Syst. Sandia Natl. Lab., USA (1997)
9. Poggi, A., Tomaiuolo, M., Turci, P.: An agent-based service oriented architecture. In: WOA 2007, Genova, pp. 157–165 (2007)
10. Christoph, U., Krempels, K.-H., Wilden, A.: JAMOCHAAGENT a rule-based programmable agent. In: ICAART 2009 Proceedings of the International Conference on Agents and Artificial Intelligence, pp. 447–454 (2009)
11. Giarratano, J.C.: CLIPS 6.4 user's guide (2014)
12. Niemeyer, P.: Beanshell - The Lightweight scripting for Java
13. Bassiliades, N.: Agents and knowledge interoperability in the semantic web era. In: Proceedings of the 2nd International Conference on Web Intelligence, Mining and Semantics- WIMS 2012, vol. 1 (2012)
14. Paschke, A., Boley, H.: Rule responder: rule-based agents for the semantic-pragmatic web. Int. J. Artif. Intell. Tools **20**, 1043–1081 (2011)
15. Aguayo, F.J., García, I., Alaiz-Moretón, H., Benavides, C.: Techniques and utilities to improve the design, development and debugging of multiagent applications with agile principles. In: Advances in Intelligent Systems and Computing (2018)

A Chaotic Cryptographic Solution for Low-Range Wireless Communications in Industry 4.0

Pilar Mareca and Borja Bordel$^{(\boxtimes)}$

Universidad Politécnica de Madrid, Madrid, Spain
mpmareca@fis.upm.es, bbordel@dit.upm.es

Abstract. Industry 4.0 refers the use of Cyber-Physical Systems as main technological component in industrial systems, but also in home and personal applications. Industrial systems tend to be critical, so privacy, trust and security are essential in this new revolutionary era. However, up to this moment, these deployments are most considered unsecure, especially when local-range communication technologies are employed. Very common systems such as the remote controls opening private areas in manufacturing companies are totally unprotected. Although this problem could be not prioritary in traditional production systems, in Industry 4.0 is critical, because of the great interdependency among components and the need of protecting the critical core managing the entire production system of a company. Therefore, in this paper it is proposed a lightweight cryptographic solution to protect low-range wireless communications. The proposed solution is based on chaotic masking technologies, which may be easily implemented in resource constraint embedded devices (as those employed in Industry 4.0). A real implementation of an infrared device including our solution is also described. An experimental validation is performed using the implemented real device.

Keywords: Industry 4.0 · Chaos · Smart devices · Masking · Wireless communications · Cryptography · Security

1 Introduction

Industry 4.0, or forth industrial revolution, refers a new technological society based on the use of Cyber-Physical Systems (CPS) [1]. Among other benefits, these integrations of cybernetic and physical processes will enable a new and more efficient economy and engineering [2]. Its applications, besides, is envisioned to be transversal. In particular, applications to Enhanced Living Environments (ELE) [3], future mobile networks [4] or education [5] have been reported. However, nowadays, industrial scenarios are still the focus of researchers around Industry 4.0.

The use of CPS in critical infrastructures, such as manufacturing and production systems, is probably the most relevant pending challenge on this topic [6]. Usually, discussions about this problem investigate the economy effects: changes in the work market, impact in the business models, etc. [7]. Nevertheless, many technological issues must be addressed before applying these future systems to industrial scenarios [8].

© Springer Nature Switzerland AG 2019
Á. Rocha et al. (Eds.): WorldCIST'19 2019, AISC 931, pp. 134–144, 2019.
https://doi.org/10.1007/978-3-030-16184-2_14

In particular, one of the most important key parameters in industry is security [9]. In this context, this word refers a double reality: on the one hand the traditional industrial safety, and, on the other hand, the new and emerging cybersecurity. Any case, with the introduction of CPS into industrial solutions, both ideas are going to merge, as industrial safety will be automatically guaranteed by cyber-protected CPS.

Thus, cybersecurity arises as the most important topic in Industry 4.0. However, nowadays, Industry 4.0 systems mostly remain unsecure [10]. Although a deep explanation about this situation would require an analysis of several different scenarios, in many cases that is because managers and workers feel secure in the local context of their company. This is the case of very common industrial devices such as remote controls opening private areas. In fact, low-range wireless communications, such as infrared communications, are understood as secure by default, as it would be impossible for any attacker to get enough close to capture the private information [11].

CPS, however, are highly interdependent solutions, and trust, privacy and security must be ensured in every single physical device or software component. Indirect attacks and other complex infection schemes are possible in those system [12]. The challenge, then, is to protect communication links and devices which are natively designed in an insecure manner and no resources are available for implement almost any standard cryptographic scheme.

Therefore, the objective of this paper is to describe a new cryptographic solution for low-range communication in the context of Industry 4.0. The proposed scheme is based on chaotic masking and considers a lightweight dynamic system to create the required numerical or electrical signals. The described scheme is computationally very low-cost and fits perfectly the requirements of Industry 4.0 solutions.

The rest of the paper is organized as follows: Sect. 2 describes the state of the art on Industry 4.0 security and chaotic cryptography; Sect. 3 describes the proposed solution, including a mathematical framework and a real implementation using microcontrollers, electronic circuits and infrared sensors; Sect. 4 presents an experimental validation using the described real implementation; and Sect. 5 concludes the paper.

2   State of the Art

Security solutions for Industry 4.0 are sparse nowadays. Most authors doing research on this topic present different reports about the security issues to be solved, the future research opportunities or the business opportunities opened in this new era [8, 13].

In a more general analysis, authors addressing security in the context of Industry 4.0 are focused on cyber-physical security. Works on cyber-physical security also describe pending challenges [14], but technical proposals have been also reported. Most solution, in fact, are specialized on critical infrastructures [15], with a special interest on smart grids [16].

Other authors address the problem from a more abstract point of view. Different architectures with a native support for security have been reported [17]. Besides, algorithms and mechanism to detect and identify attacks may be found [18]. Taxonomies and intelligent schemes to create a dynamic protection policy according to the current situation in the system have been also reported [12].

Furthermore, although they are indirectly related to Industry 4.0, different encryption schemes for future mobile communications (5G) have been also reported. In particular, symmetric key schemes have been proposed for intra-slice applications [19]. Besides, chaotic security technologies based on masking solutions have been also reported for these same applications [4].

Very complex schemes of chaotic cryptography have been defined [20, 21]. Discrete dynamics have been employed as pseudo-aleatory code [22], unidimensional maps have been integrated into spread spectrum techniques [23] and other solutions based on external keys have been described [24]. Additionally, digital and analog systems have been described [25]. However, all these proposals are based on complicated software algorithms. Thus, for small Industry 4.0 embedded devices, simpler solutions are required. In this sense, Cuomo and Oppenheim [26] propose a couple of synchronized chaotic circuits as cryptosystem (based on Lorenz dynamics), capable of hiding the transmitted information. Moreover, Kokarev [27] has demonstrated the viability of chaotic masking solutions for other dynamics, such as the Chuas's circuit [28].

3   Chaotic Cryptographic Solution for Industry 4.0

Emerging Industry 4.0 systems are envisioned to be composed of small embedded devices with a very limited computational power. These devices cannot implement complex or heavy security solutions, so new and innovative proposals must be investigated. Specifically, in this paper it is proposed a mechanism based on chaotic masking.

The proposed architecture is showed on Fig. 1. As can be seen, two modules are identified: a transmitter and a receptor. Both modules are provided with a chaotic signal generator (chaotic dynamics) which may be an electronic circuit (then an electric signal is generated) or a numerical generator based on a numerical differential equation solver.

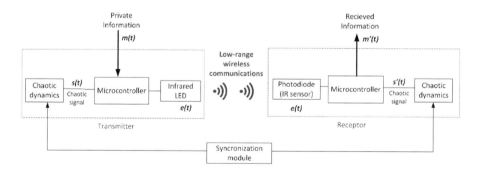

Fig. 1. Proposed architecture

In the first case, microcontrollers in both modules should sample and quantify the generated electronic signal. In the second case, the chaos generator will be a software routine (based on Runge-Kutta algorithms) embedded into the microcontroller programming. This chaotic signal, once sampled and quantified (if necessary), will be added to the private information (thus, it is encrypted in the transmitter's microcontroller). Then, due to the mathematical and physical characteristics of chaotic signals, encrypted private information will be only recovered if exactly the same synchronized chaotic signal is employed in the receptor's decryption process (in this case the decryption scheme consists of a subtraction instead of an addition, as in the encryption process). This chaotic signal is, then, the key of our system. This global encryption scheme is known as "chaotic masking". To communicate the encrypted information between transmitter and receptor, any low-range unsecure wireless communication technology (such as infrared signals) may be employed.

A final detail must be taken into account. As this chaotic masking scheme has as symmetric private key a temporal signal, both modules (transmitter and receptor) must be perfectly synchronized. A module focused on this objective is, therefore, also included. This module may take different forms: from a simple communication link to connect both dynamics and make both chaotic generators evolve together [28]; to a dual alignment module similar to those employed in CDMA (Code Division Multiple Access) solutions [29].

3.1   Mathematical Framework

The mathematical formalization of a chaotic masking encryption scheme is, actually, very simple (1–2). Basically, the encryption process consists of an addition; and, in the same way, the decryption mechanism is a simple subtraction (see Fig. 2). Despite the simplicity of this design, it is a very powerful solution. Because of the erratic behavior of chaos, only using exactly the same synchronized chaotic signal private information may be encrypted and decrypted.

$$e(t) = m(t) + s(t) \tag{1}$$

$$d(t) = m'(t) = e(t) - s'(t) \tag{2}$$

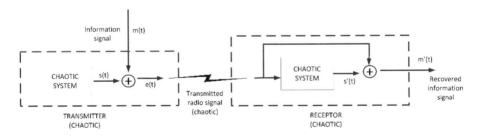

Fig. 2. Basic scheme for a chaotic masking solution

The key element in this technology, then, is the generation of a complex and highly erratic chaotic signal in the most efficient and lightweight manner. Many different chaotic dynamics have been reported to create masking solution, although they barely meet the requirements of Industry 4.0. Therefore, in this paper it is proposed a new chaotic three-dimensional dynamic based on simple mathematical operations and quadratic non-linearities. This new dynamic is named as PV1 (3).

As only simple mathematical operations are considered, it is guaranteed that Industry 4.0 embedded are able to operate these expressions; as well as very reduced electronic circuits may be employed to implement the dynamic.

$$\dot{x} = -\alpha y$$
$$\dot{y} = \beta x + z^2 \qquad (3)$$
$$\dot{z} = 1 + x + 2y - \gamma z$$

Three control parameters are identified in this new dynamic: α, β and γ. Varying the values for these parameters, different topologies and signals may be generated, including two different chaotic behavior. Figure 3 shows the associated attractors in the phase space to these two chaotic topologies. As can be seen, two equilibrium points are described in the PV1 dynamic, and different signals are generated depending on the fact if signal only oscillates around one or around both points. Any case, the Kaplan-York dimension, evaluating the complexity of the generated chaos, is around $D_L \approx 2.1$.

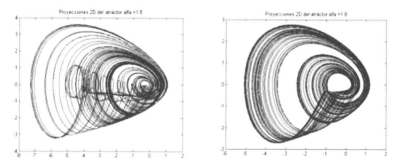

Fig. 3. Obtained attractors in the XY space, **(a)** $\alpha = 1.8, \beta = 1.5, \gamma = 1.6$ **(b)** $\alpha = 1.8,$ $\beta = 1.5, \gamma = 2.15$

The challenge, then, is to get two dynamics generating exactly the same chaotic signals. To address this problem, the synchronization theory proposes different schemes. However, studying the conditional Lyapunov exponents of this systems shows that total synchronization is not possible. In particular, being $\gamma > 0$, we obtained that $\{\lambda_1 < 0, \lambda_2 = 0\}$. In these conditions only *marginal synchronization type I* is reachable [30].

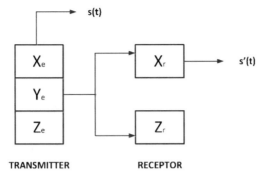

Fig. 4. Synchronization scheme for PV1 dynamic

Figure 4 shows the resulting and only possible synchronization scheme for this new dynamic. With this solution, between chaotic signals in the receptor and the transmitter there is a constant synchronization error (4) which must be corrected through software mechanisms.

$$e_1 = s(t) - s'(t) = K \tag{4}$$

3.2 Implementation

The proposed cryptographic solution is implemented in a real Arduino microcontroller, provided with low-range infrared communication modules. Figure 5 shows the proposed hardware implementation. Two built-in infrared modules were employed, so no additional amplifier or adaptation circuit was required. Microcontrollers were selected to be Arduino Uno boards.

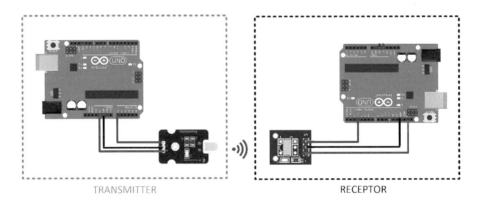

Fig. 5. Proposed real implementation

In this prototype, private information to be transmitted was based on a fixed alphanumerical message, as most remote controls do. Using the "dtostrf" function, this message is converted into a numerical sequence which may be easily combined with the chaotic signals. Then, using the inverse function, message may be transmitted as a numerical sequence or as a traditional encrypted alphanumerical message.

On the other hand, for this initial study, the PV1 dynamic was generated by numerical methods. A four-order Runge-Kutta (5) numerical algorithm was implemented in Arduino to solve the corresponding system of differential equations (3). No additional hardware component was, then, required for the chaos generator. Figure 6 shows the resulting software as a block diagram.

$$\overline{X_{n+1}} = \begin{pmatrix} x_{n+1} \\ y_{n+1} \\ z_{n+1} \end{pmatrix} = \overline{X_n} + \frac{1}{6}h\left(\overline{k_1} + 2\overline{k_2} + 2\overline{k_3} + \overline{k_4}\right)$$

$$\overline{k_1} = F\left(\overline{X_n}\right)$$

$$\overline{k_2} = F\left(\overline{X_n} + \frac{1}{2}h\overline{k_1}\right) \tag{5}$$

$$\overline{k_3} = F\left(\overline{X_n} + \frac{1}{2}h\overline{k_2}\right)$$

$$\overline{k_4} = F\left(\overline{X_n} + h\overline{k_3}\right)$$

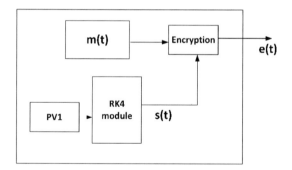

Fig. 6. Proposed real implementation

4 Experimental Validation. Methods and Results

In order to evaluate the performance of the proposed solution, an experimental validation based on the implemented real hardware device (see Sect. 3.2) was carried out.

Although a security study would be required to ensure the validity of the proposal in different scenarios, for this initial work we are only evaluating the end-to-end performance in terms of communication indicators. Besides some computational indicators will be also evaluated to provide first proves of the lightweight character of the proposed approach.

The previously described device was configured to create a complex chaotic signal, forcing the following values in the configuration parameters: $\alpha = 1.8, \beta = 1.5, \gamma = 1.6$. Other values could be employed, while preserving the chaotic behavior of the dynamic. To discover other possible configurations, the bifurcation diagram should be calculated. However, using the proposed values, the generated chaos presents a high complexity level which ensures the protection of the private information. Besides, a random 256-character message was configured to be translated. Specifically, a paragraph from the well-known "Lore ipsum" text was selected. Figure 7 shows the message before and after the encryption process.

Lorem ipsum dolor sit amet, consectetur adipiscing elit, sed eiusmod tempor incidunt ut labore et dolore magna aliqua. Ut enim ad minim veniam, quis nostrud exercitation ullamco laboris nisi ut aliquid ex ea commodi consequat. Quis aute iure reprehenderit in voluptate velit esse cillum dolore eu fugiat nulla pariatur.

Txanu qybdu mxtxa bqc junc, lxvbnlcncda jmqyqblqvo ntqc, bnm nqdbuxm cnuyxa qvlqmdvc dc tjkxan nc mxtxan ujovj jtqzdj. Dc nvqu jm uqvqu envqju, zdqb vxbcadm ngnalqcjcqxv dttjulx tjkxaqb vqbq dc jtqzdqm ng nj lxuuxmq lxvbnzdjc. Zdqb jdcn qdan anyanpnvmnaqc qv extdycjcn entqc nbbn lqttdu mxtxan nd ñdoqjc vdttj yjaqjcda

Fig. 7. Proposed clear message and its encrypted form

The experiment was designed as follows. The Arduino programming was uploaded to the microcontrollers, and its performance in terms of required processing time per transmitted character and consumed memory. On the other hand, the system was configured to operate for a long time, and the end-to-end character error rate was also evaluated.

All measurements were repeated twelve times, and Table 1 shows the obtained mean results. As can be seen, the percentage of consumed memory is in the average for a microcontroller program, and character error rate is below 0.1%. These results are a first evidence of the validity of the proposed cryptographic scheme.

Table 1. Performance of the proposed solution. Key performance indicators: results

Module	Memory consumption	Processing time		End-to-end character error rate
		No encryption	With encryption	
Transmitter	7,8%		9 ms	$7.3 \cdot 10^{-2}\%$
Receptor	8,3%		7 ms	
Total	8%		16 ms	

5   Conclusions and Future Works

In this paper we present a lightweight cryptographic solution to protect low-range wireless communications in the context of Industry 4.0. The proposed solution is based on chaotic masking technologies, which may be easily implemented in resource constraint embedded devices (as those employed in Industry 4.0).

In particular, a new simple chaotic dynamic is proposed, only based on simple mathematical operations, which can be easily executed by small embedded devices. Besides, the synchronization properties of the designed dynamic system are also evaluated.

All this information is employed to build a chaotic masking encryption scheme for low-range wireless communications. In order to validate this proposal a real prototype based on infrared communications has been also implemented.

Using this real device, an experimental validation was conducted to evaluate the end-to-end error rate in the proposed cryptography scheme. Results show this error is below 0.1%.

Future works will implement more complicated working schemes, such as chaotic generators based on electronic circuits. Moreover, other synchronization schemes and messages to be transmitted will be evaluated and tested.

Acknowledgments. Borja Bordel has received funding from the Ministry of Economy and Competitiveness through SEMOLA project (TEC2015-68284-R).

References

1. Bordel, B., Alcarria, R., Robles, T., Martín, D.: Cyber–physical systems: extending pervasive sensing from control theory to the Internet of Things. Pervasive Mob. Comput. **40**, 156–184 (2017)
2. Lasi, H., Fettke, P., Kemper, H.G., Feld, T., Hoffmann, M.: Industry 4.0. Bus. Inform. Syst. Eng. **6**(4), 239–242 (2014)
3. Bordel Sánchez, B., Pérez Jiménez, M., Sánchez de Rivera, D.: Recognition of activities of daily living in Enhanced Living Environments. IT CoNvergence PRActice (INPRA) **4**(4), 18–31 (2017)
4. Mareca, P., Bordel, B.: An intra-slice chaotic-based security solution for privacy preservation in future 5G systems. In: World Conference on Information Systems and Technologies, pp. 144–154. Springer, Cham, March 2018
5. Bordel, B., Alcarria, R., Robles, T., Martín, D.: Employing extrinsic motivation techniques and optional evaluation activities to promote student learning. In: 10th International Conference on Education and New Learning Technologies
6. Wang, L., Törngren, M., Onori, M.: Current status and advancement of cyber-physical systems in manufacturing. J. Manuf. Syst. **37**, 517–527 (2015)
7. Stock, T., Seliger, G.: Opportunities of sustainable manufacturing in industry 4.0. Procedia Cirp **40**, 536–541 (2016)
8. Lu, Y.: Industry 4.0: a survey on technologies, applications and open research issues. J. Ind. Inform. Integr. **6**, 1–10 (2017)

9. Bordel, B., Alcarria, R., Sánchez-de-Rivera, D., Robles, T.: Protecting industry 4.0 systems against the malicious effects of cyber-physical attacks. In: International Conference on Ubiquitous Computing and Ambient Intelligence, pp. 161–171. Springer, Cham, November 2017

10. Bordel, B., Alcarria, R., De Andrés, D.M., You, I.: Securing Internet-of-Things systems through implicit and explicit reputation models. IEEE Access **6**, 47472–47488 (2018)

11. Lanotte, R., Merro, M., Muradore, R., Viganò, L.: A formal approach to cyber-physical attacks. In: 2017 IEEE 30th Computer Security Foundations Symposium (CSF), pp. 436–450. IEEE, August 2017

12. Bordel, B., Alcarria, R., Robles, T., Sánchez-Picot, Á.: Stochastic and information theory techniques to reduce large datasets and detect cyberattacks in Ambient Intelligence Environments. IEEE Access **6**, 34896–34910 (2018)

13. Liao, Y., Deschamps, F., Loures, E.D.F.R., Ramos, L.F.P.: Past, present and future of Industry 4.0-a systematic literature review and research agenda proposal. Int. J. Prod. Res. **55**(12), 3609–3629 (2017)

14. Wells, L.J., Camelio, J.A., Williams, C.B., White, J.: Cyber-physical security challenges in manufacturing systems. Manuf. Lett. **2**(2), 74–77 (2014)

15. Banerjee, A., Venkatasubramanian, K.K., Mukherjee, T., Gupta, S.K.: Ensuring safety, security, and sustainability of mission-critical cyber-physical systems. Proc. IEEE **100**(1), 283–299 (2012)

16. Hahn, A., Ashok, A., Sridhar, S., Govindarasu, M.: Cyber-physical security testbeds: Architecture, application, and evaluation for smart grid. IEEE Trans. Smart Grid **4**(2), 847–855 (2013)

17. Ning, H., Liu, H.: Cyber-physical-social based security architecture for future internet of things. Adv. Internet Things **2**(01), 1 (2012)

18. Pasqualetti, F., Dörfler, F., Bullo, F.: Attack detection and identification in cyber-physical systems. IEEE Trans. Autom. Control **58**(11), 2715–2729 (2013)

19. Bordel, B., Orúe, A.B., Alcarria, R., Sánchez-de-Rivera, D.: An intra-slice security solution for emerging 5G networks based on pseudo-random number generators. IEEE Access **6**, 16149–16164 (2018)

20. Vaidya, P.G., Angadi, S.: Decoding chaotic cryptography without access to the superkey. Chaos, Solitons Fractals **17**(2), 379–386 (2003)

21. Wong, K.W., Ho, S.W., Yung, C.K.: A chaotic cryptography scheme for generating short ciphertext. Phys. Lett. A **310**(1), 67–73 (2003)

22. Li, S., Li, Q., Li, W., Mou, X., Cai, Y.: Statistical properties of digital piecewise linear chaotic maps and their roles in cryptography and pseudo-random coding. In: IMA International Conference on Cryptography and Coding, pp. 205–221. Springer, Heidelberg (2001)

23. Pareek, N.K., Patidar, V., Sud, K.K.: Cryptography using multiple one-dimensional chaotic maps. Commun. Nonlinear Sci. Numer. Simul. **10**(7), 715–723 (2005)

24. Pareek, N.K., Patidar, V., Sud, K.K.: Discrete chaotic cryptography using external key. Phys. Lett. A **309**(1), 75–82 (2003)

25. Amigó, J.M., Kocarev, L., Szczepanski, J.: Theory and practice of chaotic cryptography. Phys. Lett. A **366**(3), 211–216 (2007)

26. Cuomo, K.M., Oppenheim, A.V., Strogatz, S.H.: Synchronization of Lorenz-based chaotic circuits with applications to communications. IEEE Trans. Circuits Syst. II Analog Digital Signal Proc. **40**(10), 626–633 (1993)

27. Kocarev, L., Halle, K., Eckert, K., Chua, L.: Experimental demonstration of secure communications via chaotic synchronization. Int. J. Bifurcat. Chaos **2**, 709–713 (1992)

28. Mareca, P., Bordel, B.: Robust hardware-supported chaotic cryptosystems for streaming commutations among reduced computing power nodes. Analog Integrated Circuits and Signal Processing, pp. 1–16
29. Liu, T., Yang, C.: Signal alignment for multicarrier code division multiple user two-way relay systems. IEEE Trans. Wireless Commun. **10**(11), 3700–3710 (2011)
30. Atabakzadeh, M.H.: Synchronization-Like Behaviors of Dynamical Systems with Non Negative Lyapunov Exponents. Appl. Math. Sci. **6**(23), 1113–1120 (2012)

Yet Another Virtual Butler to Bridge the Gap Between Users and Ambient Assisted Living

Marta Carvalho[1], Ricardo Domingues[1], Rodrigo Alves[1],
António Pereira[2] ⓘ, and Nuno Costa[2](✉) ⓘ

[1] School of Technology and Management,
Polytechnic Institute of Leiria, Leiria, Portugal
[2] Computer Science and Communications Research Centre,
School of Technology and Management,
Polytechnic Institute of Leiria, Leiria, Portugal
nuno.costa@ipleiria.pt

Abstract. Ambient Intelligence promises to transform current spaces into electronic environments that will be responsive, assistive and sensitive to human presence and where all happens beyond user perception. Current computing interfaces, such as keyboards, mice, etc. will be removed at all or replaced by human ways of interaction like voice, gesture, emotions, etc. In order to materialize this new way of interaction between computing and people, we proposed a virtual butler able to make the link between ambient intelligence in general or smart spaces in particular and people. In fact, this is our second attempt to achieve that. Although tailored for smart homes, this virtual butler could be applied to other types of smart spaces and environments, such as shopping malls, airports, industry, etc. The developed virtual butler is able to interact with users by voice and it is sensitive to user location (and other context information).

Keywords: Virtual butler · Ambient assisted living · Smart spaces

1 Introduction

Ambient Intelligence (AmI) promises to transform current spaces into responsive, assistive and sensitive electronic environments (smart spaces). These electronic environments will be build upon lots of miniaturized computers, working behind user perception and hence a new way of interaction is required, as traditional computer peripherals are not suitable any more. If it is true that the major part of decisions will be taken automatically by surrounding computing, there will be some decisions and actions that will rely on user's will. In this perspective, our research team, published an article [1] at some years ago, describing an attempt to turn a regular helder's home into a smart home, where a virtual butler was used in order to support user-system interaction. The system evaluation showed very promising results however, the virtual butler exhibited some weaknesses that are being solved in these days.

© Springer Nature Switzerland AG 2019
Á. Rocha et al. (Eds.): WorldCIST'19 2019, AISC 931, pp. 145–154, 2019.
https://doi.org/10.1007/978-3-030-16184-2_15

1.1   Contextualization

During the year of 2014, part of our research team was dedicated to evaluate an attempt to transform an helder's regular home into a smart home [1], where interaction between computing environment and the inhabitant was supported by a mobile virtual butler. Although evaluation results showed very promising results, the virtual butler exhibited two (structural) weaknesses. The first one is related with the collected data which could result in a very "fat" agent and the second one is the inability to run in mobile devices as is. Specially, this second feature is very important for us as it allows virtual butler to "jump" into user smart phone and hence, virtual butler can assist user even outside his/her smart home. These two known weaknesses have been addressed in the last months and two solutions came up. The solution for the second weakness and part of the first weakness was reported in paper [2] where a new mobile agent middleware and framework was proposed in order to support really multi-platform mobile agents with the help of the JavaScript programming language (as opposed to Java language), while the "fat" problem still exists and is addressed in this paper.

1.2   Motivation

Our work reported in [2] addresses the inability to have mobile agents (virtual butlers in this context) running in mobile platforms, however the problem of having a fat mobile agent could have a big impact during its mobility, mainly when mobile devices are used. It is clear that while mobile agent paradigm is very suitable for the "follow me" [1] types of mobile agents (e.g. a virtual butler), it has the disadvantage of carrying collected data (e.g. knowledge base) with it. Obviously, this could be solved by relying in a central data repository (e.g. central knowledge database) where virtual butler should access it every time it needs some knowledge to face up a specific demand. But this possible centralized solution would approach mobile agent paradigm to the traditional client/server side. Hence, it was decided to try the client/server model for virtual butler instead of mobile agent paradigm and this paper reports that effort.

2   Related Work

This paper describes a virtual butler prototype that follows the client/server paradigm, able to interact with users by using human natural language. Hence, this section surveys the existing virtual butler projects as well as computing voice support.

2.1   Virtual Butlers

The virtual butler described in [1] is the most similar project to the one being presented in this paper. This virtual butler follows the mobile agent paradigm in order to implement the "follow me" feature. It was implemented using JADE [3] framework (Java) and it is voice and location aware like the one presented in this paper. However, it can't run in mobile devices (only a simple "clone" of it can) and it can become very "fat" as it stores knowledge is local memory.

The authors of [4] proposed an multi agent based domotics system targeted for disable people. They propose the usage of an agent for each system module (health care, nourishing, stock, etc.) and the existence of a virtual butler that is just the interface between the user and the system. However, this project seem to be in its beginning as authors did not provide enough information about the virtual butler and the targeted house is being virtually designed and domotics devices are being simulated.

Similarly, the authors of [5] presented a prof of concept where JADE agents are used for devices, human and blackboard representation. Due to the use of JADE, each device must be tied to a JADE capable device (e.g. raspberry pi) in order to run the agent. Unlike this work, our original work uses JADE mobile agents to implement the virtual butler follow-me approach and OSGI bundles (instead of agents) to digitally represent devices inside the interoperable and flexible OSGI platform.

The work described in [6] defines a speech recognition and speech synthesis for a smart home virtual butler, similar to our work but, to best of our knowledge, the virtual butler was not implemented yet and the speech prototype only reacts to orders (it does not takes the initiative yet). The work reported in [7] is probably the most similar to ours. Authors have proposed an avatar-based virtual butler to attend disabled people in a home environment. However, the evaluation was just done in a simulated "small-world" and it seems that virtual butler was not implemented yet and nothing is said about its mobility in terms of mobile devices.

2.2 Speech Recognition and Speech Synthesis

The speech recognition process can be defined as the ability of a machine to recognize and translate spoken language into text, while speech synthesis is the opposite process where written information is translated into speech.

Google Speech API. Google Speech API [8] is a Google service for speech recognition. It relies on trained neural networks and exhibits a high rate of success even in noisy environments. At first sight, this service seems to be very suitable for our system, however, at that time google was in the process of updating the API and some sporadic errors were found.

CMUSphinx. The CMUSphinx [9] project was started by Carniege Mellon University with the aim of creating a open-source system for speech recognition apart from a lighter solution to run in pocket devices. There is already some results for English language but lack of ready-to-use support for other languages, specially the Portuguese. For example, a Brazilian research group have been training the system for the Brazilian Portuguese but to the best of our knowledge no satisfying results exist yet. So, the lack of support for native Portuguese and the amount of effort needed to stablish it turned this system useless for us.

eSpeak. eSpeak [10] is an open-source and offline software for speech synthesis, available on sourceforge, that supports many languages and operating systems and it is easy driven from a third-party software. However, in the perspective of our system, this solution presents two major weaknesses. The first one is that speech synthesis still sounds like a "robotic spoken language" and the second is the lack for speech recognition.

Microsoft Speech Platform. The Microsoft Speech Platform [11] (SDK 11) is a library for speech recognition and speech synthesis that does not rely on an Internet Server. This library was selected to bring speech recognition and speech synthesis to our project mainly due to the reason present above and also due to its suitability for . NET framework development.

3 System Architecture

The developed solution is made of three main components as shown in Fig. 1: the server or controller, the client or virtual butler and the console.

3.1 The Virtual Butler

The virtual butler component is the module with the responsibility to interact with the inhabitant by showing a butler avatar. The virtual butler must be able of receiving commands by voice and answering those commands by speech synthesis. The virtual butler avatar moves its mouth when it is speaking to give a better experience to the user and is sensitive to user location. This means that if the user (inhabitant) moves to the bedroom, the virtual butler is ready to receive a stimulus in order to "move" itself to the specific bedroom.

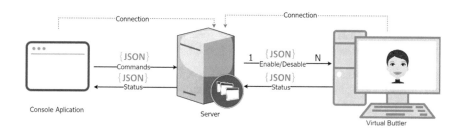

Fig. 1. System architecture.

When a (new) virtual butler application is started, it connects itself to the well-known controller (presented next) using web sockets and waits for commands for state changing. These commands are sent by controller following a JSON message format.

3.2 The Controller (Server)

The controller component is an application server that knows all virtual butler instances and is able to control or define which instance should be active. In normal conditions, only one virtual butler should be active once the user is only at one room at a time. In order to control the virtual butler activity, the controller must know all the rooms and which virtual butler instance is in each room. In order to control virtual butlers, the controller is able to receive commands (e.g. controller is the virtual butler interface). In a real scenario, those commands are sent by Smart Home Server as shown in the following example. Imagine you have a smart home that knows where the inhabitant is, probably using a presence sensor or similar. Hence, when the smart home system detects that the inhabitant has moved to other room, it sends a command to the controller in order to "move" the virtual butler to the room where the user is at that moment.

3.3   The Console Application

The console application was created to test the virtual butler solution and to show how and which commands should be sent to the controller in order to move virtual butler to another room. Hence, it is not mandatory for the system well-functioning.

4   Implementation

The solution described in the system architecture section was implemented using .NET framework mainly due to use of the Microsoft Speech Platform.

4.1   Connection Process

In connection process, controller is the well-known device, which means that both virtual butler's instances and console application connect to the defined IP where controller is listening to, as shown in Fig. 2.

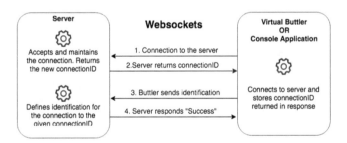

Fig. 2. Connection process.

When virtual butler connects to controller, it receives a connection ID (which should be used in next communication transactions) and then it identifies itself by referencing the room where it belongs to. After identification step, the controller is ready to send control commands to virtual butler instances. For now, the controller supports three different commands:

(a) Virtual butler listing: this command is used to get the list of the existing virtual butler instances *which is very useful for debugging purposes.*
(b) Active virtual butler: this command is used to get the identification of the active virtual butler.

(c) Activate virtual butler: this command is used to activate a specific virtual butler and, of course, deactivate the previous active one.

In the perspective of the console application or smart home server, the commands are represented by simple message format (Fig. 3).

Action	Room (optional)

Fig. 3. The controller commands' message format (for requesting)

The action field can assume one of the following operations (Fig. 4).

move	Moves (activates) virtual butler in a specific room

current	Returns the current activated virtual butler (room)

list	Returns the list of all connected virtual butlers

Fig. 4. Actions supported by controller

Although based in multiple instances of virtual butler application, the system exposes the perception that there is only one active virtual butler avatar at a time.

4.2   Speech Recognition and Synthesis

The speech recognition and speech synthesis capabilities are embedded in the virtual butler application in order to avoid network delays associated with a centralized approach. This was achieved by installing the following software packages in the device operating system.

- Microsoft Speech Platform – Runtime;
- Microsoft Speech SDK;
- ScanSoft Text To Speech Madalena - Portuguese Portugal.

During the virtual butler initialization process the language is selected (in this case Portuguese, but there is support for multiple languages) apart from the grammar dictionary that includes the sentences that are expected to be recognized. For instance, one of the sentences that must be recognized is "What time is it?" and virtual butler must answer with the current time using voice synthesis. Hence, every time the virtual butler detects voice, it tries to choose the most approximated sentence from the sentences' dictionary. However, this simple approach could identify a sentence when only part of that sentence is verbalized by the user. In order to solve this weakness, it was established a trusting level for the chosen sentence. Only when trusting level is above a specified threshold (70%) the sentence is selected. When the trusting level is below the specified threshold, the virtual butler is instructed to ask the user to repeat sentence, having a maximum number of requests for user question repetition. Of course, only one Speech recognition and synthesis engine is active at a time.

4.3   Virtual Butler Configuration

The virtual butler application includes a configuration file that must be filled with:

- Controller IP and port – as it is the virtual butler that starts the connection to the controller, it must knows the controller's socket endpoint (IP address and port where controller is listening to).
- Room name – this parameter is sent to the controller and defines in which room this virtual butler instance is.
- List of available rooms – the list of available rooms is the way the controller knows which rooms are available, or better, to each room the virtual butler could be moved/activated.

4.4   Virtual Butler Avatar

The virtual butler application includes an avatar in order to be perceived by inhabitant (Fig. 5). Hence, every time the user moves to another room, the avatar is shown in that room. Every time a sentence is detected (using the speech recognition system), the virtual butler application answers to that sentence by voice synthesis and by animating the avatar's mouth in a synchronized way. In this first version of new centralized virtual butler, the animation was implemented as a proof of concept. However, we consider this feature extremely important and in future avatar's versions it must be able to express emotions too.

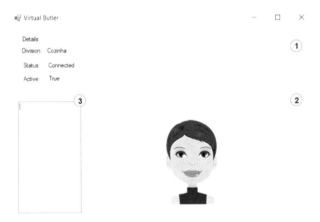

Fig. 5. The virtual butler avatar

5   Tests and Results

This section describes the tests carried on and the related evaluation as far as system objectives are concerned.

5.1   Test Bed

In order to test the virtual butler capabilities, it was established the infrastructure presented in Fig. 6.

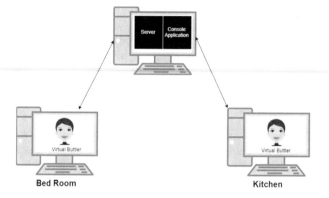

Fig. 6. The test bed

The test bed infrastructure is composed by three computers. One computer hosts the controller and console applications and the other two host one instance of virtual butler each one. Both computers have an external microphone and there is ambient music in both rooms of about 50 dB loud. The virtual butler application was configured with two rooms, the bedroom and the kitchen.

5.2   Test's Table

The following table summarizes some of the tests carried on.

5.3   Evaluation

The tests reported above were done using the console application to send actions to the controller and an external microphone to interact with active virtual butler instance. Note that it is very simple to make virtual butler react to presence sensors as demonstrated in [1], by simply connecting a presence sensor to an Arduino or similar development board and then the Arduino sends the move command to the controller module, using sockets.

According to Table 1, all virtual butler test cases were executed with success which means that the virtual butler objectives were met. Comparing this virtual butler with the virtual butler described in [1], it could be said that this new one (i) relies less in network and (ii) its size does not have impact on performance (there is no mobile code). Another advantage is that this centralized model also solves the lack of support in mobile platforms as demonstrated in [1] because a tailored virtual butler could be developed for each type of mobile platform.

Table 1. Tests listing

Test	Description	Result
Move to kitchen	The console application sent move action and kitchen as room to the controller	The avatar was shown in the kitchen computer display
Show active room	The console application sent the current action to the controller	The console application received kitchen
Show available rooms	The console application sent the list action to the controller	The console application received bedroom, kitchen
What time is it (in the bedroom)	The user said "What time is it" in the bedroom	Nothing happened
What time is it (in the kitchen)	The user said "What time is it" in the kitchen	The kitchen computer answered with current time and the avatar's mouth opened and closed during the time of speech synthesis
What time (in the kitchen)	The user said "What time" in the kitchen	The kitchen computer answered "Could you repeat please" and the avatar's mouth opened and closed during the time of speech synthesis
Turn off kitchen computer	The user turned off kitchen computer abruptly	Nothing perceived, but controller detected loss connection event
Show active room	The console application sent the current action to the controller	The console application received none
Move to bedroom	The console application sent move action and the bedroom as room to the controller	The avatar was shown in the bedroom computer

6   Conclusions and Future Work

Our research team have argued that the best way to bridge the gap between ambient intelligence and user is through virtual butlers/assistants, by having an up-to-date knowledge base where the virtual butler can get facts and contribute to build new facts. In order to reach the maximum outcome from virtual butlers usage we argue that it must be always present even when inhabitant leaves his/her home. So, the employed virtual butler must be very flexible in terms of target operating system and very lightweight. Our previous work showed the first attempt to build such a virtual butler [1] (mobile agent-based) and a new enhancement to that virtual butler [2] (also mobile agent-based) and in this paper we are proposing a new one following a new model (client-server) in the form of prof-of-concept. The evaluation has shown very promising results and puts this 3rd virtual butler generation as the most effective one to be incorporated in our next ambient intelligence project. For future work we consider that the most strategic feature to be added to this virtual butler is the ability to show emotions. Another enhancement is to improve the sentences' dictionary in order to improve the sentences it can recognize using speech recognition.

References

1. Costa, N., Domingues, P., Fdez-Riverola, F., Pereira, A.: A mobile virtual butler to bridge the gap between users and ambient assisted living: a smart home case study. Sensors **14**(8), 14302–14329 (2014)
2. Silva, C., Costa, N., Grilo, C., Veloz, J.: JavaScript middleware for mobile agents support on desktop and mobile platforms. In: International Conference on Information Theoretic Security, pp. 745–755. Springer, Cham (2018)
3. Bellifemine, F., Poggi, A., Rimassa, G.: JADE: a FIPA2000 compliant agent development environment. In: Proceedings of the Fifth International Conference on Autonomous Agents, Montreal, QC, Canada, 28 May–1 June 2001, pp. 216–217 (2001)
4. Muñoz, C., Arellano, D., Perales, F.J., Fontanet, G.: Perceptual and intelligent domotic system for disabled people. In: Proceedings of the 6th IASTED International Conference on Visualization, Imaging and Image Processing, pp. 70–75 (2006)
5. van Moergestel, L., Langerak, W., Meerstra, G., van Nieuwenburg, N., Pape, F., Telgen, D., Puik, E., Meyer, J.J.: Agents in domestic environments. In: 2013 19th International Conference on Control Systems and Computer Science (CSCS), pp. 487–494. IEEE (2013)
6. Uria, A., Ortega, A., Torres, M.I., Miguel, A., Guijarrubia, V., Buera, L., Garmendia, J., Lleida, E., Aizpuru, O., Varona, A., Alonso, E.: A virtual butler controlled by speech. In: Proceedings of the III Jornadas en Tecnologias del Habla, Zaragoza, Spain (2006)
7. Fiol-Roig, G., Arellano, D., Perales, F.J., Bassa, P., Zanlongo, M.: The intelligent butler: a virtual agent for disabled and elderly people assistance. Advances in Soft Computing, vol. 50, pp. 375–384 (2009)
8. Cloud Speech-to-Text Documentation. https://cloud.google.com/speech-to-text/docs. Accessed 16 June 2018
9. CMUSphinx Homepage. https://cmusphinx.github.io. Accessed 16 June 2018
10. eSpeak Homepage. http://espeak.sourceforge.net. Accessed 16 June 2018
11. Microsoft Speech Platform SDK 11 Requirements and Installation Homepage. https://docs.microsoft.com/en-us/previous-versions/office/developer/speech-technologies/hh362873(v=office.14). Accessed 16 June 2018

Usage of HTTPS by Municipal Websites in Portugal

Hélder Gomes[1,3(✉)], André Zúquete[2,3], Gonçalo Paiva Dias[1,4], and Fábio Marques[1,3]

[1] ESTGA, Universidade de Aveiro, 3750-127 Águeda, Portugal
{helder.gomes,gpd,fabio}@ua.pt
[2] DETI, Universidade de Aveiro, 3810-193 Aveiro, Portugal
andre.zuquete@ua.pt
[3] IEETA, Universidade de Aveiro, 3810-193 Aveiro, Portugal
[4] GOVCOPP, Universidade de Aveiro, 3810-193 Aveiro, Portugal

Abstract. This paper presents a study on the adoption of HTTPS in the official websites of all (308) Portuguese municipalities. Automated and, whenever needed, manual analysis were used to investigate its entry pages. Specifically, the pages were checked for the existence of an HTTPS site; the correctness of the certificates and their certification chain; coherence between contents of the HTTP and HTTPS sites; redirection between HTTP and HTTPS, HTTP resources fetched through HTTPS, and exploitation of HSTS. A final classification of municipalities was produced and possible determinants for the results were investigated. The general conclusion is that there is still much to be done in order to assure that citizens can communicate securely with all the Portuguese municipalities. Indeed, only 3.6% of the municipalities were considered good in this regard, while 46.1% do not guarantee the minimum conditions. These results seem to be associated with the dimension of the municipalities, although it was also identified the need for additional explanatory factors.

Keywords: E-government · Local government · HTTPS adoption

1 Introduction

All Portuguese municipalities have an official website which is the main reference point both for the provision of municipal services to the citizens and for the promotion of the municipality. Among the provided services, there are informational services, with information regarding the municipal government bodies and municipal regulations and services, and transactional services that municipal citizens may obtain electronically. Trust is fundamental for the citizen's adherence to electronic services in general and in particular to those provided by municipal websites. For that reason, municipal websites should authenticate to citizens, to guaranty that the citizen is contacting a legitimate, official website, and the access to the information and services in municipal websites is confidential and correct. Note that confidentiality in communications was one of the top concerns pointed by European citizens in a public consultation on the ePrivacy Directive Review [1].

© Springer Nature Switzerland AG 2019
Á. Rocha et al. (Eds.): WorldCIST'19 2019, AISC 931, pp. 155–164, 2019.
https://doi.org/10.1007/978-3-030-16184-2_16

The content of websites has been traditionally provided using the HTTP protocol. However, this does not include any security feature, thus being prone to interception attacks (aka Man-in-the-middle attacks) and to service impersonation attacks. Consequently, in the last years we assisted to a migration from HTTP to HTTPS [2], which is backed by standard bodies, like World Wide Web Consortium (W3C) [3] and Internet Architecture Board (IAB) [4], and by major browsers [5, 6].

Despite some vulnerabilities [7], HTTPS aims to provide authentication, integrity and confidentiality to the communication between a browser and a website, which is considered fundamental for a trustworthy interaction between municipal websites and citizens. Furthermore, the protection provided by HTTPS is in line with the European Proposal for an ePrivacy Regulation [8] - a proposal to adapt the privacy in electronic communications to the new European General Data Protection Regulation [9].

In this paper we present a survey on the adoption of HTTPS in the official websites of all (308) Portuguese municipalities. Results are presented and possible determinants for those results are investigated.

Previous studies have addressed the usage of HTTPS, namely regarding the increase of HTTPS traffic [10, 11], its adoption [10, 12–14] and the quality of the implementation at server side [12, 14, 15]. However, only one of these studies is targeted to the specific domain of local e-government, in Sweden [14], and none of them explores the determinants of the adoption.

2   Methods

The URLs of all the 308 municipalities were collected from the website of Associação Nacional de Municípios Portugueses (ANMP[1]), the national association of Portuguese municipalities. Five of them were found to be wrong, and the correct URL was searched with Google. All URLs have an HTTP anchor, instead of an HTTPS.

Most URLs contain a DNS name which follows a homogeneous and straightforward structure: www.cm-name.pt, where *name* is a diacritic-free abbreviation of the municipality name (e.g. *fozcoa* for Vila Nova de Foz Côa, *vrsa* for Vila Real de Santo António). There is no fixed rule for the abbreviations used, but they are not unnatural.

There are 38 exceptions to the default naming strategy (12.3%). Among them, two municipalities whose DNS names are under the *.com* domain, instead of *.pt*: Santana (www.cm-santana.com) and Oliveira de Frades (www.cm-ofrades.com).

Between 2018-11-12 and 2018-11-20, we conducted a series of automated analysis to all those URLs, using several tools and resorting to manual analysis with a browser only when necessary. Only the entry pages of municipal websites were accessed during this evaluation. No crawling was made throughout other pages accessible from those. Also, we did not assess the quality or the strength of the cryptographic algorithms, the presence of well-known SSL/TLS vulnerabilities, or even websites' vulnerabilities, since that would introduce many more layers of entropy in an assessment that we tried to keep as simple as possible.

---

[1] https://www.anmp.pt.

We used the *gnutls-cli*[2] tool to check the presence of HTTPS sites (on the default 443 TCP port) and to extract their certificate and certification chain. We used the result of *gnutls-cli* certificate validation and we also developed a small Java program to validate the certificates, which was run using OpenJDK root trust anchor certificates and using Debian Linux root trust anchor certificates. Both certification validation strategies produced the same result.

We used the *wget*[3] tool to learn the HTTP redirections (HTTP codes 3xx) returned within the access to those Web pages, both using HTTP and HTTPS. Finally, this tool enabled us to assert the usage of HSTS (HTTP Strict Transport Security) by the municipalities' main Web page.

Finally, we used the *OpenWPM* tool [16] to confirm the HTTP redirections and the usage of HSTS, to obtain snapshots of entry web pages, both provided through HTTP and HTTPS, and to trace the resources accessed by the main Web page of all municipalities in order to assess the uniform protection of all the resources accessed through those pages (namely, if there were resources accessed through HTTP from a Web page accessed through HTTPS). This tool also enabled us to clarify the access to some Web pages that refused to provide resources to *wget*.

To facilitate readability, the results of the municipalities were resumed in a high-level, overall quality indicator (Sect. 3.2) with four classification levels: good; reasonable; elementary; and bad. Possible socio-economic determinants for those classifications were then investigated using multiple T-tests and logistic regression analysis.

3 Results and Discussion

3.1 Descriptive Analysis

In this section, we will describe the analysis that we did to the provision of the entry Web page of Portuguese municipalities through HTTPS.

Municipalities with an HTTPS Web Server

Through an automated analysis with *gnutls-cli* and *wget* we concluded that 259 municipalities (84.1%) have a server on port TCP 443 that is able to initiate a TLS session (Fig. 1) and, possibly, provide the main Web page through HTTPS. However, not all those municipalities provide their main Web page through that server. The analysis of the characteristics of the service provided on TCP port 443 is the focus of this Descriptive analysis section.

---

[2] https://gnutls.org/.

[3] https://www.gnu.org/software/wget/.

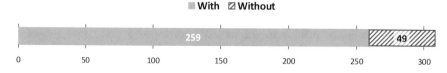

Fig. 1. Number of municipalities with and without a TLS server on port 443

Certificate Provided by Municipal Web Servers

Through an automated analysis with *gnutls-cli* we extracted the certificate of all municipal HTTPS servers and analyzed their correctness. By correctness, we mean:

- A server certificate must have in the Subject field or in the Subject Alternate Name field, the DNS name of the website or a wildcard DNS name (e.g. *.cm-albufeira.pt for all DNS names belonging to the DNS domain *cm-albufeira.pt*);
- A server certificate must be provided together with a complete and correct certification chain (excluding the root certificate);
- A server certificate cannot be self-certified;
- A server certificate or any certificate of its certification chain cannot be expired.

In all these cases, a browser blocks the access to the server and presents an error message that gives some details about the problem encountered. However, a typical user would not be able to understand the error and browsers typically suggest users not to proceed in the access to the problematic server in order to avoid problems.

Only 178 municipalities (57.8%) presented a correct certificate, and 81 (26.3%) presented a wrong certificate (Fig. 2). The breakdown of the problems found in those 81 municipalities is the following (some servers have more than one problem):

- 71 (87.7%) do not have the website DNS name in the server certificate;
- 7 (8.6%) do not provide a complete certification chain;
- 27 (33.3%) have a self-certified certificate;
- 20 (24.7%) have an expired certificate (16 of them are self-certified as well).

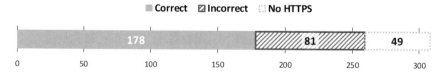

Fig. 2. Number of municipalities with correct and incorrect certificates for their HTTPS server

Content Provided by HTTPS Servers

The content provided by HTTPS servers was compared against the content provided by the corresponding HTTP servers using the *wget* tool and screenshots captured with OpenWPM. Where it was not possible, or a mismatch was detected, manual observation was used. For this analysis, we specifically required to ignore the correctness of servers' certificates, in order to get to the contents provided by HTTPS servers with a defective certificate setup. Also, to obtain HTTPS pages correctly rendered we disabled the insecure contents protection in the browser.

From the 259 HTTPS Web servers, 91 of them (35.1%) provide a wrong content (Fig. 3). Some notable cases are one municipality which provides the contents of the Amigos de Deus (God's Friends) website and two municipalities which provide the contents of the Centro Hospitalar do Porto (Porto's Health Care Center) Web page.

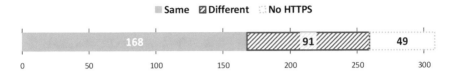

Fig. 3. Municipal websites that provide the same and different content with HTTP and HTTPS

Exploitation of HTTP Within Resources Fetched Through HTTPS

HTTP resources, namely HTML pages, use external resources to help to compose the contents presented to a user (Javascript code, CSS styles, images, movies, audio, etc.). Resource accessed through HTTPS should not use other resources fetched with HTTP, as this may create a security breach (as those resources may be tampered by an attacker with the power to intercept the communication between the browser and the server providing the unprotected resource). Only 115 (68.5%) of the 168 HTTPS Web servers of municipalities that provide correct contents do not provide resources referring other resources accessible through an HTTP URL (Fig. 4).

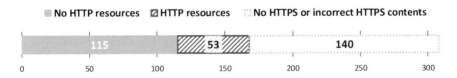

Fig. 4. Use of HTTP resources in websites with correct HTTPS content

Redirection Between HTTP and HTTPS

Web servers of municipalities should use only HTTPS and should redirect all HTTP accesses to a corresponding HTTPS access (something known as HTTPS-only access). Redirections can be made with different HTTP redirection codes (3xx codes), but the advice is to use the 301 (*Moved Permanently*), because it gives browsers an indication to cache the redirection for future use, thus saving redirections.

There are, however, some other redirection mechanisms, such as using a refresh operation on a meta tag or using Javascript code within an HTML web page. These methods may lead to the exactly same result of an HTTP redirection, but they are more difficult to assert, because they imply the parsing of downloaded HTML pages and Javascript files. Since we restricted our analysis to the meta information of munici- palities' entry Web pages, we were not able to get an accurate account of those redirections. Thus, for this analysis, those municipalities were by default accounted as not having an HTTP to HTTPS redirection. However, from other evidences, such as network traffic created by the entry pages, we found one municipality that uses content-embedded redirections and we counted it as having a redirection.

Only 103 (61.3%) of the 168 HTTPS servers that provide correct contents redirect HTTP accesses to their main Web page to HTTPS (Fig. 5). Within these (Fig. 6), 76 (73.8%) use the HTTP code 301 and 27 use other codes: 302 (*Found*), 303 (*See Other*) or 307 (*Temporary Redirect*).

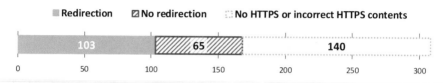

Fig. 5. Redirection of HTTP entry page to HTTPS

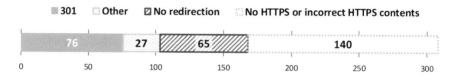

Fig. 6. Redirection codes used for the redirection of HTTP entry page to HTTPS

On the other hand, HTTPS to HTTP redirection constitutes a security downgrade and is not justifiable, from a security point of view. However, 13 of all HTTPS servers (5.0%) redirect HTTPS to HTTP, and 5 of them use the HTTP code 301 to do so.

Exploitation of HSTS

HSTS is a mechanism that HTTPS servers can use to force the usage of HTTPS by default to access all resources they provide. Servers convey this indication to browsers by means of an HTTP response header field (*Strict-Transport-Security*). Only 14 (8.3%) of the 168 HTTPS servers that provide correct contents use HSTS (Fig. 7).

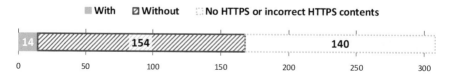

Fig. 7. Usage of HSTS in websites with correct HTTPS content

3.2   High-Level, Overall Quality Indicator

The high-level overall quality indicator provides an easy way to classify and rank the quality of the implementation of HTTPS in the web sites. For computing this indicator, we considered the following cases:

- **Good:** the municipality has a correct HTTPS Web entry page that does not use HTTP resources, HTTP access is redirected to HTTPS using code 301 and HSTS is enforced.
- **Reasonable:** the municipality has a correct HTTPS Web entry page that does not use HTTP resources.
- **Minimum:** the municipality has a correct HTTPS Web entry page.
- **Bad:** None of the previous cases.

In the previous classification, by correct HTTPS Web page we mean that the Web server certificate is valid, the content fetched through HTTPS is equal to the one obtained through HTTP and there is no HTTPS to HTTP redirection.

Given this classification, the Portuguese municipalities rank as follows (Fig. 8): 11 good (3.6%), 102 reasonable (33.1%), 53 minimum (17.21%) and 142 bad (46.1%).

Fig. 8. Rank of the Portuguese municipalities regarding HTTPS usage

3.3   Paradigmatic Cases

There are some paradigmatic cases that illustrate how confusion can create abnormal and dangerous situations. One municipality (here referred as *mname*) uses two different domains, *mname.pt* and *cm-mname.pt*. All HTTP accesses to both domains (using either the domains' name or its *www* host) redirects to an HTTPS server at *mname.pt*. However, direct accesses to the *www* HTTPS server in both domains (www.cm-mname.pt or www.mname.pt) or even to *cm-mname.pt* yield a defective server configuration (with a wrong certificate owned by another municipality).

Other paradigmatic cases are municipalities that have an HTTPS Web server but redirect the incoming requests to HTTP. This is incomprehensible, since they voluntarily use a protected access to abandon it in favor of unsecure accesses. Furthermore, some servers use the code 301 (Permanent Redirection) to perform the downgrade, which indicates browsers that HTTPS is not to be used in future accesses to them.

Some municipalities classify as bad because of a small, but relevant, technical detail: the HTTPS server does not send the complete certification chain for its certificate. This can create a problem on client browsers as it may raise the "unknown certificate issuer" error, on the absence of the intermediate certificates. Such absence depends on the popularity of the intermediate certificates in other certification chains, so this problem may occur or not depending on the past actions of the client browser.

3.4   Determinants of Results

In order to identify possible determinants for the classification of the municipalities according to the overall quality indicator (Sect. 3.2), we run multiple independent T-tests to identify significant mean differences between groups of municipalities. The tests were performed for three dichotomous dependent variables that are based on combinations of the indicator's classes: 'good'; 'good or reasonable'; and 'good, reasonable or elementary'. As independent variables we selected four indicators:

- The variables *'municipal taxes'* (logarithmized) and *'school dropout rate'*, because they were identified in [17] as relevant predictors of local e-government sophistication in Portugal;
- The variables *'total population'* (logarithmized) and *'population density'* (logarithmized), because they were identified as relevant predictors of local e-government sophistication in several international studies (for example [18, 19]);

Additionally, *'local e-government maturity'* was also selected to be tested as an independent variable envisioning to investigate a possible relation between the level of security offered and local e-government sophistication. Data from a previous study on e-government maturity of all Portuguese municipalities was used for this purpose [20]. The other indicators were obtained from the National Statistics Institute and are relative to 2017.

Table 1 shows the results of the significant T-tests (equal variance not assumed based on the previous Levene test). Results for the non-significant tests were omitted for simplicity. It turned out that there are significant mean differences only between the group of municipalities that have a classification of 'good or reasonable" and those that have a classification of 'elementary or bad'. The significant mean differences for these two groups were obtained for the variables *'municipal taxes'* and 'total population' (both logarithmized). In other words, the group of 'good or reasonable' municipalities is statistically different from the group of 'elementary or bad' municipalities when *'municipal taxes'* and *'total population'* are used as indicators.

Table 1. Results of the T-tests (only significant tests are presented)

Dependent variable	good or reasonable		
Independent variables	Sig.[***]	Mean dif.	Std. err. dif.
Municipal taxes (log)	0.030[*]	−0.1623	0.0742
Total population (log)	0.007[**]	−0.1607	0.0592

[*]$p < 0.05$; [**]$p < 0.01$; [***]Two extremities

To further investigate if those two independent variables can be used as predictors for a municipality to be classified as 'good or reasonable' we run a binary logistic regression using 'good or reasonable' as the dichotomous dependent variable and 'municipal taxes' (logarithmized) and 'total population' (logarithmized) as independent variables. The results are shown in Table 2. The model is statistically significant ($p < 0.05$) but it explains only 3.9% (Nagelkerke R^2) of the variance in the classification of municipalities and correctly classifies only 64.0% of the cases. The indicator 'total population' (logarithmized) is the only significant predictor in the model ($p < 0.05$). This suggests the existence of other relevant explanatory factors besides the ones that were tested in this study.

Table 2. Results of the binary logistic regression

Dependent variable	good or reasonable			
Independent variables	B	Std. err.	Sig.	Exp (B)
Municipal taxes (log)	0.818	0.651	0.209	0.441
Total population (log)	1.697	0.813	0.048[*]	4.989
Constant	−4.487	1.442	0.002[**]	0.021

[**] $p < 0.05$

4   Conclusions and Future Work

The results of this study allow the conclusion that there is much room for improvement in order to ensure that all the websites of the Portuguese municipalities offer the necessary conditions for citizens to communicate securely with them. This is especially important because such websites are typically used to provide transactional services to citizens and citizens need to trust that the confidentiality of the information they submit is assured in the communication process. In a global classification, only 3.6% of the municipal webpages offer good security conditions, 33.1% offer reasonable conditions, 17.1% offer elementary conditions, and 46.1% offer bad conditions (meaning that they do not offer an HTTPS webpage with a valid certificate).

These results seem to be associated with the fact that most Portuguese municipalities are rather small and consequently might not have the critical mass to correctly implement and maintain their websites. This was indirectly observed by finding that the group of 'good or reasonable' municipalities is statistically different from the group of 'elementary or bad' municipalities when 'municipal taxes' and 'total population' are used as indicators. Nevertheless, the fact that those indicators are not good predictors for the overall results (explaining only 3.7% of the phenomenon) suggests the existence of other relevant explanatory factors not identified in this study.

Due to their nature, the weaknesses presented here cannot be exploited to jeopardize the servers of the municipalities or their contents. Even so, their disclosure raises some ethical issues since they can be exploited by third parties to deceive citizens.

Acknowledgments. This work was partially funded by National Funds through the FCT - Foundation for Science and Technology, in the context of the project UID/CEC/00127/2019.

References

1. European Commission: ePrivacy: consultations show confidentiality of communications and the challenge of new technologies are key questions (2016). https://ec.europa.eu/digital-single-market/en/news/eprivacy-consultations-show-confidentiality-communications-and-challenge-new-technologies-are. Accessed 27 Nov 2018
2. Gupta, C.: The market's law of privacy: case studies in privacy and security adoption. IEEE Secur. Priv. 15(3), 78–83 (2017)
3. W3C Technical Architecture Group (TAG): Securing the Web. W3C (2015). http://www.w3.org/2001/tag/doc/web-https-2015-01-22. Accessed 27 Nov 2018
4. Morgan, C.: IAB Statement on Internet Confidentiality. IAB (2014). https://www.iab.org/2014/11/14/iab-statement-on-internet-confidentiality. Accessed 27 Nov 2018
5. Vyas, T., Dolanjski, P.: Communicating the Dangers of Non-Secure HTTP. Mozilla Security Blog (2017). https://blog.mozilla.org/security/2017/01/20/communicating-the-dangers-of-non-secure-http. Accessed 27 Nov 2018
6. Schechter, E.: A secure web is here to stay. Google Security Blog (2018). https://security.googleblog.com/2018/02/a-secure-web-is-here-to-stay.html. Accessed 27 Nov 2018
7. Ouvrier, G., Laterman, M., Arlitt, M., Carlsson, N.: Characterizing the HTTPS trust landscape: a passive view from the edge. IEEE Commun. Mag. 55(7), 36–42 (2017)
8. European Commission: Proposal for a Regulation on Privacy and Electronic Communications (2017)
9. The European Parliament and the Council of the European Union: General Data Protection Regulation. Official Journal of the European Union (2016)
10. Felt, A.P., Barnes, R., King, A., Palmer, C., Bentzel, C., Tabriz, P.: Measuring HTTPS adoption on the web. In: 26th Usenix Security Symposium, pp. 1323–1338 (2017)
11. Chan, C., Fontugne, R., Cho, K., Goto, S.: Monitoring TLS adoption using backbone and edge traffic. In: IEEE INFOCOM 2018, pp. 208–213 (2018)
12. Vumo, A.P., Spillner, J., Köpsell, S.: Analysis of Mozambican websites: how do they protect their users? In: Information Security for South Africa (ISSA), pp. 90–97 (2017)
13. Wullink, M., Moura, G.C.M., Hesselman, C.: Automating domain name ecosystem measurements and applications. In: 2018 Network Traffic Measurement and Analysis Conference (TMA), pp. 1–8. IEEE (2018)
14. Andersdotter, A., Jensen-Urstad, A.: Evaluating websites and their adherence to data protection principles: tools and experiences. In: IFIP International Summer School on Privacy and Identity Management, pp. 39–51. Springer (2016)
15. Buchanan, W.J., Woodward, A., Helme, S.: Cryptography across industry sectors. J. Cyber Secur. Technol. 1(3–4), 145–162 (2017)
16. Englehardt, S., Narayanan, A.: Online tracking: a 1-million-site measurement and analysis. In: Proceedings of ACM CCS 2016 (2016)
17. Dias, G.P., Costa, M.: Significant socio-economic factors for local e-government development in Portugal. Electron. Gov. Int. J. 10(3–4), 284–309 (2013)
18. Pina, V., Torres, L., Royo, S.: E-government evolution in EU local governments: a comparative perspective. Online Inf. Rev. 28(4), 1137–1168 (2009)
19. Chen, Y.: Citizen-centric e-government services: understanding integrated citizen service information systems. Soc. Sci. Comput. Rev. 28(4), 427–442 (2010)
20. Dias, G.P., Gomes, H.: Evolution of local e-government maturity in Portugal. In: 9th Iberian Conference on Information Systems and Technologies (CISTI), pp. 1–5 (2014)

Lego Methodology Approach for Common Criteria Certification of IoT Telemetry

George Suciu[1,2(✉)], Cristiana Istrate[1], Ioana Petre[1],
and Andrei Scheianu[1]

[1] R&D Department, BEIA Consult International,
Peroni 16, 041386 Bucharest, Romania
{george, cristiana.istrate, ioana.petre,
andrei.scheianu}@beia.ro
[2] Telecommunications Department, University POLITEHNICA of Bucharest,
Bd. Iuliu Maniu 1-3, 061071 Bucharest, Romania
george@radio.pub.ro

Abstract. In our days, almost every business relays on medium-to-high performance computer systems which presents the possibility of being the target of different threats that can exploit the vulnerable software, respectively hardware components. The concept of "security risk" can be described as a specific threat that using a specific type of attack presents the ability to exploit system vulnerabilities, action which will affect the entire integrity of the targeted systems. From this point of view, the main idea of this paper is to present a Lego methodology approach for Common Criteria certification that can be applied to IoT Telemetry systems. Furthermore, we present scenarios of implementation of our approach to increase robustness level applied for agro-telemetry system.

Keywords: Lego methodology · Common Criteria · IoT · Telemetry · Security

1 Introduction

Security certification for heterogenous Internet of Things (IoT) systems poses several challenges, as IoT devices become more ubiquitous. To discover the impact that security risks can have over a System Under Analysis (SUA), there must be specified very clearly the systems' description, considering two essential perspectives which present certain criteria, such as:

(a) **Information and functionality:**

- Evaluation of the security risks impact. An appropriate example can be the affected parts of systems' services, including data processing functions. This perspective is also known as **Aim of refinement**;
- **Criteria of partition:** Specific sections of the data or functions can require distinct security requirements which present unique criticality levels.

© Springer Nature Switzerland AG 2019
Á. Rocha et al. (Eds.): WorldCIST'19 2019, AISC 931, pp. 165–174, 2019.
https://doi.org/10.1007/978-3-030-16184-2_17

(b) **System architecture:**

- **Aim of refinement:** Every level of each component of a system can present a security risk potential. These levels must be detected in order to improve the security level of the entire system;
- **Criteria of partition:** Every component, such as software and hardware components or even an entire network, presents several security risks that can be traced. Because of this characteristic, the architecture of potentially vulnerable systems is divided into components, the specific risks of each component being identified.

In the computer security field, the threat concept is defined as an imminent danger situation that can exploit the vulnerabilities of a system in order to breach its security mechanisms. Threats can be divided into three main categories, as follows [1]:

- **Intentional threats** – represent premeditated invasive actions. The best example of intentional threats consists of computer crimes, including software attacks, theft, espionage, sabotage, etc.;
- **Unintentional threats** – represent accidental software modifications due to human mistakes or technical misunderstandings;
- **Natural threats** – represent several threats that can damage the physical equipment, such as: natural disasters, power failures, fires and floods.

The main idea of this article is the presentation of a Lego methodology approach for Common Criteria that can be applied to IoT Telemetry systems [2]. This methodology allows to select security functions according to a specific use case, integrate various evaluated and non-evaluated components, plug-and-play (exchange, add, remove and update) new components, perform additional tests to demonstrate a higher robustness level of some security functions. As it was previously specified, the natural and unintentional threats will be excluded from this work activity. The attention will be focused on the intentional threats. These threats can also be classified into 3 categories, as follows [1]:

- **Depletion:** the targeted sectors of these threats are data and resource availability. As an effect, a resource will be consumed faster than it can be replenished. After a period of time, a resource which is being attacked becomes depleted, fact that affects the target functionality;
- **Alteration:** in this case, threats occur when an unauthorized entity tries to modify private code or data, affecting in this manner the integrity of the target;
- **Disclosure:** when talking about disclosure, threats will affect only the information confidentiality, especially specific system information, such as backup and temporary files, patch levels, version numbers software distribution, etc.

The article will be structured in 5 Sections. The current Section represents the introduction, having the main purpose of providing sufficient information for a general idea of the security risks impact. In Sect. 2 will be analyzed the present work activities in this field, being mentioned only the most notable research efforts. The methodology approach for Common Criteria can be seen in Sect. 3, while several scenarios of implementation will be presented in Sect. 4. The last section of this paper draws the conclusions and envisions future work.

2 Related Work

The innovative developments in the Information and Communication Technology (ICT) field produce various embedded things/devices which contain sensors and have the property to transfer data among other objects [3].

In the paper [4], a study has been conducted regarding the methods through which the equipment from a use case on a smart service in the industrial maintenance domain are securely transmitting their data. The paper examines two solutions focused on isolation and execution environment security, namely a Security Controller and the ARM TrustZone. For their comparison, a system that takes a device's snapshot authentication has been designed. The results demonstrate increased flexibility when using the TrustZone technology and a much secure physical environment for the Security Controller. Finally, it is conducted that the best approach is only based on the desired use case. The article also proposes a hybrid solution that increases the security of industrial applications, which could bring a contribution to the future Industrial Internet of Things (IIoT).

Furthermore, there is a recent survey about technical approaches, functional requirements, and on overall, the security status based on the OpenFog's [5] architecture. The OpenFog Consortium designed the Fog Nodes functional security requirements by adopting the Common Criteria standard. The Fog Nodes bolster a trusted computing environment as well as a secure service provisioning and also the hardware virtualization. Therefore, by using these entities, the multi-tier omnipresent communication-computing infrastructure which encompasses a multitude of devices and covers sophisticated hierarchies of application areas and administration, will have a more trusted environment due to increased security levels.

The IoT paradigm emerged into the car tracking technology, is being described in [6]. The paper describes the IoT's role in designing a car tracking system as well as the needed standards, principles in order to have an increased quality factor. Reliability engineering is the concept behind these standards. Common criteria plays an important role by assuring consistency and error management, these being the basic fundamental features a system must have to guarantee it matches the function for which it was designed.

3 Methodology Approach for Common Criteria - Lego Methodology Concept

The Common Criteria (CC) represents an international standard regarding computer security information (ISO/IEC 15408) [7]. Furthermore, Common Criteria is a framework in which SFRs (Security Functional Requirements) and SARs (Security Assurance Requirements) can be foreseen in an ST (Security Target) [8].

The CC assessment methods are performed on systems and computer security items. The product or the system which represents the subject of the evaluation is called Target of Evaluation (TOE). The evaluation has to examine if the security requirements are accomplished; this is performed through Protection Profile (PP), Security Target

(ST) and Security Functional Requirements (SFRs) [9]. The PP is a document created most often by the user or by the community and its major role is to detect security requirements for a class of security devices significant for a specific target. Retailers have the possibility to implement items that submit with one or more PPs, but they also can have their products evaluated against those PPs. In this particular case, a PP can be considered a template for the ST products. ST represents the document that identifies the security features of the main goal of the evaluation. The TOE is evaluated against the SFRs settled in its ST; this allows the retailers to suit the evaluation in order to match with the abilities of their item. SFRs stipulate the functions that can be provided by an item. Common Criteria has a specific list consisting of these functions, and the list can be different depending on the type of the evaluation even if the targets of the evaluation are the same type [10].

The evaluation procedure makes an attempt in the process of determining the degree of reliability of the product's security characteristics by quality assurance procedures, represented by SARs and Evaluation Assurance Level (EAL). SARs represent depictions of the actions carried out through the product`s evolution and evaluation to provide the supposed security functionality. The requirements from the CC catalog are documented in PP and ST. EAL is the numerical evaluation which describes the depth and the harshness of an evaluation. Each EAL is correlated with a package of SARs which includes the full development of an item. Common Criteria stipulates 7 levels of EAL, EAL1 being the most fundamental with low costs, while EAL7 is the most rigorous and the most expensive. Higher EAL does not suppose a more precise security, it only intends to assure that the TOE has been verified with higher accuracy [11].

3.1 Lego Methodology Concept

The 'Lego methodology' was developed to resolve the limitations introduced by the current composition approaches and concentrates on the evaluation of a list set in advance of Security Functions (SF) necessary in a use-case already existent [12]. This solution allows a global and unique Evaluation level (EAL LEGO) of the composite Platform with different robustness levels (Vulnerability ANalysis (VAN)-HIGH, VAN-Moderate, VAN-LOW) among security functions within the Platform (see Fig. 1).

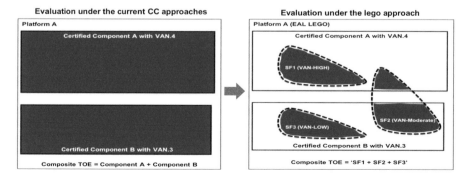

Fig. 1. Evaluation under current CC approaches vs. evaluation under Lego methodology approach

This methodology permits the selection of certain security functions related to a use case, the integration of different evaluated and non-evaluated parts, the installation and utilization of new components, and the execution of additional tests to demonstrate an increased robustness level of the security functions. This methodology leads to the following benefits:

- Evaluation with decreased effort, time and at low cost by reducing the perimeter of the evaluation, making the evaluation of the platform fast and easy;
- Dynamic plug-and-play integration by updating, adding, removing or replacing a Component with less effort;
- Possibility of obtain different robustness levels and increasing the level of robustness of required security functions within the system;

The following roles are considered in the Lego methodology:

- Component Developer: Entity developing the Component; it might also be the sponsor of the Component evaluation;
- Component Evaluator: Entity performing the Component evaluation;
- Component Certification Body: Entity performing the Component certification;
- Composite Platform Integrator: Entity integrating the Components in the Platform; it might also be the sponsor of the Composite Platform evaluation;
- Composite Platform Evaluator: Entity performing the composite Platform evaluation;
- Composite Platform Certification Body: Entity performing the composite Platform certification;
- Composite Product Evaluation Sponsor: Entity in charge of contracting the composite product evaluation.

In the concept of Lego Methodology, several suppositions are taken into consideration, such as: each Component is certified and completely specified, the objectives required by the working environment of each Component are well-defined, the Integrator should know the components functionality, the Platform has a fixed number of components in order to be certified, etc.

The described Lego methodology can be employed for demonstrating the compositional evaluation within a platform designed for security through isolation, called ODSI (On Demand Secure Isolation) [13]. The ODSI Platform combines the following independent certified components with different security levels (see Fig. 2):

(a) Configuration Manager: Certified Component was providing the Isolation security function (SF-ISO) between memory partitions. It is the base component on which the security of SFs DISP, AUTH, COMM and KEYM is built;
(b) Administration Manager: Certified Component providing the Dispatch security function (SF-DISP) of commands between partitions;
(c) Network Manager: Certified Component providing the Authentication (SF-AUTH) and the Communication (SF-COMM) security functions;
(d) Keyring Manager: Certified Component providing the key storage security function (SF-KEYM).

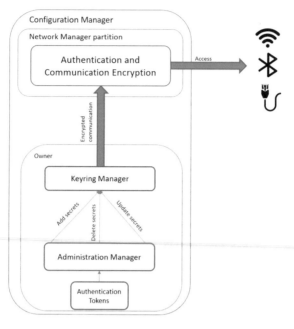

Fig. 2. The architecture behind the Network Manager, Keyring Manager and Administration Manager interaction

4   Scenarios of Implementation

In this section, there will be presented two scenarios of implementation, one being more of a general overview of the Lego methodology approach on increasing the robustness level of a SF within the CC Component and the other presenting the Agro-Telemetry use case that utilizes the CC compliant ODSI platform.

4.1   Scenario 1 - Increase the Robustness Level of a Given SF Within the Component

In this scenario, the re-evaluation effort will focus on testing that the SF1 can resist at a security level higher than VAN.3 and that no other part of the Component A1 can affect or decrease this security level.

A. Actors

The following Actors are considered in this scenario:

- Same Evaluator of the Component A1 and the SF1;
- Component Developer;
- Product Evaluation Sponsor:
  - This actor may be the Component Developer.

B. Assumptions

The following assumptions are considered for the evaluation under this scenario:

- Component A1 is CC-certified;
- The re-evaluation of SF1 and the evaluation of A1 will both be conducted by the same evaluation facility. The benefits of this assumption are the following:
  - a full evaluation of the Component A1 conducted by the new evaluation facility is not required;
  - delivery of all evaluation evidences (documents, source code, samples) to the new evaluation facility is not required;
  - the base evaluation results can be reused.
- The source code of the Component A1 is identical between the base certification and the reevaluation of SF1.

C. Inputs and Required Evidences

The following inputs are required prior to the evaluation under this scenario:

- ST, TDS (Target-of-Evaluation Design), FSP (Functional SPecification), ATE (Assurance TEsts) of the Component;
- SFRs that fulfill the SF1 are clearly identified and the rationale is provided in the Security Target;
- The SF1-Interfaces and Modules that implement the SF1 are described in the TDS and FSP Documents;
- Interactions between SF1-Interfaces and the other interfaces of the Components A1 are described in the TDS (TOE Design) document;
- SF1-Interfaces and Interactions are tested, and the conducted tests are described in the ATE Document.

D. Additional Requirements during Re-evaluation

The goal of the Lego certification within this scenario may be achieved by the following means:

- Requirements to be fulfilled by the Component Developer or the Product Evaluation Sponsor:
  - Additional information of each SF1-Interfaces may be required. This information will provide the evaluator with a better understanding of how this security function is performed. These additional details of SF1-Interfaces can be included in the TDS and the FSP documentation if the SF1-Interface is an external one (TSFI);
  - Additional information of the interactions between SF1-Interfaces and the other interfaces of the Component A1 may be required. This information will provide the evaluator a better understanding of how the SF1 interacts with the other parts of the component. These additional details about interactions can be included in the TDS;
  - The evaluator may require a characterization of the SF1-Interfaces at the implementation level (e.g. description of parameters passed from a SF1-Interface to another, variables, data identified for this SF1-Interface that are going to be used by other interfaces, return values from those interfaces, etc.). Such complete characterizations of SF1-Interfaces are meant to allow their exercise during reviewing and testing. These additional details can be included in the TDS and the FSP documentation if the SF1-Interface is an external one (TSFI);

- Since they play no role in testing, it is not mandatory to describe at an implementation level the SF1-Interfaces that have no interaction with the other interfaces of the component;
- Additional functional depth tests by the developer may be required to determine if all SF1-Interfaces and Interactions are completely tested. These additional tests can be provided in ATE.
- Requirements to be addressed by the Evaluator:
  - Compliance analysis of the updated evidences;
  - Additional penetration and/or independent tests by the evaluator may be conducted to determine if the SF1 is resistant at a level higher than VAN.3.

4.2   Scenario 2 – Agro-Telemetry System

In this scenario, it is presented the implemented Agro-Telemetry System. This system is used for precision agriculture and has two functions: data acquirement and data transmission along with processing. The first function allows sensors to collect data on temperature, humidity sunlight and actuators to execute commands in order to activate mechanic systems, such as irrigating. The data transmission and processing function illustrate the technique of sensor data transportation from the gateway to server; information is processed by the server and is presented to users through Web interface. This use-case handles two types of data: business and security data. Business data consists of raw information collected from sensors, representing the processed data produced by the server, while security data is residing on user credentials, log data, system configuration (see Fig. 3).

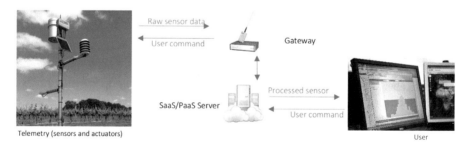

Fig. 3. Agro-Telemetry system

Table 1 concretizes the impacts of feared events.

Table 1. Impacts of all feared events

Severity	Example of feared event	Possible impact
Critical	Alteration of Business data	User commands are modified, causing over-irrigating
Critical	Disclosure of Security data	Hacker obtains user credentials to intrude Agro-Telemetry system by masquerading as Admin
Important	Unavailability of Data Processing function	Agro-Telemetry system cannot process sensor data, causing abnormal situation in the farm undetected

Regarding the threats analysis, there will be described some examples of contra measures to each threat that relates to a feared event. When alteration of Business data is involved, a man-in-the-middle attack may interfere on the network. As a contra measure, a deployment of an ODSI token verification service for user authentication will be performed. This action is provided by the Network Manager whose Certified Component supplies the Authentication (SF-AUTH) and the Communication (SF-COMM) security functions. Disclosure of Security data implies threats on the gateway such as the side-channel attack which can be prevented by deploying a technique called ODSI end-to-end encryption that embodies the Certified Component of the Administration Manager which provides the Dispatch security function (SF-DISP) of commands between partitions. Considering the unavailability of the Data Processing function, a possible threat accomplished on the server is the Denial of Service (DOS) attack which can be countered when an ODSI isolation BIP technique is deployed on the server through the Configuration Manager's Certified Component which provides the Isolation security function (SF-ISO) between memory partitions.

5 Conclusions

As shown in this paper, Lego Methodology offers an encouraging and feasible way in order to extend the current CC approach to support and facilitate composition and allow the evaluation of IoT platforms. The Lego Methodology applied on ODSI platform allows a secure communication with a remote entity, end-to-end encryption, key management and secure storage without the limitations brought in by the present composition approaches. As future work we envision developing an application in a real-world context and perform the compositional evaluation.

Acknowledgements. This work has been supported in part by UEFISCDI Romania through projects ODSI, ToR-SIM and PARFAIT, funded in part by European Union's Horizon 2020 research and innovation program under grant agreement No. 777996 (SealedGRID project) and No. 787002 (SAFECARE project).

References

1. Jouini, M., Rabai, L.B.A., Aissa, A.B.: Classification of security threats in information systems. Procedia Comput. Sci. **32**, 489–496 (2014)
2. da Cruz, M.A., Rodrigues, J.J., Paradello, E.S., Lorenz, P., Solic, P., Albuquerque, V.H.C.: A proposal for bridging the message queuing telemetry transport protocol to HTTP on IoT solutions. In: 3rd International Conference on Smart and Sustainable Technologies (SpliTech), pp. 1–5. IEEE (2018)
3. ETSI France, Orange France: Internet of Things Global Standardisation-State of Play (2018)
4. Lesjak, C., Hein, D., Winter, J.: Hardware-security technologies for industrial IoT: TrustZone and security controller. In: IECON 2015-41st Annual Conference of the IEEE Industrial Electronics Society, pp. 002589–002595. IEEE (2015)

5. Martin, B.A., Michaud, F., Banks, D., Mosenia, A., Zolfonoon, R., Irwan, S., Zao, J.K.: OpenFog security requirements and approaches. In: IEEE Fog World Congress (FWC), pp. 1–6. IEEE (2017)
6. Thomas, M.O., Rad, B.B.: Reliability evaluation metrics for internet of things, car tracking system: a review. Int. J. Inf. Technol. Comput. Sci. (IJITCS) **9**(2), 1–10 (2017)
7. Bialas, A.: Common criteria IT security evaluation methodology–an ontological approach. In: International Conference on Dependability and Complex Systems, pp. 23–34. Springer, Cham (2018)
8. Communications Security Establishment. https://www.cse-cst.gc.ca/en/canadian-common-criteria-scheme/main. Accessed 01 Oct 2018
9. Common Criteria for Information Technology Security Evaluation, Part 1: Introduction and general model, Version 3.1, Revision 4 (2012)
10. Common Criteria for IT security evaluation. https://www.commoncriteriaportal.org/files/epfiles/anssi-cible-cc-2017_50en.pdf.pdf. Accessed 01 Oct 2018
11. Common Criteria for Information Technology Security Evaluation, Part 3: Security assurance components, Version 3.1, Revision 5 (2017)
12. Chae, H., Lee, D.H., Park, J., In, H.P.: The partitioning methodology in hardware/software co-design using extreme programming: evaluation through the lego robot project, pp. 187. IEEE (2006)
13. Suciu, G., Istrate, C., Petrache, A., Schlachet, D., Buteau, T.: On demand secure isolation using security models for different system management platforms. In: Advanced Topics in Optoelectronics, Microelectronics, and Nanotechnologies IX, vol. 10977, p. 109770R (2019)

A Two-Phase Algorithm for Recognizing Human Activities in the Context of Industry 4.0 and Human-Driven Processes

Borja Bordel$^{(\boxtimes)}$, Ramón Alcarria, and Diego Sánchez-de-Rivera

Universidad Politécnica de Madrid, Madrid, Spain
{bbordel,diegosanchez}@dit.upm.es,
ramon.alcarria@upm.es

Abstract. Future industrial systems, a revolution known as Industry 4.0, are envisioned to integrate people into cyber world as prosumers (service providers and consumers). In this context, human-driven processes appear as an essential reality and instruments to create feedback information loops between the social subsystem (people) and the cyber subsystem (technological components) are required. Although many different instruments have been proposed, nowadays pattern recognition techniques are the most promising ones. However, these solutions present some important pending problems. For example, they are dependent on the selected hardware to acquire information from users; or they present a limit on the precision of the recognition process. To address this situation, in this paper it is proposed a two-phase algorithm to integrate people in Industry 4.0 systems and human-driven processes. The algorithm defines complex actions as compositions of simple movements. Complex actions are recognized using Hidden Markov Models, and simple movements are recognized using Dynamic Time Warping. In that way, only movements are dependent on the employed hardware devices to capture information, and the precision of complex action recognition gets greatly increased. A real experimental validation is also carried out to evaluate and compare the performance of the proposed solution.

Keywords: Industry 4.0 · Pattern recognition · Dynamic Time Warping · Artificial Intelligence · Hidden Markov Models

1 Introduction

Industry 4.0 [1] refers the use of Cyber-Physical Systems (unions of physical and cybernetic processes) [2] as main technological component in future digital solutions, manly (but not only) in industrial scenarios. Typically, digitalization has caused, at the end, the replacement of traditional work mechanisms by new digital instruments. For example, workers in the assembly lines were substituted by robots during the third industrial revolution.

However, some industrial applications cannot be based on technological solutions, being human work still essential [3]. Hand-made products are an example of applications where the presence of human works is essential. These industrial sectors, any

© Springer Nature Switzerland AG 2019
Á. Rocha et al. (Eds.): WorldCIST'19 2019, AISC 931, pp. 175–185, 2019.
https://doi.org/10.1007/978-3-030-16184-2_18

case, must be also integrated into fourth industrial revolution. From the union of Cyber-Physical Systems (CPS) and humans acting as service providers (active works), humanized CPS arise [4]. In these new systems, human-driven processes are allowed [5]; i.e. processes which are known, executed and managed by people (although they may be watched over by digital mechanisms).

To create a real integration between people and technology, and move the process execution from the social subsystem (humans) to the cyber world (hardware and software components), techniques for information extraction are needed. Many different solutions and approaches have been reported during the last years, but nowadays pattern recognition techniques are the most promising one.

The use of Artificial Intelligence, statistical models and other similar instruments have allowed a real and incredible development of pattern recognition solutions, but some challenges are still pending.

First, pattern recognition techniques are dependent on the underlying hardware device for information capture. The structure and learning process changes if (for example) instead of accelerometers we consider infrared sensors. This is very problematic as hardware technologies evolve much faster than software solutions.

And, second, there is a limit to the precision in the recognition process. In fact, as human actions turn more complicated, more variables and more complex models are required to recognize them. This approach generates large optimization problems whose residual error is higher as the number of variables increases; which causes a decreasing in success recognition rate [6]. In conclusion, mathematics (not software, thus, not dependent on the implementation) force a certain precision for the pattern recognition process given the actions to be studied. To avoid this situation, a lower number of variables should be considered, but this also reduces the complexity of actions that may be analyzed; a solution which is not acceptable in industrial scenarios where complex production activities are developed.

Therefore, the objective of this paper is to describe a new pattern recognition algorithm addressing these two basic problems. The proposed mechanism defines actions as a composition of simple movements. Simple movements are recognized using Dynamic Time Warping (DTW) techniques [7]. This process is dependent on the selected hardware for information capture; but DTW are very flexible and updating the pattern repository is enough to reconfigure the entire algorithm. Then, complex actions are recognized as combinations of simple movements through Hidden Markov Models (HMM) [8]. These models are totally independent from hardware technologies, as they only consider simple actions. This two-phase approach also reduces the complexity of models, increasing the precision and success rate in the recognition process.

The rest of the paper is organized as follows: Sect. 2 describes the state of the art on pattern recognition for human activities; Sect. 3 describes the proposed solution, including the two defined phases; Sect. 4 presents an experimental validation using a real scenario and final users; and Sect. 5 concludes the paper.

2 State of the Art on Pattern Recognition

Many different pattern recognition techniques for human activities have been reported. However, most common proposal many be classified into five basic categories [9]: (i) Hidden Markov Models; (ii) the Skip Chain Conditional Random Field; (iii) Emerging Patterns; (iv) the Conditional Random Field; and (v) Bayesian classifiers.

In fact, most authors propose the use of Hidden Markov Models (HMM) to model human activities. HMM allow modeling actions as Markov chains [10, 11]. Basically, HMM generate hidden states from observable data. In particular, the final objective of this technique is to construct the sequence of hidden states that fits with a certain data sequence. To finally define the whole model, HMM must deduct from data the model parameters in a reliable manner. Figure 1 shows a schematic representation about how HMM work. When human activities are recognized, the actions composing the activities are the hidden states, and sensor outputs are data under study. HMM, besides, allow the use of training techniques considering prior knowledge about the model. This training is sometimes essential to "induce" all possible data sequences required to calculate the HMM. Finally, it is very important to note that simple isolated HMM can be combined to create larger and more complex models.

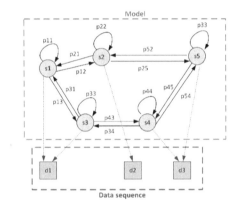

Fig. 1. Graphical representation of an HMM

HMM, nevertheless, are useless to model certain concurrent activities, so other authors have reported a new technique named, Conditional Random Field (CRF). CRF are employed to model those activities that present concurrent actions or, in general, multiple interacting actions [12, 13]. Besides, HMM requires a great effort on training to discover all possible hidden states. To solve these problems, Conditional Random Field (CRF) employs conditional probabilities instead of joint probability distributions. In that way, activities whose actions are developed in any order may be easily modeled. Contrary to chains in HMM, CRF employs acyclic graphs, and enables the integration conditional hidden states (states that depend on past and/or future observations).

CRF, on the other hand, are still useless to model certain behaviors, so some proposals generalize this concept and propose the Skip Chain Conditional Random

Field (SCCRF). SCCRF is a pattern recognition technique, more general than CRF, that enables modeling activities that are not sequence of actions in nature [14]. This technique tries to capture long-range (skip chain) dependencies; and may be understood as the product of different linear chains. However, calculating this product is quite heavy and complicated, so this technique is usually too computationally expensive to be implemented in small embedded systems.

Other proposals employ higher level description techniques such as Emerging Patterns (EP). For most authors, EP is a technique describing activities as vectors of parameters and their corresponding values (location, object, etc.) [15]. Using distances between vectors it is possible to calculate and recognize actions developed by people. Finally, other authors have successfully employed secondary techniques such as Bayesian classifiers [16], which identify activities making a correspondence between human activities and the most probable sensor outputs while these actions are performed, considering all sensors are independent. Decision trees [17], HMM extensions [18], and other similar technologies have been also studied in the literature, although these proposals are sparse.

Among all described technologies, HMM is not the most powerful one. However, it fits perfectly with Industry 4.0, where actions are very complex but very structured and ordered (according to company protocols, efficiency policies, etc.). Besides, fast feedback is required (sometimes even real-time), in order to guarantee human-driven processes operate correctly before a global critical fail occurs. Thus, computationally expensive solutions are not a valid approach, and we are selecting HMM as main base technology. In order to preserve its lightweight character and, at the same time, being able to model complex activities, we introduce a two-phase recognition scheme which enable dividing complex actions in two simpler steps.

3   A Two-Phase Pattern Recognition Algorithm

In order to (i) make independent the pattern recognition process from employed hardware devices to capture information, (ii) enable the recognition of complex actions, and (iii) preserve the lightweight character of the selected models, the proposed solution presents an architecture with three different layers (see Fig. 2).

The lowest layer includes the hardware platform. Monitoring devices such as accelerometers, smartphones, infrared sensors, RFID tags, etc., are deployed to capture information about the people behavior. The outputs of these devices create physical data sequences whose format, dynamic range, etc., are totally dependent on the selected hardware technologies.

These physical data sequences are then processed in the middle layer using DTW techniques. As result, for each physical data sequence, a simple movement or action is recognized. These simple actions are represented using a binary data format to make the solution as lightweight as possible. Software at this level must be modified each time the hardware platform is updated, but DTW technologies do not require a heavy actualization process, and refreshing the pattern repository is enough to configure the algorithm at this level.

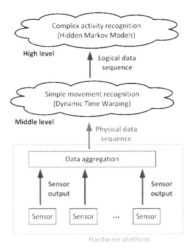

Fig. 2. Architecture of the proposed pattern recognition solution

Recognized simple movements, then, are grouped to create logical data sequences. These sequences feed a high-level pattern recognition system based on Hidden Markov Models. At this level, software components require a heavy training process, but middle layer makes totally independent the hardware platform and high-level models. Thus, any change in the hardware platform does not enforce an actualization in the HMM, which would be extremely computationally costly. By the analysis of the sequence of simple movements, complex actions are recognized.

Next subsection describes both proposed pattern recognition phases in detail.

3.1 Simple Movement Recognition: Dynamic Time Warping

In order to recognize simple gestures or movements, a Dynamic Time Warping solution is selected. DTW technologies fulfill the requirements of middle-level software components as they adapt to the underlying hardware platform's characteristics very easily and are quite fast and efficient (so small embedded devices may implement them).

In our solution, human behavior is monitored through a family of sensors S containing N_s components (1).

$$S = \{s_i, i = 1, \ldots, N_s\} \tag{1}$$

The outputs of these sensors are periodically sampled each T_s seconds; obtaining for each time instant, t, a vector of N_s values (each value from each sensor). This vector Y_t is called "a multidimensional sample" (2).

$$Y_t = \{y_t^i, i = 1, \ldots, N_s\} \tag{2}$$

Then, a simple movement Y will a have a duration of T_m seconds and will be described by the temporal sequence of N_m multidimensional samples collected during this time (3). In order to later recognize movements, a pattern repository $\mathcal{R}$ is created containing the corresponding temporal sequences for each one of the K simple actions to be recognized (4).

$$Y = \{Y_t, t = 1, \ldots, T_m\} = \{Y^i, i = 1, \ldots, N_m\} \tag{3}$$

$$\mathcal{R} = \{R_i, i = 1, \ldots, K\} \tag{4}$$

In general, people perform movements in similar but different manners. Thus, transitions may be slower or faster, some elemental actions may be added or removed, etc. Therefore, given a sequence X with N_x samples, representing a movement to be recognized, it must be located the pattern $R_i \in \mathcal{R}$ closer to X; so R_i is recognized as the performed action. To do that it is defined a distance function (5). This distance function may be applied to calculate a cost matrix, required as samples usually don't have the same length neither they are aligned (6).

$$d : \mathcal{F} \times \mathcal{F} \rightarrow \mathbb{R}^+, X^i, r^i_j \in \mathcal{F} \tag{5}$$

$$C \in \mathbb{R}^{N_x \times N_m} \ C(n, m) = d\left(X^n, R^m_j\right) \tag{6}$$

In positional sensors (accelerometers, infrared devices, etc.) distance function is applied directly to the sensors' outputs (contrary to, for example, microphones whose outputs must be evaluated in the power domain). Although other distance functions can be employed (the symmetric Kullback–Leibler divergence or the Manhattan distance), for this first work we are employing the standard Euclidian distance (7)

$$d\left(X^n, R^m_j\right) = \sqrt{\sum_{i=1}^{N_s} \left(x^n_i - r^{m,j}_i\right)^2} \tag{7}$$

Then, it is defined a warping path $p = (p_1, p_2, \ldots, p_L)$ as a sequence of pairs (n_ℓ, m_ℓ) with $(n_\ell, m_\ell) \in [1, N_x] \times [1, N_m]$ and $\ell \in [1, L]$, satisfying three conditions: (i) the boundary condition, i.e. $p_1 = [1, 1]$ and $p_L = [N_x, N_m]$; (ii) the monotonicity condition, i.e. $n_1 \leq n_2 \leq \ldots \leq n_L$ and $m_1 \leq m_2 \leq \ldots \leq m_L$; and (iii) the step size condition, i.e. $p_\ell - p_{\ell-1} \in \{(1, 0), (0, 1), (1, 1)\}$ with $\ell \in [1, L-1]$.

Then, the total cost of a warping path p_i is calculated adding all the partial costs or distances (8). With all this, the distance between two data sequences R_i and X is defined as the cost (distance) of the optimum warping path p^* (9).

$$d_{p_i}\left(X, R_j\right) = \sum_{\ell=1}^{L} d\left(X^{n_\ell}, R^{m_\ell}_j\right) \tag{8}$$

$$d_{DTW}(X, R_j) = d_{p^*}(X, R_j)$$
$$= min\{d_{p_i}(X, R_j), being\ p_i\ a\ warping\ path\} \tag{9}$$

Finally, the simple movement recognized from the data sequence X is that whose pattern R_i has the smallest distance (is the closest) to X. The use of this definition is tolerant to speed variations in the movement execution, to the introduction of new micro-gestures, etc. Besides, as can be seen, when a different hardware technology is deployed, it is enough to update the patter repository R to reconfigure the entire pattern recognition solution (as no training is required).

3.2   Complex Action Recognition: Hidden Markov Models

Previously proposed mechanism is very useful to recognize simple actions, but complex activities involve a huge number of variables and require much more time. Thus, DTW tend to become imprecise, and probabilistic models are required. Among all existing models, HMM is the most adequate for industrial scenarios and human-driven processes.

From the previous phase, the universe of possible simple movements to be recognized is $\mathcal{M} = \{m_i, i = 1, \ldots, K\}$. Besides, it is defined a state universe $\mathcal{U} = \{u_i, i = 1, \ldots, Q\}$, describing all the states that people may cross while performing any of the actions under study.

Then, a set of observations $\mathcal{O} = \{o_i, i = 1, \ldots, Z_o\}$ (simple movements recognized in the previous phase) is also considered, as well as the sequence of states $V = \{o_i, i = 1, \ldots, Z_v\}$ describing the action to by modeled by HMM. In this initial case, we are assuming each observation corresponds to a new state, so $Z_v = Z_o$ Then, three matrices are calculated: (i) the transitory matrix A (10) describing the probability of state u_j following state u_i; (ii) the observation matrix (11) describing the probability of observation o_i caused by state u_j independently from k; and (iii) the initial probability matrix (12).

$$A = [a_{i,j}] \quad a_{i,j} = P(v_k = u_j \mid v_{k-1} = u_i) \tag{10}$$

$$B = [b_j(o_i)] \quad b_j(o_i) = P(x_k = o_i \mid v_k = u_j) \tag{11}$$

$$\Pi = [\pi_i] \qquad \pi_i = P(v_1 = u_i) \tag{12}$$

Then, the HMM for each complex activity λ_i to be recognized is described by these previous three elements (13).

$$\lambda_i = \{A_i,\ B_i,\ \Pi_i\} \tag{13}$$

Two assumptions are, besides, made: (i) the Markov assumption (14) showing that any state is only dependent on the previous one; and (ii) the independency assumption (15) stating that any observation sequence depends only on the present state not on previous states or observations.

$$P(v_k \mid v_1, \ldots, v_{k-1}) = P(v_k \mid v_{k-1}) \tag{14}$$

$$P(o_k \mid o_1, \ldots, o_{k-1}, v_1, \ldots, v_k) = P(o_k \mid v_k) \tag{15}$$

To evaluate the model and recognize the activity being performed by users, in this paper we are using a traditional approach (16). Although forward algorithms have been proved to be more efficient, for this initial work we are directly implementing the evaluation expression in its traditional form.

$$
\begin{aligned}
P(\mathcal{O} \mid \lambda) &= \sum_V P(\mathcal{O} \mid V, \lambda) P(V \mid \lambda) \\
&= \sum_V \left(\prod_{i=1}^{Z_o} P(o_i \mid v_i, \lambda) \right) \left(\pi_{v1} \cdot a_{v1v2} \cdot \ldots \cdot a_{v_{zv-1}v_{zv}} \right) \\
&= \sum_{v1,v2,\ldots,v_{zv}} \pi_{v1} \cdot b_{v1}(o_1) \cdot a_{v1v2} \cdot b_{v2}(o_2) \cdot \ldots \cdot a_{v_{zv-1}v_{zv}} \cdot b_{vzv}(o_{zo})
\end{aligned}
\tag{16}
$$

The learning process was also implemented in its simplest way. Statistical definitions were employed for transitory matrix, observation matrix and initial probability matrix. In particular, the Laplace definition of probability was employed to estimate these three matrices from statistics about the activities under study (17–19). The operator $count(\cdot)$ indicates the number of times an event occurs.

$$a_{i,j} = P(u_j \mid u_i) = \frac{count(u_j \ follows \ u_i)}{count(u_j)} \tag{17}$$

$$b_j(o_i) = P(o_i \mid u_j) = \frac{count(o_i \ is \ observed \ in \ the \ state \ u_j)}{count(u_j)} \tag{18}$$

$$\pi_i = P(v_1 = u_i) = \frac{count(v_1 = u_i)}{count(v_1)} \tag{19}$$

4   Experimental Validation: Implementation and Results

In order to evaluate the performance of the proposed solution, an experimental validation was designed and carried out. An industrial scenario was emulated in some large rooms in Universidad Politécnica de Madrid. The scenario represented a traditional company manufacturing handmade products. In particular, a small PCB (printed circuit boards) manufacturer was emulated.

In order to capture information about people behavior, participants were provided with a cybernetic glove, including accelerometers and a RFID reader [19]. Objects around the scenarios were identified with an RFID tag, so the proposed hardware platform may identify the hand position (gesture) and the objects people interact with.

A list of twelve different complex activities where defined and recognized using the proposed technology. Table 1 describes the twelve defined activities, including a brief description about them.

Table 1. Complex activities' description

Activity	Description
Draw the circuit's paths	The circuit to be printed is designed using a specific software PC program
Print the circuit design using a plotter	Using plastic sheet and a special printer called plotter, the circuit design is printed
Clean the copper-sided laminate of boards	Using a special product all dust and particles are removed from the cooper-sided laminate
Copy the circuit design on cooper boards	The circuit design in the plastic sheet is copied into the cooper laminate using a blast of UV light
Immerse the boards in the acid pool	To remove all unwanted cooper, the printed board is immersed in an acid bath
Wash off the cooper using a dissolvent bath	After the acid bath, the remaining cooper surface is washed off in a dissolvent bath
Layer alignment	PCB are composed of several layer; stacked and aligned during this phase
Optical inspection	Using a laser, the layer alignment is checked
Join outer layers with the substrate	Using an epoxy glue, the final and outer layer in the board are joined
Bond the board	The bonding occurs on a heavy steel table with metal clamps
Drill the required holes	Holes for components, etc., are bored into the stack board
Plating	In an oven, the board is finished

Eighteen people (18) were involved in the experiment. People were requested to perform the activities in a random number. The real order, as well as the order the activities are recognized were stored by a supervisory software process. The global success rate for the whole solution was evaluated, identifying (moreover), the same rate for each one of the existing phases.

In order to evaluate the obtained improvement in comparison to existing similar solutions, the same physical data sequences were employed to feed a standard pattern recognition solution based only on HMM. Using statistical data processing software, some relevant results are extracted.

Figure 3 represents the mean success rate for three cases: the global solution, the first phase (DTW), and the second phase (HMM). Besides, the success rate for the traditional HMM-based approach is also included. As can be seen, the proposed technology is, globally, around 9% better than traditional pattern recognition techniques based on HMM exclusively. Besides, first phase (based on DTW) is around 20% worse than the second phase (HMM) which is meaningful as Dynamic Time Warping techniques are weaker by default.

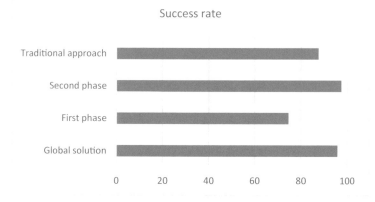

Fig. 3. Mean success rate for the proposed solution

5   Conclusions and Future Works

In this paper we present a new pattern recognition algorithm to integrate people in Industry 4.0 systems and human-driven processes. The algorithm defines complex activities as compositions of simple movements. Complex activities are recognized using Hidden Markov Models, and simple movements are recognized using Dynamic Time Warping. In order to enable the implementation of this algorithm in small embedded devices, lightweight configurations are selected. An experimental validation is also carried out, and results show a global improvement in the success rate around 9%.

Future works will consider most complex methodologies for data processing, and comparison for different configurations of the proposed algorithm will be evaluated. Besides, the proposal will be analyzed in different scenarios.

Acknowledgments. The research leading to these results has received funding from the Ministry of Economy and Competitiveness through SEMOLA (TEC2015-68284-R) project and the Ministry of Science, Innovation and Universities through VACADENA (RTC-2017-6031-2) project.

References

1. Bordel, B., Alcarria, R., Sánchez-de-Rivera, D., Robles, T.: Protecting industry 4.0 systems against the malicious effects of cyber-physical attacks. In: International Conference on Ubiquitous Computing and Ambient Intelligence, pp. 161–171. Springer, Cham (2017
2. Bordel, B., Alcarria, R., Robles, T., Martín, D.: Cyber–physical systems: extending pervasive sensing from control theory to the Internet of Things. Pervasive Mobile Comput. **40**, 156–184 (2017)
3. Neff, W.: Work and Human Behavior. Routledge, London (2017)
4. Bordel, B., Alcarria, R., Martín, D., Robles, T., de Rivera, D.S.: Self-configuration in humanized cyber-physical systems. J. Ambient Intell. Hum. Comput. **8**(4), 485–496 (2017)

5. Bordel, B., de Rivera, D.S., Sánchez-Picot, Á., Robles, T.: Physical processes control in industry 4.0-based systems: a focus on cyber-physical systems. In: Ubiquitous Computing and Ambient Intelligence, pp. 257–262. Springer, Cham (2016)

6. Pal, S.K., Wang, P.P.: Genetic Algorithms for Pattern Recognition. CRC Press, Boca Raton (2017)

7. Müller, M.: Dynamic time warping. In: Information Retrieval for Music and Motion, pp. 69–84 (2007)

8. Eddy, S.R.: Hidden Markov models. Current Opinion Struct. Biol. **6**(3), 361–365 (1996)

9. Kim, E., Helal, S., Cook, D.: Human activity recognition and pattern discovery. IEEE Pervasive Comput./IEEE Comput. Soc. [and] IEEE Commun. Soc. **9**(1), 48 (2010)

10. Li, Z., Wei, Z., Yue, Y., Wang, H., Jia, W., Burke, L.E., Sun, M.: An adaptive hidden markov model for activity recognition based on a wearable multi-sensor device. J. Med. Syst. **39**(5), 57 (2015)

11. Ordonez, F.J., Englebienne, G., De Toledo, P., Van Kasteren, T., Sanchis, A., Krose, B.: In-home activity recognition: Bayesian inference for hidden Markov models. IEEE Pervasive Comput. **13**(3), 67–75 (2014)

12. Zhan, K., Faux, S., Ramos, F.: Multi-scale conditional random fields for first-person activity recognition on elders and disabled patients. Pervasive Mob. Comput. **16**, 251–267 (2015)

13. Liu, A.A., Nie, W.Z., Su, Y.T., Ma, L., Hao, T., Yang, Z.X.: Coupled hidden conditional random fields for RGB-D human action recognition. Signal Process. **112**, 74–82 (2015)

14. Liu, J., Huang, M., Zhu, X.: Recognizing biomedical named entities using skip-chain conditional random fields. In: Proceedings of the 2010 Workshop on Biomedical Natural Language Processing, pp. 10–18. Association for Computational Linguistics (2010)

15. Gu, T., Wu, Z., Tao, X., Pung, H.K., Lu, J.: epSICAR: an emerging patterns based approach to sequential, interleaved and concurrent activity recognition. In: IEEE International Conference on Pervasive Computing and Communications, PerCom 2009, pp. 1–9. IEEE (2009)

16. Hu, B.G.: What are the differences between Bayesian classifiers and mutual-information classifiers? IEEE Trans. Neural Netw. Learn. Syst. **25**(2), 249–264 (2014)

17. Wang, X., Liu, X., Pedrycz, W., Zhang, L.: Fuzzy rule based decision trees. Pattern Recognit. **48**(1), 50–59 (2015)

18. Davis, M.H.: Markov Models & Optimization. Routledge, New York (2018)

19. Bordel Sánchez, B., Alcarria, R., Martín, D., Robles, T.: TF4SM: a framework for developing traceability solutions in small manufacturing companies. Sensors **15**(11), 29478–29510 (2015)

Kitchen Robots: The Importance and Impact of Technology on People's Quality of Life

Ema Fonseca[1], Inês Oliveira[1], Joana Lobo[1], Tânia Mota[1],
José Martins[2,4], and Manuel Au-Yong-Oliveira[1,3(✉)]

[1] Department of Economics, Management, Industrial Engineering and Tourism,
University of Aveiro, 3810-193 Aveiro, Portugal
{emafonseca,i.oliveira,joanalobo,taniamota,mao}@ua.pt
[2] INESC TEC, University of Trás-os-Montes e Alto Douro, Vila Real, Portugal
jmartins@utad.pt
[3] GOVCOPP, Aveiro, Portugal
[4] Polytechnic Institute of Bragança - EsACT, Mirandela, Portugal

Abstract. The interest in technology allied to household chores has been growing exponentially. Robots like Bimby have revolutionized the way of cooking, since they perform several functions, which were once done manually. How do users of kitchen robots see this continuous evolution and what is the impact on their routines? What are the main advantages associated with this technology and how do non-users see them? This study is a focus on the variables gender, quality of life and technological evolution, as a way to determine if women and men use kitchen robots on an equal scale, if the potentialities of these machines contribute to real improvements in the lives of their users and if, in a near future, this technology will replace the human element in the preparation of meals. To answer these questions, two methodological approaches were followed: quantitative (via questionnaires) and qualitative (via interviews and a focus group). The first approach allowed us to conclude on the profile of kitchen robots and their impact on people's quality of life. The second approach led us to understand the interest of suppliers, and whether the interest is to help human beings or to replace their role altogether in the kitchen. With this study we conclude that kitchen robots have effectively brought improvements in terms of time spent with household tasks, the typical user of this technology is indeed a woman and, finally, that it will be difficult for a robot to replace humans entirely, since anyone who really likes to cook will never stop doing it.

Keywords: Innovation · Domestic technologies · Kitchen robots · Gender ·
Technological evolution · Cooking

1 Introduction

Nowadays, relationships and interactions are highly dependent on existing technological mechanisms. Even people reluctant to adhere to technological evolution are confronted with it, for example, in a simple phone call. [7] argue that we - humans - love to develop electronic gadgets to help us with common tasks. Considering that food and meal preparation are essential and common issues in our routines, this function is increasingly facilitated by small home appliances.

© Springer Nature Switzerland AG 2019
Á. Rocha et al. (Eds.): WorldCIST'19 2019, AISC 931, pp. 186–197, 2019.
https://doi.org/10.1007/978-3-030-16184-2_19

Kitchen robots are just one example of a device created to help prepare meals and these have been gaining importance in Portuguese homes [2]. It is important to state that when a kitchen robot is mentioned, it refers to an equipment that has the autonomy and ability to prepare several pre-programmed meals, according to the information that the user provides, it only being necessary to put the ingredients inside the equipment. Based on this technology, this study aims to discover if this technology has improved quality of life in terms of time spent and resources necessary and if it is used in equal scale by the two genders, or rather whether women are the gender most associated with these domestic tasks [3]. Herein, we intend also to research if the concern of kitchen robot suppliers is to replace the human role in the kitchen on the one hand or, on the other hand, whether the objective is to facilitate the lives of its users. In this context, three research questions were defined:

1. *Has the technology associated with cooking (robots) improved the quality of life of families?*
2. *Are kitchen robots used on an equal scale by both genders?*
3. *Did robots revolutionize the traditional way of cooking? That is, is the focus of suppliers of this type of technology the development of technology capable of replacing the human in the kitchen, or is the goal just to facilitate the lives of its users?*

2 Kitchen Robots - Literature Review

This study focuses on a particular home appliance, namely multifunctional kitchen robots, which allow weighing, baking, grinding and even kneading [1]. According to [10] – a Portuguese brand of kitchen robots – this technology should allow the user to only need to program the time, speed and temperature in order to prepare a meal. The most known robot in the world is Bimby or Thermomix, created by Vorwerk. The first robot came to Portugal in 1971, and Vorwerk has been investing and developing new models over the years, always aiming to improve the performance of its robots while working also to increase the number of functionalities available. Like other techno-logical inventions, Bimby is not the only kitchen robot in the market. Several companies around the world have tried to conquer the market with a "copying" strategy, emulating existing machines, creating these few examples of multifunctional devices: Yämmi, Moulinex, Ladymaxx, Philips, KitchenAid, Evolution Mix, Kenwood, MyCook, Bosch, Cooksy, Thermochef Natura, Mamy Gourmet, Monsieur Cuisine and Chef Express. These robots appear in the market at a price considerably lower than that of Bimby, which still remains the most expensive robot in the market (at around 1,000€). Bimby is also the only robot adapted to the latest technological evolution, namely allowing for an Internet connection to thus make possible access to recipes [2], for example.

Focusing attention on home space and on the family, housework is increasingly facilitated by equipment that allows people to have more time to perform other tasks – such as to live life more, to its fullest, or simply providing for more time to rest. It is even argued that robots or equipment that have been developed over time, are vital in the execution of domestic tasks, especially for middle-class women, who have their jobs throughout the day, and who are always confronted with traditional household chores [3]. [3] also considers that the kitchen is a special space in the sense that it is the place where, especially women, work on the food themselves and, eventually, also to support other people. [4] even argues that the kitchen is designed by men and used by women.

The great advantage attributed to this type of multifunctional kitchen robots is that they can be used by all people. It is an inclusive technological invention, in the sense that it can be used by seniors, by people who do not understand the different cooking times, by single men and/or divorced people who do not know how to cook, by single mothers and even by people seriously concerned about healthy eating. However, [5] argue that men have a different view to that of women in relation to technological objects. Women are more receptive to innovations that help them in some way in social tasks, and men are more affectionate to objects that give them social status. The second comparative advantage of robots is the possibility of preparing diversified meals in a short time: there is "more time to do other things, rather than spending time in the kitchen waiting or even preparing, […] cooking incredible food at home has never been easier, leading to the ultimate new cooking experience" [2, p. 18]. In addition to diversified meals, it is also possible to prepare healthy foods, which could counter the massive consumption of ready-made food and fast food: "makes for easy and fast preparation of meals and encourages the taking up of healthy eating habits through the use of fresh and raw ingredients at the expense of convenience foods" [1, p. 45]. This author also argues that this type of robots save space, replacing numerous tools that would be necessary to produce each type of meal.

Finally, multifunctional kitchen robots can improve quality of life. However, this concept is very subjective and, in many cases, embracing. According to [8], quality of life can be defined as "the level of satisfaction and comfort that a person or group enjoys", realizing that this will differ from person to person – according to their ideals, tastes and goals. In agreement with [9], quality of life is "the standard of health, comfort, and happiness experienced by an individual or group". For [6, p. 59], quality of life is a more comprehensive term: "[it] is an individual's perception of their position in life in the context of the culture and value systems in which they live and in relation to their goals, expectations, values and concerns, incorporating physical health, psychological state, level of independence, social relations, personal beliefs and their relationship to salient features of the environment - quality of life refers to a subjective evaluation which is embedded in a cultural, social and environmental context" [12, 13]. In this sense, [3] and [2] argue that robots allow people to save time and resources, and this can be a measurable factor for quality of life. In this study quality of life will be considered as the possibility to save economic resources, to save time and even to improve people's living conditions, in the sense that multifunctional robots perform many tasks that once had to be performed manually. All in all, it is considered that the positive characteristics attributed to this type of robots have impacts in all segments of society. At the industrial level, it allows to cook a meal more efficiently. At the level of the health of users, it is possible to cook a healthy meal in a short period of time. And finally, economically, it allows for savings in the quantity of food used (since the different brands provide a wealth of detailed recipes) and in economic terms, it allows for savings in goods such as gas, electricity and water [1]. However, as with everything else, authors also attribute disadvantages to this type of machine, such as the loss of the traditional way of cooking and talent in producing a meal and lessening the contact with the handling of food: "[…] people have less physical contact with food due to technology mediation, they use less perceptive skills to deal with the aesthetics and sensorial elements of cooking practices" [1, p. 49]. Therefore, there is a loss of the need for senses such as touch, taste, and even the talent of conjugating them to produce a good meal.

In order to be able to answer the selected research questions and thus confirm or refute the theories presented in this chapter, the present study will explore the vision and reality of people who use robots and of those who do not, to understand if the literature corresponds to the reality in Portugal.

3 Methodology

Before defining the methodology of this study, a SWOT analysis (Table 1) was developed to understand the dynamics of kitchen robots and also to use it as support in the elaboration of the two types of analysis chosen to respond to the proposed research questions.

Table 1. A SWOT analysis of kitchen robots

Strengths	Weaknesses
• Product technology	• High price
• Assistent in the confection of meals	• People who do not like/work with technology, do not work with it
• Autonomy of the robot	• Possible loss of traditional taste
Opportunities	**Threats**
• Product not explored by the market yet	• Take away/cold meals
• Innovator product	• There may be other types of robots that replace these
	• There is no consumption pattern

Through the SWOT analysis, it becomes possible to perceive that kitchen robots have the greatest strength in the fact that they are a technological object, an ally in the confection of meals, in the sense that they perform numerous functions independently. Kitchen robots have as opportunities those presented in a vast market defined by the technological evolution, facing such threats as those that competitors offer namely cold meals, and the lack of a pattern of consumption.

The selected methodology intends to ascertain the current impact of kitchen robots on the quality of life of its users and the expectation, about this technology, of those who do not have it. The option of analyzing customers and "non-customers" from a comparative perspective is due to the greater likelihood of obtaining credible and unbiased data on current advantages or potentialities and possible interest in acquiring this product. After the research method was defined a study sample was chosen. In order to gather enough information to answer the research questions, two types of methodological approaches were chosen: a quantitative one - which aims to respond to the gender issue (who uses this robots the most?) and the impact on quality of life - and a qualitative one, which aims to respond to the revolution brought to the kitchen and to understand how suppliers see these products and the needs of their consumers; that is, that these products intend to be perceived as facilitators of daily activity, or as an increasingly technological product, capable of replacing the role of the human being.

It is important to note that the quantitative analysis was carried out through two surveys, with the aim of collecting demographic and socioeconomic data, as well as data involving perspectives on the use of this technology. The sample of this survey was composed of 300 respondents, of which 150 have a kitchen robot and 150 do not (the latter do not have any contact with this technology). Both surveys were distributed online, because it is considered that in this way, anonymity is guaranteed, the environmental impact is reduced and there is no influence on the respondent's behaviour. Regarding the age group of the selected sample, the respondents are at least 18 years old, and a maximum age was not stipulated. This choice is based on the fact that until the age of 18, there is usually still a great dependence on parents, which could lead to a bias in the data if these people answered that they have a kitchen robot, when in fact the users/owners were their parents. Both surveys were launched on October 8, 2018 (Monday) at 10:00 pm, and remained accessible until October 10, 2018 (Wednesday) at 10:00 pm, that is, they were available to receive responses for 48 h. The numerous responses received by women was not a purposeful choice of the group of authors. The questionnaires were distributed randomly on the internet. When the group started to collect responses, we realized that women were in fact the main users of kitchen robots, explaining the percentage of 95% of women who have the technology. The questionnaires were not delivered to more women than men. Women were the gender, however, which offered to respond, much more so than men.

Both questionnaires (for users and non-users) were distributed online – via the social network Facebook and the communication tool Messenger. The questionnaire of the users of kitchen robots was also made available in two Facebook groups that are frequented, mainly, by people who have the technology and who share experiences, recipes, etc.: "Recipes of Yämmi" (https://www.facebook.com/groups/706607436034 598/) and "Bimby, Without Limits" (https://www.facebook.com/groups/BimbySem Limites/). The group also sent the questionnaire to people who actually owned a robot, since before being shared, those interested in responding were questioned about whether or not they possess this technology.

The qualitative analysis was based on a focus group and two semi-structured interviews with clients, professional users and a robot supplier. The focus group was held on October 12, 2018, at 6:00 pm, in Porto, and lasted approximately for one hour. This action brought together in the same room four people who use kitchen robots to develop much of their work online (bloggers or digital influencers) or in their own homes: Dulce Salvador (client), Teresa Abreu (*Healthy Bites* blog, nominated for blogs of the year 2018 in the category of "Culinary"), Rui Ribeiro (*Faz e Come* blog) and a representative of the commercial department of a brand of kitchen robots: Patrícia Cayolla (Yämmi: leading market brand in Portugal) - Table 2. This action aimed to get experts on this technology to talk about the importance of these robots and also to complete with the perspective of a company, to see if what is really offered in the market is what the customer seeks most.

Table 2. Profile of the participants in the focus group

	Gender	Profession	Age
Dulce Salvador	Female	Senior Technician Communication and Image – client of kitchen robots	39 years
Teresa Abreu	Female	Naval Officer - *Healthy Bites* blog	34 years
Rui Ribeiro	Male	Trainer - *Faz e Come* blog	35 years
Patrícia Cayolla	Female	Commercial Department – Yämmi – a brand of kitchen robots	45 years

The two interviews were held on October 11 and on October 14 (2018), to reinforce the results of the focus group, but with the difference being that they were non face-to-face (and, rather, done at a distance) due to the impossibility of traveling and scheduling by the interviewees. This activity was also developed with a client and a digital influencer: Mariana Teixeira (client) and Sílvia Martins (*Bocadinhos de Açúcar* blog) - Table 3.

Table 3. Profile of the interviewees

	Gender	Profession	Age
Mariana Teixeira	Female	Housewife	47 years
Sílvia Martins	Female	Pharmaceutical and *Bocadinhos de Açúcar* blog	40 years

The methodology chosen did present some challenges, but it was important to gather all the information necessary for the subsequent analysis of the results.

4   Discussion of the Field Work

4.1   Quantitative Analysis

A data analysis was performed based on the 300 answers obtained (150 users and 150 non-robot users) in the survey.

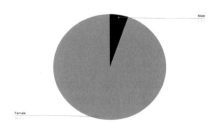

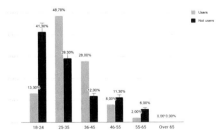

Fig. 1. Users divided by gender

Fig. 2. Age of users and non-users of kitchen robots

As can be seen in Fig. 1, about 95% of kitchen robot users are female, corresponding to most of the sample. This result confirms the paradigm initially exposed in the assumptions, that those who use this technology in the kitchen are women. Regarding the age range, Fig. 2 shows that about 49% of kitchen robot users are between 25 and 35 years of age, followed by 28% of users between the ages of 36 and 45 years. The age group above 65 years was the one that represented the lowest percentage of users of this technology, which can be justified by the distribution of online surveys, which were not as easy to access. Regarding socio-economic data, 56% of the users of kitchen robots in our sample have a bachelor's degree or higher, followed by 37% who have only completed high school. It is also important to note that 47% of the non-users in our sample also have a bachelor's degree or higher. So, it isn't possible to establish with this study a relationship between literacy and the use of kitchen robots. On the other hand, it can be concluded that it is people aged between 25–35 years and 36–45 years who have the most access to this technology in their homes because they have a more stable life and a higher salary, as well as being more receptive to new technologies. Younger people between 18 and 24 years of age have just entered the labor market and may not have the economic possibilities to acquire this technology yet. From the non-users' perspective, this same fact can be verified, since 41.3% of the people surveyed correspond to young people aged between 18 and 24 years. Concerning marital status, it was verified that the majority of people who use kitchen robots in their home are married or are living together in a union, reaching a percentage of about 75%, followed by 21% who are single. Therefore, it was noticed that people who do not live alone are those who use the kitchen robot more, and this can be explained by the fact that they have to cook in greater quantity, which sometimes makes the task of preparing meals more difficult.

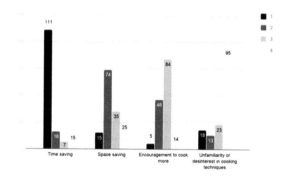

Fig. 3. Advantages of owning a kitchen robot, mentioned by users.

Robot users were asked to order from 1 to 4 (with 1 being the most important and 4 being the least) the advantages of owning such a device - Fig. 3. The main advantage attributed by the people surveyed was the possibility of saving time. The advantage pointed as the second most important is the saving of space, because the robot allows the replacement of several small appliances such as choppers and crushers. There were other advantages, such as the incentive to cook more often and the fact that it doesn't require much cooking practice or knowledge to prepare meals. Regarding the time that the robot saves in preparation, it is perceptible that 40% of the users in the sample believe that this technology allows to save between 1 to 3 h weekly. About 29% of the people surveyed consider that they save between 3 to 5 h a week and about 7% believe that they save more than 7 h a week. This shows that about 40% save, by using the robot every day of the week, between 8 and a half and 25 min a day in the preparation of their meals. For 7% of the users who consider that they save more than 7 h a week, it can be said that they save more than 1 h a day. It was possible to note that the meal for which users most use the robot is for soups, with a percentage of 45%, followed by 21% of the users who use it for steaming, and 16% to make desserts. Only 12% use this technology to cook side dishes and entrees. So, the most prepared meal by kitchen robots is soup, possibly because it is the meal that implies fewer procedures and, consequently, the one that allows to save more time.

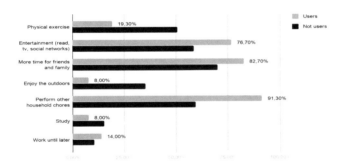

Fig. 4. Use of the time saved by the kitchen robot to dedicate themselves to these activities (by the users and non-users alike, considering they had this technology)

The question posed and visualized in Fig. 4 was one of the most important ones for this study, because it allowed us to associate the use of the robot with quality of life. It is important to mention that they were asked to choose three options about what they do or intend to do while the robot prepares meals. Therefore, when confronted with the question "use this free time for which of the following activities?", about 91% answered "perform other household chores", 83% answered "more time for friends and family" and 77% answered "entertainment". The activities that users accomplish less in the time they save are: "enjoying the outdoors", "studying", "physical exercise" and "work until later". This means that the kitchen robot allows them to save time that is mostly used to perform other tasks at home, but also to enjoy their time with family and friends and for their entertainment, which denotes an improvement in the quality of people's lives. In fact, 98% of the people surveyed considered that that happened. Regarding the 150 non-users, the same question was asked so that people could reflect on the advantages of having a robot. 70% felt that they would really benefit to spend more time with friends and family and 59.3% would do other tasks. It is noticeable that both agree with the main improvements in their routine.

The non-users have shown either that the main reason for not purchasing a kitchen robot is that they are not interested and do not feel it is a necessity, followed by no financial resources. Another important indicator for assessing quality of life is related to people's beliefs. 69% of the people who use this technology in their kitchen believe that it is more sustainable and that they have reduced food waste, which is beneficial at the ecological level. 17% of the respondents still think that the robot has a greater impact on gas savings because they do not have the need to use the stove so often.

4.2   Qualitative Analysis

In the qualitative analysis performed, the main ideas drawn up mention that kitchen robots should serve to facilitate the work of their users, be intuitive and easy to use, provide support in recipes that are difficult to produce and make the flavor come near to that of traditional cooking. In the brand's representative perspective (focus group participant), when users think about kitchen robots they associate them to trust, security, ease of use and good after-sales support service, so they should be seen as a complement to the client's life. Saving time and the fact that they could avoid other small appliances were also some of the advantages mentioned in the discussion. When questioned about the paradigm men versus women, participants were unanimous and responded that this technology is - and will continue to be - more used by women than by men, because women are the gender that spend more time on similar tasks (for example: shopping for cooking). They added in fact that man can even cook more often and be a bigger help to women, but society will not feel the change in reality, since it will not be the robot which puts the man in the kitchen, but the situation that leads him to need it, such as in the cases of living alone or enjoying the topic. This is confirmed by Fig. 2 relating to the quantitative analysis, where in a sample of 150 people, about 95% correspond to women. In terms of the potentialities of this technology, the interviewees mentioned that the technology allows to facilitate the routine in the kitchen (such as cutting, crushing, kneading and weighing), which is the most valued aspect. When considering a future perspective, the idea was common that those who like to cook will not want to be replaced by a robot, that is, the passivity of the user will only exist if there is not an interest in cooking. The kitchen is increasingly seen, according to the participants, as a time to escape the stress of everyday life, a therapy for those who like to spend time cooking. Meal confection requires creativity and the sharing with family or friends and this is not something that they can find in technology. Everyone has assumed that they are aware robots will effectively begin to possess many potentialities that never have been imagined before, but the human factor will always play a dominant role in the kitchen. Thus, the interest is not in having an independent robot, but rather in having a facilitator, since the most important thing is the active participation in the kitchen, both for those who like it or for those who want to learn. In Table 4 it is possible to see some of the main and synthesized ideas of the focus group and interviews.

Table 4. Main ideas of the focus group and interviews

	Focus group	Interviews
Technology advantages	Creation of digital communities for recipe sharing; preparing meals that were not prepared before the purchase or helping with a baby/children	Makes cooking easier while saving time
Buying decision	Weighted decision between the high cost of investment and the real need felt	Assessment of the advantages and disadvantages, without regret after the purchase
Males vs. females	Will always be more used by women than by men, because this is the reality associated with the tasks in the kitchen	It is an asset to both genders and men may feel more curious, but they will not play the prime role in the kitchen
Future perspective	There will actually be a technological evolution, but humans will never be a passive agent; the passion for cooking is growing, it is seen as an escape from the routine	Kitchen robots will evolve in their functions, but creativity will be the connecting point with humans; the robot should be a facilitator

5   Conclusions and Suggestions for Future Research

Nowadays people face innumerous tasks, associated with time management and high levels of stress. Therefore, there is a growing interest in technological gadgets that facilitate everyday activities. A simple meal is a necessity for which technology proves to be a powerful ally, in terms of time spent and simplification of the task.

In this article the main interest was to perceive the effective impact of kitchen robots on people's daily lives (starting with the task of producing a meal) and their extension to improved quality of life. Thus, in relation to the first proposed research question: "*Has kitchen technology (robots) improved the quality life of families?*", it was possible to perceive, by way of the quantitative analysis, that kitchen robots effectively brought improvements in terms of time spent with household tasks and with space that was occupied by other home appliances. As mentioned above, the concept of quality of life is associated with the level of satisfaction and comfort of people, and according to our study, our results show that robots can allow people to perform other domestic tasks or to simply spend more time together, since there is a real saving of time. Thus, as [3] and [2] argue, saving time and space may indeed be the major advantages of this technology. Our study also confirms that kitchen robots have improved quality of life from both an environmental and economic point of view. A small part of the sample admitted that they have started saving more gas and food, in line with the statements of [1]. It is also important to note the health issue. Most of the users surveyed use the robots to make soups and stews, which are healthy meals that can counteract the consumption of fast food and ready meals sold in big shopping malls. Considering the second research question "*Are kitchen robots used in equal scale by both genders?*", the bibliography used in the theoretical framework associates the use of household gadgets to women and, in our study, it is possible to confirm this – with the survey, the focus group and the interviews, that it is in fact women who most use kitchen robots. We recall also that in a sample of 150 users, 95% were female. We also conclude that the typical user of this technology is a woman, who is married or living with a companion, between the ages of 25 and 35 years. Regarding the third (and fourth) research questions *"Have robots revolutionized the traditional way of cooking? Should the focus of suppliers of this type of technology be the development of technology capable of replacing human beings, or rather should the focus be to facilitate the lives of its users?"*, the answers from the focus group and interviews are illustrative. Robots revolutionized the traditional way of cooking in the sense that they proved to be an ally when making a simple meal. All participants in the qualitative analysis agreed that it will be difficult for a robot to replace human beings, since anyone who really likes to cook will never stop doing it. [11] also provided arguments along the same line of thought of our research. The robot will inevitably evolve, but the main interest is to facilitate people's lives. Here it is also important to emphasize the strategies of the brands which sell kitchen robots studied herein, and we conclude that what is offered in the market is what the customer seeks most. Suppliers may have to improve (if the robots are too complex then seniors, for example, may not use them) or maintain their technology (if it is at an appropriate level), since the public of this technology seeks a "friend" and a support to facilitate the production of fast meals and spare the use of more utensils. Although technological evolution is a great opportunity for companies that want to stand out against their competitors, in this market, it is still necessary to evaluate the potential failures of others to understand where they can compensate recent users and captivate potential customers.

Since it is not possible to present in this article all the information collected in the questionnaires, focus group and interviews, due to the limitation of pages and time, it is important to mention that only a small part of the graphics which we possess have been used in the article. If there was more space, the study would reinforce variables such as the cost of kitchen robots, the importance of the price-quality ratio, the most important functionalities and possibilities (cut, grind, knead, digital weighing and balancing or the possibility to see step-by-step recipes) and the sense of fairness/injustice felt in the price paid for the robot – which is knowledge present in the data collected in the questionnaires. As suggestions for future studies, it would be interesting to explore the same variables in a country where women do not assume such a predominant role in the kitchen, as for example in northern Europe. A study in another country would also be interesting to see how people view the type of robots studied herein. It would be interesting to also analyze senior citizens, to understand if this group of people are averse to technology or if they see it as an ally.

Compliance with Ethical Standards

The present study would not have been possible without the intervention of all those involved. Therefore, we would like to thank the participants we surveyed, as well as the interviewees and the participants in the focus group, for volunteering their time to help us and for sharing some of their important knowledge with us.

References

1. Truninger, M.: Cooking with Bimby in a moment of recruitment: exploring conventions and practice perspectives. J. Consum. Cult. **11**, 37 (2011)
2. Reffóios, A.: Bimby, The Game Changing Innovative Technology. University Católica Portuguesa (2018)
3. Treusch, P.: Robotic companionship, the making of anthropomatic kitchen robots. Linköping Studies in Arts and Science, number 649, Linköping (2015)
4. Søraa, R.: Mechanical genders, how do humans gender robots? Genderm Technol. Dev. **21**(1–2), 99–115 (2017)
5. Badaloni, S., Perini, L.: The Influence of the Gender Dimension in Human-Robot Interaction. University of Padova (2017)
6. Susniene, D., Jurkauskas, A.: The Concepts of Quality of Life and Happiness – Correlation and Differences. Kaunas University of Technology, Panevezys Institute (2009)
7. Rusu, R., Gerkey, B., Beetz, M.: Robots in the kitchen: exploiting ubiquitous sensing and actuation. Robot. Auton. Syst. **56**, 844 (2008)
8. Dictionary of Cambridge. https://dictionary.cambridge.org/dictionary/english/quality-of-life. Accessed 16 Oct 2018
9. English Oxford Living Dictionaries. https://en.oxforddictionaries.co/definition/quality_of_life. Accessed 16 Oct 2018
10. Official Website of Yämmi. https://www.yammi.pt. Accessed 16 Oct 2018
11. Yamazaki, T.: Beyond the smart home. In: 2006 International Conference on Hybrid Information Technology, vol. 2, pp. 350–355 (2006)
12. Au-Yong-Oliveira, M., Gonçalves, R., Martins, J., Branco, F.: The social impact of technology on millennials and consequences for higher education and leadership. Telemat. Inform. **35**, 954–963 (2018)
13. Gonçalves, R., Martins, J., Rocha, Á.: Internet e redes sociais como instrumentos potenciadores de negócio. RISTI-Revista Ibérica de Sistemas e Tecnologias de Informação 09–11 (2016)

Vehicle-Pedestrian Interaction in SUMO and Unity3D

Leyre Artal-Villa[1] and Cristina Olaverri-Monreal[2]($\boxtimes$) ⓘ

[1] Department Electrical and Electronic Engineering,
Public University of Navarra, Campus de Arrosadía s/n, 31006 Pamplona, Spain
leireartalvilla@gmail.com
[2] Chair for Sustainable Transport Logistics 4.0,
Johannes Kepler University Linz, Altenberger Straße 69, 4040 Linz, Austria
cristina.olaverri-monreal@jku.at

Abstract. Road fatalities that involve Vulnerable Road Users (VRU) outnumber in some countries and regions ones that involve vehicular drivers and passengers. As most vulnerable road user fatalities happen in urban areas, where the traffic conditions are more demanding and an increased pedestrian interaction can result in unpredictable scenarios, it is imperative to study solutions to reduce the high rate of accidents in which pedestrians are involved. To this end, we present in this paper a simulation framework that provides a framework to generate a variety of pedestrian demands to simulate vehicle-pedestrian interaction and vice versa. A Transmission Control Protocol (TCP) connection combines the game engine Unity 3D with the Simulation of Urban Mobility (SUMO) open source traffic simulator. After creating the 2D scenario SUMO was connected with Unity 3D by using the Traffic Control Interface (TraCI) Protocol and TraCI as a Service (TraaS) library. The motion in Unity took place after instantiating the pedestrians retrieved from SUMO. The system was evaluated by detecting and visualizing pedestrians and vehicles that were within a specific range.

Keywords: VRU · P2V · V2P · SUMO · Unity 3D

1 Introduction

According to the World Health Organization, 85 000 people die annually from road traffic injuries in the European Region [1]. Urban areas have more complex intersections and a higher number of pedestrians and cyclists that might be overseen by drivers [2], particularly if they are involved in non-driving tasks. Due to the increased complexity in urban road and traffic scenarios, vulnerable road users (VRUs) are involved in more fatalities.

Pedestrians are especially vulnerable as road users because an armored vehicle does not protect them, nor do they wear any protective helmets [3]. Addressing the risk of death in road traffic is fundamental to achieve the Sustainable Development Goals (SDGs), by targeting efficient transportation services that do not affect health security. Therefore, it is essential to reduce mortality and injuries derived from road crashes [4]. In this context, P2V (Pedestrian-to-Vehicle) and V2P (Vehicle-to-Pedestrian) communication technologies have become crucial.

© Springer Nature Switzerland AG 2019
Á. Rocha et al. (Eds.): WorldCIST'19 2019, AISC 931, pp. 198–207, 2019.
https://doi.org/10.1007/978-3-030-16184-2_20

We present a platform to evaluate P2V and a V2P communication in a simulated 3D environment that includes the microscopic modeling of vehicles and pedestrians relying on the recent release of the last version of Simulation of Urban Mobility (SUMO) and TraaS (TraCI as a Service).

Next section outlines related literature in the field. Section 3 delineates the proposed work and describes the pedestrian modeling, creation of the scenario in SUMO and the connection of SUMO with Unity 3D. Section 4 describes the 3D simulation in Unity 3D and Sect. 5 presents the process to evaluate the implemented system. Finally, Sect. 6 concludes the work.

2 Related Literature

Communication targeting VRU protection addresses the use of smart devices to send Personal Safety Messages (PSM), over a wireless communication channel. This approach is also known as General Packet Radio Service (GPRS)-based, since the smart devices use the latitude and longitude coordinates, which are then transformed into local coordinates to estimate the relative position of the communicating pedestrians and vehicles [5]. Vehicle-to-Pedestrian (V2P) and Pedestrian-to-Vehicle (P2V) communication or a combination of them rely on GPRS.

Pedestrian detection has been the focus of research in many works. For example, by using AdaBoost and support vector machine algorithms [6] or by detecting and tracking pedestrians using cameras, (i.e. from a moving vehicle using both Histogram of Oriented Gradients (HOG) and Kalman filter [7] or by using the back-camera of a mobile device using image-processing techniques [8, 9]).

By using both P2V and V2P communication, the authors in [10] developed an application based on a collision-prediction algorithm. The proposed application broadcast the device's position to the vehicles nearby, and reciprocally broadcasts the vehicular position to the pedestrians nearby.

As far as simulation tools are concerned, PARAMICS, VISSIM, AIMSUM and SUMO stand among the most recognized simulators, which use microscopic models and allow the inclusion of pedestrian flow in the simulation [11]. For example, PARAMICS has its own software to simulate pedestrian behavior in real word environments. VISSIM allows making a 3D simulation with pedestrians, but it fails in the calibration process of certain parameters and it is a difficult pro-gram to handle [12]. AIMSUM enables pedestrian-vehicle interactions at uncontrolled, actuated-controlled or fixed-controlled intersections, but it is not convenient for navigating between different periods and it does not provide background maps [13].

SUMO is an open source, highly portable, microscopic and continuous road traffic simulation package designed to handle large road networks [18]. The recent release from December 2017 of the last version (0.32.0) together with the last version of TraaS from August 2017 reveal the possibility of performing a remote-controlled realistic 3D traffic scenario, which includes pedestrians. The aforementioned version of SUMO implements Traffic Control Interface (TraCI) which provides the necessary commands for both remotely retrieving and changing the state of pedestrian objects.

By means of TraaS library and the above mentioned methods, authors in [14] retrieved the information (vertices, length, width, type) of lanes and cars (speed, position, angle) from SUMO and instantiated them as game objects in Unity 3D. Authors in [15, 16] reused the translated library and connected the traffic light system data from SUMO into Unity 3D.

As the integration of pedestrians in current implemented simulators add realism to the scenarios and use cases, we focus on developing a 3D visualization of traffic, which includes both pedestrians and vehicles from a driver centric perspective simultaneously.

3   Framework Implementation

Relying on the work presented in [10] we implemented a scenario in order to establish connections between vehicles and pedestrians and vice versa as part of the SUMO OSM environment, an option in the CoAutoSim3D simulation platform [17]. The CoAutoSim3D simulator is independent from the operating system and always up-to-date with the latest version of Unity 3D. When running 3DCoAutoSim the user can choose among several environments. One of them is the SUMO OSM environment: a dynamic and configurable environment by the user. It allows for one to take advantage of the benefits provided by a microscopic 2D traffic-modeling tool (SUMO) and a powerful game-engine (Unity 3D). In fact, due to the flexibility provided by this option, the user can perform a 3D realistic simulation of any area of interest. Furthermore, the 3D reconstructed map presented in [14] was used to maintain the high quality visualization provided by the implemented environment.

When the user selects the SUMO OSM environment, the simulator accesses a SUMO configuration file (.cfg or .SUMOcfg) that contains all the necessary information to conduct the simulation in SUMO and performs the Transmission Control Protocol (TCP) connection to SUMO. SUMO acts as a server and Unity 3D as a client. The simulation can be performed by loading the configuration file in Unity 3D. To implement the 3D simulation with the pedestrian-detection system we proceeded as follows.

3.1   Pedestrians Modeling in SUMO

Pedestrians in SUMO need dedicated lanes and areas, which differ from the rest of the elements of a scenario in terms of attributes that determine their behavior.

For example, when walking along an edge, pedestrians use sidewalks if they are available. With respect to zebra crossings, there are two possible cases:

- If the network contains walking areas, pedestrians may only cross a street whenever there is a pedestrian crossing.
- However, if the network does not include walking areas, pedestrians will move between any two edges that allow pedestrians at an intersection.

Pedestrian crossing behavior does not only depend on the type of roads, but also on the vehicles. SUMO contains rules to mimic ideal traffic scenarios by targeting the avoidance of collisions between vehicles and pedestrians. Therefore, pedestrians will only use a crossing if the whole length of the crossing is free of vehicles for the whole time needed to cross. This behavior is not expected and not occurring in real life. If a vehicle occupies the whole width of the lane and gets too close to a pedestrian, the pedestrian may briefly move to the side of the lane in order to let the vehicle pass.

The last version of SUMO provides a framework to generate a variety of pedestrian demands both in an explicit and random fashion, defined as follows:

- Explicit: manually defining the pedestrian movement in an .xml file.
- Random: using the tool `randomTrips.py` with the option —pedestrian --pedestrian that supports generating random pedestrian demand.

3.2   Simulation in SUMO: 2D Scenario

The approach taken for explaining the generation of the 2D scenario consists of four steps:

- Generation of a realistic network from OSM data
- Edition of the network with `netedit` and `netconvert`
- Generation of the vehicle routes and the pedestrian demand
- Executing a simulation with pedestrians in SUMO-GUI

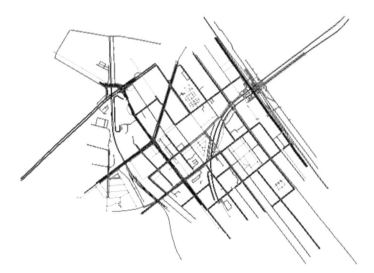

Fig. 1. .net file obtained from OSM WebWizard before editing.

In order to have a realistic scenario, a network was imported from Open Street Map (OSM). Since the conversion process included some imperfections that made the simulation differ from reality, the network was edited with `netedit`, which is a Graphical User Interface (GUI) application for editing traffic networks. It can be used to create networks from scratch and to modify all aspects of existing networks. Figure 1 shows the `.net` file obtained from the OSM WebWizard before editing. Figure 2 represents an example of pedestrian topology in `netedit` before and after edition.

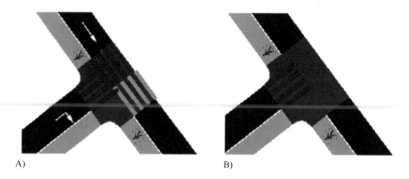

A)   B)

Fig. 2. Pedestrian topology in `netedit` before (A) and after (B) edition.

`Netconvert` is a network generator that provides SUMO-format networks. After obtaining the network file (`.net` file) and generating both the vehicle routes and the pedestrian demand, the simulation was executed in SUMO-GUI, which is the same application as SUMO, just extended by a graphical user interface.

3.3   Connecting SUMO with Unity 3D: TraCI Protocol and TraaS Library

As previously mentioned 3DCoAutoSim connects both a microscopic traffic simulator, SUMO, and a powerful 3D Graphic Engine, Unity 3D, by setting a TCP connection between them. TraCI is the Application Programming Interface (API) used for that purpose. It is based on a client/server architecture and it enables the retrieval of values from simulated objects from the server (SUMO) to the client (Unity 3D) and their behavior manipulation on-line, as well as to execute the simulation from the client side.

A TraaS library with a class for pedestrians that contained the necessary TraCI methods and which could be imported in Unity 3D was required for retrieving and changing information about pedestrian objects between SUMO and Unity 3D without changing the internal logics and structure of the already implemented and evaluated 3DCoAutoSim simulator. The implemented class made it possible to use the available commands to retrieve the values of each person from TraCI.

4   Simulation in Unity

The 3D simulation contains buildings created with the 3D city editing and visualization tool CityEngine. In addition it consist of the following objects that were retrieved from SUMO: car lanes, vehicles, sidewalks, pedestrian crossings, pedestrians, manually placed traffic lights and road markings, a user-controlled vehicle, a bidirectional communication system between cars and pedestrians using P2V and V2P and as many GPS as the number of pedestrians in the simulation. The motion of the pedestrians in the simulation occurred after their instantiation as follows.

The SUMO-retrieved-objects are instantiated in Unity 3D as Game-Objects. In order to have each SUMO pedestrian correctly classified in Unity, a class Pedestrian is created, which has the following attributes: position, id, speed and angle. The class Pedestrian also contains some getters that allow to get both X and Z coordinates of the position of each pedestrian (getX() and getZ()), the angle (getAngle ()), the speed (getSpeed()) and the ID (getId()).

When the connection between both programs has been established and all the lanes retrieved from SUMO have been read, the function ReadPedestrians() is called (every 5 frames in Unity in order to match the SUMO timestamp), which is in charge of creating a list of pedestrians that will be further printed and moved in the environment (in PrintPedestrians()). The code below depicts which parameters are received from SUMO and how information of every pedestrian in the Unity list is updated each time-step.

Thus, every time there is a change in SUMO, the list is updated and filled with the pedestrians existing in that time-step. The position, the ID, the speed and the angle for every pedestrian is retrieved from SUMO using TraCi commands contained in the so-called "person value retrieval".

Then, a pedestrian-object is created in Unity 3D for each simulated pedestrian in SUMO and its attributes position, id, speed and angle are set to the SUMO 2D position, the SUMO id, the SUMO speed and the SUMO 2D angle (in degrees, with respect to the Y axes) of every simulated pedestrian respectively. The created pedestrian object is added to the list of pedestrians in the last position. It should be noted that until now, nothing has been displayed in Unity 3D, as the information about the simulated pedestrians is in Unity, but it cannot be seen in the game view or in the scene view until all the pedestrians from the list are printed in Unity and moved in PrintPedestrians().

```
Function ReadPedestrians() SUMOmanager.cs
//-Reading the retrieved pedestrians from SUMO-
public void ReadPedestrians()
{
//Start with an empty list every SUMO time-step
pedlist.Clear();
//Read the ID-s of the pedestrian in the current time-
step
pedIDs = (SumoStringList)sumoTraciConnection.do_job_get
(Person.getIDList());
//For every pedestrian in the list
foreach (string perid in pedIDs)
{
//--Retrieving information from SUMO-
//SUMO 2D position of this pedestrian
positionperson =
(SumoPosition2D)sumoTraciConnection.do_job_get
(Person.getPosition(perid));
//SUMO speed of this pedestrian
speedperson=
((java.lang.Double)sumoTraciConnection.do_job_get
(Person.getSpeed(perid))).doubleValue();
//SUMO angle of this pedestrian
angleperson =
((java.lang.Double)sumoTraciConnection.do_job_get
(Person.getAngle(perid))).doubleValue();
//Create the pedestrian object with the SUMO information
of this pedestrian
pers = new Pedestrian(positionperson, perid, speedperson,
angleperson);
//add this pedestrian object to the list
pedlist.Add(pers);
}
//Create the pedestrian game-object in Unity 3D
// if it has not been previously created or update
// the game-object if it has been previously created
PrintPedestrians();
//Perform another time-step in SUMO
sumoTraciConnection.do_timestep();
}
```

5 System Evaluation Procedure

To evaluate the framework we implemented a bidirectional communication between vehicles and pedestrians and detected whether they were within the specific range of 40 m.

To visualize that the V2P communication had been achieved in the simulated 3D environment, the pedestrians were color-coded, pedestrians' detection (P2V communication) by the vehicle was visualized through an object-bounding box. Figure 3 shows an example for the pedestrians' detection visualization.

The list of all the simulated pedestrians was iterated and an array of renderers was created for every person-object. A renderer is what makes an object appear on the screen. In this case, renderers were used to access and modify the person-objects. The array stored all renderers of each person-object as a component in its children elements. Then, the difference between the position of each pedestrian and the position of the user-controlled car was obtained. After that, the list of all the simulated cars was iterated.

For each pedestrian i, a subtraction was calculated and vice versa between the position of that pedestrian and the position of each simulated car. At this point, if the distance of pedestrian i and any simulated car or the distance of the pedestrian i and the user-controlled car was less than 40 m, the color of the pedestrian object i, which represents the pedestrian with id = i in SUMO, was changed in the 3D environment from the initial color (white) to red. If the distance was more than 40 m and the i_{th} pedestrian had been previously red colored, the pedestrian object was reversed to white. This procedure helped to represent the pedestrians that were aware of the presence of vehicles in the surroundings.

Fig. 3. Example of pedestrians' detection by using an object-bounding box.

To establish a P2V communication, a pedestrian list was iterated. Since the function OnGUI() is called every frame, the variable pedlist contained all the pedestrians that existed at a given time in SUMO and therefore all the pedestrians that had been instantiated as Game-Objects in Unity. For every pedestrian the difference between its position and the position of the user-controlled vehicle was obtained by subtracting both 3D positions and computing the norm of the three dimensional vector obtained because of the mentioned subtraction.

The above describe strategy has been proved to be effective to establish the bidirectional communication between the pedestrians and the vehicles in the surroundings.

6 Conclusion and Future Work

We proposed in this work a system to detect VRU and vehicles in the vicinity as a means to enhance road safety. The system has the potential to assist both drivers and pedestrians in a variety of situations in which technology used by both parties is evaluated. The implementation of a framework that relies on the linking through a TCP connection of the game engine (Unity 3D) to access data from the SUMO open source traffic simulator has been proved to be capable of simulating vehicle-pedestrian and pedestrian-vehicle interaction. Future work will be pursued by validating the simulation framework in a variety of use cases.

Acknowledgments. This work was supported by the BMVIT endowed Professorship Sustainable Transport Logistics 4.0 and the Erasmus Program, code A WIEN 20.

References

1. Jackish, J., Sethi, D., Mitis, M., Szymański, T., Arra, I.: European facts and the Global status report on road safety 2015. WHO Regional Office for Europe, Copenhagen (2015). http://www.euro.who.int/__data/assets/pdf_file/0006/293082/European-facts-Global-Status-Report-road-safety-en.pdf?ua=1. Accessed 13 Oct 2018
2. BrainonBoard.ca Vulnerable Road Users: Pedestrians and Cyclists. http://brainonboard.ca/program_resources/VulnerableRoadUsersPedestriansandCyclists_Fact_Sheet_Eng_4.pdf. Accessed 13 Oct 2018
3. Olaverri-Monreal, C., Pichler, M., Krizek, G.C., Naumann, S.: Shadow as route quality parameter in a pedestrian-tailored mobile application. IEEE Intell. Transp. Syst. Mag. **8**(4), 15–27 (2016)
4. World Health Organization Europe Road Safety: Fact sheets on sustainable development goals: health targets. http://www.euro.who.int/__data/assets/pdf_file/0003/351444/3.6-Factsheet-SDG-Road-safety-FINAL-10-10-2017.pdf?ua=1. Accessed 2 Oct 2018
5. Rostami, A., Cheng, B., Lu, H., Gruteser, M., Kenney, J.B.: Reducing unnecessary pedestrian-to-vehicle transmissions using a contextual policy. In: Proceedings of the 2nd ACM International Workshop on Smart, Autonomous, and Connected Vehicular Systems and Services - CarSys 2017, pp. 3–10 (2017)
6. Guo, L., Ge, P.S., Zhang, M.H., Li, L.H., Zhao, Y.B.: Pedestrian detection for intelligent transportation systems combining AdaBoost algorithm and support vector machine. Expert Syst. Appl. **39**(4), 4274–4286 (2012)

7. Nkosi, M.P., Hancke, G.P., dos Santos, R.M.A.: Autonomous pedestrian detection. In: AFRICON IEEE (2015). https://doi.org/10.1109/AFRCON.2015.7332014
8. Allamehzadeh, A., Olaverri-Monreal, C.: Automatic and manual driving paradigms: cost-efficient mobile application for the assessment of driver inattentiveness and detection of road conditions. In: IEEE Intelligent Vehicles Symposium Proceedings, pp. 26–31 (2016)
9. Allamehzadeh, A., Urdiales de la Parra, J., Garcia, F., Hussein, A., Olaverri-Monreal, C.: Cost-efficient driver state and road conditions monitoring system for conditional automation. In: Proceedings IEEE Intelligent Vehicles Symposium, Los Angeles, USA, pp. 1497–1502 (2017)
10. Hussein, A., García, F., Armingol, J.M., Olaverri-Monreal, C.: P2V and V2P communication for pedestrian warning on the basis of autonomous vehicles. In: IEEE Proceedings Intelligent Transportation Systems Conference (ITSC), pp. 2034–2039 (2016)
11. Kokkinogenis, Z., Sanchez Passos, L., Rossetti, R., Gabriel, J.: Towards the next-generation traffic simulation tools: a first evaluation. In: 6th Iberian Conference on Information Systems and Technologies, pp. 15–18 (2011)
12. Doina, K.S.Y., Chin, H.C.: Traffic Simulation Modelling: VISSIM. https://docplayer.net/9916680-Traffic-simulation-modeling-vissim-koh-s-y-doina-1-and-chin-h-c-2.html. Accessed 14 Nov 2018
13. Salgado, D., Jolovic, D., Martin, P.T., Aldrete, R.M.: Traffic microsimulation models assessment - a case study of international land port of entry. Procedia Comput. Sci. **83**, 441–448 (2016)
14. Biurrun-Quel, C., Serrano-Arriezu, L., Olaverri-Monreal, C.: Microscopic driver-centric simulator: linking unity 3D and SUMO. In: Rocha, Á., Correia, A., Adeli, H., Reis, L., Costanzo, S. (eds.) Recent Advances in Information Systems and Technologies. AISC, vol. 569, pp. 851–860. Springer, Cham (2017)
15. Olaverri-Monreal, C., Errea-Moreno, J., Díaz-Álvarez, A.: Implementation and evaluation of a traffic light assistance system in a simulation framework based on V2I communication. J. Adv. Transp. **2018**, 11 (2018). https://doi.org/10.1155/2018/3785957. Article ID 3785957
16. Olaverri-Monreal, C., Errea-Moreno, J., Díaz-Álvarez, A., Biurrun-Quel, C., Serrano-Arriezu, L., Kuba, M.: Connection of the SUMO microscopic traffic simulator and the unity 3D graphic engine to evaluate V2X communication-based systems. Sens. J. **18**(12), 439 (2018). https://doi.org/10.3390/s18124399
17. Hussein, A., Diaz-Alvarez, A., Armingol, J.M., Olaverri-Monreal, C.: 3DCoAutoSim: simulator for cooperative ADAS and automated vehicles. In: Proceedings 21st International IEEE Conference on Intelligent Transportation Systems, ITSC2018, Hawaii, pp. 3014–3019 (2018)
18. SUMO - Simulation of Urban Mobility. http://sumo.dlr.de/index.html. Accessed 2 Dec 2018

Electronic System for the Detection of Chicken Eggs Suitable for Incubation Through Image Processing

Ángel Fernández-S.[1], Franklin Salazar-L.[1] , Marco Jurado[1],
Esteban X. Castellanos[2], Rodrigo Moreno-P.[3],
and Jorge Buele[1,2(✉)]

[1] Universidad Técnica de Ambato, Ambato 180103, Ecuador
{afernandez8448, fw.salazar, marcoajurado,
jl.buele}@uta.edu.ec
[2] Universidad de las Fuerzas Armadas ESPE, Latacunga 050104, Ecuador
excastellanos@espe.edu.ec
[3] Escuela Superior Politécnica de Chimborazo, Riobamba 060155, Ecuador
rodrigo.moreno@espoch.edu.ec

Abstract. In order to increase quality control in products and services provided by the poultry industry, the automation of its processes is required. Therefore, this research work presents a system for identifying chicken eggs suitable for incubation, using an image processing algorithm. The design of the prototype is based on a metal bucket with a soft-coated surface, where the units are housed safety. This bucket is located in a structural base, which in its lower part has a LED matrix to position the eggs by transillumination. In the upper part, a mechanical system has been implemented that allows a digital camera to be slide to perform the respective analysis. This system can be modified, since it has been developed using hardware and free software and is a low-cost proposal for the use of Raspberry Pi 3, as a central control unit. In this embedded board, web servers and a database have been implemented, for the storage of the obtained data. To validate this work, the respective experimental tests have been developed in manual mode (controlled by the user) and automatic mode.

Keywords: Image processing · Embedded device · Open source software · Servers · Quality management

1 Introduction

The agricultural industry has been responsible for the commercial and economic growth of societies throughout history [1–3]. Thus, using automation was able to modernize the processes, performing them in a shorter time and with a lower investment [4–7]. Latin America has become a benchmark in terms of export and domestic consumption of animal meat and its derivatives [8]. Local population prefers the consumption of chicken, pork and beef, where more than 50% choose for intake the poultry products [9]. In Ecuador, the consumption of chickens per person registers an annual value of 35 kg, it means 14 birds each person. According to data from the

© Springer Nature Switzerland AG 2019
Á. Rocha et al. (Eds.): WorldCIST'19 2019, AISC 931, pp. 208–218, 2019.
https://doi.org/10.1007/978-3-030-16184-2_21

National Institute of Statistics and Census (INEC), the annual consumption per family is 50'692,246.5 and the weekly production of chicken eggs is 55,427,095 units. Therefore, 91.84% of national production goes to sale, 8.04% to incubation and 0.12% to self-consumption [10].

To develop this research, a preliminary study was made of some related works to this topic. The work proposed in [11] is based in a system to detect defects in poultry eggs. The prototype consists of a transport module, an image processing and texture collection module and a human-machine interaction module (HCI). Through the experimental tests it can be identified cracks and spots, with a high rate of accuracy. In [12] an automatic method of quality control which can be established and customized by the user is presented. By the analysis of the image characteristics of the contour of a chicken egg it can be determined which are suitable for human consumption.

In this context, after making an analysis of the current situation of management and quality control of these products in local industries, the need to automate these processes have been raised [13–15]. In addition to evidencing the use of image processing for the detection of low-quality products in the manufacturing area, as in [16] where it is used to identify cracks and other faults in tire styling. This work is a contribution to the poultry sector, to present an automated low-cost system with hardware architectures and free software. The prototype contributes to the fulfillment of national and international standards of companies dedicated to the production, from the incubation, of chickens. This way it can be verified the internal and external physical conditions of chickens, using a special lighting system and acquisition of images made by a digital camera. In addition, a user-friendly interface is incorporated that allows the consumer to choose whether the process will be manual or automatic. The data obtained is stored in a database, available through web servers that allow easy access. Figure 1, describes the general diagram of the proposed system.

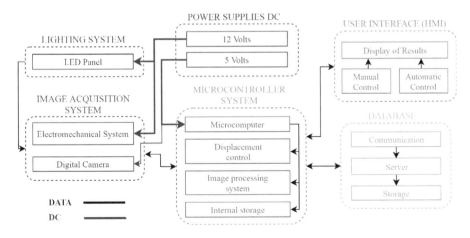

Fig. 1. General diagram of the proposed system.

Therefore, this paper is organized as follows: the introduction is shown in Sect. 1, in Sect. 2 the hardware elements are exposed. In Sect. 3, the software elements are described and in Sect. 4 the setup of necessary parameters for the prototype implementation. Experimental tests are performed and showed the results in Sect. 5. Finally, conclusions are described in Sect. 6.

2   Hardware

Figure 2 shows up the electronic system of this prototype, whose central control unit is the Raspberry Pi 3 Model B. It consists of two stages that are presented below:

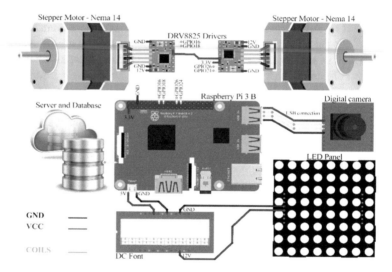

Fig. 2. Schematic diagram of the prototype electronic system.

2.1   Lower Structural Design

This design is made based on a metal bucket capable of housing 36 chicken eggs, it has a soft-coated surface to prevent damage to them. This bucket can be easily inserted and removed from a base in the shape of a rectangular prism, which at the bottom has a light matrix for the eggs position by transillumination. Table 1 details the technical characteristics of light bulbs for lighting. The design does not include a system to flip the eggs since the process could break them due to eggs fragility.

In the lighting system a panel of 216 bulbs is used. It has been designed to concentrate the light energy in each of the units to be analyzed. LED technology has been selected for its low cost, longer life, low energy consumption and low thermal emission, avoiding alterations in the incubation process. It is connected directly to a 12 V source, obtaining the maximum power of its elements.

Table 1. Characteristics of several devices for illumination.

	Incandescent bulb	Halogen bulb	Compact fluorescent lamp	LED bulb
Power (watts)	60	43	14	13
Useful life (years)	1	1–3	6–10	15–25
Heat emission	High	High	Middle	Low
Approximate cost (USD)	2	7,5	3	4

2.2   Superior Structural Design

Through the GPIO pins of the Raspberry Pi 3 board, the control actions corresponding to the DRV8825 drivers are issued. By means of these, the movement of two Nema 14 DC stepper motors is controlled. The amount of micropasses that the motors must provide with a full-pass configuration is controlled in order to locate the mechanism in the established positions. A Genius FaceCam 1000x digital camera is adapted to this structure, which moves in 2 dimensions (x, y) automatically. The distance of the camera on the bucket (z axis), is established manually by holding this mechanism to the fixed base using some nuts. This structure is shown in Fig. 3.

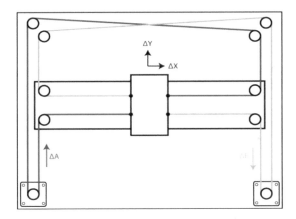

Fig. 3. Mechanic design of the CoreXY system (Top view).

3   Software

3.1   Image Processing

The previously described camera performs the image acquisition, for further processing on the Raspberry Pi 3 B embedded board. The Raspbian operating system is installed in it and, for the development of image processing algorithms the Python language is used. Based on some methods of the OpenCV library, the manipulation of visual information is carried out under the following guidelines:

Gray Scale Conversion. Using the cvtColor() method which receives the image to be converted and the color conversion code as parameters, an image with the same grayscale dimensions is obtained. To determine the gray scale value, each pixel is analyzed, establishing an output matrix (GS) of dimensions UxV that corresponds to the image number of pixels. Each pixel value that is stored in the matrix is calculated with the expressed formula in (1).

$$GS_{xy} = \frac{R_{xy} + G_{xy} + B_{xy}}{3} \tag{1}$$

Where the variables x, y fixes the pixel position, R is the red color value of the pixel, G the green color value and B the blue color value. The sum of these quantities is divided for 3, in order to determine a single value for the resulting matrix in each position.

Noise Reduction by Gaussian Filter: After obtaining the grayscale image, the Gaussian filter which is the most robust for image analysis is applied. Through the GaussianBlur() method, which receives as parameters the gray scale image, kernel size and the method of pixel extrapolation. When applying the Gaussian filter, a new matrix (GF) is obtained based on the standard deviation (σ), as shown in (2).

$$GF_{xy} = \frac{1}{2\pi\sigma^2} e^{\frac{x^2 + y^2}{2\sigma^2}} \tag{2}$$

Contours Detection: The method findContours() is used, it receives the filtered or binary image, the hierarchy or relationship of contours and the approach method as parameters. The presence of discontinuities between pixels is determined by applying the gradient where a two-dimensional function is obtained, as evidenced in (3).

$$G[GF_{xy}] = \begin{bmatrix} G_x \\ G_y \end{bmatrix} = \begin{bmatrix} \frac{\partial}{\partial x} GF_{xy} \\ \frac{\partial}{\partial y} GF_{xy} \end{bmatrix} \tag{3}$$

The module of vector G is set in (4) and its angle in (5).

$$|G| = \sqrt{G_x^2 + G_y^2} \tag{4}$$

$$\theta_{xy} = \tan^{-1} \frac{G_y}{G_x} \tag{5}$$

Segmentation: For independent analysis it is necessary to segment or cut the processed image. A new vector (C) of dimensions RxS is established and the initial and final points of the 2 dimensions of the previous image, as described in (6).

$$C_{ij} = G[y_0 : y_n, x_0 : x_n] \tag{6}$$

Comparison of Shape Descriptors: Image descriptors are obtained using the Canny () algorithm, which receives as parameters the filtered or grayscale image and the minimum and maximum limits for the hysteresis procedure. After obtaining the descriptors, the matchShapes() method is applied. At this point, two images to be compared in grayscale or contours are entered as parameters and the comparison method that returns the metric or similarity value is expressed in (7).

$$M = \sum_{i=1...7} \left| \frac{1}{m_i^A} - \frac{1}{m_i^B} \right| \tag{7}$$

Where A denotes the image 1 and B, image 2 respectively, h_i^A and h_i^B are the *Hu* moments of each image.

3.2 Storage of Analysis Results

The information acquired after analyzing the product is recorded in a database managed by phpMyADmin and MySQL. Centralized web servers are also implemented. Everything is loaded on the Raspberry Pi 3 board.

3.3 Development of Prototype User Interface

The interface is developed in Python 3.5 language, whose process is discussed in the flow diagram described in Fig. 4. This language gives the user full control of the system, through two methods of analysis. The manual method allows manipulation of the camera position with defined buttons, selecting at personal criteria, chicken eggs suitable for incubation. In the automatic method, the movement of the mechanism is executed directly and the analysis is carried out according to the established parameters.

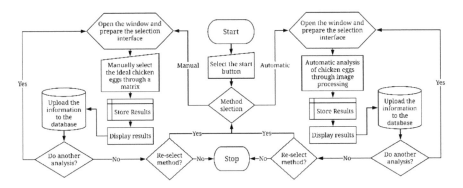

Fig. 4. Control system flowchart.

Through scripting programming, the automatic storage of information is managed in a database and in the report files with time, date and type of analysis. At the end of the process, several options are presented involving the return to the main screen, perform another analysis under the same method or exit completely from the program. In Fig. 5 the main screen of the presented interface is shown.

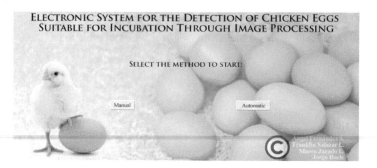

Fig. 5. Main screen of the developed interface.

4   Implementation

4.1   Image Acquisition Parameters

For the mechanism position, the scope of the camera was determined, performing image acquisition tests based. Thus, a cube segmentation was determined in 4 quadrants, which generates 4 3x3 egg matrices. Placing the camera at 31.5 cm on the z-axis captures the desired matrix. According to the prototype configuration, the displacement of the camera on the cuvette was determined. On the x-axis, 1160 steps of the motors were determined in full-pass mode and, 918 steps on the y-axis. When crossing the four quadrants, a total time of 34.08 s is used, determined through the use of the library "time", specifically the time() method in Python 3.5. Figure 6 shows the system in execution. It must be pointed that chicken eggs are being analyzed just from one side, the dark part which is located directly over the illumination system does not have any incidence in the realized analysis. It has been determined taking into account that, *e.g.* if the product has excess pores or very thin skin on the analyzed part, it must be discarded, since they are conditions that affect it globally because it is a unique entity. Similarly, if any fissure is detected, an air chamber at the end of the shortly section is generated so it could be visualized in any position.

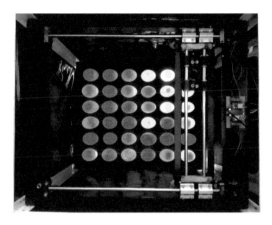

Fig. 6. Control prototype in operation.

4.2 Parametrization of Ranges for the Comparison of Product Quality Characteristics

In order to perform an individual analysis of chicken eggs, by comparing descriptors, the silhouette of an egg that shows no damage (pores, cracks or other types of marks) is taken as reference. This image approaches an ellipse with a 4:3 ratio of A with respect to B.

5 Results

To carry out the tests, 108 units were obtained in stores, commercial and field premises. A sample of 36 was chosen to perform the tests.

5.1 Manual Method

Through the personal criteria of the user, the units adequate for incubation are determined. In Fig. 7 the respective results are displayed, where 9 have been chosen as fit

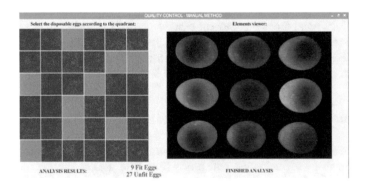

Fig. 7. Graphic interface – Manual Method Results.

(gray boxes) and the rest will be used for consumption (red boxes). After the process ends, the options to perform another analysis or to exit the interface are presented.

5.2   Automatic Method

The problematic characteristics that prevent a satisfactory hatching are analyzed through the processing of the acquired images, as shown in Fig. 8a. With these parameters using the cv2.matchShapes() method of OpenCV, the threshold value of the comparison metric is obtained which is equal to 0,1. Those eggs that are over than the said value is discarded and marked with an "X", while the 7 that fulfill the respective standards do not have any signal. The operation of the interface can be evidenced in Fig. 8b.

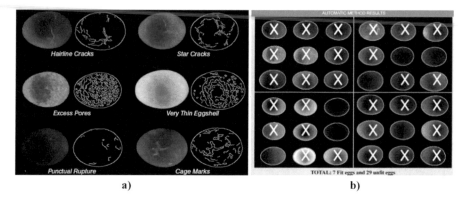

a)   b)

Fig. 8. (a) Defective characteristics detected in chicken eggs by image processing. (b) Graphic Interface – Automatic Method Results.

The results of the analysis performed by the automatic method with the 108 chicken eggs are presented below. 23 (21,3%) of them are determined as fit and 85 (78,7%) that can't be incubated is subdivide determined as unsuitable for incubation are subdivided into two groups, with a metric of equivalent similarity, those that have excess pores or a fine skin (33,8%) and those that present fissures or slight porosity (44,9%). In the one hand, the average time it takes to analyze a chicken egg using the developed prototype is 1.72 s, determined by the time() method. In the other hand, the maximum RAM consumption of the embedded board is 440 MB.

6   Conclusions

The focus and close up of the camera are determinant for the movement of the prototype mechanism, segmentation and processing. The greater the distance from the bucket, the greater the amount of chicken eggs that are visualized, but the image quality will be reduced and the matrix distribution will lose symmetry, negatively influencing the results of the individual analysis. As an initialization parameter for the system analysis, a base image must be established to perform the comparison and obtain a similarity value. In this case of study, the silhouette of an egg that does not present any disturbance or internal damage was used, ensuring that in each analysis the obtained value presents high reliability and allows discarding the product that moves away from the threshold value determined through the scientific method. Since the analysis has not been affected as detailed above, within the general considerations of structural design has avoided the egg flip due to its fragility.

Tests validate that the prototype is capable of detecting problems in the composition of products such as cracks, excessive porosity, very thin shell and marks which can affect the incubation process. Results can strongly show the achieve of the goal of controlling the quality of the chicken eggs. Using the automatic mode, the average analysis time is 1.72 s per egg. When carrying out the analysis, it is determined that the percentage of eggs unsuitable for incubation is higher, since most of them have been prepared for human consumption and are exposed to constant handling and transfer. These are more prone to damage such as cracks, sudden changes in temperature and even very thin shells, because of their producers do not necessarily focus on adequate feeding for birds. In addition, the number of suitable units determined by the manual method is greater than those of the automatic method, since the operator overlooks some characteristics that image processing surely show.

References

1. Zhanga, G., McAdams, D.A., Shankar, V., Darani, M.M.: Modeling the evolution of system technology performance when component and system technology performances interact: commensalism and amensalism. Technol. Forecast. Soc. Chang. **125**, 116–124 (2017)
2. Singh, P., Pawar, P., Takwale, P., Karia, D.: Agriculture monitoring system AMS. In: 2017 International Conference on Energy, Communication, Data Analytics and Soft Computing (ICECDS), pp. 742–745. IEEE, Chennai (2017)
3. Abdelsalam, H.M., et al.: Web-based knowledge evolution for thinking green transportation using expert systems. Int. J. Eng. Technol. **7**(2), 837–839 (2018)
4. Oh, H.S., Lee, Ch.H.: Origin and evolution of quorum quenching technology for biofouling control in MBRs for wastewater treatment. J. Membr. Sci. **554**, 331–345 (2018)
5. Almousa, H., Alenezi, M.: Measuring software architecture stability evolution in object-oriented open source systems. J. Eng. Appl. Sci. **12**(2), 353–362 (2017)
6. Mettler, T., Sprenger, M., Winter, R.: Service robots in hospitals: new perspectives on niche evolution and technology affordances. Eur. J. Inf. Syst. **26**(5), 451–468 (2017)
7. Johnson, R.D., Lukaszewski, K.M., Stone, D.L.: The evolution of the field of human resource information systems: co-evolution of technology and HR processes. Commun. Assoc. Inf. Syst. **38**, 533–553 (2016)
8. Brand, U., Boos, T., Brad, A.: Degrowth and post-extractivism: two debates with suggestions for the inclusive development framework. Curr. Opin. Environ. Sustain. **24**, 36–41 (2017)
9. White, R.R., Hall, M.B.: Nutritional and greenhouse gas impacts of removing animals from US agriculture. In: Proceedings of the National Academy of Sciences of the United States of America, pp. 10301–10308 (2017)

10. INEC. http://www.ecuadorencifras.gob.ec/institucional/home/
11. Yao, Y., Xuesong, S., Zhigang, X.: Poultry egg body defect detecting and sorting system based on image processing. Int. J. Simul. Syst. Sci. Technol. **17**(13), 191–194 (2016)
12. dos Santos, C.D.F., do Vale Nascimento, D., dos Santos, J.C.M.: A software for selection of eggs using digital image processing with customization between profits and quality. In: 2013 8th Iberian Conference on Information Systems and Technologies (CISTI), pp. 1–7. IEEE, Lisboa (2013)
13. Vasileva, A.V., Gorbunova, E.V., Vasilev, A.S., Peretyagin, V.S., Chertov, A.N., Korotaev, V.V.: Assessing exterior egg quality indicators using machine vision. Br. Poult. Sci, 1–10 (2018)
14. Stadelman, W.J.: The preservation of quality in shell eggs. In: Egg Science and Technology, pp. 67–79. CRC Press (2017)
15. Shimizu, R., Yanagawa, S., Shimizu, T., Hamada, M., Kuroda, T.: Convolutional neural network for industrial egg classification. In: 2017 International SoC Design Conference (ISOCC), pp. 67–68. IEEE, Seoul (2017)
16. Wiseman, Y.: Take a picture of your tire! In: Proceedings of 2010 IEEE International Conference on Vehicular Electronics and Safety, pp. 151–156. IEEE, Qingdao (2010)

Mobile Application Based on Dijkstra's Algorithm, to Improve the Inclusion of People with Motor Disabilities Within Urban Areas

Luis Arellano$^{(\boxtimes)}$, Daniel Del Castillo$^{(\boxtimes)}$, Graciela Guerrero$^{(\boxtimes)}$ [iD],
and Freddy Tapia$^{(\boxtimes)}$

Universidad de las Fuerzas Armadas ESPE,
Av. General Rumiñahui S/N, 171-5-231B, Sangolquí, Ecuador
{learellano1, dadel2, rgguerrero, fmtapia}@espe.edu.ec

Abstract. The technology trend is to become an inseparable part of the users daily experiences. In this article will be discussed about the implementation of an application that can be a fundamental tool to guide disability people. The application is based on the geolocation of the user and it destination, this generates the optimal path from a database that has the coordinates of the access ramps for the disabled people. For the route development it use the Dijkstra algorithm that includes the value or "weight" of each path.

The results shows that the application helps to improve the time of mobilization from one point to another. Finally, the developed application accomplishes the objective of reducing time and efforts in the mobility of people with motor disabilities in the center of the city of Quito.

Keywords: Disabled people · Mobile application · Dijkstra algorithm · Mobility · Geolocation

1 Introduction

One of the purposes that technology seeks is to improve the quality of life of people and with this become a part of the human being, currently there are applications that can make different things, starting with improve the way in which the user plan his day, prepare his meals; to automate processes inside companies. Gleeson [1], says that most of these applications have forgotten critical social issues such as disability. Currently it is estimated that more than a billion people, including children (or about 15% of the world population) lives with disability [2, 17]. The lack of support services can make disabled people too dependent of their families, what prevents them from being economically active and socially included [3], for this reason this research focus on the development and implementation of a navigation system for people with limited mobility, allowing them to mobilize better and in less time, inside the historic center of the City of Quito – Ecuador; this application integrates several technologies such as mobile phones, global positioning system GPS, database, where the coordinates of tourist places of the city and the points with specially designed access for people with disabilities are stored; and Dijkstra algorithm with which an optimal route can be calculated.

© Springer Nature Switzerland AG 2019
Á. Rocha et al. (Eds.): WorldCIST'19 2019, AISC 931, pp. 219–229, 2019.
https://doi.org/10.1007/978-3-030-16184-2_22

There are a few works in this area, the most relevant one call's GAWA [9], this application is related to the use a smartphone to mobilize trough different places, using Dijkstra algorithm. The main contribution of this application is that is focus on give disable people a useful tool when they are moving around a city that has an irregular geography, because it has been considered that disable people can only mobilize trough several paths in which ramps are present, this causes a huge restriction when they try to move to different places.

The present work has been organized as follows: (i) Sect. 2 describes the theoretical foundation obtained through the selection of relevant related works, according to the theme to be treated; (ii) Sect. 3 discusses the methodological and experimental configuration of the proposal, this includes the description of the used architecture, the coordinates extraction process, the development tests; (iii) Sect. 4 presents the results analysis after the validation of two proposed scenarios; (iv) Finally, Sect. 5 presents the conclusions of the research and the future work section.

2   Related Work

Chen et al. [4] says that the Dijkstra algorithm is known to be the best algorithm for finding the shortest or optimal path. This algorithm depends of a graph obtained by an ordered pair (V, E) formed by set V of vertices or nodes and a set E of edges that connects the nodes. For the correct implementation of the shortest path algorithm, the edges of the graph must have a value, this means that each nodes connections must have their own weights. The algorithm analyze all the possible paths between the initial and the final node in order to find the route in which the sum of weights is the lowest (minimum cost) [5].

There are many applications and some of them are relevant in different knowledge areas, such as: traffic control, transport route planifications and applications that involve trajectories [6].

For the development of this study, the use of Google API's has been essential, API's are based on web pages formed by JavaScript, XHTML and CSS [7]. Elevations API was the main, this API belongs to Google Maps, which helps to define nodes.

Users who has wireless devices with location ability can recognize their environment at any time at anyplace [8], to use a wireless smartphone service, it is necessary to have at least one of the following two options enabled: (i) the first one depends on the Communication Network for Mobile Devices, that is, the network of the telephone service provider; (ii) the second is the GPS, which is mainly based on a satellite that send the location, this option is highly accurate and works no matter weather, place and time. In recent times, many people use smartphones to navigate to different parts of the world [9], a lot of these users use applications that provide addresses, places or other information that is based on current location.

This research is based on access improvement and social equality, for people with disabilities in a more profitable way. Taking advantage of the assistance and conventional technologies, this allow to increase the independence of people with disabilities creating a social and individual benefit [10]. Is a State policy to promote equality, inclusion and the improvement of people's live, mobility problems have a big impact on the quality of life of the citizens and on the sustainability of cities. For example, travel time shows a strong relationship between satisfaction and life in small cities, but that relationship is non-existent in large cities, due to the costs of traffic congestion [11], this leads to consider the importance of improving the access of people with disabilities, because the percentage of people with disabilities is big around the world [12]. It is estimated that more than one billion people live with some type of disability, according to the World Health Organization [13].

It is necessary to consider the importance of understanding that current technological devices are designed for average users, therefore, users with motor disabilities must adapt to the use of this devices [14]. Rodriguez et al. [9] says that there are four challenges that must be overcome in order to have an application with an universal access, and they are: (i) external navigation, (ii) internal navigation, (iii) accessible use of applications and (iv) an accessible database which contains information about points and routes of interest. Finally, Cledou et al. [15], suggest that applications must have a high precision and these must be tested using emulators, and in a real environment with physical devices.

3 Experimental Configuration and Methodology

The application has been develop for mobile devices with Android OS, this has been done in relation to the high number of Android users on Ecuador.

3.1 Proposed Architecture

The proposed system consists of an API that allows the use of a geographic map provided by Google Maps. It has a MySQL relational that is hosted on a cloud server. All this, to ensure the availability of the application through any type of mobile network or Wi-Fi (Fig. 1).

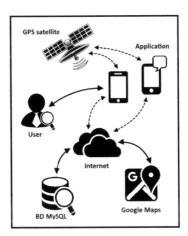

Fig. 1. General proposed architecture

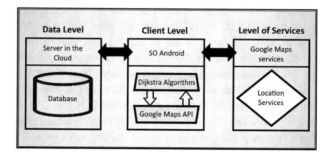

Fig. 2. Architecture represented by level

The application has been represented by a three-level model, which are shown in Fig. 2, describing each level: (i) The Data Level, contains a cloud server that stored in a relational database the coordinates and places for the correct deployment of the optimal path. (ii) The Client Level, that allows the visualization and localization of the user, through the mobile device, in order to process the information that will be shown to the user. (iii) The Service Level, which provides the application with the localization services through Google Maps.

3.2   Algorithm Application

In order to obtain the shortest path, we consider the Dijkstra Algorithm because is a search algorithm that uses a graph with weights, it algorithmic functioning is [16]:

a. A graph with nodes and distances is established. **b.** One node for origin and other for destination is defined, considering that these nodes must have a link. **c.** The algorithm label each node with its route value, it previous node and the iteration number. **d.** The minor route values are compared, in order to convert this node into the main (greedy). **e.** The optimal path is found, considering the lowest route value.

The Dijkstra algorithm will allow the correct selection of the shortest path, and is going to be a fundamental part while the application display the route that the user must follow.

Additionally, the elements that were used for the correct implementation of the algorithm from Google Maps API are exposed:

Google Maps API: An API is a group of methods and tools that can be used in the creation of software applications.

Elevation API: It allows to make locations queries on the terrestrial surface and as results the elevation data is obtained.

Directions API: The Directions API is a service that calculates directions between locations. You can search for directions for several modes of transportation, including transit, driving, walking, or cycling.

3.2.1   Coordinates Validation

To perform the validation of the obtained coordinates, two evaluations were made.

1. Each node is checked obtaining the coordinates in decimal degrees (DD) (Fig. 3a), using Google Maps.
2. The second validation is done in a practical way, by using the GPS of a smartphone (Fig. 3b). For this verification the sample size of the total universe (800 nodes) (Fig. 7) was considered, the nodes were considered in Zone 1 (260 nodes), with a confidence level of 95% and a margin of error of 5%.

(a)   **(b)**

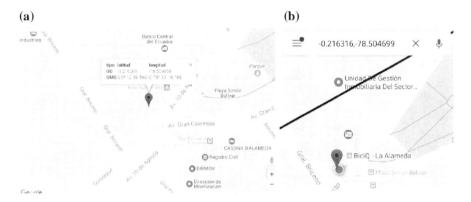

Fig. 3. (a) Obtaining coordinates in DD. (b) Obtaining coordinates on GPS

Once the comparison between the two scenarios was made, an error of 0.01% was obtained, concluding that the obtained coordinates have an acceptable margin.

To obtain the weight of each existing link between nodes, the following must be considered: (i) Calculation of distance and speed. (ii) Testing the weights with Grafos Software. (iii) Deployment of the optimal route.

3.2.2   Distance and Speed Calculation

At the beginning it was planned to use the time that Google's API indicates, so this time was validate with a practical test (Table 1), getting the following results.

In this test we obtain an average error of 20.54 s, that is a minor error, but we want to be more precise, because this time is calculated for an average user.

Then we proceed to use the Directions API, which requires two coordinates (origin and destination) to return the distance (meters) and the travel time of that link (minutes) (Fig. 4). With the obtained distances, we proceed to save all of them on the database, then we start to calculate the speed of disable people.

Table 1. Google's API time versus practical time

Slope	Google's API (seconds)	Practical test (seconds)	Error (seconds)
−3,196218906	50	67	17
3,196218906	35	76	41
−6,594321534	41	60	19
−1,816844617	59	62	3
1,816844617	49	72	23
6,594321534	37	70	33
−6,813464957	69	74	5
6,813464957	54	84	30
−2,055899664	56	63	7
2,055899664	44	68	24
		Average	20,54

```
{
  "destination_addresses" : [ "Manabi, Quito 170401, Ecuador" ],
  "origin_addresses" : [ "Manabi, Quito 170401, Ecuador" ],
  "rows" : [
    {
      "elements" : [
        {
          "distance" : {
            "text" : "1 m",
            "value" : 1
          },
          "duration" : {
            "text" : "1 min",
            "value" : 0
          },
          "status" : "OK"
        }
      ]
    },
  ],
  "status" : "OK"
}
```

Fig. 4. Directions API.

For the speed calculation, there were three stages types: (i) A street with an angle of 0° and 10°, (ii) A street with an angle of 10°–20°, and finally (iii) A street with and angle of 30° or more. For this test around 30 people with motor disabilities were needed, we divided them on three groups of 10 people each, with this test we could obtain the time of each travel, then we calculate the average speed, with the following formula.

$$Speed = \frac{Distance\,Travelled}{Time\,Travelled} \tag{1}$$

With the average speed and the node link distance, the approximate travel time is calculated inside the application. With this method we obtain an average error of 11,80 s, getting a reduce of 8,74 s.

3.2.3   Grafos Software Test

For the evaluation with Grafos software, a map of one of the proposed zones (Zone 1) was implemented, to proceed with the insertion of the location of each node (Fig. 5a); considering that the software does not use coordinates, it only use weighted points.

When the links between nodes were made, the weight was considered as the distance in meters between each of them. After making the insertion of the nodes to which the test is intended to run, we proceed to start the simulation of the shortest path, in this way 2 nodes are selected (Fig. 5b) and the software shows the shortest path. The use of Grafos software allowed to create the graph with it nodes, and then we proceed to integrate this graph in the Android application.

(a)   **(b)**

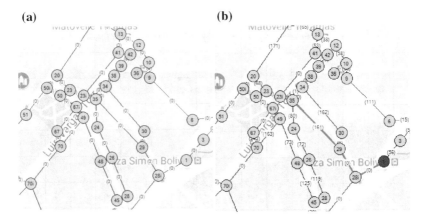

Fig. 5. (a) Location of nodes in Grafos software. (b) Use of Dijkstra in Grafos software

3.2.4   Deployment of the Optimal Route

To observe the deployment of the optimal route, the user selects a route defining the origin and the destination on the developed application, which accesses the Database and checks the links for each node inside the path, this is where the Dijkstra algorithm is applied. It proceeds to compare the weights (distances) respect to each link, so that way it shows the route that this person must take, indicating the ramps that he must take to reach his destination. The application uses features from the Google Maps application, it aims to correct the times shown to the users when they mobilize. Providing a guide for the correct deployment of people with motor disabilities through the city of Quito that has an irregular geography (Fig. 6).

Fig. 6. Deployment of the optimal route on the developed application

The developed application does not use Google paths, because these are based on streets or roads, therefore, they are unusable to guide and reach a destination.

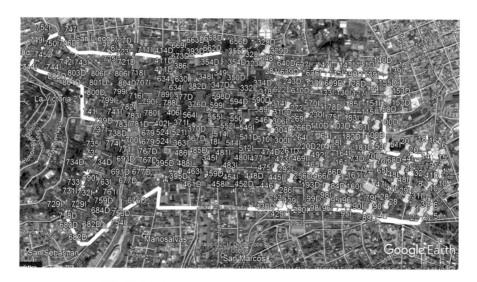

Fig. 7. Nodes taken for the application of the algorithm

4   Evaluation and Results

The evaluation of the mobile application was made with staff and family members of Comando Conjunto de las Fuerzas Armadas del Ecuador, that have mobility problems and that use a wheelchair.

Three practical evaluations were made, considering the following routes:

Route 1: PLAZA GRANDE (Plaza Grande – Iglesia Santa Bárbara),
Route 2: CIRCULO MILITAR (Círculo Militar – Plaza Grande) and
Route 3: SAN FRANCISCO (San Francisco – Plaza Grande).

For the evaluation, the average information was taken from 45 participants with mobilization problems and who mobilized on wheelchairs.

Two stages were planned that consisted on (a) Verify the average time displayed by the application. (b) Verify the real average time when people executes the routes using the application.

Table 2. Routes time – application and test comparative

Route	Average time stage 1	Average time stage 2	Absolute time error	Percentage error between stages
Plaza Grande	08:47:00	09:15:00	00:28:00	5,05%
Círculo Militar	04:26:00	04:23:00	00:03:00	1,13%
San Francisco	04:21:00	03:51:00	00:30:00	11,49%

In the evaluation and analysis of the obtained results from time measurements (Table 2), it is observed that the mobility and inclusion application for people with motor disabilities provides a similar time to the tested time, being: (i) The estimated time of the application in Route 1 is less than 5% of the actual execution when the journey with the users is make. (ii) The estimated time by the application in Route 2 is higher by 1.13% compared to the actual execution of the second scenario. Finally, (iii) the estimated time for the application in Route 3 is higher with 11, 49% in comparison to scenario 2. Once the average time values of the two scenarios and the percentage of differences between each route have been obtained, it is demonstrated that the use of the proposed application, for the selection of the most optimal route allows to obtain similar information of the actual time of arrival at the destination.

5   Conclusions, Recommendations and Future Works

This present study has focused on the design and development of a mobile application for Android smartphones that has been implemented using the Google Maps API for the map deployment, and the calculation of the optimal route using the Dijkstra algorithm. To this is added, the storage of the nodes in a database, capable of being accessed by any mobile network or Wi-Fi.

When the exploratory study was carried out to obtain nodes (ramps), an amount greater than 800 units was obtained, these were restructured and now there is an amount no greater than 400 units.

The correct segmentation of zones within the map, allowed to have a better manipulation of the data. In the extraction of nodes it was noticed that in certain tourist areas of the city of Quito there is a deficit and an excess of ramps.

The values obtained in the evaluation were obtained through the execution of two scenarios with three routes each, the first scenario was about the prediction of the average time of 45 users who mobilize in wheelchairs, while the second scenario adjusted the average time in real execution data to the prediction, concluding that there is no average error of 5.89% in the three proposed routes.

The calculation for the predicted time does not count the traffic lights, this is why the real time can vary in greater or lesser amounts.

When the tests were made, it was possible to obtain that the application can be used not only for disable people, because they are not the only people who needs ramps, it can also be used by fathers with baby strollers or people who have suffered an accident and need to use ramps to move around the city.

As a complementary study, it has been proposed to expand the coverage area, within the city of Quito, this would include the use of a new algorithm for the calculation of the optimal route. In the same way, it is proposed to carry out the evaluation with a greater number of users in different scenarios.

References

1. Gleeson, B.: A geography for disabled people? Trans. Inst. Br. Geogr. **21**(2), 387–396 (1996)
2. Domingo, M.C.: An overview of the internet of things for people with disabilities. J. Netw. Comput. Appl. **35**(2), 584–596 (2012)
3. Karimanzira, D., Otto, P., Wernstedt, J.: Application of machine learning methods to route planning and navigation for disabled people. In: Modelling, Identification, and Control, Germany (2006)
4. Chen, Y.-Z.: Path optimization study for vehicles evacuation based on Dijkstra algorithm. In: 2013 International Conference on Performance-Based Fire and Fire Protection Engineering, Wuhan (ICPFFPE) (2013)
5. Szücs, G.: Decision support for route search and optimum finding in transport networks under uncertainty. J. Appl. Res. Technol. **13**(1), 125–134 (2015)
6. Cardona, M., Castrillón, O., Tinoco, H.: Determinación del Método Óptimo de Operaciones de Ensamble Bimanual con el Algoritmo de Dijkstra (o de Caminos Mínimos). Información tecnológica **28**(4), 28 (2017)
7. Hu, S., Dai, T.: Online map application development using google maps API, SQL database, and ASP.NET. Int. J. Inf. Commun. Technol. Res. **3**(3), 102–110 (2013)
8. Ibrahim, O., Mohsen, K.: Design and implementation an online location based services using google maps for android mobile. Int. J. Comput. Netw. Commun. Secur. **2**(3), 113–118 (2014)
9. Rodriguez, M., Martínez, J.: GAWA – manager for accessibility wayfinding apps. Int. J. Inf. Manage. **37**(6), 505–519 (2017)
10. Agree, E.: The potential for technology to enhance independence for those aging with a disability. Disabil. Health J. **7**(1), S33–S39 (2014)
11. Li, K.: Test for mobility. Revista Antioqueña de las Ciencias Computacionales y la Ingeniería de Software **4**(1), 7–10 (2014)
12. Pomboza, M.D.R., Cloquell, V.: Anthropometric determination for school furniture designed to children with motor disabilities in Ecuador. Ciencia y trabajo **1**(53), 154–158 (2014)

13. OMS, Informe Mundial sobre la Discapacidad 2011, Organización Mundial de la Salud (2015). http://www.who.int/disabilities/world_report/2011/accessible_es.pdf. Accessed 15 Sept 2018
14. Sánchez, J., Zapata, C., Jiménez, J.: Evaluación heurística de la usabilidad de software para facilitar el uso del computador a personas en situación de discapacidad motriz. Revista EIA **14**(14), 63–72 (2017)
15. Cledou, G., Estevez, E., Soares, L.: A taxonomy for planning and designing smart mobility services. Gov. Inf. Quart. **35**(1), 61–76 (2018)
16. Bautista, E.: Simulación de los Algoritmos Dijkstra y Bellman-Ford para ruteo de paquetes en redes de comunicaciones (2005). http://catarina.udlap.mx/u_dl_a/tales/documentos/lem/bautista_h_e/. Accessed 17 Dec 2018
17. Vázquez, M.Y., Sexto, C.F., Rocha, Á., Aguilera, A.: Mobile phones and psychosocial therapies with vulnerable people: a first state of the art. J. Med. Syst. **40**(6), 157 (2016)

Design and Construction of a Semi-quantitative Estimator for Red Blood Cells in Blood as Support the Health Sector

Yadira Quiñonez[1(✉)], Said Almeraya[2(✉)], Selin Almeraya[2],
Jorge Reyna[2], and Jezreel Mejía[3]

[1] Universidad Autónoma de Sinaloa, Mazatlán, Mexico
yadiraqui@uas.edu.mx
[2] Instituto Tecnológico de Mazatlán, Mazatlán, Mexico
{141000013, reynajr}@itmazatlan.edu.mx,
almerayamorales@hotmail.com
[3] Centro de Investigación en Matemáticas, Zacatecas, Mexico
jmejia@cimat.mx

Abstract. This paper shows the steps involved in the creation of a device capable of giving an approximate amount of red cells in blood, without the need of laboratory analysis tests. This device relies on the use of photo sensitive sensors such as a phototransistor and a red LED. As the intensity of the light varies, so does the amount of current flowing through the sensor. In consequence, this variation in electric current will cause a variation on the voltage drop across the connections of a resistor. Subsequently, this voltage drop will be read by a microcontroller that will calculate an approximate number for the amount of red cells. To achieved this some formulas were established that represent the relationship among the extreme points of a data group obtained during a sampling process in patients requiring hematic biometry studies.

Keywords: Extreme points · LED · Light · Microcontroller · Phototransistor · Red cells · Voltage

1 Introduction

The beginning of the methods or techniques for counting the elements that form the blood started 1658 when Jan Swammerdam first observed under a microscope, and the Red Blood Cell (RBC) count was performed in 1852 by Professor Karl Vierordt [1]. The first devices whose objective was to measure the thickness of the blood based on its opacity began to appear around that time. According to the Leukemia & Lymphoma Society [2] the normal range for the amount of red blood cells in for men is between 4.7 to 6.1 million of cells per microliter. Similarly, the normal range for the amount of red blood cells in women encompasses from 4.2 to 5.4 million of cells per microliter.

There are several related works to counting of red blood cell using different methods. Mazalan et al. [3] proposed an automated of RBC counting in microscopic image using circular Hough transform. The authors in [4] proposed a system based on the radon transform to automated marker identification in grey-scale images obtained

© Springer Nature Switzerland AG 2019
Á. Rocha et al. (Eds.): WorldCIST'19 2019, AISC 931, pp. 230–239, 2019.
https://doi.org/10.1007/978-3-030-16184-2_23

by optical light microscopy of blood smears. In [5] a review of the different methodologies used for the counting of blood cells through the segmentation of images was presented. In this research the authors described different techniques for processing of images such as filtering, histogram equalization, thresholds, masking, morphological operations, contrast enhancement, among others.

Recently, Christy et al. [6] have presented a system to classify the blood cell count through the blood smear using microscopic images with high resolution cameras, to perform image processing using different techniques (gray scale conversion, image enhancement, contrast stretching, and histogram equalization, thresholds). In another work, Acharya et al. [7] have proposed the technique of image processing to carry out the red blood cell count through the blood smear using the labeling algorithm and circular Hough transform, according to the results, better results are obtained with the circular Hough transform.

According to the related works analysis, at the moment, there does not exist a device or non-invasive method for the estimation of red blood cells. The present work deals with the design and construction of a semi-quantitative estimator for the amount of red blood cells in blood. In addition, this device can be used to make a faster diagnosis when dealing with a big population. Since that a higher cell density, in this case red blood cells, will reduce the amount of light coming through a person's finger, a photosensitive sensor will, in consequence, be less stimulated resulting in a lower voltage at its output. Conversely, if the sensor is stimulated by a higher amount of light, the result will be a higher voltage level at its output. This photosensitive device consists of a phototransistor with a transparent casing. A transparent casing will allow the red light coming from a red LED to penetrate the sensor. Therefore, the main goal of this paper is to show a proposed device that estimate, to a certain degree of accuracy, the amount of red cells in blood. This is based on the amount of light passing through the patient's index finger.

This paper is structured as follows: section two presents the material and methods; section three presents the data analysis; section four describes the software design to calculate the red blood cells, section five shows the overall results, and finally, section six shows the conclusions.

2 Material and Methods

The development of the project required its fragmentation into six steps. These steps are: (a) design of the internal circuitry, (b) soldering of the required components, (c) data obtainment through a sampling process, (d) data analysis (e) software design and (f) evaluation of the equations developed during the data analysis step with the data obtained during the sampling process.

2.1 Design of the Internal Circuitry

The whole circuit is composed of four parts: (a) a photosensitive element and a red LED, (b) an amplifier stage for the signal that comes from the phototransistor, (c) a circuit whose work is to adjust the voltage signal to a level that is compatible with a microcontroller and (d) a digital circuit whose work is the processing of the voltage signal and control of a LCD display. Adjoined to this stage are all the necessary elements that help the microcontroller to work properly. These elements are a quartz crystal resonator, two ceramic capacitors, pull-up resistors for the reset pin and other buttons. Finally, the whole circuitry is fed by a voltage regulating stage that delivers the necessary voltage levels for each stage of the device.

The element that produces the light that will pass through the finger is composed of a light emitting diode. This diode is forward biased through a resistor. This resistor ensures the LED to work within the safe current ranges while granting the maximum level of brightness.

The current that circulates through the diode is given by the following equation:

$$I_{LED} = \frac{V_s - V_{LED}}{R_{LED}} \tag{1}$$

where I_{LED} denotes the current flowing through the series circuit formed by the resistor and the LED. V_S indicates the voltage of the power source, V_{LED} represents the voltage drop through the LED, and R_{LED} designates the resistor in series with the LED.

$$I_{LED} = \frac{8\,V - 1.8\,V}{330\,\Omega} = 18.7\,mA \tag{2}$$

The maximum current intensity for a common red LED is 20 mA. The brightness will increment according to the current, but it should be taken into account the maximum current that a given LED can withstand (Universidad de Granada, 2018). The aforementioned amplifier circuit helps increase the current coming out from the photosensitive sensor. This task is accomplished by a general purpose NPN transistor. In relation to the microcontroller has 13 general input/output ports. Six of these ports can be configured as PWM signal outputs. Besides, the device incorporates a UART port and a six channel analog to digital converter [8].

2.2   Final Design

This step involves the construction of the circuit board as well as the choosing of an outer casing for the device. For the printed circuit board, a factory holed, with a track distribution similar to a prototype board, printed circuit board was used. The optical receptacle used for taking in the index finger is made of a PVC coupling pipe with three openings. The placement of the optical elements inside the PVC coupling must be done in such a way so that the maximum amount of light can be caught by the phototransistor. This means, the phototransistor is placed in front of the LED, the source of light. This space will be occupied by the patient's finger at the moment of the sampling. Given the fact that these optical elements must remain fixed at all moments, two cork caps were used for this task. They were drilled in the middle with an adequate size for the optical elements. Once having assembled the whole internal circuitry alongside the required wiring for the other components, it is needed to place all these objects inside a plastic casing. This way the device will be ready to begin the sampling process. Figure 1 depicts the final facade of the device.

Fig. 1. Final design of the red blood cell estimator.

2.3 Data Obtainment Through a Sampling Process

In order to create a program for the device, data must be obtained and processed. The first part involves the retrieval of laboratory test results in relation to the amount of voltage measured by the semi-quantitative estimator of red blood cells (see Fig. 2). Initially, the device was programmed to show the voltage signal coming from the amplifier stage. A list of 17 males from different ages was made. This list also showed the specific voltage drop for each person in particular. The laboratory test that contained the data of interest, in this case, amount of red cells per microliter is called hematic biometry.

Fig. 2. Semi-quantitative estimator of red blood cells.

3 Data Analysis

After having finished the sampling process, the new task at hand involved the creation of a set of equations that satisfy the relationship between the amount of red blood cells and the voltage obtained for each one of the individuals involved in the sampling process. Table 1, it is shown the relationship between the number of red blood cells and the level of voltage measured with the device.

Table 1. Relationship among hemoglobin levels, amount of red cells, voltage, age, and gender of the patients.

Gender	Age	Hemoglobin	# Red cells	Voltage
M	73	13.4	4.65×10^6/uL	0.23 V
M	43	11.7	4.20×10^6/uL	0.29 V
M	29	15.8	5.33×10^6/uL	0.29 V
M	57	14.0	5.15×10^6/uL	0.29 V
M	37	16.2	5.13×10^6/uL	0.14 V
M	16	17.2	6.11×10^6/uL	0.11 V
M	84	11.1	3.45×10^6/uL	0.43 V
M	49	10.6	3.55×10^6/uL	0.40 V
M	80	12.9	4.46×10^6/uL	0.14 V
M	52	11.5	4.24×10^6/uL	0.43 V
M	30	13.0	5.30×10^6/uL	0.27 V
M	70	12.9	4.64×10^6/uL	0.23 V
M	39	11.7	4.20×10^6/uL	0.20 V
M	50	12.0	4.16×10^6/uL	0.40 V
M	27	13.0	5.12×10^6/uL	0.16 V
M	76	11.0	3.54×10^6/uL	0.14 V
M	58	14.3	4.59×10^6/uL	0.37 V

Figure 3, shows the relationship between the amount of hemoglobin and red blood cells, it can be seen that there is a linear relationship between the two basic parameters of hematic biometry. In relation to the voltage as a function of red blood cells, a general tendency can be observed that shows a decrease in the amount of voltage when the number of red blood cells increases.

Figure 4, it is shown the percentage amounts of the male population falling on different ranges expressed in millions of red blood cells per microliter.

Based on those percentage amounts shown above, it is possible to determine which part of the population falls within a specific blood cell range. In reference to Fig. 3, where the percentage amounts for the male population are located, it is observed that 47% of them fall within the 4 to 5 million of red blood cells per microliter range.

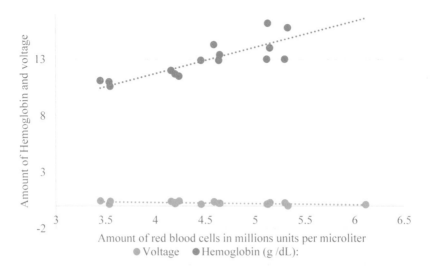

Fig. 3. Hemoglobin and voltage depending on the amount of red blood cells.

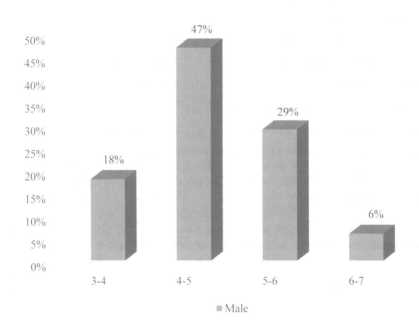

Fig. 4. Percentage amounts of the male population falling on different ranges of number of blood cells.

These percentages will serve to set the required parameters to fit a straight line that will allow for the calculation, at a certain extent, of an amount of red blood cells within a specific range. Definite extreme points for the male populations will serve for this purpose. In this sense, for the male population, these points are: 6.11 million of red blood cells per microliter at 0.11 V and 3.45 million of red blood cells per microliter at 0.43 V. Thus, the coordinates are: (6.11, 0.11) and (3.45, 0.43). According to the formula used to find the inclination of a straight line, it is stated that:

$$m = \frac{y_2 - y_1}{x_2 - x_1} \tag{3}$$

If $y_2 = 0.11$, $y_1 = 0.43$, $x_2 = 6.11$ and $x_1 = 3.45$; the inclination for this line is:

$$m = \frac{y_2 - y_1}{x_2 - x_1} = \frac{0.11 - 0.43}{6.11 - 3.45} = -0.12 \tag{4}$$

The general form for a straight line equation is:

$$y = mx + b \tag{5}$$

Finally, the straight line equation that relates the amount of red blood cells with voltage levels is:

$$y = -0.120x + 0.85 \tag{6}$$

4   Design to Calculate the Red Blood Cells

The next step consists of the creation of an Arduino program that allows the micro-controller ATMEGA328P to calculate an approximate of the amount of red blood cells based on the voltage levels it is acquiring through its analog input. First off, the microcontroller executes a reading operation of the voltage seen at its pin A5 (analog input).

After that, the microcontroller executes a scanning operation to determine the state of the pin where the "Select Gender" switch is located. That is, if the person who is going to undertake the reading is a woman, then the switch must be put in the open position. Therefore, a high state will be read. With that, depending on the voltage level measured from the optic sensor, the following messages will be displayed on the LCD: (a) If there is no finger placed inside the receptacle, a "Insert Finger" message will be shown, (b) If the reading operation is zero because there is a total blockage on the photosensitive sensor, a "Wrong reading" message will be displayed and (c) if the former two conditions are not satisfied, then the microcontroller will proceed with the calculation of the amount of red blood cells (see Fig. 5).

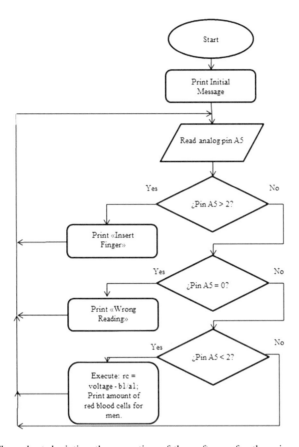

Fig. 5. Flow chart depicting the operation of the software for the microcontroller.

4.1   Evaluation of the Equations

Finally, an evaluation process was undertaken. For this, those equations developed in the data analysis step were tested using some values of voltage presented on Table 1. The values chosen for this step had to be within the range of red blood cells for which these equations were created. For the male population, the range is 4 to 5, and the required formula is:

$$x = \frac{(y - 0.85)}{-0.120} \tag{10}$$

where, for both equations: x = amount of red blood cells and y = voltage level.

5 Results

Following the process of evaluation of the equations, the final action involves the comparison among the data obtained with laboratory tests and the results given by the evaluated equations. On Table 2, it is shown a list with the amount of red blood cells per microliter of three different male patients.

Table 2. Relationship between mount of red blood cells obtained in the laboratory and on the device.

Men	
Laboratory	Device
$6.11 \times 10^6/\mu L$	$6.16 \times 10^6/\mu L$
$5.33 \times 10^6/\mu L$	$4.83 \times 10^6/\mu L$
$3.45 \times 10^6/\mu L$	$3.5 \times 10^6/\mu L$

As it can be appreciated on Table 2, there are variations between the results given by the laboratory tests and the amounts that would be calculated by the semi-quantitative estimator. However, this is to be expected since it is a non-invasive device that relies on optic sensors. Variations in light intensity due to physical traits such as nail thickness, finger thickness, nail polish and the amount of pressure exerted by the person on the sensor will cause a change in the results.

In spite of all the above mentioned, there is a tendency that indicates a behavior that, when the red blood cells level is elevated, the voltage seen by the machine is low. On the contrary, if the concentration of red blood cells is low, a higher amount of light will be able to pass through the finger causing an increment on the voltage seen by the device.

6 Conclusions

Due to the fixed structure of the receptacle that allows the device to take a given sample, patients with finger sizes differing of those established will experience variations in the measurement.

Every patient presents variation in nail thickness, and as a result, the amount of light that the photosensitive sensor receives will vary depending on it. In women, another important factor that interferes with the measuring is the presence of nail polish. During the sampling process, measurements in patients with colored nails had to be discarded.

It must be added that it is possible to improve the design of the receptacle where the optic devices are located. One possible way to improve the design involves the change of the receptacle for a device more versatile, similar to an oximeter pulse probe.

References

1. Verso, M.L.: The evolution of blood-counting techniques. Med. Hist. **8**(2), 149–158 (1964)
2. Leukemia & Lymphoma Society Homepage. https://www.lls.org/. Accessed 20 Nov 2018
3. Mazalan, S.M., Mahmood, N.H., Razak, M.A.A.: Automated red blood cells counting in peripheral blood smear image using circular hough transform. In: 1st International Conference on Artificial Intelligence, Modelling and Simulation, pp. 320–324. IEEE, Kota Kinabalu (2013)
4. Kolhatkar, D., Wankhade, N.: Detection and counting of blood cells using image segmentation: a review. In: World Conference on Futuristic Trends in Research and Innovation for Social Welfare (Startup Conclave), pp. 1–5. IEEE, Coimbatore (2016)
5. González-Betancourt, A., Rodríguez-Ribalta, P., Meneses-Marcel, A., Sifontes-Rodríguez, S., Lorenzo-Ginori, J.V., Orozco-Morales, R.: Automated marker identification using the Radon transform for watershed segmentation. IET Image Proc. **11**(3), 183–189 (2017)
6. Christy, N., Annalatha, M.: Computer aided system for human blood cell identification, classification and counting. In: Fourth International Conference on Biosignals, Images and Instrumentation, pp. 206–212. IEEE, Chennai (2018)
7. Acharya, V., Kumar, P.: Identification and red blood cell automated counting from blood smear images using computer-aided system. Med. Biol. Eng. Comput. **56**(3), 483–489 (2018)
8. Pathak, M., Rawat, R., Singh, S., Thakur, R.: Fire figthing robot remotely controlled by android application. Int. J. Sci. Eng. Res. **5**(5), 106–109 (2017)

Automatic Forest Fire Detection Based on a Machine Learning and Image Analysis Pipeline

João Alves[1], Christophe Soares[1,2], José M. Torres[1,2(✉)], Pedro Sobral[1,2], and Rui S. Moreira[1,2,3]

[1] ISUS Unit, University Fernando Pessoa, Porto, Portugal
{joao.alves,csoares,jtorres,pmsobral,rmoreira}@ufp.edu.pt
[2] LIACC, University of Porto, Porto, Portugal
[3] INESC-TEC, FEUP - University of Porto, Porto, Portugal
http://isus.ufp.pt

Abstract. Forest fires can have devastating consequences if not detected and fought before they spread. This paper presents an automatic fire detection system designed to identify forest fires, preferably, in their early stages. The system pipeline processes images of the forest environment and is able to detect the presence of smoke or flames. Additionally, the system is able to produce an estimation of the area under ignition so that its size can be evaluated. In the process of classification of a fire image, one *Deep Convolutional Neural Network* was used to extract, from the images, the descriptors which are then applied to a *Logistic Regression* classifier. At a later stage of the pipeline, image analysis and processing techniques at color level were applied to assess the area under ignition. In order to better understand the influence of specific image features in the classification task, the organized *dataset*, composed by 882 images, was associated with relevant image metadata (eg presence of flames, smoke, fog, clouds, human elements). In the tests, the system obtained a classification accuracy of 94.1% in 695 images of daytime scenarios and 94.8% in 187 images of nighttime scenarios. It presents good accuracy in estimating the flame area when compared with other approaches in the literature, substantially reducing the number of false positives and nearly keeping the same false negatives stats.

Keywords: Forest fire detection · Computer vision · Deep learning · Feature extraction · Classification · Machine learning

1 Introduction

In southern Europe, and more specifically in the Mediterranean basin, the forest fires are a top concern as the summer approaches. In fact, the risk period of forest fires has been increasing in Portugal [1], regardless of the year season, as a result of climate change fueled by global warming [2]. The consequences

© Springer Nature Switzerland AG 2019
Á. Rocha et al. (Eds.): WorldCIST'19 2019, AISC 931, pp. 240–251, 2019.
https://doi.org/10.1007/978-3-030-16184-2_24

of forest fires can be devastating if not detected, combated and extinguished in the initial phase [3]. According to [2], the prospect of an overall increase in temperature by $2\,°C$ accompanied by the decrease in humidity levels will cause higher levels of dryness in the environment. As a result, the risk of climate-related forest fires is expected to increase in the future, particularly around the Mediterranean. The use of an automatic forest fire detection system, targeted for its initial phase, would be a desirable solution to mitigate this problem. Such system could complement and enhance the current work of human observers on fire detection. The use of automatic surveillance technologies is a very promising option, backed-up by a significant number of studies presented in [4].

Among the available fire detection techniques, we emphasize the use of video cameras as input source for automatic detection systems. The optical systems may mimic the human eye looking for apparent signs of fire. However, the forest does not present uniform landscape scenarios, which makes fire detection a very challenging process. Dynamic phenomena such as reflections, cloud shadows and fog, or even the presence of human apparatus or activities translate into more demanding problems and increase the possibility of false alarms [5]. With these problems in mind, the present work was carried out in order to implement a low cost solution able to tackle these inherent challenges of forest fire detection with the use of video cameras feeding machine learning algorithms.

The automatic fire detection process described in this article aims identifying a fire situation. This ability requires two initial steps, first the extraction of image descriptors which then feeds a posterior classification. Together, these extraction and classification phases form the pipeline of a *Deep Learning* (DL) model. The extraction of descriptors returns unique properties that characterize an image by means of a numerical vector. This vector is the output of a *Deep Convolutional Neural Network* (DCNN) module. The obtained descriptors are then used to train a *Machine Learning* (ML) classifier. The aim is to carefully plan the training process of the classification model, addressing particular situations in a variety of scenarios, so that the model is as best fitted as possible to detect fires in real situations. On the final stage, after obtaining a fire classification, the pipeline focuses on spotting the image areas with flames, through the application of Computational Vision (CV) techniques.

2   State of the Art

In this section a more focused study is conducted for projects that propose the use of mobile or static video cameras in fire detection. The systems presented are divided in two classes: one based on CV techniques and other using DL techniques. CV systems mostly seek to detect only flames or only smoke, while in DL systems there is no such restriction.

2.1   Deep Learning Systems in Fire Detection

In [6] the best choice of optimizers, reduction functions and learning rates for the DCNN Inception-V3 is analyzed. By default, Inception-V3 is implemented according to certain parameters, however, the existing alternatives have been analyzed. They conclude that the best combination of parameters encompasses the *Adam* optimizer, the function *reduce max* and a learning rate of 0.1. As demonstrated in [7], it is proposed to use two DCNNs whose goal is to improve detection accuracy by fine-tuning the fully connected layer of both. The two DCNNs involved are VGG16 and Resnet50. The results demonstrate that by adding Fully Connected layers at the end of networks, an accuracy improvement of only 1% occurs. In [8], the best DCNN for the problem of detection of forest fires is analyzed. As such, five DCNNs are studied: AlexNet, GoogLeNet, modified GoogLeNet, VGG13 and modified VGG13. After training, its concluded that GoogLeNet (version prior to Inception-V3), has the best result, with 99% accuracy, following the modified VGG13. On the other hand, the original VGG13 has the worst accuracy and the longest training time due to overfitting and the large number of parameters.

Table 1. Comparison between DCNN based fire detection systems

System	DCNN	Flame	Smoke	Night	Precision	*Dataset*
[6]	Inception-V3	✓	✓	✗	99.7%	++
[7]	Resnet50 modified	✓	✓	✓	92.15%	++
[7]	VGG16 modified	✓	✓	✓	91.18%	++
[8]	GoogLeNet	✓	✓	n/a	99%	++
[8]	AlexNet	✓	✓	n/a	94.8%	++
[8]	VGG13 modified	✓	✓	n/a	96.2%	++

Most of the systems presented in Table 1 detect both fire flames and the presence of smoke, however, some do not address nocturnal scenarios. All the systems exhibit high precision (above 90%) which proves the ability of the DCNN in the detection of forest fires. It is noted that all systems only mention the adoption of a large *dataset*. They do not perform any analysis of the image set, in order to understand what types of situations are used and which ones should be added in order to make the *dataset* more comprehensive. Given that the systems presented used different *datasets*, it is not fair to compare their relative precision. Still, the best results in terms of accuracy were obtained in [6], which is based on the DCNN Inception-V3. However, the *dataset* used does not address several forest scenarios, small fires and night scenarios.

2.2   Computer Vision Systems in Fire Detection

Flame Detection: A method for detecting flames and smoke is proposed in [9]. It is suitable for daytime scenarios and is intended to be applied to drones. The method consists in the use of a new color index called, FFDI, whose algorithm only involves the manipulation of the RGB color space through the application of a set of color indexes. A method for flame detection based on image processing is shown in [10]. Here, flame detection is achieved by manipulating RGB and YCrCb color spaces. The algorithm consists of seven rules to be applied sequentially. Another method is proposed in [11] which seeks to improve the accuracy of flames detection through fixed surveillance cameras. The system uses two parallel image streams, one for the detection of the flame and the other for confirmation of the results. The analysis of the HSI, YCbCr and RGB color spaces of dynamic textures and spatial-temporal frequency are used in the first stream, while in the second stream a *Background Subtraction* algorithm is used to detect and trace moving objects.

Smoke Detection: A method for smoke detection based on color and motion analysis is presented in [12]. This system provides rapid detection in daytime scenarios and is designed to allow multi-camera signal processing without large computational complexities. The detection process involves signal preprocessing, feature extraction and classification. In [13] a method is proposed based on the segmentation of zones with movement, extraction of characteristics of the smoke and, finally, a classifier. For the segmentation of motion zones, the *Visual Background extractor* method is used. Next, the detection is based on certain characteristics, such as the high frequency energy based on the wave transformation, the consistency and the direction of the smoke movement. Finally, in [14], an approach is proposed to detect smoke through static surveillance cameras. The *Support Vector Machine* classification template is used based on the previously obtained data. They perform a preprocessing to reduce the noise and to divide the image into small blocks. The idea is to determine the suspected regions of smoke based on the detection of color and slow movements and, finally, to analyze the regions through a co-occurrence matrix. The co-occurrence matrix is aimed at detecting three characteristics that are representative of the presence of smoke, such as contrast, energy and correlation.

Table 2. Comparison between CV based fire detection systems

System	Camera	Flame	Smoke	Night	Cost	Precision	Dataset
[9]	Mobile	✓	✗	✗	+	96.82%	+++
[10]	Mobile	✓	✗	✗	+	99%	+
[11]	Static	✓	✗	✗	+++	91.1%	++
[12]	Static	✗	✓	✗	++	90%	+++
[13]	Static	✗	✓	✗	+++	92.5%	+
[14]	Static	✗	✓	✗	++	87%	+++

Table 2 summarises several analysed systems where the detection accuracy is very good, some of them above 95%, as is the case of [9, 10]. Again, given that each system uses a different *dataset*, it is not fair to directly compare their precision, but other features such as their cost of implementation or the quality of applied techniques can be evaluated. Daytime flame detection systems [9–11], in general, have a lower implementation cost than those that perform smoke detection [12–14]. This is because the process of identifying the smoke characteristics is usually more complex than simply applying color rules, as is the case in some systems [9, 10] in the detection of flames.

Regarding the quality of the used *dataset*, that is, if it is vast and varied in terms of situations, the [9, 12, 14] systems are the ones that best meet these requirements.

3 Implementation of the Fire Detection Pipeline

The proposed fire detection system aims to fulfill, in sequence, two specific phases over the collected images. First, to correctly recognize and classify a fire situation. Second, to detect the flame areas present in the image classified with fire. The whole process of detection is carried out autonomously from the moment it receives an image until its final decision.

3.1 Dataset Description and Organization

Before deploying the fire detection system, a varied set of images was collected and selected to cover a range of real world scenarios. The image selection process was carried out carefully to train and test the classifier, and maximize the correct classification of future image samples, containing images from a multitude of landscape regions and seasons of the year. The set of relevant images was selected from the Internet, and contained a total of 882 images organized into four classes, as described in Table 3.

Table 3. Dataset classes statistics

Classes	Number of images
Day + Fire	247
Day + No fire	448
Night + Fire	101
Night + No fire	86
	882

In the construction of the *dataset*, specific image content characteristics were taken into consideration in order to obtain a *dataset* that is most representative of reality. This aimed also to increase the training set variability and, hopefully, reduce the number of future false positives classified by the system. The selected images contained several types of forest zones, daytime and nighttime scenarios and fires with several dimensions, among other features. In order to keep track of all these variants, a metadata set of information was created, containing all the characteristics associated to each image. Table 4 depicts the variables and associated values considered in the collected metadata information. The advantage of using this metadata is the possibility of performing various tests depending on the desired characteristics. With this, one could verify, to what extent, a certain characteristic negatively influences the performance of the classifier.

Table 4. Metadata associate with images

Variables	Values
Mode	Day; Night
Flames	0; 1
Smoke	0; 1
Clouds	0; 1
Fog	0; 1
Human elements	0; 1
Terrestrial surface type	0 (other); 1 (forest)
Vegetation color	0 (other); 1 (green)
Fire stage	None; Initial; Advanced

The images have mainly, but not exclusively, an aerial perspective and were taken away from the area of interest. Due to the adoption of this research criterion, the images available for the collection were more restricted. In addition, it was a concern to collect as many images as possible with initial fire, both for day and night scenarios, since the main purpose is that the system is able to detect small spots of fire. None of the images used were submitted to any preprocessing, and therefore they possess their original varied resolutions.

3.2   System Pipeline

The automatic fire detection system proposed in this paper, uses a pipe and filter architecture with three sequential modules, as illustrated in Fig. 1. Initially, the images are acquired and then analyzed in sequence by the features extraction module, followed by the classification module. At this point, a correct decision is expected on the existence or not of a fire. In the last module, all images previously classified as fire are subjected to the flame detection module. However, flames are not always visible and detectable, which in such cases, leads to the conclusion that the fire was detected due to the presence of smoke (Fig. 2).

3.3   Features Extraction

The extraction of descriptors from an image can be performed through a DCNN. The chosen DCNN for this module was Inception-V3 which output vector serves

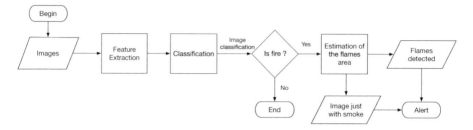

Fig. 1. Diagram of the fire detection system pipeline

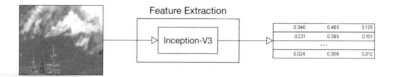

Fig. 2. Flowchart of the features extraction module

as the input basis for the next classification module. All DCNNs have a descriptor extraction layer and a classification layer at the end of their network. However, it is possible to disassociate the classification layer in order to test various learning algorithms, to find the best solution. Briefly, this module extracts from each image a numerical vector of features. Two different images will originate two different vectors. To acquire a numerical representation of the images, they are fed to the DCNN Inception-V3, which in this case will discard the classification layer and consider only the output of the penultimate layer of the network. This layer returns a vector with 2048 descriptive values of each image (cf. base extracted features) (Fig. 3).

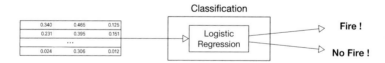

Fig. 3. Flowchart of the classification module

3.4 Classification

The classification module follows the descriptor extraction module, receiving as input the 2048 descriptors of each image. The goal is now to obtain a binary response, whether or not there is a fire in the image, provided the descriptors supplied as input to the classifier. The classifier model was tuned by a training stage. Generically speaking, the training followed a supervised learning model by using a set of input images, for which, the output associated class (of each image) was known (cf. training set). After testing several classifiers, it was concluded that the Logistic Regression (LR) model offered the better performance for our purpose. However, since the set of images has both daytime and nighttime situations, and taking into account the significant differences between them, it was decided to use separated classifiers for day and night images.

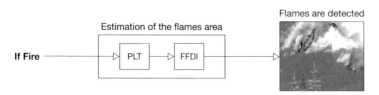

Fig. 4. Flowchart of the flame area estimation module

3.5 Flame Area Estimation

The flame area estimation module builds on the previously developed CAFE project [15]. In order to achieve a more efficient detection and, at the same time, reduce the number of false positive situations, instead of using only the FFDI color index, the CAFE approach introduces a preprocessing module, in the Lab color space, which provides better results. This module, generically, aims to estimate the area of visible flames in an image. Since the classification module does not specify whether a fire was detected with smoke, with flames, or both, in this module, such omission can be solved through the following rule: if flames are detected in the image, then the image is composed of smoke and flames; if no flames are detected then it may be presumed that the module is facing an image where the fire signal is just smoke (Fig. 4).

4 Evaluation

With respect to the classification task of the system, two different evaluations are carried out. On the one hand we intend to analyze the performance of the classification model when trained with the whole set of images. On the other hand, a second evaluation is carried out to study the influence of each characteristic on the classification task, with the help of metadata. Finally, the system's ability to correctly detect flaming areas is analyzed.

4.1 Performance of Trained Models with the Entire Set of Images

The two classification models (day and night classifiers), were trained and evaluated with the whole set of images, by following the *Cross Validation* evaluation methodology.

The evaluation of the day and night models culminated in an accuracy of 94.1% and 94.8% respectively. Given the number of images used in the training stage, the results obtained are very promising and clearly demonstrate the model capability to distinguish fire situations (Fig. 5).

		Predicted						Predicted		
		FIRE	NO FIRE	Σ				FIRE	NO FIRE	Σ
Actual	FIRE	221	26	247		Actual	FIRE	94	7	101
	NO FIRE	15	433	448			NO FIRE	3	83	86
	Σ	236	459	695			Σ	97	90	187
		(a) Daily Model						(b) Nightly Model		

Fig. 5. Confusion matrices

4.2 Influence of Image Characteristics in the Classification Process

In order to analyze the influence of the characteristics present in the images, in the classification process, a test was performed for each of the metadata represented in Table 4. The process consisted in, for each characteristic, to separate on a 70/30 proportion, the images characterized by having or not that specific characteristic.

For example, 70% of the fog images are separated for training while the remaining 30% are reserved for testing purposes. Thus, the training set is composed by the union of those 70% with the remaining *dataset*. In the end, the 30% of the fog images previously reserved, are used to test the performance of the model with the evaluation of a specific feature in mind (Table 5).

Table 5. Influence of image characteristics in the classification process

Tested characteristic	Negative influence?
Fog	YES
Clouds	NO
Humanized elements (day)	NO
Open field landscapes	NO
Varied tonality vegetation	NO
Initial fire	YES
Advanced fire	NO
Humanized elements (night)	YES

Through the tests performed with the metadata variables, it was verified that, for daily scenarios, FP situations are especially due to the presence of fog. FN situations, however, occur mainly in cases of very small initial fires. During the night, it has been proven that artificial lighting contributes to the increase in the number of FP cases.

4.3   Estimation of the Area of the Flames

Considering that the CAFE approach was used to detect the flaming zones, the tests performed attempted to compare the CAFE solution, that is, the PLT component followed by the application of the FFDI block, versus the use of FFDI only (FFDI [9] approach). The images used come from 4 daytime scenarios with different characteristics, as shown in the Table 6, which allows to evaluate the CAFE approach in different situations. The detailed methodology used is described in a previous work [15].

Table 6. Characteristics of the considered daytime scenarios

A	Little vegetation, rocky area
B	Burned vegetation
C	Dry brownish surface
D	Housing area

Table 7. CAFE versus FDDI approach in daily scenarios (TP and FP results)

	Flame area in source IMG	TP FFDI	TP CAFE	FP FFDI	FP CAFE
A	28	28	27	3	0
B	7	7	5	241	0
C	25	25	23	106	0
D	6	6	5	114	4

In Table 7 it is verified that in the scenarios presented we obtain similar results as to the identified TPs, however, there is a noticeable difference in the number of detected FPs. In this way we have been able to improve our flame detection mechanism and the respective regions affected by the fire.

5   Conclusion

This work presents a system of automatic detection of forest fires, oriented to operate continuously, day and night, and in different forest scenarios. Its development took into account the possibility of being applied in mobile cameras, as in drones, and static, as in watchtowers.

The architecture of the detection system starts by extracting descriptors from an image. The choice of DCNN Inception-V3 was based on being the better one, in terms of precision, when compared with others. In the classification process, the supervised learning model LR demonstrated, when trained with the whole set of images, a high detection accuracy in both daytime (94.1%) and nighttime (94.8%) scenarios. However, there are cases of false negatives that should be avoided because they can not occur when the system is actually used.

The results obtained after tests with the metadata, clarify that the presence of fog during the day, and of artificial illumination lights at night, are the characteristics with greater responsibility for the occurrence of false positives by the classification model. They further clarify that detecting small fire spots on an image is a challenging task and requires a more voluminous training set.

The estimation of the area of the flames is the last step to be applied in this process of detection of forest fires. The proposal used, the CAFE, is an improvement of an index used in the detection of flames, the great difference being translated into a smaller number of false positives detected.

Acknowledgements. This work was partially funded by: FCT-Fundação para a Ciência e Tecnologia in the scope of the strategic project LIACC-Artificial Intelligence and Computer Science Laboratory (PEst-UID/CEC/00027/2013); and by Fundação Ensino e Cultura Fernando Pessoa.

References

1. Gabinete do Secretário de Estado das Florestas e do Desenvolvimento Rural, Despacho n.o 9081-E/2017 de 13 de Outubro, Portugal (2017)
2. de Rigo, D., Libertá, G., Houston Durrant, T., Artés Vivancos, T., San-Miguel-Ayanz, J.: Forest fire danger extremes in Europe under climate change: variability and uncertainty, Luxembourg (2017)
3. EUFOFINET: Project Detection Synthesis of Good Practices, Zvolen
4. Li, M., Xu, W., Xu, K.E., Fan, J., Hou, D.: Review of fire detection technologies based on video image. J. Theor. Appl. Inf. Technol. **49**(2), 700–707 (2013)
5. Alkhatib, A.A.A.: A review on forest fire detection techniques. Int. J. Distrib. Sens. Netw. **10**(3), 597368 (2014)
6. Huttner, V., Steffens, C.R., da Costa Botelho, S.S.: First response fire combat: deep leaning based visible fire detection. In: 2017 Latin American Robotics Symposium (LARS) and 2017 Brazilian Symposium on Robotics (SBR), pp. 1–6 (2017)
7. Sharma, J., Granmo, O., Goodwin, M., Fidje, J.T.: Deep convolutional neural networks for fire detection in images. In: Engineering Applications of Neural Networks, vol. 744, pp. 183–193. Springer International Publishing (2017)
8. Lee, W., Kim, S., Lee, Y.-T., Lee, H.-W., Choi, M.: Deep neural networks for wild fire detection with unmanned aerial vehicle. In: 2017 IEEE International Conference on Consumer Electronics (ICCE), vol. 10132 LNCS, pp. 252–253 (2017)
9. Cruz, H., Eckert, M., Meneses, J., Martínez, J.-F.: Efficient forest fire detection index for application in unmanned aerial systems (UASs). Sensors **16**(6), 893 (2016)
10. Vipin, V.: Image processing based forest fire detection. Int. J. Emerg. Technol. Adv. Eng. **2**(2), 87–95 (2012)
11. Gomes, P., Santana, P., Barata, J.: A vision-based approach to fire detection. Int. J. Adv. Rob. Syst. **11**, 1–12 (2014)
12. Sedlak, V., Stopjakova, V., Brenkus, J.: A real-time method for smoke detection in monitored forest areas. In: International Conference on Applied Electronics (AE), pp. 1–6 (2017)

13. Cai, M., Lu, X., Wu, X., Feng, Y.: Intelligent video analysis-based forest fires smoke detection algorithms. In: 2016 12th International Conference on Natural Computation, Fuzzy Systems and Knowledge Discovery (ICNC-FSKD), pp. 1504–1508 (2016)
14. Luu-Duc, H., Vo, D.T., Do-Hong, T.: Wildfire smoke detection based on co-occurrence matrix and dynamic feature. In: International Conference on Advanced Technologies for Communications (ATC), pp. 277–281 (2016)
15. Alves, J., Soares, C., Torres, J., Sobral, P., Moreira, R.S.: Color algorithm for flame exposure (CAFE). In: 2018 13th Iberian Conference on Information Systems and Technologies (CISTI), pp. 1–7 (2018)

Cybersecurity Threats Analysis for Airports

George Suciu[1,2(✉)], Andrei Scheianu[1], Ioana Petre[1],
Loredana Chiva[1], and Cristina Sabina Bosoc[1]

[1] R&D Department, BEIA Consult International,
Peroni 16, 041386 Bucharest, Romania
{george, andrei.scheianu, ioana.petre, loredana.chiva,
cristina.bosoc}@beia.ro
[2] Telecommunication Department, University Politehnica of Bucharest,
Bd. Iuliu Maniu 1-3, 061071 Bucharest, Romania

Abstract. In today's society, the cyber-security subject raises a lot of uncomfortable questions and gives no alternative to cyber-attacks. Thus, for a better understanding of the problem, people should know that such attacks are deliberate malicious acts undertaken via cyber-space to compromise the system directly and extract valuable information. In terms of airport informational security, there are a lot to take into consideration, from digitalized processes (e.g. flight display, check-in process) to all human activities (e.g. ensuring guidance to the gates, welcoming people on board). Due to the latest trend of BYOD (Bring Your Own Device), the introduction of the new generation e-Enabled aircraft, such as Boeing 787 (including all the technologies they support and wireless connectivity) contributed to the growth of the cyber-security breaches. In order to reduce these events, people should understand the danger they can easily generate using, for instance, mobile terminals during a flight. For the prevention of malicious attacks, the authorized personnel have to conduct an in-depth analysis of all the security interconnections which were established between different networks. The main idea of this paper is to describe several use-cases where the cyber-security field related to airports is brought to the public attention.

Keywords: Cyber-security · Cyber-attacks · Airport · Threats · Vulnerabilities

1 Introduction

The aviation segment is one of the basic foundation frameworks that is not only defenseless against physical dangers, yet in addition digital dangers, particularly with the expanded use of BYOD at the airport terminals [1]. It has been perceived that there is currently no cyber-security standards set up for airplane terminals in the United States as the current benchmarks have primarily centered around aircraft CS (Control System).

As the aviation domain develops and new technologies rises, the cyber-threat will adjust. Cyber-security refers to the protection of data systems (hardware, software and the associated infrastructure), the data on them, but also on the services they provide, from unauthorized access, damage or misusage [2].

© Springer Nature Switzerland AG 2019
Á. Rocha et al. (Eds.): WorldCIST'19 2019, AISC 931, pp. 252–262, 2019.
https://doi.org/10.1007/978-3-030-16184-2_25

Airport cyber-security can be characterized as the avoidance of the reaction to several purposeful activities taken through cyber-resources to bargain an airport terminal's framework specifically or indirectly where those networks assume an important role in the wider aviation system [3].

In spite of the considerable number of activities carried on to accomplish the security of the international airports, the flight area represents an issue which must be settled quickly [4]. The attack that occurred against the United States of America on September 11, 201 by which terrorists took control of the planes, crashing two of them into the Twin Towers of the World Trade Center in New York, killing all the people on board and many other people represented the start of an overall assaults arrangement, such as body bombs (2009 and other endeavors which happen even today), endeavors to demolish airplanes by using extemporized hazardous gadgets (observed mainly in 2010), etc. Other significant terrorist attacks on airports which have caused death of a high number of the passengers or just injured them, occurred in Moscow, Russia (Domodedovo Airport) and Sofia, Bulgaria (Burgas Airport) on 2011, respectively 2012 [5].

Cyber-security has become a national essential and a government priority. Improving cyber-security policies will help the authorities to protect consumers and business. Protection refers to any activity that aims to ensure the functionality, continuity and integrity of critical infrastructure to diminish and neutralize a threat or vulnerability. Today, cyber-security threats can be performed via USB, large-scale, targeted botnet, social networks, and cross-site scripting Web, phishing from "trusted" third parties and data exfiltration and also insider threats [6].

The fast advances in aircraft and network connectivity technologies raise concerns about cyber-security risks for aircraft. The external connectivity of new aircraft models and current aircraft changes add the risks of cyber-security attacks to flight operations and also to the airline's services and businesses [7].

Rapidly expanding data communication services and widely available mobile devices allow the airlines to provide internet connectivity to passengers. While the passengers are allowed to access the aircraft networks, the aircraft becomes vulnerable to cyber-security threats during flight. Since aircrafts do not have on-board IT assistance, monitoring, detection and reporting malicious activities in real time could be a viable alternative to providing the necessary protection [8].

The RMS Cyber Loss Experience Database has pointed out that cyber losses take place in affairs of all dimensions and activities, from a large number of causes [9]. The EU General Data Protection Regulation (GDPR) came into force on May 25, 2018. This represents a major change when speaking of the results of leaks in the personal data of the European citizens. GDPR aims to prevent the growing tendencies in data exfiltration by standardizing data protection laws across Europe.

The article consists of three sections, as follows: Sect. 2 of this paper provides a general description of related work to address the cyber-security issues, while Sect. 3 analyzes use cases and proposes solutions to control cyber-security attacks. The conclusions of the paper are presented in Sect. 5.

2 Related Work

This section illustrates a series of results from the selected literature to emphasize the importance of cyber-security awareness among people and companies and, also, several methods used to prevent the global attempts of not respecting legislation related to cyber-security and personal safety in airports.

With the increase of cyber-attack amongst civil aviation sector in 2014 CANSO (Civil Air Navigation Services Organization) established a research group for improving the implementing provisions of sectoral methods between the ANSP (Air Navigation Service Provider) providers [10].

With all the available technologies there are still breaches in the security of an airport, especially on international airport computer systems, in-flight aircraft control systems and air traffic management systems [11] which need the attention of authorities.

Nowadays, the process of digitalization and the innovative devices, but also the systems regularly associated with the Web, establish solid risks for flight cyber-security. The study presented in paper [12] is only a portion of a wide debate regarding the bundle between critical infrastructures and cyber-security.

In "A taxonomy and comparison of computer security incidents from the commercial and government sectors" [13] are discussed and classified several incidents of cyber-physical attacks based on sectors, sources, targets and impact of the incidents. This article presents an example of how the classification of cyber-incidents information can help the victims and aids in understanding the threats and the consequences.

Also, in paper [14] are presented four dimensions of the taxonomy which can help researchers to understand better the previous attacks and find solutions for the future cyber-threats. The first dimension of the taxonomy describes the attack behavior and the second is responsible with the target of the attack. Last but not least, the third and the fourth dimensions are treating vulnerabilities and payloads.

The framework in [15] describes kernel components in cyber-terrorism and defines cyber-terrorism from six perspectives: target, motivation, domain, attack method, perpetrator action and the impact of the attack. Also, the proposed framework provides a dynamic way to define cyber-terrorism and describe its significances.

Incident analysis security ontology research presented in [16] also provides a taxonomy which has some similarities with the framework presented in [15] but includes aspects such as action and unauthorized results added in their classification.

The taxonomy proposed by Giraldo et al. classifies cyber-physical systems by focusing on CPS specific features such as: domains, defenses, attacks, network security, research trends, security-level implementation and computational strategies. In terms of types of attacks there are numerous ways of initiating it such as worm, trojan, virus, DDos, whistleblower, denial of service or account hijacking. When it comes to the targeted sector, the most common attacks are redirected to the Government, private sector, industries or single individuals. The are many intentions when launching a cyber-attack, including service delay, extracting sensitive data, political repercussions or even death [17].

Several studies directed throughout the years have demonstrated that the number of cyber-attacks is rising [18]. For example, Accenture Security iDefense has noticed cyber-criminal use of deception policies (including anti-analysis code, steganography, and expendable command and control servers used for secrecy).

On Friday, 12 May 2017, it was launched a large cyber-attack using WannaCry [19]; in just a few days, WannaCry ransomware virus targeting Microsoft Windows operating systems infected more than 230.000 computers. The way this virus works is quite simplistic: once activated, it claims ransom payments in order to unlock the infected system [20].

A monitoring system to prevent cyber-security attacks on aircrafts was illustrated in [21]. The security monitoring system is an application running in the ACD (Aircraft Control Domain) and it is configured to gather event logs from all the applications in the ACD. Logs contain the harshness level of the security event occurred. High harshness logs are processed in order to find out if a security alert needs to be sent to the system or to the responsible person. For example, a log aroused when the system receives an unasked-for or a sudden ACARS (Aircraft Communications Addressing and Reporting System) message indicating a potential malicious communication represents a high severity security event log.

3   Airports Cyber-Security Attacks – Description and Prevention

In this section are analyzed several use cases in which are identified potential vulnerabilities of the APOC (Airport Operations Centre). There are also simulated situations where is identified if a communication system is vulnerable or not to different cyber-attacks.

3.1   Use Case 1: Slow and Steady Infiltration to Steal Data

A team of motivated and skilled hackers are willing to infiltrate through an airport security in order to steal sensitive data. Apart from gathering information, their purpose is to delete their tracks by destroying parts of the airport IT system [22].

The attack begins with the well-known phishing method in which the computers are compromised via an attachment or URL redirecting to a website hosting the malware. Once the attack is launched, the harmful program will start credential escalation in order to take control over hosts (see Fig. 1).

Fig. 1. Description of the stealing data method

In order to avoid detection, the data exchanged among infected equipment will be multiple-encrypted so that it acts like a proxy server but with no ability to decode the information.

With full access to the Active Directory, cyber-criminals can access, or worst, can sell all the data related to flights and intelligence about airport stakeholders.

Impact on Airport: The stolen data could refer to basic information about flights or sensitive details about delays, as well as passenger's personal information. The data leakage, due to the new GDPR, could easily bring up to 5% of annual profit.

One attack can also bring along multiple future attacks in order to damage the whole IT infrastructure, which will cause long term disruption of the airport.

Vulnerabilities: An airport network consists in a lot of members, from employees to devices, that can become a vulnerability in terms of an attack. The entities vulnerable to threats are presented below:

- Hardware: Lack of periodic replacement schemes;
- Software: Well-known flaws in the software, Poor password management;
- Network: Unprotected communication lines;
- Personnel: Insufficient security training;
- Organization: Lack of formal process for access right review (supervision) or lack of records in administrator and operator logs.

Mitigation: In an airport context this means first identifying the critical IT assets that have to be secured and then evaluating the associated costs. If the airport loses people' confidence and the revenue is higher than the cost of mitigation, the resources that have been compromised must be replaced. However, this is an imprecise judgement given the difficulty of anticipating the eventual costs of any future attack. The measures that can be taken in case of a cyber-attack include:

- Using white-lists and identification as well as scanning of networks;
- Traffic profiling and understanding what traffic is expected on the network;
- Educating the staff to be aware of cyber-security issues;
- Establishing relations with cyber-security authorities that would send Indicators of compromise (IoC) to be searched for on the network.

3.2   Use Case 2: Major Integrity Loss

A team of highly motivated people intend to disrupt operations at any European airports. To accomplish their purpose, they send incorrect flight information to the desired airport using a worldwide messaging service used by airlines, airports and other aviation related companies. For a hacker, it's relatively easy to gain both physical and digital access to a connection by compromising one of these legitimate businesses [22].

In the next step, they are willing to send false information to the right target. When searching for the target and the messaging service names on a proprietary search engine, the address of the target can be easily found.

Flight information sent to an airport is formatted according to IATA (International Air Transport Association) specifications. Attackers need to know how to write the wrong messages they want to send. This task can be very easily done because there are public servers around the world, where all relevant information is illegally stored. These documents define the MVT (Aircraft Movement Message) specification, which is enough to disturb an airport (see Fig. 2).

Fig. 2. Description of the scenario

In the end, a list of incoming and outgoing flights must be created that helps the attacker to know details about departures and arrivals. The useful information can be found either on the airport's own site or on the other freely available sites such as FlightRadar24. The attackers now have everything, even they can write a script that sends information related to flights in a right IATA syntax.

Once the attack begins, the AOP (Airport Operations Plan) will receive inconsistent updates but will not be able to check the sender's list because it is not required to enter a sender's address in a message. It is also possible to use another stakeholder address.

Impact on Airport: The message sent by the attackers will introduce false data into the AOP. The following effects will be the result of this:

- Flights will be postponed since planed parking spot will be disturbed by messages indicating that airplanes need to stay longer. Furthermore, the departing flights sequence will be disturbed by these delays;
- Meanwhile, real data will be handled by IT systems and will be interleaved with incorrect data send by the hackers. All this could cause a significant AOP slowdown due to the information that needs to be processed.

Vulnerabilities: There is a List of Possible Threats:

- Network: Single point of failure; Lack of identification and authentication of sender and receiver;
- Personnel: Lack of monitoring mechanisms.

Mitigation: A solution of mitigating such an attack could be to confirm all messages with operational impact by secondary means. Anyway, such duplication could be significantly expensive and should be carefully implemented with a sufficient separation so that if the endpoint is compromised then the primary and secondary resources are not compromised.

Also, XML messages could be used instead of plain text because the XML structure contains several fields, one of which could contain a certificate confirming that a message has been emitted from a permitted source. If it is known that the endpoint has been digitally compromised, it may be included in the blacklist, as its identity is known through XML structure. It could take some time until someone identifies the source. If plain text is used, the messaging provider will be contacted to detect the sender and block messages. This could take several hours.

In case of a major integrity loss, a temporary mitigation would be to reduce the connection to the messaging service. This means that the AOP will not be updated by legitimate messages.

3.3   Use Case 3: Security Model for Cyber-Attacks Prevention

In this study [23], the main focus is on airports Critical Infrastructure (CI) as Cyber-Physical-Systems (CPS). Considering the effect of the cyber-attacks on the airport, there is necessary to elaborate a model capable to portend impact of the cyber-attacks or blackouts regarding the airport performance.

In order to create a deductive modeling approach [24], it must be created the qualitative model, which in this case, it is divided in five categories interconnected with each other: energy, ICT (Information and Communications Technology), economy, traffic and ecology (Fig. 3).

Even though aspects such as: traffic, economic and ecologic are significant for Total Airport Management (TAM), for airport operation are also required ICT systems and power supplies. The ICT categories and the energy are the base for an airport model.

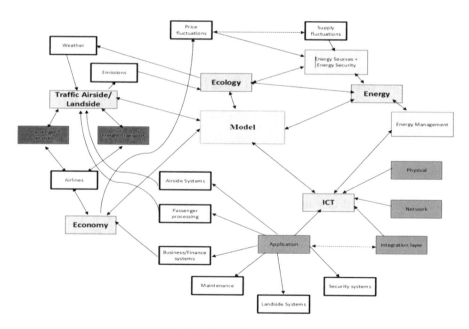

Fig. 3. The qualitative model

In a CPS, the properties of the components are various, and for that, it was applied a method named call network component objectification, which enables the visualization of partial properties and structures just for a limited number of essential properties. Thus, the method is simplified using object-based programming languages. In the model are included external factors (such as cyber-attacks) that might spread through the system timed. Two control feedback loops are used to monitor the working conditions of the objects [25, 26]. The object properties were divided into input and output parameters connected by a mathematical system description. The power-flow analysis method is used to determine unknown values for electric network stability control [27–29].

In Fig. 4 the working scheme of the dynamic grid stability analysis is presented. With blue it is presented the phase when the simulation is prepared, while the input of the control feedback loop is calculated via the power-flow analysis. While checking the grid stability, if the result is positive (the grid is stable), the dynamic load conditions are updated before starting the new time step. If the check fails, processes like load shedding/adding loads are started and then the power-flow analysis and stability check are restored. The green color marks the unaffected working conditions, while red color includes the event-based system changes. With yellow are represented the if-statements (path decisions).

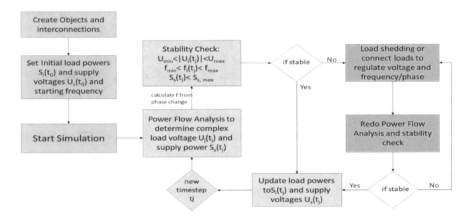

Fig. 4. Working scheme of the dynamic grid stability analysis

4 Security System Architecture

In order to create a security system architecture, two coercions were taken into consideration: ease of detection and ease of mitigation [30]. These two coercions compose a hierarchy of duties, as presented in the architecture of the system (Fig. 5). The checking module consists of several encryption and firewall tests to check information. If the result is positive the system is compromised, otherwise the system is secured.

The mitigation module is designed knowing several solutions or mitigation techniques. The module also divides the threats hierarchically and operates in order to fix them. The main purpose of the team who worked on this project is to implement the system on a networked aircraft.

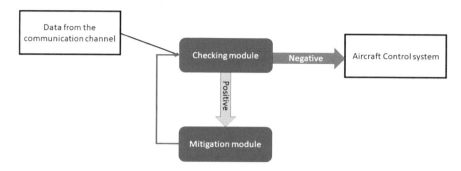

Fig. 5. The system architecture

5 Conclusions

Taking everything into consideration, attacks can have a big impact on the airplane flight operations and/or the business of the airline. Also, the external connectivity in new aircraft designs and legacy aircraft modifications exposes the aircraft to potential cyber-security threats from within and outside the aircraft. This paper has presented several use cases in which were described potential cyber-attacks and the methods to prevent such attacks. As future work, the first supposition would be to develop a prototype on how to track the main source of the cyber-events on an airport.

Acknowledgements. This paper has been supported in part by UEFISCDI Romania through projects ALADIN, ToR-SIM and ODSI, and funded in part by European Union's Horizon 2020 research and innovation program under grant agreement No. 777996 (SealedGRID project) and No. 787002 (SAFECARE project).

References

1. Gopalakrishnan, K., Govindarasu, M., Jacobson, D.W., Phares, B.M.: Cyber security for airports. Int. J. Traffic Transp. Eng. **3**(4), 365–376 (2013)
2. Aviation Cyber Security Strategy. https://assets.publishing.service.gov.uk/government/uploads/system/uploads/attachment_data/file/726561/aviation-cyber-security-strategy.pdf. Accessed 1 Oct 2018
3. National Safe Skies Alliance (2018). https://www.sskies.org/images/uploads/subpage/PARAS_0007.CybersecurityQuickGuide.FinalReport.pdf. Accessed 1 Oct 2018
4. Suciu, G., Scheianu, A., et al.: Cyber-attacks – the impact over airports security and prevention modalities. In: Trends and Advances in Information Systems and Technologies, WorldCIST 2018, pp. 1–8 (2018)
5. Bernard, L.: Emerging threats from cyber security in aviation – challenges and mitigations. J. Aviat. Manag. (2014)
6. Nojeim, G.T.: Cybersecurity and freedom on the internet. J. Natl. Secur. Law Policy **4**, 119 (2010)

7. Duchamp, H., Bayram, I., Korhani, R.: Cyber-security, a new challenge for the aviation and automotive industries. In: Seminar in Information Systems: Applied Cybersecurity Strategy for Managers, pp. 1–4 (2016)
8. Strohmeier, M., Schäfer, M., Smith, M., Lenders, V., Martinovic, I.: Assessing the impact of aviation security on cyber power. In: 8th International Conference on Cyber Conflict, pp. 1–4 (2016)
9. Cyber Risk Outlook 2018: Centre for Risk Studies, University of Cambridge (2018). https://www.jbs.cam.ac.uk/fileadmin/user_upload/research/centres/risk/downloads/crs-cyber-risk-outlook-2018.pdf
10. Lal, K., Fitzhardinge, N., et al.: Global air navigation services performance report 2014 (2014)
11. ICAO: Aviation Unites on Cyber Threat. http://www.icao.int/Newsroom/Pages/aviation-unites-on-cyberthreat.aspx. Accessed 1 Oct 2018
12. Tommaso, D.Z., Fabrizio, A., Federica, C.: The Defence of Civilian Air Traffic Systems from Cyber Threats. IAI (2016). http://www.iai.it/sites/default/files/iai1523e.pdf
13. Kjaerland, M.: A taxonomy and comparison of computer security incidents from the commercial and government sectors. Comput. Secur. 25(7), 522–538 (2006)
14. Hansman, S., Hunt, R.: A taxonomy of network and computer attacks. Comput. Secur. 24(1), 31–43 (2005)
15. Rabiah, A., Zahari, Y.: A dynamic cyber terrorism framework. Int. J. Comput. Sci. Inf. Secur. 10(2), 1–4 (2012)
16. Blackwell, C.: A security ontology for incident analysis. In: Proceedings of the Sixth Annual Workshop on CyberSecurity and Information Intelligence Research, pp. 1–4 (2010)
17. Giraldo, J., Sarkar, E., et al.: Security and privacy in cyber-physical systems: a survey of surveys. IEEE Des. Test 34(4), 7–17 (2017)
18. Accenture: Midyear cybersecurity risk review forecast and remediations (2017). https://www.accenture.com/t20171010T121722Z__w__/us-en/_acnmedia/PDF-63/Accenture-Cyber-Threatscape-Report.pdf
19. Mohurle, S., Patil, M.: A brief study of WannaCry Threat: Ransomware Attack 2017. Int. J. Adv. Res. Comput. Sci. 8(5), 1938–1940 (2017)
20. Ehrenfeld, J.M.: WannaCry, cybersecurity and health information technology: a time to act. J. Med. Syst. 41, 104 (2017)
21. Waheed, M., Cheng, M.: A System for real-time monitoring of cybersecurity events on aircraft. In: IEEE/AIAA 36th Digital Avionics Systems Conference (DASC), pp. 1–4 (2017)
22. Delain, O., Ruhlmann, O., Vaultier, E., Johnson, C., Shreeve, M., Sirko, P., Prozserin, V.: Addressing airport cyber-security – Final report (2016)
23. Nasser Al-Mhiqani, M., Ahmad, R., Mohamed, W.: Cyber-security incidents: a review cases in cyber-physical systems. Int. J. Adv. Comput. Sci. Appl. 9(1), 499–508 (2018)
24. Koch, T., Moller, D.P.F., Deutschmann, A.: Model-based airport security analysis in case of blackouts or cyber-attacks. In: 2017 IEEE International Conference on Electro Information Technology (EIT), Lincoln, NE (2017)
25. Moller, D.P.F.: Guide to Computing Fundamentals in Cyber-Physical Systems. Springer, Heidelberg (2016)
26. Astrom, K.J., Murray, R.M.: Feedback Systems - An Introduction for Scientists and Engineers. Princeton University Press, Princeton (2009)
27. Crow, M.L.: Computational methods for electric power systems. In: Grigsby, L.L. (ed.) Electric Power Engineering, 2nd edn. CRC Press, Boca Raton (2010)

28. Ilic, M.D., Zaborsky, J.: Dynamics and Control of Large Electric Power Systems. Wiley Interscience, New York (2000)
29. Grainger, J.J., Stevenson Jr., W.D.: Power system analysis. In: Electrical and Computer Engineering, 1st edn. McGraw-Hill Education (1994)
30. Kanlapuli Rajasekaran, R., Frew, E.: Cyber security challenges for networked aircraft. In: International Conference on Neuron Scattering (ICNS) (2017)

Player Engagement Enhancement with Video Games

Simão Reis[1][✉], Luís Paulo Reis[1], and Nuno Lau[2]

[1] LIACC/FEUP, Artificial Intelligence and Computer Science Laboratory,
Faculty of Engineering, University of Porto, Porto, Portugal
simao.reis@outlook.pt, lpreis@fe.up.pt
[2] DETI/IEETA, Institute of Electronics and Informatics Engineering of Aveiro,
University of Aveiro, Aveiro, Portugal
nunolau@ua.pt

Abstract. This work aims to present and summarize the identified main research fields about player engagement enhancement with video games. The expansion of video game diversity, complexity and applicability increased development costs. New approaches aim to automatize the design process by developing algorithms that can understand players requirements and redesign games on the fly. Multiplayer games have the added benefit of socially engage all involved parties through gameplay. But balancing becomes more important as feeling overwhelmed by a stronger opponent may be demotivating, as feeling underwhelmed by a weaker adversary that cannot provide enough challenge and stimulation. Our research concludes that there is still lack of research effort in the identified fields. This may be due to the lack of academy incentive on the subject. The entertainment industry depends on game quality to increase their revenue, but lack interest on sharing their knowledge. We identify potential application on Serious Games.

Keywords: Computer games · Dynamic Difficulty Adjustment · Procedural Content Generation

1 Introduction

Videos games have become one of main social devices and their consumption has steadily increased over the last few years. The produced content become more generic and the video game player base has become diverse. It is now more difficult then ever for games to offer ideal experience to players [28]. Content development now represents a major percentage of the development budget and time of production [16]. There is more incentive than ever to find ways to mitigate video games development costs [28]. Classical approaches for game difficulty design go from allowing the player to manually choose the challenge level (e.g. easy, medium or hard) to the conception of a difficulty curve by the game designer where the challenge increases as the player progresses through game levels [39].

© Springer Nature Switzerland AG 2019
Á. Rocha et al. (Eds.): WorldCIST'19 2019, AISC 931, pp. 263–272, 2019.
https://doi.org/10.1007/978-3-030-16184-2_26

These static configurations do not take into account player performance. This makes classic game design game-centered instead of user-centered [39].

Challenge can be adapted through game balancing [6]. The balance between challenge and skill of the player is considered one of the most important variables that affects players enjoyment [5]. Game balancing consists on adjustments that makes the activity fun, not easy as to bore players, nor too challenging as not to frustrate them [27]. There is the risk that players may find the adjustments unfair, decreasing their engagement [2]. It is not easy for game designers and developers to map a complex game state into one variable.

Balancing is a non-trivial task. In our opinion, current video game complexity justifies the use of Machine Learning (ML) to solve balancing problems. As first step to solve such challenge, we aimed to review existing literature in player engagement through automatic adaptive techniques. We present multiple works, analyze its strengths and weaknesses. In the end we try to understand the limitations of the analyzed work.

The rest of this work is organized as follows. In Sect. 2 we introduce Game-Flow Theory, the foundation to task engagement with games. In Sect. 3 we review the fields that explore engagement enhancement in video games from literature. We finish with Sect. 4 where we discuss our overall conclusions and present future work.

2   GameFlow

It is widely accepted that games have a wide array of purposes, from entertainment to training, but regardless of which, players must have fun, must retain their attention on the playing tasks for as long as possible. GameFlow is a model developed that explains what enhancement through game-play aims to achieve, its the application of the Flow Theory to games [32].

Flow consists on a experience so gratifying that a person would be willing to do for its own sake, as much difficult or dangerous it is for him or her [13]. Flow happens when performing a task that is goal oriented, bounded by a set of rules and require appropriated skills [32]. This setting is comparable to games, where a player is required to have sufficient skill, must receive immediate feedback on his task, and has a clear comprehension of the goal. This allows for a direct mapping between Flow and Game attributes. GameFlow maps eight different game elements to Flow Theory elements [32]:

1. The Game $\implies$ A task that can be completed.
2. Concentration $\implies$ Ability to concentrate on the task.
3. Challenge Player Skills $\implies$ Perceived skills should match challenges and both must exceed a certain threshold.
4. Control $\implies$ Allowed to exercise a sense of control over actions.
5. Clear goals $\implies$ The task has clear goals.
6. Feedback $\implies$ The task provides immediate feedback.
7. Immersion $\implies$ Deep but effortless involvement, reduced concern for self and sense of time.

3 Automatic Game Balancing and Adaptability

In this section we summarize and discuss the main fields identified in the literature that explores player engagement through automatic adaptive techniques.

3.1 Dynamic Difficulty Adjustment

Dynamic Difficulty Adjustment (DDA) is a set of real-time techniques that aims to adapt to players during video game-play. DDA tecniques react to the player performance by providing a good match between player ability and challenge [15]. It may use pre-game data and adapt the game preemptively, but it also must take into consideration real-time playing data. Personalized and adaptive experience helps to sustain the player attention for longer period of times, making DDA an interesting approach [32]. In video games, balancing can be achieved, for example, with assistance techniques at the software level [1]. The designers can focus only on controlling the development process, leaving adaptation to the DDA algorithm [14]. DDA techniques have two primary goals [36]:

- Adjust behaviour rules, numerically [38] or structurally [9].
- Generate adaptive game contents.

Behaviour-wise and content-wise approaches are not used together often [36].

Some authors already explored how ML techniques can be employed in DDA. The authors of [22] explore ranking of available actions. It is considered that a player has a set of available action/strategies and its playing against an AI agent. The actions have a natural ordering or raking, there are better strategies that the player can choose against the opponent AI, and same for the AI agent. This means the AI agent can evaluate the player performance by the actions taken by him and adapt its own actions to the player.

In [23,25] is explored modelling of playing types based on a supervised learning method. Data is provided trough game-play during testing stage. It is assumed that there is a finite number of player types. The methodology consists on three steps: cluster the recorded game traces; average the supervision over each cluster; and learn to predict the right cluster from short period of game-play.

In [24] online learning is explored. The DDA problem is formalized as a meta-game between the player and the game master, where the game master tries to predict the most appropriate setting playing over a partially ordered relation of 'more difficult than' over actions. The game is proceeded in the three steps:

1. Game Master chooses game state with a certain difficulty;
2. Player plays one episode of the game;
3. Master receives feedback and adapts its choice in the first step.

The previous three works are driven by the following three properties:

- Universality - should be applicable in the most possible range of games;

- Non-intrusiveness - should work in a seamless way, transparent to the players;
- Feasibility - should have performance guarantees.

In [36], a goal oriented system called Multi-Level Swarm Model (MLSM) that uses Particle Swarm is introduced. MLSM uses a Particle Swarm Optimization (PSO) algorithm and ecosystem mechanism to achieve game balance on content generation and behavioral control with some degree of diversity. Goal-oriented means that it ignores the relation between action and content, only focusing on the end goal.

From these works is clear that there is some attempt on using ML and optimization algorithms for DDA. These works are leaning towards general purpose solutions, where most literature leans to game specific solutions. Feasibility is where in our opinion, most solutions fail. For example, evaluating the difficulty of game states in [24].

Reinforcement Learning (RL) has been reassuring over the last years using deep learning to solve complex human tasks RL agents learning by self play were able to beat professional Dota 2 players [3]. AlphaZero [31] mastered the games of Chess, Shogi and Go. We conclude that RL is an interesting avenue for game balancing and DDA which is also a complex human task.

3.2   Multiplayer Dynamic Difficulty Adjustment

One of the most positive contributions of multiplayer games is the social interaction with friends and the challenge to beat others. On the other hand, one of the most prejudicial aspects that contributes to a decline in engagement on playing with others is mismatched skills. Weaker opponents provide little challenge and competing with more stronger opponents causes frustration [12].

Fairness in multiplayer games can be defined as a game where for all players or teams the winning ratio is statistically equal or nearly [35]. Multiplayer Dynamic Difficulty Adjustment (MDDA) explores to adapt in real-time a game taking into account the playing performance between two or more players. MDDA techniques try to adapt to the skill level of all involved players. Furthermore, giving some sort of disadvantage to the highest performing player may been seen as punishment and may affect negatively that player's engagement, and improving, or giving some sort of advantage to the less skill player may encourage the player without removing the rewarding feeling of the highest skillful player [5].

Multiplayer Online Battle Arena (MOBA) games make use of multiple fairness mechanisms [35]. One of the important steps of this class of games is character selection, where each has different abilities that contribute for clearing the game. Ban and Pick is a mechanism where the player choose characters to not be used for both teams during the next game. The players should maximize their odds by leaving characters they may want to use and ban characters they think the opponent will want to choose. Other mechanism is the balancing of characters which involves re-configuring their parameters like: disable abilities, increase stats, etc. The third set of mechanisms are on the map, environment or level. While previous mechanisms are offline (before the game starts), MOBA is a real time game, so online mechanisms need to be employed.

In [30], the authors explore dynamic adaptation in MOBA games, one of most popular online multiplayer game genres. Its a sub-genre of Real Time Strategy (RTS) games. In these games a player should choose their hero (or unit) and conquer the enemy base with its teammates. To evaluate a player's performance they identify three game features and build a heuristic performance function. The three features are: the hero's level, which represents its evolution along the a game; hero's death, the number of times the hero fall from battle and towers destroyed; how many enemy bases were conquered by the player and its teammates.

Recently in [17], agents achieve human-level capacity in Quake III Arena Capture the Flag, with only pixels and game points as input.

In [5], the authors develop a framework to describe the components of MDDA instances. The components allow for the abstraction of MDDA instances regardless of genre or mechanics. The MDDA instances identified by the authors can be described as a 7-tuple:

- Determination decides whether to use the MDDA instance using past performance or actual performance;
- Automation identifies if it is applied manually or automatically;
- Recipient indicates if the target of the instance is an individual of a team;
- Skill dependency indicates if it is required for a player to play with a certain degree of performance to activate the instance;
- Action required indicates if the instance must be initiated trough some kind of interaction from the player;
- Duration identifies the number or timing of use of an instance. Can be used once per game, multiple times per game or periodically;
- Visibility indicates which players are informed about the occurrence or use of an instance.

Dancing or racing games may have players perform without interaction or influence from the other players. These are called parallel games [28]. In non parallel games the opponents serve as obstacle to the player. Score adjustment can be used for balancing a two player game in parallel games, but it might fail in providing a better interaction between players.

From these works we conclude that MOBA games are a primary target for multiplayer game balancing and fairness. There have been recent works using RL algorithms to achieve human level behaviour in multiplayer games, were teams compete against other teams. These are interesting because incentive player interaction. As such RL is still an interesting avenue to achieve game balancing, even in multiplayer games.

3.3   Procedural Content Generation

Procedural Content Generation (PCG) aims to automatically generate game content through an algorithm [33]. PCG application to game balancing has been limited [34]. But more recently, PCG evolved to generate personalized content to players, motivated by the previous production of low-quality game content that leaded to player disengagement [37]. PCG may be divided as experience-driven and search-based [36]. Experience-driven PCG does a more controllable generation process [37]. Search-based tries to use an evaluation function of content and gradually choose the most appropriate content [33].

In [29], the authors try to tackle two main issues on PCG: of quality assurance and DDA. They develop an algorithm to address both issues by using the classic Super Mario Bros[1] platform game as a test bed. They proposed a hybrid approach for generating Constructive Primitives (CPs) using simple rules and a learning-based method. The learning method is divided in three steps: sampling; lustering; and active learning. In the sampling step a tractable data set with 19.000 entries is generated from the tailored content space. In the clustering step the sample data-set is structured. In the active learning step, 800 game segments are manually validated by one of the authors, which is an expert player. Real-time DDA is done in response to the players performance in the last played CP and affects the next CP accordingly.

Match balancing is one of the main design issues on multiplayer shooting games [20]. An algorithm for matchmaking is usually used to form balanced teams taking into account their skill, but not in all games this can be applied, for example, when players form their teams. As such, in [20] is proposed a technique based on search-based PCG with a genetic algorithm. They use a open-source shooter game, Cube 2[2] as a test-bed. The map is encoded in such away that all spaces by default are walls and only the empty spaces are encoded. It is used a fitness function that evaluates the balance between score of the players in function to the number of deaths and self-deaths. Two bots were used in a series of experiments using the same or different shooting weapon (which have different characteristics). Each bot had a different skill level. In different scenarios new wall passages were opened (walls removed).

In [18] is presented an algorithm that can classify a match as balanced or favorable to one of the teams by using convolutional neural networks, represented by an image using players weapons as parameters. 39 level generators were used to distribute small and large object trough the map to generate sufficient data for ML. Also for the data volume needed for deep learning, AI agents with pre-built behaviours were used.

3.4   Player Modeling

Player modeling consists on measuring the skill level of each individual player and represent it in a data structure. Contemporary games are being played with increase interest of game aesthetic and narrative besides the challenge of mastering a game [40]. Traditionally evaluation of user experience was based on

---

[1] http://mario.nintendo.com/.
[2] http://sauerbraten.org/.

Human-Computer Interaction techniques that do not take account game-specific characteristics like enjoyment or its purpose [7].

In [28], the authors aimed to develop a new methodology that helps developers to better understand the players. The work considers four major concepts: Developers, Game, Player and Player Context. Player model is divided in player characteristics, which describes information about the player, containing information about the way that the player interacts with the game. For example, the information could answer if the player is casual or hardcore. Context model stores information about how the game is played. For example, player location in recent mobile augmented reality games. Game model has information about how the player plays the game, what records he or she have achieved.

In [40] is explored tensor factorization in Action Role-Playing Game, developing a temporal player model for Challenge Tailoring (CT). CT generalizes DDA which in contrast performs online and offline optimization. A three dimension tensor is decomposed in player factors, spell type factors and time factors. Similar to recommender systems, where user preference is stored for each item, here the users are the players and the items are the spell types.

In [8], the authors present a model for in-game recognition of four playing styles (Competitor, Dreamer, Logician, and Strategist) based on the Kolb's experiential learning theory [19]. They use linear regression to classify players game play style. The model coefficients are found by means of heuristics and optimization, and is effective to recognize play styles with high accuracy.

3.5   Affective Computing

Affective computing is a research field that studies how video games should react to players emotions and also how it should provoke certain emotions on player [21]. The three main steps are: Measure human emotions, how to adapt the game to those emotions, and to finally provoke them. A player model or profile can be used to predict the affective state of the player [4].

In [21], the authors propose a taxonomy for affective games based on the type of feedback, direct and indirect and the scope. Direct feedback is provided from biometric sensors. Indirect feedback is inferred by how the player is maneuvering the game controllers or how is playing the game. First task is to detect and measure the affective state of the player. The second task is the adaptation of games to the affective state of the player.

4   Conclusion

Research on these subjects are yet not very wide, perhaps for the lack of incentive and application. We identified potential interest in balancing of Serious Games. Cognitive rehabilitation of patients with some disability must be early, intensive and repetitive, and maintain patient motivation and interest during rehabilitation tasks is difficult, as exercises have a (very) repetitive nature [10,11,26]. Games require cognitive and motor activity, engaging a person's attention.

With multiplayer games a social component is added. But the skill gap between players may be a demotivating factor, as such game balancing becomes crucial for the engagement enhancement of a patient.

From these work we conclude that current solutions in literature are divided in two main paradigms: DDA and PCG. These become interesting as they do not depend of specific peripherals, increasing their practical use. Modern player modeling is based on game content and leads to better evaluation of players skill. Finally, research on affective computing is still active. However, we believe that ML based solutions are more interesting avenue of research.

For future work, we aim to explore a MDDA solution scheme by modeling the game balancing problem as a meta-game, where a RL agent must learn to play a game that affects another multiplayer game's rules, assuming the role of a game master. It would learn how to adjust during game-play of a video games.

Acknowledgments. This work is supported by: Portuguese Foundation for Science and Technology (FCT) under grant SFRH/BD/129445/2017; LIACC (PEst-UID/CEC/00027/2013); IEETA (UID/CEC/00127/2013);

References

1. Altimira, D., Mueller, F.F., Clarke, J., Lee, G., Billinghurst, M., Bartneck, C.: Enhancing player engagement through game balancing in digitally augmented physical games. Int. J. Hum.-Comput. Stud. **103**(C), 35–47 (2017)
2. Altimira, D., Mueller, F.F., Lee, G., Clarke, J., Billinghurst, M.: Towards understanding balancing in exertion games. In: Proceedings of the 11th Conference on Advances in Computer Entertainment Technology, ACE 2014, pp. 10:1–10:8. ACM, New York (2014)
3. Andersen, P., Goodwin, M., Granmo, O.: Deep RTS: a game environment for deep reinforcement learning in real-time strategy games. In: 2018 IEEE Conference on Computational Intelligence and Games (CIG), pp. 1–8 (2018)
4. Bakkes, S., Whiteson, S., Li, G., Vişniuc, G.V., Charitos, E., Heijne, N., Swellengrebel, A.: Challenge balancing for personalised game spaces. In: 2014 IEEE Games Media Entertainment, pp. 1–8 (2014)
5. Baldwin, A., Johnson, D., Wyeth, P., Sweetser, P.: A framework of dynamic difficulty adjustment in competitive multiplayer video games. In: 2013 IEEE International Games Innovation Conference (IGIC), pp. 16–19, September 2013
6. Bateman, S., Mandryk, R.L., Stach, T., Gutwin, C.: Target assistance for subtly balancing competitive play. In: Proceedings of the SIGCHI Conference on Human Factors in Computing Systems, CHI 2011, pp. 2355–2364. ACM, New York (2011)
7. Bernhaupt, R., Ijsselsteijn, W., Mueller, F.F., Tscheligi, M., Wixon, D.: Evaluating user experiences in games. In: CHI 2008 Extended Abstracts on Human Factors in Computing Systems, CHI EA 2008, pp. 3905–3908. ACM, New York (2008)
8. Bontchev, B., Georgieva, O.: Playing style recognition through an adaptive video game. Comput. Hum. Behav. **82**, 136–147 (2018)
9. Booth, M.: The AI systems of left 4 dead. In: Artificial Intelligence and Interactive Digital Entertainment Conference at Stanford (2009)
10. Burke, J.W., McNeill, M.D.J., Charles, D.K., Morrow, P.J., Crosbie, J.H., McDonough, S.M.: Optimising engagement for stroke rehabilitation using serious games. Vis. Comput. **25**(12), 1085 (2009)

11. Burke, J.W., McNeill, M.D.J., Charles, D.K., Morrow, P.J., Crosbie, J.H., McDonough, S.M.: Augmented reality games for upper-limb stroke rehabilitation. In: 2010 Second International Conference on Games and Virtual Worlds for Serious Applications, pp. 75–78 (2010)
12. Clarke, D., Duimering, P.R.: How computer gamers experience the game situation: a behavioral study. Comput. Entertain. **4**(3) (2006)
13. Csikszentmihalyi, M.: Flow: The Psychology of Optimal Experience. Harper Perennial, New York, NY (1991)
14. Hendrikx, M., Meijer, S., Van Der Velden, J., Iosup, A.: Procedural content generation for games: a survey. ACM Trans. Multimed. Comput. Commun. Appl. **9**(1), 1:1–1:22 (2013)
15. Hunicke, R.: The case for dynamic difficulty adjustment in games. In: Proceedings of the 2005 ACM SIGCHI International Conference on Advances in Computer Entertainment Technology, ACE 2005, pp. 429–433. ACM, New York (2005)
16. Iosup, A.: Poggi: puzzle-based online games on grid infrastructures. In: Sips, H.J., Epema, D.H.J., Lin, H.-X., (eds.) Proceedings of the 15th International Euro-Par Conference on Parallel Processing (Euro-Par). LNCS, vol. 5704, pp. 390–403. Springer, Berlin (2009)
17. Jaderberg, M., Czarnecki, W.M., Dunning, I., Marris, L., Lever, G., Castañeda, A.G., Beattie, C., Rabinowitz, N.C., Morcos, A.S., Ruderman, A., Sonnerat, N., Green, T., Deason, L., Leibo, J.Z., Silver, D., Hassabis, D., Kavukcuoglu, K., Graepel, T.: Human-level performance in first-person multiplayer games with population-based deep reinforcement learning. CoRR, abs/1807.01281 (2018)
18. Karavolos, D., Liapis, A., Yannakakis, G.: Learning the patterns of balance in a multi-player shooter game. In: Proceedings of the 12th International Conference on the Foundations of Digital Games, FDG 2017, pp. 70:1–70:10. ACM, New York (2017)
19. Kolb, D.A.: Experiential Learning: Experience as the Source of Learning and Development. Prentice-Hall PTR, Englewood Cliffs, New Jersey (1984)
20. Lanzi, P.L., Loiacono, D., Stucchi, R.: Evolving maps for match balancing in first person shooters. In: 2014 IEEE Conference on Computational Intelligence and Games, pp. 1–8 (2014)
21. Lara-Cabrera, R., Camacho, D.: A taxonomy and state of the art revision on affective games. Future Generation Computer Systems (2018)
22. Missura, O., Gaertner, T.: Online adaptive agent for connect four. In: Proceedings of the 4th International Conference on Games Research and Development CyberGames, pp. 1–8 (2008)
23. Missura, O., Gärtner, T.: Player modeling for intelligent difficulty adjustment. In: Gama, J., Costa, V.S., Jorge, A.M., Brazdil, P.B., (eds.) Discovery Science, pp. 197–211. Springer, Heidelberg (2009)
24. Missura, O., Gärtner, T.: Predicting dynamic difficulty. In: Shawe-Taylor, J., Zemel, R.S., Bartlett, P.L., Pereira, F., Weinberger, K.Q. (eds.) Advances in Neural Information Processing Systems 24, pp. 2007–2015. Curran Associates, Inc. (2011)
25. Mladenov, M., Missura, O.: Offline learning for online difficulty prediction. In: Workshop on Machine Learning and Games at ICML (2010)
26. Moya, S., Grau, S., Tost, D., Campeny, R., Ruiz, R.: Animation of 3D avatars for rehabilitation of the upper limbs. In: 2011 Third International Conference on Games and Virtual Worlds for Serious Applications, pp. 168–171, May 2011

27. Mueller, F., Vetere, F., Gibbs, M., Edge, D., Agamanolis, S., Sheridan, J., Heer, J.: Balancing exertion experiences. In: Proceedings of the SIGCHI Conference on Human Factors in Computing Systems, CHI 2012, pp. 1853–1862. ACM, New York (2012)
28. Oliveira, S., Magalhães, L.: Adaptive content generation for games. In: Encontro Português de Computação Gráfica e Interação (EPCGI), pp. 1–8, October 2017
29. Shi, P., Chen, K.: Learning constructive primitives for real-time dynamic difficulty adjustment insuper mario bros. IEEE Trans. Games **10**(2), 155–169 (2018)
30. Silva, M.P., do Nascimento Silva, V., Chaimowicz, L.: Dynamic difficulty adjustment on MOBA games. CoRR, abs/1706.02796 (2017)
31. Silver, D., Hubert, T., Schrittwieser, J., Antonoglou, I., Lai, M., Guez, A., Lanctot, M., Sifre, L., Kumaran, D., Graepel, T., et al.: Mastering chess and shogi by self-play with a general reinforcement learning algorithm. arXiv preprint arXiv:1712.01815 (2017)
32. Sweetser, P., Wyeth, P.: Gameflow: a model for evaluating player enjoyment in games. Comput. Entertain. **3**(3), 3–3 (2005)
33. Togelius, J., Yannakakis, G.N., Stanley, K.O., Browne, C.: Search-based procedural content generation: a taxonomy and survey. IEEE Trans. Comput. Intell. AI Games **3**(3), 172–186 (2011)
34. Togelius, J., Preuss, M., Beume, N., Wessing, S., Hagelbäck, J., Yannakakis, G.N., Grappiolo, C.: Controllable procedural map generation via multiobjective evolution. Genet. Program. Evolvable Mach. **14**(2), 245–277 (2013)
35. Wu, M., Xiong, S., Iida, M.: Fairness mechanism in multiplayer online battle arena games. In: 2016 3rd International Conference on Systems and Informatics (ICSAI), pp. 387–392, November 2016
36. Xia, W., Anand, M.: Game balancing with ecosystem mechanism. In: 2016 International Conference on Data Mining and Advanced Computing (SAPIENCE), pp. 317–324, March 2016
37. Yannakakis, G.N., Togelius, J.: Experience-driven procedural content generation. IEEE Trans. Affective Computing **2**(3), 147–161 (2011)
38. Yannakakis, G.N., Maragoudakis, M.: Player modeling impact on player's entertainment in computer games. In: Ardissono, L., Brna, P., Mitrovic, A. (eds.) User Modeling 2005, pp. 74–78. Springer, Heidelberg (2005)
39. Yun, C., Shastri, D., Pavlidis, I., Deng, Z.: O' game, can you feel my frustration?: Improving user's gaming experience via stresscam. In: Proceedings of the SIGCHI Conference on Human Factors in Computing Systems, CHI 2009, pp. 2195–2204. ACM, New York (2009)
40. Zook, A.E., Riedl, M.O.: A temporal data-driven player model for dynamic difficulty adjustment. In: Proceedings of the Eighth AAAI Conference on Artificial Intelligence and Interactive Digital Entertainment, AIIDE 2012, pp. 93–98. AAAI Press (2012)

Multimedia Systems and Applications

VR Macintosh Museum: Case Study of a WebVR Application

Antoni Oliver$^{(\boxtimes)}$ ⓘ, Javier del Molino ⓘ, Maria Cañellas,
Albert Clar, and Antoni Bibiloni ⓘ

Multimedia Information Technologies Laboratory (LTIM),
Department of Mathematics and Computer Science,
University of the Balearic Islands, Palma, Spain
{antoni.oliver,m.canellas,albert.clar,
toni.bibiloni}@uib.es, j.delmolinol@estudiant.uib.es

Abstract. Virtual reality (VR) applications have spread in the entertainment sector and are being introduced in other areas. Popularity of VR headsets as well as mobile headsets has created a lot of opportunities for VR applications, which can also be played in web pages thanks to HTML5's WebVR. We propose a case study of a VR museum of Macintosh computers, which can be played using VR and non-VR interfaces as a web application. Participants are invited to explore the virtual world and interact with historic Apple Macintosh computers in an immersive, virtual world. The system has been implemented using A-Frame, a Mozilla-backed WebVR library that enabled us to create a multiplatform interface. User behavior is recorded in real time leveraging the experience API (xAPI) and may be analyzed both in real time and offline, so we can understand it better. This data, plus feedback received as part of the case study are discussed in the end of this article.

Keywords: Virtual reality · Virtual museum · WebVR · xAPI · VR analytics

1 Introduction

The use of interactive multimedia technologies in 3D, including virtual reality (VR) and augmented reality (AR), has been possible thanks to notable advances in the performance of consumer hardware, spread of affordable, immersive and non-immersive interfaces, as well as a rapid growth in the available bandwidth, enough to transmit large volumes of data, required for net-based interactive multimedia applications.

This technological situation has popularized the concept of cyberspace, as a new reality of space-time, where the users are ready to change 2D to 3D interfaces. Popularity of 3D videogames, Internet communities and 360-degree videos have increased the users' familiarity, expectations and applications with three-dimensional techniques. Nowadays, these applications have become common in entertainment, and similar experiences are a reality in other areas, such as education, cultural promotion, formation, tourism or electronic commerce, offering improved interfaces in contrast to conventional 2D environments.

© Springer Nature Switzerland AG 2019
Á. Rocha et al. (Eds.): WorldCIST'19 2019, AISC 931, pp. 275–284, 2019.
https://doi.org/10.1007/978-3-030-16184-2_27

This article covers an immersive VR multimedia application, the Mac Museum, in which users are invited to interact with antique Apple Macintosh computers in a virtual world, while their actions are recorded to be analyzed. The system was demonstrated in *Ciència per a tothom* (Science for all), an event organized by the University to promote the use of the Science among primary and secondary education students.

This section includes a revision of the current state of the art in Virtual Reality interfaces, libraries and applications. Section 2 describes the implementation of the Mac Museum. Section 3 explains the case study held in *Ciència per a tothom*, including an overview of their actions. Finally, the conclusion can be found in Sect. 4, containing the feedback collected from the demonstration and the future steps for this project.

1.1   State of the Art

Current dedicated 3D devices include headsets like the Oculus Rift[1], HTC Vive[2], PlayStation VR[3] and Google Daydream[4] or mobile headsets like the Samsung Gear VR[5], or even the low-cost Google Cardboard[6].

Typically, VR systems are native applications for the platform which will play them, introducing a differentiated marked for every technology and different development workflows. Web applications are easily spread to the public and, thanks to the advances in HTML5, the current technologies enable them to address a wide range of use cases.

WebVR is an HTML5 specification[7] that enables browsers to use virtual reality (VR) devices, including sensors and head-mounted displays on the Web, so they can be played not only using the VR hardware listed above but also directly on computers and smartphones without any kind of VR hardware. Multiple libraries and frameworks wrap WebVR and the graphics specification WebGL[8], like A-Frame[9], React 360[10], BabylonJS[11] or Primrose[12]. WebVR is being used in the literature for 360-degree videos [1], virtual reality [2], social experiences in VR [3] and interactive visualization of complex data [4].

Results obtained in literature show that sensorial immersion in VR multimedia scenes, having 2D and 3D objects to interact with, is a resource understood by the

---

[1] https://www.oculus.com/rift/.

[2] https://www.vive.com/.

[3] https://www.playstation.com/explore/playstation-vr/.

[4] https://vr.google.com/daydream/.

[5] https://www.samsung.com/global/galaxy/gear-vr/.

[6] https://vr.google.com/cardboard/.

[7] https://immersive-web.github.io/webvr/spec/1.1/.

[8] https://www.khronos.org/webgl/.

[9] https://aframe.io/.

[10] https://facebook.github.io/react-360/.

[11] https://www.babylonjs.com/.

[12] http://www.primrosevr.com/doc/.

users. An early study of the technology as a learning tool is given in [5], as well as more modern applications in distance learning [6] and gamification [7].

The concept of a museum in virtual reality has already been explored in [8], where the authors present a system that allows museums to build and manage VR and AR exhibitions. Other projects include a touch museum that features additional information in VR [9], a consumer technologies museum implemented in JanusVR [10] and even a location-based VR museum that uses Samsung Gear VR [11]. Other projects revolve around bringing the VR inside the museum [12–14]. We seek to provide a modern, multidevice, web-based VR museum experience that can be enjoyed at home.

2 The Macintosh Museum System

The purpose of this project was to create a virtual museum displaying the evolution of the Apple Macintosh computers as an immersive multimedia resource that could be accessed from a wide range of devices, including personal computers, mobile devices and VR headsets.

To make these computers look as real as possible in the virtual world, we displayed real-world photographs and videos in the synthetic world. The user is able to interact with 39 computers released between the eighties and the 2000's thanks to the display of images in the virtual scene, which are replaced with a video that shows the advertisement when they were released when they are touched. While we grabbed these videos from the Internet, pictures of the computers were taken by us.

Following Fig. 1, the Mac Museum system includes a VR Player based on web technologies that can be played on multiple devices. This application contains all the interactive experience of the system. Being a web application, the VR Player is delivered by a web server to the users' web browsers. This application also transmits data back to the app server, via the WebSocket protocol[13]. This data contains what the user is currently doing now and is set to the analysis module. The Monitor application is another web application that displays in real time the position and orientation of all the users that are interacting with the system.

The analysis module is a generic development we first created to collect usage data in 360-degree videos [15, 16]. It uses the Experience API (xAPI) to provide a standard interface to collect usage data from the application.

2.1 The VR Player

The VR player is the web application that displays the virtual museum of Macintosh computers. Users can play the experience in four different interfaces, to move around the gallery and interact with the computers.

In all the devices, the looks of the application and its virtual world are very similar: the player stands in a first-person view in the middle of a large room, surrounded by Macintosh computers that have their name below them. A virtual cursor appears in the

---

[13] https://html.spec.whatwg.org/multipage/web-sockets.html.

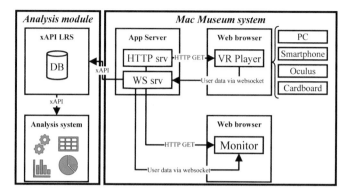

Fig. 1. General architecture.

middle of the screen, used to select those elements to interact with them; the cursor will change its size to notify the user that they are on a valid target. When selected, the computer will boot, show the TV commercial that featured this computer in its screen and a summary of the specification of the machine will appear in the scene (see Fig. 2). On a corner of the screen, the "Enter VR" button appears, used to display a stereoscopic version of the scene so it can be viewed with a VR headset, like the Oculus Rift or the Cardboard.

When accessing the application with a personal computer, no additional hardware is required: the user can look around by moving the mouse, since the application captures the pointer, turning the player in an experience similar to a first-person shooter videogame; movement is performed using the keyboard WASD or arrow keys; and selection is issued by clicking when a target is aligned with the virtual cursor that is displayed in the middle of the screen, displaying the technical information next to the computer.

When using a smartphone, the hardware changes drastically; the user can benefit from the gyroscope of the device to look around the scene; their movement is limited, though, and is performed through the selection of a target: when the user looks at a

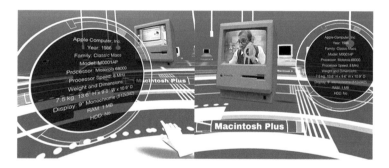

Fig. 2. Macintosh computer selected, with a video playing and technical information. On the left, it has been selected using the Oculus interface, so the text appears on the opposite hand's palm; on the right, as it is shown in the other interfaces.

computer and taps the screen, it will boot and the player will automatically move next to their target, so they can watch the video and read the technical information.

To use the Oculus interface, the user needs to use their computer, as in the first interface and then push the "Enter VR" button. This time the user enjoys a true VR experience and is able to look around by moving their head; the movement is performed with the joysticks of the Oculus Touch controllers; and the user can turn on a Macintosh computer by touching them with their virtual hands using the "point gesture" (see Fig. 3), that eliminate the need of a virtual cursor, and are used to display the computer specifications: when touching a computer with a hand, the text will appear on the palm of the other hand, as a piece of paper we would read (see Fig. 2-left).

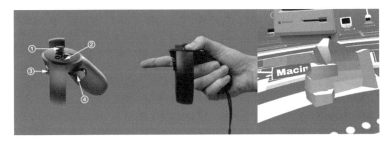

Fig. 3. Oculus Touch controller and hand gesture. The thumbstick (1) is used to move around; to make the "point gesture", one must place their thumb on the thumbrest (2), not press the trigger (3), and press the grip button (4) with their middle finger.

Finally, to use a Cardboard or another mobile VR headset, one should use their phone as in the second interface, push the "Enter VR" button and slide their phone into the headset. Following that second interface, the user still controls the orientation thanks to the gyroscope of the device, that is now attached to their head; movement is still linked to the selection of a target; but this time the user cannot touch the screen to select a computer: for this reason, we introduced the "gaze" control, in which a target will be selected when the user stares at it during two seconds.

We imposed ourselves the requisites that the application had to be multiplatform and as easy to obtain as possible, so we leveraged WebVR and other Web technologies to create a web application that could be accessed from browsers in a wide range of devices. The first decision was to choose the WebVR framework: at that moment, while A-Frame was already relatively mature, with a rich example set, documentation and community support, React 360 (then React VR) was just released. For this reason, we chose A-Frame as the WebVR framework (which internally uses Three.js to interface with WebGL). Despite this, the application was written as a React[14] application, using the aframe-react[15] package to interface these two libraries. React enabled us to follow a more modular design and to create a modern web application, thanks to

---

[14] https://reactjs.org/.

[15] https://github.com/ngokevin/aframe-react.

the set of tools that come with its ecosystem: WebPack[16], Babel[17], and all their extensions. The communication between the web applications and the application server is via the WebSocket protocol, leveraged by the Socket.IO library[18]. The application server was also coded in Javascript, using Node.js[19], the express module[20] to serve the application via HTTP, and the Socket.IO library again to handle Web-Socket connections. The A-Frame scene included specific components[21] to provide appearance, behavior and/or logic, some of which were created especially for this application.

2.2   Integration with the Analysis Module

The Mac Museum system leverages our existing interactive video analysis module to record user behavior. A Websocket connection is stablished with the application server and is used to report whenever a client starts a session; moves or looks around; the video in a computer starts, stops or finishes playing; and enters or leaves the VR Mode via the "Enter VR" button.

The following information is sent and stored via xAPI into the Learning Record Store[22] (LRS), the entity responsible for storing this information: type of interaction, unique session id, device type, position and orientation. When applicable, the following information is also collected: accept-language, referrer, user-agent HTTP headers (when the user starts a session), trigger (was the computer touched by the virtual hand, clicked/tapped or *fused* via gaze?) current video playback time (when the user interacts with a computer).

Once stored, this information is available via the standard xAPI; in our case, we use the tool we developed previously to obtain an overview of what did the users do.

2.3   The Monitor Application

The idea of the Monitor application is to serve as a map that shows where different users are in the 3D scene. Its purpose was to be a first step of live usage analysis, so the route of users can be observed in real time. Also an A-Frame application, it borrows much of the features from the VR Player, but this time the view is static from the top and triangles are displayed to represent the current location and orientation of the players (see Fig. 4).

The Monitor registers as such with the websocket server and receives information about the players when they emit changes (see the previous subsection). The triangles on the scene and the HTML list on the right are updated accordingly.

---

[16] https://webpack.js.org/.

[17] https://babeljs.io/.

[18] https://socket.io/.

[19] https://nodejs.org/.

[20] https://expressjs.com/.

[21] https://aframe.io/docs/0.8.0/core/component.html.

[22] https://xapi.com/learning-record-store/.

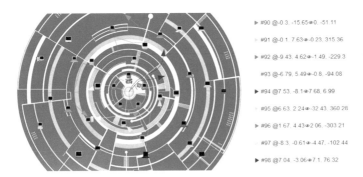

Fig. 4. The Monitor application, displaying all players in the virtual scene.

3 Case Study: *Ciència per a tothom*

The system was demonstrated at the science fair at the University, *Ciència per a tothom*[23] (Science for all). Hundreds of school and high school students (ages 6–16) attended the sessions on May 10th, 11th and 12th, and many of them visited our venue (Fig. 5).

Fig. 5. The setup at *Ciència per a tothom*: on the left, a player uses the Oculus Rift + Touch interface, what she sees is mirrored on the left screen and the right one displays the Monitor application; on the right, a group of students use the Cardboard interface.

Our stand contained two of the four interfaces explained previously: we set up the Oculus Rift + Touch hardware in a table, so a single user could play with the system using the virtual hands while their colleagues saw what the player saw in a monitor; and in another table we asked the attendants to use their own smartphones to browse to the web application and slide their phones in a plastic version of the google cardboard, enabling them to enjoy the virtual world in groups of around 6 people. Participants

---

[23] http://seras.uib.cat/ciencia/2018/.

were given a short training before using the system but were not given a specific task to perform more than to "explore the virtual world".

185 mobile sessions were recorded during the case study that interacted with at least a computer. Due to the overwhelming success of the Oculus interface, individual sessions per user could not be recorded and the data is aggregated. In total, Macintosh computers were selected 1436 times from smartphones and 995 times from the Oculus interface.

Computers that were in front of the player (when they were facing the table) who was using the Oculus interface are the most selected (Fig. 6a, bottom part), their paths are clearly observed in Fig. 6b, as they even tried to walk outside the allowed area. When using a mobile headset, all the users started in the center of the scene and spread in all directions, although computers in the center were more visited. Notice the circles of red points in Fig. 6a, they are the result of the moving mechanism when using a mobile device, as the users can only move by teleporting to a computer, at a fixed distance.

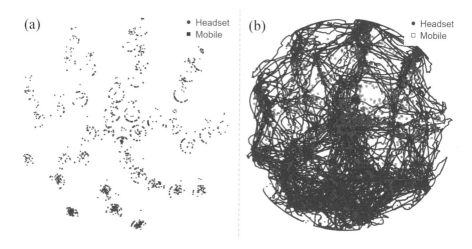

Fig. 6. Spatial representation of the data collected in the case study. (a) computers selected grouped by interface type; (b) paths walked by the users grouped by interface type.

More analysis could be performed from our tracked data, of course. We only give some insights in this article, so that its potential may be perceived.

4   Conclusion and Future Work

First, the use of A-frame and their components allowed us to create a VR web application that can be displayed in four different user interfaces. The system proved to be suitable to provide this experience to a group of players simultaneously in the case study. The data collected in the case study and the feedback collected on-site shows that the virtual reality is a new concept that is understood by young users.

Despite the very good reception of the demonstration, participants showed difficulties when performing the "pointing gesture" in the Oculus interface: they were required to place their thumb on the thumbrest (see Fig. 3) and most did not understand that even though there was no button there, it was necessary. Others tried to move and touch the computers with the same hand, so their thumb was on the joystick. After the experiment, we are certain that the "pointing gesture" should be more flexible.

Since there was a physical table with screens in the area, many players thought that walking towards that place was the "right direction" and ignored the rest of the scene. Also, due to the success of the experiment, participants rushed to play the system and continued from the state of the previous player, trying to advance even further.

The entities in the scene, when wearing the Oculus interface, are on the plane $y = 1.5$ m. While may seem reasonable, the youngest players could not reach the objects because of not being tall enough. In the case of adolescent players, a group often formed around the now unable to see player, annoying him or her.

Being an immersive activity, the Virtual Reality isolates the player from the real world, losing track of space and time. Some users were not aware of the group of people around them and hit them when trying to reach the computers or walked into the table. Some users forgot they were in a shared environment and would have kept playing unless they were forced to stop.

Most players asked what the goal of the game was. When they were told that there was not a goal, that they just had to explore the museum, they seemed confused. Some players were also very rigid and expressed their concern of not seeing anymore the real world distressed them. This complicated the tasks of selecting a computer, since they did not raise their arms nor moved their fingers correctly.

When using the Cardboard interface, users were relatively more relaxed. They were also confused for the lack of an objective, but they turned the experiment into a social experience, since the group was playing at the same time, they talked among themselves and tried to find a specific computer and even tried to find themselves –which was impossible–. They understood that they had to stare at a computer to select it and move next to it, but they also tried to move physically to move in the virtual world, to no avail.

In the future, besides fixing these small issues, we will introduce a short tutorial to introduce the user to the virtual reality and how to perform the actions in the interface they are using. Since users demanded an objective to play, we will add easy instructions that will direct players to find a specific target. Finally, we will assign an avatar to the player, that is going to be added to the scene, so users will be able to see each other in the virtual world, augmenting the social component of the experiment.

Finally, a formal usability evaluation will be performed to compare the interfaces supported by the system.

References

1. Pakkanen, T., Hakulinen, J., Jokela, T., Rakkolainen, I., Kangas, J., Piippo, P., Raisamo, R., Salmimaa, M.: Interaction with WebVR 360° video player: comparing three interaction paradigms. In: 2017 IEEE Virtual Reality (VR), pp. 279–280. IEEE (2017)
2. Dibbern, C., Uhr, M., Krupke, D., Steinicke, F.: Can WebVR further the adoption of Virtual Reality? In: Hess, S., Fischer, H. (eds.) Mensch und Computer 2018 - Usability Professionals, pp. 377–384. Gesellschaft für Informatik e.V. Und German UPA e.V., Bonn (2018)
3. Gunkel, S., Prins, M., Stokking, H., Niamut, O.: WebVR meets WebRTC: towards 360-degree social VR experiences. In: 2017 IEEE Virtual Reality (VR), pp. 457–458. IEEE (2017)
4. Hadjar, H., Meziane, A., Gherbi, R., Setitra, I., Aouaa, N.: WebVR based interactive visualization of open health data. In: Proceedings of the 2nd International Conference on Web Studies - WS.2 2018, pp. 56–63. ACM Press, New York (2018)
5. Chittaro, L., Ranon, R.: Web3D technologies in learning, education and training: motivations, issues, opportunities. Comput. Educ. **49**, 3–18 (2007)
6. Coyne, L., Takemoto, J.K., Parmentier, B.L., Merritt, T., Sharpton, R.A.: Exploring virtual reality as a platform for distance team-based learning. Curr. Pharm. Teach. Learn. **10**, 1384–1390 (2018)
7. Lam, J.T., Gutierrez, M.A., Goad, J.A., Odessky, L., Bock, J.: Use of virtual games for interactive learning in a pharmacy curriculum. Curr. Pharm. Teach. Learn. **11**, 51–57 (2018)
8. Wojciechowski, R., Walczak, K., White, M., Cellary, W.: Building virtual and augmented reality museum exhibitions. In: Proceedings of the Ninth International Conference on 3D Web Technology - Web3D 2004, p. 135. ACM Press, New York (2004)
9. Zhao, Y., Forte, M., Kopper, R.: VR touch museum. In: 2018 IEEE Conference on Virtual Reality and 3D User Interfaces (VR), pp. 741–742. IEEE (2018)
10. Subramanian, A., Barnes, J., Vemulapalli, N., Chhawri, S.: Virtual reality museum of consumer technologies. In: Advances in Intelligent Systems and Computing, pp. 549–560. Springer, Cham (2017)
11. Lugrin, J.-L., Kern, F., Schmidt, R., Kleinbeck, C., Roth, D., Daxer, C., Feigl, T., Mutschler, C., Latoschik, M.E.: A location-based VR museum. In: 2018 10th International Conference on Virtual Worlds and Games for Serious Applications (VS-Games), pp. 1–2. IEEE (2018)
12. Carrozzino, M., Bergamasco, M.: Beyond virtual museums: experiencing immersive virtual reality in real museums. J. Cult. Herit. **11**, 452–458 (2010)
13. Izzo, F.: Museum customer experience and virtual reality: H.BOSCH exhibition case study. Mod. Econ. **8**, 531–536 (2017)
14. Chang, H.-L., Shih, Y.-C., Wang, K., Tsaih, R.-H., Lin, Z.: Using virtual reality for museum exhibitions: the effects of attention and engagement for National Palace Museum. In: PACIS 2018 Proceedings (2018)
15. Bibiloni, T., Oliver, A., del Molino, J.: Automatic collection of user behavior in 360° multimedia. Multimed. Tools Appl. **77**, 20597–20614 (2018)
16. Oliver, A., del Molino, J., Bibiloni, A.: Automatic view tracking in 360° multimedia using xAPI. Communications in Computer and Information Science, pp. 117–131. Springer, Cham (2018)

Web-Based Executive Dashboard Reports for Public Works Clients in Construction Industry

Alvansazyazdi Mohammadfarid[1,4],
Nelson Esteban Salgado Reyes[2(✉)], Amir Hossein Borghei[3],
Alejandro Miguel Camino Solórzano[4],
Maria Susana Guzmán Rodríguez[1],
and Mario Augusto Rivera Valenzuela[1]

[1] Universidad Central del Ecuador, Av. Universitaria, Quito 170129, Ecuador
alvansaz@gmail.com, {sguzman,mariverav}@uce.edu.ec
[2] Pontifical Catholic University of Ecuador, Quito 170135, Ecuador
nesalgado@puce.edu.ec
[3] Project Estimator at Reputable Construction and Fitout Company Based
in Sydney, Sydney, Australia
[4] Universidad Laica Eloy Alfaro de Manabí, Manta 130803, Ecuador
camino2h@yahoo.es

Abstract. Public sector clients are responsible for multitude of construction projects, and each of these organizations deal with a multitude of projects every year. Hence, executives and top-managers are overloaded with projects' information and this issue possibly reduce their efficiency in the process of monitoring and control as well as their decision making. By adopting an executive dashboarding approach from business into construction, busy managers can have a quick view of projects' performance and their up-to-date status with adequate attention to more sensitive or problematic ones. An extensive literature review was conducted to study the dashboards in business and to identify the information needs of executives in client organizations. By means of interviews with executives and conducting questionnaire-based surveys, the important categories and subcategories of information required for executive dashboards were distinguished and importance of them were identified. Finally, based on the previous researches as well as the results of the interviews and questionnaire surveys, a conceptual framework, a standard input form and a schematic view of the executive dashboard system has been proposed.

Keywords: Executive dashboard · Project status · Performance reporting · Construction clients

1 Introduction

Construction industry is one the most important industries in the economy of any country. This industry has many stakeholders involved in every project from cradle to grave. One of the important stakeholders in Iranian construction industry are city municipalities who has the authority to work as a client for public work construction,

© Springer Nature Switzerland AG 2019
Á. Rocha et al. (Eds.): WorldCIST'19 2019, AISC 931, pp. 285–294, 2019.
https://doi.org/10.1007/978-3-030-16184-2_28

similar to JKR in Malaysia. Hence a city municipality is the client for numerous construction projects in their jurisdiction. According to annual statistics of municipality of Tehran (2009) more than 2000 construction projects were under construction in 2009 and this figure is increasing. These projects include highways, roads, tunnels, bridges, buildings, parks, sewerage system, sport complexes, and so forth. As it is obvious a huge amount of reporting is needed for top managers and executives in this organisation to monitor and control several projects, however most of the time because of lacking a systematic approach for monitoring and control some projects are neglected. Managers in these large organisations are often overwhelmed with reports and information produced from a multitude of projects and people which all compete for managers' attention. This phenomenon is generally known as information overload [6]. The problem is further worsened when reports are poorly designed with respect to how information is presented, which often distract than guide decision makers' attention. This issue happens in business world as well, and similarly executives and managers in a large corporation are flooded with information from numerous sources. In contrast, if the managers are provided all relevant data, expressed in a visual exhibit that makes organised displays of useful information, they can observe and monitor progress and performance of all projects with special attention to more sensitive or problematic projects. In the early years of the 2000s with software prices coming down, new ideas such as business intelligence (BI) technologies hitting the market, and data sources opening up, dashboarding suddenly became a mainstream word in corporations and governmental organisations worldwide. For managers to make the best possible decisions in the shortest amount of time, it is essential to turn data into structured information and then present this information to them in a format that is easy to read and that supports effective monitoring and control [1, 3, 4]. In recent years, software vendors have embraced this need, and now numerous solutions, commonly referred to as Dashboards, have emerged on the market, however the majority of these solutions are suitable for business organisations but not for construction industry. According to Few [2] dashboard measures are intended to give management a quick view of projects' performance, or in other words performance of the projects at-a-glance. An effective deployment of dashboards within an organisation can dramatically reduce the need for multiple ad-hoc reports. It will also support better decision making and ultimately help improve the process of monitoring and control [5, 7]. Because most dashboard tools are highly graphical, dynamic, and easy to use, with simple training, users across an organisation can be empowered to monitor and analyse the information relevant to their areas of responsibility and to make informed decisions [6]. With this tool, top management finally has the information needed to ensure they are focused on only the most critical project issues. Hence, dashboarding can work as an effective monitoring tool for busy managers in large organizations such as municipalities. Thus, the goal of this study is to adopt a dashboarding system from corporate world into construction industry to improve the process of reporting and performance monitoring and control for managers and executives in construction organizations who deal with several projects and multitude of reports.

1.1   Research Methodology

This study employed both the quantitative and qualitative research methodology. As Fig. 1 suggests, to achieve the objectives of this research, after gathering information among literature review, Primary Data were collected from interview with a panel of top-managers in client organisations. This interview aimed to identify the current system of project progress monitoring and reporting in target organisations and the information needs of top-managers. After that a questionnaire based on the literature review and also responses from interview was developed. The questionnaire was distributed among respondents who were involved in the public sector construction projects in Tehran, Iran and they were among top-managers and executives, including deputy regional mayors, district mayors, and regional project control audits. They were introduced with the concept of dashboard and its benefits for their organisations and they were asked to state the importance of each category and sub-category of information they may need in an executive dashboard by choosing a scale of 1 for "Unimportant" to 5 for "Very Important" based on Likert Scale. In addition, they were asked to recommend a reasonable time interval for the dashboard information to be updated. In addition, Secondary Data was collected from books, journals, articles and websites to complement the results of interviews and questionnaires. After gathering information through questionnaires and interviews, the complete data analysis is presented. Finally, based on the results of interviews and questionnaire survey as well as literature review, a conceptual framework, a standard form for gathering information needs in dashboard and a schematic graphical presentation of suggested executive dashboard were developed. In addition, conclusions and recommendations has been conducted.

1.2   Interview with Expert Panels

Semi-structured interviews were used as method for data collection. The semi structured interview is a powerful research tool, widely used in research in many fields, and is capable of producing rich and valuable data [8]. In this study, interview was conducted with five respondents in a managerial position in the target organization to identify the current system of monitoring and reporting, as well as their information need for effective monitoring. They were asked to state their information need from projects and their opinion for the current system of reporting as well as their comment on the appropriate performance criteria that should be included in a dashboard report.

1.3   Questionnaire Survey

The questionnaire survey is a powerful method to uncover current conditions in many fields of research and used to gather resourceful data [9]. The questionnaire has been created based on the related published work on the subject of "Executive Dashboards and Lean Methodology", results of the interview with Executive managers in the construction and related books and websites, own authors knowledge in the construction industry.

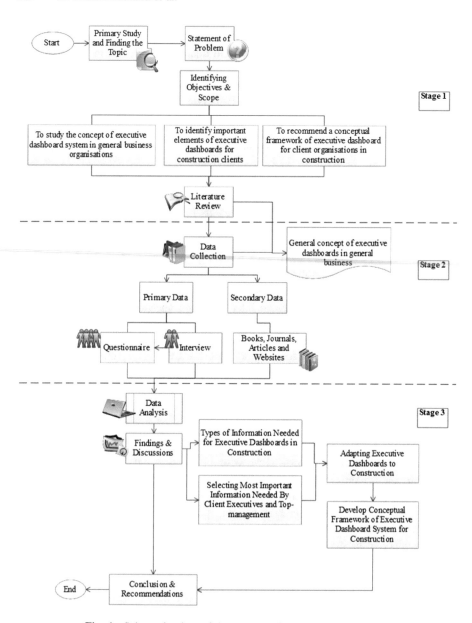

Fig. 1. Schematic view of the process of research methodology

Questions in the questionnaire survey purposed to collect information on the followings:

i. The important categories of information and performance measures in an Executive Dashboard for top manager in construction
ii. The importance of each sub-category of information and performance measures in an Executive Dashboard
iii. The suggested update interval for the Dashboard information to be updated.

1.4   Data Analysis

To identify current system of performance reporting and to find out important categories of information in the viewpoint of executives in client organisations semi-structured interviews with five respondents in top-managerial levels were conducted. The respondents were among "District Mayors" in Municipality of Tehran.

The collected responses were analyzed by "SPSS 18" software Package which is the recommended software for statistical analysis in academia. The data collected from the responses of the questionnaire survey were analyzed using the Average Index methodology based on the following formula [10]. In this method, the frequency of respondents determined the classification of each project information and performance criteria. The average index achieved by each criterion would determine the level of importance for the criteria in an executive dashboard. Importance of each category and sub-category of information in executive dashboards were identified according to the [10] Average Index as Table 1:

Table 1. Average importance index.

Average Index (AI)	Importance criteria of performance level
$1.00 \leq AI \leq 1.50$	*Not Important*
$1.50 < AI \leq 2.50$	*Less Important*
$2.50 < AI \leq 3.50$	*Moderately Important*
$3.50 < AI \leq 4.50$	*Important*
$4.50 < AI \leq 5.00$	*Very Important*

From the interviews the following results can be inferred:

- Current system of reporting of project performance status in target organisation is mainly based on ad-hoc paper-based reports which do not follow a regular pattern of time.
- The most important information that top-managers want to know about the status of projects in their area of responsibility consist of last status of projects, information about cost, time and quality, major problems that interrupt normal progress of project, satisfaction of supervision team on the performance of project and contractor, and non-compliances to specification, standards, and contract provisions.

- Depending on the importance and sensitivity of project the update interval for information in a dashboard can be from daily to weekly or even monthly.
- Most respondents in the executive managerial level do not need to go to details. In most cases they only want important and summarised information about the performance status of projects with problematic areas to be highlighted.
- The respondents suggested some features for an executive dashboard drill-down capability and history keeping that has been taken into consideration in the proposed executive dashboard.

1.5   Development of Conceptual Framework of Dashboard Report for Construction

Based on the dashboard samples in business organizations and literature on the conceptual framework for "Web-based Reporting Systems" as in the work of [12] and "Performance Monitoring and Reporting Systems" in an article from [13] a conceptual framework for the web-based executive dashboard report system has been adopted which is based on the need of construction industry.

2   Developing a Conceptual Framework

This conceptual executive dashboard was derived from multitude of samples of executive dashboards in business which was illustrated and discussed in the books of Few [2], Malik [6], Kerzner [15] and Watson [16]. In addition, results from the interview and the questionnaire survey were used to identify the important pieces of information on performance measure that need to be presented in an executive dashboard report meant for top-managers in public sector client organizations.

The first page in the dashboard Fig. 2 is designed in a way that gives the management a quick view of the performance of all projects that are in progress in their district. A colored matrix is used to indicate overall performance of projects in each category of dashboard reports. As mentioned earlier, each color has special meaning, i.e. green means that the performance of that category is satisfactory, while yellow means that project has some minor problems, and red color means that an urgent attention to that performance measure is required. Managers can click on the problematic project to view further information on the status of that project in more detailed view. The Project Dashboard page provides a useful "dashboard overview" of the current status of the project. By viewing a managerial dashboard chart, any user can quickly see the status of key items, metrics and Key Performance Indicators (KPIs) on a project. Conceptual web-based executive dashboard was derived, which allows top-managers to monitor the up-to-date status of projects and their performance.

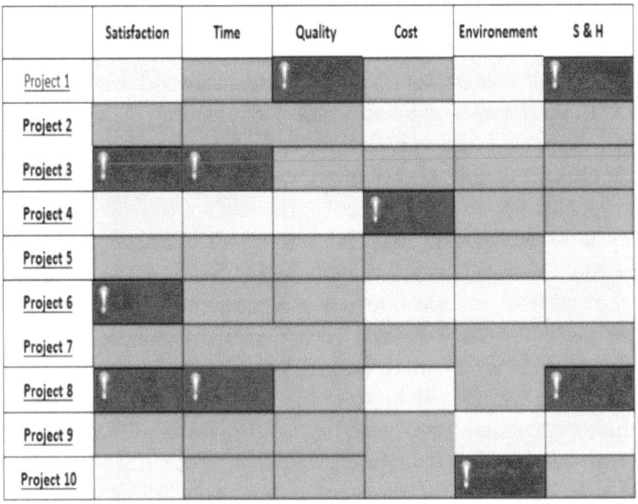

	Satisfaction	Time	Quality	Cost	Environement	S & H
Project 1			!			!
Project 2						
Project 3	!	!				
Project 4			!			
Project 5						
Project 6	!					
Project 7						
Project 8	!	!				!
Project 9						
Project 10					!	

Fig. 2. First page of executive dashboard - overview of all projects

3   Results and Discussion

To achieve the objectives of this study, an exhaustive literature review was conducted. In addition, the questionnaires were distributed. The required categories and sub-categories of required information and its optimal update interval for an executive dashboard were identified. A total of 50 sets of questionnaires were distributed to the respondents by Google Docs package and 18 completed questionnaires were returned. According to analysis of results, there are seven categories of information which are required by top-managers in an executive dashboard which consist the following and sorted according to their importance in the perspective of respondents including Last Status of Projects, Client Satisfaction, Time Information, Information Regarding Quality, Cost Information, Information Regarding Environment, and Information on Safety and Health. Based on the definition of executive dashboard [11], they should only contain important pieces of information to be presented to top-managers. Therefore, sub-categories of information with average importance index of higher than 3.5 was selected to be included in the executive dashboards. Based on the analysis of results from the questionnaire survey, the categories of information in an executive dashboard and their sub-categories that should be included in executive dashboard are shown in Fig. 1. The results according to Fig. 2 show that top-managers need some categories in the dashboard report to be updated more regularly such as "Last Status of Project", "Supervisors' Satisfaction" and "Time Information". In contrast, some other information on "Safety and Health" and "Information Regarding Environment" may be updated less frequently. "Cost Information" was the only category of information in the perspective of top-management in target organization which requires monthly updates to get a meaningful results in. To establish the third objective a conceptual framework was recommended that was based on previous research studies [12, 14], and the major elements of the proposed system were identified. Surveys, interviews and literature searches were conducted to establish the structure and data requirements of the conceptual framework. The conceptual framework of the web-based executive dashboard system consists of three key components such as Recording the required information in standard forms, Appraisal and verification, and Dissemination and presentation. The information recording module allows engineers, supervisors and other client's representatives or consultants to submit up-to-date data on each category of required information by top-managers through the web-based interface. The appraisal module aims to ensure that the data is valid, reliable and accurate. In this module, all raw data will first be checked and authenticated by the superintendent as the system administrator, and the verified data will then be the raw information for visualizing. The outcomes of this module are the performance scores and rankings and other qualitative and quantitative data, which will be stored in a database. The dissemination module makes the required processed and visualized information available to respective registered users (top-managers) via the web–based interface. This information is mostly graphical and easy to digest for top-level of management. They can click on each category of information on the application to view the most important pieces of information pertinent to the status of the project. The application also has a summary of all data in first page which allows the user to drilldown and view performance measures and supporting details for problematic projects or more sensitive ones.

4 Conclusion

The aim of this research was to formulate a professional executive dashboard reporting system for construction organizations' top managers who work as public-sector client. All the three objectives set for the research have been successfully achieved and the findings are summarized based on the objectives of the research as follows:

i. To study the concept of executive dashboard systems used in general business organization. As a conclusion for this objective, it can be said that academic literature and books suggest that executive dashboards are gaining considerable attention in recent years and business users especially in executive and top-managerial levels of medium sized or large enterprises are interested in implementing the dashboards in their organizations because of its numerous benefits in monitoring and control, decision making, etc. So in recent years many researchers and also software vendors tried to develop this concept for business. However, very limited number of them has considered this concept for construction industry.

ii. To identify important elements of executive dashboard report systems for construction. As a result of this objective, the most important and important pieces of information which top managers usually require knowing about the projects in their area of responsibility was identified. The categories of information according to their importance in the top-manager's perspective were: 1. Last Status of Projects, 2. Client Satisfaction, 3. Time Information, 4. Information Regarding Quality, 5. Cost Information, 6. Information Regarding Environment, and 7. Information on Safety and Health.

iii. To recommend a conceptual framework of executive dashboard for construction client organizations. For this objective a conceptual framework for an executive dashboard was recommended which consist of web-based standard forms as inputs of the system which should be filled by relevant users, appraisal and verification of data that requires superintendents to review and authenticate the data, and dissemination and appraisal module which present the data in a graphical and easy to digest format to top-managers to inform them about the latest status and performance of multiple projects in their district.

References

1. Tokola, H., Gröger, C., Järvenpää, E., Niemi, E.: Designing manufacturing dashboards on the basis of a key performance indicator survey. Procedia CIRP, 619–624 (2016). https://doi.org/10.1016/j.procir.2016.11.107
2. Few, S.: Information Dashboard Design. The Effective Visual Communication of Data, vol. 3 (2006). https://doi.org/10.1017/s0021849904040334
3. Järvenpää, E., Lanz, M., Tokola, H., Salonen, T., Koho, M.: Production planning and control in Finnish manufacturing companies Current state and challenges (2015). https://research.aalto.fi/en/publications/production-planning-and-control-in-finnish-manufacturing-companies-current-state-and-challenges(f6749e41-8407-40d5-a8ee-6b3724a7c6c9)/export.html. Accessed 13 Feb 2018
4. Nadoveza, D., Kiritsis, D.: Concept for context-aware manufacturing dashboard applications. IFAC Proc. Vol., 204–209 (2013). https://doi.org/10.3182/20130619-3-ru-3018.00103
5. Mazumdar, S., Varga, A., Lanfranchi, V., Ciravegna, F.: A knowledge dashboard for manufacturing industries. In: CEUR Workshop Proceedings, pp. 51–63 (2011). https://doi.org/10.1007/978-3-642-25953-1_10
6. Malik, S.: Enterprise Dashboards: Design and Best Practices for IT. Wiley, Hoboken (2005)
7. Freese, W.: You cannot manage, what you cannot measure. In: Praxishandbuch Corporate Magazines, pp. 202–217. Gabler Verlag, Wiesbaden (2012). https://doi.org/10.1007/978-3-8349-3702-5_18
8. Myers, M.D., Newman, M.: The qualitative interview in IS research: examining the craft. Inf. Organ. 17, 2–26 (2007). https://doi.org/10.1016/j.infoandorg.2006.11.001
9. Chern, T.Y., Te Chen, W.: A study on the satisfaction with the service quality by land administration agent in Taiwan. In: Proceedings of the IEEE International Conference on Advanced Materials for Science and Engineering, IEEE-ICAMSE 2016, pp. 35–38 (2017). https://doi.org/10.1109/icamse.2016.7840224
10. Al-Hammad, A., Assaf, S.: Assessment of work performance of maintenance contractors in Saudi Arabia. J. Manag. Eng. 12, 44–49 (1996). https://doi.org/10.1061/(ASCE)0742-597X(1996)12:2(44)
11. DeBusk, G.K., Brown, R.M., Killough, L.N.: Components and relative weights in utilization of dashboard measurement systems like the balanced scorecard. Br. Account. Rev. 35, 215–231 (2003). https://doi.org/10.1016/S0890-8389(03)00026-X
12. Chassiakos, A.P., Sakellaropoulos, S.P.: A web-based system for managing construction information. Adv. Eng. Softw. 39, 865–876 (2008). https://doi.org/10.1016/j.advengsoft.2008.05.006

13. Cheung, S.O., Suen, H.C.H., Cheung, K.K.W.: PPMS: a web-based construction project performance monitoring system. Autom. Constr. **13**, 361–376 (2004). https://doi.org/10.1016/j.autcon.2003.12.001
14. Ng, S.T., Palaneeswaran, E., Kumaraswamy, M.M.: A dynamic e-Reporting system for contractor's performance appraisal. Adv. Eng. Softw. **33**, 339–349 (2002). https://doi.org/10.1016/S0965-9978(02)00042-X
15. Kerzner, H.: Project management metrics, KPIs, and dashboards: a guide to measuring and monitoring project performance (2015)
16. Watson, H.J., Frolick, M.N.: Determining information requirements for an EIS. MISQ **17**, 255 (1993). https://doi.org/10.2307/249771

Novel Robust Digital Watermarking in Mid-Rank Co-efficient Based on DWT and RT Transform

Zahoor Jan[1(✉)], Inayat Ullah[1], Faryal Tahir[1], Naveed Islam[1],
and Babar Shah[2]

[1] Department of Computer Science,
Islamia College University, Peshawar 25000, Pakistan
{Zahoor.jan,naveed.islam}@icp.edu.pk,
inayatbcs2012@yahoo.com, faryal.tahirkhan@gmail.com
[2] College of Technological Innovation,
Zayed University Abu Dhabi, Abu Dhabi, United Arab Emirates
babar.shah@zu.ac.ae

Abstract. Recent advancement in multimedia is facilitating humanity in data manipulation, transportation, and transmission. All at once, this improvement in multimedia data has exposed sensitive data to various threats, like threats of piracy and copyright materials. The most important application of digital image watermarking scheme are ownership protection of copyrighted materials and authorization of multimedia data. Digital watermark is an imperceptible mark embedded in digital images for various objectives including image captioning, authentication, authorization, copyright protection and proof of ownership of multimedia contents. Watermarking is a procedure of embedding secret information such as logo, number, text and image in multimedia data. The proposed watermarking scheme is based on Discrete Wavelet (DWT) and Ranklets Transform (RT) domain in which the mid-rank coefficients in low frequency sub-band (LL3) of DWT are selected for random number watermark embedding. Experimental results shows that no visible distortion among the cover and watermarked image, representing high level of imperceptibility. The scheme has been tested against various set of malicious attacks, proving that the proposed scheme computationally surpasses other state of the arts schemes.

Keywords: Watermark · Discrete Wavelet Transform (DWT) ·
Ranklets Transform (RT) · PSNR · MSE · SSIM · RMSE

1 Introduction

The effectiveness of watermarking technique can be judge through in terms of imperceptibility, robustness, security, efficiency and capacity [1]. Recently watermarking techniques gain importance and increased security demand of copyrighted materials contents and proof of ownership data [2]. It's essential to safe user data from illegal and unauthorized access, different techniques, such as steganography, cryptography and digital robust watermarking techniques are used for this purpose [3].

© Springer Nature Switzerland AG 2019
Á. Rocha et al. (Eds.): WorldCIST'19 2019, AISC 931, pp. 295–302, 2019.
https://doi.org/10.1007/978-3-030-16184-2_29

Digital watermarking techniques are used for data authentication, prevent illegal and authorized access of multimedia products [4].

2 Related Work

Invisible digital watermarking schemes can be categories into binary grouped, one is spatial domain and other is frequency domain. A dual mask is matched with the Least Significant Bits of the image pixel values, if binary mask is identical to Least Significant Bits, then pseudo-random binary numbers are added to the image values, if not equal, and then subtracted. Tai et al. [10] suggested a method, which is based on visual secrete sharing. This method was not robust and not withstand against some attacks, such as scaling, translation and rotation. These methods are not resisting against image processing attacks, liked lossy JPEG compression, adding noises, low/high pass filters, cropping, and resizing. These kinds of attacks easily damage the watermark data in the watermarked image. Cox et al. [11] proposed a non-blind watermarking technique used a spread spectrum for embedding pseudo-random number watermark data in the Discrete CosineTransform domain. The proposed watermarking scheme is withstand against image manipulation operations, such as high pass filter, low pass filter and lossy JPEG compression etc. and geometric attacks, liked scaling, translation and rotation. A DFT based digital watermarking technique is proposed by [12] and used two watermarks to embed in the lower and higher frequencies. This kind of watermark embedding technique known as circular watermarking process. The advantage of this type of watermark embedding is to resist various attacks like JPEG compression, resizing, scaling, rotation and translation etc. Khan et al. [13] presented another DCT-based watermarking technique in which contrast masking, luminance sensitivity and frequency are used to form the watermark perceptually for the host image to gain high level imperceptibility. Wang et al. [14] suggested wavelet-based digital watermarking technique, the watermark bits are inserted in different frequency sub-bands. DWT transform domain divide the cover image into four quadrant as shown in Fig. 1(a) and (b).

(a)   (b)

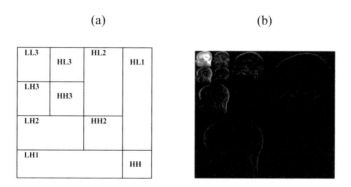

Fig. 1. Three level wavelet decomposition of brain image

3   Ranklet Transform (RT) Domain

RT domain is a multi-resolution image processing algorithm alike to that of dyadic wavelet transform [17]. The variance between these two domains as, it describes the rank of pixels values and not defined the gray intensity values. The positioning selective response is gained by suitable choice of Treatment T and Control C sets of intensity pixels values that are matched by the assessment, extra precisely in the circumstance of RT domain. These groups are well-defined based on three Haar wavelets $hi^{(\vec{x})}$, i = 1, 2, 3 as shown in Fig. 2(a), (b) and (c).

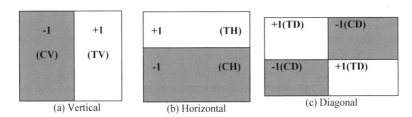

Fig. 2. The bi-dimensional haar wavelet, the letter in parenthesis shows the Treatment (T) and Control (C) region.

These T and C regions of image pixels of any orientation are first sorted in ascending order and then assigned rank values, which is basically performed to pick out the mid-ranks co-efficient from the selected image window and then these co-efficient are used for random number sequence watermark embedding. W_s^i is the function that represent the brightness of Ti [14]. The Ranklets Intensity are defined by Eq. (1).

$$R_w^i = 2W_{VW}^i/n^2 - 1 \tag{1}$$

So Ranklets response replicate the positioning selectivity features of the three Haar wavelets and Ti and Ci sets are resultant [13] and then the watermark coefficients were inserted in the low frequency sub-band LL3 of the original transformed image.

4   Proposed Watermarking Scheme

Our proposed watermarking technique is based on RT domain is a new research and area for digital images authentication and copyright protection [9]. In contrast if the watermark data are inserted in the most insignificant location, it will be very simple to conceal the watermark data but the proposed method are more robust against attacks [10].

4.1   Proposed Scheme

Let $X(i, j)$ signify the original cover image of dimension $M \times N$ and W is the watermark pattern with size $W(i \times j)$ to be inserted using additive embedding technique. The proposed watermark inserting scheme is as follows:

Step 1:   Initial read the cover image.

Step 2:   By applying DWT to divided the cover image into four quadrants: LL1, LH1, HL1 and HH1.

Step 3:   Again applied DWT on LL1 sub-band, further divided it into four multi-resolution sub-bands, i.e. LH2, HL2 and HH2.

Step 4:   Used again DWT domain on LL2 to decompose it further into four sub-band sets: LH3, HL3 and HH3.

Step 5:   Now applied RT domain on LL3 sub-band to divide it into two blocks as T and C regions of dimension are 64×64.

Step 6:   Then assigned ranks to the co-efficient of these T and C regions.

Step 7:   After that, applied Ranklets computation on both regions to find out mid-ranks coefficients from the brighter region (T) in the LL3 sub-band. The mid-ranks coefficients of the brighter region are selected for random number watermark embedding.

Step 8:   Then the Wilcoxon rank sum test statistics used to compute the Ranklets values in Treatment T region.

Step 9:   Then embed the random binary sequence watermark in the selected mid-rank coefficients LL3 sub-bands.

The following is the watermark embedding Eq. (2)

$$E = LL3 \pm K \times W(i) \tag{2}$$

Where E is the watermarked image, LL3 is the approximation sub-band of the transform image, $W(i)$ is the pseudo random number watermark, and K is the secrete key which control the watermark embedding strength in the embedding scheme. The strength of watermark K kept as a constant value.

Step 10:   Taken inverse DWT to reconstruct the watermarked image.

The watermark embedding algorithm of the proposed techniques is shown in Fig. 3.

4.2   Watermark Extraction Process

The block diagram of watermark extraction is shown in Fig. 4. The watermark embedding strength factors αi used for approximation sub-band image is set to 0.45 and 0.1 for the remaining sub-band images. The following Eq. (3) are used for watermark detection as

$$T = 1/M \sum x i . t . \alpha \, i \tag{3}$$

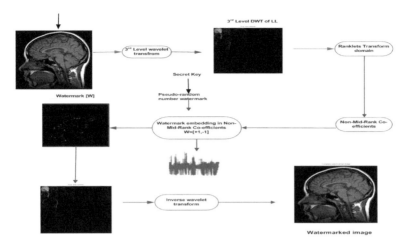

Fig. 3. The proposed watermark embedding scheme architecture.

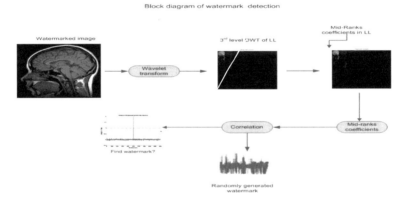

Fig. 4. The general architecture of watermark detection and extraction.

Where T is the sum of frequencies to be marked, xi is the selected frequencies; t is the watermark and αi shows the watermarking level for specific frequency values of the selected coefficients.

5 Experimental Result and Discussion

5.1 Dataset

To shows the robustness of the proposed watermarking scheme standard images has been taken from the database. Figure 5(a) shows the cover image of brain; Fig. 5(b) shows the watermarked image that shown no visual distortion to the watermarked images (Fig. 6).

Fig. 5. (a) Cover image of brain; (b) watermarked image of brain

Fig. 6. (a) Cover image of brain; (b) watermarked image of brain; and (c) difference between two images.

5.2 Imperceptibility and Robustness Results of the Proposed Watermarking Scheme

MATLAB R2012a is used for the implementation of our proposed scheme the performance, perceptual invisibility and imperceptibility of the proposed scheme can be tested and evaluated on the basis of PSNR, MSE, RMSE and SSIM. Table 1 contains the imperceptibility results of the proposed digital watermarking scheme. The robustness of the proposed robust watermarking scheme has been experienced against various attacks. Table 2 shows the robustness of the proposed watermarking scheme after various attacks.

Table 1. Imperceptibility of our proposed work

Serial#	Test images	PSNR	MSE	RMSE	SSIM
1	Bird.gif	39.3487	7.5546	2.7486	0.9909
2	Forest.gif	39.3677	7.5217	2.7426	0.9962
3	Map.gif	39.2906	7.6563	2.767	0.9990
4	Pretty_girls.gif	39.3401	7.5695	2.7513	0.9974
5	Ship.gif	39.4761	7.3362	2.7085	0.9936
6	World.gif	39.4285	7.4170	2.7234	0.9981
7	Lena.jpg	40.0750	6.3912	2.8195	0.9840
8	Brain.jpg	40.7272	5.4999	2.3452	0.9929
9	Bridge.tiff	39.3247	7.5964	2.7562	0.9983
10	Livingroom.tif	39.3659	7.5247	2.7431	0.9962
11	Barbara.tiff	39.4566	7.3691	2.7146	0.9919
12	Boat.tiff	39.3659	7.5247	2.7431	0.9941
13	Tank.tiff	39.3333	7.5815	2.8008	0.9967
14	Tank1.tiff	39.3008	7.6383	2.7637	0.9961
15	Chest.jpg	39.9370	6.5975	2.7352	0.9933
16	Peppers.png	39.3642	7.5276	2.7437	0.9957
Average 16 images		39.53139375	7.2691375	2.72540625	0.99465

Table 2. Robustness of the proposed scheme.

Kind of attacks	Attack value	PSNR	MSE	RMSE	SSIM
Low pass filter	2 × 2 window	31.9579	41.4278	6.4364	0.9671
	3 × 3 window	35.1546	19.8437	4.4546	0.9750
	4 × 4 window	31.0938	50.5476	7.1097	0.9392
	5 × 5 window	32.1131	39.9735	6.3225	0.9334
	6 × 6 window	29.4283	74.1732	8.6124	0.9008
High pass filter	2 × 2 window	31.9579	41.4278	6.4364	0.9671
	3 × 3 window	35.1546	19.8437	4.4546	0.9750
	4 × 4 window	31.0938	50.5476	7.1097	0.9392
	5 × 5 window	32.1131	39.9735	6.3225	0.9334
	6 × 6 window	29.4283	74.1732	8.6124	0.9008
Gaussian noise	G = 0.001	30.0296	64.5831	8.03	0.8548
Salt & pepper noise	S = 0.001	33.3231	30.2534	5.4151	0.9716
Speckle noise	S = 0.001	36.8496	13.4313	3.6569	0.9858
Wiener JPEG Compression	J = 10	28.7128	87.4587	9.3519	0.8737
	N = 10				
JPEG Compression	100%	39.6002	7.1294	2.6701	0.9908
	75%	39.5390	7.2306	2.689	0.9907
	50%	39.5475	7.2166	2.6864	0.9907
	25%	39.0983	8.0029	2.8289	0.9870
	10%	62.7704	0.0344	0.1854	1.0000
Median filter	2 × 2 window	31.8561	42.4106	6.5123	0.9668
	3 × 3 window	36.7789	13.6519	3.6948	0.9853
	4 × 4 window	31.3905	47.2098	6.8709	0.9548
	5 × 5 window	33.9425	26.2317	5.1217	0.9554
	6 × 6 window	30.4518	58.6000	7.6551	0.9265

6   Conclusion

DWT divided the cover image into four quadrants up to level three and applied RT domain on LL3 to divide it into two blocks as Treatment and Control C regions to extract the mid-rank coefficients for random number watermark embedding to achieve good results of imperceptibility and robustness. There is also the option to include a subheading within the Appendix if you wish.

References

1. Jabeen, F., Jan, Z., Jahangir, F.: Energy-based coefficient selection for digital watermarking in wavelet domain. Int. J. Innov. Comput. Appl. 5(1), 18–25 (2013)
2. Jan, Z., Khalid, M., Mirza, A.M., Jabeen, F., Abbasi, A., Tarin, Z., Durrani, A.: Digital image watermarking using multilevel wavelet decomposition and human visual system. In: International Conference on Information Science and Applications (ICISA), pp. 1–5. IEEE (2012)

3. Pradhan, C., Rath, S., Bisoi, A.K.: Non blind digital watermarking technique using DWT and cross chaos. Procedia Technol. **6**, 897–904 (2012)
4. Jan, Z., Mirza, A.M.: Genetic programming-based perceptual shaping of a digital watermark in the wavelet domain using Morton scanning. J. Chin. Inst. Eng. **35**(1), 85–99 (2012)
5. Rafat, K.F.: Enhanced text steganography in SMS. In: 2nd International Conference on Computer, Control and Communication, IC4 2009, pp. 1–6. IEEE, 17 February 2009
6. Kutter, M., Jordan, F.D., Bossen, F.: Digital signature of color images using amplitude modulation. In: Electronic Imaging 1997, pp. 518–526. International Society for Optics and Photonics, 15 January 1997
7. Macq, B.M., Quisquater, J.J.: Cryptology for digital TV broadcasting. Proc. IEEE **83**(6), 944–957 (1995)
8. Hoffberg, S.M.: Intelligent electronic appliance system and method. United States patent US 6,850,252, 1 February 2005
9. Pitas, I.: A method for signature casting on digital images. In: Proceedings of the International Conference on Image Processing, vol. 3, pp. 215–218. IEEE, 16 September 1996
10. Tai, S.C., Wang, C.C., Yu, C.S.: Digital image watermarking based on VSS in BTC domain. J. Chin. Inst. Eng. **26**(5), 703–707 (2003)
11. Cox, I.J., Kilian, J., Leighton, F.T., Shamoon, T.: Secure spread spectrum watermarking for multimedia. IEEE Trans. Image Process. **6**(12), 1673–1687 (1997)
12. Shieh, J.M., Lou, D.C., Chang, M.C.: A semi-blind digital watermarking scheme based on singular value decomposition. Comput. Stand. Interfaces **28**(4), 428–440 (2006)
13. Khan, A., Mirza, A.M., Majid, A.: Intelligent perceptual shaping of a digital watermark: exploiting characteristics of human visual system. Int. J. Knowl. Based Intell. Eng. Syst. **10**(3), 213–223 (2006)
14. Wang, S.H., Lin, Y.P.: Wavelet tree quantization for copyright protection watermarking. IEEE Trans. Image Process. **13**(2), 154–165 (2004)
15. Hussein, J., Mohammed, A.: Robust video watermarking using multi-band wavelet transform, 9 December 2009. arXiv preprint: arXiv:0912.1826
16. Mehmood, I., Ejaz, N., Sajjad, M., Baik, S.W.: Prioritization of brain MRI volumes using medical image perception model and tumor region segmentation. Comput. Biol. Med. **43**(10), 1471–1483 (2013)
17. Smeraldi, F.: Ranklets: orientation selective non-parametric features applied to face detection. In: Proceedings of the 16th International Conference on Pattern Recognition, vol. 3, pp. 379–382. IEEE (2002)

Multiple Moving Vehicle Speed Estimation Using Blob Analysis

Muhammad Khan, Muhammad Nawaz[✉], Qazi Nida-Ur-Rehman,
Ghulam Masood, Awais Adnan, Sajid Anwar, and John Cosmas

Centre for Excellence in IT, Institute of Management Sciences,
Peshawar, Pakistan
m.nawaz@imsciences.edu.pk

Abstract. Nowadays, traffic problems including vehicles monitoring for their speed is the need of this modern world. The video-based surveillance system is used to monitor and identifying the moving vehicles on road for safety measurement of public. The proposed work is aimed to find out the over-speeding of multiple vehicles through an automatic vehicle video tracking system. The method build for the automatic vehicles tracking systems is based on a couple of steps; initially the frame differencing is applied for the object detection. In the frame differencing, the background model is computed and the current frame is subtract from previous frame which is computed during background modeling for the feature extraction. The blob analysis is then used to track the detected object or the region of interest. Ultimately the speed is estimated to compute and find out the over speeding of vehicles. The Euclidian distance determines the centroid of every region in both horizontal and vertical direction. The overall accuracy achieved by our proposed system is 90% for the detection and measurement of speeds of multiple vehicles.

Keywords: Video acquisition · Reference frame · Blob analysis · Euclidian distance

1 Introduction

These days auto collision is the fundamental issue because of which many people pass on. To tackle this issue one way is to implement video analysis systems in light of the premise of activity base observation. It comprises of irregular action acknowledgement, checking of movement and well-being driving. The video-based analysis framework is utilized to find and distinguishing the moving item in a region. It comprises of anomalous action acknowledgement, observing of activity and security driving. The vehicle discovery and following framework can play a vital commitment for expressway movement observation control administration and the arranging of activity. The vehicle speed can be resolved to control activity issue and additionally for security driving.

Nowadays, there is too much traffic on roads and most of the time collision and accident occur because of irresponsible driving and speed. To handle this kind of

© Springer Nature Switzerland AG 2019
Á. Rocha et al. (Eds.): WorldCIST'19 2019, AISC 931, pp. 303–314, 2019.
https://doi.org/10.1007/978-3-030-16184-2_30

situation there is a need of intelligent system to control the flow of the traffic and especially fast moving vehicles. Moreover, the use of this specific intelligent traffic surveillance system will help in real-time moving vehicle detection, speed estimation, proficiencies.

In the proposed work, precise speed estimation problem was identified. One of the major reasons of this work is to estimate the speed of multiple vehicles to determine the vehicle which breaks down the rule of speed violation. The check will apply on slow, medium or over speeding to avoid the incident, theft of a vehicle, traffic jam, violation of rule, providing secure transportation, traffic monitoring. For this purpose, the background modelling through frame differencing with Blob Analysis which is used to track the vehicles and after the tracking a speed estimation algorithm is activated to estimate the speed of the vehicles.

The rest of the paper is organised as follow. In Sect. 2 the most relevant literature is presented. In Sect. 3 the methodology developed in this research work is explained in detail. In Sect. 4 the experimental results are presented. In Sect. 5 the final conclusion of this research work is explained.

2   Literature

In this part, the most significant and most recent research work connected to our study is discussed in detail.

Bodkhe et al. [1] provides a review of methods used for the purpose of vehicles and human objects detection.

Yabo et al. [3] develop a real-time system for the vehicles speed estimation using the camera install on the highways. The system is based on SURF features detection and used those features for classification of the vehicles. El Jaafari et al. [5] proposed a new and novel system for vehicle detection and tracking. The sliding window is used to detect the vehicles for features descriptor for classification, in the second phase the tracking is perform using canny edge detector. Sun et al. [11] develop an automatic system to extract six different types of features for the classification of vehicles from the rare view of the camera. The proposed model is based on two main ideas; one is hypothesis generation from multiscale driven and in the second verification of appearance based hypothesis. The different features considered in this research work are PCA, Wavelet, Quantized Wavelet, Gabor, and Combination of Wavelet and Gabor features. The evaluation of the features is performed through NN and SVM.

Sofia Janet et al. [2] presents an image processing technique for vehicles detection based on background differencing using a road image without vehicles and to subtract from the current frame, 2d median filters used to reduce the noise and then Otsu's thresholding is applied with morphological operations to find the vehicles. Uppaal [6] presents a vision base system for the detection and tracking of vehicles which is based on the infrared thermographic camera.

Mu et al. [4] present a descriptive system for video surveillance using roadside cameras to detect and track vehicles. The system is based on SIFT features matching occurrence and to find out the geometrical points of the two images to relate the two images. Krutak [7] develops a counter base computer vision system for the purpose of collision and prediction of vehicles. The system is based on to find the centroid using counter based tracking of vehicles on centroid vehicle detection and tracking system and constructed a rectangle on the vehicle for collision prediction. Satzoda et al. [8] develop a system based on Vehicle Detection using Active learning and Symmetry. The system is working on Haar-like features and Adaboost classifiers which work in both conditions fully and partially visible vehicles.

Kim [9] a real-time night and daytime vision based vehicles detection system are developed utilising background extraction and automatic vanishing detection. Kurz et al. [10] develops a long-term vehicles detection system utilising the properties of sensors. The Cell Transmission Model (CTM) is deploying to track the vehicles in the longer term in a complex environment, for this purpose the airborne images sequences. Daigavane et al. [13] present a system based on morphological operations for the detection and counting of vehicles on road The algorithm developed for this purposed consists of four major steps. First, divide the given video into multiple frames, then the registered image is produced from the frame subtraction. In the third stage, post processing is applied to segment the foreground and in the final and fourth step, the detected objects are count.

Brendan et al. [12] present two different approaches for the monitoring of highways and analysis. The first approach is Visual Vehicle Classifier and Traffic Flow Analyzer (VECTOR) in which the vehicles are classified into different types and generate different highway analysis. The second approach is build to monitor the behaviour of the traffic flow and to maintain the lane on the highway, this is basically used to detect any kind of anomaly during the traffic flow. Sepehr et al. [14] present Horn-Schunk based optical flow algorithm for the object detection and tracking from an ariel or stationary view. The median filter is used to remove extra noise and morphology operation is applied to shape the segmented object. In the final stage blob analysis is applied to track the object.

3   Methodology

Initially, the video is acquired for processing. After the video acquisition, a reference image is formed which helps in determining the position of the moving vehicle. The reference frame is used to compare the current frame with the previous frame for the feature extraction. The purpose of blob analysis in the proposed algorithm is to detect those points or regions within the frames that are different from the rest to track the moving vehicles. Speed analysis can be helpful in making runtime decisions for the concerned authority. In our work, we have applied a red, green and yellow label to distinguish slow, medium and fast speed. The detail of the proposed methodology is shown in the below Fig. 1.

3.1   Frame Differencing

After video data is acquired, the background model is computed using frame averaging in order to detect foreground pixels within the video frame. Frame averaging is a statistical method to create a background frame by the combination of multiple frames.

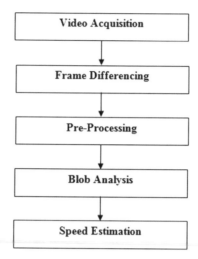

Fig. 1. Proposed methodology

For background modelling greater the number of frames for averaging, the better the results will be as the noise is decreased. In our case, we have considered 16 initial frames to the model background.

After successfully creating the background model the active or current frame is then subtract from the background model frame/image consider as a reference frame to detect the object. The basic steps for the frame differencing are as follow:

First, the background models from the initial frames are computed and get the background image/frame. In the following Eq. (1), x and y represent the coordinates of the frames/images and t represents the time interval of every frame during background modelling, and M represent the final background image computed.

$$M = (a, b, t) \tag{1}$$

Now the current frame is considered to subtract from the previous frame which is the background model image computed in the previous Eq. 1.

The current frame is:

$$C = (a, b, t) \tag{2}$$

After computing the background model image and find the current frame in the video sequence, the C is then subtracted from M and apply a threshold value to get the final image with detected object. The threshold value is applied in the preprocessing steps. The F is the final image acquired after frame differencing.

$$F = M - C \tag{3}$$

3.2 Pre-processing

Before going to functionalize blob analysis, the video frame/image generated during frame differencing is converted to a grayscale and apply a threshold value to get the binaries image with moving objects detected in the frame/image. The threshold value is selected manually to get the required result of binaries frame. The T represents the threshold value to get the binaries image applied on the final frame/image get in the frame differencing.

$$F > T \tag{4}$$

3.3 Blob Analysis

The main idea behind the use of blob analysis is that blob analysis detects those features in binary video frames which can be different in terms of area or brightness etc. [15]. In this research work blob analysis is used to find out the connected region of pixels and to determine which of these pixels can be grouped together as "blobs" or connected components to represent distinct objects.

Two matrices one consist of centroid coordinates of the detected blobs while the other consists of bounding box coordinate are returned through blob analysis. Blob analysis guarantees sure criteria in detecting blobs. Blob analysis reduces computation time to get rid of some blobs on the base of a spatial characteristic which have an only essential blob for more analysis [17].

Blob analysis is also used to find statistical information such as the size number and location of the blobs and also for the existence of blob regions. The tool of blob analysis parameter "MinimiumBlobArea" which sets the smallest size for the threshold value to a group of pixels to estimate a connected component [16]. We have tested with large parameter values but in this situation, the vehicles are not correctly detected because a higher threshold values show fewer or no detected blobs. On the other hand, a low threshold leads to a bundle of detected blobs because the blob analysis tool would consider every vehicle a non-apparent blob. The blob analysis process scans the entire image and detects all the blobs on the base of this it makes a full report of each particle of the vehicle. The reports in general contact information of shape size location and route of the vehicle. These whole perimeters are used for the detection and categorization of the blobs [17]. Once the object is detected then the features are calculated.

1. Find the area of the object, centroid and bounding box parameter.
2. Locate the Bounding box and centroid on the object.

After object detection, vehicle tracking is carrying out through Blob Analysis. The blob analysis is used to calculate the area of every detected object for tracking and speed estimation. To calculate the speed of an object through blob analysis the centroid of the objects is find out. Blob analysis is a region of connected pixels and is used for the identification and investigation of the connected regions in a video frame. The main reason behind the use of blob analysis is to detect the area and centroid of the object which is detected. The calculating area of the object is generating the stats which define the area for every vehicle.

3.4   Area

The total number of pixels in a video frame is measured and consider as a blob area. Area of connected pixels with the identical logic state is considered as a blob. For this purpose, the characteristic is frequently used to get rid of blobs that are very small or big from a frame [18].

3.5   Bounding Box

Bounding Box of a Blob is a minimum rectangle which consists of a number of regions or blobs. It is well defined by indicating all the pixels of a blob to find the four coordinates with the minimum value of x and y and the maximum value of x and y values too. Height divided by the width of the bounding box is the bounding box ratio of a Blob. The ratio of blobs area to the bounding box is the compactness of a blob [18].

$$Compactness = blob\ area/width * height \qquad (5)$$

3.6   Centroid

The centroid of a binary object is calculated by averaging the x and y positions used for a binaries video frame. The sum of the x coordinates of every pixel in the Blob dividing by all the numbers of pixels correspondingly for y value. The Centroid (X_c, Y_c) is calculated mathematically as follows using (6) and (7).

$$Xc = 1/N \sum_{i=1}^{N} xi \qquad (6)$$

$$Yc = 1/N \sum_{i=1}^{N} yi \qquad (7)$$

In the above equations of centroid calculation, N represents a total number of pixels in the blob xi where $yi = x$ and y coordinates of N pixels.

3.7   Speed Estimation

The velocity of moving the vehicle is computed the distance travelled with regard to the time taken. The Euclidian distance formula is used to compute the distance travelled between sequences of frames. The velocity of the vehicle is determined by using the distance formula with respect to frame rate [19].

3.8   Velocity Calculation

The velocity of moving object is finding out by the distance covered by the centroid to the frame rate of the video. Algorithm for computing velocity is given below.

1. First read the distance travelled by the object

$$Velocity = distance\ travelled/frame\ rate \qquad (8)$$

2. The array is used to store the value.
3. The velocity of moving object in the consecutive frames is defined in pixels/second.

The velocity of the moving vehicle objects can also be calculated as in the following equation. The distance between vehicle and starting point measured in kilometre:

$$Distance = Df * (D/Dx) * (Pn - Po) \qquad (9)$$

3.9 Speed Measurement

Speed measurement is the basic idea of our research. Speed estimation of the vehicle is very important to detect the vehicle which moves fast on the road. For this purpose, blobs are used to represent the detected vehicle. Centroid of every blob characterises a single centre pixel of the blob.

The location of the centroid in every consecutive frame will change as the vehicle travels. The displacement of the centroid in each successive frame can be adjusted to represent the speed of the vehicle. The Euclidean distance between the centroid of the blob in two consecutive frames will give the distance travelled in terms of pixels [20]. The aim of this research work is that we need to track multiple vehicles, firstly in the video and then we need to estimate the speed of those vehicles. After we have the vehicles along with their speeds we simply need to place a check which ensures that if any of the tracked vehicles goes beyond the range specified, and then it will raise an alert message. For this purpose, the Euclidian distance is applied to compute the distance of the vehicle between different frames and hence dividing total distance cover by the total time taken.

$$d = \sqrt{(x_2 - x_1)^2 + (y_2 - y_1)^2} \qquad (10)$$

The following steps of the algorithm describe different parameters of vehicles regarding speeds. First, we will assume different parameters for speed to define slow, average, and fast moving vehicles.

ALGORITHM FOR SPEED ESTIMATION

IF

   SMV < 100

    Indicator → Slow Moving Vehicles (SMV):

Else IF

    100<AMV<110

    Indicator → Average Moving Vehicles (AMV):

Else IF

    MV > 110

    Indicator →Fast Moving Vehicles (FMV)/Over Speeding

4   Results and Discussion

In this paper, we analyse multiple vehicles detection. The proposed method is applied for slow motion, medium motion and fast motion.

4.1   Slow Motion

On roads, there is a range of vehicles speed which should be strictly followed. If the vehicles move slowly from this range this is not a violation of rules but may cause problems for other vehicles. There is a certain range of vehicles speed which should be followed. In the following Fig. 2, shows the object detected with the green bounding box are the slow moving vehicles causing problems for the rest of the vehicles on the road. This kind of vehicles needs to move on one side of the road.

(a)   (b)

Fig. 2. Shows the detection of slow moving vehicles, In (a) shows frame with moving vehicles on road, In (b) slow vehicle detected and represents with the green rectangle.

4.2   Medium/Average Motion

If the movement of any vehicles is in between this kind of motion status then will be considered as a normal and average speed for any kind of vehicles. It is a normal range which can be helpful to avoid the incident and other criminal activities. The detection of this kind of vehicles can be seen in Fig. 3 surrounding by a blue rectangle, which shows that there are couples of vehicles moving in this range.

(a)   (b)

Fig. 3. Shows the detection of average moving vehicles in video frames. In (a) the vehicles can be seen moving in a video frame, In (b) the vehicles are detected and highlighted with the blue rectangle representing average moving vehicles.

4.3   Fast Motion

This is the kind of motion that need to be detected as wrongdoing speed of vehicles and a violation of traffic rules and regulations. If any vehicle cross this specific limit set by the authority then it will consider the rule of violation which may causes problem on the road for others. Figure 4 shows the detection of fast moving vehicles in a video frame.

(a)   (b)

Fig. 4. In (a) the vehicles can be seen moving in a video frame. In (b) the fast moving vehicles are detected and highlighted through red rectangle.

The main objective of the proposed system is to mention the efficiency of multiple vehicle detection for overspeeding. It has been tested on a standard dataset [21]. The results are based on 50 videos taken from the mentioned datasets. Table 1 shows the overall video samples obtained from the benchmark dataset.

Table 1. Videos details

Motion type	No. of videos	Total frame	Size	fps
Slow	50	2400	6 MB	30
Medium	40	1500	5.2 MB	30
Fast	30	2300	4.4 MB	30
Total	120	4200	5.2 MB	30

Table 2 shows the overall results for slow medium and fast motion with the accuracy rate achieved by the proposed method.

Table 2. Experimental result

Motion type	Accuracy rate	Computation time	False detection
Slow	89%	85 MS	11%
Medium	90%	87 MS	10%
Fast	91%	88 MS	9%
Overall result	90%	87 MS	4.7%

5   Conclusion

Nowadays, over speeding of vehicles on roads and highways are amongst the main issues that causes traffic problems. The video-based surveillance system is used to monitor and identifying the moving object on the roads. Blob analysis is used to estimate various speed ranges of multiple vehicles. The main contribution of this research is that various ranges of speeds like (slow medium and fast) are estimated accurately using blob analysis. This research work can help to estimate the speed of multiple vehicles, which subsequently prevent over speeding. The proposed work is also helpful to control and find vehicles breaking traffic rules on highways or roads. These days auto collision is the fundamental issue, because of which many people pass on. To tackle this huge issue, one way is to implement video analysis systems in light of the premise of activity base observation. It comprises of irregular action acknowledgement, checking of movement and well-being driving etc. We have developed a video-based automatic surveillance system to control the over speeding of multiple vehicles to avoid collision and other criminal activities. In this research, the object is detected through frame differencing. For the frame differencing, first the reference image is computed through background modelling and then the current image is subtracted from the reference image. The blob analysis is used to track the object after detection and finally, the speed is estimated of the vehicles in three categories slow, average and fast moving vehicles. The proposed system achieved an overall of 90% result. The works is based on stationary camera and further expand for moving the camera and make it more dynamic regarding the movement of the vehicles and cameras.

References

1. Bodkhe, A.P., Nirmal, S.A., Thakre, S.A.: A literature review on different models for human and vehicle tracking. Int. J. Innov. Res. Comput. Commun. Eng. **3**(1), 322–327 (2015)
2. Sofia Janet, R., Bagyamani, J.: Traffic analysis on highways based on image processing
3. Yabo, A., Arroyo, S., Safar, F., Oliva, D.: Vehicle classification and speed estimation using computer vision techniques. In: XXV Congreso Argentino de Control Automático (AADECA 2016), Buenos Aires (2016)
4. Mu, K., Hui, F., Zhao, X.: Multiple vehicle detection and tracking in highway traffic surveillance video based on SIFT feature matching. J. Inf. Process. Syst. **12**(2), 183–195 (2016)
5. El Jaafari, I., El Ansari, M., Koutti, L., Ellahyani, A., Charfi, S.: A novel approach for on-road vehicle detection and tracking. Int. J. Adv. Comput. Sci. Appl. **1**(7), 594–601 (2016)
6. Uppaal, R.: Exploring robust vehicle detection and tracking methods for various illumination settings using infrared thermographic images. Int. J. Res. Advent Technol. (2016). E-ISSN: 2321-9637, Special Issue National Conference "NCPCI-2016", 19 March 2016
7. Krutak, M.: Multiple Vehicle Detection and Tracking from Surveillance Camera with Collision Prediction. Brano University of Technology, Faculty of Information Technology (2016)
8. Satzoda, R.K., Trivedi, M.M.: Multipart vehicle detection using symmetry-derived analysis and active learning. IEEE Trans. Intell. Transp. Syst. **17**(4), 926–937 (2016)
9. Kim, H.: Tracking and counting vehicles in nighttime. Int. J. Web Sci. Eng. Smart Device **3**(2), 13–18 (2016)
10. Kurz, F., Rosenbaum, D., Runge, H., Cerra, D., Mattyus, G., Reinartz, P.: Long-term tracking of a specific vehicle using airborne optical camera systems. ISPRS Int. Arch. Photogramm. Remote Sens. Spat. Inf. Sci. **41**, 521–525 (2016)
11. Sun, Z., Bebis, G., Miller, R.: Monocular precrash vehicle detection: features and classifiers. IEEE Trans. Image Process. **15**(7), 2019–2034 (2006)
12. Morris, B.T., Trivedi, M.M.: Learning, modeling, and classification of vehicle track patterns from live video. IEEE Trans. Intell. Transp. Syst. **9**(3), 425–437 (2008)
13. Daigavane, P.M., Bajaj, P.R.: Real time vehicle detection and counting method for real time vehicle detection and counting method for unsupervised traffic video on highways unsupervised traffic video on highways. IJCSNS **10**(8), 112 (2010)
14. Aslani, S., Mahdavi-Nasab, H.: Optical flow based moving object detection and tracking for traffic surveillance. Int. J. Electr. Electron. Commun. Energy Sci. Eng. **7**(9), 789–793 (2013)
15. Khan, S.D., Porta, F., Vizzari, G., Bandini, S.: Estimating speeds of pedestrians in real-world using computer vision. In: International Conference on Cellular Automata. Springer, Cham (2014)
16. Judd, D., Wu, R., Taylor, C.J.: Squash sport analytics & image processing. In: MIT Sloan Sports Analytics Conference (2014)
17. Hari Hara Santosh, D., Venkatesh, P., Poornesh, P., Narayana Rao, L., Arun Kumar, N.: Tracking multiple moving objects using gaussian mixture model. Int. J. Soft Comput. Eng. (IJSCE) **3**(2), 114–119 (2013)
18. Manasa, J., Pramod, J.T., Jilani, S.A.K., Javeed Hussain, S.: Real time object counting using Raspberry pi. Int. J. Adv. Res. Comput. Commun. Eng. (IJARCCE) **4**(7), 540–544 (2015)

19. Upadhyay, M., Patel, J.R.: Real time movable object detection, tracking and velocity estimation using image processing. In: National Conference on Recent Research in Engineering and Technology (NCRRET 2015) (2015). E-ISSN: 2348-4470, PRINT-ISSN: 2348-6406
20. Gokule, R., Kulkarni, A.: Video based vehicle speed estimation and stationary vehicle detection. Int. J. Adv. Found. Res. Comput. (IJAFRC) **1**(11), 93–99 (2014)
21. Papageorgiou, C.P., Poggio, T.: A trainable object detection system: Car detection in static images (1999). http://cbcl.mit.edu

Computer Networks, Mobility and Pervasive Systems

New SIEM System for the Internet of Things

Natalia Miloslavskaya$^{(\boxtimes)}$ and Alexander Tolstoy

National Research Nuclear University MEPhI
(Moscow Engineering Physics Institute),
31 Kashirskoye shosse, Moscow, Russia
{NGMiloslavskaya, aitolstoj}@mephi.ru

Abstract. Based on the available standards, the generalized architecture and the reference model of the IoT as a security object to be protected are presented. On the IoT layers, different security controls collecting data for further detection of security-related events are located. The security incident management process for the IoT needs automation, for which Security Information and Event Management (SIEM) systems are the best applicable solutions. But modern challenges dictate the need to modify these systems for the IoT. A new blockchain-based SIEM system for the IoT is proposed.

Keywords: Internet of Things ·
Security Information and Event Management (SIEM) system ·
Blockchain technologies · Security incident

1 Introduction

The Internet of Things (IoT) term has been coined in 1999 by British K. Ashton, executive director of the Auto-ID Center at the Massachusetts Institute of Technology. After 20 years the IoT is still a new concept in its early stages. No single definition has been agreed to yet. Here are a few examples from the well-known organizations:

- "A global infrastructure for the information society, enabling advanced services by interconnecting (physical and virtual) things based on existing and evolving inter-operable information and communication technologies...Through the exploitation of identification, data capture, processing and communication capabilities, the IoT makes full use of things to offer services to all kinds of applications, whilst ensuring that security and privacy requirements are fulfilled..." [ITU-T Y.4000] [1];
- "An infrastructure of interconnected objects, people, systems and information resources together with intelligent services to allow them to process information of the physical and the virtual world and react" [ISO/IEC 20924] [2];
- "A cyber-physical ecosystem of interconnected sensors and actuators, which enable decision making" [ENISA, Baseline Security Recommendations for IoT in the context of Critical Information Infrastructures, 2017] [3].

In addition, in [4] the IoT system term as "a system that is comprised of functions that provide the system the capabilities for identification, sensing, actuation, communication, and management, and applications and services to a user" was introduced.

© Springer Nature Switzerland AG 2019
Á. Rocha et al. (Eds.): WorldCIST'19 2019, AISC 931, pp. 317–327, 2019.
https://doi.org/10.1007/978-3-030-16184-2_31

The IoT scenarios cover a broad range of applications for consumer electronics, healthcare, automotive technology, agriculture, retail, industrial control systems, wildlife monitoring, smart cities, etc. All IoT devices (like smart phones, home TVs, refrigerators, door locks, medicine droppers, traffic lights, connected cars) are inter-connected in a unified IoT network and send or receive data from various mobile applications and cloud services. To work efficiently, the IoT requires new applications and complementary technologies, where information of both real and virtual worlds will be gathered, identified, normalized, correlated, analyzed and distributed in a per-vasive way. It is imperative that the IoT handles heterogeneous systems of hardware, software and services that enable various objects to connect each other using public and private wired and wireless network connections. In this landscape, the integration of the conventional Internet, mobile communication, cross-sensing platforms and social networks lead to an exponential increase in attacks against all IoT elements.

When creating, storing, updating and sharing large volumes of data in various applications, securing these data against intruders is of a prime concern in order to provide intended IoT services to authorized users in due course. Due to multiple vulnerabilities in the communication protocols connecting IoT devices with each other and networks, the physical, management and application interfaces for transmitting data to and from cloud-based resources, IoT hardware and devices' firmware, many targeted attacks such as malicious modification, brute force password cracking, Dis-tributed Denial-of-Service (DDoS), Main-in-the-Middle (MitM), buffer overflow and others are continuing to evolve [5]. Undoubtedly, security and privacy of the IoT as a whole, as well as sensitive data circulating in it, and continuous delivery of services become a real challenge to its providers and customers. Other important issues are the accountability of all operations with IoT devices, data, applications and services while millions of things are exchanging their data for collaborative work. In addition, IoT devices often store Personally Identifiable Information (PII) of their users, which is of great value for malicious persons (one note: privacy issues are out of the scope of this paper). Of course, that is not the complete list of all security issues, which require security management. Thus, it is obvious that for the IoT it is vital to know which threats exist at the moment, and how they could grow into an incident (unwanted or unexpected events that have a significant probability of compromising IoT' security).

The entire IoT requires two complementary information security management approaches as "information lies at the heart of IoT, feeding into a continuous cycle of sensing, decision making, and actions" [3]. Specific security capabilities depend on IoT-specific requirements, while generic ones include traditional security controls such as anti-virus and anti-malware systems, launching with minimal privileges, access control based on authentication and authorization, installing updates and others. Among the traditional tools, there are Security Information and Event Management (SIEM) systems designed for monitoring security in networks. Two generations of SIEM systems exist, but modern challenges dictate the need to modify them for the IoT. The goal of the paper is to present our innovative idea to apply the advanced blockchain technologies for this purpose.

Thus, the remainder of the paper is organized as follows. The section with the related work is intentionally missed as there are no publications of other authors in open access proposing the same idea of SIEM 3.0 systems. Section 2 is devoted to standardization of security-related issues for the IoT. Section 3 copies the generalized IoT architecture and the reference model from ISO/IEC 30141:2018 and ITU-T Y.2068 to show the IoT as a security object to be protected and IoT layers, on which different security controls collect data for further detection of security-related events. Section 4 briefly introduces security incident management for the IoT and gives some examples of these incidents. Section 5 describes a new blockchain-based SIEM system proposed for the IoT. The main areas of further research conclude the paper.

2   Security Standardization for the IoT

There are only a few adopted standards or standard under development for IoT security or security-related issues. The following is almost a complete list of November 2018: ISO/IEC 29181-5:2014 (Future Network – Problem statement and requirements – Part 5: Security), 27030 (Guidelines for security and privacy in IoT), 30149 (IoT Trustworthiness framework) and 30147 (Methodology for trustworthiness of IoT system/service), ITU Y.4806 (Security capabilities supporting safety of the IoT), X.1362 (Simple encryption procedure for IoT environments), Y.4102/Y.2074 (Requirements for IoT devices and operation of IoT applications during disasters), Y.4455 (Reference architecture for IoT network service capability exposure), Y.4118 (IoT requirements and technical capabilities for support of accounting and charging), Y.4702 (Common requirements and capabilities of device management in the IoT), Q.3952 (The architecture and facilities of a model network for IoT testing), and Q.3913 (Set of parameters for monitoring IoT devices).

The European Union Agency For Network And Information Security (ENISA) has published in 2017 the "Baseline Security Recommendations for IoT in the context of Critical Information Infrastructures" to map assets and relevant threats, assess possible attacks and identify good practices and security measures for the IoT systems [3].

In the other standards some important notes on the IoT security are also given. For example, three key security properties of information, namely confidentiality, integrity and availability, are defined in relation to the IoT in ISO/IEC 30141:2018 [6]:

- The confidentiality protection supports prohibiting people or systems from reading data or control messages (e.g., secret tokens, financial information, PII) when they are not authorized to do so. For example, IoT motion detection sensors and smart meters could show thieves whether a property is occupied or not;
- Data integrity ensures that the data used for decision-making in the system and executable software has not been altered by faulty or unauthorized devices or by malicious actors. For example, an intermediate node may increase the temperature in a room, but this should not cause to increase in cooling;
- Availability of a device is related to its correct operation over time and to its network connectivity. Availability of data means the ability of the system to get the requested data from a system. Availability of services is the system's ability to provide the requested service to users with a pre-defined Quality of Service (QoS).

At that moment the National Institute of Standards and Technology (NIST) has no special publications for the IoT with only two draft internal reports (NISTIR) 8222 (IoT Trust Concerns) and 8228 (Considerations for IoT cybersecurity and privacy risks). Hence, ensuring security for the IoT remains an active research area.

The standards mentioned above can be taken as a basis to formulate security requirements for the IoT and further to use them for wording security incidents to be detected in the IoT.

3 Generalized IoT Architecture and Reference Model

To describe the IoT as a security object to be protected from various incidents, it is reasonable to use the entity-based IoT's reference model presented in ISO/IEC 30141:2018 (Fig. 1) [6]. This figure illustrates the interactions between the major entities using arrowhead lines, corresponding to data streams. These entities, as well as data streams, first require ensuring their confidentiality, integrity, and availability.

More detailed implementation view of the IoT functional framework is given in ITU-T Y.2068 (Fig. 2) [7]. The framework contains security controls, which collect data for further analysis and selection of those data that are suspicious or accurately indicate the incident occurred. Thus, Information Protection Tools (IPTs) are the security-related data sources.

4 Security Incident Management in the IoT

Following the main principles of ISO/IEC 27035-1:2016 [9] and NIST SP 800-61 [10], an effective security incident management (SIM) process in place is the essential part of security management processes for the IoT. SIM for the IoT can be considered as a part of the cyclic process, consisting of a set of targeted actions to achieve the IoT's objectives by ensuring its security and including continuous assessment of security risks, planning, implementation and evaluation of the effectiveness of appropriate security controls for their treatment, as well as immediate response to unwanted security-related events (further security events) violating security policies or practices. According to these policies, some of the security events are classified as incidents. In ISO/IEC 27000:2018, information security (IS) incident refers to a single or a series of unwanted or unexpected IS events that have a significant probability of compromising business operations and threatening IS [8]. In turn, IS event is an identified occurrence of a system, service or network state indicating a possible breach of IS policy or failure of controls or a previously unknown situation that may be security relevant. IS event can be a part of one IS incident, while IS incident can be a set of IS events. The same definitions are completely applicable to the IoT's security violation.

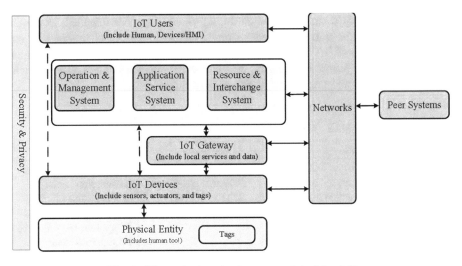

Fig. 1. The entity-based reference model of the IoT

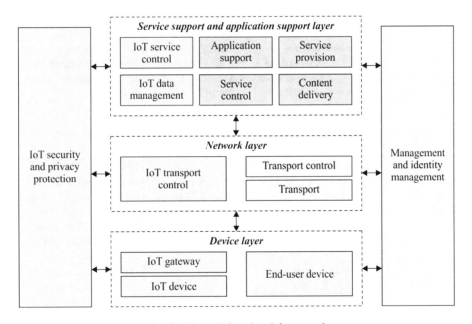

Fig. 2. The IoT functional framework

SIM for the IoT should be based on a structured approach to detect, assess, and respond to security incidents, including the activation of adequate security controls to prevent, reduce, and recover from impacts on one hand, to report vulnerabilities for dealing with them appropriately on the other hand and, at least, to learn from these incidents and vulnerabilities, and make improvements to the process.

Initial data for SIM can be collected during monitoring. It is intended for continuous control over the normal operations mode in the constantly changing IoT, and analyzing the status of IoT devices, network channels, systems, services, applications, information assets, processes, activity, etc. The events indicate non-compliance with the requirements from policies and appointed regulations.

Table 1. Examples of security-related events for the IoT

Event type	Events	Data sources	Further actions
Suspicious single (atomic) event 1	To access one of the IoTs services the authorized user enters his credentials three times, first two times with errors	Access Control System (ACS) of the IoT service	The block is formed for transmission to the sidechain; the event is stored in the individual ACS's DB; the reference to the initial event is inserted in the block
Suspicious event 2 – Compound event	• To access one of the IoT services the authorized user quickly and without any mistakes types his/her credentials • The user initializes an update of a firmware of a few IoT devices (maybe to fix bugs and security vulnerabilities) • The user checks that the updates were installed successfully	• ACS of the IoT service • Task manager, user's actions recording tool (UART) • Task manager, UART, logs of the IoT devices	The block is formed for transmission to the sidechain; the separate events are stored in the individual DBs of ACS, UART, and IoT devices' logs; the references to the initial events are inserted in the block
Pure single (atomic) security event 1	In the absence of the owner, one of the sensors was replaced by a fake product	Video Surveillance System (VSS), visual inspection of the sensor	The block is formed for transmission to the main chain; the event is stored in the VSS's DB; the reference to the initial event is inserted in the block
Pure compound security event 2	• In the absence of the owner, one of the sensors was replaced by a fake one • The fake sensor generates incorrect measurement results • The fake sensor sends a virus (attaches malicious payloads) to other sensors	• Video Surveillance System (VSS) • Automated and visual checks of applications' outputs • Antivirus software on this and other sensors, task manager	The block is formed for transmission to the main chain; the separate events are stored in the individual DBs of VSS, antivirus software, sensor's ACS, etc.; the references to the initial events are inserted in the block

According to the IoT's security policy, an event can be classified as a non-security event, suspicious event or security event/incident (Table 1). For example, log in to start a service with the right credentials is only an event. Any login attempt with wrong credentials is a security event, and it can be either a user's error or the beginning of an attack against the IoT. In the latter case, an actual unauthorized activity, leading to a security incident in the future, may be detected after this event.

So-called SIEM systems are used to solve the problem of real-time flow control over large volumes of security-related data to computerize this process, in particular, to find specific data, characterizing attacks' phases and their typical consequence [11]. First-generation SIEM 1.0 systems of the late 1990's were log-centric, detected events through preset rules and simple correlation techniques mostly for IP addresses, used relational DBs with little semantic richness. They could not detect non-standard network traffic, protocol anomalies, malicious malware's activity, etc. Since the late 2000's, SIEM 2.0 systems perform behavioral and contextual analysis and implement detection and centralized collection of data from distributed sources, correlation of ports, protocols and users, accumulation of statistics on previous and current users' and applications' activity, tracking the entire lifecycle of each security event, automated generation of reports and recommendations to handle these events, etc. But modern challenges dictate the need to modify these systems for the IoT. From the other side, all SIEM systems' data and its subsystems should be protected as they are vital for collecting and analyzing sensitive data, for example, for digital forensics.

5   New SIEM System for Managing Security Events in the IoT

In recent times, the BC (BC) technologies (BCT) for creating verifiable digital records have shown notable success in many application domains. The BC is a secure distributed data structure (database, DB) that maintains a constantly expanding list of non-editable time-stamped blocks without centralized administration and data storage and sets rules about transactions tied to these blocks [12]. This DB is shared by a group of participants with the right to submit new blocks for inclusion in it. In the IoT, transactions reflect events happening in it and gathered from different data sources – IPTs. All the events in a block as parts of one incident are grouped together along with a cryptographic hash of the previous block. Any block will be included in the BC only after the consensus of a majority of participants who will agree that any event from the block under consideration looks valid. This block cannot be removed or changed after published. All the blocks are linked to each other in one chain in a chronological order of their creation. The public history of all blocks' inclusion is securely stored in the BC. It can be shared, immutable and verifiable for recording the history of events relying on peer-to-peer, consensus and data storage protocols.

After exploring the BCT features, it was concluded that they are well applicable for designing a new BC-based SIEM 3.0 system for the IoT. It can provide the following significant benefits for this purpose: interoperability of data from all sources presented in a structural format for further processing; proof of data source's identity prohibiting any anonymity; ensuring data integrity for transmissions to, from and within the BC, rather than keeping it confidential; real-time event recording with time-stamping and an event history for traceability; an opportunity to investigate a consequence of events that lead to an incident; automatic updates every time a new event occurs; the possibility of inserting in a block some additional data attributes; independence of the BC from the type and number of data sources; a sufficient capability to register, validate, process and transmit to the BC billions of events per second in hard real-time; hard to attack because of the multiple shared copies of the same BC, etc.

In the BC-based SIEM 3.0, the BC will streamline all the events taking place in the IoT. It is a unidirectional chain with a linked by hash pointers blocks, each one referring back to its predecessor. Blocks will have unique identifiers (IDs) and hashes and will contain data on security events collected by agents – IPTs. Agents should be deployed on each IoT asset to be monitored. They will preprocess the collected events to reduce their number before transmitting them to the BC by discarding those, which do not affect the IoT's security.

Rules are set for the events tied to a specific incident. Security events, belonging to the same incident as its consequent steps, will be combined in one block. Creating a block with a content similar to one of the preceding blocks (so-called repeated block) but with a larger ID means that some repeated event has appeared in the IoT that characterize, for example, a DoS attack.

The SIEM 3.0 system is proposed to utilize a permissioned ledger. Firstly, it is used in a closed IoT community. Secondly, all trusted data sources (IPTs) collect raw events, discard those not related to IoT's security and send all remaining events to a single centralized DB of pending events (PEDB) shared by all IPTs (this idea falls well in the BCT). The sources should be able to view in the PEDB only their collected data. Events from the PEDB after validation are used for insertion into blocks by "miners". Using events' timestamps, the miner decides on this block to become the next block in the BC and creates its unique hash. A special type of nodes called "peers" will reduce the workload of miners: they do peer exchange and share information on other nodes (e.g., they can tell them that "block N+1 has parent N, has event 102, its hash is HASH N+1, etc.") and their status (e.g., out-of-service because of compromise), as well as perform read/write operations to the BC.

IPTs collecting raw events store them themselves. All preprocessed results on suspicious events and pure security events are transmitted to the PEDB. After the initial preprocessing, these events can be considered later as the steps of one incident.

One block will contain one atomic suspicious, atomic security or compound from several atomic security events (Table 1). To formulate the events typical attacks against the IoT targets described in [5] can be used. Besides that, each block will include references to all initial raw events for additional analysis required in the future.

To illustrate how to form a block, let us look at the compound security event 2 from Table 1. During non-working hours in the absence of an owner, somebody replaced one of the sensors by a fake product that was due to not properly control for physical access to it. This sensor was located in a zone being in sight of one Video Surveillance System (VSS), which recorded this action. After that, the fake sensor has begun to generate incorrect results of some measurements and to send viruses and malware. The day after somebody logged remotely in it without any authentication. No other actions have been recorded yet. A new block will contain all these events one by one, excluding cases of a repeated description of the same event by different data sources (event 4 was fixed on three infected sensors), and the references to all sources with initial raw data on these events and their timestamps (Fig. 3).

To guarantee that the same incident is documented from different perspectives, all data sources transmitting event data to PEDB, should be synchronized through a reconciliation process. Blocks with "pure" events will come to the main chain. Blocks with suspicious events will come to the sidechain for additional analysis or waiting for the arrival of blocks with events into both chains that will confirm their "pure security events" status. If so, these blocks will be transferred to the main chain.

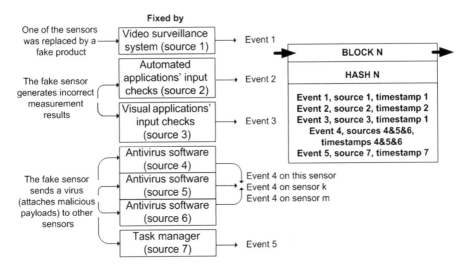

Fig. 3. An example of forming a block in SIEM 3.0 system for the IoT

Further, when implementing the BC-based SIEM 3.0 system, the typical atomic security events for the whole IoT can be inserted in it in the form of reusable patterns.

6 Conclusion

It is obvious that the BC in the SIEM 3.0 system will be updated every time a new block will come. Its size will grow very quickly, and the system will run over time slower. Hence, a technique to support its flexibility and scalability is very much appreciated. The idea of a convolution (pruning) can be applied for this purpose in two ways: automatic meaning the deletion of all blocks older than a pre-set lifetime or selective when a designated peer (operator/administrator) scans manually or using a specialized application the BC to find and remove "expired" blocks. An expire block is a block with obsolete security events or suspicious events not resulted in any security incident or not being its part for a long time period, for example, one year. But when removing a block, the main issue is what to do with its cryptographic links to the previous one. As it was found later, this idea has come to us in parallel to Dr. G. Ateniese and the Accenture company. They proposed so-called "redactable BC" [13] with modified chameleon hashes that allow designated authorities to edit and remove previous blocks without breaking the BC by means of building a virtual padlock on this link. After that, the chameleon hash key is used to unlock the link and to substitute the block with a new one without breaking the hash chain.

Further, the resulting BC-based SIEM 3.0 system should be tested for satisfying the requirements for the SIEM systems and its efficiency (performance). Although this system is not yet implemented, a methodology should be chosen in advance for proving that it is realizable, complete, consistent, unambiguous, and verifiable. A tool-supported B method, which is the most popular in safety-critical system applications and allows for highly accurate expressions of the properties required by specifications and models systems in their environment, was chosen [14]. The main efficiency indicators are typical for these systems: downtime, response time, work in case of temporary limited available resources, operating modes, data loss, false positives/negatives, the ability to interface with various types of IPTs and to update incident classifications and so on, as well it SIEM's self-protection against attacks like blocking of data transmission channels from the IPTs, erasing data on events, etc.

Summing up, it can be concluded that the BC concept is fully applicable to the creation of the next-generation SIEM 3.0 systems that are designed to detect security incidents in a modern, fully interconnected network environment like the IoT. In comparison with SIEM 2.0 systems, a BC-based 3.0 system has two main advantages as it stores data on all IS-related events in a secure manner (in respect to its integrity) for further computer forensics investigation and in the form of connected blocks, each of which describes one separate IS incident consisting of a few (or only one) IS events with their timestamps and sources. Our further work is aimed at bringing to life the ideas proposed.

Acknowledgement. This work was supported by the MEPhI Academic Excellence Project (agreement with the Ministry of Education and Science of the Russian Federation of August 27, 2013, project no. 02.a03.21.0005) and possibly the Russian Foundation of Basic Research (its decision on funding as of 04.03.2019 is not known).

References

1. ITU-T Y.4000/Y.2060 Overview of the Internet of Things
2. ISO/IEC 20924 Information technology - Internet of Things - Definition and Vocabulary
3. European Union Agency for Network and Information Security (ENISA). Baseline Security Recommendations for IoT in the context of Critical Information Infrastructures (2017). https://www.enisa.europa.eu/publications/ (AD: 10.01. 2019)
4. Bahga, A., Madisetti, V.: Internet of Things: A Hands-on-Approach, 443 p. (2014)
5. Miloslavskaya, N., Tolstoy, A.: Internet of Things: information security challenges and solutions. Cluster Comput. (2018). https://doi.org/10.1007/s10586-018-2823-6
6. ISO/IEC 30141:2018 Internet of Things (IoT) – Reference Architecture
7. ITU-T Y.2068 Functional framework and capabilities of the Internet of Things

8. ISO/IEC 27000:2018 Information technology – Security techniques – Information security management systems – Overview and vocabulary
9. ISO/IEC 27035-1:2016 Information technology – Security techniques – Information security incident management – Part 1: Principles of incident management
10. NIST SP 800-61 Rev. 2. Computer Security Incident Handling Guide (2012)
11. Miloslavskaya, N.: Security operations centers for information security incident management. In: Proceedings of the 4th International Conference on Future Internet of Things and Cloud (FiCloud 2016), Vienna, Austria, pp. 131–138 (2016)
12. Miloslavskaya, N.: Designing blockchain-based SIEM 3.0 system. Inf. Comput. Secur. **26**(4), 491–512 (2018). https://doi.org/10.1108/ics-10-2017-0075
13. Ateniese, G., Magri, B., Venturi, D., Andrade, E.: Redactable Blockchain – or – Rewriting History in Bitcoin and Friends. https://eprint.iacr.org/2016/757.pdf (AD: 10.01. 2019)
14. Lano, K.: The B Language and Method: A Guide to Practical Formal Development. Springer-Verlag New York, Inc., Secaucus (1996)

Standardization Issues for the Internet of Things

Natalia Miloslavskaya$^{(\boxtimes)}$, Andrey Nikiforov, Kirill Plaksiy,
and Alexander Tolstoy

The National Research Nuclear University MEPhI (Moscow Engineering Physics
Institute), 31 Kashirskoye Shosse, Moscow, Russia
{NGMiloslavskaya,aitolstoj}@mephi.ru,
andreinikiforov993@gmail.com, kirillplaksiy@mail.ru

Abstract. The development of the Internet of Things (IoT) entails the emergence of new security threats and risks. The daily promotion of using more devices in more areas of life requires the development of new security standards. Interactions with and within the IoT have to be regulated by the documents of leading international organizations. At the same time, the problems of ensuring IoT security are not yet fully worked out because of the constantly expanding methods, tools, and devices involved in these processes. The paper discusses existing standards in the field of IoT's information security (IS). This research is focused on how the IS issues are addressed in these standards. An assumption on the significance and prospects of the progress in this field is made on the basis of the analysis performed.

Keywords: Internet of Things · Standards · Information security

1 Introduction

The Internet of Things (IoT) has densely entered billions of people lives around the world. However, the increase in the number of connected IoT devices leads to an increase in risks: from causing physical harm to people to downtime and equipment damage. Since many IoT systems have already been attacked with considerable damage, their protection comes to the foreground [1].

The IoT is a relatively new concept. It refers to a computer network of physical objects (things) equipped with embedded technologies for interacting with each other or with the external environment [2]. The IoT considers the organization of such networks as a phenomenon to restructure economic and social processes that exclude the need for human participation from the part of actions and operations [3]. The IoT has actively developed thanks to the ubiquitous spread of wireless networks and cloud computing, the development of machine-to-machine technology and software-defined networks, and the active transition to IPv6. It is necessary to develop generally accepted standards that will allow the same methods and tools to be assessed adequately to ensure a safe environment for effective work within the IoT framework. The IoT is developing at a tremendous speed, but all the innovations introduced add serious

© Springer Nature Switzerland AG 2019
Á. Rocha et al. (Eds.): WorldCIST'19 2019, AISC 931, pp. 328–338, 2019.
https://doi.org/10.1007/978-3-030-16184-2_32

information security (IS) challenges. The Business Insider news website published in 2013 a study, which shows IS as the biggest problem when implementing the IoT [4].

The IoT market is currently experiencing a period of rapid growth. According to Ericsson [5], in 2018 the number of IoT sensors and devices will exceed the number of mobile phones and will become the largest category of connected devices. The total annual growth rate of this segment for the 2015–2021 period will be 23%. The company predicts that there will be approximately 28 billion connected devices worldwide by 2021 and about 16 billion will be connected to the IoT. The Russian market for the IoT is also actively developing. According to IDC [6], the total investment in it in 2017 amounted to more than 800 billion US dollars. The forecast is 1.4 trillion by 2021. The Direct INFO [7] estimated the total size of this market in 2016 in 17.9 million devices (42% more in comparison to 2015). By 2021, the total number of IoT devices will increase to 79.5 million and by 2026 to 164.7 million. The total potential of the Russian market was estimated at 0.5 billion devices.

At this rate of growth, the issue of ensuring privacy and IS for the IoT is very critical [8]. For example, all the information that IoT devices store is highly demanded because it shows a complete picture of users' everyday activities and habits. The availability of databases with this content is useful for companies that can direct their resources to the production of goods and services focused on the habits and preferences of the masses. Special protection systems, including encryption, for downloading and storing data in clouds can help to minimize these problems [9].

In 2015, Hewlett Packard conducted a large-scale study, which reported that 70% of IoT devices have vulnerabilities in their passwords, data encryption, and access control, and 50% of applications for mobile devices do not exchange data [10]. Kaspersky Lab did another interesting study among the Russian companies. It conducted tests showing that video monitors and coffee machines can be hacked to intercept video and to transmit information in an unencrypted form. The device can also save Wi-Fi network password, to which it was connected [11]. IoT devices, which do not have adequate protection, can be attacked remotely over the Internet or consciously infected with malicious code to create a botnet [12].

In order to review the state of addressing the IoT's IS issues in international standards, the rest of the paper is organized as follows. Section 2 studies related work. Section 3 reviews existing standards. In Sect. 4, a comparison of these standards is carried out according to the selected criteria. In conclusion, the main results of this research are summarized, and directions for possible further work are outlined.

2 Related Work

There are a lot of papers devoted to the problems and peculiarities of the IoT. The work of Russian scientists [13] contains a detailed analysis of governing organizations that are engaged in standardization in this area as well as examines cyber-physical systems. In [14] the IoT is considered from IS point of view; the paper examines major security problems and their solutions and discusses typical vulnerabilities and attacks against the IoT and key areas of further development.

There are a few specialized books on IoT security [15–18]. The book [15] explores how attackers can abuse popular IoT-based devices like wireless LED light bulbs, electronic door lockers, smart TVs, connected cars, etc., and their tactics. In [16], some guidelines to architect and deploy a secure enterprise IoT, namely how to build a security program, to select individual components affecting the security posture of the entire system, to design a secure IoT using Systems Security Engineering and Privacy-by-design principles, and to leverage cloud-based systems to support the IoT are provided. The author of [17] presents attack models of IoTs with countering principles, details the security design of sensors and devices linked into IoTs, discusses new IoT network security protocols, examines IoT back-end security issues (trust and authentication), and analyzes privacy preservation schemes. The small-volume book [18] explains the IoT security concept from theoretical and practical viewpoints, which take into account end-node resource limitations, IoT hybrid network architecture, communication protocols, and applications characteristics.

Prospects for the emergence of the IoT, as well as measures taken worldwide to implement this concept, are considered in [19]; standardization, addressing, and security are studied. There are some sections dedicated to the problems of standardization in [20]. Papers [21, 22] offer a brief overview of the state of standardization issues in general. A comparative analysis of the IoT standards on security issues is presented in [23].

3   The IoT Standardisation Issues

With the development of the IoT, its users and manufacturers become more and more concerned with ensuring the safety of people, systems, devices, data transmission channels, etc. In addition to physical protection, it is necessary to ensure IS for the entire IoT. To do this, it is required to develop standards in this area and bring all security requirements to a single universal form. Standards provide people and organizations with a basis for a mutual understanding of the IoT. The strongest contributors to the field are the International Standardization Organization (ISO), the International Electrotechnical Commission (IEC), the International Telecommunication Union (ITU), and the IEEE Standards Association (IEEE-SA). The IoT creation and use from different viewpoints, including ensuring IoT's IS, are presented in the adopted in 2012–2018 standards shown below.

3.1   ITU Standards

Our review begins with the organization, which is most interested in the IoT topic.

Standard Y.4000/Y.2060 (Overview of the IoT) [24] gives an overview of the IoT with the main objective of highlighting this important area for future standardization. The interrelation of various components of the most complex systems leads to significant security threats, such as threats towards confidentiality, authenticity, and integrity of both data and services. The document describes succinctly generic (at application, network and device layer) and specific security capabilities in general.

Document Y.4050/Y.2069 (Terms and definitions for the IoT) [25] specifies the terms and definitions relevant to the IoT from an ITU-T perspective, in order to clarify the IoT and IoT-related activities.

Specification Y.4100/Y.2066 (Common requirements of the IoT) [26] classifies demands of the IoT systems into the categories and security and privacy protection requirements are among them. The document contains the functional requirements for data capturing, storing, transferring, aggregating and processing and provision of services, describes the different roles of participants and shows some use cases.

Y.4111/Y.2076 (Semantics-based requirements and framework of the IoT) [27] contains requirements for security capabilities with the use of technologies for security-related decision making. Semantic annotation, security policy management, and access control are proposed for efficient use of semantic technologies.

Document Y.4113 (Requirements of the network for the IoT) [28] introduces a basic model of the network for the IoT, general characteristics of smart meters and sensors, and general issues of this network. The document gives a good explanation of the suitable networks, but security issues are not considered.

Y.4103/F.748.0 (Common requirements for IoT applications) [29] has much in common with Y.4100/Y.2066, but focuses on IoT applications. In the IoT, all things are connected that results in significant security threats (like threats to confidentiality, authenticity, integrity of both data and services) and the need to integrate different security policies and techniques. The proper threat analysis is required to pay attention to the IoT applications' characteristics and to chose corresponding measures for them. The document also considers cases, when security and privacy requirements are imposed by laws and regulations, application maintenance and support issues.

Document Y.4552/Y.2078 (Application support models of the IoT) [30] provides three (configurable, adaptable, reliable) application support models with their basis. This document and the previous one overlap a little and complement each other. List of capabilities and components and use cases for the models are also of interest.

Standard Y.4453 (Adaptive software framework for IoT devices) [31] addresses the adaptive software framework (ASF) concept, identifies high-level requirements and provides a reference functional architecture for IoT devices. There are some security capabilities provided for ASF secure execution and a proper example use case and workflow.

Y.4101/Y.2067 (Common requirements and capabilities of a gateway for IoT applications) [32] has more issues of interest from the IS perspective. The common requirements and capabilities provided are intended to be generally applicable to gateways' application scenarios. The document gives a brief introduction to gateways for IoT applications, their general requirements, common capabilities, reference technical framework, typical high-level flows, and use cases. These topics are more sensitive from the IoT's IS point of view.

Specification Y.4112/Y.2077 (Requirements of the plug and play capability of the IoT) [33] contains the requirements for the plug and play (PnP) capability of the IoT, as a basis for further standardization work related to this issue. It describes the concept and the purpose of this capability and then provides its components as well as its requirements. It also describes Device and Gateway PnP capability, security protection from a counterfeit device, firewall protection, PnP authorization, and access control. PnP use cases are worth exploring as they will be useful for future use.

Y.4401/Y.2068 (Functional framework and capabilities of the IoT) [34] provides a description of the key IoT capabilities based on the functional, implementation and deployment view of the IoT functional framework (FF) described in order to fulfill the common requirements from Y.2066. Here one can see concepts of the IoT FF, IoT FF in functional, implementation and deployment view with proper structures, capabilities for integration of cloud computing and big data technologies and security considerations. At the end, the list of management capabilities and components, as well as the list of security and privacy protection capabilities are given.

Standard Y.4806 (Security capabilities supporting safety of the IoT) [35] identifies threats that may affect safety and security capabilities based on [34]. It determines these threats and assigns security capabilities which can be applied to mitigate them. The IoT poses specific security challenges, which may not be covered by existing security objectives (like confidentiality, integrity, availability – CIA) completely. Traditionally security threats are considered as issues that arise in the virtual environment and target the data handling process. This leads to the interpretation of information technology (IT) security as the ensuring of its CIA. Improper IT system behavior (e.g., software bugs, backdoors, Trojan programs) is also considered as a source of problems that affect these aspects and therefore cause only data handling concerns. The document suggests two general universal methods for keeping the system in a secure state. It also provides a list of IoT's security threats, security capabilities and some cases for further use.

3.2 ISO/IEC Standards

Other organizations who study different IoT-related issues are ISO and IEC. They were the first who decided to cooperate and assemble best practices with the aim to create IoT standards. Established in 2012, ISO/IEC JTC 1/SWG 5 Internet of Things is a standardization special working group of the Joint Technical Committee ISO/IEC JTC 1 of ISO and IEC which develops and facilitates the development of standards for the IoT. Unlike ITU standards, all documents of these organizations are not laid out in open access.

First of all, it should be noted that work in this field does not stop. Standards such as ISO/IEC 20924 (IoT definition and vocabulary), ISO/IEC 21823 (Interoperability for IoT systems – Part 1: Framework; Part 2: Network connectivity; Part 3: Semantic interoperability), ISO/IEC 23093-1 (Internet of media things - Part 1: Architecture), ISO/IEC 27030 (Guidelines for security and privacy in IoT), ISO/IEC 30149 (IoT Trustworthiness framework) and ISO/IEC 30147 (Methodology for trustworthiness of IoT system/service) are under development or at the revision stage (November 2018).

ISO/IEC 22417:2017 (IoT use cases) [36] suggests using ISO/IEC terms and consists of IoT use case scenarios. They provide a practical context for considerations on interoperability and standards based on user experience, clarify when these standards can be applied and highlight where standardization work is needed.

ISO/IEC 29161:2016 (Unique identification for the IoT) [37] establishes a unique identification scheme for the IoT, based on existing and evolving data structures. This standard specifies the common rules applicable to unique identification to ensure full compatibility across different identities. The unique identification is a universal construct for any physical or virtual object or person. It is used in IoT information systems, which need to track or otherwise refer to entities. It is intended for use with any IoT media.

ISO/IEC 29181-9:2017 (Future Network – Problem statement and requirements, Part 9: Networking of everything) [38] describes the general characteristics of Networking of Everything (NoE), which can be applied to Future Networks, especially from an IoT perspective. This standard specifies a conceptual NoE model and its definition, problem statements in conventional networking, standardization activities of other standards-development organizations, requirements for NoE from an IoT perspective, and technical aspects.

ISO/IEC 30141:2018 (loT reference architecture) [39] provides a standardized IoT Reference Architecture using a common vocabulary, reusable designs and industry best practices. It uses a top-down approach, beginning with collecting the most important characteristics of the IoT, abstracting those into a generic IoT Conceptual Model, deriving a high-level system based reference with subsequent dissection of that model into five architecture views from different perspectives.

The last two standards are interesting for consideration from the IS point of view. A more detailed study of ISO/IEC standards is planned in our future research.

3.3   IEEE Standards

P2413 (Architectural framework for the IoT) [40] defines an architectural framework (AF) for the IoT, including descriptions of various IoT domains, definitions of IoT domain abstractions, and identification of commonalities between different IoT domains. The IoT's AF provides a reference model that defines relationships among various IoT verticals (e.g., transportation, healthcare, etc.) and common architectural elements. It also provides a blueprint for data abstraction and the quality "quadruple" trust that includes protection, security, privacy, and safety. Furthermore, this standard shows a reference architecture that builds upon the reference model. The reference architecture covers the definition of basic architectural building blocks and their ability to be integrated into multi-tiered systems. This architecture also addresses how to document and, if strived for, mitigate architecture divergence. This standard leverages existing applicable standards and identifies planned or ongoing projects with a similar or overlapping scope.

P1451-99 (Harmonization of IoT devices and systems) [41] defines a method for data sharing, interoperability, and security of messages over a network, where sensors, actuators, and other devices can interoperate, regardless of underlying communication technology. This standard does not cover Application Programming Interfaces (APIs) for existing IoT or legacy protocols. But it utilizes the advanced capabilities of the Extensible Messaging and Presence Protocol (XMPP), such as providing globally authenticated identities, authorization, presence, lifecycle management, interoperable communication, IoT discovery, and provisioning. Descriptive meta-data about devices and operations provides sufficient information for infrastructural components, services, and end-users to dynamically adapt to a changing environment. For a successful Smart City infrastructure, key components and needs are identified and addressed.

P1931.1 (Architectural framework for real-time onsite operations facilitation for the IoT) [42] defines an AF, protocols, and APIs for providing Real-time Onsite Operations Facilitation (ROOF). ROOF computing and networking for the data and IoT devices include next-hop connectivity for the devices, real-time context building and decision triggers, efficient backhaul connectivity to the cloud, and security and privacy. This standard covers interoperability, collaboration and autonomous operation of an IoT system with computing required for context building, security, access control, data storage, data aggregation and ability to choose different cloud and application service providers. It defines how an end user is able to securely provision, commission/decommission the devices, as well as leverages existing applicable standards and is complementary to AFs defined in broader IoT environments.

P2668 (Maturity index of IoT: evaluation, grading and ranking) [43] gives the basis for measuring the maturity of objects in IoT environment, namely things, devices or the entire IoT. The standard defines the mechanism and specifications for evaluation, grading and ranking of the performance of IoT objects by using an indicator value IoT Index (IDex). IDex classifies the objects into multiple levels of performance, gives a quantitative performance representation and indication and manifests guidance on blending of IoT objects to evolve into better performance.

There are some more standards to consider in future.

3.4 IoT Security Standards

There are far fewer standards for IoT security or security-related issues. The following is almost a complete list: ISO/IEC 29181-5:2014 (Future Network – Problem statement and requirements – Part 5: Security), X.1362 (Simple encryption procedure for IoT environments), Y.4102/Y.2074 (Requirements for IoT devices and operation of IoT applications during disasters), Y.4455 (Reference architecture for IoT network service capability exposure), Y.4118 (IoT requirements and technical capabilities for support of accounting and charging), Q.3952 (The architecture and facilities of a model network for IoT testing), Y.4702 (Common requirements and capabilities of device management in the IoT), and Q.3913 (Set of parameters for monitoring IoT devices).

Three very important projects under development must be added to this list: ISO/IEC 27030 (Guidelines for security and privacy in IoT), 30149 (IoT Trustworthiness framework) and 30147 (Methodology for trustworthiness of IoT system/service).

In 2017, the European Union Agency For Network And Information Security (ENISA) has published the "Baseline Security Recommendations for IoT in the context of Critical Information Infrastructures" [44] to map critical assets and relevant threats, assess possible attacks and identify potential good practices and security measures for protecting the IoT systems.

At that moment (November 2018) the National Institute of Standards and Technology has no special publications for the IoT with only NISTIR 8228 (Considerations for IoT cybersecurity and privacy risks) [45] and Draft NIST Cybersecurity White Paper, Internet of Things (IoT) Trust Concerns [46].

The issues of ensuring IoT's IS look of less importance in other organizations' documents, and their consideration is often limited to recommendations about contacting the support service of the organization that oversees the project for further research and problem solution.

From all documents mentioned above it can be concluded that the IoT standardization remains a prospective research area.

4   Comparison of Standards Considered

According to the full content of available documents studied, one can define several comparison criteria in case. Using these criteria, common and unique parts of these documents could be identified. Table 1 shows the results of this comparison.

Table 1. Comparison of IS issues in standards considered

IS issues/Name of the standard	ITU	ISO/IEC	IEEE
Terms and definitions	Own relevant terms and definitions	Own relevant terms and definitions	Own relevant terms and definitions
IoT requirements	General and specific, different from others	General and specific, different from others	General and specific, different from others
IoT capabilities	Disclosed in detail in general and particular cases	More focused on general cases	More focused on particular cases
IoT specifics	Reviewed from different points of view	Reviewed from different points of view	Reviewed from the perspective of certain issues
IoT threat classification	A generic and concrete version with examples	A generic version with examples	Not defined yet

The results demonstrate that each of the organizations developing the IoT standards has its own vision of the situation and its priorities. It cannot be disclaimed that problems they are studying are relevant. With the results shown, one can conclude that all the organizations pay great attention to the IoT issues. The requirements are described in sufficient detail, specific points are noted, and valuable comments are made. Specific applications of these standards are of interest, since they may give a different vision of problems, initial bases, and areas of knowledge. From the organizations discussed, it can be seen that the ITU standards contain more general and specific recommendations than others, with appropriate examples. Security issues are considered from different points of view, although IS problems are not paid increased attention. Despite this fact and the pace of development in this area, it can be said that these gaps will be filled in the next few years.

5   Conclusion

Although the IoT concept and all its aspects are widely discussed, there are some issues which require further detailed research. Development of new principles for working in such networks, as well as the requirements for participants, is really relevant. The main problem of IoT's IS standardization today is that the IoT requires a different approach than a regular network. These issues are now under consideration. Most of the standards are focused either on general problems or consider their specific subjects. The general nature of recommendations for ensuring IS can be looked upon as a shortcoming of the existing standards lacking (although not everywhere) of disclosure of security issues not only for the IoT software but also hardware. It is expected that the situation will be improved as the best practices will be collected.

Further work in this area lies in the development of detailed recommendations, which can be sent to international organizations involved in the standardization process to verify their correctness and relevance, as well as subsequent incorporation into documents. The second direction is the detailed study of IoT's IS issues and assessment of the applicability and correctness of the developed recommendations.

Acknowledgement. This work was supported by the MEPhI Academic Excellence Project (agreement with the Ministry of Education and Science of the Russian Federation of August 27, 2013, project no. 02.a03.21.0005) and possibly the Russian Foundation of Basic Research (its decision on funding as of 04.03.2019 is not known).

References

1. Reference Architecture of the Internet of Things (IoT) (2017). https://www.anti-malware.ru/practice/solutions/iot-the-reference-security-architecture-part-1. (AD: 10.01.2019), (in Russian)
2. Gartner IT glossary. https://www.gartner.com/it-glossary/internet-of-things/. (AD: 10.01.2019)
3. Ashton, K.: That 'Internet of Things' Thing. In the real world, things matter more than ideas. RFID Journal, 22 June 2009
4. We Asked Executives About The Internet of Things and Their Answers Reveal That Security Remains A Huge Concern. Business Insider (2015). https://www.businessinsider.com/internet-of-things-survey-and-statistics-2015-1. (AD: 10.01.2019)
5. Ericsson. Internet of Things forecast. https://www.ericsson.com/en/mobility-report/internet-of-things-forecast. (AD: 10.01.2019)
6. Lagutenkov, A.: Silent expansion of the Internet of Things. Sci. Life **5**, 38–42 (2018). (in Russian)
7. Internet of Things Market in Russia and the World. Direct INFO Inc. (2017). (in Russian)
8. Panda Security: You don't realize how many devices in your workplace are connected to the Internet. 2016. https://www.pandasecurity.com/mediacenter/mobile-security/shadow-iot-workplace-internet. (AD: 10.01.2019)
9. Hewlett Packard Enterprise. Protecting data into and throughout the cloud (2016). https://www.hpe.com. (AD: 10.01.2019)
10. Smith, C.: HPE Fortify and the Internet of Things (2016). http://go.saas.hpe.com/fod/internet-of-things. (AD: 10.01.2019)

11. Do not expose children to the threat of a cyber attack (2016). https://www.kaspersky.ru/blog/kid-safety-iot/10587/. (AD: 10.01.2019), (in Russian)
12. Greenberg, A.: Hackers Remotely Kill a Jeep on the Highway—With Me in It, Wired (2015). https://www.wired.com/2015/07/hackers-remotely-kill-jeep-highway/. (AD: 10.01.2019)
13. Kupriyanovskij, V.P., Namiot, D.E., Kupriyanovskij, P.V.: Standardization of Smart Cities, Internet of Things and Big Data. Considerations for practical use in Russia. Int. J. Open Inform. Technol. **4**(2), 34–40 (2016). (in Russian)
14. Miloslavskaya, N., Tolstoy, A.: Internet of Things: information security challenges and solutions. Springer, Cluster Computing (2018)
15. Dhanjani, N.: Abusing the Internet of Things: Blackouts, Freakouts, and Stakeouts. O'Reilly Media, Sebastopol (2015)
16. Russell, B., Van Duren, D.: Practical Internet of Things Security. Packt Publishing, UK (2016)
17. Hu, F.: Security and Privacy in Internet of Things: Models, Algorithms, and Implementations. CRC Press, Boca Raton (2016)
18. Li, S., Xu, L.D.: Securing the Internet of Things. Elsevier, Amsterdam (2017)
19. Alguliev, R., Mahmudov, R.: Internet of Things. Inform. Soc. **3**, 42–48 (2013). (in Russian)
20. Roslyakov, A.V., et al.: Internet of Things. Samara: Povolzhskiy State University of Telecommunications and Informatics, vol. 340. As Guard Publishing (2014). (in Russian)
21. Saryan, V.K., et al.: The Past, Present, and Future of Internet of Things standardization. Proc. Radio Res. Inst. **1**, 2–7 (2014). (in Russian)
22. Kess, P., et al.: Standardization with Iot (Internet-of-Things). In: Managing Innovation and Diversity in Knowledge Society Through Turbulent Time: Proceedings of the MakeLearn and TIIM Joint International Conference 2016, pp. 1069–1076. ToKnowPress (2016)
23. Hwang, I., Kim, Y.G.: Analysis of security standardization for the Internet of Things. In: 2017 IEEE International Conference on Platform Technology and Service (PlatCon), pp. 1–6 (2017)
24. ITU Y.4000/Y.2060 Overview of the IoT
25. ITU Y.4050/Y.2069 Terms and definitions for the IoT
26. ITU Y.4100/Y.2066 Common requirements of the IoT
27. ITU Y.4111/Y.2076 Semantics-based requirements and framework of the IoT
28. ITU Y.4113 Requirements of the network for the IoT
29. ITU Y.4103/F.748.0 Common requirements for IoT applications
30. ITU Y.4552/Y.2078 Application support models of the IoT
31. ITU Y.4453 Adaptive software framework for IoT devices
32. ITU Y.4101/Y.2067 Common requirements and capabilities of a gateway for IoT applications
33. ITU Y.4112/Y.2077 Requirements of the plug and play capability of the IoT
34. ITU Y.4401/Y.2068 Functional framework and capabilities of the IoT
35. ITU Y.4806 Security capabilities supporting safety of the IoT
36. ISO/IEC TR 22417:2017 Information technology – Internet of things (IoT) use cases
37. ISO/IEC 29161:2016 Information technology – Data structure – Unique identification for the Internet of Things
38. ISO/IEC TR 29181-9:2017 Information technology – Future Network – Problem statement and requirements – Part 9: Networking of everything
39. ISO/IEC 30141:2018 Internet of Things (IoT) – Reference Architecture
40. P2413 Standard for an Architectural Framework for the Internet of Things (IoT)
41. P1451-99 Standard for Harmonization of Internet of Things (IoT) Devices and Systems

42. P1931.1 Standard for an Architectural Framework for Real-time Onsite Operations Facilitation (ROOF) for the Internet of Things
43. P2668 Standard for Maturity Index of Internet-of-things: Evaluation, Grading and Ranking
44. European Union Agency For Network And Information Security (ENISA). Baseline Security Recommendations for IoT in the context of Critical Information Infrastructures (2017). https://www.enisa.europa.eu/publications/baseline-security-recommendations-for-iot. (AD: 10.01.2019)
45. Boeckl, K., et al.: Draft NIST IR 8228 Considerations for Managing Internet of Things (IoT). Cybersecurity and Privacy Risks (2018) https://nvlpubs.nist.gov/nistpubs/ir/2018/NIST.IR.8228-draft.pdf. (AD: 10.01.2019)
46. Voas, J., et al.: Internet of Things (IoT) Trust Concerns. 2018. https://csrc.nist.gov/CSRC/media/Publications/white-paper/2018/10/17/iot-trust-concerns/draft/documents/iot-trust-concerns-draft.pdf. (AD: 10.01.2019)

An Efficient Sybil Attack Detection for Internet of Things

Sohail Abbas[✉]

Department of Computer Science, University of Sharjah, Sharjah, UAE
sabbas@sharjah.ac.ae

Abstract. The Internet of Things paradigm is about to emerge in full scale but various security vulnerabilities are still to be addressed. One of these is the threat of Sybil attacks. A Sybil attacker creates and controls more than one identity on its physical device. These illegitimate identities of the Sybil attacker may be used for numerous malicious activities without the fear of being detected and hence accountable for committed malign actions. One of the promising countermeasures of Sybil attacks is received signal strength based localization and detection systems. However, these schemes detect only the direct Sybil attackers, where no collusion among the identities is assumed; and these schemes also incur overhead in the form of periodic and persistent localization. In this paper, we propose a detection system that detects both direct and indirect Sybil identities using one-time localization without causing overhead in the form of period localization information dissemination. The analysis of our scheme shows that the incurred overhead is significantly low in terms of communication, storage, and computation.

Keywords: Sybil attack · Collusion · RSSI based localization ·
Internet of Things · Internet of Vehicles

1 Introduction

The Internet of Things (IoT) paradigm has already emerged on the horizon. The futurists are working on to divert the existing infrastructures to adapt to IoT based platforms. Due to the recent development in wireless communication technologies; such as short-range wireless and WiFi, the IoT enables amalgamation and integration of numerous "things". Each of these "things" [1] become network entities that may act as "smart" objects and may enjoy the ubiquitous connectivity to the Internet enabling the integration of humans, things, sensors, drones, vehicles, networks, and infrastructures. The aim is to establish an intelligent network that will support various services for big cities or even for a country; such as, intelligent transport system for a city [2], Internet of Vehicles (IoT) for road conditions, safety services, Unmanned Aerial Vehicles (UAVs) networks for aerial surveillance [3], and so on.

In wireless networks like IoT, mobile hosts and access points are usually represented by a unique identity. Each identity implicitly denotes one network entity, i.e. a mobile host or an access point, portraying a one-to-one mapping of identity and entity.

© Springer Nature Switzerland AG 2019
Á. Rocha et al. (Eds.): WorldCIST'19 2019, AISC 931, pp. 339–349, 2019.
https://doi.org/10.1007/978-3-030-16184-2_33

In infrastructureless domains of IoT, where nodes join and leave at any time, due to the absence of identity management and the distributed architecture, the one-to-one mapping of identity and entity is really difficult to impose. Sybil attackers violate the one-to-one identity and entity mapping thereby forging more than one identity using their physical devices [4].

Sybil attacks can cause detrimental damages to the IoT environment in various forms. For instance, the Sybil attacker can disrupt routing process thereby injecting the forged identities into various routing paths posing the false impression of being distinct nodes on various locations. This scenario can then be used for various malicious activities, such as Denial of Service (DoS) attack. The participation of a Sybil attacker in multiple paths may grow the scale of the attack. Sybil attackers can also cause disruption to the reputation and trust based schemes, i.e. increase their own ratings or defame other good nodes in the network. In voting based protocols and data aggregation protocols, Sybil attackers may exploit their fake identities in order to manipulate the final outcome. In vehicular networks, Sybil attackers can use their forged identities to portray false traffic congestion for malicious traffic diversions. These are very few examples of the damages they can cause.

In the literature various countermeasures for Sybil attacks have been proposed, among those techniques, RSSI (Received Signal Strength Indicator) based position verification is considered to be the most promising one [5] owing to its intrinsic and lightweight nature. In these schemes, each location is considered to be bound by only one identity; transmissions received from identical location tagging more than one identity are deemed to be a Sybil attack. There are two main problems in this category of techniques. First, most of the solutions have been proposed to detect the direct Sybil attacks, such as [6–9]. They do not detect the collusion among Sybil attacker or their identities. Second, the detection schemes, such as [10, 11], usually require periodic and persistent localization based broadcasts or periodic beaconing. Our contribution in this paper is that we propose an efficient and lightweight RSSI based technique in order to detect and isolate not only direct Sybil attacks, but also the colluded or indirect communication based Sybil attacks. In our proposed scheme, we use one-time localization for the detection of each identity. The analysis demonstrates that the incurred overhead is significantly low in terms of communication, storage, and computation.

The rest of the article is organised as follows. In Sect. 2, we discuss the related literature; Sect. 3 is regarding our detection rationale, methodology, and analysis. The paper is concluded in Sect. 4, highlighting the future work.

2   Related Work

Newsome *et al.* [5] categorized the countermeasures of the Sybil attacks into three broad classes: cryptographic, resource testing, and position verification based solutions. In the cryptographic based solutions, each identity is tagged with one digital certificate obtained from centralized or semi centralized Trusted Third Party (TTP) [12]. The cryptographic based solutions, a better preventive countermeasure though, got various problems;

such as it requires heavy cryptographic computations and strong reliance on TTP makes it less efficient and not scalable. The TTP will also be needed as all-time online and accessible to all network nodes. Other issues include proper and secure certificate issuance, revocation, management, and its distribution. The resource testing based solutions are not practical; these schemes usually cast a line to plummet the users' resources beneath that line, such as the assumption that each node must contain only one network interface card, etc. Some of the resources now-a-days are very cheap to buy. The last category of countermeasures, called position verification based detection, is considered to be a promising one. According to this category, each identity is assumed to be bound by a distinct location, i.e. messages received from two identities belonging to the same location would imply Sybil attack.

In the literature various attempts have been made to make the localization lightweight and precise for the improved detection of Sybil attacks. Various authors, such as [13–15], employed extra hardware in the form of Geographical Positioning System (GPS) and directional antennae in order to improve the detection process. Whereas some authors, such as [10, 11, 16], use fixed anchor nodes and periodic beaconing for the improved accuracy. The RSSI based countermeasures proposed for Sybil attacks, on the other hand, are good alternative to the above proposed schemes for various reasons; such as, RSSI is lightweight and it is an intrinsic property of nodes: nodes don't need to be equipped with extra hardware. Though the RSSI varies with time and it is susceptible to various environmental factors, it is still an attractive choice for static networks. In case of mobility, still it is a good choice if few-meter accuracy is ok to be compromised or the node size is large, such as in the Internet of Vehicles scenario a vehicle can occupy few meter space.

In the RSSI based Sybil attack detection schemes, most of the schemes [6–9] detect the direct Sybil nodes and do not detect the collusion among Sybil attacker or their identities. Similarly, the detection schemes [10, 11, 16] usually require periodic and persistent localization based broadcasts or periodic beaconing; no matter if the same nodes stay in the network for long time. In the coming section, we will discuss our scheme in order to address both of these issues.

3   The Proposed Detection System

3.1   RSSI Based Localization

According to Frii's free space radio model, if a transmitter node t transmits with power P_t, the signal strength P_r at the receiver r can be represented by the following equation.

$$P_r = P_t \frac{G_t G_r \lambda^2}{(4\pi d)^\alpha} \tag{1}$$

where G_t and G_r are antenna gain at the transmitter t and the receiver r, respectively; λ is wavelength of the signal, d is separation between t and r. The exponent of d is α, which is usually called path-loss exponent or distance-power gradient, its value varies depending upon on the environment in which signal travels. For example, for free space (Line-of-Sight outdoor environment) its value is 2; similarly, for indoor communications, its value is greater than 2 [17]. The P_r is usually called Received Signal Strength Indicator (RSSI). The above can also be written in concise form as

$$P_r = P_t \frac{K}{d^\alpha}, \quad K = \frac{G_t G_r \lambda^2}{(4\pi)^\alpha} \tag{2}$$

If the transmitted power, i.e. P_t, of nodes is known, any receiver of the signal can compute the distance between them which is then used to localize the transmitter node using simple geometric triangulation. However, RSSI is considered to be unreliable due to its time varying and non-isotropic nature [18].

Given the above discussion, Sheng et al. [19] proposed a localization algorithm and demonstrated that a node can efficiently be localized if monitored by at least four nodes even in the presence of RSSI fluctuation and RSSI being non-isotropic. In order to demonstrate the scenario, suppose a node S is monitored by four nodes named as A, B, C, and D, as shown in Fig. 1(a). After receiving the first RSSI from S, each monitoring node will record the RSSI as per Eq. (2), which can be written as

$$\left(P_{r(A)} = \frac{P_{t(S)} K}{d_A^\alpha}\right), \left(P_{r(B)} = \frac{P_{t(S)} K}{d_B^\alpha}\right), \left(P_{r(C)} = \frac{P_{t(S)} K}{d_C^\alpha}\right), \left(P_{r(D)} = \frac{P_{t(S)} K}{d_D^\alpha}\right) \tag{3}$$

Now suppose node A acts as a primary monitor that collects the RSSI from its subordinate nodes, i.e. B, C, and D, and computes the ratios of RSSIs as follows.

$$\left(\left(\frac{R_A}{R_B}\right) = \left(\frac{d_B}{d_A}\right)^\alpha\right), \left(\left(\frac{R_A}{R_C}\right) = \left(\frac{d_C}{d_A}\right)^\alpha\right), \left(\left(\frac{R_A}{R_D}\right) = \left(\frac{d_D}{d_A}\right)^\alpha\right) \tag{4}$$

The interesting point in Eq. (4) is that it is independent of the transmit power P_t. For the sake of simplicity, let the locations are considered to be in 2-dimensional Cartesian coordinates, node A can easily compute the location of node S (x, y) by solving the following equation.

$$\begin{aligned}
((x - x_A) &+ (y - y_A))^2 \\
&= (R_B/R_A)^{\frac{1}{2}}((x - x_B) + (y - y_B))^2 \\
&= (R_C/R_A)^{\frac{1}{2}}((x - x_C) + (y - y_C))^2 \\
&= (R_D/R_A)^{\frac{1}{2}}((x - x_D) + (y - y_D))^2.
\end{aligned} \tag{5}$$

The Sheng's algorithm can be used to detect Sybil attackers. As shown in Fig. 1(b), the Sybil node has two Sybil identities named as S_1 and S_2. Since, these two identities have been generated on the same physical device; both must refer to the same location (x, y). Suppose that the Sybil attacker exploits its S_1 identity and transmits data at time t_1; which will be received by the monitoring nodes A, B, C, and D accordingly as per

Eq. (5) and associating the resulted location with S_1. Later in time, suppose the Sybil attacker uses its second identity, i.e. S_2, to transmit data. All the monitoring nodes again will compute the location and tagging it with S_2. If the two identities refer to the same location, S_1 and S_2 will be detected as Sybil identities.

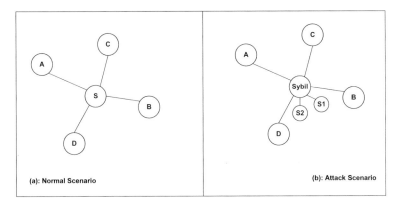

(a): Normal Scenario   (b): Attack Scenario

Fig. 1. (a) RSSI collection without attackers and (b) with Sybil attackers

Sheng's algorithm required at least four collaborating nodes, Demirbas *et al.* [8] proposed an enhanced version of this scheme thereby using ratios of RSSIs. Let at time t_1, S_1 transmits data and all the neighboring nodes calculate the RSSI and then share it with the primary detector, i.e. node A, as shown by Eq. (6). Then the same is repeated for S_2 when it transmits at time t_2, the gathered RSSI at node A is given as the following.

$$\frac{R_A^{S1}}{R_B^{S1}}, \frac{R_A^{S1}}{R_C^{S1}}, \frac{R_A^{S1}}{R_D^{S1}} \quad and \quad \frac{R_A^{S2}}{R_B^{S2}}, \frac{R_A^{S2}}{R_C^{S2}}, \frac{R_A^{S2}}{R_D^{S2}} \tag{6}$$

After collecting RSSI related data for S_1 and S_2 at time t_1 and t_2, respectively, the primary detector A may detect Sybil identities thereby comparing the ratios of RSSIs at time t_1 and t_2, as shown by Eq. (7). If the difference of ratios happens to be zero or close to zero, the event will be considered as Sybil attack; and no attack otherwise.

$$\frac{R_A^{S1}}{R_B^{S1}} = \frac{R_A^{S2}}{R_B^{S2}}, \quad \frac{R_A^{S1}}{R_C^{S1}} = \frac{R_A^{S2}}{R_C^{S2}}, \quad \frac{R_A^{S1}}{R_D^{S1}} = \frac{R_A^{S2}}{R_D^{S2}} \tag{7}$$

Demirbas *et al.* [8] experimentally verified that using the ratios of RSSIs, at least two nodes' collaboration is required as opposed to the Sheng's algorithm.

Most of the proposed Sybil attack detection systems including Demirbas *et al.* [6–9] detect the Sybil attackers that claim to be in one-hop vicinity of the detectors nodes. In other words, the Sybil attackers directly communicate with the detector nodes. However, the already proposed schemes, may not detect the smart Sybil attackers that use its original identity (that acts as benign or clean identity) to communicate with the neighbors and the rest of the Sybil identities then indirectly communicate with the neighbors via the original identity. Newsome *et al.* [5] termed it as Sybil attack with indirect communication. More specifically, the original identity introduces the Sybil identities, such as S_1 and S_2 in Fig. 1(b), to its neighbors, i.e. the detectors A, B, C, and D. The detectors consider S_1 and S_2 to be indirect nodes, i.e. at 2-hop distance from the detectors.

In the following section, in our proposed scheme, we use the ratios of RSSI based localization algorithm.

3.2   The Proposed Scheme

We assume an IoT based scenario that is in the form of a wireless network consisting of different types of nodes. Nodes may be mobile or static of different kind, such as smart phones, tablets, laptops, or vehicles. We propose a distributed detection scheme that will work for the scenarios shown in Fig. 2. In both of these scenarios, the communication with the gateway node is performed through multi-hopping. The first scenario is Internet of Vehicles (IoV), i.e. vehicles are connected to a nearby communication tower via multihop communication that provides Internet connectivity, Fig. 2(a). The scenario shown in Fig. 2(b) is a wireless network having no centralized control other than the gateway node that also provides Internet connectivity.

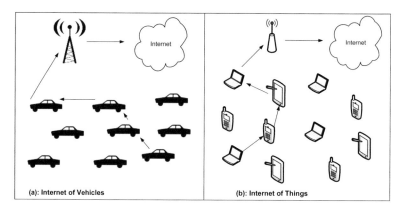

Fig. 2. Two different mobile networks scenarios: (a) IoV and (b) IoT

We assume that each node will work in two phases. First, each node will collect RSSI information from its vicinity. Second, after collection of RSSI related information, each node will share this information further with its 1-hop neighbors. We assume that these messages will be encrypted by symmetric or asymmetric cryptographic system in order to impose message authentication. After these two phases, each node must have the appropriate information to compute ratios of RSSI for each neighboring node according to the formulation given in Sect. 3.1 in order to detect Sybil attackers.

Following the above procedure, once it has been verified that an identity is benign or Sybil, we believe that there is no need for further localization process, until a new identity emerges. Each node will record and share those benign identities in the form of a list, called Benign Identity List (BIL), in the network. The BIL will not contain Sybil identities in order to prevent malicious nodes from blackmailing attack or to prevent negative information propagation in the network. In short, each new node will be confirmed to be Sybil or non-Sybil by its 1-hop neighbors, at its entry into the network. The 1-hop neighbors will then inform their neighbors about the newly registered or rejected node, and so on. We believe that once a node has been verified at its entry time, re-localization of it later when it moves will be just an overhead. Logically, this process can be perceived as an entrance door, as shown in the Fig. 3. As a result, nodes will communicate and provide services, such as packet forwarding, etc., only with the identities present in the BIL. Unregistered identities will be considered as untrusted; hence, no communication. This is what we call one-time or one-off localization.

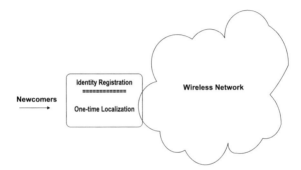

Fig. 3. Global view of the one-time localization

Since, our proposed detection system is distributed in nature, 1-hop neighbors play a vital role to detect and isolate Sybil attackers. Once it is confirmed that a new identity is not a Sybil one, 1-hop neighbors will simply broadcast this information in the form of BIL. The result of this broadcast is twofold. One, all the neighbors that did not take part in the localization process (those neighbors are usually lies in 2-hop distance) will be informed of this new identity and they will update their BILs accordingly. Two, by receiving BILs, new node will also build its own BIL.

In static networks this sort of detection system works fine. However, in mobile networks the above mentioned scheme poses some predicaments. For instance, a new node registration is reflected to 2-hops, what is going to happen if the newly registered node moves out of its 2-hop range? In order to address this issue, we need to closely study the problem, for instance, when a registered node moves out of its 2-hop range, then at the edge of that region some of its 2-hop neighbors would discover a change in hop count for this node in their BILs. For example, the stored hop count for the node at the neighboring nodes will become inconsistent. Please note that the contact between the nodes may be direct or indirect. In the indirect case, due to the broadcast nature of wireless medium nodes collect information through overhearing in promiscuous mode. Whenever, the hop count change is observed by the 2-hop neighbors, they will update their BILs to reflect the hop count change (i.e. from 2-hop to 1-hop conversion) and will re-broadcast their BILs. The result of this second broadcast will update BILs in the 2-hop vicinity of the mobile node. As shown in Fig. 4, in node B's BIL, node A's hop count is 2. Now if due to mobility A moves and enters into B's radio range. Node B, from direct or indirect communication would observe this change in the hop count and will update its BIL accordingly and further broadcast it, so that D, I, and F update their BILs. Node C, E, and m will discard the B's broadcast because it is the same hop count for them.

The above detection is based on one-off localization, i.e. each identity is localized for the detection only once as long as it stays in the network.

In the above proposed scheme the threat for indirect communication version of Sybil attack is still there. For example, a registered malicious node m, as shown in Fig. 4, can exploit these broadcast messages to collude with either its own created Sybil node s or with an unregistered node u. The malicious node m can secretly provide services either to another unregistered malicious node u or to its own created Sybil identity without being detected. As a result, the Sybil attacker along with its colluded identities may evade the detection process and therefore may not be deem accountable for their malicious actions. For instance, m may broadcast the identity of u or s to present a false impression to its neighbors that a new identity has just been registered at 1-hop or a mobile node has induced a hop count change of in its BIL. In both the cases, neighbors (such as B, C, E) receiving these broadcasts would update their BILs with a Sybil or unregistered node identity without being aware of the reality.

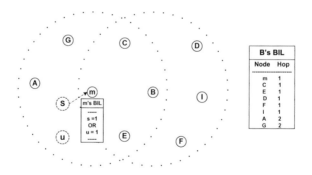

Fig. 4. Collusion or indirect communication attack scenario

In order to mitigate the above mentioned collusion attack we assume that the number of good nodes in any neighborhood is greater than that of the Sybil identities. Then we propose that nodes would accept the update of BIL for an identity only when the message for the same identity is validated by at least one other registered neighbor within time t. For instance, as shown in Fig. 4, when a node, let's say B, receives a first update message from the neighbor m, B must not accept it without being validated. So node B will wait for time t to receive the same update message from other neighbors.

After the timeout, node B will send an explicit identity confirmation request message to the neighbors and upon receipt of the confirmation reply from the neighbors (other than m) the update will be accepted; otherwise, it will be simply ignored.

3.3   Overhead Analysis

Our proposed detection method causes low overhead because BILs are broadcasted only when a new node registers into the network or in case of mobility when a node moves in the network beyond its 2-hop. However, overhead will be increased in the environments where the join and leave rate of nodes is high.

Due to space limitation we consider a very simple scenario of 2-hop network, connected further with a gateway node. Assuming that there are N network nodes uniformly distributed in area $A = X * Y$. Let $\mathcal{D}(q, R)$ denotes a disk surface of radius R representing the radio range of node q that lies at the origin of the disk. The average number of 1-hop nodes of node q, denoted by $n(q)$ or simply n, in terms of radio range and density is given as

$$n = \frac{N}{A} \cdot \pi R^2$$

Nodes are using IEEE 802.11 protocol and every new node will cast a *join request* upon its emergence to the 1-hop neighbors. The 1-hop neighbors will then capture the RSSI of its request-to-send (rts) and *data* control frames (i.e. 802.11 four-way handshake) in order to localize the new identity; hence two broadcasts from the new node. The receiving nodes will collect the RSSI of these two messages and further broadcast them in order to reflect it to 2-hop vicinity. The total communication overhead in order to localize the new identity is given below.

Communication Overhead $= 2 + n - 1 \rightarrow n - 1$

Each node will store the record of $n - 1$ nodes, i.e.

Memory Overhead $= n - 1$

The processing overhead as per the localization algorithm is that each node performs n division operations twice and n comparisons as well, i.e.

Computation Overhead $= 2n + n \rightarrow 3n$

In Fig. 5, the overhead for network with 500-meter dimension having three different radio ranges, i.e. $R1$, $R2$, and $R3$, with respect to the number of nodes is shown. It is evident from the figure that our proposed technique incurs minor (linear) overhead in terms of communication, storage, and computation.

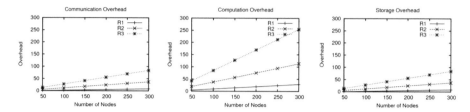

Fig. 5. Communication, computation, and storage overhead w.r.t. range and number of nodes

4   Conclusion and Future Work

Sybil attacks pose detrimental threats to the Internet of Things and Internet of Vehicles based architectures in various forms. In this paper we have proposed an RSSI based detection mechanism in order to safeguard the network against Sybil attacks. The novel aspect of our scheme was twofold: detection of colluded Sybil attacks using lightweight detection mechanism, i.e. one-time localization. We analytically analysed our scheme for overhead. In our future work, we aim at extensive simulation based evaluation of our scheme as well as the test-bed conduction for more realistic approach. Our scheme in its current form is applicable to situations where users may not change their wireless transmit powers. However, the capabilities of an attacker may not be underestimated. We will also focus on extending our work in that direction. Our scheme is also susceptible to spoofing attacks which we will try to address in our future work.

References

1. Qiu, T., Chen, N., Li, K., Atiquzzaman, M., Zhao, W.: How can heterogeneous Internet of Things build our future: a survey. IEEE Commun. Surv. Tut. **20**(3), 2011–2027 (2018). Early access
2. Paul, A., Daniel, A., Ahmad, A., Rho, S.: Cooperative cognitive intelligence for Internet of Vehicles. IEEE Syst. J. **11**(3), 1249–1258 (2015)
3. Lin, X., Yajnanarayana, V., Muruganathan, S.D., Gao, S., Asplund, H., Maattanen, H.-L., Bergstrom, M., Euler, S., Wang, Y.-P.E.: The sky is not the limit: LTE for unmanned aerial vehicles. IEEE Commun. Mag. **56**(4), 204–210 (2018)
4. Douceur, J.R.: The Sybil attack. In: First International Workshop on Peer-to-Peer Systems, pp. 251–260 (2002)
5. Newsome, J., Shi, E., Song, D., Perrig, A.: The Sybil attack in sensor networks: analysis & defences. Presented at the Third International Symposium on Information Processing in Sensor Networks (IPSN 2004) (2004)
6. Han, S., Ban, D., Park, W., Gerla, M.: Localization of Sybil nodes with electro-acoustic positioning in VANETs. In: IEEE Global Communications Conference GLOBECOM 2017, pp. 1–6 (2017)
7. Yao, Y., Xiao, B., Wu, G., Liu, X., Yu, Z., Zhang, K., Zhou, X.: Multi-channel based Sybil attack detection in vehicular ad hoc networks using RSSI. IEEE Trans. Mob. Comput. (2018). Early Access

8. Demirbas, M., Song, Y.: An RSSI-based scheme for Sybil attack detection in wireless sensor networks. In: Proceedings of the International Symposium on World of Wireless, Mobile and Multimedia Networks (2006)
9. Shaohe, L., Xiaodong, W.F., Xin, Z., Xingming, Z.: Detecting the Sybil attack cooperatively in wireless sensor networks. In: International Conference on Computational Intelligence and Security, pp. 442–446 (2008)
10. Amini, F., Misic, J., Pourreza, H.: Detection of Sybil attack in beacon enabled IEEE 802. 15.4 networks. In: International Wireless Communications and Mobile Computing Conference, pp. 1058–1063 (2008)
11. Xiao, B., Yu, B., Gao, C.: Detection and localization of Sybil nodes in VANETs. Presented at the Proceedings of the 2006 Workshop on Dependability Issues in Wireless Ad Hoc Networks and Sensor Networks, Los Angeles, CA, USA (2006)
12. Hashmi, S., Brooke, J.: Towards Sybil resistant authentication in mobile ad hoc networks. In: Fourth International Conference on Emerging Security Information Systems and Technologies (SECURWARE), pp. 17–24 (2010)
13. Yuan, Y., Huo, L., Wang, Z., Hogrefe, D.: Secure APIT localization scheme against Sybil attacks in distributed wireless sensor networks. IEEE Access 6, 27629–27636 (2018)
14. Iwendi, C., Uddin, M., Ansere, J.A., Nkurunziza, P., Anajemba, J., Bashir, A.K.: On detection of Sybil attack in large-scale VANETs using spider-monkey technique. IEEE Access 6, 47258–47267 (2018)
15. Tangpong, A., Kesidis, G., Hsu, H.-Y., Hurson, A.: Robust Sybil detection for MANETs. In: Proceedings of the 18th International Conference on Computer Communications and Networks, ICCCN 2009, pp. 1–6 (2009)
16. Chen, Y., Yang, J., Trappe, W., Martin, R.P.: Detecting and localizing identity-based attacks in wireless and sensor networks. IEEE Trans. Veh. Technol. 59(5), 2418–2434 (2010)
17. Garcia-Naya, J.A., Heath, R., Kaltenberger, F., Rupp, M., Via, J.: Experimental evaluation in wireless communications. EURASIP J. Wirel. Commun. Netw. 2017(1), 59 (2017)
18. Newport, C., Kotz, D., Yuan, Y., Gray, R., Liu, J., Elliott, C.: Experimental evaluation of wireless simulation assumptions. Simul. Trans. Soc. Mod. Sim. Int. 83(9), 643–661 (2007)
19. Sheng, Z., Li, L., Yanbin, L., Richard, Y.: Privacy-preserving location based services for mobile users in wireless networks. Department of Computer Science, Yale University, Technical report ALEU/DCS/TR-1297 (2004)

Li-Fi Embedded Wireless Integrated Medical Assistance System

Simona Riurean[1](✉) ⓘ, Tatiana Antipova[2] ⓘ, Alvaro Rocha[3] ⓘ,
Monica Leba[1] ⓘ, and Andreea Ionica[1] ⓘ

[1] University of Petrosani, 332006 Petrosani, Romania
{simonariurean,monicaleba,andreeaionica}@upet.ro
[2] Institute of Certified Specialists, Perm, Russia
antipovatatianav@gmail.com
[3] University of Coimbra, Coimbra, Portugal
amrocha@dei.uc.pt

Abstract. The hybrid Wireless Integrated Medical Assistance System (WIMAS) addressed in this paper relay both on optical wireless communication (Light-Fidelity - Li-Fi, Visible Light Communication - VLC, infrared - IR) and conventional Radio Frequency (RF) wireless communication. The medical system consists of two Wireless Medical Body Area Networks (WMBAN) based on VLC (an insulin wearable kit and ECG test device) and an Emergency Remote Medical Assistance (ERMA) with Li-Fi wireless communication technology embedded. Using RF in medical facilities is subject of strict regulations due to interferences with other RF devices, negative effects on human health and lack of security. On the other hand, both Li-Fi and VLC are suitable to be set as wireless communication technologies in medical environments and by the patients with wearable WMBAN. Research on VLC and IR transdermal communication for implantable medical devices has also been demonstrated as feasible with promising future and the Li-Fi technology recently deployed on the market is mature enough to be integrated in the emergency remote medical assistance system presented here.

Keywords: Li-Fi · Visible Light Communication · WMBAN ·
Health-monitoring · Medical devices · Transdermal communication ·
Remote medical assistance system

1 Introduction

Different scenarios of a Wireless Integrated Medical Assistance System (WIMAS) have been already discussed and implemented as reliable systems, most of them based on the RF communication [1–3]. The WIMAS we propose, consists of two different modules (Fig. 1) connected by optical wireless communication. The first module is a health monitoring integrated system, WMBAN type and the second one is an Emergency Remote Medical Assistance module (ERMA).

The health monitoring system refers to, for example, an insulin device with wireless transmission between sensor, glucose monitor and insulin pump and/or an

© Springer Nature Switzerland AG 2019
Á. Rocha et al. (Eds.): WorldCIST'19 2019, AISC 931, pp. 350–360, 2019.
https://doi.org/10.1007/978-3-030-16184-2_34

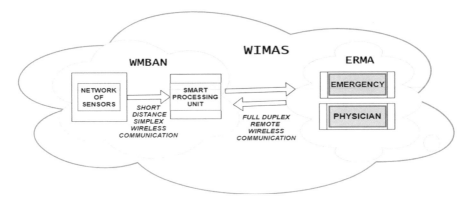

Fig. 1. A general view of a Wireless Integrated Medical Assistance System (WIMAS)

electrocardiogram (ECG) test system. All data acquired are sent wireless to a smart processing unit (smart phone, laptop, tablet, desktop) and then remotely forwarded to a dedicated server, an emergency room or a physician for emergency medical assistance. The smart processing unit plays a major role in collecting all the information gathered by the sensors, storing and/or sending data acquired via an external gateway to the "outside world".

2   A WIMAS Based on RF Transmission

2.1   General Considerations on WMBAN

Body Area Network (BAN) technology use tiny, low-power smart devices that can be carried on or embedded inside the human body. The main two activities where BAN are used aim to monitor human general or specific health condition or wellness.

Wireless BAN (WBAN) consists in ultra-low-power, self-sustained, intelligent and light weight body-borne sensors, designed to work on/in or around the human body and mainly operates in the license-free industrial, scientific, and medical overcrowded radio band centred at 2.45 GHz (IEEE 802.15.6 standard, 402–405, 420–450, 863–870, 902–928, 95–956, 2,60–2400, 2400–2438.5 MHz) [4]. WMBAN has a network of on-body wireless medical sensor devices and/or dedicated medical devices (MDs) that can communicate with a monitoring device placed on/around the human body. Such monitoring devices, in their role of MBAN hub, display and process physiological parameters from WMBAN devices and may also forward them using different communication technologies. Wireless Medical BAN (WMBAN) technology (IEEE 802.15.4j standard, 2360–2490 MHz) has been designed and developed for human body in order to acquire and send several vital signals such as heart rate, blood pressure, oxygen or insulin level, temperature, ECG and so on. Since the network of sensor nodes sense, process and send the physiological signals in real time, WMBANs allow early detection, diagnosis, prevention, monitoring, treatment or alleviation of disease or injury and fast aid in the medical field, thus offering a reliable monitoring of different human diseases and fast interventions in case of emergency situations [5].

In order to be uniquely identified, each node of the MBAN has assigned a logical IP address. The IEEE 802.15.6. Task Group 6 develops dedicated applications and energy-efficient devices, while operating at maximum frequency and highest modulation order by considering all the overheads of the Media Access Control (MAC) and Physical (PHY) layers [6] providing capability to perform carrier-sense multiple access with collision avoidance (CSMA/CA) [7]. The RF based wireless communication technologies that enable WIMASs are short-range, such as WLAN aka Wi-Fi (IEEE 802.11a/b/g/n standard, RF from 2,4 GHz and 5 GHz), Bluetooth and Bluetooth Low Energy (BLE) (IEEE 802.15.1 standard, RF 2.4 GHz), ZigBee (IEEE 802.15.4 standard 868/915, 2400 MHz) or ultra-wideband (UWB) (IEEE 802.15.4a standard, 75–724, 3000–5000, 6000–10000 MHz) [8].

Fig. 2. Different wireless communication technologies for a reliable WIMSS

Wireless technologies for a real-time health monitoring that support short range communication within the surrounding area of the human body (BLE and VLC), indoor communication based on both RF (BLE, Wi-Fi, Bluetooth) and optical transmission (VLC, Li-Fi), as well as remote transmission (Wi-Fi, WLAN, GSM, UMTS - Universal Mobile Telecommunications System) are presented in Fig. 2.

2.2 Medical Applications of WMBAN

The European Telecommunication Standardization Institute (ETSI) supervises, analyses and carry out measures in order to determine the specific requirements for WMBAN for three types of indoor communication (home, hospital and office) [9]. ETSI has developed updated System Reference Documents for the many medical applications, as it can be seen in Table 1.

2.3 Weaknesses of Wireless Data Communication Based on RF

The most important issues regarding WMBAN refers to monitoring and sensing, power efficient protocols, system architectures, routing, and security. WMBAN requires specific features for a high data rate communication and security, enhanced robustness

Table 1. System reference documents of ETSI

MBAN	Medical application	Frequency range
SRD ULPWMCE	Short Range Devices Ultra Low Power Wireless Medical Capsule Endoscopy	UHF wideband 430 MHz–440 MHz
ULP-AMMI	Ultra Low Power Active Medical Membrane Implants and Accessories	30–37.5 MHz
ULP-AID	Ultra Low Power Animal Implant Devices	12.5 MHz–20 MHz 315 kHz–600 kHz
ULP AMI	Ultra Low Power Active Medical Implants	401–402 MHz 405–406 MHz
LP-CIS	Low Power Cochlear Implant Systems	2,483.5 MHz–2,500 MHz
MBANSs	Medical Body Area Network Systems	2483.5 MHz–2500 MHz 1 785 MHz–2 500 MHz
ULP-AMI	Ultra Low Power Active Medical Implants	very low duty cycles < 0.01%
ERM RSM SRD	EM compatibility and Radio Spectrum Matters; Short Range Devices Radio equipment	315 kHz–600 kHz

in the existence of interferences, low energy consumption and low complexity of MAC layer to allow an extended autonomy [10].

RF Negative Effects on Human Health

Potential health effects of exposure to electromagnetic fields (EMF) has constantly been under evaluation by the Scientific Committee on Emerging and Newly Identified Health Risks (SCENIHR). SCENIHR deals with queries related to emerging or newly identified health and environmental risks, for example physical hazards such as noise and EM fields from cell phones, transmitters and electronically controlled indoor environments as well as procedures to evaluate new risks [11].

Most RF safety limits defined by American National Standards Institute/Institute of Electrical and Electronics Engineers (ANSI/IEEE) are in terms of both the electric and magnetic field strengths and in terms of power density. The most restrictive limits on exposure are in the frequency range of 30–300 MHz where the human body absorbs RF energy most efficiently when exposed in the far field of an RF transmitting source. The limit for human exposure to RF emission from hand held (e.g. cell phone) devices in terms of specific absorption rate (SAR) is between 1.6 and 2 (W/kg) averaged over 10 g of tissue [12].

Although general perception about RF wireless communication is that it negatively affects human health, in 2015, SCENIHR concluded that, on cell phone, RF, EM radiation exposure do not indicate an increased risk of brain tumours or other cancers of the neck region and head and studies do not show increased risk for other malignant diseases, including childhood cancer [13]. However, taking into account that the number of cell phone calls per day, the length of each call, the amount of data communication and long period of time when people use cell phones have grown, as well as the number of base stations for transmitting wireless, enforce researchers and responsible governmental bodies to keep under observation this subject [14].

RF Interference Issue

There are other concerns regarding the signal interferences between electronic implantable devices (EID) (such as pacemakers) and other MDs.

Since any electronic device is susceptible to EM signals, possible EID malfunctions can be expected. According to U.S. FDA regulation, EID manufacturers have to test their products for susceptibility to electromagnetic interference (EMI) over a wide range of frequencies, design the EID to incorporate EM shields to prevent RF signals to interfere with the electronic circuitry and submit the results as a prerequisite for market approval. Smart devices, such as cell phones, using wireless communication based on RF have adverse effects on the MD used in critical care setup [15] and therefore they are usually limited or forbidden to be used in some specific places of the healthcare facilities.

Magnetic Resonance Imaging (MRI) scanners, for example, emit large amounts of EM radiation, therefore special measures have to be taken in their proximity: patients must remove jewelleries before using the scanner and hospitals must use RF shielding to prevent EM radiation from causing interference in nearby of MDs. EM interference (EMI) causes power flows and spikes with undesirable effect on both data collected by medical equipment and the way MDs interact with each other. When decisions are being made on the basis of incorrect data given by the equipment, the patient health or even life can be put in danger. To avoid this kind of problems, insulation transformers (MITs) can be used to allow most Class I equipment to pass the electrical side of the EN60601 regulation to become safe for use in hospitals [16].

Both EN 60601 (European) and IEC 60601 (U.S.) standards provide requirements for EM compatibility for MDs, setting limits on the amount of EM radiation a MD can produce and provide test criteria to ensure that the MD continues to operate safely in the presence of external EMI [17]. They refer to MDs such as external pacemaker (2–31), nerve and muscle stimulators (2–10), ECG (2–27), EMG (2–40), hearing aids (2–66), alarm systems (8), patient monitoring (2–49). Moreover, since there are increased pervasive wireless interference from Wi-Fi, Bluetooth, etc., and the increased likelihood of using a wireless device, such as a cell phone or tablet, very close to the MDs, new regulation has entered into force since May 26th 2017 and will replace the current EU's MD Directive (93/42/EEC) and the EU's directive on active implantable MD (90/385/EEC). All manufacturing companies (both from EU and U.S) that produce currently approved MDs will have a transition time until May 26th 2020 to meet the requirements of the new regulation [18].

RF Security Issue

The medical wireless data communication from sensors to the medical device, general processing unit (smart phone, PC, laptop or tablet) as well as the remote service by internet must be strictly private and confidential and therefore encrypted in order to guarantee security and protect the patient's privacy and hence data confidentiality. On the other hand, the medical staff should receive non altered data (data integrity requirement) due to high secure management of authentication and authorization processes. Furthermore, the network should be accessible in case of emergency, e.g. trauma situations when the user is not capable to give the password.

RF communication already implemented in MD has proved to be easily accessible by unauthorized persons. In 2009, Kevin Fu, an assistant professor hacked into a defibrillator. He reprogramed the device, so it would give a shock to a patient's heart and was also able to disable the defibrillator's power-saving mode, causing the battery to run down in hours rather than years. At the Black Hat security conference in 2011, Jay Radcliffe, an IBM security expert, demonstrated how easy a hacking operation can be done when the insulin pump operates in RF wireless communication. From a distance of 30–50 m, the insulin pump can be remotely manipulated to disperse a lethal amount of insulin [19].

3 Hybrid WIMAS Based on VLC, Li-Fi and RF

The three arguments showed above, are strong enough to implement the Li-Fi, optical wireless technology in MDs and medical facilities also. Optical wireless communication (VLC and infrared IR) does not have negative effects on health, there is no interference with other MDs based on RF and there are no security issues since light cannot penetrate solid objects (Fig. 3).

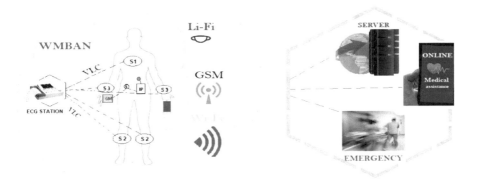

Fig. 3. The medical assistance system proposal with optical WMBAN for ECG test (S1, S2, S3 – sensors for ECG) and insulin MD (IP – insulin pump, GM – glucose monitor, S – sensor) having optical wireless communication technology embedded.

VLC uses the light intensity of a light emitting diode (LED) that is modulated by a message signal. The light signal propagates through the optical wireless channel, is detected by a photodiode (PD) that converts optical signal into electrical signal for further processing. The most important characteristics of the Optical Transmitter (oTx) and Receiver (oRx) refer to the electrical modulation bandwidth, optical spectral response, patterns of both radiation and detection, LED optical power, the PDs active area, as well as the PDs sum of noises [20].

The optical channel modelling depends on the impulse response of a finite duration [18] and the path loss. Path loss (dB) is linear over logarithmic distance, from 27 dB to 80 dB in case of both line-of-sight (LoS) and non-LoS communication scenarios [21]. The indoor light distribution (channel characterisation) for a specific setup can be determined either with recursive methods or by means of a Monte Carlo (or modified Monte Carlo) ray-tracing (MCRT) method. Chanel impulse response describes the optical delay spread in a dispersive optical wireless channel [22]. Root mean squared delay spread (RMSDS) for a LoS setup is between 1.3 ns and 12 ns and in case of a non-LoS setup, RMSDS' values are between 7 and 13 ns [20]. The single-carrier pulse modulated signals such as, On Off Keying (OOK), multi-level Pulse Position Modulation (M-PPM) and multi-level Pulse Amplitude Modulation (M-PAM) are typically applied in a VLC system, but at high speed modulation, Inter Symbol Interference (ISI) appears due to the frequency selective optical wireless channel [23]. A modified Orthogonal Frequency Division Multiplexing (OFDM), suitable for VLC and Li-Fi (purely unipolar - in frequency and time domain also) can be used with a single-tap equalizer in the frequency domain. Some appropriate solutions are: asymmetrically clipped optical OFDM (ACO-OFDM) [24], DC biased optical OFDM (DCO-OFDM) [25], PAM discrete multi-tone (PAM-DMT) [26], flip-OFDM [27], unipolar OFDM (U-OFDM) [28] and so on.

3.1   VLC Embedded in WMBAN

The wireless technology mainly avoids the risk of infection or disconnection that arises from wired connections. ECG test can be done using 3, 4, 5 or 10 sensors, in order to provide increased detailed information of the electrical activity of the heart. A wireless communication platform for ECG (operating in the unlicensed 2.4–2.4835 GHz band) was proposed to maintain reliable performance while considerably reduce sensor transmission power. Data were captured and monitored by a wearable board, and is then transmitted wireless to a nearby listening smart device [29]. An ECG monitoring using VLC has been tested with maximum 3% error in received signal at a distance of 2 m [30]. Therefore, the proposed WMBAN for the ECG system with VLC embedded, consists of optical transmission between sensors (with an array of LEDs as oTx) and the ECG equipment (with an array of PDs as oRx). Effectiveness of LED index modulation and non-DC biased OFDM for optical wireless communication with and array of LEDs and PDs have been demonstrated as optically power efficient [31].

Simulation of prototypes with low cost off-the-shelf optical devices with VLC technology embedded in MDs, proved to meet the quality requirements and therefore have ubiquitous availability in medical environments (both facilities and MDs) for safe and fast data communication [32].

There are alternative studies aiming to find minimal invasive techniques to measure blood glucose level with different types of sensors (IR, photoacoustic, ultrasound and fluorescence). Instead of extracting blood, different other fluids (saliva, sweat or tears) or body tissues (skin, tongue, oral mucosa) has been proposed to be used as an alternative method to measure glucose concentration [33].

Youngseok et al. propose a new type of sensor spoofing attack based on saturation using two medical infusion pumps equipped with IR drop sensors to control precisely the amount of medicine injected into a patients' body [34].

On the other hand, implantable medical devices (IMDs) have been subject of simulation and extended research, too. Early studies regarding transdermal optical links have been done since 1991 (neuromuscular stimulators) [35], artificial hearts as well as implanted cardiac assist devices [36]. The first duplex transdermal optical transmission system for monitoring and control an artificial heart at data rate of 9600 bps was demonstrated in 2005 [37]. The optical transdermal channel model considers transmittance (or attenuation) of the skin, the misalignment between the oTx and oRx, and the background noise. Three types of misalignment (longitudinal, lateral, and angular) have direct effect on the power level at the PD active area.

When oTx and oRx are perfectly aligned, the power of the PD on the radial direction (P_{oRx} (rd)) is defined by expression:

$$P_{oRx}(rd) = P_{oTx}(rd)\left(1 - e^{\frac{-2d}{w^2(t)}}\right) \tag{1}$$

Where P_{oTx} is power of the LED, d is diameter of the PD's active area and $w(t)$ is the optical beam radius [38].

Transdermal simulations with oTx outside the body and the oRx inside, immediately under the skin showed that solar lighting as well as artificial lights highly affect the quality of the data signal and the quality of communication is acceptable for an ideal emission wavelength of 1100 nm and a skin thickness of 4 mm [39]. Different optical communication measurements through the porcine skin, tissue and bone were made. Although the transmission intensity through bone is ≈75% less than the tissue in between, there is more than sufficient signal to obtain threshold conditions for accurate 115 Kbps optical communication with an oTx LED (880 nm, 80°) and infrared IR sensor (PIN diode with 850 nm peak sensitivity, 140°) as oRx [39]. Even though, transdermal IR communication has been demonstrated as a reliable transmission, the optical VLC (from 375 nm to 780 nm) is not in this moment an efficient way of collecting data for IMDs. However, due to fast technical developments, as well as intensive research on the area, the VLC short range transdermal transmission has promising future.

3.2 Li-Fi Fast Indoor Data Wireless Communication

Li-Fi technology (coined by Haas in 2011 during a TED conference [40]) enables full duplex, multiple-in multiple-out (MIMO) high speed optical data wireless communication between mobile devices and other networked devices via LED lights fixtures used to carry data, being a promising candidate for the upcoming ultra-high capacity indoor data transmission. A Li-Fi dongle (VLC download and IR upload enabled) connected to the mobile device receives and converts light signals into data, providing data communication by modulating the light emitted by an LED light being imperceptible by the eye. Li-Fi data wireless communication is a suitable replacement of WLAN technologies indoor, such as medical facilities, taking into account that some video streaming available prototypes have already been launched on the market [41, 42].

4 Conclusions

The hybrid WIMAS proposed here consists of a WMBAN and a remote emergency medical assistance based on both RF and optical transmission. Since RF has adverse effects on life support MDs, such as negative effects on human health, RF interference and security issues, Li-Fi is a suitable solution and reliable alternative as wireless indoor data transmission technology for WMBAN. Two different WMBAN are considered here as examples. One of them, refers to ECG test, consisting in VLC communication between sensors (placed on human body) and the ECG equipment with oRx embedded. The other WMBAN intended to replace the RF transmission with simplex optical transmission is embedded in MD worn by patients with type 1 diabetes. Optical transdermal communication for IMDs, as subject of numerous research projects where the channel model (the skin optical properties) has been considered (in both LoS and non LoS scenarios with longitudinal, lateral, and angular misalignment, skin attenuation and additive noise), has encouraging future, also.

Li-Fi has been already deployed as indoor, full duplex, MIMO, mobile wireless communication technology based on VLC (as download data path) and IR (as upload data path), being therefore a proper solution for indoor wireless data communication in medical facilities and embedded in MDs, also. The appropriate remote wireless communication for the emergency medical assistance is based on classical Ethernet, WLA networks or the GSM system.

References

1. Panescu, D.: Emerging technologies [wireless communication systems for implantable medical devices]. IEEE Eng. Med. Biol. Mag. **27**(2), 96–101 (2008). https://doi.org/10.1109/emb.2008.915488
2. Baig, M.M., Gholamhosseini, H.: Smart health monitoring systems: an overview of design and modelling. J. Med. Syst. **37**(2), 9898 (2013). https://doi.org/10.1007/s10916-012-9898-z
3. Uddin, A., Barakah, D.M.: A survey of challenges and applications of Wireless Body Area Network (WBAN) and role of a virtual doctor server in existing architecture. In: Third International Conference on Intelligent Systems, Modelling and Simulation (ISMS) (2012)
4. IEEE Standard for Local and Metropolitan Area Networks—Part 15.6: Wireless Body Area Networks; IEEE Std 802.15.6-2012, pp. 1–271. IEEE, Piscataway (2012)
5. Arefin, T., Hanif, M., Haque, A.K., Fazlul, M.: Wireless body area network: an overview and various applications. J. Comput. Commun. **5**(7), 53–64 (2017)
6. Liu, B., Yan, Z., Chen, C.W.: MAC protocol in wireless body area networks for E-health: challenges and a context-aware design. IEEE Wirel. Commun. **20**, 64–72 (2013). https://doi.org/10.1109/MWC.2013.6590052
7. Waheed, M., Ahmad, R., Waqas, A., Drieberg, M., Mahtab Alam, M.: Towards efficient wireless body area network using two-way relay cooperation. Sensors (Basel) **18**(2), 565 (2018)

8. Sudhir, D., Ramjee, P.: Human Bond Communication: The Holy Grail of Holistic Communication and Immersive Experience. John Wiley and Sons Inc., Hoboken (2017)
9. https://www.etsi.org/technologies-clusters/technologies/. Accessed 15 Oct 2018
10. Liu, R., Wang, Y., Shu, M., Wu, S.: Throughput assurance of wireless body area networks coexistence based on stochastic geometry. PLoS ONE **12**(1), e0171123 (2017)
11. Potential health effects of exposure to electromagnetic fields (EMF). SCENIHR adopted this Opinion at the 9th plenary meeting on 27 January 2015. Accessed 11 Oct 2018
12. International Standards on Absorbed Radiation (SAR), Jabra, White Paper (2018)
13. Signals, the exposure from cell phone use-power output-has changed. https://www.cancer.gov/about-cancer/causes-prevention/risk/radiation/cell-phones-fact-sheet. Accessed 11 Oct 2018
14. Cleveland R.F., Ulcek Jr., J.L.: Questions and answers about biological effects and potential hazards of radio frequency electromagnetic fields. In: OET Bulletin 56 Fourth Edition (1999)
15. Hans, N., Kapadia, F.N.: Effects of mobile phone use on specific intensive care unit devices. Indian J. Crit. Care Med. **12**(4), 170–173 (2008). https://doi.org/10.4103/0972-5229.45077
16. EMI in hospitals – and what can be done to reduce it. http://www.epdtonthenet.net/article/133645. Accessed 08 Oct 2018
17. http://ec.europa.eu/health/scientific_committees/emerging/docs/. Accessed 11 Oct 2018
18. https://www.tuv-sud.com/industries/medical-devices-healthcare. Accessed 15 Oct 2018
19. https://www.pcworld.com/article/255841/wireless_tech_makes_health_care_security_a_major_concern.html. Accessed 11 Oct 2018
20. Dimitrov, S., Haas, H.: Principles of LED Light Communications: Towards Networked Li-Fi. Cambridge University Press, Cambridge (2015)
21. Kahn, J.M., Barry, J.R.: Wireless infrared communications. Proc. IEEE **85**(2), 265–298 (1997)
22. Dimitrov, S., Mesleh, R., Haas, H., Cappitelli, M., Olbert, M., Bassow, E.: On the SIR of a cellular infrared optical wireless system for an aircraft. IEEE J. Sel. Areas Commun. (IEEE JSAC) **27**(9), 1623–1638 (2009)
23. Le Bas, C., Sahuguede, S., Julien-Vergonjanne, A., Combeau, P.: Infrared and Visible links for medical body sensor networks. In: IEEE Global LIFI Congress, Paris, France (2018)
24. Armstrong, J., Lowery, A.: Power efficient optical OFDM. Electron. Lett. **42**(6), 370–372 (2006)
25. Carruthers, J.B., Kahn, J.M.: Multiple-subcarrier modulation for non-directed wireless infrared communication. IEEE J. Sel. Areas Commun. **14**(3), 538–546 (1996). https://doi.org/10.1109/49.490239
26. Lee, S.C.J., Randel, S., Breyer, F., Koonen, A.M.J.: PAM-DMT for intensity - modulated and direct-detection optical communication systems. IEEE Photonics Technol. Lett. **21**(23), 1749–1751 (2009)
27. Fernando, N., Hong, Y., Viterbo, E.: Flip-OFDM for optical wireless communications. In: Information Theory Workshop (ITW), pp. 5–9. IEEE, Paraty (2011)
28. Tsonev, D., Sinanović, S., Haas, H.: Novel unipolar orthogonal frequency division multiplexing (U-OFDM) for optical wireless. In: Proceedings of the Vehicular Technology Conference (VTC Spring), Yokohama. IEEE, Japan (2012)
29. Tsouri, G.R.: Low-power body sensor network for wireless ECG based on relaying of creeping waves at 2.4 GHz. In: Proceedings of the International Conference on Body Sensor Networks, BSN 2010, pp. 167–173 (2010)

30. Al-Qahtani, A., Al-hajri, H., Al-kuwari, S., et al.: A non-invasive remote health monitoring system using visible light communication. In: Future Information and Communication Technologies for Ubiquitous HealthCare (Ubi-HealthTech) (2015). https://www.researchgate. net/publication/281814516_A_non-invasive_remote_health_monitoring_system_using_visible_light_communication
31. Khan, M.I., Mondal, M.R.H.: Effectiveness of LED index modulation and non-DC biased OFDM for optical wireless communication. In: IEEE International Conference on Telecommunications and Photonics (ICTP), Dhaka, pp. 227–231 (2017)
32. Riurean, S.M., Leba, M., Ionica, A.: VLC embedded medical system architecture based on medical devices quality' requirements. In: International Multidisciplinary Symposium "Universitaria SIMPRO 2018" Petrosani, Romania (2018). Quality-Access to Success Journal
33. Matzeu, G., Florea, L., Diamond, D.: Advances in wearable chemical sensor design for monitoring biological fluids. Sens. Actuators B Chemical **211**, 403–418 (2015)
34. Youngseok, P., Yunmok, S., Hocheol, S., Dohyun, K., Yongdae, K.: This ain't your dose: sensor spoofing attack on medical infusion pump. In: The 10th USENIX Workshop on Offensive Technologies, Austin, TX, USA (2016)
35. Jarvis, J.C., Salmons, S.: A family of neuromuscular stimulators with optical transcutaneous control. J. Med. Eng. Technol. **15**(2), 53–57 (1991)
36. Miller, J.A., Belanger, G., Song, I., Johnson, F.: Transcutaneous optical telemetry system for an implantable electrical ventricular heart assist device. Med. Biol. Eng. Comput. **30**(3), 370–372 (1992)
37. Okamoto, E., Yamamoto, Y., Inoue, Y., Makino, T., Mitamura, Y.: Development of a bidirectional transcutaneous optical data transmission system for artificial hearts allowing long-distance data communication with low electric power consumption. J. Artif. Organs **8**(3), 149–153 (2005)
38. Ghassemlooy, Z., Alves, L.N., Zvanovec, S., Khalighi, M.A.: Visible Light Communications: Theory and Applications. CRC Press, Boca Raton (2017)
39. Abita, J.L., Schneider, W.: Transdermal Optical Communications. Johns Hopkins APL Tech. Dig. **25**(3), 261–268 (2004)
40. (2011). https://www.ted.com/talks/harald_haas_wireless_data_from_every_light_bulb. Accessed 11 Oct 2018
41. Pure LiFi. https://purelifi.com/. Accessed 11 Oct 2018
42. https://www.eldoled.com/led-drivers/product-news/visible-light-communication–vlc-/. Accessed 11 Oct 2018

Cloud in Mobile Platforms: Managing Authentication/Authorization

Vicenza Carchiolo[1], Alessandro Longheu[2(✉)], Michele Malgeri[2],
Stefano Iannello[3,4], Mario Marroccia[3,4], and Angelo Randazzo[3,4]

[1] Dip. di Matematica ed Informatica, Università degli studi di Catania, Catania, Italy
vincenza.carchiolo@unict.it

[2] Dip. di Ingegneria Elettrica, Elettronica e Informatica,
Università degli studi di Catania, Catania, Italy
{alessandro.longheu,michele.malgeri}@dieei.unict.it

[3] ICT Technical Excellence Center, ST Microelectronics, Catania, Italy
stefano.ianniello@gmail.com,
{mario.marroccia,angelo.randazzo}@st.com

[4] ICT Technical Excellence Center, ST Microelectronics, Naples, Italy

Abstract. This work presents the solution used within STMicroelectronics to grant security in Mobile Cloud computing (MCC), where Cloud computing paradigma meets the mobile environment, the enabling technology to provide access network services anyplace, anytime and anywhere. We first illustrate the scenario, then MCC main issues are introduced, and the corresponding solution in the STMicroelectronics case is shown and briefly discussed.

Keywords: Mobile devices · Mobile platform · Cloud computing · Security · Authentication · Authorization

1 Introduction

The Mobile Cloud Computing (MCC) enhances classical mobile environment with the Cloud computing paradigm [3]; it consists of an infrastructure where both data storage and the processing occur outside of the mobile device; this shifting allows applications and mobile computing being accessed by a much broader range of subscribers than conventional smartphone users.

Joining Cloud computing and mobile networks poses many questions, as low bandwidth, availability, heterogeneity, computing offloads, data accessing, security, privacy, and trust, all enforced by the more and more significant increase in the use of smartphones.

In this paper, authentication and authorization issues in mobile Cloud computing security are addressed; in other works, e.g. [24], authors either focus on specific mechanisms provided by products or services, or they concentrate on the theory behind user authentication with no consideration to industry solutions. Here we illustrate the MCC security solution developed and applied within the

© Springer Nature Switzerland AG 2019
Á. Rocha et al. (Eds.): WorldCIST'19 2019, AISC 931, pp. 361–372, 2019.
https://doi.org/10.1007/978-3-030-16184-2_35

STMicroelectronics IC manufacturer plants [23]; the system allows users authentication and authorization according to the security standards and lies in the Cloud and mobile Cloud environment.

Our proposal aims to (1) reduce the need to store multiple passwords and multiple username for different services, (2) reduce time spent re-entering passwords and username for the same identity and (3) simplify the definition and management of security policies; ICT and related security management indeed play a key role in ST overall success and it is crucial in process transformation, systems development and innovation promotion tools, and for revenue growth, and higher productivity. In addition, the solution also considers geographical and organizational company extension, and all the set of interconnected ERP activities as supply chain, factory automation, finance, sales/purchases and marketing (compliant with recent Cloud platforms as [20]) and business intelligence.

In Sect. 2 we briefly introduce MCC and related security issues, in particular authentication and authorization. In Sect. 3 the solution adopted within STMicroelectronics is shown, together with a description of the tools and technologies used. Our conclusions are finally discussed in Sect. 4.

2   An Overview of MCC

MCC is an extension of cloud computing and incorporates mobile computing and wireless networking aiming to provide typical cloud services to the mobile consumers [21].

Main difference with the classic cloud computing service is that in MCC execution time and energy consumption matters and both are enhanced by shifting the execution of resource-intensive application from the hosting mobile device to the cloud-based resource, therefore with MCC even low performance mobile devices can work. In addition, MCC also shift data storage to cloud servers. Migrating these two workloads made MCC an useful solution to obtain higher-level service performance on mobile devices despite their (local) resources.

When considering MCC security issues, people's perception is stronger than in Cloud Computing because of the problems inherent in the mobile environment. Moreover, most security problems in MCC are related to privacy issues, i.e. potential sensitive information leakage during the mobile cloud service processes. Privacy issues are important concerns of customers and has a tight relationship with security as the security vulnerabilities can potentially result in the privacy leakage, due to the unexpected information disclosure caused by data security weaknesses.

While different measures can be implemented to improve security of the services offered in the MCC its vulnerabilities are strictly related to technologies it is based on, so hardening security of mobile devices, wireless networks and cloud computing is important, in particular for what concerns:

- Regulatory compliance: Cloud service providers should have external audits and security certification

- Privileged user access: When sensitive data get offloaded to the cloud, it may appear that the data are no longer under the direct physical, logical, and personal control of the user owner of data.
- Data location: Exact physical location of users data is not transparent, and this may result in confusion in particular authorities and commitments on local privacy needed.
- Recovery: Cloud providers should provide proper recovery management schemes for data and services when a technical fault or disaster arises.
- Data segregation: As cloud data are usually stored in a shared space in a multi-tenant environment, each users data should be separated and isolated from the others with efficient encryption methodologies.
- Long-term viability: It must be ensured that users data would be safe and accessible even in the event the cloud company itself goes out of business.
- Investigative support: For multiple customers, logging and data may be co-located. Thus, it may be vital, but hard, to predict any inappropriate or illegal activity.

2.1 Users Authentication Issues

User authentication in MCC is the process of identifying and validating the mobile user to ensure that it is legitimate to access mobile cloud resources. Both mobile device and cloud server should authenticate each other in order to secure the communication when the user with its device accesses the cloud from any geographical location in the world, using different networks and various mobile devices. Due to limited resources of mobile devices, optimal MCC authentication mechanisms should be lightweight with the least possible computing, memory, and storage requirements.

Authentication mechanisms and strategies in MCC are different from classical cloud computing. The most relevant causes of discrepancy are low resources and mobility of users devices:

1. *Resource limitations.* Among mobile devices refers to incapacitation in computational power, battery lifetime, and storage capacity in comparison to typical computers in cloud networks. Computational performance and functionalities of mobile devices are significantly hindered by such as incapacitation. Consequently, most of the mobile devices are not able of efficiently executing sophisticated resource-intensive encryption algorithms (e.g. RSA 2048 bits). Therefore, mobile device requires robust but lightweight authentication mechanism that can ensure the authenticity of users without draining local resources.
2. *Mobile and device sensors.* Mobile device sensors such as touch screen, gyroscope, fingerprint reader, accelerometer, camera, digital compass, and microphone, give to developer and researcher a great chance to add other authentication factors, particularity biometrics to improve the level of security of MCC authentication mechanisms. Although the authentication methods in cloud computing can benefit from peripheral equipment on end-user computers, additional costs can create a hurdle.

3. *High mobility.* Latency in mobility environment is inevitable due to WAN architecture that is intensified by signal handoff in the presence of heterogeneous networks and devices. Moreover, the miniature nature of mobile devices increases the possibility of robbery and loss leading to the high probability of user privacy and security violation in the absence of a reliable authentication solution. In addition, fast authentication procedure is desired to protect seamless connectivity for mobile devices in roaming.
4. *Network heterogeneity.* The mobile devices connect to heterogeneous network over time (Wi-Fi, WiMAX, 2G, 3G, 4G, and LTE to accomplish the data traffic demand in MCC). An authentication method must be designed based on security and performance requirements of each network technology that is a challenging aspect in MCC.

As shown in Fig. 1, authentication methods can be classified into two main categories,*cloud-side*, and *user-side*, each splitted into two sub-categories based on types of authentication credentials, i.e. *identity-based* or *context-based*.

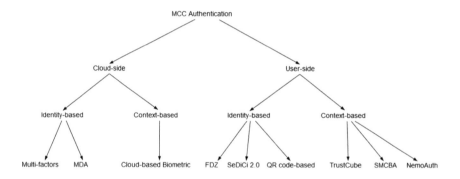

Fig. 1. Taxonomy of the state-of-the-art authentication in MCC

In Cloud-side authentication approaches, most of the authentication steps are processed in the cloud server. The cloud resources bring some advantages to improve the performance of authentication method by providing elastic processing and storage capabilities. Furthermore, different kinds of authentication procedure can be incorporated into authentication procedure based on security requirements of the user. In addition, the cloud-based authentication methods are more flexible, efficient, and adjustable compared to other authentication methods because of using the unlimited resources of cloud servers.

Although, the cloud-based authentication introduces some benefits in case of performance and usability, it introduces some security and privacy issues. The user's private authentication information (e.g. passwords, and biometrics) are highly exposed to risk. Therefore, robust authentication method is a critical requirement for mobile-cloud environment.

Table 1. Usability of existing MCC authentication schemes

Scheme	Pros	Cons
Multi-factors [10]	Using voice and face recognition, which is easy to provide and hard to replicate instead of complex password	User should memorize complicated password Using biometrics and password at the same time, demands more computational resources
MDA [7]	Using just one factor for authentication simplifies the authentication procedure	User should memorize complicated password to achieve a high level of security
Cloud-based biometrics [18]	Inserting password using handwriting is more intuitive than using tiny keyboard of the mobile device to enter password 4-digit handwritten password which is easy to remember	Any other authentication factor is not available if the system fails to identify the handwritten image
FDZ [22]	Using graphical password, which is easier to remember	Memorizing some secrets makes the procedure more difficult for a mobile user
QR code-based [17]	No need to memorize all images	The authentication procedure is complicated to the user
TrustCube [6]	Using the fine-grained method that can be customized based on user preferences Using implicit authentication method that does not need the user interferes	Using specific mobile data patterns is required to improve the accuracy of this method, which is difficult for the users
SMCBA [19]	Easy procedure for user is provided by no longer needing to memorize passwords	Using error-prone biometric reduces the usability of the authentication mechanism

In *Identity-based authentication methods*, users are authenticated through user identification attributes such as unique ID, password, and biometrics. In this type of authentication users attributes are generally fixed. The authenticator entity checks the user attributes directly without extra analyzing indirect procedures such as the user behavioral analysis. Besides, the authentication service provider is the responsible entity for identity management and performing primary user authentication.

Table 2. Security and robustness of existing MCC authentication schemes

Scheme	Pros	Cons
Multi-factors [10]	Using TLS and SSL for communication between network entities Utilizing several authentication factors	Mutual authentication is not applied
MDA [7]	Providing mutual authentication procedure Hashing ID and password before transmission	User untraceability is not considered
Cloud-based biometrics [18]	The uniqueness of handwriting style prevents an attacker to do any actions, even if he knows the user password	Mutual authentication is no applied Lack of security attack-resistant mechanism
FDZ [22]	Providing secure channel between mobile device and server using Diffie-Hellmann key exchange Resistant against device loss, impersonation, and MITM attack Using AES encryption algorithm	The password space of proposed graphical password is not large enough Mutual authentication is not applied
QR code-based [17]	Using Diffie-Hellman key exchange to prevent reply attack Preventing an impersonation attack using graphical password	Lack of security attack-resistant mechanisms The password space of proposed graphical password is not large enough Mutual authentication is not applied
TrustCube [6]	It is difficult to impersonate the user, because an attacker must access to different user information such as calling pattern, SMS, website access, and location	Mutual authentication is not applied
SMCBA [19]	Uniqueness of user fingerprint authentication	Mutual authentication is not applied User untraceability and unlinkability mechanism are not considered

In *Context-based authentication methods*, the users are authenticated by analyzing multiple passive user information such as IP address, device location, user biometrics, signal to noise ratio, and behavioral features of users. Unlike identity-based, context-based authentication method is more autonomous because the authentication procedure requires minimal user interaction. However, the accuracy of context-based method is lower than identity-based method because the authentication procedure depends on the accuracy of the results of analyzing users pattern information. Furthermore, analyzing the MCC user private information such as location, biometrics, calling patterns, and web history, expose and increase the privacy issue. In contrast to the accuracy and privacy issues, context-base methods can get some benefits from the various smartphone sensors and peripherals for both capturing and processing required user information to authenticate itself.

When considering user-side authentication method, most of the authentication steps are processed in mobile devices. Considering that in recent years the mobile devices capabilities have been improved deeply and rapidly, these mobile devices are capable of running resource-intensive applications. Moreover, the mobile devices are equipped with different input technologies that can be used to capture and analyze different patterns. Thus, such technologies, supported by device performance, are utilized in user-side authentication methods to check authenticity and authorization of mobile users.

Compared to the conventional PCs, mobile devices suffer several weakness dues to using of a heterogeneous network, additionally there is the real issue in case of losing or misplacing the mobile device, which involves the loss of valuable user confidential information. This approach contrasts with one of the main goals in MCC environment that is transferring resource-intensive processing tasks to the cloud, in this way, user-side authentication approach is less efficient and secure for cloud-connected mobile devices compared to cloud-side methods.

Like identity-based within cloud-side, in *Identity-based authentication methods* it uses user identity information to authenticate the user, but here is the mobile device that processes and analyzes user attributes to check user authentication instead of cloud servers. The private user identities such as biometrics are stored locally in the mobile device during authentication procedure, which increase the privacy and security issues.

In *Context-based authentication methods* in the user-side analyze user behavior features, similar to the cloud-side methods. The only difference between cloud-side and user-side context-based authentication method is that the mobile device processes and evaluates user information instead of cloud server.

Generally, a context-based authentication mechanism needs more computation resource compared to the identity-based methods, introducing, in this way, performance issue due to resource limitations of mobile devices. Therefore, context-based user-side authentication methods are less appropriate in MCC compared to cloud-based methods. In addition, various types of users' sensitive information are stored in mobile devices, increasing users' privacy risk due to device loss compared to the more reliable cloud environment.

Tables 1 and 2 illustrate pros and cons of several works for what concerns usability, security and robustness.

3   STMicroelectronics Authentication Services

This section discusses the authentication mechanism developed inside a STMicroelectronics test site respecting the constraints and specifications derived from the complex corporate context. A major requirement is the usability of the mechanism both with mobile devices and with PCs desktop using different architecture. Moreover the authenticator must be reachable from both the intranet network and the extranet traversing the corporate firewalls. We use the CMS-style web application, based on the PHP framework *Laravel* [11] for the part of the application and OAuth 2 for the authentication and authorization part, since it can be used effectively via mobile and desktop devices. Moreover the use of https allows us to traverse firewall minimizing the configuration issues.

The first part of this section covers the study of the framework and the realization of the relational entities for the subsequent implementation of the database. After that, paper focuses on the whole infrastructure. The infrastructure is based on cloud technologies offered by *Amazon Web Services* (AWS), in particular the application host server runs on a micro-type ec2 platform with an *Apache* server on board.

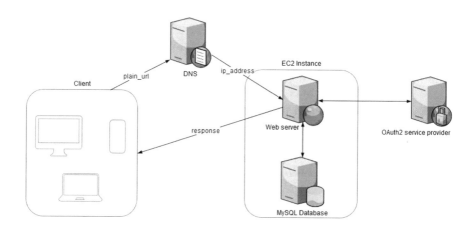

Fig. 2. Application's architecture

As the Fig. 2 shows, the client may be any device with browsing capabilities. Clients' requests are forwarded through a dynamic DNS service provided by *No-IP* [15] to the Amazon EC2 instance.

3.1   Laravel

In order to speed the development of the platform one of the requirements was to built the web-application using one of the many available development frameworks, such as, Rails, Sinatra, that use Ruby; Django and Pyramid, using Python; CodeIgniter, Symphony, and Laravel, written in PHP. The selected framework Lavarel join simple development and support for several authentication methods.

Laravel is a free, open-source PHP web framework intended for the development of web applications following the model–view–controller (MVC) architectural pattern as most of modern web application development frameworks where the is logically divided into three templates: *Models* accessing and modifying data, *Views* that present data and *Controllers*, which coordinate everything. In a web application another fundamental module is the *Routing module* that has the task to examine the URL path in each incoming request and to pass the control to the correct *controller method*.

3.2   Socialite

Lavarel provides the developers with a simple, convenient way to authenticate with OAuth providers thanks to *Laravel Socialite* [12] that handles almost all of the boilerplate social authentication code you are dreading writing. Indeed, it is possible to use *Facebook*, *Twitter*, *Google*, *LinkedIn*, *GitHub* and *Bitbucket* and any system based on OAuth.

3.3   Routing

The core of HTTP request handling is Laravel routing. All Laravel routes are defined in route files that are automatically loaded by the framework.

To fulfill the authentication purposes two routes at least are necessary: one for redirecting the user to the OAuth provider, and another for receiving the callback from the provider after authentication:

```
Route::get('login/{provider}', '
    AuthController@redirectToProvider');
Route::get('login/{provider}/callback','
    AuthController@handleProviderCallback');
```

Listing 1.1. web.php

To access Socialite, the Socialite facade is needed; the Auth facade is also required to access various Auth methods.

The Fig. 3 shows the sequence diagram implemented according with the modules and the standard.

From this point, onwards the application can simply fetch all of this data and operate with them as it prefers.

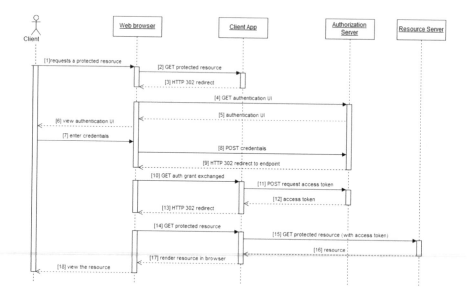

Fig. 3. OAuth2 workflow sequence diagram

3.4   OAuth2 Server with Node.js

The second step was creating an OAuth2 server allowing access to API end-points for the authorized user or authorized applications. The application so far described integrates *OAuth2orize* [16] into the application. Node.js[1] was adopted for its content richness and broad support community, in addition, it provides a large number of modules for interoperability with various services. The authorization server uses as a back-end MongoDB [1], a non-relational database since it is supported thanks to *Mongoose* [2] library and it exploit benefits as high performance/low delay, simplicity, horizontal scalability and rapid object-relational mapping.

4   Conclusions

The authentication and authorization approach used within MCC STMicro-electronics has been presented, addressing cloud computing and MCC issues. No multiple credentials for different services and easy management of security policies is achieved, addressing geographical/organizational company extensions. Further works of this proposal include performance assessment, integration with existing company's architectures and effective exploitation in other domains, e.g. in the Dispersed Computing [9] and related paradigma [8,13], as well as in different scenarios [4,5].

---

[1] Node.js is an open-source, cross-platform JavaScript run-time environment for executing JavaScript code server-side. See [14].

Acknowledgement. This work was partially supported by the 'Piano per la Ricerca 2016/2018 DIEEI Università degli Studi di Catania'.

References

1. MongoDB. https://www.mongodb.com. Accessed 29 Aug 2017
2. Mongoose. http://mongoosejs.com/. Accessed 15 Sept 2017
3. Bokhari, M.U., Makki, Q., Tamandani, Y.K.: A survey on cloud computing. In: Aggarwal, V.B., Bhatnagar, V., Mishra, D.K. (eds.) Big Data Analytics, pp. 149–164. Springer, Singapore (2018)
4. Buzzanca, M., Carchiolo, V., Longheu, A., Malgeri, M., Mangioni, G.: Direct trust assignment using social reputation and aging. J. Ambient Intell. Hum. Comput. **8**(2), 167–175 (2017). https://doi.org/10.1007/s12652-016-0413-0
5. Carchiolo, V., Longheu, A., Malgeri, M., Mangioni, G.: Gain the best reputation in trust networks. In: Brazier, F.M.T., Nieuwenhuis, K., Pavlin, G., Warnier, M., Badica, C. (eds.) Intelligent Distributed Computing V, pp. 213–218. Springer, Heidelberg (2012)
6. Chow, R., Jakobsson, M., Masuoka, R., Molina, J., Niu, Y., Shi, E., Song, Z.: Authentication in the clouds: a framework and its application to mobile users. In: Proceedings of the CCSW 2010, pp. 1–6. ACM, New York (2010). http://doi.acm.org/10.1145/1866835.1866837
7. Dey, S., Sampalli, S., Ye, Q.: Message digest as authentication entity for mobile cloud computing. In: 2013 IEEE 32nd International Performance Computing and Communications Conference (IPCCC), pp. 1–6, December 2013
8. Dolui, K., Datta, S.K.: Comparison of edge computing implementations: fog computing, cloudlet and mobile edge computing. In: 2017 Global Internet of Things Summit (GIoTS), pp. 1–6, June 2017
9. García-Valls, M., Dubey, A., Botti, V.: Introducing the new paradigm of social dispersed computing: applications, technologies and challenges. J. Syst. Archit. **91**, 83–102 (2018). http://www.sciencedirect.com/science/article/pii/S1383762118301036
10. Jeong, Y.S., Park, J.S., Park, J.H.: An efficient authentication system of smart device using multi factors in mobile cloud service architecture. Int. J. Commun. Syst. **28**(4), 659–674 (2015). http://dx.doi.org/10.1002/dac.2694
11. Laravel. https://laravel.com/. Accessed 30 Aug 2017
12. Laravel Socialite. https://github.com/laravel/socialite. Accessed 7 July 2017
13. Laszka, A., Dubey, A., Walker, M., Schmidt, D.: Providing privacy, safety, and security in IoT-based transactive energy systems using distributed ledgers. In: Proceedings of the Seventh International Conference on the Internet of Things, IoT 2017, pp. 13:1–13:8. ACM, New York (2017). http://doi.acm.org/10.1145/3131542.3131562
14. Node.js. https://nodejs.org. Accessed 5 Sept 2017
15. Noip. https://www.noip.com/. Accessed 15 Sept 2017
16. OAuth2orize. https://github.com/jaredhanson/oauth2orize. Accessed 5 Sept 2017
17. Oh, D.S., Kim, B.H., Lee, J.K.: A study on authentication system using QR code for mobile cloud computing environment, pp. 500–507. Springer, Heidelberg (2011). https://doi.org/10.1007/978-3-642-22333-4_65
18. Omri, F., Hamila, R., Foufou, S., Jarraya, M.: Cloud-ready biometric system for mobile security access, pp. 192–200. Springer, Heidelberg (2012). https://doi.org/10.1007/978-3-642-30567-2_16

19. Rassan, I.A., AlShaher, H.: Securing mobile cloud computing using biometric authentication (SMCBA). In: Proceedings of CSCI 2014, pp. 157–161. IEEE Computer Society, Washington, DC (2014). http://dx.doi.org/10.1109/CSCI.2014.33
20. Salesforce.com, inc., SalesForce. https://www.salesforce.com. Accessed 5 July 2017
21. Sanaei, Z., Abolfazli, S., Gani, A., Buyya, R.: Heterogeneity in mobile cloud computing: taxonomy and open challenges. IEEE Commun. Surv. Tutor. **16**(1), 369–392 (2014)
22. Schwab, D., Yang, L.: Entity authentication in a mobile-cloud environment. In: Proceedings of CSIIRW '13 (CSIIRW 2013), pp. 42:1–42:4. ACM, New York (2013). http://doi.acm.org/10.1145/2459976.2460024
23. STMicroelectronics. http://www.st.com/
24. Todorov, D.: Mechanics of User Identification and Authentication: Fundamentals of Identity Management. CRC Press, Boca Raton (2007)

An Ontology-Based Recommendation System for Context-Aware Network Monitoring

Ricardo F. Silva[1], Paulo Carvalho[1($\boxtimes$)], Solange Rito Lima[1],
Luis Álvarez Sabucedo[2], Juan M. Santos Gago[2], and João Marco C. Silva[3]

[1] Centro Algoritmi, Universidade do Minho, 4710-057 Braga, Portugal
{pmc,solange}@di.uminho.pt
[2] Department of Telematics, University of Vigo, Vigo, Spain
{Luis.Sabucedo,Juan.Gago}@det.uvigo.es
[3] HASLab, INESC TEC, Universidade do Minho, Braga, Portugal
joao.marco@inesctec.pt

Abstract. Current network management systems urge for a context-aware perspective of the provided network services and the underlying infrastructure usage. This need results from the heterogeneity of services and technologies in place, and from the massive traffic volumes traversing today's networks. To reduce complexity and improve interoperability, monitoring systems need to be flexible, context-aware, and able to self-configure measurement points (MPs) according to network monitoring tasks requirements. In addition, the use of sampling techniques in MPs to reduce the amount of traffic collected, analysed and stored has become mandatory and, currently, distinct sampling schemes are available for use in operational environments.

In this context, the main objective of this paper is the ontological definition of measurement requirements and components in sampling-based monitoring environments, with the aim of supporting an expert recommendation system able to understand context and identify the appropriate configuration rules to apply to a selection of MPs. In this way, the ontology, defining management needs, network measurement topology and sampling techniques, is described and explored considering several network management activities. A use case focusing on traffic accounting as monitoring task is also provided, demonstrating the expressiveness of the ontology and the role of the recommendation system in assisting context-aware network monitoring based on traffic sampling.

Keywords: Ontology · Network monitoring · Traffic sampling

1 Introduction

The semantic support for services, in a broad sense, is a common feature in many platforms nowadays. The features unleashed by semantic tools are achieving a

© Springer Nature Switzerland AG 2019
Á. Rocha et al. (Eds.): WorldCIST'19 2019, AISC 931, pp. 373–384, 2019.
https://doi.org/10.1007/978-3-030-16184-2_36

high-maturity level, and their support for add-value services are the reasons for its broad adoption in a large number of scopes. Nevertheless, its adoption in traffic monitoring scenarios has yet a substantial way to evolve, especially, in heterogeneous and context-dependent environments, such as Smart Cities.

As a crucial task supporting network management activities, network monitoring must attend to each specific context requiring traffic measurements. This means that a system can use important context information to provide customised and optimised measuring services to meet the needs of network users/administrators. Furthermore, context-aware monitoring enables saving computational and communication resources, thus fostering the provision of more relevant and agile services. Due to the constant increase in data volumes, it is also essential to enhance the monitoring process without compromising its efficiency. For this purpose, the use of traffic sampling techniques is mandatory to enable capturing the behaviour of services and networks resorting uniquely to a subset of traffic [1].

In terms of context-representation requirements, several considerations must be taken into account when selecting the technologies that can satisfy a context middleware. The choice of how to store, represent and infer context in a context processor are important requirements for the operation of a context-based system. The use of ontological support allows to formalise semantically the interaction between the users and the machine, allowing reutilisation principles which facilitate the process of representing the information [2].

This paper handles the issue of defining an ontology to assist context-aware monitoring environments which resort to traffic sampling to improve monitoring efficiency. Using a highly practical approach, the present work contributes for the identified objective through: (i) the definition of a context-aware monitoring system architecture and associated recommendation system, capable of mapping measurement needs into a set of rules to configure measurement points (MPs); and (ii) the definition of the corresponding ontology, expressing relevant classes, relations and attributes of a monitoring task, the underlying network measurement topology and the available sampling techniques, which supports the configuration of the sampling-based monitoring environment. The semantic validation of this proposal is achieved through the application of the ontology in several management competence queries and, in particular, when supporting traffic accounting. For this purpose, the ontology has been previously populated with real data collected from the University of Minho campus network.

The first step to achieve the high-level design goals mentioned above is the provision of ontological support. Thus, after discussing related work in Sect. 2, a context-aware monitoring architecture is proposed in Sect. 3. The ontological layer, fully described in Sect. 4, must endow the maximum possible level of interoperability among components, enabling the deployment of autonomic network monitoring. This section formalises the domain concepts in terms of their relations and attributes in order to assist the deployment of added-value monitoring services in this context. A description of technologies involved in the process is also included. This semantic layer is intended to cope with requirements of sampling-based monitoring. The latter are presented under the form of competence questions in a case study, described in Sect. 5, as an initial validation of the semantic model of the system. Finally, the main conclusions are presented to the user in Sect. 6, where the reader can access insight information about the proposed model, its current and future features in the scope of semantic-based expert systems in the domain of network monitoring.

2   Related Work

Mapping network measurement requirements into the most suitable MP, traffic sampling technique, and underlying operation parameters have been topics of research along the last years. Globally, such efforts have identified different strategies able to provide high accurate results in manifold tasks, such as traffic classification and characterisation [1,3,4], SLA compliance [5,6], QoS monitoring [7,8], and network security [9,10]. However, as network traffic is heavily dynamic and heterogeneous, a sampling solution used to estimate a particular parameter correctly may not be adequate for a different parameter or traffic type [11]. This creates the need of having previous and detailed knowledge about the monitored traffic as well as direct intervention of network administrators for tuning the measurement process according to monitoring requirements.

High-level recommendation systems based on ontologies emerge as a new mechanism to improve the semantic expressiveness of management activities, being a valuable approach to overcome the challenges mentioned above. In this way, some works have been addressing interoperability issues, where ontologies are used to map network managed objects defined in information models, such as SMI, GMDO, MIF, and IPFIX [12,13].

Considering network management activities, related research is exploring ontological representation as a mechanism for supporting autonomic networks, in particular, for automated configuration [14,15]. More specific works which require traffic measurements are mainly focused on QoS monitoring [16,17] and network security [18,19].

Although being considered a key enabler within network semantic management, exploiting ontology's capabilities to face the challenges of selecting the most suitable sampling-based monitoring strategy in context-dependent network environments is still an open issue.

3   Context-Aware Monitoring Architecture

The proposed context-aware monitoring architecture, represented in Fig. 1, illustrates the expert recommendation system as a key component for assisting network management tasks. At upper level, from a management plane perspective, each service or network management task is expected to specify, and subsequently meet, particular measuring requirements. These requirements specificity will be handled within the control plane, where the expert recommendation system will act downstream to suggest adequate configuration parameters for a

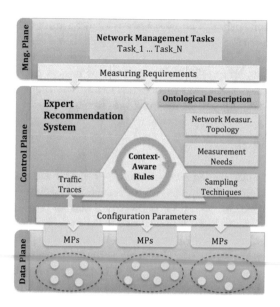

Fig. 1. Monitoring architecture

set of measurement points (MPs). At lower level, MPs will collect and report traffic descriptors as traffic traces or datasets. Thus, the recommendation system receives as main inputs the specification of measurement requirements and traffic data, and will include the required ontological support and reasoning components.

Depending on the monitoring context and on network traffic variability, the system is expected to suggest a configuration profile so that the monitoring task can be efficiently accomplished. Efficiency is here understood as a trade-off between measurements accuracy and overhead.

The main modules of the semantic recommendation system are illustrated in Fig. 2. The module called ML incorporates machine learning techniques for identifying traffic behaviours from real data. The *Traffic variation detection* module detects traffic fluctuations and assesses when these may require changes on monitoring configuration. The module called *Rules for changes* is a repository of semantic rules (e.g., specified in SWRL) that indicate what changes must be accomplished in the process of monitoring a task. The module *Parameters to optimize* handles the parameters to be changed for a given monitoring configuration. As output, the system produces a specific configuration profile to be applied to a set of MPs.

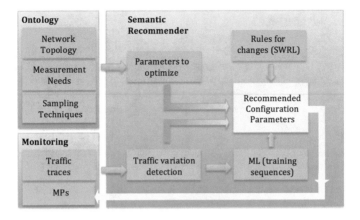

Fig. 2. Semantic recommender

4   Ontology Definition

This section will explain the building steps of the ontology sustaining the monitoring architecture.

4.1   Conceptual Model

Prior to the ontology definition, the preliminary step is the identification of the application domain and the specification of the questions of competence, i.e., the questions to which the ontological system is expected to answer. The purpose of the ontology is to serve as the basis for implementing a monitoring service. Therefore, the next step is to identify the concepts in the domain that should be represented in the ontology taking into account the specifications of the monitoring architecture.

4.2   Competence Questions

Competence questions play a very important role both in the creation of an ontology since they allow to justify the existence of the ontology, and in the consequent evaluation of the ontology. When creating the ontology, the questions of competence must be verified so that the development of the ontology does not deviate from the purpose initially defined. Examples of competence questions are: (a) how much memory and CPU are available at a particular MP?; (b) which MPs are border routers?; (c) what are the characteristics of a particular MP?; (d) what is the id of the existing monitoring tasks? (e) which MPs are available in the network?; (f) what are the sampling techniques supported by a particular MP?; (g) what are the MPs capable of supporting a specific sampling technique?, or (h) what is the active technique and setup parameters at a particular MP?

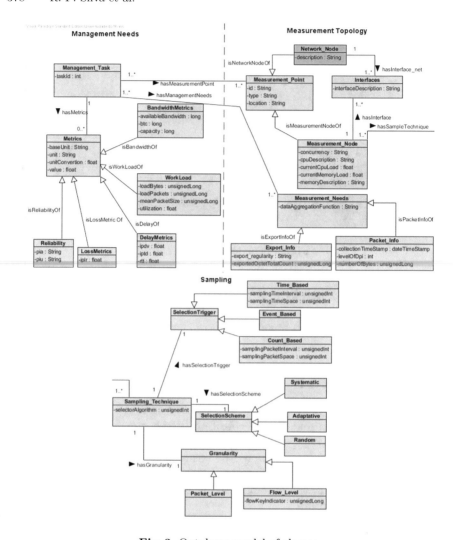

Fig. 3. Ontology: model of classes

4.3   Ontology Representation

Figure 3 illustrates the class model developed for the ontology, along with the corresponding object properties.

The monitoring architecture components *Management Needs*, *Measurement Topology* and *Sampling* are defined as classes in the ontology. This ontological representation allows a comprehensive understanding of the domain, establishing how these classes and the object properties are interrelated in a service monitoring context based on traffic sampling. As illustrated, a `Management_Task`, such as accounting or traffic classification, determines the `Measurement_Needs`, in terms of `Metrics` to be measured, and a set of `Measurement_Point`. An MP can be a general purpose `Network_Node` (e.g., router or switch) performing network measurements or a dedicated `Measurement_Node`. Each `Measurement_Node` determines the sampling techniques supported in that node. These techniques are defined through the classes `SelectionTrigger`, `SelectionScheme` and `Granularity`, which determine the time and space characteristics regulating traffic sampling. In sampling techniques, timers, packet counters or events may trigger the sampling process, ruling the intervals in which packets are collected. Sampling intervals can be systematic, random or adaptive according to the current network load.

The model of classes also includes the attributes (data properties) of each object, e.g. `loadBytes`, `loadPackets`. Defining a management task, e.g. accounting, corresponds to an object instantiation (individual), which has object properties such as `hasMeasurementPoint` and `taskID`.

5   Case Studies - Semantic Validation

In this section, as a semantic validation of the devised ontology expressiveness, examples of real uses of the ontology in the context of network management are presented, as well as a study on the requirements of a specific management task, i.e., *traffic accounting*. The data used to populate the ontology was collected from the University of Minho network during a working day.

5.1   Querying the Ontology

As mentioned, the ontology must be able to answer a list of competence questions. These questions ground the existence of the ontology and allow to evaluate whether the ontology responds to the defined purposes. To achieve this, the SPARQL Protocol and RDF Query Language (SPARQL), based on the Resource Description Framework (RDF), is used.

As an example, we only include a competence question from Sect. 4.2, and the corresponding answer. Other queries are covered within traffic accounting task.

– How much memory and CPU are available at a particular MP?
  Input of Query 1:

```
1   PREFIX rdf: <http://www.w3.org/1999/02/22-rdf-syntax-ns#>
2   PREFIX owl: <http://www.w3.org/2002/07/owl#>
3   PREFIX xsd: <http://www.w3.org/2001/XMLSchema#>
4   PREFIX rdfs: <http://www.w3.org/2000/01/rdf-schema#>
5   PREFIX sm: <http://www.semanticweb.org/kaiser/ontologies/2016/3/
6   service-monitoring#>
7   SELECT ?nome ?memoria ?cpu
8   WHERE{
9   ?mp rdf:type sm:Measurement_Point.
10  ?mn rdf:type sm:Measurement_Node.
11  ?mp sm:id ?nome.
12  ?mp sm:hasMeasurementNode ?mn.
13  ?mn sm:currentMemoryLoad ?memoria.
14  ?mn sm:currentCpuLoad ?cpu
15  }
16  ORDER BY ?nome
```

name	memory	cpu
"Point1"@	"76566"^^<http://www.w	"5.03"^^<http://www.w3
"Point2"@	"96410"^^<http://www.w	"17.95"^^<http://www.w
"Point3"@	"86163"^^<http://www.w	"18.26"^^<http://www.w
"Point4"@	"80765"^^<http://www.w	"10.76"^^<http://www.w
"Point5"@	"85551"^^<http://www.w	"97.27"^^<http://www.w

Fig. 4. Available memory and CPU at MPs

In the example of Query 1, five variables are created: mp of type
Measurement_Point (line 9); mn of type Measurement_Node (line 10); name that
collects the values of the id attribute associated with each Measurement_Point
(line 11); memory that collect the values of the currentMemoryLoad attribute
associated with each Measurement_Node (line 13); and cpu that collects the values of the currentCpuLoad attribute associated with each Measurement_Node
(line 14). The variables selected for the query output are name, memory and cpu,
and corresponding values are shown in Fig. 4.

5.2   Traffic Accounting

Traffic accounting, a vital network management task for service providers, was
selected as case study for testing the ontological system. In its simplest form,
this task keeps track of users traffic volumes, including the amount of time spent
per session, the amount of data transferred during a session, and the type of
services accessed.

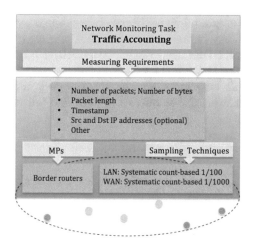

Fig. 5. Accounting task specification

Figure 5 illustrates essential aspects to consider when performing traffic
accounting, namely: the measuring requirements; the MPs involved in traffic

accounting; and the most adequate sampling techniques to use, i.e., the techniques that achieve a better trade-off between accuracy and overhead (allowing to obtain accurate results without interfering with normal network operation). As shown, common requirements of traffic accounting include: the amount of traffic monitored (in packets and bytes); the size of the collected packets; a time instant when a sample is collected; the source and destination IP addresses; and other specific header (or payload) fields under observation. The selection of the most suitable MPs for the execution of accounting are, usually, the border routers of the network domain, but can also be dedicated measurement nodes, as defined in Sect. 4.

The selection of the sampling technique that best suits a monitoring task depends on several factors, such as the memory and CPU available at the MP, the amount of data collected, the network performance, etc. Sampling techniques with best experimental results in performing accounting are: systematic count-based 1/100 for local area networks, and systematic count-based 1/1000 for non-local networks [20].

Next, examples of queries and results obtained when performing traffic accounting using sampling techniques are presented.

In Query 2, four variables are created: mp of type Measurement_Point (line 4); name that collects the values of the id attribute of each Measurement_Point (line 5); technique of type Sampling_Technique that collects the techniques associated to Measurement_Points through the object property hasSampleTechnique (line 6); and type that collects the types of Measurement_Points through type attribute (line 7). The variables selected for output are name, technique and type (line 2). Figure 6 shows the obtained values of this query.

- Which are the type and name of each MP, and the corresponding sampling techniques in use;

  Input of Query 2:

```
1    PREFIX ...
2    SELECT ? name ? technique ? type
3    WHERE {
4      ?mp rdf: type sm: Measurement_Point .
5      ?mp sm: id ? name .
6      ?mp sm: hasSampleTechnique ? technique .
7      ?mp sm: type ? type
8    }
```

name	technique	type
"Point4"@	MUST	"core"@
"Point2"@	SystT	"Border Router"@
"Point1"@	SystC	"Border Router"@
"Point3"@	RandC	"core"@
"Point5"@	LP	"Border Router"@

Fig. 6. Output of Query 2

Other example listing the characteristics of a specific MP (MP3) - Query 3 and Fig. 7, is provided below.

– Listing the characteristics of a particular MP

Input of Query 3:

```
1    PREFIX  . . .
2    SELECT ? name ? cpuDescription ? cpu ? memory
3    WHERE {
4    ?mp rdf: type sm: Measurement_Point  .
5    ?mp sm:id ? name  .
6    FILTER REGEX (? name ," Point3 ").
7    ?mp sm: hasMeasurementNode ?mn.
8    ?mn sm: cpuDescription ? cpuDescription  .
9    ?mn sm: currentMemoryLoad ? memory  .
10   ?mn sm: currentCpuLoad ?cpu
11   }
```

name	cpuDescription	cpu	memory
"Point3"@	"ARM V7 900MHZ"@	"18.26"^^<http://v	"86163"^^<http://

Fig. 7. Output of Query 3

As a final note, once the semantic layer defined as proposed is applied to all nodes in the network under management, monitoring the entire network would be simplified. The provision of an interoperable layer (common ontology), would benefit the development of network monitoring solutions in an objective, systematic and automatic manner.

6   Conclusions

This work has proposed the construction of a semantic model to assist the development of solutions based on expert agents for context-aware network monitoring. That is why an ontology has been proposed to describe the domain and a software architecture has been described that provides these high added-value services for network monitoring in a highly automated way. The proposed system was also validated by verifying the competency questions identified and the support of a crucial monitoring task - *traffic accounting.* The experimental validation of the system in a realistic network environment is planned as future work. The authors hope to develop a collaborative system that can anticipate the needs of the system with rules deduced from the behaviour of the system under realistic loading conditions.

Acknowledgments. This work has been supported by FCT – *Fundação para a Ciência e Tecnologia* within the Project Scope: UID/CEC/00319/2019.

References

1. Silva, J.M.C., Carvalho, P., Lima, S.R.: Inside packet sampling techniques: exploring modularity to enhance network measurements. Int. J. Commun. Syst. **30**(6), e3135 (2017)
2. Grüninger, M., Fox, M.: Methodology for the design and evaluation of ontologies. In: IJCAI 1995, Workshop on Basic Ontological Issues in Knowledge Sharing, 13 April 1995. http://citeseer.ist.psu.edu/grninger95methodology.html
3. Lin, R., Li, O., Li, Q., Dai, K.: Exploiting adaptive packet-sampling measurements for multimedia traffic classification. J. Commun. **9**(12), 971–979 (2014)
4. Tammaro, D., Valenti, S., Rossi, D., Pescapé, A.: Exploiting packet-sampling measurements for traffic characterization and classification. Int. J. Netw. Manag. **22**(6), 451–476 (2012)
5. Zseby, T., Hirsch, T., Claise, B.: Packet sampling for flow accounting: challenges and limitations. In: International Conference on Passive and Active Network Measurement, pp. 61–71. Springer, Heidelberg (2008)
6. Hu, C., Wang, S., Tian, J., Liu, B., Cheng, Y., Chen, Y.: Accurate and efficient traffic monitoring using adaptive non-linear sampling method. In: IEEE INFOCOM 2008: The 27th Conference on Computer Communications, pp. 26–30. IEEE (2008)
7. Mahmood, A.N., Hu, J., Tari, Z., Leckie, C.: Critical infrastructure protection: resource efficient sampling to improve detection of less frequent patterns in network traffic. J. Netw. Comput. Appl. **33**(4), 491–502 (2010)
8. Gu, Y., Breslau, L., Duffield, N., Sen, S.: On passive one-way loss measurements using sampled flow statistics. In: IEEE INFOCOM 2009, pp. 2946–2950. IEEE (2009)
9. Yoon, S., Ha, T., Kim, S., Lim, H.: Scalable traffic sampling using centrality measure on SDNs. IEEE Commun. Mag. **55**(7), 43–49 (2017)
10. Jun, J.-H., Ahn, C.-W., Kim, S.-H.: DDoS attack detection by using packet sampling and flow features. In: Proceedings of the 29th Annual ACM Symposium on Applied Computing, pp. 711–712. ACM (2014)
11. Duffield, N., et al.: Sampling for passive internet measurement: a review. Stat. Sci. **19**(3), 472–498 (2004)
12. Martinez, A., Yannuzzi, M., López, V., López, D., Ramírez, W., Serral-Gracià, R., Masip-Bruin, X., Maciejewski, M., Altmann, J.: Network management challenges and trends in multi-layer and multi-vendor settings for carrier-grade networks. IEEE Commun. Surv. Tutor. **16**(4), 2207–2230 (2014)
13. Wong, A.K.Y., Ray, P., Parameswaran, N., Strassner, J.: Ontology mapping for the interoperability problem in network management. IEEE J. Sel. Areas Commun. **23**(10), 2058–2068 (2005)
14. Martinez, A., Yannuzzi, M., de Vergara, J.E.L., Serral-Gracià, R. Ramírez, W.: An ontology-based information extraction system for bridging the configuration gap in hybrid SDN environments. In: 2015 IFIP/IEEE International Symposium on Integrated Network Management (IM), pp. 441–449. Ottawa (2015). https://doi.org/10.1109/INM.2015.7140321
15. Xu, H., Xiao, D.: Applying semantic web services to automate network management. In: 2nd IEEE Conference on Industrial Electronics and Applications, ICIEA 2007, pp. 461–466. IEEE (2007)
16. Rodrigues, C., Lima, S.R., Sabucedo, L.M.Á., Carvalho, P.: An ontology for managing network services quality. Expert Syst. App. **39**(9), 7938–7946 (2012)

17. Moraes, P.S., Sampaio, L.N., Monteiro, J.A., Portnoi, M.: Mononto: a domain ontology for network monitoring and recommendation for advanced internet applications users. In: IEEE Network Operations and Management Symposium Workshops: NOMS Workshops 2008, pp. 116–123. IEEE (2008)
18. Simmonds, A., Sandilands, P., Van Ekert, L.: An ontology for network security attacks. In: Asian Applied Computing Conference, pp. 317–323. Springer, Heidelberg (2004)
19. Silva, D.V., Rafael, G.R.: Ontologies for network security and future challenges. In: International Conference on Cyber Warfare and Security, p. 541. Academic Conferences International Limited (2017)
20. Silva, J.M.C., Carvalho, P., Lima, S.R.: Computational weight of network traffic sampling techniques. In: 2014 IEEE Symposium on Computers and Communications (ISCC), pp. 1–6. IEEE (2014)

Analysis of Performance Degradation of TCP Reno over WLAN Caused by Access Point Scanning

Toshihiko Kato[✉], Sota Tasaki, Ryo Yamamoto,
and Satoshi Ohzahata

University of Electro-Communications, Chofu, Tokyo 182-8585, Japan
{kato, t.souta, ryo_yamamoto,
ohzahata}@net.lab.uec.ac.jp

Abstract. In wireless LANs, terminals regularly perform the access point scanning in order to obtain up-to-data information on access points located nearby. During the access point scanning, a terminal sends a packet with the power management bit set to suppress the data transfer, and this operation sometimes causes packet losses in the terminal. If TCP Reno is used as a congestion control algorithm and the transmission delay is large, the increase of the congestion window size is suppressed and the performance may be deteriorated. This paper shows an experimental result of TCP Reno with access point scanning and proposes a method to prevent the throughput degradation.

Keywords: IEEE802.11 wireless LAN · TCP Reno · Access point scanning · Congestion window size

1 Introduction

Nowadays, high speed wireless LANs (WLANs) such as IEEE 802.11n and 802.11ac [1] are used widely. They establish high speed data transfer using the higher data rate support, the frame aggregation in Aggregation MAC Protocol Data Unit (A-MPDU), and the Block Acknowledgment mechanism. Although stations can leverage the high speed data transfer when the radio conditions between them and the access point with which they communicate are good. In order avoid to communicate with access points under bad radio conditions, IEEE 802.11 standard defines a procedure which allows stations to look for access points around them. This procedure is called an *access point scanning* [1]. It has two types of procedures; active scanning and passive scanning. In the active access point scanning, a station broadcasts Probe request frames and waits for the responses from multiple access points. In the passive access point scanning, a station just waits for receiving Beacon frames from access points. In either case, the station needs to refrain from receiving data frames from the access point with which the station is associated at that time [2]. For this purpose, stations use the power management function by sending frames with the power management bit set in the WLAN header [3].

© Springer Nature Switzerland AG 2019
Á. Rocha et al. (Eds.): WorldCIST'19 2019, AISC 931, pp. 385–395, 2019.
https://doi.org/10.1007/978-3-030-16184-2_37

In the previous papers [4, 5], we analyzed how the access point scanning affects the TCP throughput. Since the access point scanning is performed independently of TCP, and the shift to the power save mode stops transmitting TCP data segments, the TCP module keeps sending data segments during an access point scanning and the data segments sent may be lost due to the buffer overflow in the transmission queue installed in the WLAN device driver. Our previous papers examined the TCP performance degradation invoked by the access point scanning, focusing on the uploading or downloading data transfer, the types of congestion control algorithms, and whether the TCP small queues [6] are adopted or not. Our previous papers concluded that, in the uploading data transfer from a station to an access point, the TCP throughput is largely affected by whether the TCP small queues are used or not, not by the congestion control algorithm that TCP connections use.

Our previous papers [4, 5] used the network configuration where the delay between communicating nodes is very small, that is, the communicating nodes are connected via WLAN and Ethernet. However, in the actual Internet access, there is some amount of delay between a client and a server, in the order of 10 ms and 100 ms. In this situation, the TCP throughput is dependent on the congestion control algorithm. For example, the throughput of TCP Reno (or NewReno) [7], which is traditional but used widely even nowadays, may be low if the round-trip time (RTT) is large and several packets are lost contiguously. In this paper, we evaluate again the performance of TCP during an access point scanning under the condition that a data sending station uses TCP Reno, and that the RTT between the station and the server is relatively large.

This paper also evaluates the results of performance evaluation when CUBIC TCP [8], an example of high speed congestion control algorithms, is used in the uploading data transfer. Moreover, we propose a method to stop access point scanning while data transfer is being performed. We implement the proposed method over the *NetworkManager* software module [9] running over the Linux operating system. This paper also describes the overview of proposed method and its performance evaluation.

The rest of this paper consists the following sections. Section 2 shows the technologies relevant to this paper. Section 3 explains the experimental settings. Section 4 gives the detailed analysis of uploading TCP data transfer together with access point scanning. Section 5 proposes a method to protect the throughput degradation by the access point scanning. In the end, Sect. 6 gives the conclusions of this paper.

2   Relevant Technologies

2.1   Access Point Scanning and Power Management Function

IEEE 802.11 standards provide an access point scanning procedure. This is introduced for multiple purposes. One is to obtain an up-to-data information on the surrounding access points of a station, even when the station is communicating through a specific access point. Another purpose is to select another access point in a better radio condition with which a station maintains the roaming relationship. As described above, 802.11 standards define active and passive access point scanning, both of which refrain from data transmission by use of the power management function.

In the WLAN frame format depicted in Fig. 1(a), bit 12 in the Frame Control field is the Power Management bit (shown in Fig. 1(b)). By setting this bit to 1, a station informs the associated access point that it is going to the *power save mode*, in which the station goes to sleep and wakes up only when the access point sends beacon frames. By setting the bit to 0, it informs the access point that it goes back to the *active mode*.

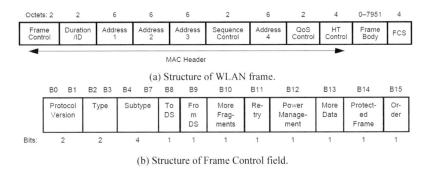

(a) Structure of WLAN frame.

(b) Structure of Frame Control field.

Fig. 1. IEEE 802.11 WLAN frame format [1].

This function is used for several purposes. One example is the case that a station is actually going to sleep to save its power consumption for a while. In this case, a station wakes up only at the timing of receiving beacon frames from the access point. If the access point has data frames to deliver to the sleeping station, it indicates this fact in the Traffic Indication Map element in a beacon frame. In response to this information, the station requests the delivery of data frames by use of a PS-Poll frame.

Another example is the access point scanning described above. During the scanning, a station asks the current access point to stop sending data frames to it, by sending a frame with the Power Management bit set to 1.

In order to inform access points of the shift to the power save mode or to the active mode, stations use several methods. One is to use Null data frames [10]. This method is shown in our previous papers [4, 5]. A Null data frame is a frame that contains no data (Frame Body in Fig. 1(a)). It is different from an ordinary data frame. While an ordinary data frame has three Address fields, a Null data frame has only two Address fields; the transmitter is a station MAC address and the receiver is an access point MAC address. By using Null data frames with the Power Management bit set to 0 or 1, stations can request the power management function for access points.

Another method is to use RTS (Request to send) and CTS (Clear to send) frames, without sending any data frames. RTS and CTS frames are control frames prepared to prevent the hidden terminal problem. This method is explained in this paper.

2.2   TCP Reno and CUBIC TCP

TCP Reno is a classical congestion control method which adopts an additive increase and multiplicative decrease (AIMD) algorithm. Here, congestion window size (*cwnd*) is increased each time the TCP sender receives an ACK segment acknowledging new data. In the slow start phase, cwnd increases one packet for one new ACK segment. During the congestion avoidance phase, the increase is $1/cwnd$ packet, and as a result, cwnd is expected to be increased by one packet during one RTT. The slow start phase shifts to the congestion avoidance phase when cwnd exceeds the slow start threshold (*ssthresh*). ssthresh maintains half of the cwnd value at the time of last retransmission. When multiple packets are lost, ssthresh contiguously decreases to half of the previous value, with the limitation that the minimum value is two packets. That is, multiple packet losses make the congestion avoidance phase start with a small value of cwnd.

CUBIC TCP is designed for high speed and long delay networks. It is a default congestion control algorithm in the Linux operating system. It defines cwnd as a cubic function of elapsed time T since the last congestion event [8]. Specifically, it defines cwnd by the following equation.

$$cwnd = C\left(T \cdot \sqrt[3]{\beta - \frac{cwnd_{max}}{C}}\right)^3 + cwnd_{max}$$

Here, C is a predefined constant, β is the decrease parameter, and $cwnd_{max}$ is the value of *cwnd* just before the loss detection in the last congestion event. Comparing with TCP Reno, cwnd increases faster in CUBIC TCP.

3   Experimental Settings

Figure 2 shows the network configuration of the performance evaluation experiment we conducted. There are two stations conforming to 802.11n with 5 GHz band and one access point connected to a server through 1 Gbps Ethernet. One station called STA1 is associated with the access point, and communicates with the server through the access point. The other station called STA2 is used just to monitor WLAN frames exchanged between STA1 and the access point. We use commercially available personal computers (PCs) for the stations and the server. The access point used is also an off-the-shelf product. The detailed specification of the PCs and the access point is shown in Table 1. It should be noted that the version of Linux operating system in STA1 is Ubuntu 12.04 LTS, which is an older version developed at April, 2012, because we wanted to suppress the TCP small queues [6] in the uploading data transfer.

In the experiment, the data is generated by iperf [11] for 300 s in the direction from STA1 to the server. We used TCP NewReno or CUBIC TCP as a congestion control mechanism. During the data transmissions, the following detailed performance metrics are collected for the detailed analysis of the communication;

- the packet trace at the server and STA2, by use of *Wireshark* [12],
- the TCP throughput for every second, calculated from packet trace at TCP receiver (the server), and
- the cwnd and ssthresh values at STA1, by use of *tcpprobe* [13] (an average during one second, calculated from the values obtained for every segment reception at STA1).

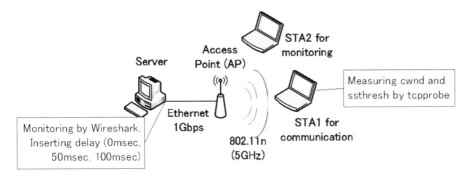

Fig. 2. Network configuration.

Table 1. Specification of equipment used in experiment.

STA1	Manufacturer/model	DELL Inspiron 15
	Operating system	Ubuntu 10.04LTS (kernel 3.2)
	WLAN driver	ath9k
SRV	Manufacturer/model	DELL Inspiron 15
	Operating system	Ubuntu 10.04LTS (kernel 3.1.3)
AP	Manufacturer/model	BUFFALO AirStation WZR–HP–AG300H
	WLAN chip	Atheros AR7161
	WLAN driver	ath9k

In order to emulate the Internet access environment, delay is inserted at the server (receiving side). The delay value is 0 ms (no delay), 50 ms, and 100 ms. This delay is inserted by use of tc (traffic control) command with the netem filter.

4 Analysis of Uploading TCP Data Transfer with Access Point Scanning

4.1 Results Using TCP NewReno

Figures 3 through Fig. 5 show the results of performance evaluation when STA1 uses TCP NewReno, with changing delay between STA1 and the server. Each of the figures contains the time variation of throughput, calculated in each second, and the time variation of cwnd and ssthresh.

When the delay is 0 ms, the access point scanning is performed around 120 s and 240 s. At this timing, the throughput drops sharply from 150 Mbps to 20 Mbps. At the same timing, cwnd and ssthresh decrease largely. The sharp drop of ssthresh means that there are multiple packet losses. As a result, cwnd increases relatively slowly in the congestion avoidance phase. Since the delay is very small, however, the throughput recovers quickly to a stationary value (Fig. 3).

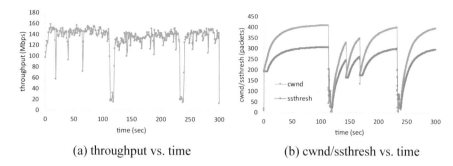

(a) throughput vs. time   (b) cwnd/ssthresh vs. time

Fig. 3. Time variation of throughput and TCP parameters (Reno, no delay).

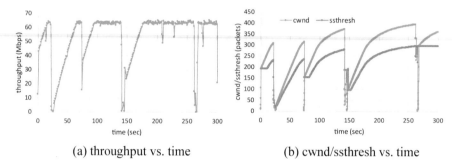

(a) throughput vs. time   (b) cwnd/ssthresh vs. time

Fig. 4. Time variation of throughput and TCP parameters (Reno, 50 ms delay)

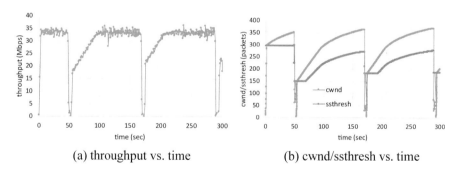

(a) throughput vs. time   (b) cwnd/ssthresh vs. time

Fig. 5. Time variation of throughput and TCP parameters (Reno, 100 ms delay)

Figure 6 shows the WLAN frames exchanged at the access point scanning around 120 s. At 120.936082 s, STA1 sends an RTS frame with the Power Management bit (shown as PM in the figure) set to 1. This means that STA1 goes to the power save mode. The Duration field in this frame is 164 μs, and that in the following CTS frame is 104 μs. This means that there are no data frames following the RTS and CTS frames. The RTS frame here is transmitted just for requesting the shift to the power save mode. After that, STA1 receives a Beacon frame from the access point (blue part in the figure), and it sent a Block Ack frame with the Power Management bit set to 0 (red part). This indicates that STA1 goes back to the active mode. In the experiments in our previous paper [4, 5], Null data frames are used to request the mode changing. The reason is that we changed a PC used as STA1. In our previous papers, a PC for STA1 uses a Lite-on Technology's WLAN chip, and in this paper, STA1 uses a SparkLan Communication's chip. It considered that how to request the mode changing is an implementation dependent matter.

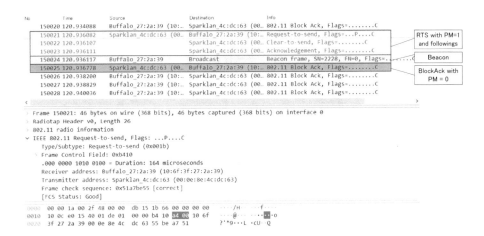

Fig. 6. Frame capture of access point scanning.

When the inserted delay is 50 ms, the throughput degradation becomes a problem. At around 25 s, the access point scanning is performed and the throughput goes to almost zero at this point. Both cwnd and ssthresh also goes to small values. As a result, the cwnd value increases slowly according to the congestion avoidance. In order to recover the throughput to a stationary value (65 Mbps), it takes about 40 s. This is caused by the inserted delay 50 ms. At around 150 s, there is another access point scanning. Although the ssthresh value is 100 packets and cwnd grows sharply by the slow start until this value, it also takes about 30 s so that the throughput is recovered. In contrast, the access point scanning performed around 270 s does not give a large effect on the performance.

When the inserted delay is 100 ms, the throughput is also degraded similarly with the case of 50 ms delay. There are three invocations of access point scanning at around 50 s, 170 s, and 290 s. In all the cases in this measurement run, the ssthresh value does not go down to 2 packets, but stays at the values between 150 and 175 packets. As a result, cwnd increases sharply until it reaches this value. However, it takes about 40 s until the throughput recovers to a stationary value (35 Mbps). In this case, the throughput degradation caused by the access point scanning is an issue for TCP NewReno.

We conducted ten measurement runs for individual values of inserted delay. Table 2 shows the average throughput and its standard deviation for 300 s data transfer. In the case of inserted delay of 50 ms and 100 ms, the average throughput is smaller than the corresponding stationary throughput, about 65 Mbps and 35 Mbps. From those results, it can be said that the access point scanning gives a bad influence to uploading TCP Reno (NewReno) communication through the Internet.

Table 2. Results of ten measurement runs

Inserted delay	Average (Mbps)	Standard deviation (Mbps)
0 ms	150.6	37.6
50 ms	53.1	17.5
100 ms	23.3	9.0

4.2 Results Using CUBIC TCP

For the purpose of comparison, we conducted a similar experiment using CUBIC TCP as the congestion control algorithm in STA1. Figures 7 and 8 show the time variations of the throughput measured at individual second, and the cwnd and ssthresh values.

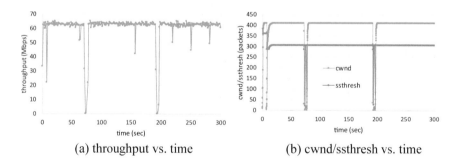

(a) throughput vs. time   (b) cwnd/ssthresh vs. time

Fig. 7. Time variation of throughput and TCP parameters (CUBIC, 50 ms delay).

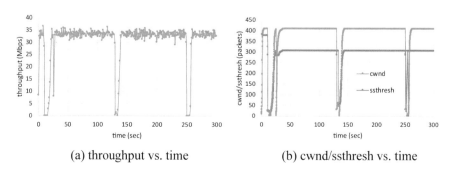

(a) throughput vs. time   (b) cwnd/ssthresh vs. time

Fig. 8. Time variation of throughput and TCP parameters (CUBIC, 100 ms delay)

In Fig. 7 with the inserted delay of 50 ms, the access point scanning is performed at around 70 s and 190 s. In either scanning, all of the throughput, cwnd and ssthresh go to close to zero, but the throughput and cwnd recover very quickly. Similarly, in Fig. 8 with the inserted delay of 100 ms, the access point scanning is performed at around 10 s, 130 s and 250 s. In this case, the throughput also recovers quickly. Specifically, the recovery is slower than the case of 50 ms inserted delay, but the instantaneous throughput degradation caused by the access point scanning does not affect the overall throughput.

From those results, it can be said that CUBIC TCP can prevent serious throughput degradation in the uploading data transfer, although there are several packet losses during the access point scanning. But, TCP Reno (NewReno) is widely used in various Internet terminals, which cannot change their congestion control algorithm. The throughput degradation caused by the access point scanning is a serious problem.

5   Approach to Stop Access Point Scanning During Good Radio Condition

5.1   Proposal

The duration between invocations of access point scanning is specified in the NetworkManager source program `mn-device-wifi.c`. The minimum and maximum values of the duration are 3 s and 120 s. In the experiments described above, the maximum duration seems to be adopted. The access point scanning procedure itself is realized by the C language functions described in Fig. 9 [5].

Fig. 9.   Functions for access point scanning.

- In the function `schedule_scan()`, the function `request_wireless_scan ()` is called periodically by use of `g_timeout_add_seconds()`.
- The function `g_timeout_add_seconds()` is a function to make another function called at regular intervals, which is defined in the framework of GTK+ (the GIMP Toolkit) [14].
- The function `request_wireless_scan()` calls `nm_suplicant_interface_request_scan()`, which performs access point scanning actually.

In this paper, we modified the function `request_wireless_scan()` in the way that, if the received radio strength is large enough, the actual scanning is skipped.

The received radio strength can be obtained from/sys/class/net/wlan0/wireless/link. The purpose of this modification is that the large received radio strength indicates a stable data transfer at the WiFi level, resulting the use of high data rate.

5.2   Performance Evaluation

Figure 10 shows the time variation of throughput when the proposed method is applied to the TCP NewReno uploading data transfer with 50 ms and 100 ms delay. There are no drops in the throughput. We conducted ten measurement runs and obtained the average throughput for 300 s data transfer such that 63.3 Mbps (4.9 Mbps standard deviation) for 50 ms delay, and 33.6 Mbps (2.5 Mbps standard deviation) for 100 ms delay. The results were much improved compared with those described above.

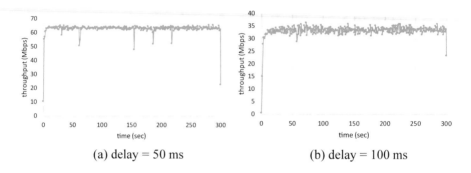

(a) delay = 50 ms   (b) delay = 100 ms

Fig. 10.   Time variation of throughput using proposed method.

6   Conclusions

This paper described the throughput degradation caused by the access point scanning in WLAN when TCP Reno (NewReno) is used in an uploading data transfer. This degradation is serious, 30% decrease when RTT is 100 ms. In order to avoid this degradation, this paper also proposed a method to skip scanning when the radio condition is good, and showed that the proposed method recover the degradation.

References

1. IEEE Standard for Information technology: Local and metropolitan area networks Part 11: Wireless Medium Access Control (MAC) and Physical Layer (PHY) Specifications (2016)
2. ath9k-devel@lists.ath9 k.org: disable dynamic power save in AR9280, January 2011. http://comments.gmane.org/gmane.linux.drivers.ath9k.devel/5199. Accessed 13 Nov 2018
3. My80211.com: 802.11: Null Data Frames. http://www.my80211.com/home/2009/12/5/80211-null-data-frames.html. Accessed 12 Nov 2018

4. Kobayashi, K., Hashimoto, Y., Nomoto, M., Yamamoto, R., Ohzahata, S., Kato, T.: Experimental analysis on access point scanning impacts on TCP throughput over IEEE 802.11n wireless LAN. In: Proceedings of the 12th International Conference on Wireless and Mobile Communications (ICWMC 2016), Barcelona, pp. 115–120 (2016)

5. Kato, T., Kobayashi, K., Tasaki, S., Nomoto, M., Yamamoto, R., Ohzahata, S.: Experimental analysis and resolution proposal on performance degradation of TCP over IEEE 802.11n wireless LAN caused by access point scanning. Int. J. Adv. Telecommun. **11** (1&2), 10–19 (2018)

6. Eric Dumazet: [PATCHv2 net-next] tcp: TCP Small Queues (2012). http://article.gmane.org/gmane.network.routing.codel/68. Accessed 13 Nov 2018

7. Henderson, T., Floyd, S., Gurtov, A., Nishida, Y.: Then NewReno Modification to TCP's Fast Recovery Algorithm. IETF RFC 6582 (2012)

8. Ha, S., Rhee, I., Xu, L.: CUBIC: a new TCP-friendly high-speed TCP variant. ACM SIGOPS Oper. Syst. Rev. **42**(5), 64–74 (2008)

9. archLinux: NetworkManager. https://wiki.archlinux.org/index.php/NetworkManager. Accessed 13 Nov 2018

10. Gu, W., Yang, Z., Xuan, D., Jia, W.: Null data frame: a double-edged sword in IEEE 802.11 WLANs. IEEE Trans. Parallel Distrib. Syst. **21**(7), 897–910 (2010)

11. iperf. http://iperf.sourceforge.net/. Accessed 29 Nov 2018

12. Wireshark. https://www.wireshark.org/. Accessed 29 Nov 2018

13. Linux foundation: tcpprobe. http://www.linuxfoundation.org/collaborate/workgroups/networking/tcpprobe. Accessed 29 Nov 2018

14. The GTK+ Project: What is GTK+, and how can I use it? https://www.gtk.org/. Accessed 29 Nov 2018

Stratifying Measuring Requirements and Tools for Cloud Services Monitoring

Paulo Carvalho[1]([✉]), Solange Rito Lima[1], and Kalil Araujo Bispo[2]

[1] Centro Algoritmi, Universidade do Minho, Campus de Gualtar,
4710 057 Braga, Portugal
{pmc,solange}@di.uminho.pt
[2] Departamento de Computação, Universidade Federal de Sergipe,
São Cristóvão, SE 49100-000, Brazil
kalil@dcomp.ufs.br

Abstract. Monitoring plays an essential role in the management and engineering of today's communication networks. Facing paradigms such as cloud computing and network virtualization, the challenges of monitoring are even more demanding, attending to the variety and dynamics of services and resources involved. The lack of standards regarding cloud services monitoring makes even more difficult to reach a common ground on how to assess these services. In this context, this paper takes a high-level stratified view of the problem to clarify and systematise the involved pieces in cloud services monitoring, from the physical/virtual infrastructure to the customer/provider layer. In each layer, relevant measuring requirements and metrics are discussed, following existing guidelines for the evaluation of performance, reliability, security, and service level agreements fulfilment. In addition, representative cloud services in the market are aggregated per service model and relevant monitoring tools are surveyed and mapped into the proposed functional layers. The present study is a step forward contributing to achieve a more modular, flexible and consensual view to cloud services monitoring.

Keywords: Cloud Computing · Service monitoring · Monitoring tools

1 Introduction

Currently, there is a clear move toward Cloud Computing paradigm as result of the significant advantages it brings both in economic and technological terms. The adoption of cloud technology decreases the capital and operational costs (CapEx and OpEx) [1,2] when deploying and maintaining cloud services, as consequence of multiple economy of scale effects (e.g., increased productivity with reduced spending on infrastructure resources, workforce, maintenance, software licensing, etc.). The possibility of deploying services on demand in a cost-effective and flexible way also contributes to hide technology details and understanding.

© Springer Nature Switzerland AG 2019
Á. Rocha et al. (Eds.): WorldCIST'19 2019, AISC 931, pp. 396–406, 2019.
https://doi.org/10.1007/978-3-030-16184-2_38

Conversely, this flexibility and variety of provided cloud services represent a huge challenge for the supporting monitoring systems, which urge for a systematization of the measuring requirements involved in the multitude of resources and services requiring monitoring. The pool of virtualized hardware and software resources (e.g., storage devices, communication devices, integrated development environments, user service platforms, etc.), dynamically assigned, is a further demand to the usual service and physical infrastructure monitoring.

Moreover, providers of public clouds still need to provide monitoring information to customers, which poses additional flexibility, security, customization and service quality concerns. The customer/provider relation, settled through the negotiation of Service Level Agreements/Service Level Specification (SLAs/SLSs), involves the verification of the corresponding quality of service (QoS) compromises, i.e., the fulfilment of the contractualized cloud services. Ensuring other relevant requirements such as security and quality of experience (QoE) are issues that lead to a complex monitoring process due to the use of virtual environments and, eventually, outsourcing infrastructures. In the event of conflicts of interest that may arise between providers and customers, the adoption of a neutral third-party responsible for monitoring service performance seems to be the best solution [3,4].

Facing the above concerns, this paper aims at clarifying and layering the predominant measuring requirements in cloud services monitoring, in order to sustain and promote the development of *comprehensive and flexible* monitoring systems. This layered perspective for monitoring encompasses requirements from physical/virtual resources to high-level customer/provider relation. Structuring and measuring relevant metrics in a layered approach allows to identify and interrelate anomalies in the provided cloud services, for instance, by establishing a relation between performance degradation in high-level service metrics with infrastructural causes. Studying this cross-layer interrelation is an important step to improve cloud monitoring systems' efficiency. This paper also provides a survey perspective of: (i) representative cloud services, aggregating them into the corresponding cloud services model; and (ii) relevant monitoring tools, mapping their functionality is the proposed monitoring layers.

This article is organized as follows: the related work identifying cloud service models, platforms and tools for cloud monitoring is discussed in Sect. 2; the layered monitoring approach for cloud services and related metrics are presented in Sect. 3; the relation between monitoring layers and the main service models and monitoring tools is discussed in Sect. 4; and finally, the main conclusions are included in Sect. 5.

2   Related Work

This section presents briefly the main cloud service models, discusses cloud monitoring issues, and reviews the main cloud measurement methodologies.

Table 1. Examples of cloud services aggregated per service model

Model	Examples	Sellers	Buyers
IaaS	Amazon Web Services (EC2), Google Computing Engineering Microsoft Hyper-V, GoGrid, Proofpoint, Rackspace, RightScale, SoftLayer/IBM (Blue Cloud), Windows Azure, Sun (Project Caroline), HP Adaptive IaaS, EMC, VMWare VCloud	Datacenter Providers	Enterprises
PaaS	Google App. Engine, Windows Azure, dotCloud, Salesforce, Redhat, Oracle, Cloudera, Cloud Foundry	Plataform Service Providers	Developers Software Companies
SaaS	Office 365, Salesforce, Google Apps., Yahoo, (Zimbra/VMware), Concur, Taleo, Netsuite, Proofpoint, Dropbox, Workday, Hotmail	Software Companies	End Users

2.1 Service Models

Cloud computing services are divided into service models according to their nature. Well-known service models are broadly defined in the cloud architecture as IaaS (Infrastructure as a Service), PaaS (Platform as a Service) and SaaS (Software as a Service). According to this classification, Table 1 includes a list of representative examples of cloud services.

2.2 Monitoring Systems

Cloud monitoring is not just about monitoring servers hosted. It is also relevant monitoring cloud-based services that enterprises consume. There is a need to identify, anticipate and report failures in order to enhance service provision and customer satisfaction [5].

The design of a monitoring system involves defining aspects such as: (i) the architecture upon which the measurements are performed, identifying the main components of the monitoring system, their functionality and how they interact; (ii) the timeliness requirements of measurement data, identifying the timing issues regarding the objectives that measurements intent to fulfill, e.g., whether real-time measures (online measures) are required or not (offline measures); (iii) the measurement methodology, defining if measurements are achieved based on passive or active methodologies, and whether additional techniques are needed to improve measurement efficiency (e.g., traffic sampling techniques); and (iv) the spatial distribution of measurement points - this involves deciding locality and cardinality issues, e.g., if all or a subset of specific network nodes are involved in the measurement process and where they are located in the network.

In cloud environments where monitoring has to be carried out in multiple dimensions, ranging from physical/virtual resources on single nodes to end-to-end service delivery, passive and active measurements are usually combined to achieve more scalable monitoring systems. According to [6], in cloud environments, distributed monitoring systems should exhibit properties such as scalability, elasticity, adaptability, timeliness, autonomicity, comprehensiveness, extensibility, intrusiveness, resilience, reliability, availability and accuracy of measurements.

Table 2. Examples of monitoring tools.

Monitoring	Examples
Local	Sysstat (Isag, Ksars)
Remote	*Open source*
	Cacti, Cloud Stack Zenpack, Dargus, Ganglia, Nimbus, Open Nebula, Sensu, VMware Hyperic HQ Open Source Ed.
	Commercial
	3Tera, Amazon CloudWatch, Aneka, AppDynamics, Aternity, AzureWatch, CA Nimsoft, CloudKick, Enstratus, Gomez APM, GridICE, GroundWork, Kaavo, Kaseya Traverse, Keynote, Landscape, Logic Monitor, MonALISA, Monitis, Nagios, NewRelic, Rackspace Cloud Monitoring Tool, RevealCloud, RightScale, ScienceLogic, Tap in systems, TechOut, VMware Hyperic/Cloud Status, YLastic, Zenoss, PCMONS

Security is the most barrier to companies e consumers to adopt cloud as a environment for applications or business, considering privacy and data protection [7]. In this case, cloud monitoring is a essential service for helping secure cloud environments [8].

2.3   Cloud Monitoring Tools

Currently, there is a lot of tools for cloud monitoring. As an effort of systematizing the present offer, Table 2 includes most of open-source and commercial monitoring tools available.

According [9], important features of an efficient cloud monitoring tool should be: scalability, portability, non-intrusiveness, robustness, multi-tenancy, interoperability, customizability, extensibility, usability, affordability and archivability.

The literature review has shown that there is no consensus on the monitoring solutions definition that satisfy requirements of complex cloud environments. This reality has motivated the present study in pursuing a structured approach to envision monitoring in these environments. Viewing the problems through layers, this study aims to contribute to the development of efficient and flexible solutions for managing and optimizing cloud services.

3   Cloud Service Monitoring Model

This section presents the proposed layered approach for cloud services monitoring. For each layer, relevant measuring requirements are identified to satisfy the corresponding monitoring dimension.

In the analysis, four main monitoring layers have been identified, being divided into specific categories. These layers are presented in Fig. 1 which also illustrates the high-level correspondence between the suggested monitoring layers and the typical cloud service models. This aspect is further explored in Sect. 4.1.

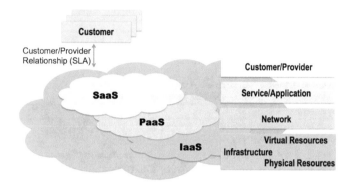

Fig. 1. Stratified view for monitoring cloud services.

Taking a bottom-up approach to the layered overview of the measuring requirements presented in Table 3, the proposed monitoring layers are discussed in more detail below.

Infrastructure - Monitoring cloud infrastructures (physical and virtual) involves the systematic collection and analysis of data to track the performance and availability of the cloud infrastructure components. Focusing on physical resources, three categories of metrics are considered: Components, Energy and Security. Components includes from processing and storage devices to network equipment. The percentage of CPU utilization, RAM, storage memory, and statistics of physical network interfaces are the most consensual metrics considered in this category [5,10]. In Energy, the main purpose is to assess and optimize the use of energy, to control temperature and to reduce the emission of carbon monoxide. Therefore, besides power consumption, metrics for temperature control and backup power systems (generators, UPS) are also proposed [9,11]. In Security, considering the seven-layer security model proposed by the Cloud Security Alliance (CSA) [12,13], the proposed metrics are aligned with the first three levels - Facility, Network and Hardware - of this model.

Still within the Infrastructure layer, focusing now on virtual resources, two categories of metrics are considered: Components and Security. Facing the dynamic operations involved in virtualization processes (start, stop, suspend, resume, migration of Virtual Machines), monitoring virtual resources across VMs-oriented metrics assume a crucial role in a cloud computing environment, increasing transparency, dynamics and scalability [14,15]. The requirements of measuring the security of virtual resources can be associated with the OS and Middleware CSA layers discussed in [16,17], involving metrics extracted from the monitoring of OS-level events and system calls between VMs and hardware.

Table 3. Monitoring layers overview

Layer	Category	Measuring requirements
Customer/ Provider	ServiceLevel	Availability; Turn-around time; Mean Time to Recover; Uptime; User Experience and QoS
	Auditing	SLA violations; SLA nonfulfillment and associated penalties
	Accounting/Pricing	Monitoring of usage; Money cost and revenue evaluation
	Other	Security, privacy; Helpdesk/Trouble tickets; Specific metrics affecting the business
Service/ Application	Availability	Service uptime/downtime; Service availability/unavailability
	Performance	Service contention ratio; Response time (average/maximum) Service ratio (responses vs. requests); Number of transactions per time unit
	Reliability	Mean time to repair upon failure; Mean time between failures; Bug and software vulnerabilities
	Security	Access patterns; login processes; Password management Number of security vulnerabilities
	Other	Login times; IP access records (historic); Specific service/app. metrics
Network	Throughput	Traffic volume per time unit; Link capacity; Bandwidth used and available
	Performance	Packet duplication; Packet loss (OWPL, OWLP, IPLR); Packet delay (OWD, RTT, IPTD, IPDV); Response time (average/maximum)
	Availability	Uptime/downtime; Availability/unavailability of the network; Connectivity (one or two-way)
	Reliability	Mean time to repair in case of failure; Mean time between failures
	Security	Firewalls; IDS and IPS monitoring
Infrastructure Virtual Resources	Components	CPU (usage, n. of cores, speed, time per execution); RAM (usage); Memory storage (usage, r/w speed, throughput, page _exch/s and /exec); Number of active instance; VM statistics (interfaces, startup time, acquisition/release time, exec/access time, uptime);
	Security	Monitoring events and OS calls between VMs and hardware
Infrastructure Physical Resources	Components	CPU (usage, n. of cores, speed, time per execution); RAM (usage, r/w speed); Network interface statistics Memory storage (usage, r/w speed, throughput); Topology connectivity
	Energy	Energy consumption; temperature; Generator and UPS state
	Security	Fire alarms/sensors; Surveillance; Access control; Authentication systems

Network - Aiming at assessing the network infrastructure and the internetworking service, this layer considers the following categories: Throughput, Performance, Availability, Reliability and Security. In this layer, both ITU-T and IETF (e.g., IP Performance Metrics workgroup - IPPM) have devoted substantial efforts to define metrics and methodologies for IP performance measuring.

Throughput-related metrics are considered essential in cloud monitoring both for traffic engineering and SLSs verification [6,18,19]. For bandwidth estimation, several useful tools are accessible at CAIDA. Delay and loss-related metrics are the most common way to access network performance [20], existing a rich set of measures proposed for the effect, mostly with an one-way context. Regarding availability and reliability assessment, the quantification of (un)availability of a network and connectivity, affected by failures in network components or routing misconfigurations, and the efficient recover to network failures are common metrics. Within the security context, the CSA inputs at network level are considered.

Service/Application - This is a context-aware layer, as the relevant parameters to consider are tight related with the software/service being monitored. Despite that, there are metrics that span distinct contexts, being categorized as: Availability, Performance, Reliability and Security.

Measuring the availability of a service involves registering the periods of time during which the service is running or is unavailable, which is also critical for assessing SLSs violations and corresponding penalties. Aspects impacting on performance are typically associated with service ratios and time responsiveness [7,18]. Similarly to the Network layer, the metrics related with service failures and recover times are considered within the reliability category, being also measures of the efficiency of a service. Security is a fundamental aspect for this layer, being relevant aspects to control the number of security vulnerabilities/violations, digital certificates, private keys, Domain Name System Security Extensions (DNSSEC), login processes, access patterns, user passwords format/renew [21]. Regarding the CSA model, Application and User are the top layers to consider as reference. In addition to the metrics mentioned above, specific metrics for each Service/Application type should be evaluated.

Customer/Provider - The negotiation of SLAs/SLSs, specifying the administrative and technological aspects of services, is the basis of a Customer/Provider service offer relation. It is also the ground for an effective service monitoring, acting as an auditing instrument for both the enrolled partners.

The measuring requirements at this layer are defined in three major categories: Service Level, Auditing and Accounting/Pricing. Considering the negotiated SLA/SLS, a multilevel description and evaluation of each service can be obtained, for instance regarding QoS, uptime, security, privacy, backup procedures, responsibilities and compensation of both parties, geographic location of resources (e.g., datacenters) [22]. In addition, as a cross-layer task spanning all the layers defined in Table 3, SLS violations assessment can rely on metrics obtained from lower levels, e.g., average of CPU usage [5]. QoS evaluation usually relies on metrics evaluated at network layer. Service accounting is also an important issue to control, helping defining the cloud business model according to the market tendencies [18,20].

Table 4. Relation to service models

	IaaS	PaaS	SaaS
Infrastructure	✓		
Network	✓	✓	✓
Service/Application			✓
Customer/Provider	✓	✓	✓

4 Discussing Monitoring Layers

The objective of this section is twofold: (i) to establish a relation between the proposed monitoring layers and the main cloud service models; and (ii) to identify the monitoring layer(s) in which representative cloud monitoring tools operate.

4.1 Service Models

Due to significant differences between the existing cloud service models, there is consensus that a generic solution for cloud management will be hard to attain in practice. In fact, each type of service has specific objectives, characteristics and management requirements, therefore, a tailored monitoring system needs to be planned and developed in order to meet the goals of management. Apart from considering the multiple components of an entire cloud environment, monitoring should also target aspects such as QoS and QoE, SLAs and features such as security, robustness, scalability, elasticity, among other [23].

In this context, it is useful to relate the proposed layered approach for monitoring cloud services with the service models. This is illustrated in Table 4, taking into account service models (IaaS, PaaS and SaaS) and the identified layers Infrastructure, Network, Service/Application and Customer/Provider.

The layer Infrastructure is essentially associated with the service model IaaS. In fact, the Infrastructure layer includes components of physical and virtual resources for the processing, storage and network communications, i.e., aspects that also characterize the IaaS service model.

As regards the Network layer, if considering an end-to-end monitoring perspective, this level is directly or indirectly related to the three service models. For instance, in SaaS, Network aspects (such as Throughput and Performance) also need to be monitored from the source (infrastructure, datacenters) to the service access point. The same occurs with PaaS, where monitoring Network layer components is required.

In turn, the Service/Application layer is associated with the service model SaaS as the parameters under monitoring/measurement and how they are collected depend mainly on the software and not the infrastructure.

Finally, the Customer/Provider layer, addressing aspects such as accounting and auditing, is directly related to the three service models. Due to the complexity of a cloud environment and the existence of multiple stakeholders, it is

Table 5. Classifying cloud monitoring tools

Monitoring Tools	Infrastructure		Net.	Service/App.	Customer/Provider
	Physical	Virtual			
AppDynamics				✓	
Aternity				✓	✓
Azurewatch		✓		✓	
CA Nimsoft	✓	✓		✓	
Ganglia	✓				
Gomez APM				✓	✓
Hyperic/Cloud status	✓	✓			
Kaseya Traverse	✓	✓	✓	✓	✓
LogicMonitor	✓	✓		✓	
Monitis	✓			✓	
Nagios	✓	✓	✓	✓	
New Relic	✓			✓	
Rackspace Cloud Mon.	✓	✓		✓	
RevealCloud	✓	✓		✓	
Zenoss	✓	✓		✓	

normal to have business relations between providers and customers at all levels. Providers can provide both datacenters, as service platforms or software, while customers may either be technology companies or final users.

4.2   Cloud Monitoring Tools

Taking into account the discussion in Sect. 2, the characteristics and popularity of off-the-shelf cloud monitoring tools, Table 5 summarizes popular tools (top 15), relating the main target of their operation to the cloud monitoring layers.

As illustrated, most of monitoring tools are focused on measuring infrastructure resources (physical and/or virtual) and service/application metrics. Although few tools also focus on the network layer, the research and development of tools for monitoring IP metrics is rather vast and precedes the cloud hype. The customer/provider layer, as shown, evinces the need for improving monitoring facilities at this layer. As consequence, currently, achieving an encompassing cloud monitoring solution may involve considering complementary tools.

5   Conclusions

Despite diversity, the cloud paradigm embraces a panoply of services and resources which management imposes similar monitoring requirements. Several tools exist for this purpose, nevertheless, the lack of standards and guidelines for a systematic and multilevel evaluation of cloud services is still evident.

Taking a customer and service provider perspective, this study has identified and suggested measuring requirements, metrics and best practices in the context of cloud monitoring. Through a layered view of the problem, we believe to have undertaken the right steps toward modular monitoring and optimization of cloud services deployment. This layered view has spanned measuring requirements from the underlying physical/virtual infrastructure to the customer/provider contractual relation. In addition, to better frame the problem in the actual monitoring panorama, representative cloud services in the market have been positioned per service model and existing monitoring tools have been surveyed and mapped into the suggested functional layers. Future work includes going further exploring cross-layer dependencies of cloud monitoring metrics.

Acknowledgments. This work has been supported by FCT *Fundação para a Ciência e Tecnologia* within the Project Scope: UID/CEC/00319/2019.

References

1. Ferraz, F., Ribeiro, F., Lima, W., Sampaio, C.: A disturbing question: what is the economical impact of cloud computing? A systematic mapping. In: 2018 IEEE 11th International Conference on Cloud Computing (CLOUD), vol. 00, pp. 853–856 (2018). https://doi.org/10.1109/CLOUD.2018.00120
2. Rafique, K., Tareen, A.W., Saeed, M., Wu, J., Qureshi, S.S.: Cloud computing economics opportunities and challenges. In: 2011 4th IEEE International Conference on Broadband Network and Multimedia Technology, pp. 401–406 (2011). https://doi.org/10.1109/ICBNMT.2011.6155965
3. Noor, T.H., Sheng, Q.Z., Zeadally, S., Yu, J.: Trust management of services in cloud environments: obstacles and solutions. ACM Comput. Surv. **46**(1), 12:1–12:30 (2013). https://doi.org/10.1145/2522968.2522980
4. Hussain, M., Al-Mourad, M.: Effective third party auditing in cloud computing. In: 2014 28th International Conference on Advanced Information Networking and Applications Workshops (WAINA), pp. 91–95 (2014). https://doi.org/10.1109/WAINA.2014.158
5. Hauser, C.B., Wesner, S.: Reviewing cloud monitoring: towards cloud resource profiling. In: 2018 IEEE 11th International Conference on Cloud Computing (CLOUD), pp. 678–685 (2018). https://doi.org/10.1109/CLOUD.2018.00093
6. Aceto, G., Botta, A., De Donato, W., Pescapè, A.: Cloud monitoring: a survey. Comput. Netw. **57**(9), 2093–2115 (2013). https://doi.org/10.1016/j.comnet.2013.04.001
7. Subashini, S., Kavitha, V.: A survey on security issues in service delivery models of cloud computing. J. Netw. Comput. Appl. **34**(1), 1–11 (2011). https://doi.org/10.1016/j.jnca.2010.07.006
8. Zhang, H., Luna, J., Suri, N., Trapero, R.: Flashlight: a novel monitoring path identification schema for securing cloud services. In: Proceedings of the 13th International Conference on Availability, Reliability and Security, ARES 2018, pp. 5:1–5:10. ACM, New York (2018). https://doi.org/10.1145/3230833.3230860
9. Fatema, K., Emeakaroha, V.C., Healy, P.D., Morrison, J.P., Lynn, T.: A survey of cloud monitoring tools: taxonomy, capabilities and objectives. J. Parallel Distrib. Comput. **74**(10), 2918–2933 (2014). https://doi.org/10.1016/j.jpdc.2014.06.007

10. Palhares, N., Lima, S.R., Carvalho, P.: A multidimensional model for monitoring cloud services. In: Rocha, Á., Correia, A.M., Wilson, T., Stroetmann, K.A. (eds.) Advances in Information Systems and Technologies, pp. 931–938. Springer, Heidelberg (2013)

11. Sheikhalishahi, M., Grandinetti, L.: Revising resource management and scheduling systems. In: Proceedings of the 2nd International Conference on Cloud Computing and Services Science, pp. 121–126 (2012). https://doi.org/10.5220/0003955401210126

12. Khan, M.A.: A survey of security issues for cloud computing. J. Netw. Comput. Appl. **71**, 11–29 (2016). https://doi.org/10.1016/j.jnca.2016.05.010

13. Singh, S., Jeong, Y.S., Park, J.H.: A survey on cloud computing security: issues, threats, and solutions. J. Netw. Comput. Appl. **75**, 200–222 (2016). https://doi.org/10.1016/j.jnca.2016.09.002

14. AbdElRahem, O., Bahaa-Eldin, A.M., Taha, A.: Virtualization security: a survey. In: 2016 11th International Conference on Computer Engineering Systems (ICCES), pp. 32–40 (2016). https://doi.org/10.1109/ICCES.2016.7821971

15. Jain, R., Paul, S.: Network virtualization and software defined networking for cloud computing: a survey. IEEE Commun. Mag. **51**(11), 24–31 (2013)

16. Spring, J.: Monitoring cloud computing by layer, part 1. IEEE Secur. Priv. **9**(2), 66–68 (2011). http://doi.ieeecomputersociety.org/10.1109/MSP.2011.33

17. Spring, J.: Monitoring cloud computing by layer, part 2. IEEE Secur. Priv. **9**(3), 52–55 (2011). http://doi.ieeecomputersociety.org/10.1109/MSP.2011.57

18. Wieder, P., Butler, J.M., Theilmann, W., Yahyapour, R.: Service Level Agreements for Cloud Computing. Springer Publishing Company, Incorporated, New York (2011)

19. European Commission C-SIG SLA Subgroup: Cloud Service Level Agreement Standardization Guidelines (2014)

20. Choi, T., Kodirov, N., Lee, T.H., Kim, D., Lee, J.: Autonomic management framework for cloud-based virtual networks. In: 2011 13th Asia-Pacific Network Operations and Management Symposium (APNOMS), pp. 1–7 (2011). https://doi.org/10.1109/APNOMS.2011.6077011

21. de Chaves, S., Westphall, C., Lamin, F.: SLA perspective in security management for cloud computing. In: 2010 Sixth International Conference on Networking and Services (ICNS), pp. 212–217 (2010). https://doi.org/10.1109/ICNS.2010.36

22. Stamou, K., Morin, J.H., Gâteau, B., Aubert, J.: Service level agreements as a service - towards security risks aware SLA management. In: CLOSER, pp. 663–669 (2012)

23. Clayman, S., Galis, A., Chapman, C., Toffetti, G., Rodero-Merino, L., Vaquero, L.M., Nagin, K., Rochwerger, B.: Monitoring service clouds in the future internet. In: Tselentis, G., Galis, A., Gavras, A., Krco, S., Lotz, V., Simperl, E.P.B., Stiller, B., Zahariadis, T. (eds.) Future Internet Assembly, pp. 115–126. IOS Press (2010)

Consensus Problem with the Existence of an Adversary Nanomachine

Athraa Juhi Jani[✉]

Department of Computer Science, College of Science,
Al-Mustansiriyah University, Baghdad, Iraq
athraa.jj@gmail.com

Abstract. A nanomachine is considered to be the basic functional unit in nanotechnology. The rapid evolution in nanotechnology has provided appropriate development in miniaturization and fabrication of nanomachines with simple sensing, computation, data storing, communication and action capability. Further capabilities and applications can be enabled if multiple nanomachines communicate to perform collaborative and synchronous functions in a distributed manner to form a nanonetwork. In this paper the consensus problem in diffusion based molecular communication is considered in a network of n nanomachines. A nanomachine $node_c$ that can control and direct processes in the network is one of the n nanomachines. The considered model is time slotted. An adversary nanomachine $node_A$ is located within the transmission range of the network, where $node_A$ aims to jam the communication among these n nanomachines. The adversary nanomachine $node_A$ is assumed to follow Poisson probabilistic distribution in diffusing its jamming molecules. The n nanomachines need to estimate the concentration of the jamming molecules, in order to improve the possibility of reaching consensus, taking into account the additional jamming molecules in the environment. Thus, during the first k time slots, each nanomachine from n senses (listens to) the jamming molecules diffused by $node_A$, and stores the sensed molecular concentration during each time slot in a vector of length k. Based on the stored molecular concentration in its vector, each nanomachine from n attempts to estimate the average of diffused jamming units. After estimating this value, the processes to reach consensus start. Each nanomachine n has an initial value, the initial values of all nanomachines are diffused to $node_c$. Then $node_c$ computes the average of all initial values. However, $node_c$ also needs to take into account the jamming molecules when it computes the average of initial values. Thus, same as nanomachines n, nod_c is assumed to estimate the jamming molecular concentration during the first k time slot. Then, $node_c$ diffuses the average of the initial values to the other nanomachines n.

Keywords: Nanonetworks · Consensus problem · Diffusion ·
Molecular communication · Adversary · Poisson distribution

© Springer Nature Switzerland AG 2019
Á. Rocha et al. (Eds.): WorldCIST'19 2019, AISC 931, pp. 407–419, 2019.
https://doi.org/10.1007/978-3-030-16184-2_39

1   Introduction

Nanonetwork is considered a new research branch, which derived from applying nanotechnology in the digital communication field [2]. The communication between nanoscale devices expands the possible applications, and moreover, it increases the complexity and range of operation of the system [1]. Many options for communications in nanoscale have been revealed and studied, several of which use natural mechanisms and processes as a model, by directly applying different elements from nature to serve their purposes [13]. Molecular communication is a bio-inspired communication mechanism, where information is exchanged through transmitting, propagating and receiving molecules between two nanometer-scale devices [15]. The characteristics and rules that govern molecular communication are motivated by the communication in biological systems.

1.1   Previous Work

Security in nanonetworks has been explored in few researches, as it is considered a serious challenge. The authors in [4] discussed security in nano communication exploring the forms of possible threats and attacks, through studying sensor networks, in order to derive security requirements and to check the possibility of applying the available security solutions to the nano communication. Suggesting that new security solutions needed for the bio-inspired nano communication, as it have different functionalities, different environmental rules, therefore the existing security protocols won't be applicable. This led to form a new security field known as *biochemical cryptography*, which can be considered as new research direction. It can be used to provide a secure bio-inspired nano communication, as the protection mechanism is based on biological and molecular processes. In [16] the idea of using the characteristics in human immune system as basis to establish security in nanonetwork is presented. The authors in [10] gave an overview of the conventional ways to tackle attacks in wireless networks. Then, an inclusive general characterization of security and privacy in molecular communication is presented. The authors state that a collaboration between many diverse disciplines is required in order to efficiently understand the security in molecular communication based systems.

1.2   Related Work

Quorum sensing process [9,14] is an example of signalling between bacteria, where bacteria can use it to estimate the density of their population in the environment through estimating the concentration of a certain type of molecules. Consensus problem in diffusion based molecular communication has been studied in related research [5–7]. Mainly, [5] trying to map the Quorum Sensing to consensus problem under diffusion based molecular communication. Where their goal is to study consensus problem by spreading information about an event or any variation through a diffusion based network. Through communication all nanomachines try to obtain the best estimate of this random variable.

1.3   The Work in This Paper

In this paper, an attempt to explore security issues in molecular communication, consensus problem in diffusion based network with the existence of an adversary nanomachine is presented. The adversary nanomachine aim is to jam the communication between the network nanomachines. The adversary nanomachine is following Poisson random distribution in diffusing its jamming molecules. The network nanomachines attempt to estimate the concentration of the jamming molecules. Thus, through k time slots, each nanomachine senses the molecular concentration and stores it in a vector. Then, each nanomachine attempts to estimate the average of the jamming molecular concentration, based on the stored molecular concentration during k time slots. After estimating the jamming molecular concentration, the processes to reach consensus start, where each nanomachine in the network has an initial value. Each nanomachine diffuses its initial values to a special node in the network. This special node computes the average of all initial values and diffuses it to the network nanomachines. The special node is assumed to estimate the jamming molecular concentration in the same way and during the same interval that the network nanomachine attempted to estimate it. Thus, the special node takes in consideration the jamming molecular concentration when it computes the average of all initial values.

1.4   Structure of the Paper

This paper is organized as follows, In Sect. 2 the proposed model is described. Section 3 discusses the process of estimating parameter λ (which is related to the jamming molecules distribution) by all n nanomachines in the network. The steps of the proposed consensus processes are presented in Sect. 4. Finally, Sect. 5 combines conclusions and prospective future work.

2   Model

Network Environment: In this paper a system of n nanomachines is considered, communicating according to diffusion based molecular communication Fig. 1. Each nanomachine $n(i)$ (where, $i \in \{1, 2, ..N\}$) has the ability to sense the concentration of molecules from the environment and to emit molecules at a particular rate into the environment, and one of these n nanomachines is considered to be a special node (leader node) $node_c$ (which has some responsibilities for directing and controlling processes of the consensus protocol), and other nanomachines in different positions from $node_c$ within its transmission range distance d_{max}. Information molecules are encoded based on the variation in the concentration of molecules in the communication medium. The communication medium might contain residual molecules from previous diffusion (as it is not necessary that all molecules to be received by the other nanomachines), and also contains molecules from other nanomachines (that are not among n) and these molecules can be considered as noise. The nanomachines are assumed to be located within a close range, so that if one nanomachine diffuses then all other nanomachines can receive some molecules.

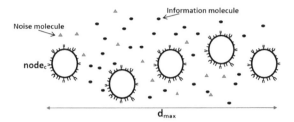

Fig. 1. Model representation

Communication among Nanomachines. Communication between nanomachines is based on diffusing and sensing molecules. Each nanomachine i from n can decide to diffuse, for example, a unit u of molecules at time t, and any other nanomachine j at distance d from the nanomachine i can sense the impulse of the released unit u of molecules within the interval $[t, t + T]$, through the following [11]

$$c(u, d, T) = \int_0^T u \cdot \frac{1}{(4\pi Dt)^{\frac{2}{3}}} \cdot \exp(\frac{-d^2}{4Dt}) \tag{1}$$

where D is the diffusion coefficient of the communication medium. If more than one nanomachine diffuses a unit u molecules, a receiver nanomachine j accumulates the sensed molecules through the summation of the values of $c(u, d, T)$ over diffusing nanomachines i, i.e., nanomachine j senses molecules in total during the interval $[t, t + T]$

$$\sum_i c(u_i, d_{(i,j)}, T_i) - c(u_i, d_{(i,j)}, [T_i - T]_+) \tag{2}$$

where, $d_{(i,j)}$ is the distance between nanomachines i and j, T_i is the time that passed from the diffusion of nanomachine i up to time $t + T$, i.e., nanomachine i diffused at time $(t + T - T_i)$, u_i is the unit of molecules by nanomachine i at that time, and $[T_i - T]_+$ equals to $\max\{T_i - T, 0\}$. In other words, the receiver nanomachine senses the total amount of molecules that have been in its nearest proximity in the time interval $[t, t + T]$ without being able to distinguish which molecules come from which transmitter. If the amount of sensed molecules is greater than or equal to threshold τ, it will be considered as 1, otherwise it is 0.

Adversary Nanomachine: The adversary nanomachine $node_A$ continues to diffusing jamming molecules. The distribution of the jamming molecules is stochastic. All the other n nanomachines are assumed to be placed in such a way that it can sense the same distribution range of the jamming molecules; thus, the distance between $node_A$ and the other nanomachines n is d_A. The adversary nanomachine is assumed to follow Poisson probabilistic distribution in diffusing its jamming molecules as shown in 3.

$$P(j) = \frac{\lambda^j e^{-\lambda}}{j!} \tag{3}$$

The number of molecules diffused by adversary nanomachine $node_A$ can vary in each time it diffuses; thus, j represents the observed units of molecular concentration from $node_A$, i.e., $j \in \{0, 1, 2, \cdots, N\}$, while λ represents the average of diffused molecular concentration (units of molecules) by $node_A$. $P(j)$ represents the diffusion rate of $node_A$ and the sensed molecular concentration by the other nanomachines n.

The other nanomachines n in the network need to estimate the value of λ in order to improve the possibility of reaching consensus, taking into account the additional jamming molecules in the environment. Thus, during the first part of the model, the nanomachines n including $node_c$ are assumed to sense the diffused molecular concentration for k time slots (each one with the length T_0), and store these molecular concentration in a vector. The statistics to estimate the value of λ show that it is the arithmetic mean of the samples [3].

$$\lambda = \frac{\text{summation of samples}}{\text{number of samples}} \tag{4}$$

Consensus Problem. Each nanomachine $n(i)$ has an initial value. Through k time slots, each nanomachine $n(i)$ diffuses its initial value to nod_c. The accumulative molecular concentration of the initial values and the jamming molecules from $node_A$ are sensed by $node_c$. Thus, $node_c$ stores the sensed molecular concentration in a vector of length k. Then, $node_c$ computes the average of the sensed molecular concentration during the k time slots, while attempting to exclude the jamming molecules from the sensed initial values. Then, $node_c$ computes the average of the estimated initial values and diffuses it to the other nanomachines $n(i)$. The initial values of each nanomachine from $n(i)$ are assumed to be relatively close to the value of a known global parameter u in the network. Thus, consensus is reached, if the average of the initial values sensed by each nanomachine from $n(i)$ has a deviation of ϵ compared to the global parameter u. Although only the sum of the sensed jamming molecules is needed; however, the sensed jamming molecules in each time slot are stored in a vector. This can enable nanomachines to perform more specific computations that require a comparison between the received molecular concentration for example.

Time Slots Length: All the nanomachines n in the network are assumed to be synchronized, and can communicate in a predefined time slot T_0. Where T_0 is a system parameter and its length depends on the network's geometric properties, such that $T_0 = v \frac{d_{max}^2}{D}$, where v is a constant that can be equal 1; d_{max} is the transmission range distance of $node_c$, and D is the diffusion coefficient. During the first k time slots, all nanomachines sense molecular concentration from $node_A$ to estimate λ. Then, after estimating λ, the process of consensus and the diffusion of the initial values to $node_c$ begins, each node $n(i)$ diffuses its initial value to nod_c through k time slots. However, in order to avoid Inter Symbol Interference, each nanomachine $n(i)$ waits for at least $v \frac{d_{max}^2}{D}$ before diffusing its initial value again. Where the Inter Symbol Interference (ISI), means the residue molecules from the previous diffused symbol which can affect the current diffused symbol.

The ISI in [11] is assumed to come from a sufficient number of interfering sources (nanomachines which are diffusing), in a way that these residue molecules follow a normal distribution.

3   Estimating λ by Nanomachines n

Each nanomachine from n is assumed to recognize that the distribution of the jamming molecules from $node_A$ is based on Poisson distribution. Thus, each nanomachine from n aims to sense the jamming molecules through k time slots and stores the sensed molecular concentration in a vector, in order to estimate λ. Through the observation of the sensed molecular concentration, each nanomachine from n attempts to estimate λ. The following steps show the sensed molecular concentration by a nanomachine from n when an adversary nanomachine $node_A$ diffuses units of molecules u_A, during each time slot ts from the observed k time slots. The molecular concentration is stored in a *vector* of the nanomachine $n(i)$.

---

1   $ts \leftarrow 0$;
2   $t \leftarrow 0.1$;
3   **while** $ts \leq k$ **do**
4   **while** $(c(u_A, d_A, T_0) \geq \tau)$ & $(t \leq T_0)$ **do**
5           $c(u_A, d_A, T_0) = \int_t^{T_0} u_A \cdot \frac{1}{(4\pi Dt)^{\frac{2}{3}}} \cdot \exp(\frac{-d_A^2}{4Dt})$
6       $vector[ts] = c(u_A, d_A, T_0)$;
7       $ts = ts + 1$;

As Eq. (4) shows that λ is the mean of samples. The maximum likelihood of λ estimation can be as follows [8]:

$$P(u_A) = \frac{e^{-\lambda}\lambda^{u_A}}{u_A!}$$

Where $u_A = \{0, 1, 2, \cdots, N\}$, $j = \{1, 2, \cdots, k\}$. Thus, using the log likelihood, the function would be:

$$l(\lambda) = \log \prod_{j=0}^{k} P(u_A(j), \lambda)$$

$$= \sum_{j=1}^{k} \log \frac{e^{-\lambda}\lambda^{u_A(j)}}{u_A(j)!}$$

$$= \sum_{j=1}^{k} \log e^{-\lambda} + \sum_{j=1}^{k} u_A(j) \log \lambda - \sum_{j=1}^{k} \log u_A(j)!$$

Then, the first derivative is used to find the maximum likelihood of λ estimation:

$$\frac{\partial l(\lambda)}{\partial \lambda} = 0$$

$$\frac{\partial \sum_{j=1}^{k} \log e^{-\lambda} + \sum_{j=1}^{k} u_A(j) \log \lambda - \sum_{j=1}^{k} \log u_A(j)!}{\partial \lambda} = 0$$

$$-k + \frac{1}{\lambda} \sum_{j=1}^{k} u_A(j) = 0$$

$$\lambda = \frac{1}{k} \sum_{j=1}^{k} u_A(j)$$

i.e.

$$\lambda = \frac{\sum_{j=1}^{k} u_A(j)}{k} \tag{5}$$

Thus, each nanomachine from n computes the average of the sensed molecular concentration during k time slots, and considers it to be an estimation of λ. To compute the sensed molecular concentration from other n nanomachines diffusions, each nanomachine $n(i)$ deducts the estimated value of λ from the total amount of the sensed molecular concentration during a certain time slot T_0. Thus, in the case of communication between nanomachines n, each nanomachine $n(i)$ can sense units of molecular concentration u_n diffused by the other nanomachines n, plus the jamming molecules from the adversary nanomachine u_A. If the total amount of the sensed molecular concentration by a nanomachine $n(i)$ during a certain time slot T_0 is R_i, then nanomachine $n(i)$ can assume that:

$$u_{n(i)} \approx \begin{cases} R_i - (\lambda + \varepsilon), & \text{if } \varepsilon \leq \frac{1}{k} \\ R_i - (\lambda - \varepsilon), & \text{if } \varepsilon \geq 1 - \frac{1}{k^2} \end{cases} \tag{6}$$

where ε is also estimated by each nanomachine based on the number of nanomachines n in the network, during estimating λ, in such a way that:

$$\varepsilon \begin{cases} \leq \frac{1}{k}, & \text{if } k \leq n \\ \geq 1 - \frac{1}{k^2}, & \text{otherwise} \end{cases} \tag{7}$$

In [12], Chernoff bound is used for Poisson random variable estimation in **Theorem 5.4**, which shows that:

1. If $u_A > \lambda$ then

$$Pr(X \geq u_A) \leq \frac{e^{\lambda}(e\lambda)^{u_A}}{u_A^{u_A}};$$

2. If $u_A < \lambda$, then

$$Pr(X \leq u_A) \leq \frac{e^{\lambda}(e\lambda)^{u_A}}{u_A^{u_A}}.$$

4   Consensus Protocol

After waiting for k time slots to estimate λ (so that the jamming molecules from $node_A$ are taken into account in the communication among nanomachines n), the process to reach consensus begins.

4.1   Estimate the Number of Nanomachines

If it is required to estimate the number of nanomachine in the network before the consensus protocol starts. Then, $node_c$ diffuses a unit u of molecular concentration, which is assumed to equal a global parameter known for all the nanomachines in the network. The other nanomachines $n(i)$ should respond to this by diffusing the same amount of molecular concentration. Then, $node_c$ accumulates the total diffused molecules and divides it by the value of u, to estimate the number of nanomachines.

Thus, during the round $k+1$, $node_c$ diffuses a general parameter u, and the other nanomachine $n(i)$ sense the diffused parameter u through Eq. 1. The steps during the round $k+1$ are described in the following algorithm.

Algorithm 1. Sensed molecular concentration after $node_c$ diffuses u

---

Input: Initial values of 1 parameters
Output: Molecular concentration sensed by each of $n(i)$
1 $t \leftarrow 0.03$; // time consumed from the round in diffusing by $node_c$
2 **while** $(c(u, d_i, T_0) \leq \tau)$ & $(t \leq T_0)$ **do**
3 $\quad c(u, d_i, T_0) = \int_{t=0.3}^{T} u \cdot \frac{1}{(4 \pi D t)^{\frac{2}{3}}} \cdot \exp(\frac{-d_i^2}{4Dt})$;
4 $\quad t = t + \delta$

During the round $k+1$, it is assumed that $node_c$ continues to diffuse unit u of molecules during the first 0.3 ms of the length of the round. Thus, in the Algorithm 1 the initial value of t is assumed to equal 0.3. During the rest of the round $k+1$, each nanomachine $n(i)$ at distance d_i attempts to sense the diffused unit u, and continues sensing as long as the amount of molecules is larger or equal to the threshold τ and the value of t is less than the length of T_0. Thus, d_i represents the distance of each specific nanomachine $n(i)$ from $node_c$.

The sensed molecular concentration $c(u, d_i, T_0)$ of each nanomachine $n(i)$ is considered to be an estimation of the diffused unite u molecules, i.e., $u_{estimate}$. Each nanomachine from $n(i)$ should diffuse its estimation $u_{estimate}$ to $node_c$, during the next round; however, $node_c$ senses the accumulative molecular concentration of all the diffused $u_{estimate}$. During the round $K+1$, $node_c$ diffused u and the other nanomachines $n(i)$ compute $u_{estimate}$. The next round is assumed to be a waiting round, so that the molecules diffuse away from the communication environment. Thus, during the round $k+3$, the other nanomachines $n(i)$ diffuse their estimations $u_{estimate}$ to $node_c$ based on the following steps:

Algorithm 2. Sensed $u_{estimate}$ by $node_c$

---

1 $t \leftarrow 0.03$;   // time consumed from the round in diffusing by $n(i)$
2 **while** $(t \leq T_0)$ **do**
3      $u_{estimate} = u_{estimate} + (c(u_{estimate}, d_i, t) - c(u_{estimate}, d_i, [t - T_0]_+))$
4      $t = t + \delta;$

Each nanomachine of $n(i)$ is assumed to diffuse its $u_{estimate}$ during the first 0.3 ms of the length of the round $k + 3$. In order to avoid the Inter Symbol Interference, round $k + 2$ is assumed as a waiting round. During the rest of $k + 3$ round length, nod_c attempts to sense the accumulative molecular concentration of all the diffused $u_{estimate}$. The value of d_i in the algorithm represents the distance of each specific $n(i)$ from $node_c$.

During the round $k + 1$, depending on the distance from $node_c$, the other nanomachines $n(i)$ might sense molecular concentration less than u. Thus, when the other nanomachines $n(i)$ diffuse their estimations $u_{estimate}$ during the round $k + 3$, $node_c$ might not receive the same amount of molecular concentration it diffused, i.e., u. Eventually, $node_c$ can't estimate the number of nanomachines.

Therefore, each nanomachine from $n(i)$ attempts to compute the effects of its distance from $node_c$, to ensure that, when it diffuses u molecular concentration, $node_c$ senses approximately u molecules concentration. Thus, before diffusing back to $node_c$, each nanomachine $n(i)$ aims to increase the amount of its diffused molecules with a specific value $\rho(i)$.

Where:

$$\frac{1}{\rho(i)} = \frac{u_{estimate}(i)}{u} \tag{8}$$

Thus, the amount of molecules that the other nanomachines would diffuse depends on its ρ. The value of ρ can vary from one nanomachine to another depending on its distance from $node_c$. This means, each nanomachine from $n(i)$ diffuses:

$$u_{n(i)} = \rho(i) \times u \tag{9}$$

i.e. $u_{n(i)}$ is approximately u when it reaches $node_c$, $\rho \times u \approx u$. Thus, during the round $k+3$, each nanomachine $n(i)$ diffuses it $u_{n(i)}$ to $node_c$, through the similar steps described in Algorithm 1.

Meanwhile, $node_c$ accumulates the sensed molecules from the diffused $u_{n(i)}$, in order to get u_{total} by the end of the round. Then, $node_c$ estimates N which equals the number of nanomachines $n(i)$, by dividing the value of the total units it has received u_{total} by the value of unit u, where: $N = \frac{u_{total}}{u}$.

4.2   Consensus Protocol Steps

After $node_c$ has estimated N, the consensus protocol steps can be initiated. The consensus protocol includes a number of steps in different time rounds. Each nanomachine $n(i)$ has an initial value, which represents an estimation of a certain parameter in the environment. In order to avoid the Inter Symbol Interference, it is assumed that there are waiting rounds. These waiting rounds are utilized so that molecules diffused a way from the communication environment. Thus, the $k + 4$ round is a waiting round. The consensus protocol steps start from round $k + 5$, can be explained as follows:

Each nanomachine from $n(i)$ diffuses its initial value to $node_c$ during the $k + 5$ round, in such a way, that $node_c$ senses almost the same diffused molecules concentration amount (i.e. the initial value of each $n(i)$). Thus, if $\zeta(i)$ is the initial value of a nanomachine from $n(i)$, this means, it should diffuse:

$$\zeta(i) \times \rho(i) \tag{10}$$

to ensure that at least $\zeta(i)$ molecular concentration reaches $node_c$.

In the meantime, $node_c$ follows the steps described in Algorithm 2 in order to sense the accumulative molecular concentration of initial values from each $n(i)$ plus the jamming molecular concentration from the adversary nanomachine $node_A$. Each nanomachine $n(i)$ keeps diffusing its initial value to nod_c through k time slots, and $node_c$ keeps storing the sensed molecular concentration in its vector. However, these k time slots are not consecutive, as a nanomachine $n(i)$ diffuses its initial value during a time slot, then waits for at least T_0, before diffusing its initial value again, where, $node_c$ does not store the sensed molecular concentration during the waiting time slot as it is mainly jamming units. Thus, the sensed molecular concentration by $node_c$ during k time slots represents R_c:

$$R_c = (\sum_{i=0}^{n} \zeta(i) + u_A 1) + (\sum_{i=0}^{n} \zeta(i) + u_A 2) + \cdots + (\sum_{i=0}^{n} \zeta(i) + u_A k) \tag{11}$$

$$R_c = \sum_{j=1}^{k} (\sum_{i=0}^{n} (\zeta(i) + u_A j)) \tag{12}$$

where $(\zeta(i))$ is the initial value of nanomachine $n(i)$, and u_A is the amount of the jamming molecules from $node_A$. After waiting for $2k$ time slots, and sensing molecular concentration during one of these k time slots, nod_c estimates the initial values from the total stored molecular concentration in its vector that has been computed in R_c.

$$InitR_c = \frac{R_c - \sum_{i=1}^{k} u_A i}{k} \tag{13}$$

i.e.

$$InitR_c \approx \begin{cases} \frac{R_c - (k\lambda + \varepsilon)}{k}, & \text{if } \varepsilon \leq \frac{1}{k} \\ \frac{R_c - (k\lambda - \varepsilon)}{k}, & \text{if } \varepsilon \geq 1 - \frac{1}{k^2} \end{cases} \tag{14}$$

Then, during $3k + 6$ slot, $node_c$ computes the average of the estimated initial values $InitR_c$, assuming that $node_c$ knows N the number of nanomachines in the network based on the explained process in Subsect. 4.1. Thus, the average of the estimated initial values can be:

$$Init_{av} R_c = \frac{InitR_c}{N} \tag{15}$$

During the next time slot, $node_c$ diffuses $Init_{av}R_c$ to the other nanomachines, through the same steps described in Algorithm 1.

In the meantime, the other nanomachines attempt to sense the molecules diffused by $node_c$ (through the steps described in Algorithm 2), in order to recognize the average of all initial values. As explained in Subsect. 4.1, each nanomachine from $n(i)$, identified the effects of distance on the received concentration from $node_c$. Thus, each nanomachine $n(i)$ counts on its $\rho(i)$ to distinguish the right value of $Init_{av}R_c$. However, besides $\rho(i)$, here each nanomachine from $n(i)$ takes into consideration the jamming molecular concentration and the estimated value of λ, as shown in the following:

$$Init_{av}R_c \approx \begin{cases} \rho(i) \times Init_{av}R_c(i)_{received} - (\lambda + \varepsilon), & \text{if } \varepsilon \leq \frac{1}{k} \\ \rho(i) \times Init_{av}R_c(i)_{received} - (\lambda - \varepsilon), & \text{if } \varepsilon \geq 1 - \frac{1}{k^2} \end{cases} \tag{16}$$

where, $Init_{av}R_c(i)_{Received}$ is the sensed (received) molecular concentration by a nanomachine $n(i)$ when $node_c$ diffuses the estimated average of initial values.

Reaching Consensus: The model assumes that consensus is reached when the average of the initial values $Init_{av}R_c(i)$ approaches to the value of u with ϵ deviation, i.e., the deviation of average of the initial values from the assumed global parameter u is:

$$|u - Init_{av}R_c(i)| \leqslant \epsilon \tag{17}$$

where, $\epsilon > 0$. As the initial values of each nanomachine from $n(i)$ are assumed to be relatively close to the value of u, in order to increase the possibility to reach an agreement among the networks' nanomachines.

Time to Reach Consensus: In order to compute the time needed to compute the average of all initial values $Init_{av}R_c$, the time needed to estimate λ and the time required for the consensus processes, should be taken into consideration. Thus, the first k time slots needed to estimate λ are multiplied by the length of the time slot, i.e., T_0. Besides this, the next $2k$ time slots needed for diffusing the initial values of nanomachines $n(i)$ with the two time slots required to compute the average of the initial values and then diffusing it by $node_c$, are also multiplied by T_0. In addition to the waiting time slots and the needed slots to compute N the number of $n(i)$ by $node_c$.

$$t_{Init_{av}R_c} = (T_0 \times k) + (T_0 \times (2k + 2)) + (T_0 \times 5) \tag{18}$$

5   Conclusions

In this work, consensus problem in diffusion based network is explored with the existence of an adversary nanomachine. The adversary nanomachine follows Poisson random distribution in diffusing molecules to jam the communication among the network's n nanomachines. The network's nanomachines attempt to estimate the jamming molecular concentration, through sensing molecular concentration for k time slots and storing this value in a vector. Then, the network nanomachines compute the average of the sensed jamming molecular concentration during k time slots, and take in consideration estimation error ε. The consensus processes start after estimating the jamming molecular concentration. The process of computing the number of nanomachines in the network is demonstrated. Besides that, the needed time to reach consensus in the model is computed.

References

1. Abadal, S., Akyildiz, I.F.: Bio-inspired synchronization for nanocommunication networks. In: 2011 IEEE Global Telecommunications Conference (GLOBECOM 2011), pp. 1–5. IEEE (2011)
2. Akyildiz, I.F., Brunetti, F., Blázquez, C.: Nanonetworks: a new communication paradigm. Comput. Networks **52**(12), 2260–2279 (2008)
3. Chang, A.: Statistics toolkit: Poisson distribution: explained. https://www.statstodo.com/Poisson_Exp.php. Accessed 09 Jan 2017
4. Dressler, F., Kargl, F.: Towards security in nano-communication: challenges and opportunities. Nano Commun. Networks **3**(3), 151–160 (2012)
5. Einolghozati, A., Sardari, M., Beirami, A., Fekri, F.: Consensus problem under diffusion-based molecular communication. In: 2011 45th Annual Conference on Information Sciences and Systems (CISS), pp. 1–6. IEEE (2011)
6. Einolghozati, A., Sardari, M., Beirami, A., Fekri, F.: Data gathering in networks of bacteria colonies: Collective sensing and relaying using molecular communication. In: 2012 IEEE Conference on Computer Communications Workshops (INFOCOM WKSHPS), pp. 256–261. IEEE (2012)
7. Einolghozati, A., Sardari, M., Fekri, F.: Networks of bacteria colonies: a new framework for reliable molecular communication networking. Nano Commun. Networks **7**, 17–26 (2016)
8. Gourieroux, C., Monfort, A., Trognon, A.: Pseudo maximum likelihood methods: applications to poisson models. Econometrica J. Econometric Soc. **52**, 701–720 (1984)
9. Hammer, B.K., Bassler, B.L.: Quorum sensing controls biofilm formation in vibrio cholerae. Mol. Microbiol. **50**(1), 101–104 (2003)
10. Loscri, V., Marchal, C., Mitton, N., Fortino, G., Vasilakos, A.V.: Security and privacy in molecular communication and networking: opportunities and challenges. IEEE Trans. Nanobiosci. **13**(3), 198–207 (2014)
11. Meng, L.S., Yeh, P.C., Chen, K.C., Akyildiz, I.F.: Mimo communications based on molecular diffusion. In: 2012 IEEE Global Communications Conference (GLOBECOM), pp. 5380–5385. IEEE (2012)
12. Mitzenmacher, M., Upfal, E.: Probability and Computing: Randomization and Probabilistic Techniques in Algorithms and Data Analysis. Cambridge University Press, Cambridge (2017)
13. Nakano, T., Eckford, A.W., Haraguchi, T.: Molecular Communication. Cambridge University Press, Cambridge (2013)
14. Ng, W.L., Bassler, B.L.: Bacterial quorum-sensing network architectures. Annu. Rev. Genet. **43**, 197–222 (2009)

15. Pierobon, M., Akyildiz, I.F.: Information capacity of diffusion-based molecular communication in nanonetworks. In: 2011 Proceedings IEEE INFOCOM, pp. 506–510. IEEE (2011)
16. Vassiliou, V.: Security issues in nanoscale communication networks. In: 3rd NaNoNetworking Summit, pp. 1–53. Network Research Laboratory (2011)

Flow Monitoring System for IoT Networks

Leonel Santos[1,2,3]($\boxtimes$), Carlos Rabadão[1,2], and Ramiro Gonçalves[3,4]

[1] School of Technology and Management, Polytechnic Institute of Leiria,
Leiria, Portugal
leonel.santos@ipleiria.pt
[2] Computer Science and Communication Research Centre, Leiria, Portugal
[3] Universidade de Trás-os-Montes e Alto Douro, Vila Real, Portugal
[4] INESC TEC (Formerly INESC Porto), Porto, Portugal

Abstract. The big number of Internet of Things (IoT) devices, the lack of interoperability and the low accessibility of many of them in a vast heterogenous landscape will make it very hard to design specific monitor, manage and security measures and apply specific mechanism to IoT networks. Administration tasks like reporting, performance analysis, and anomaly detection also depend on monitoring for decision making. For that purpose, a solution used in IoT networks must be scalable and interoperable.

In this work, we are concerned with the design of a real time monitoring system for IoT networks. To do this, after studying the various traditional network monitoring solutions, we concluded that there are still several developments to be made to this type of mechanism.

The design proposed will consider the specific architecture of an IoT network, the scalability and heterogeneity of this type of environment, and the minimization of the use of resources. To do so, we considered the various network monitoring methods available and select a flow monitoring solution in an IoT network. After the presentation of a workflow for flow monitoring on IoT networks, the workflow was tested. By doing analysis of flows, rather than packets, we concluded that this type of solution could be more scalable and interoperable than traditional packet-based network monitoring, make it suitable in an IoT environment.

Keywords: Internet of Things · Network monitoring · Flow monitoring · IPFIX

1 Introduction

With the rapid growth of the number of devices being connected to the internet, network administration responsibilities like devices management, traffic monitoring and security is a concern that must be on everyone's mind. According to various reports, by 2020 there will be fifty billion devices connected to the internet, and each person will own around seven devices.

Network traffic monitoring and analyses represents a key component for network administration as it allows the development of several types of mechanisms, such as flow analysis and, threats and anomalies detection, and performance monitoring.

© Springer Nature Switzerland AG 2019
Á. Rocha et al. (Eds.): WorldCIST'19 2019, AISC 931, pp. 420–430, 2019.
https://doi.org/10.1007/978-3-030-16184-2_40

Network traffic monitoring approaches have been proposed and developed throughout the years. They can be classified into two categories: active and passive. Active approaches, such as implemented by tools like Ping, Traceroute, SNMP, and NETCONF, inject traffic into a network to perform different types of measurements and to perform analysis. Passive approaches observe existing traffic as it passes by a measurement point and therefore observe and collect traffic generated by users and systems for being analyzed.

Packet capture and flow export are common passive monitoring approaches. In the first, complete packets are captured providing deep insight into the traffic. This approach requires expensive hardware and infrastructure for storage and analysis.

Flow export aggregate packet into flows and export them to a collector for storage and analysis. In our work, we follow the revised definition of flow proposed by [1]. His revised definition is based on the definition published in [2] as following:

A flow is defined as a sequence of packets passing an observation point in the network during a certain time interval. All packets that belong to a particular flow have a set of common properties derived from the data contained in the packet, previous packets of the same flow, and from the packet treatment at the observation point.

If used in high-speed networks, this approach is more scalable and less costly than packet capture due to the integration of flow export protocols into network devices, such as routers, switches and firewalls. Other advantages are: the reduction of the amount of data stored, the possible use for forensic investigation and the achievement of the privacy of the traffic data captured since traditionally only packet headers are considered.

As known, IoT is used for applications in different domains such as home automation, industrial process, human health and environmental monitoring and the privacy preserving of data collected is a constant concern. IoT networks and devices are resource constrained, with very restrictions in computational capability and power consumption [3]. Therefore, the use of a network traffic monitoring passive approach like flow export must be considered and analyzed.

Considering the above mentioned, the research team decided to undergo an analysis on the feasibility of application of a flow-based network traffic monitoring system on IoT networks. As result of our work, we present a workflow for flow-based network traffic monitoring on IoT networks, along with its prototype implementation.

The rest of this paper is organized as follows. Section 2 details the basic notions of network monitoring and flow-based network traffic monitoring systems, providing a revision of relevant terms and applications. Section 3 presents the proposed workflow in detail and, at the same time, explains how technologies, approaches, and protocols can be used to implement it. In Sect. 4 presents a proof-of-concept implementation in a testbed using IPFIX protocol and other analysis tools, with the aim of validate the workflow's ability to monitor network traffic on an IoT network environment. Finally, in Sect. 5, we present a brief set of conclusions complemented with a discussion of open issues and future work considerations.

2 Background and Related Work

As the goal of this work is to analyze the viability of using a flow monitoring system on IoT networks, this section briefly reviews some basic notions of flow network monitoring.

2.1 Flow Monitoring Architecture

The typical architecture of flow monitoring setups consists of several steps [5], each of which is explained bellow and are represented in Fig. 1 [6].

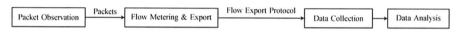

Fig. 1. Architecture of a typical flow monitoring setup [6].

2.2 Packet Observation

Packet observation is the process of capturing packets from the line and pre-processing them for further use.

Packet observation involves capturing packets from an observation point that is part of an observation domain. This step **architecture can be resumed in five stages**: Packet Capture, Timestamping, Truncation, Packet sampling, and Packet filtering.

The **installation of packet capture devices** can be positioned in-line and in mirroring mode, which may have a significant impact on capture and network operation. In terms of **packet capture technologies**, there are libraries available to be used in most of operating systems (OS), such as *libpcap* [7] and *libtrace* [8]. Several methods have been proposed to speed up this process [9], as following: Interrupt mitigation and packet throttling (Linux NAPI), Network stack bypass techniques, such as PF_RING, Memory-map techniques.

Packet sampling and filtering strategies aim to reduce the number of packets that are being sent to the Flow Metering and Export stage thus reducing the consumption of bandwidth, memory and computational power. Zseby et al. [10] defines several sampling strategies. The role of packet filtering is to separate the packets that have a specific characteristic from those who don't [6].

2.3 Flow Metering and Export

The Flow Metering & Export stage is where packets are aggregated into flows and flow records are exported. Its architecture is composed by 3 steps, as follows: Metering Process; optional Flow sampling and filtering; and Exporting Process.

Flow records are defined in [2] as "information about a specific flow that was observed at an observation point", which include flow keys, such as characteristic properties of a flow, e.g., IP addresses and port numbers, and measured properties, e.g., packet and byte counters.

The packet aggregation is performed within the **Metering Process**, based on a list of Information Elements (IEs) [11] that define the layout of a flow that can be considered unidirectional or bidirectional [12]. In terms of IEs, flow collectors are always informed by flow exporters by means of templates, which are used to define which IEs are used for every flow. After aggregation, an entry per flow is stored in a flow cache, until a flow is considered to have terminated and the entry is expired. To set a new flow cache it should be considered features such the layout, type [13] and size.

After the Metering Process, one or more optional **flow-based sampling and filtering functions** could be applied to select a subset of flow records.

The last step is the **Exporting Process**. This step consists in encapsulation of selected flow records in IPFIX messages and transport them to Flow Collector using a selected transport protocol [2] like TCP (Transmission Control Protocol), UDP (User Datagram Protocol) or SCTP (Stream Control Transmission Protocol).

When selecting a flow exporter for implementation, it is important to verify the following criteria: Throughput, Flow cache size, Supported IEs, and Application awareness. There are various open-source flow exporters solutions that support IPFIX, such as YAF [14], nProbe [15], QoF [16], and Vermont [17]. Besides the open-source flow exporters listed below, there are also commercial solutions from various vendors, such as Cisco, Juniper, nTopng, and Plixer Scrutinizer.

2.4   Data Collection

Flow collectors are an important step of flow monitoring setups, as they receive, store, and pre-process flow data from one or more flow exporters.

The **flow data storage format** is an important characteristic of flow data collecting step because it defines the performance and functionality level of the flow collectors. Flow data storage can be stored in two different formats [6]: *Volatile* and *Persistent*.

Regarding the persistent storage, the different solutions can be distinguished between the following types: Flat files, Row-oriented databases, and Column-oriented databases.

When selecting a flow collector for deployment, it is important to verify the following criteria: Performance, Storage format, Export protocol features, Processing delay, Flow record deduplication, and Integration with other systems.

There is various open-source flow collector solutions that support IPFIX, such as IPFIXCOL [18], nProbe [15], SiLK [19], and Vermont [17]. Besides the open-source flow collectors listed below, there are also commercial solutions from various vendors, such as Cisco Stealthwatch, nTopng, and Plixer Scrutinizer.

2.5   Data Analysis

Data Analysis is the final step in a flow monitoring setup. There are three main areas where analyses of flow data can be applied [6]: Flow analysis & reporting, Threat detection, and Performance monitoring.

The applicability statement of IPFIX issued by the IETF in [20] and the survey on network flow applications provided in [21] should also be considered for exploring more applications examples of flow data analysis.

Flow analysis & reporting is the most basic functionality provided by flow analysis applications and offers possibility for diverse actions: browsing and filtering flow data; statistics overview; and reporting and alerting. As examples of applications that provide this functionality are NfSen [22], nTopng [23] and SiLK [19]. Another works in this field could be mentioned, in [24] the authors use flow data statistics to characterize and categorize network traffic on five different networks of Masaryk University. In [25] the authors also propose an IP flow analysis solution to analyze large volume of IP flows to report scalability issues.

When flow data is used for **threat detection**, it can be used for analyzing which host has communicated with which each other, potentially including summaries of the number of packet and bytes involved, the number of connections, etc. Many techniques have been proposed in recent years, as surveyed in [5, 26].

Performance monitoring is used to observe the status of the services running in the network in order to verify if the SLA (Service Level Agreement) is being fulfilled and see how the end-user experience is going. Metrics that are being observed include the response time, the delay, the jitter, and the bandwidth usage.

3   Proposed Workflow for Flow Monitoring on IoT Networks

A proposed workflow for monitoring flows in IoT networks is presented to verify the feasibility of using a solution of this kind to monitor all the network traffic that exists in a type of networks so specific and special as the IoT networks, as shown in Fig. 2.

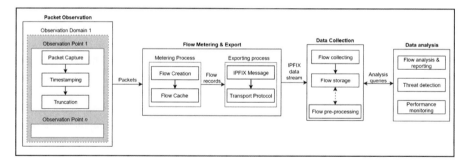

Fig. 2.   Workflow for flow monitoring on IoT networks

As the traditional workflow of flow monitoring, this workflow consists in four main steps, namely: packet observation, flow metering & export, data collection, and data analysis.

3.1   Packet Observation

In this step our proposal is to define one or more domains of observation. These observation domains can be organized by application domain of IoT networks, or could be through communication technology, i.e. an observation domain for the Wi-Fi network (802.11) and for Bluetooth Low Energy (BLE).

In each observation domain there could be one or more observation points. These observation points, also known as *probes*, can have two types of installation: *mirroring mode* and *in-line mode* and will have to be installed in order to capture the network packets at strategic points in the IoT network.

For reasons of performance of the capture process, this should be preferably done it using the PF_RING library.

In order to ensure the temporal synchronization of the devices, the observation point should ensure that it keeps the system date and time synchronized with other devices for a correct timestamping using the Network Time Protocol (NTP) service.

3.2   Flow Metering and Export

In this step, the packets captured in the previous step will be aggregated into flow records. The most important IEs are the usual known as 5-tuple fields of transport headers, but other IEs should be selected in order to increase the amount of information about each flow to achieve a better data analysis.

Our proposal is to use bidirectional flows, since it is possible to obtain data on the reverse direction. As an example, we propose that the selection of the IEs for the start and end timestamps of the flow, source and destination IP and MAC address, source and destination ports, total bytes, total packets, TCP statistics (example: TCP flags, sequence numbers, inter arrival times, etc.), and flow end reason.

This selection of IEs defines the layout of the flow cache in the flow exporter that will be used. In terms of type, the flow cache must be of the permanent cache type.

These flow records will be integrated into IPFIX messages that will be exported using the TCP as transport protocol over TLS in order to ensure the confidentiality.

3.3   Data Collection

For this step, we propose that the flow collector should be scalable and placed on a Cloud Service, regarding the computational limitations of node devices and border router utilized in IoT networks.

The flow collector should develop the collection of flow data using TCP over TLS.

The collected data flow must be stored in a persistent type storage.

Regarding the type of persistent storage, we propose the use of flat files storage due to its optimized use of disk space and data insertion performance, despite some limitations regarding query flexibility and query performance.

3.4   Data Analysis

In this step, our proposal is to use applications with the ability to query the flat file-based database created and maintained in the data collector stage. The applications used need to be capable to query flow data to, for example, make available reports and analysis about the traffic of the IoT networks. Another important feature is flow data analysis to detect threats or intrusions in IoT networks.

4   Validation and Discussion

In this section, we describe the validation of this work, starting by describing the validation setup, as well as specifics regarding the deployment. Finally, we discuss the results.

4.1   Setup and Deployment

The workflow described in Sect. 3 has been validated using an IoT network scenario as show in Fig. 3.

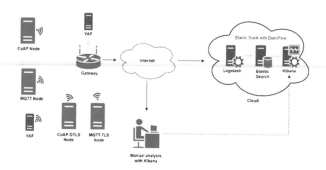

Fig. 3. IoT network prototype diagram

The IoT network scenario was configured using Wi-Fi (802.11n) and ethernet links. For the gateway, that act also as wireless access point, we used a Raspberry Pi 3 Model B+ platform with two network interfaces. Ethernet interface acting as WAN interface connected to the Internet, and the wireless interface acting as a wi-fi access point, allowing connection of the IoT nodes to the network and the Internet. The gateway has the latest Raspbian operating system (OS) installed. The address used was IPv4 and a DHCP service is configured at the gateway.

All IoT nodes are Raspberry Pi 3 Model B+ and have installed Raspbian OS. We configured most of these nodes with protocols typically used in IoT networks such as Message Queuing Telemetry Transport (MQTT) and Constrained Application Protocol (CoAP), with and without security [4].

On the IoT nodes that act as MQTT servers or clients, the *Eclipse Mosquitto* [27] package is installed and configured. In the IoT nodes that act as CoAP servers and clients, the *libcoap* [28] and *coapthon* [29] packages are installed and configured. The objective of the use of these types of protocols in the IoT nodes is to simulate traffic typical of IoT networks, namely simulate the sensing and the actuation.

In terms of deployment, there is only one observation domain with two observation points. One of these observation points is installed in *mirroring mode* and is configured in the gateway, in order to capture the incoming and outcoming traffic to the IoT network. The other observation point is installed *in-line mode* and is configured in an IoT node, in order to capture internal packets of the IoT network.

In both observation points, the Yet Another Flowmeter (YAF) [14].

Considering the limitations of the devices usually used in IoT networks, the flow collector must be installed on a host located in a Cloud Service. In our prototype, the host used has an Intel quad-core processor, 8 GB RAM, 500 GB HDD, a gigabit ethernet port, and Ubuntu Server 18.04 LTS as OS.

For flow collecting and analysis solution, we selected the Elasticstack (ELK) stack from Elasticsearch [30]. The ELK stack consists of three projects:

- Elasticsearch - a RESTful search and analytics engine.
- Logstash - processing pipeline that ingests data from sources, transforms it and sends it to Elasticsearch.
- Kibana - lets you visualize Elasticsearch data and navigate through it.

As a flow collector we use Logstach with an IPFIX decoder. After decoding IPFIX messages, it will send the flow data decoded to the Elasticsearch to be stored in documents. Elasticsearch use indexes based on the IEs used in the flow data collected.

Kibana will be used as a flow analysis application, because it provides a graphical interface for visualizing statistics of the flows stored in the Elasticsearch through dashboards. The dashboards use queries that can be customized. In order to improve the use of the ELK stack, we use the Elastiflow [31] project that provides a specific IPFIX Flow index for Elasticsearch. In addition, it provides specific dashboards for visualization statistics and flow monitoring to add to Kibana.

4.2   Results

The setup presented in the previous section demonstrate the implementation and setup of a flow monitoring system on an IoT network.

As can be seen in the following Figs. 4, 5 and 6, it is possible to develop flow analysis of an IoT network traffic, as well as analysis of the application protocols that are usually used in this type of networks. With the use of Elastiflow dashboards, we have been able to improve the results presented by the queries made to the flow storage.

Fig. 4. Elastiflow - top-services monitored

Based on the information gathered from this setup, the fields of the flows existing in the flow storage can be used to produce mechanisms to detect intrusions and threats, as well as to evaluate the performance of the application protocols used in IoT networks, for example MQTT and CoAP.

Fig. 5. Elastiflow – CoAP over DTLS records

Fig. 6. Elastiflow – MQTT over TLS records

In terms of scalability and interoperability, the use of the IPFIX protocol to build and export the network flows assure the interoperability between devices and technologies used in the IoT networks. The placement in a Cloud environment of the data collect and analysis solution guarantee the scalability of this solution, since it allows the connection of several flow exporters and can support the integration of new ones.

5   Conclusions

The goal of this work was to develop an analysis on the feasibility of application of a flow-based network traffic monitoring system on IoT networks.

We start to present some background and related work existing in this subject with the objective to introduce some awareness around the difference between the usual passive packet monitoring systems and the flow monitoring systems.

Then, we introduce and propose a workflow for flow-based network traffic monitoring on IoT networks and present all the important requirements and characteristics that should be considered in the implementation of a flow monitoring system on a network so specific like IoT networks are.

Yes, it is feasible to setup a scalable and interoperable flow monitoring system to monitor and analyze the traffic on an IoT network, which has been proven by the setup and deployment of a prototype on our laboratory.

Our results show that the proposed workflow and the requirements presented could be an important output to considerer in the implementation of a flow monitoring system, so it could lead to an efficient system.

As result of the analysis that could be visualized in the reports, we could see various kinds of flow statistics such as the IP sources and destinations addresses, the flow end reason, the total number of packets and bytes of a flow, top-talkers, etc.

Future works includes investigating flow analysis techniques that cloud be applied in intrusion and threat detection in IoT networks.

Acknowledgements. This work was supported by Portuguese national funds through the FCT - Foundation for Science and Technology, I.P., under the project UID/CEC/04524/2019.

References

1. Velan, P.: Improving network flow definition: formalization and applicability. In: NOMS 2018 - 2018 IEEE/IFIP Network Operations and Management Symposium, Taipei, pp. 1–5 (2018)
2. Claise, B., Trammell, B., Aitken, P.: Specification of the IP flow information export (IPFIX) protocol for the exchange of flow information. In: RFC 7011 (Internet Standard), pp. 1–76, September 2013
3. Santos, L., Rabadao, C., Gonçalves, R.: Intrusion detection systems in Internet of Things: a literature review. In: 13th Iberian Conference on Information Systems and Technologies (CISTI), pp. 1–7 (2018)
4. Al-Fuqaha, A., Guizani, M., Mohammadi, M., Aledhari, M., Ayyash, M.: Internet of things: a survey on enabling technologies, protocols, and applications. IEEE Commun. Surv. Tutorials **17**(4), 2347–2376 (2015)
5. Sperotto, A., Schaffrath, G., Sadre, R., Morariu, C., Pras, A., Stiller, B.: An overview of IP flow-based intrusion detection. IEEE Commun. Surv. Tutorials **12**(3), 343–356 (2010). Third Quarter
6. Hofstede, R., Čeleda, P., Trammell, B., Drago, I., Sadre, R., Sperotto, A., Pras, A.: Flow monitoring explained: from packet capture to data analysis with NetFlow and IPFIX. Commun. Surv. Tutorials IEEE **PP**(99), 2037–2064 (2014)
7. Jacobson, V., Leres, C., McCanne, S.: libpcap, Lawrence Berkeley Laboratory, Berkeley, CA. Initial public release June (1994)
8. Alcock, S., Lorier, P., Nelson, R.: Libtrace: a packet capture and analysis library. SIGCOMM Comput. Commun. Rev. **42**(2), 42–48 (2012)
9. Braun, L., Didebulidze, A., Kammenhuber, N., Carle, G.: Comparing and improving current packet capturing solutions based on commodity hardware. In: Proceedings of 10th ACM SIGCOMM IMC, pp. 206–217 (2010)
10. Zseby, T., Molina, M., Duffield, N., Niccolini, S., Raspall, F.: rfc5475 @ Tools.Ietf.Org, March 2009. https://tools.ietf.org/html/rfc5475
11. IEs @ www.iana.org (2007). https://www.iana.org/assignments/ipfix/ipfix.xhtml. Accessed 04 Nov 2018
12. Trammell, B., Boschi, E.: Bidirectional flow export using IP Flow Information Export (IPFIX), RFC 5103 (Standards Track) Internet Engineering Task Force, January 2008
13. Sadasivan, G., Brownlee, N., Claise, B., Quittek, J.: Architecture for IP Flow Information Export, RFC 5470 (Informational) Internet Engineering Task Force, March 2009
14. Inacio, C.M., Trammell, B.: YAF: yet another flowmeter. In: Proceedings of 24th International Conference on LISA, pp. 1–16 (2010)
15. Deri, L.: nProbe: an open source NetFlow probe for gigabit networks. In: Proceedings of TNC, pp. 1–4 (2003)
16. QoF. https://www.ict-mplane.eu/public/qof. Accessed 11 Nov 2018
17. Lampert, R.T., Sommer, C., Munz, G., Dressler, F.: Vermont - a versatile monitoring toolkit for IPFIX and PSAMP. In: Proceedings of IEEE/IST Workshop MonAM, pp. 1–4 (2006)

18. Velan, P., Krejčí, R.: Flow information storage assessment using IPFIXcol. In: Proceedings of 6th International Conference AIMS, vol. 7279, pp. 155–158 (2012)
19. Gates, C., Collins, M., Duggan, M., Kompanek, A., Thomas, M.: More NetFlow tools for performance and security. In: Proceedings 18th International Conference on LISA, pp. 121–132 (2004)
20. Zseby, T., Boschi, E., Brownlee, N., Claise, B.: IP Flow Information Export (IPFIX) applicability. RFC 5472 (Informational) Internet Engineering Task Force, March 2009
21. Li, B., Springer, J., Bebis, G., Gunes, M.H.: A survey of network flow applications. J. Netw. Comput. Appl. **36**(2), 567–581 (2013)
22. Haag, P.: Watch your flows with NfSen and NFDUMP. In: Proceedings of 50th RIPE Meeting, pp. 1–32 (2005)
23. Deri, L., Suin, S.: Ntop: beyond ping and traceroute. In: Proceedings of 10th IFIP/IEEE International Workshop DSOM, vol. 1700, pp. 271–283 (1999)
24. Velan, P., Medková, J., Jirsík, T., Čeleda, P.: Network traffic characterisation using flow-based statistics. In: Network Operations and Management Symposium (NOMS), 2016 IEEE/IFIP, pp. 907–912. IEEE (2016)
25. Jirsik, T., Cermak, M., Tovarnak, D., Celeda, P.: Toward stream-based IP flow analysis. IEEE Commun. Mag. **55**(7), 70–76 (2017)
26. Umer, M.F., Sher, M., Bi, Y.: Flow-based intrusion detection: techniques and challenges. Comput. Secur. **70**, 238–254 (2017)
27. Mosquitto. https://projects.eclipse.org/projects/technology.mosquitto. Accessed 11 Nov 2018
28. LibCoAP. https://libcoap.net/. Accessed 11 Nov 2018
29. Tanganelli, G., Vallati, C., Mingozzi, E.: CoAPthon: easy development of CoAP-based IoT applications with Python. In: 2015 IEEE 2nd World Forum on Internet of Things (WF-IoT), pp. 63–68. IEEE (2015)
30. Elasticstack. https://www.elastic.co/elk-stack. Accessed 11 Nov 2018
31. ElastiFlow Project. https://github.com/robcowart/elastiflow. Accessed 11 Nov 2018

The User's Attitude and Security of Personal Information Depending on the Category of IoT

Basma Taieb[1]([✉]) [iD] and Jean-Éric Pelet[2] [iD]

[1] Léonard de Vinci Pôle Universitaire,
Research Center, Paris-La Défense, France
basma.taieb@devinci.fr
[2] ESCE International Business School,
10 rue Sextius Michel, 75015 Paris, France

Abstract. Internet of Things (IoT) is already part of our everyday life and is expected to grow considerably in the coming years. Such an invasion is not without risk, in particular on the privacy level. This study highlights the issue related to the security of personal information as perceived by users. Through an exploratory study conducted with 14 interviewees followed by a confirmatory study with 1006 people, we strive to look for needs and concerns of consumers about the three different categories of IoT. Results reveal that users are willing to use IoT for daily tasks, in particular, smart home devices that they consider relevant for safety purposes or to save the energy of their house and finally to manage their time in a better manner. They are willing to provide their personal information when they feel that they control these smart devices. Hence, the user's attitude and the security of perceived personal information vary according to the category of IoT. Results also show significant differences between men and women regarding the category of IoT they would prefer to buy as well as the main criteria they consider when purchasing smart devices.

Keywords: Internet of Things · User's attitude ·
Security of personal information

1 Introduction

In recent years, IoT (Internet of Things) and wearable devices are rapidly gaining ground in various landscapes such as healthcare industry, smart homes, smart community and smart city. They bring different kinds of devices into the market which collect, process and distribute data such as sport watches with GPS for fitness aids, security cameras with face recognition technology, vocal assistant for cars and smart homes. Users of smart home devices exceed 1 billion in 2017 and 30 to 212 billion smart devices could be sold by 2020 according to Gartner (2015)[1]. The real stake for businesses is to anticipate tomorrow's new trends and indispensable smart objects in order to foresee future interactions consumers can have with IoT (Hoffman and Novak 2017). For example, 'Bosch Smart Home', launched in June 2017, is present in four

---

[1] Gartner, Cisco, Idate ou IDC, 2015.

© Springer Nature Switzerland AG 2019
Á. Rocha et al. (Eds.): WorldCIST'19 2019, AISC 931, pp. 431–437, 2019.
https://doi.org/10.1007/978-3-030-16184-2_41

lines of business: mobility solutions, industrial technology, consumer goods, energy and accommodation where interactions with customers are expected to increase (Acas 2017).

With the development of smart devices, privacy assumes a new importance. The new technologies allow companies an unprecedented intrusion into privacy, because of the transparency of these IoT. The American Federal Trade Commission (US FTC) has also listed the practices linked to the respect of privacy, including informing, upstream data collection; giving individuals control over the type of information they must provide, giving them the choice whether to allow the company to share the collected information; guaranteeing that the information is secure.

This study aims to identify in which extent consumers are concerned about IoT used for daily tasks by highlighting the security of personal information as perceived by users with respect to the category of IoT.

2   Security of Personal Information and User's Attitude Towards IoT

The IoT is defined as *"a global infrastructure for the Information Society, enabling advanced services by interconnecting (physical and virtual) things based on existing and evolving, interoperable information and communication technologies"* (International Telecommunication Union 2012). IoT is divided into three main categories: smart home device (smart water meter, smart thermostat …), leisure/sports device (smart shoes, smart clothing, smart tennis racket …) and health/wellness tracking device (watch, health tracker, body scale…) (ref).

By accessing houses, IoT can provide companies with silos of information in relation to the everyday usage consumers make of the later, through big data obtained with these "smart" objects. Likewise, if we refer to the 'smart home device' category, consumers choose their preferences via the "Smart Home app", and the controller receives them to transfer the information to the different devices, on areas such as indoor climate, security and lighting. Such a system provides intelligent regulation of heating for each room, to obtain a perfect temperature, depending on the time of the day, consumers' preferences and ambient conditions. Simultaneously, the "intelligent socket" is able to switch on and off automatically and thereby, save energy, which is relevant for safety purposes and security in general. Thanks to these connected devices, the socket is able to turn on by itself, 10 min before you wake up so that the coffee machine is ready to use. An intelligent and autonomous security system offers a surveillance service to protect homes. Thereby, the "360° Indoor Camera" lets consumers have an eye on their homes at any time. It can be used anywhere around the world to check what is happening in any room whenever consumers want. Finally, the "lighting solution" provides the perfect lighting in relation to the hour of the day. Many other IoT exist, that furnish a great variety of information enabling companies to foresee usage and create new needs.

IoT can bring many benefits to users. They can also harm users' privacy without their notice, in particular wearable devices such as smart clothes, smart motorbikes helmets, smart bands and eye glasses (Morris 2015). Concerns related to how information is collected and transferred through the Internet prevail. Users might doubt the trustworthiness of wearable devices. For example, they are likely to worry about their data being shared with other parties silently. Users with high privacy concerns will have doubts regarding the use of the new technology. According to Yildirim and Ali-Eldin (2018), individuals who believe that it would be risky to use a smart device have high privacy concerns regarding collection, errors, secondary use and improper access. Individuals who consider smart devices risky or unsafe have a low intention to use them, and as the risk increases, the intention to use smart devices decreases. Using IoT and smart devices in different contexts such as healthcare can raise concerns on users' privacy (Yiwen et al. 2015). Therefore, we propose the following hypotheses:

H1. User's attitude varies according to the category of IoT
H2. Security of personal information as perceived by user varies according to the category of IoT

Previous research has found that consumers' reactions to IoT are paradoxical (Rijsdijk and Hultink 2009; Bonnin et al. 2014; Ardelet et al. 2017). Even if consumers perceive the main benefits of connected objects such as saving time and the ease of accomplishing daily tasks, nearly a third of buyers of IoT abandon their connected object within six months of purchase (Bonnin et al. 2014). They describe these objects as useless complex gadgets and they are afraid of being dependent (Rijsdijk and Hultink 2009). Therefore, it seems important to identify the relevant criteria when purchasing an IoT as well as the category of IoT that consumers prefer to buy. For this purpose, an exploratory research has been conducted followed by a confirmatory research.

3   Exploratory Study

14 individual interviews have included seven women, users and non-users as well as buyers and non-buyers of IoT and smart devices. Respondents were asked about the criteria they consider important when purchasing an IoT and the category of smart devices they would prefer to buy. They have argued that quality of the product is a key purchasing factor for IoT and smart devices. Moreover, the design of the device is an important criterion mainly for women when purchasing smart home devices. Nowadays, open kitchens are becoming more and more trendy in houses. They create a nice space for cooking and sharing moments. Household equipment needs to fit in the kitchen and becomes design objects.

Notwithstanding the price which is important but it doesn't appear as the first determining criterion for most of users. Thereby, people are willing to pay more for useful IoT, for example, a product that helps to lose weight, or enables to save time and save money (e.g. intelligent thermostat, etc.). The study shows interesting findings when coming to data security and privacy that should be considered as important factors when purchasing IoT. When being burgled, we don't have the feeling of having things stolen only, but also that someone come inside our lives. This creates an unpleasant feeling. Thus, when individuals adopt IoT, they need to have the certainty to control them. They need to be able to deactivate their objects whenever they want or need it. Users want to have the feeling they control their homes. Furthermore, most of the respondents seem to be interested by the IoT in smart home and health devices.

This study also points out the differences between men and women regarding their preference towards the category of IoT and the criteria they consider when purchasing smart devices. It is interesting to examine this finding during the confirmatory phase. The following hypotheses were therefore proposed:

H3. There is a difference between men and women regarding the main criteria they consider when purchasing an IoT
H4. There is a difference between men and women regarding the category of IoT they prefer to buy.

4   Quantitative Study

4.1   Questionnaire and Sample

An online questionnaire was developed to collect data. The questionnaire has included rank-order questions related to the category of IoT and the main criteria for purchasing smart devices that have been emerged from the exploratory research and completed by a literature review. The user's attitude towards IoT was evaluated through 7 items thanks to the adapted scale of attitude towards the website from Yoo and Donthu (2001) and the perceived security of personal information was evaluated using 9 items from Salisbury et al. (2001). (see Appendix 1). Items were measured on a 5-point Likert-type scale, ranging from strongly disagree (1) to strongly agree (5).

1006 people have been participated to this online survey. Among the respondents, 60% are users of IoT and smart devices. 57% are females and 43% are males, aged from 25 to 45 years old.

4.2   Results

The measurement scales are reliable, the Cronbach's alpha (α) is 94% for user's attitude and 89% for perceived security of personal information.

ANOVA is employed to test H1 and H2. Results indicate that the category of IoT has a significant impact on the user's attitude ($F = 2.82$, $p < .05$) and the security of personal information perceived by user ($F = 2.32$, $p < .05$). Thus, H1 and H2 are statistically supported. Users show a positive attitude towards IoT for health and smart home devices that they consider as utilitarian objects in comparison to IoT of leisure that they perceived as gadget objects. Similarly, users are willing to provide personal information when using smart devices perceived as utilitarian and they don't feel uncomfortable about the security of their personal information.

H3 and H4 have been examined using Chi-square tests. As supposed, the importance of criteria of smart devices varies between men and women when purchasing an IoT. More particularly, significant differences exist for these criteria: product/quality ($\chi^2 = 11.61$, $p < .05$), price ($\chi^2 = 14.09$, $p < .05$), design ($\chi^2 = 13.17$, $p < .05$), interoperability - which means the ability to work with other products or systems without any restrictions of access - or implementation ($\chi^2 = 10.86$, $p < .05$) and functionality/configuration ($\chi^2 = 11.14$, $p < .05$). Hence, H3 receives statistical support. The quality of the smart device, its price and its design are the most important criteria for women.

Similarly, results show significant differences regarding the preference for the category of IoT that men and women want to buy ($\chi^2 = 16.32$, $p < .05$). Hence, H4 is statistically supported. Specifically, women prefer to buy smart home devices such as electrical appliance and devices to manage light and rolling shutters, in general. Men prefer to buy smart devices to manage sound and broadcasting services and also prefer devices for autonomous irrigation.

5 Discussion, Implications and Future Ways of Research

This study shows interesting findings when coming to the security of personal information and privacy related to the use of IoT. According to the category of IoT, privacy and data security may restrain a consumer from purchasing an IoT since their use can frighten some individuals. People are willing to use smart home or health devices where the benefit of saving time will overcome the reluctance to divulge personal information or share private content. At the same time, they perceive that the security of their personal information is guaranteed through their control of these smart devices. Otherwise, privacy is closely intertwined with control. In the online setting, there is a reported consumer demand for control over their personal data, in terms of disclosure or use (Pelet et al. 2013). When consumers feel their privacy is protected in their relationship with online vendors, they become more satisfied, which can lead to increased loyalty to a vendor's website (Liu et al. 2005). On the contrary, consumers' privacy concerns about the loss of control over their information may negatively influence customer-vendor relationships and affect whether consumers want to engage in online purchasing from a website in the future (Eastlick et al. 2006). Consequently, one of the ways for companies to differentiate themselves on the IoT market seem to lie on the security of personal information by offering the absolute control for users.

Belk (2013) has argued that the meaning of everyday objects changes, depending on who is connecting them, what they are being connected to, and where they are being connected. The ability of smart objects to connect not only with the consumer and each other, but also with other virtual and physical devices on the Internet such as his/her home, will accelerate the pace of change in the meaning of these everyday objects (Verhoef et al. 2017). This will be even more possible that consumers don't feel in danger with privacy, when using IoT.

IoT are changing the landscape of the healthcare industry by offering continuous monitoring (Chan et al. 2012). Numerous interesting research avenues exist such as understanding how will consumers interact with smart devices. One can also question how much control over smart devices do consumers want? Do they trust smart products and are they willing to rely on them? How do they handle a lack of consumer autonomy? By determining which is the most practical application of the concept of Smart Home, companies could know what is the most suitable support for managing locally and remotely smart devices/IoT for their house?

Questions related to older consumers also arise and reveal potentially strategic for the IoT market. Due to the aging of the population, it becomes more and more interesting to question whether smart devices could lead to better autonomous management of daily tasks for seniors (IoT of health and well-being, management of lights, shutters, electrical appliances…). Issues related to the security of personal information are also as important, and it might be important to consider feelings about using wearable devices, and permanently request user's authorization for any request. IoT could increase chances to engage in a healthier way of living, it is of paramount importance to know more on this topic as smart devices can bring more comfort as well as greater control over health and wellness. Attitude and intention towards use also need some questioning since disclosing personal information in order to be able to use a smart device doesn't seem that obvious for users.

Appendix 1. Measurement Scales

Construct	Source
Attitude of user	
Using Internet of things/smart devices is a good idea It is useful to use Internet of things/smart devices Using Internet of things/smart devices is a wise idea Using Internet of things/smart devices is pleasant I like the idea of using Internet of things/smart devices I would use Internet of things/smart devices voluntarily I would enjoy using Internet of things/smart devices	
Security pf personal information	
I feel apprehensive about using Internet of things/smart devices I am worried about the amount of personal information you may be revealing I would hesitate to use a wearable device as in the case I make mistakes I could not correct them Internet of things/smart devices makes me feel somewhat uncomfortable I am concerned that the Internet of things/smart devices may collect too much personal information As a result of me using Internet of things/smart devices, information about me which I consider private is now more readily available to others As a result of me using Internet of things/smart devices, information about me is out there, and if used, it will invade my privacy When providing personal information in order to use the Internet of things/smart devices, I am concerned that it may be a means of getting hold of information on myself for other purposes I am concerned that Internet of things/smart devices share my personal information with other entities without requesting my authorization	

References

Acas, R.: Conference Smart Home Bosch. Published 9 December 2016. http://www.bosch-presse.de/pressportal/de/en/ces-2017-bosch-will-be-presenting-what-a-smart-home-can-do-today-81472.html. Accessed 18 Nov 2017

Ardelet, C., Veg-Sala, N., Goudey, A., Haikel-Elsabeh, M.: Entre crainte et désir pour les objets connectés: Comprendre l'ambivalence des consommateurs. Décisions Marketing 86 (avril – juin), pp. 31–46 (2017)

Belk, R.W.: Extended self in a digital World. J. Consum. Res. **40**(3), 477–500 (2013)

Bonnin, G., Goudey, A., Bakpayev, M.: Meet the robot: Nao's Chronicle. In: Advances in Consumer Research, Film Festival, Baltimore, MD, 23–25 October 2014

Chan, M., Estève, D., Fourniols, J.-Y., Escriba, C., Campo, E.: Smart wearable systems: current status and future challenges. Artif. Intell. Med. **56**(3), 137–156 (2012)

Eastlick, M.A., Lotz, S.L., Warrington, P.: Understanding online B-to-C relationships: an integrated model of privacy concerns, trust, and commitment. J. Bus. Res. **59**(8), 877–886 (2006)

Gartner, Cisco, Idate ou IDC (2015)

Hoffman, D.L., Novak, T.P.: How to market the consumer IoT: focus on experience. In: MSI Meetings Web Conference (2017)

International Telecommunication Union: New ITU standards define the internet of things and provide the blueprints for its development (2012). http://www.itu.int/ITUT/newslog/New?ITU?Standards?Define?The?Internet?Of?Things?And?Provide?Aspx. Accessed 2018

Liu, Y., Doucette, W.R., Farris, K.B., Nayakankuppam, D.: Drug information–seeking intention and behavior after exposure to direct-to-consumer advertisement of prescription drugs. Res. Soc. Adm. Pharm. **1**(2), 251–269 (2005)

Morris, R.: Wearable Technology. Function, Fit, Fashion (2015). http://www.onebeacontech.com/Technology/pages/news/detail/whitepaper.page?id=d537ede26772af84785f5a19ffb5be56. Accessed 30 June 2016

Pelet J.-É., Diallo, M.F., Papadopoulou, P.: How can Social Networks Systems be an m-commerce Strategic Weapon? Privacy Concerns based on Consumer Satisfaction. Congrès Européen de Marketing EMAC, 4–7 June 2013, Istanbul (2013)

Rijsdijk, S.A., Hultink, E.J.: How today's consumers perceive tomorrow's smart products. J. Prod. Innov. Manag. **26**(1), 24–42 (2009)

Salisbury, W.D., Pearson, R.A., Pearson, A.W., Miller, D.W.: Perceived security and World Wide Web purchase intention. Ind. Manag. Data Syst. **101**(4), 165–177 (2001)

Verhoef, P.C., Andrew, T.S., Kannan, P.K., Luo, X., Vibhanshu, A., Andrews, M., Bart, Y., Datta, H., Fong, N., Hoffman, D.L., Hu, M., Novak, T., Rand, W., Zhang, Y.: Consumer connectivity in a complex, technology-enabled, and mobile-oriented world with smart products. J. Interact. Mark. (2017). SSRN. https://ssrn.com/abstract=2912321 or http://dx.doi.org/10.2139/ssrn.2912321

Yildirim, H., Ali-Eldin, A.M.T.: A model for predicting user intention to use wearable IoT devices at the workplace (2018)

Yiwen, G., He, L., Yan, L.: An empirical study of wearable technology acceptance in healthcare. Ind. Manag. Data Syst. **115**(9), 1704–1723 (2015)

Yoo, B., Donthu, N.: Developing a scale to measure the perceived quality of an internet shopping site (SITEQUAL). Q. J. Electron. Commer. **2**(1), 31–47 (2001)

Development of Self-aware and Self-redesign Framework for Wireless Sensor Networks

Sami J. Habib[1(✉)], Paulvanna N. Marimuthu[1], Pravin Renold[2], and Balaji Ganesh Athi[2]

[1] Computer Engineering Department, Kuwait University, Kuwait City, Kuwait
sami.habib@ku.edu.kw
[2] Electronic System Design Laboratory, Velammal Engineering College, Chennai, India

Abstract. We propose a self-aware self-redesign framework (SASR), which embeds an existing computing system (CS) with awareness of abnormalities to trigger a self-adaptation process through necessary redesigning to handle the operational challenges. We view a wireless sensor network (WSN) deployed in a hostile environment, often subject to gusty winds, as a computing system to be embodied with SASR framework for a seamless transmission at the frequency band of 2.4 GHz. We classify the environment severity into four levels based on the visibility as a clear sky, dust haze, dust storm and heavy dust storm, and our framework generates a hybrid-awareness within the sensor nodes by embedding the relationship between the meteorological changes surrounding the sensing area (public awareness) and the corresponding transmission losses (private awareness). Then, by partitioning the transmission band into four channels with varying transmission frequencies possessing varying transmission powers, and by selecting event and time based channel hopping, the adaptiveness of WSN towards the environment is ensured. We have utilized Contiki's Cooja simulator, as it suits for low-power and lossy networks, and multichannel communication, to generate the data transmission with respective transmission path losses for the defined environmental conditions. We have utilized two types of WSN deployment, such as uniform and random and we have attempted the channel hopping mechanism under both uniform and random deployments to test SASR feasibility within WSN. The simulation results showed that the channel hopping is not feasible under uniform deployment for the selected environmental conditions. It is also further observed that the random deployment with multi-channel hopping based on channel idle level and interference level showed better adaptiveness towards adverse environmental conditions, where the packet delivery ratio showed a drop by 5 to 8% compared to 15 to 30% in random deployment with single channel.

Keywords: Computing system · Random deployment · Self-aware · Self-redesign · Simulator · Transmission losses · Uniform deployment · Wireless sensor networks

© Springer Nature Switzerland AG 2019
Á. Rocha et al. (Eds.): WorldCIST'19 2019, AISC 931, pp. 438–448, 2019.
https://doi.org/10.1007/978-3-030-16184-2_42

1 Introduction

Presently, the growth of industrial automations, ubiquitous computing and internet-of-things try to increase autonomy in system functioning, thereby increasing the human dependability on computing systems. However, the computing systems are not designed with complete self-aware, and the awareness is playing a vital role in managing their intended services besides maintaining a guaranteed quality-of-service (QoS). The self-awareness can be developed either through human operators or through embodied sub-systems, after perceiving and comprehending various possible internal states (situations) the system may go through within a defined time span. This leads to the development of self-aware and self-adaptive systems, which are capable of knowing their environment and the interactions of others with their environment in the selected timespan and the adaptive methods to face the challenges in their operations.

A computing system, such as a wireless sensor network (WSN) has become increasingly popular to solve highly challenging real-world problems, including industrial (indoor) and environmental (outdoor) monitoring; however, the presence of internal and external concerns may influence the sensor system to behave abnormally from its designed specifications. The abnormal behavior may be due to: (i) failure of a component, (ii) run out of power, (iii) change in environment, which triggers variation in selected component behavior, (iv) relocation of the system/part of the system due to arid and gusty environment and (v) lack of resources and so on. Hereby, the component failure and resource insufficiency are internal to the system and to some extent they are controllable, whereas adapting to harsh external environment variations are challenging, especially the presence of gusty winds in Middle Eastern countries.

Dust storms are common in Middle East during summer and spring seasons, lasting up to 6 to 8 months [1], where the wind speeds accompanying the dust storms may exceed 25 mph (40k mh) and the airborne dust causes reduced visibility, traffic accidents, airport shutdown, damaged telecommunications and health hazards [2, 3]; moreover, the rising plumes of hot air comprehended with passing 'dust whirls' introduces transmission losses in electromagnetic waves by attenuating and scattering the signal severely. In addition, it may move the transmitting antenna away from their designated directions, thus, producing a measurable impact, say loss on the transmission signal of wireless systems [4]. Transmission path loss play a vital role in the analysis and design of link budget of a communication system. Thus, developing a SASR framework is an essential for the seamless operation of WSN, which is deployed for outdoor monitoring. Moreover, testing of such framework in a simulator is a critical step to validate SASR before implementing it in a real-time network.

Many simulators, such as NS2, Qualnet, and OMnet++ are available. However, we selected the network simulator Cooja, which comes with Contiki operating system [5] and it provides a simulation environment with simple radio propagation models and follows the IEEE 802.15.4e standard [6] to support multi-channel hopping based on interference level in a single channel. The emulation level support and the availability of multi-channel hopping mechanism facilitate the selection of Cooja simulator for our simulation runs. Further, Contiki operating system supports the time synchronized channel hopping (TSCH) mechanism, which is necessary to implement channel hopping.

In this paper, we have developed a SASR system framework for a WSN equipped with ZigBee wireless transmission, deployed for an outdoor monitoring facing transmission losses under gusty environment. SASR creates a hybrid self-awareness within a sensor by embedding the transmission path losses (private awareness) it may experience during the diverse environmental conditions (public awareness) and it initiates channel hopping to adapt to the environment so as to transmit the sensing data with reduced path loss. We classified the environment into four levels based on the visibility (v_i), according to [7, 8]: (i) $v_i > 40$ km, (ii) $5 > v_i > 6$ km, (iii) 2 km $< v_i > 3$ km, (iv) $v_i < 0.5$ km and the atmospheric conditions as clear sky, dust haze, dust storm and heavy dust storm respectively. We have utilized the relation between environmental factors and the path losses, derived according to [8] and as a preliminary attempt, we have simulated the signal transmission under the classified environments in Cooja simulator with two types of deployments; namely random and uniform deployment. Then, we have tested the adaptiveness of WSN system based on event and time based channel hopping under the defined environmental conditions. The simulation results showed that the channel hopping is possible under random deployment for the defined adverse environment conditions.

2 Related Work

Our extensive literature survey revealed that the term self-aware started appearing before three decades, when the researchers [9, 10] were trying to define the self-aware hypothesis from human perspective, as the ability to understand one's own state and the behavior of others, and they described it to be evolving and adaptive. Morin [11] defined the self-awareness as the capacity to become the object of one's attention, whereas Duval and Wickland [12] classified self-awareness in living organisms into a subjective and objective based on their behaviors. Further to previous classification, Neisser [13] developed a model detailing five levels of awareness in a single organism: Ecological self, interpersonal self, private self, extended self and conceptual self.

Inspired from the fields of Psychology and Cognitive Science, Cox [14] carried out a study on meta-cognition in computing systems, and the author introduced the concept of self-awareness in computing systems. Then, Agarwal et al. [15] introduced a paradigm shift from procedural design practice to self-aware engineering systems while designing autonomic systems. Chen et al. [16] followed the classification of self-awareness in living organisms by Morin [11] and the authors categorized engineering self-awareness into two: private self-awareness and public self-awareness. The computing node having knowledge about itself and its internal phenomena is said to be possessed with a private self-awareness and the node which has the knowledge of external phenomena, is said to be possessed with a public self-awareness. In this paper, we have chosen a hybrid awareness scheme, where both the public and private awareness are utilized to generate awareness in a sensor.

We studied the research works dealing with the dust storm and its effect on transmission losses. The rain and dust are the main sources of signal attenuation in the atmosphere, which are known to cause a transmission loss in electromagnetic waves by scattering. Especially, the dust particles cause attenuation and depolarization of the electromagnetic waves propagating in the sandy desert environment [17]. Many of the research works focused on the study of path loss on radio frequencies (RF) around 40 GHz [18, 19]. We found few research works dealing with the study of RF transmission losses in outdoor environment. Rama Rao et al. [20] analyzed the received signal strength at 868, 916 and 2400 MHz in forest and Mango and Guava vegetation environment, and Mujlid [21] carried out an empirical study on transmission losses due to dust storm at RF frequencies at lower range, especially around 2.4 GHz. In this paper, we have utilized the data collected in [8, 21] and the derived path loss equations to simulate dusty environments.

3   Modeling Self-aware and Self-redesign Computing System

We have developed a self-aware and self-redesign framework for an existing computing system, to analyze the possibilities in embedding the given system with a self-aware, and a self-redesign capability to sustain with the variations in operational environment and to maintain system design goals. The existing computing system $S(X(K))$ is represented as a function of known inputs, and measured outcome, illustrated as in Fig. 1, which has a set of K input parameters $X = \{x_1, x_2, x_3, \ldots, x_K\}$ and a set of Q output parameters $Y = \{y_1, y_2, y_3, \ldots, y_Q\}$ at various time instances. Hereby, the term X represents the input attributes, the device intended to compute (sense or measure), the term Y which represents the system outputs, at any time instance.

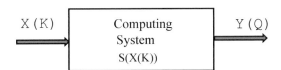

Fig. 1. A representation of a generic computing system.

In addition to the input and output (functional data) parameters of a general system in Fig. 1, we include a set of external factors $E = \{e_1, e_2, e_3, \ldots, e_L\}$ influencing system output, as a non-functional data set and a redesigned output $R(Y(Q))$, where the normal output is redesigned to improve the quality of the generated output. Here, we defined the impact of environmental variations on the system as $P(X(K))$, which are the changes observed during the monitoring period and the extent of adaptation to be implemented on the system as $A(X(K))$. We define the normal system, the system with self-aware functionalities and the system with self-aware and self-redesign functionalities mathematically, as in Eq. (1).

$$Y(Q) = \begin{cases} S(X(K)) & \text{Normal System} \\ S(X(K)) \pm P(X(K)) & \text{Self-aware} \\ S(X(K)) \pm P(X(K)) \pm A(X(K)) & \text{Self-aware \& self-redesign} \end{cases} \quad (1)$$

Here, the self-aware system is embodied with a monitoring system to observe the behavior of the designated parameters and explore the situation causing the change in system output. Then, recommendations from the monitoring system trigger the necessary redesign operation to sustain its normal behavior in a complete SASR system.

4   Modeling a Typical WSN Self-aware System

We analyze the monitoring data set utilizing a time-series analysis, as the analysis enables us to explore the trend, and abnormality over the specified time period. The set of N observations in the monitoring data set P_i for a time instance t_i would be represented as a row of N elements in a matrix P^{W_i}, and each row of data in P^{W_i} would represent the situations collected at uniformly spaced time intervals. Thus, the matrix P^{W_i} represents a log of situations, and for a time-series window of size M, the matrix would be comprised of M rows and N columns (N observations), as illustrated in Eq. (2). The log of situations is collected over a specified time span, and it occupies a specified time-series window W_1.

$$P^{W_i} = \begin{bmatrix} P_1 \\ P_2 \\ \vdots \\ P_i \\ \vdots \\ P_M \end{bmatrix} \text{ and } P_i = \begin{bmatrix} p_{i1} & p_{i2} & p_{i3} & \cdots & p_{iN} \end{bmatrix},$$

$$\text{So } P^{W_i} = \begin{bmatrix} p_{11} & p_{12} & p_{13} & \cdots & p_{1N} \\ p_{21} & p_{22} & p_{23} & \cdots & p_{2N} \\ p_{31} & p_{32} & p_{33} & \cdots & p_{3N} \\ \vdots & \vdots & \vdots & \vdots & \vdots \\ p_{M1} & p_{M2} & p_{M3} & \cdots & p_{MN} \end{bmatrix} \tag{2}$$

In this work, we have selected four different environmental conditions, such as clear sky, dust haze, dust storm and heavy dust storm, and we have utilized the empirical model derived in [8, 21] from the log of records of transmission path loss and respective environmental parameters, to quantify the impact on system, which are presented using Eqs. (3) to (6).

$$\text{Clear sky:} \quad Tl_{cl} = 28.07 \times \log(d) + 55.505 \tag{3}$$

$$\text{Dust haze:} \quad Tl_{dh} = 36.085 \times \log(d) + 54.346 \tag{4}$$

$$\text{Dust storm:} \quad Tl_{ds} = 32.399 \times \log(d) + 53.298 \tag{5}$$

Heavy dust storm: $Tl_{hds} = 37.507 \times \log(d) + 48.522$ (6)

The Eqs. (3) to (6) would be implanted in the sensor node to create self-awareness about the transmission losses due to the environment. Thus, a hybrid self-awareness is created within WSN from the defined four environmental conditions (public) and the corresponding measured transmission losses (private).

5   Wireless Sensor Network Model

We have considered a clustered WSN topology under two types of deployment model namely: uniform and random, as shown in Fig. 2(a) and (b).

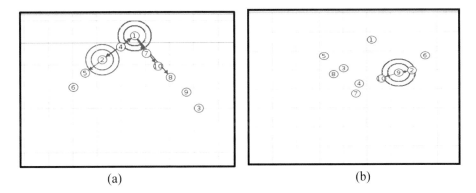

(a)   (b)

Fig. 2.   Network model: (a) uniform and (b) random.

In the case of uniform deployment, the grid of size 10 m is formed. The information of grid size and minimal distance between the sensors nodes are determined by the link budget estimation with the proper assumptions.

5.1   IEEE802.15.4e Time Slotted Channel Hopping (TSCH) for Designing Self-redesign WSN System

In order to provide the self-redesign within WSN, we have selected Contiki operating system, which followed IEEE 802.15.4e standard [6] based time synchronized channel hopping (TSCH) to partition the given transmission channel of a sensor node into four, suiting to the selected environmental conditions. TSCH supports multi-channel hopping based on the interference level in a single channel. The duration of the timeslot is fixed as in Fig. 3, and it ranges from 10 s to 1000 s. TSCH uses absolute slot number (ASN) and it is initialized to 0. ASN is incremented by 1 whenever a new network node is added. The purpose of ASN is used to determine the frequency for communication.

The nodes, which are already in the network periodically send enhanced beacons (EBs) to announce the presence of the network. When a new node joins, it listens for EBs to synchronize to the TSCH network. EB frames contain information about the timeslot length, the slot frames and timeslots the beaconing mote is listening on, and a 1-byte join priority.

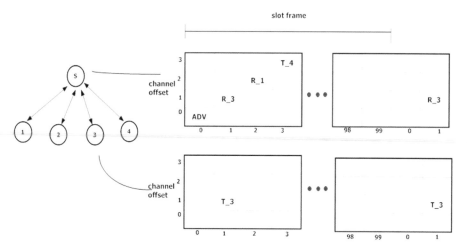

Fig. 3. The schedule table of sink and node 3 with a total time slot of 100 s with each slot length is 10 s.

The channel selection is based on:

$$frequency = (channel\,Offset + ASN)\,MOD\,nF,$$

Where, channel Offset is the total channels, ASN is the absolute slot number, and nF is the number of available channels.

We modified TSCH algorithm as in Fig. 4 to show adaptation for channel hopping. Once the network is deployed, it starts to learn the topology and the sink node announces its personal area network (PAN) identifier. The other nodes scan the channel and the process of synchronization takes place. Once the node has a data to transmit, it checks for the availability of a sink in that channel. The channel hopping happens if the slot is idle for more than a threshold (slot/2), and if the packet retransmission request is received more than two times. Then, the node tries to transmit at high power amplifier (PA) level i.e. 31 in the case of CC2420. Even if no acknowledgement is received, the channel hopping process starts to identify the interference free channel. The different power levels adopted in the work [22] are listed in Table 1.

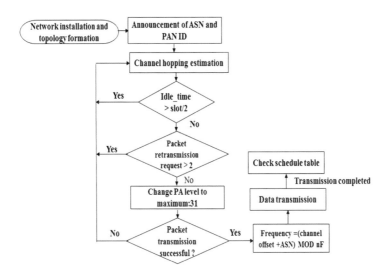

Fig. 4. Modified TSCH algorithm.

Table 1. Various power levels, according to [22].

PA-level	Output power [dBm]
31	0
27	−1
23	−3

6 Results and Discussion

We have simulated a wireless sensor networks (WSN) comprising of 10 nodes, communicating through star topology to the sink node under random and uniform deployments in Cooja network simulator. The simulation parameters are tabulated in Table 2. The simulation results are obtained as the average of five different simulation runs with different seed values. The transmission path loss Eqs. (3) to (6) are embedded within the sensor node to be aware of the environmental changes. The nature of traffic generated in the simulation is of time driven and event driven. For every 2 s, a sensor node would be triggered to transmit data towards the sink node. Whenever, the environment changes, event driven mechanism is on and TSCH selects the channel hopping mechanism, say from clear sky to dust haze, to facilitate the data transmission.

In order to evaluate the performance of wireless sensor network, the packet delivery ratio is considered as an important performance index. The study performs the packet delivery ratio (PDR) for different environmental conditions and the results are shown in Fig. 5(a). The uniform deployment of nodes showed better performance when compared to random. In the case of uniform deployment, the sensor nodes are regularly spaced, hence the nodes are able to perform the communication with minimal interference. However, the channel hopping condition based on interference worked better in random deployment scenario. The performance of the channel hopping mechanism in uniform deployment is poor due to the time taken in synchronizing sink and the source nodes. We also observed a drop in packet delivery when the nodes are switching its channels and there is change in the nature of traffic in the network. The packet delivery ratio (PDR) in channel hopping is increased by the incorporation of different control signals to improve the level of synchronization. We observed a drop in PDR by 5% to 8% in the case of random deployment with channel hopping, whereas the PDR value is dropped by 15–30% in random deployment.

Table 2. Simulation parameters.

Nodes	10
Routing protocol	Mesh; Unicast
MAC	IEEE802.15.4; IEEE 802.15.4e (TSCH enabled)
Sink node	1
Simulation time	250 s
Sensor node	TelosB
Transceiver	CC2420
Receiver sensitivity	−95 dBm
Type of traffic	Time based and event driven

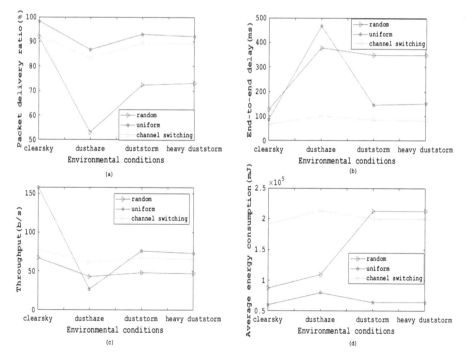

Fig. 5. Outcome of the experiment against the environments: (a) packet delivery ratio (b) end-to-end delay (c) throughput and (d) average energy consumption.

The end-to-end delay at the time of propagation of signals between the nodes and the throughput have been evaluated for different environmental conditions and the results are shown in Fig. 5(b) and (c) respectively. The end-to-end delay in channel hopping mechanism outperforms random and uniform deployment with single channel. It is due to the fact that better channel is identified based on the interference level. In sand storm condition the delay is higher, as it is not able to find better neighbor to forward packet.

The throughput of channel hopping is nearly constant in all the deployment scenarios. The random deployment achieves lesser throughput, due to more packet loss in the network. The uniform deployment achieves better throughput except in the case of sand storm. It is due to increased retransmission and additional packet loss in the network due to sand storm. Also, the number of packets successfully reached the destination is less under sand storm. The throughput may be improved by deploying more nodes, i.e., with greater neighbor density.

7 Conclusion

We proposed a self-aware and self-redesign framework (SASR) within an existing WSN computing system to explore the environmental impact on its data transmission at 2.4 GHz frequency band and to trigger a self-adaptation by hopping the transmission channel with varying transmission power. We have utilized Contiki simulator to generate four types of data transmission losses under environmental conditions, such as clear sky, dust haze, dust storm and heavy dust storm and to implement the awareness of environmental impacts under two deployment scenarios: uniform and random. We partitioned the transmission band of each sensor into four channels with varying transmission power, suited to the path loss under the selected environments so as to adapt its transmission. We selected packet delivery ratio, end-to-end delay, throughput and energy consumption as the performance metrics and the simulation results showed that the channel hopping mechanism showed better outcome in random deployment than with uniform. We also observed that the transmission with channel hopping showed a drop in packet delivery ratio by 5% to 8%, whereas in a single channel random deployment, the packet drop was 15% to 30%. We are continuing our research work to include more environmental parameters and more number of clusters with more participated network resources to ensure better propagation of signals between nodes.

References

1. Al-Awadhi, J.M., Al-Shuaibi, A.A.: Dust fallout in Kuwait city: deposition and characterization. Sci. Total Environ. **461–462**, 139–148 (2013)
2. Prakash, P., Stenchikov, G., Kalenderski, S., Osipov, S., Bangalath, H.: The impact of dust storms on the Arabian Peninsula and the Red Sea. Atmos. Chem. Phys. **15**, 199–222 (2015)
3. Notaro, M., Alkolibi, F., Fadda, E., Bakhrjy, F.: Trajectory analysis of Saudi Arabian dust storms. J. Geophys. Res. Atmos. **118**, 6028–6043 (2013)

4. Faroq-I-Azam, M., Ayyaz, M.N.: Location and position estimation in wireless sensor networks. J. Immunol. **178**(3), 1301–1311 (2016)
5. Roussel, K., Song, Y.-Q., Zendra, O.: Using Cooja for WSN simulations: some new uses and limits. In: The Proceedings of International Conference on Embedded Wireless System Networks, Graz, Austria, 15–17 February, pp. 319–324 (2016)
6. IEEE802.15.4e: IEEE standard for local and metropolitan area networks. Part 15.4: Low-rate wireless personal area networks (LRWPANs) Amendment 1: MAC Sublayer. Institute of Electrical and Electronics Engineers Std., April 2012
7. Wang, P.Z., Vuran, M., Al-Rodhaan, M., Al-Dhelaan, A., Akyildiz, I.: Topology analysis of wireless sensor networks for sandstorm monitoring. In: The Proceedings of IEEE International Conference on Communications, Kyoto, Japan, 5–9 June, pp. 1–5 (2011)
8. Mujlid, H., Kostanic, I.: Propagation path loss measurements for wireless sensor networks in sand and dust Storms. Front. Sens. **4**, 33–40 (2016)
9. Humphery, N.: Consciousness: a just-so story. New Sci. **19**, 474–477 (1982)
10. Gallup, G.G.: Self-awareness and the emergence of mind in primates. Am. J. Primatol. **2**, 237–248 (1982)
11. Morin, A.: Levels of consciousness and self-awareness: a comparison and integration of various neurocognitive views. Conscious. Cogn. **15**, 358–371 (2006)
12. Duval, S., Wicklund, R.A.: A Theory of Objective Self Awareness. Academic Press, Oxford (1972)
13. Neisser, U.: The roots of self-knowledge: perceiving self, it, and thoua. Ann. N. Y. Acad. Sci. **818**, 19–33 (1997)
14. Cox, M.T.: Metacognition in computation: a selected research review. Artif. Intell. **119**(2), 102–141 (2005)
15. Agarwal, A., Miller, J., Eastep, J., Wentziaff, D., Kasture, H.: Self-aware computing, Technical report. AFRL-RI-RS-TR-2009-161. Massachusetts Institute of Technology and DARPA, Cambridge, MA, USA (2009)
16. Chen, T., Fanivi, F., Bahsoon, R., Lewis, P.R., Yao, X., Minku, L.L., Esterle, L.: The handbook of self-aware and self-expressive systems, Technical report, EPiCS EU FP7 project consortium (2014)
17. Abuhdima, E.M., Saleh, I.M.: Effect of sand and dust storms on microwave propagation signals in southern Libya. In: The Proceedings of 15th IEEE Mediterranean Electro technical Conference, Valletta, Malta, 25–28 April, pp. 695–698 (2010)
18. Srivastava, S.K., Vishwakarma, B.R.: Study of the loss of microwave signal in sand and dust storms. IETE J. Res. **50**(2), 133–139 (2004)
19. Yan, Y.: Microwave propagation in saline dust storms. Int. J. Infrared Millim. Waves **25**(8), 1237–1243 (2004)
20. Rama Rao, T., Balachander, D., Nanda Kiran, A., Oscar, S.: RF propagation measurements in forest and plantation environments for wireless sensor networks. In: The Proceedings of 2012 International Conference on Recent Trends in Information Technology, Chennai, India, 19–21 April, pp. 308–313 (2012)
21. Mujlid, H.M.: Real-time monitoring of sand and dust storm winds using wireless sensor technology (2016). https://repository.lib.fit.edu/handle/11141/1073. Accessed 3 Oct 2018
22. De Guglielmo, D., Brienza, S., Anastasi, G.: A model-based beacon scheduling algorithm for IEEE 802.15.4e TSCH networks. In: The Proceedings of IEEE 17th International Symposium on A World of Wireless, Mobile and Multimedia Networks, Coimbra, Portugal, 21–24 June, pp. 1–9 (2016)

Intelligent and Decision Support Systems

Comparison of Evolutionary Algorithms for Coordination of Cooperative Bioinspired Multirobots

A. A. Saraiva[2,3]([⊠]), F. V. N. Silva[2], Jose Vigno M. Sousa[1,2],
N. M. Fonseca Ferreira[4,5], Antonio Valente[3,6], and Salviano Soares[3,7]

[1] University Brazil, Sao Paulo, Brazil
[2] UESPI-University of State Piaui, Piripiri, Brazil
aratasaraiva@gmail.com, {ffranciscovinicius,
josevignog}@prp.uespi.br
[3] School of Science and Technology,
University of Tras-os-Montes and Alto Douro, Vila Real, Portugal
{avalenteg, fsalblues}@utad.pt
[4] Department of Electrical Engineering, Polytechnic Institute,
Institute of Engineering of Coimbra, Coimbra, Portugal
nunomig@isec.pt
[5] Knowledge Engineering and Decision-Support Research Center (GECAD),
Institute of Engineering, Polytechnic Institute of Porto, Porto, Portugal
[6] INESC-TEC Technology and Science,
Campus da FEUP, Rua Dr. Roberto Frias, 378, 4200-465 Porto, Portugal
[7] Institute of Electronics and Informatics Engineering of Aveiro, IEETA-UA,
Aveiro, Portugal

Abstract. This paper compares optimal path planning algorithms based on a Genetic Algorithm and a Particle Swarm Optimization algorithm applied to multiple bioinspired robots in a 2D environment simulation. The planning objectives are related to the harvesting of an apple plantation in which three swarm of butterflies were run, counting the fruits on the ground to optimize the harvest in a cooperative way. Robotic swarms must travel through points on the map to count the fruits. The time for each swarm was also counted for the comparison results.

Keywords: Evolutionary multirobots bioinspiratory

1 Introduction

Robot patrolling work is developing because they can perform tasks of different complexities like flying, floating, and moving underground [12]. Typically, robot swarm control systems are built on the basis of intelligent robots capable of planning their route in unfamiliar environments [3].

In the practices, the tasks developed by the robots can be related to the search in places for detection, such as soil types, climatic conditions of temperature. [6]. It is taken into consideration the determination of the limits of the domain, all the points of

© Springer Nature Switzerland AG 2019
Á. Rocha et al. (Eds.): WorldCIST'19 2019, AISC 931, pp. 451–460, 2019.
https://doi.org/10.1007/978-3-030-16184-2_43

the which have some property in common e.g. the level of radiation where it does not exceed a value or elaborate on a map the edge of a given area [11].

This paper discusses the path of a swarm of robotic butterflies and the harvest of an orchard. Comparing the efficiency of two methods the Genetic Algorithm (GA) and Particle Swarm Optimization (PSO). The algorithms were chosen because of the swarm to be able to ensure a uniform coverage of the considered field.

2   Related Work

The field of research of robotics operates in several areas, the work of [12] demonstrating the combination of chemotactic and anemiotaxic models, known as odor-controlled reotaxy (OGR), to solve problems of location of the source of real-world odors. Chemotaxis is a basis for many algorithms such as PSO. However, combining as two orientations within a modified PSO-based algorithm, odors within a complete propagation environment can be localized and dynamic advection-diffusion problems can be solved.

In [10] discusses a method that performs gesture recognition, with the objective of extracting characteristics of the segmented hand, from dynamic images captured from a webcam and identifying signal patterns. With this method it is possible to manipulate simulated multirobots that perform specific movements. The method consists of the Continuously Adaptive Mean-SHIFT algorithm, followed by the Threshold segmentation algorithm and Deep Learning through Boltzmann restricted machines. As a result, an accuracy of 82.2

In [9] describes a group of robots for cleaning a simulated environment and proposes an efficient algorithm for navigation based on Path finding A*. No need for vision sensors. As a result it was observed that the robots can work cooperatively to clear the ground and that the navigation algorithm is effective in cleaning. In order to test its efficiency it was compared the combination of the Path finding A* algorithm and the decision algorithm proposed in this paper with Path finding A* and Euclidean distance, resulted in an improvement in time and distance traveled.

Works such as [2] use hybridization of these algorithms in the work of classifying Indian pines samples and detecting roads. The results confirm that the proposed method is able to discriminate between road and background pixels and works better than the other approaches used for comparison in terms of performance metrics.

In [6] applies a PSO update strategy to increase population diversity and use the Biogeography Graft Optimization algorithm to optimize paths in the network network for AVBN modeling. Experimental results show that the pattern is feasible and effective.

In [1] showed that continuous PSO has the highest overall fitness, and Q-learning with continuous states performs significantly better than Q-learning with discrete states. Also that in the case of the single robot, PSO and Q-learning with discrete states require a similar amount of total learning time to converge, while the time required with Q-learning with continuous states is significantly greater.

In [8] applied an GA and a PSO to deal with the complexity of the problem and to calculate feasible and quasi-ideal trajectories for fixed-wing UAVs in a complex 3D environment, considering the dynamic properties of the vehicle. The characteristics of the optimal path are represented in the form of a multiobjective cost function that we develop. The paths produced are composed of line segments, circular arcs and vertical propellers. In addition, the rigorous comparison of the two algorithms shows, with statistical significance, that the GA produces superior trajectories for the PSO.

In [7] presents a survey of recent research on BIAs that focuses on conducting multiple BIAs based on different working mechanisms and mobile robot control applications to help understand BIAs comprehensively and clearly. The research has four main parts: a classification of the BIAs from the biomimetic mechanism, a summary of several typical BIAs of different levels, an overview of the current applications of BIAs in the control of mobile robots and a description of possible future directions for research.

3   Methodology

In this section the methods used to compare the algorithms implemented in the simulation were written. In the composition of the simulation three swarms are positioned on the side of the map, their goal is to reach the other side of the map by counting the fruits found on the ground. The exit points of each swarm will occur an explosion of butterflies in to different parts around the points. The stopping criterion of the simulation is based on the arrival points of each swarm and the fruit harvest term shown in Algorithm

1. When a vector is loaded with distance and target, a test is made to the agent who finds himself in that position. As it approaches the target, the values are divided by 60. With chance to score at 0, 0.001, this value is added to the distance to decrease the chance of the division resulting in 0.

Algorithm 1 Fitness

```
distance   (position;target)
if distance == 0 then
    dist    0:001
end if
retur60=dist
```

3.1   Implementation with PSO

PSO is a stochastic optimization algorithm based on a kind of collective intelligence of a swarm based on psycho-sociological principles [7]. It can be used to demonstrate social behaviors or engineering applications. It was first described in 1995 by James Kennedy and Russell C. Eberhart [7]. Inspired by the social behavior of fish and birds, it simulates collective intelligence mechanisms in order to find a solution for the problem [4]. The Fig. 1 shows an illustrative scheme of the iterations of the particles [4].

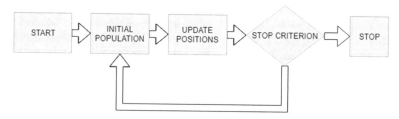

Fig. 1. PSO algorithm

For this algorithm, each solution is called a particle and represents a point D in space, if that point D is the number of parameters to be optimized [5]. Thus, the position of the particle i can be described by the vector xi:

$$x_i = [x_{i1}; \ x_{i2}; \ x_{i3} \ ::: \ x_{iD}] \tag{1}$$

And the population of N solutions constitutes the swarm:

$$X = (x_1; \ x_2; \ :: \ x_n) \tag{2}$$

To obtain the best solution, the particles define trajectories in the parameter space based on the following equation of motion:

$$x_i(t+1) = x(t) + v_i(t+1) \tag{3}$$

Where t and t + 1 indicate two successive iterations of the algorithm and vi is the vector collecting the velocity components of the particle i along the dimensions D [5].

Algorithm 2 PSO Algorithm

    for i do 1 n
        p[i]   x
        pBest   f (p)
        if pBeste < vBest then
            vBest   pBest
            g   p(i)
        end if
    end for

The algorithm 2 describes that for each traversed point is added to a variable best point (pBest) in which randomly start [5]. When this best point is reached (vBest), another point with the same value is allocated instead. This cycle repeats to a stop condition (i to n). In this case each swarm will have its ideal point, properly saying an apple located in a certain point of the terrain [5].

In the field where the plantation is located, a test of butterflies robots is simulated to harvest the apples. By setting a point on the fruits in which the swarm will pass, they must follow the path to this set fruit of the code. During the journey, the robots should collect the fruits found in their path. The path follows a line on the X-axis line as shown in Fig. 2.

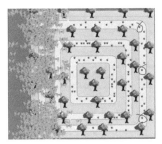

Fig. 2. Path of the butterflies at the beginning with destiny to the selected fruit

Each fruit in which the exam passes must account for it. One factor that should also be taken into account is the time in which the map will travel. The count of time will run until the butterflies reach the apple on the ground. It was considered in the simulation that the apples would be on the ground when the swarm passes through them.

3.2   Implementation with GA

GA belongs to the family of computational models inspired by the theory of the evolution of the species described by Charles Darwin Goldberg (1989) [12]. This technique is formed by algorithms inspired by the mechanisms of natural evolution and genetic recombination, so that the technique provides an adaptive search method that is based on the principle of reproduction and survival of the best individual [12]. The Fig. 3 demonstrates the flowchart of an GA.

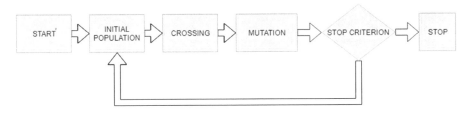

Fig. 3. Flow chart GA

Algorithm 3 describes the creation of a genome with 50 genes initially. Each gene with a random size. All this process is done during the course of the created genome.

At the intersection new objects are created from those previously selected. Different methods can be used for the crosstalk [12]. It resembles the mating of the animal kingdom in which you can cross a parent object, the percentage of each, and so on.

In the algorithm 4 one has the crossing of the genes in the father and relative vectors. Based on the chance of crossing, these vectors will be crossed generating new genes.

Algorithm 3 Genome Selection

```
sizeGenoma  50
gene   randon( 1:0; 1:0)
for i do 1 sizeGenoma
    gene[i] + +
end for
```

Algorithm 4 Crossing

```
sizeGenoma  50
gene
dad
partner
chanceCrossing  randon(0; 2)
for i do 1 sizeGenoma
    if chanceCrossing == 2 then
        dad[i] ( dad[i] + gene
    end if
    if chanceCrossing <> 2 then
        partner[i] ( partner[i] + gene
    end if
end for
```

After the crossing, based on probability, several objects are randomly mutated, each with a random value [12]. This means that you have the possibility of different solutions [8]. Algorithm 5 describes the mutation of the selected items. Based on the size of the mutation and its chance. Whenever this chance is less than or equal to the size of the mutation, a new gene will be created. All this traversing the gene vector.

Algorithm 5 Genome Mutation

```
dad    gene[i]
newMutation
sizeMutation   0:01
chanceMutation   randon(0:0; 0:1)
for i do 1 genes
    if chanceMutation <= sizeMutation then
        newMutation[i]   dad[i]
    end if
end for
```

The last step to be described is the appropriateness or criterion of evaluation of the decisions of the algorithm. If this criterion is reached, the code is stopped, otherwise a pre-selection of a new population or a new crossover can be made from the individuals resulting from the mutation. With the same terrain as the plantation, robots simulating a cloud of robotic butterflies flying over the spot behind their food were placed at different points on the map. One point to leave the swarm and another to reach. The points were placed at the ends of the map, that is, one has the exit points where the agents are initially positioned and their goal is to reach another part of the map where the points destined for arrival are located. This facilitates the locomotion of the robots to learn about the terrain in which they are working Fig. 4.

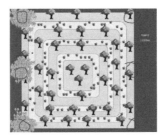

Fig. 4. Plantation with set points for recognition

Your scrolling on the map will not depend on the fruit selected on the map. Because at each iteration, a new gene is generated by crossing with the previous gene until it reaches the indicated point, which is the other end of the map. While learning the route, robots must do the same work as those who use the PSO, which will count the fruits and the time at which this count is demonstrated in Fig. 5. The time is counted during robot training, house seha the entire period in which it takes to get to the demarcated points will be counted because all the parts of the map in which they will pass will count points saved by the butterflies. Each loop started in the simulation has already saved its last point, making it possible to reach the end of the route at the other point on the map.

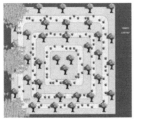 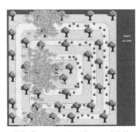 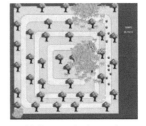

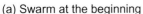
(a) Swarm at the beginning   (b) Swarm in the middle   (c) Swarm in the end

Fig. 5. Path of the swarm covered on the map

4   Results

Table 1 shows the results of the simulation made with the swarm of butterflies using the implementation of the PSO from each swarm to the arrival at the other end of the map. A total of 175 apples were used during the simulation divided into the paths of each swarm. The middle swarm detained the shortest time and the largest number of apples, this is due to its better positioning in relation to the others found in the sides. After the simulation was finished, nine apples were left that did not enter the count because the swarm had reached its set fruit.

Table 1. Quantity and time of harvest in the Terrain (PSO)

Robots	Amount harvested	Time (secs)
1^a swarm	45 apples	87.4402
2^a swarm	80 apples	85.1598
3^a swarm	43 apples	88.2398

The results of the simulation with the AG implementation are described in Table 2.

Table 2. Quantity and time of harvest in the Land (GA)

Robots	Amount harvested	Time (secs)
1^a swarm	48 apples	85.4402
2^a swarm	82 apples	84.1598
3^a swarm	45 apples	94.2398

The same number of apples were used in the simulation with the PSO. The middle swarm also obtained the largest number as well as in the previous simulation, although it had a smaller time variation. All apples used were counted in this simulation, since the three exams were able to traverse the map completely. Both obtained the same number of explosions, but with the time variation this is due to speed the particles updated after each position held by the swarm. This implementation also had the same number of explosions, but with a shorter arrival time than the last implementation.

It was observed that the examination of the medium had the largest number of maces harvested. This happened because they are positioned in the middle of the map where their explosion occurs in greater number, different from the others that are located in the sides of the terrain.

In the Table 3 the best rates of progress and the time of percussion on the ground of both implementations are present. Taking into account the best time and the highest amount of fruit harvested in which each swarm traveled in the map of the flattening and the shortest time to reach the other end of the map. It is noted that both during their course have made the same progress, presenting a difference in the number of apples counted and in the time of the simulation. The Fig. 6 contains the graph with the progression of these results.

Table 3. Best result for harvesting

Swarm	Amount harvested	Time (secs)
PSO	80 apples	84.1598
AG	82 apples	84.1598

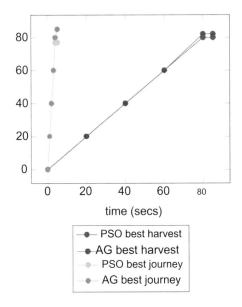

Fig. 6. Best result of each implementation

5 Conclusion

When we compare the time for each test GA proves the is more efficient. The PSO in some simulations although it realized the count of the apples left some pass or it closed the simulation. The GA made the complete count during the course of each swarm without letting any fruit pass.

The time factor of the run is important because with it was possible to say that the GA had better result in the simulation. The PSO on the other hand had a considerable time, but inferior in relation to the AG that even with the same number of explosions took a less time to perform its task.

This work aimed to compare the efficiency of two algorithms implemented in a simulation. Both met what was proposed in the context, but with a small difference in the final results the GA proved to be more efficient. The PSO performed well, but in the tests some fruits did not count. The AG also presented better results in the movement of the swarm during the course, because different from the PSO was not necessary the use of an apple in the terrain to carry out the work.

Acknowledgments. This work was funded by National Funds through the FCT - Foundation for Science and Technology, in the context of the project UID/CEC/00127/2019.

References

1. Di Mario, E., Talebpour, Z., Martinoli, A.: A comparison of PSO and reinforcement learning for multi-robot obstacle avoidance. In: 2013 IEEE Congress on Evolutionary Computation (CEC), pp. 149–156. IEEE (2013)
2. Ghamisi, P., Benediktsson, J.A.: Feature selection based on hybridization of genetic algorithm and particle swarm optimization. IEEE Geosci. Remote Sens. Lett. **12**(2), 309–313 (2015)
3. Jatmiko, W., Sekiyama, K., Fukuda, T.: A PSO-based mobile robot for odor source localization in dynamic advection-diffusion with obstacles environment: theory, simulation and measurement. IEEE Comput. Intell. Mag. **2**(2), 37–51 (2007)
4. Liu, Q., Ma, J., Zhang, Q.: PSO-based parameters optimization of multi-robot formation navigation in unknown environment. In: 2012 10th World Congress on Intelligent Control and Automation (WCICA), pp. 3571–3576. IEEE (2012)
5. Marini, F., Walczak, B.: Particle swarm optimization (PSO): a tutorial. Chemom. Intell. Lab. Syst. **149**, 153–165 (2015)
6. Mo, H., Xu, L.: Research of biogeography particle swarm optimization for robot path planning. Neurocomputing **148**, 91–99 (2015)
7. Ni, J., Wu, L., Fan, X., Yang, S.X.: Bioinspired intelligent algorithm and its applications for mobile robot control: a survey. Comput. Intell. Neurosci. **2016**, 16 (2016)
8. Roberge, V., Tarbouchi, M., Labonte, G.: Comparison of parallel genetic algorithm and particle swarm optimization for real-time UAV path planning. IEEE Trans. Industr. Inf. **9**(1), 132–141 (2013)
9. Saraiva, A.A., Costa, N., Sousa, J.V.M., De Araujo, T.P., Ferreira, N.F., Valente, A.: Scalable task cleanup assignment for multi-agents. In: Memorias de Congresos UTP, vol. 1, pp. 439–446 (2018)
10. Saraiva, A.A., Santos, D.S., Junior, F.M., Sousa, J.V.M., Ferreira, N.F., Valente, A.: Navigation of quadruped multirobots by gesture recognition using restricted boltzmann machines. Memorias de Congresos UTP **1**, 431–438 (2018)
11. Shorakaei, H., Vahdani, M., Imani, B., Gholami, A.: Optimal cooperative path planning of unmanned aerial vehicles by a parallel genetic algorithm. Robotica **34**(4), 823–836 (2016)
12. Zhu, Z., Wang, F., He, S., Sun, Y.: Global path planning of mobile robots using a memetic algorithm. Int. J. Syst. Sci. **46**(11), 1982–1993 (2015)

An Actionable Knowledge Discovery System in Regular Sports Services

Paulo Pinheiro[1](✉) and Luís Cavique[2]

[1] Universidade Aberta, Cedis, Lisbon, Portugal
ppinheiro@cedis.pt
[2] Universidade Aberta, BioISI-MAS, Lisbon, Portugal
luis.cavique@uab.pt

Abstract. This work presents an actionable knowledge discovery system for real user needs with three steps. In the first step, it extracts and transforms existing data in the databases of the ERP and CRM systems of the sports facilities and loads them into a Data Warehouse. In a second phase, predictive models are applied to identify profiles more susceptible to abandonment. Finally, in the third phase, based on the previous models, experimental planning is carried out, with test and control groups, in order to find concrete actions for customer retention.

Keywords: Actionable knowledge discovery system · Sport services · Predictive analysis · Experimental planning · Loyalty

1 Introduction

The sports services sector is characterized by a high dropout rate [1]. In Portugal, according to the Fitness Barometer of the Association of Portuguese Gymnasiums and Academies [2] overall dropout rate in 2016 was 69% corresponding to a retention rate of 31%.

Actionable Knowledge is the knowledge required to initiate changes in the operational environment in order to create value. In this customer retention work we reuse some concepts of Database Marketing [3] and IDIC model (Identify, Differentiate, Interact, Customize) [4]. We intend to use existing data to identify the regular sports users services at risk of moving out, differentiating them from the rest through machine learning techniques and to interact with them using personalized loyalty actions. The proposed model, which includes the data preparation, the profile discovery using predictive models and the loyalty actions with evaluation, can be presented in the following schematic form:

$$Data \rightarrow Models \rightarrow Loyalty$$

This document has the following structure. In Sect. 2, a brief approach is given to a related work that has been done in the area of sports service retention and in the application of predictive analysis in retention. In Sect. 3, we present the data preparation methodology. In Sect. 4, we introduce the predictive model applied to the data

© Springer Nature Switzerland AG 2019
Á. Rocha et al. (Eds.): WorldCIST'19 2019, AISC 931, pp. 461–471, 2019.
https://doi.org/10.1007/978-3-030-16184-2_44

presented in the previous section and measure the results obtained. In Sect. 5, we approach the planning of experiences and the actions of loyalty that can be introduced. In Sect. 6 we present not only a proposal that allows us to measure the results obtained, but also a conclusion on the effectiveness of the loyalty actions. Finally, in Sect. 7 we present the contributions of the work.

2   Related Work

Studies [1, 5] carried out in the area of sports service retention generally conclude that user retention and loyalty is related to the quality of the facilities, staff and the overall quality of services provided, results which have been used by sports facilities management. This type of work is always based on surveys carried out on a sample of the population, which suggests the search for other methods to measure or ensure a lesser character generic and that allows to activate mechanisms directed to users in pre-dropout phase.

In addition to two studies [6, 7] applied to fitness and regular sport services, there were no other studies in the area of data mining applied directly to data obtained in sports services. However, the problem of high drop-out rates/low retention rates in other types of services has led to such churn prediction, especially in telecommunications, where high dropout rates are also observed. However, given the large size of databases and costs involved, most studies in this area use small samples of customer records, which may result in poor reliability and validity of the results obtained [8].

In order to focus their efforts on the clients they are most likely to meet and/or those who will be most profitable, companies seek to identify patterns and needs in customer groups (segmentation process). There are simple methods that do not require predictive models are RFM and RM [9] based on properties such as the Recency (when the customer's last visit occurred), Purchasing Frequency and the overall Monetary value.

As segmentation methods that use predictive models Siegel [10] refers to the Lift and Uplift models. Lift identifies customers who are most susceptible to a particular communication or marketing action. Uplift, in order to learn how to distinguish influential clients - those who make a difference in doing some treatment - learns from customers who have been contacted and those who have not been contacted, so it is necessary to use two data sets to train the model, a group of clients who are "treated" - treatment group - and another group of clients who are not - control group. The Uplift method also uses data mining techniques to segment through Uplift trees that, similar to decision trees, use attributes to automatically identify subgroups, but in a different way, try to identify extreme segments by the difference of treatments, identifying segments that are particularly influential.

Once one tries to define profiles of behaviors that lead to abandonment, it is necessary to find characteristics or attributes that somehow allow to trace those profiles. Work related to retention in sports services [1, 5, 11–13] allows systematizing and identifying attributes necessary to characterize users and their behavior, both those who continue to use the services and those who leave, and that can be found in the databases of the computer systems of these facilities: (1) demographic attributes such as age and gender; (2) Attributes related to contracting the service such as contracted frequency, number of months of enrollment and turnover (LTV); (3) Attributes related to frequency such as actual frequency, average frequency and number of days without visiting the premises; (4) Other attributes related to quality of service such as complaints or other manifestations of dissatisfaction, contacts made, assessments of the physical condition or any other type of evaluation.

3 Preparation of Data

Considering the need to obtain records with the attributes of the type indicated in the referred groups, it was considered a Lisbon sport facility database that uses a market application (e@sport) to which Extract, Transform and Load (ETL) processes were applied, as described by Trujillo [14], considering the entire history of users who were (or still are) enrolled in aquatic or fitness activities between 01/June/2014 and 31/October/2017.

According to the first step of Database Marketing is intended to create a data warehouse with a fact table where will reside the relevant attributes that will support the predictive model. Since the performance of some Machine Learning techniques is limited to the manipulation of values of a certain type or the performance itself is influenced by the range of values [15], in addition to the attributes directly mapped from the source database, some attributes have been transformed, discretized through numeric-symbolic conversions, or created new attributes that derive from classifications and transformations made on the original data or attributes. As so the ETL process resulted in the construction of a fact table in the data warehouse with 51 relevant attributes, although only forty-five have valid data. Relevant attributes, such as those related to the quality of the service were not filled due to lack of data. The attributes considered are presented in Table 1.

Table 1. Considered attribute groups.

Group	Attributes
(1) Demographic	Age (2 attributes), Gender, References (2 attributes), Distance to the facility
(2) Service level agreement	Number of months of enrollment (3 attributes), Turnover (2 attributes), Free Use, Attended activities (10 attributes), Number of activities attended, Contracted frequency (2 attributes), Number of renewals
(3) Frequency	Number of days without attendance (3 attributes), Average frequency (3 attributes), Total number of frequencies (2 attributes), Number of classes (2 attributes), Average frequency of classes (2 attributes), Ratio (real frequency/contracted frequency) (2 attributes), Training duration (2 attributes)
(4) Service quality	Number of contacts established, Indications of dissatisfaction (3 attributes), Number of manifestations of dissatisfaction, Last response NPS, Number of assessments of physical condition, Number of prescriptions

Table 1 shows that some attributes have variations that correspond to derivations that aim to discretize the value of the original attribute, and sometimes more than one method has been used. The Hughes method [16] (classification A - attributes whose name starts by *class*) was used in attributes *Number of months of enrollment, Turnover, Number of days without attendance, Average frequency, Total number of frequencies, Number of classes* and *Average frequency of classes*.

Through the indications obtained in the referenced literature, variant attributes were also added in a second classification (classification B - attributes whose name begins with *class2*) from the base attributes: *Days without frequency, Age, Number of months of enrollment, Average frequency, Ratio (real frequency/contracted frequency)* and *Training duration*.

In addition to the referred operations on the attributes, situations of missing values were also corrected through the strategy of removing the respective records.

Since most users of the sports facility practice aquatic or fitness activities, the users were grouped in three different fact tables according to the activities that they practiced during their frequency. After the execution of all ETL processes the final number of records is presented in Table 2.

Table 2. Number of users in data warehouse fact tables

Users	In aquatic activities	In fitness activities	Total number of users
Active	1226	803	1927
Dropouts	1697	4926	6454
Total	2923	5729	8381

4 The Predictive Model

To build and validate the predictive model we used Microsoft SQL Server Analysis Services Designer Ver. 13.0.1701.8. This product provides a classification algorithm, Microsoft Decision Trees, based on decision trees that are, according to several authors [7, 10, 17], adequate and most used in studies related to retention where we need to predict a class from a nominal attribute. In this case, the *Withdrawal* attribute is the attribute that we want to predict (the target attribute). By definition we considered the value 1 to classify a Dropout user and the value 0 for an Active user. In addition, decision trees produce human readable results which is very useful in this case.

The algorithm used evaluates the available attributes by punctuating each attribute according to the information it provides and proposes the construction of models based on the most scoring attributes. Since some attributes result from different forms of classification or discretization of the same characteristic, the proposed models eventually use redundant attributes. Gama *et al.* [15] states that since the process of constructing a tree selects the attributes to use, they result in models that tend to be quite robust in relation to the addition of irrelevant and redundant attributes. However, it is desired to obtain models with significant predictive capacity and that at the same time lead to actionable profiles redundant attributes should be avoided.

The elimination of the redundant attributes of the proposed models resulted in several models in which adjustments were made to obtain shallower trees with leaves that always have a number of examples greater than fifty. Table 3 present the evaluation metrics for each model obtained with the Holdout method considering 70% of data for training and 30% for testing of the models.

Table 3. Evaluation metrics of Predictive Models with Holdout Method

Model	#Nodes	Depth	Accuracy	Sensitivity	Specificity	Precision	F-Score
Ret71	30	6	87.90%	92.69%	72.53%	91.54%	92.11%
Ret81	37	6	87.90%	92.69%	72.53%	91.54%	92.11%
Ret91	45	7	87.58%	93.11%	69.85%	90.83%	91.95%
Fit71	18	5	87.94%	92.14%	63.64%	93.61%	92.87%
Fit81	24	5	88.40%	93.30%	60.08%	93.11%	93.21%
Fit91	22	5	87.94%	92.14%	63.64%	93.61%	92.87%
Aq71	12	4	87.91%	86.36%	90.03%	92.19%	89.18%
Aq81	20	5	88.14%	91.70%	83.29%	88.21%	89.92%
Aq91	18	5	88.14%	91.70%	83.29%	88.21%	89.92%

If it is not possible to consider a substantially better model among the models created, the choice on the model to be used falls on the model Ret71 created on the basis of all the users and that presents a tree in which the attributes used are less redundant and have fewer nodes (Principle of Parsimony/Occam's Razor) since it reduces complexity, minimizes the possibility of overfitting and facilitates the creation of actions in the next phase.

Figure 1 presents the correlation diagram between the attributes used in the chosen model that results the tree shown in Fig. 2.

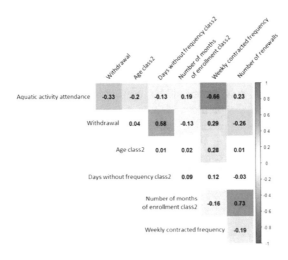

Fig. 1. Correlation diagram between the attributes used in model Ret71

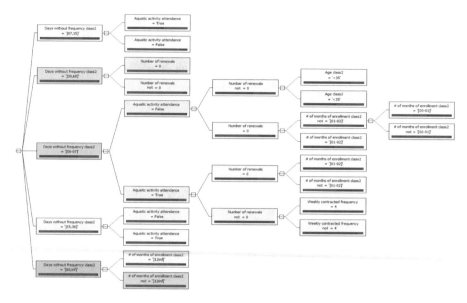

Fig. 2. Ret71 model decision tree

In each leaf of the tree there are examples that correspond to users quitting and examples that correspond to active users. The relationship between these quantities defines, on each leaf, a probability's threshold of withdrawal for the set of rules that define it. It is thus possible to draw dropout profiles from the rules on each leaf that shows a dropping rate above a considered threshold.

Considering, for example, a 90% dropout threshold, we found in the decision tree of Fig. 2 some leaves with upper thresholds, from which we can construct the profiles of Table 4.

5   Loyalty: Planning the Actions

In order to concentrate the efforts to increase the retention rate in groups more susceptible to abandonment of the sport, a bottom-up approach was adopted, according to Gorgoglione [18], and it is proposed the application of loyalty actions on the segmentation obtained with the predictive model referred to in the previous section, which obeys the segment utility criterion according to Kotler [19], which indicates that segmentation is only useful if the segments meet five criteria: they are measurable, substantial, accessible, differentiable and actionable.

For each of the actionable profiles obtained with the predictive model, a flow of conditionally sequential actions must be defined in which, at first, all the users that present this profile are the target of the first action. Secondly, only the users who did not change their behavior are the target of the second action, and finally, in a third moment, only the users who did not change their behavior after the first and second actions are the target of the third action.

Table 4. Dropout profiles obtained from the decision tree of the Ret71 model

Tree branch	Dropout profile

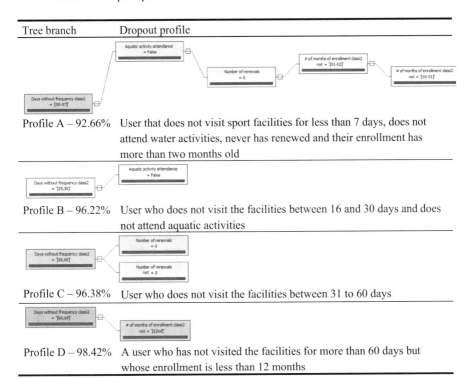

Profile A – 92.66%   User that does not visit sport facilities for less than 7 days, does not attend water activities, never has renewed and their enrollment has more than two months old

Profile B – 96.22%   User who does not visit the facilities between 16 and 30 days and does not attend aquatic activities

Profile C – 96.38%   User who does not visit the facilities between 31 to 60 days

Profile D – 98.42%   A user who has not visited the facilities for more than 60 days but whose enrollment is less than 12 months

If the change in characteristics and behavior of the user no longer places it on the leaf of the tree where it was initially placed, the user is no longer the target of this sequence or flow of actions. If this change fits into a target segment of another workflow, then the user is now framed in the flow of actions provided for the new profile.

Since it is expected that at each stage at least some of the target users of the corresponding action will change their behavior in order to not become dropouts, it is proposed to implement the workflow in three stages starting with the email, where the cost is practically zero, followed by the use of SMS and finally the personal contact, thus constructing a pyramid chaining, as shown in Fig. 3.

In addition to the cost, it is important to take into account the capacity of personalization and feedback and therefore many authors [19–21] attribute to email and phone great potential in promoting actions whose goal is to increase loyalty and retention rate.

Since the decision tree created by the algorithm may have multiple leafs where the withdrawal threshold is larger than indicated, multiple workflows with messages appropriate to the profile in which they affect can be designed and implemented.

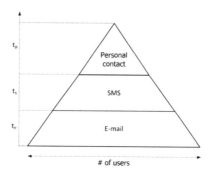

Fig. 3. Loyalty actions pyramid

The possible actions to be applied at each stage of the workflow can be grouped according to their purpose into four main groups: (a) informative actions, with personalized information about the schedules that the user can attend and about the benefits included in the contracted service of which may not be making use; (b) satisfaction perception actions, namely Net Promoter Score (NPS) surveys, reasons for absences and other quality surveys; (c) offer of benefits, namely at the level of free attendance of classes of other activities not included in the contracted service, substitution classes, participation in master classes or other events, free use of sports venues, gifts or vouchers; (d) withdrawal of subscription and/or updating of consents under the General Data Protection Regulation (GDPR).

6   Loyalty Evaluation: A/B Tests

After defining the model described above, it is necessary to evaluate its effectiveness according to the following hypotheses:

H0: After performing the loyalty actions, the number of dropouts is the same as if no loyalty actions had been taken;
H1: After performing the loyalty actions, the number of dropouts is lower than if the loyalty actions had not been carried out.

The determination of which of the hypotheses occurs in a certain confidence interval will allow us to conclude whether or not there is a causal relationship between the application of the actions to the constructed profiles and the reduction of the number of withdrawals. It is proposed to validate hypotheses by planning experiments with the configuration indicated in Table 5.

Table 5. Experimental planning

Groups		Email		SMS		Personal	
R_t	O_{t1}	X_1	O_{t2}	X_2	O_{t3}	X_3	O_{t4}
R_c	O_{c1}		O_{c2}		O_{c3}		O_{c4}

The experiments are constructed through the implementation of A/B tests and evaluated through the chi-square method, which will allow to gauge a statistical conclusion for the problem in question.

A/B tests are used by manipulating a causal variable and where it is sought to determine the impact of this manipulation on two different groups of individuals, one experiencing the experiment and the other does not. In this case, the groups are created by splitting the target users of the actions directed to a profile in two groups of users: those on whom the loyalty actions are applied, the test group (t), and another group on which they will not be applied, the control group (c).

For the creation of groups it is proposed that their constitution be made at the beginning of the experiment by dividing in equal parts the users who present the profile defined by the leaf of the tree on which the set of actions is intended to be applied. Since the users that present the profile have different dropout probability, it is proposed to sort them in descending order of that probability, alternating their placement in each of the test and control groups until exhaust the users. This process, although not purely random, creates homogeneous and equivalent groups in terms of probability of withdrawal, which allows avoiding problems with validation in terms of selection for group composition and generalization of experience results [22].

The application of the chi-square method to these groups should have two objectives so it must be done in two steps. In a first step, the application of the method must be done after each action, which will allow a performance evaluation of the action in concrete. In a second step, the application of the method must be done after all actions have been taken to evaluate the model as a whole.

The application of the method can be carried out by the construction of a matrix where the observed (O) and expected values (E) of the number of users who have withdrawn and do not give up at the beginning and after the application of each action or after all actions have been taken, in the case of overall assessment.

If the result obtained for the overall evaluation of the model has a confidence level higher than 95%, H0 can be rejected and it can be concluded with statistical relevance that after performing the loyalty actions, the number of dropouts is lower than if no loyalty actions had been taken.

7 Conclusions

With this work we try to present a valuable contribution to increase loyalty and retention rates in regular sports services through the development of a system that generates actionable knowledge based on real data.

Based in database marketing concepts this Actionable Knowledge Discovery System is based on three steps:

- Use real data from the ERP and CRM systems to extract records and attributes that characterize the behavior of the users that dropout;
- To obtain segments and profiles through the construction of predictive models that allow in a certain threshold of certainty to differentiate the users who dropout from the other users;

- To interact with users who are about to give up by implementing loyalty actions specifically directed to the characteristics of the profiles where each user fit; the value-added of loyalty actions is measured using A/B tests.

This actionable knowledge discovery system was developed in Cedis enterprise and is being implemented in two final customers before GDPR came into effect.

References

1. Avourdiadou, S., Theodorakis, N.: The development of loyalty among novice and experienced customers of sport and fitness centres. Sport Manag. Rev. **17**, 419–431 (2014)
2. AGAP: Barómetro 2016 (2016)
3. Cavique, L.: Relatório da Unidade Curricular de Database Marketing, 2005–2006, Escola Superior Comunicação Social Instituto Politécnico de Lisboa
4. Peppers, D., Rogers, M.: Managing Customer Relationships. Wiley, Hoboken (2004). (30097)
5. Howat, G., Assaker, G.: Outcome quality in participant sport and recreation service quality models: empirical results from public aquatic centres in Australia. Sport Manag. Rev. **19**, 520–535 (2016)
6. Pinheiro, P., Cavique, L.: Determinação de padrões de desistência em Ginásios. Revista de Ciências da Computação **10**, 33–60 (2015)
7. Pinheiro, P., Cavique, L.: Modelos para incremento da retenção em serviços desportivos regulares: Análise preditiva e ações de fidelização. In: 13ª Conferência Ibérica de Sistemas e Tecnologias de Informação (CISTI), pp. 1–6 (2018)
8. Mahajan, V., Misra, R., Mahajan, R.: Review of data mining techniques for churn Prediction in Telecom. J. Inf. Organ. Res. **39**, 183–197 (2015)
9. Cavique, L.: Micro-Segmentação de Clientes com Base de Dados de Consumo: Modelo RM-Similis. Revista Portuguesa e Brasileira de Gestão **2**, 72–77 (2003)
10. Siegel, E.: Predictive Analytics: The Power to Predict Who Will Click, Buy, Lie or Die. Wiley, Hoboken (2013)
11. Surujlal, J., Dhurup, M.: Establishing and maintaining customer relationships in commercial health and fitness centres in South Africa. Int. J. Trade Econ. Finance **3**, 14–18 (2012)
12. Gonçalves, C.: Variáveis internas e externas ao indivíduo que influenciam o comportamento de retenção de Sócios no Fitness. Podium Sport Leis. Tour. Rev. **1**, 28–58 (2012)
13. Frota, M.: Gestão da Retenção, Manual de Gestão de Ginásios e Health Clubs – Excelência no sector do Health & Fitness, pp. 103–148 (2011)
14. Trujillo, J., Luján-Mora, S.: A UML based approach for modeling ETL processes in data warehouses. In: Conceptual Modeling ER, vol. 2813, pp. 307–320 (2003)
15. Gama, J., Carvalho, A., Faceli, K., Lorean, A., Oliveira, M.: Extração de Conhecimento de Dados (2017)
16. Hughes, A.: Strategic Database Marketing. McGraw-Hill, New York (2011). ISBN 9780071773485
17. Tan, P., Steinbach, M., Kumar, V.: Introduction to Data Mining. Pearson, Boston (2016)
18. Gorgoglione, M.: Beyond customer churn: generating personalized actions to retain customers in a retail bank by a recommender system approach. J. Intell. Learn. Syst. Appl. **3**, 90–102 (2011)
19. Kotler, P., Keller, K.: Marketing Management, 14th edn. Pearson (2009)

20. Pousttchi, K., Wiedemann, D.: A contribution to theory building for mobile marketing: categorizing mobile marketing campaigns through case study research. In: Mobile Business, vol. 2925 (2006)
21. Merisavo, M., Raulas, M.: The impact of e-mail marketing on brand loyalty. J. Prod. Brand Manag. **13**(7), 498–505 (2004)
22. Smith, S., Albaum, G.: An Introduction to Marketing Research (2010)

Real Time Fuzzy Based Traffic Flow Estimation and Analysis

Muhammad Abbas[1,2,3], Fozia Mehboob[1,2,3], Shoab A. Khan[1],
Abdul Rauf[2(✉)], and Richard Jiang[3]

[1] Department of Computer Engineering,
National University of Science and Technology, Islamabad, Pakistan
`Fouzia.malik@ceme.nust.edu.pk`
[2] Department of Computer Science, Al-Imam Mohammed Ibn Saud Islamic
University (IMSIU), Riyadh, Saudi Arabia
`drraufmalik@gmail.com`
[3] Department of Computer Science, Northumbria University,
Newcastle upon Tyne, UK

Abstract. Real-time traffic flow analysis using road mounted surveillance cameras presents multitude of benefits. In this paper, we used surveillance videos to design optical flow based technique for robust motion analysis and estimation. Region growing method is employed for detection of objects of interest. Autonomous density estimation of vehicles is crucial for traffic congestion analysis so that countermeasures can be taken at the earliest possible opportunity. A video based data extraction scheme for traffic data is proposed to determine the right traffic conditions which alleviates the false alarms and detrimental noise effects. Evaluation of proposed system is done by applying approach on several surveillance videos obtained from different sources and scenarios. An experimental study illustrates estimation and analysis results accuracy as compared to state-of-the-art approaches.

Keywords: Traffic surveillance videos · Flow estimation · Smart city

1 Introduction

Goal of the intelligent transport system (ITS) is to enhance the public safety, provide efficient transit and travel information, reduce the congestion and detrimental impacts of the environment. The efficacy of these systems depends on the comprehensiveness and performance of vehicle detection technologies. Detection and tracking of vehicles are integral part of such detection techniques that gather all the information required in effective and intelligent transport system. Video based vehicle detection is one of most successful technologies for the implementation of advance technologies of traffic management, tracking schemes, and collecting data on large scales. For such techniques, information extraction from surveillance video based systems is vital for efficient traffic management and signal control [1]. To alleviate the above-mentioned problems i.e. surveillance video information extraction, a number of algorithms have been employed for automatic detection of incidents and vehicles.

© Springer Nature Switzerland AG 2019
Á. Rocha et al. (Eds.): WorldCIST'19 2019, AISC 931, pp. 472–482, 2019.
https://doi.org/10.1007/978-3-030-16184-2_45

In this paper, a novel optical flow based method is proposed for estimation of robust density and tracking. In this methodology: Sect. 2 presents the literature review, Sect. 3 offers the comprehensive information regarding the proposed approach. In Sect. 4, the performance of the proposed approach is discussed in experimental results. Lastly, Sect. 5 presents the concluding remarks.

2   Relevant Work

There are two main approaches which are in fashion for the detection of vehicles such as motion information and vehicle inherent features. To improve the segmentation of moving objects, background subtraction approach has been presented by different researchers [6, 7]. Several other methods for the estimation of the background were also proposed in [12, 15]. Full search block matching based method was presented in [16] which reduced the false motion by applying adaptive threshold and detecting only non-static parts. In addition, level set method was utilized and kalman filter was employed to enhance the vehicle classification accuracy [21]. A generative model was presented on the basis of edges and Markov chain model which perform vehicle detection and segmentation in static images [23]. Few of studies combined motion information and inherent features such as edge and color information [22]. Two algorithms were proposed in [2] which deal with the classification of traffic density in night-time and daylight traffic conditions. Collins [10] proposed a three-frame differencing method by incorporating adaptive background subtraction model. Subtraction algorithm was also used for detection of vehicle but it required scene information and approach was unable to handle occlusion problem [11].

3   Proposed Methodology

An efficient vehicle detection method is significant for eliminating noise; which otherwise affects object detection process. Proposed approach eradicated the detrimental noisy effects and parses only the object of interest. The Proposed adapted method comprises of subsequent steps as depicted in Fig. 1.

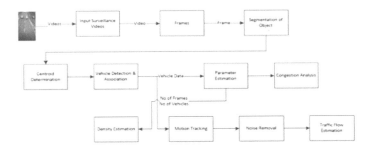

Fig. 1. Schematic diagram of proposed approach

We proposed herein a novel technique for real time estimation of traffic density. Though optical flow is computationally expensive but it offers higher detection accuracy and is appropriate for moving object analysis in complex scene of videos [18]. Moreover, robust region growing method is used for occlusion problem handling.

3.1 Vehicle Segmentation and Noise Removal

For vehicle segmentation, unknown frame (input) is segmented out into multiple objects of interest using region growing method. Conversion of grey scale images to binary images is performed using thresholding function. An appropriate threshold value was set; which reduces the false alarms effectively. In region growing method, 3X3 neighborhood is used and pixels are extracted from continuous changing region (Fig. 2).

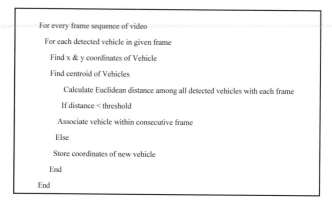

Fig. 2. Pseudo code of the vehicle tracking

Vehicle position is determined by identifying the centroid value using:

$$\sum_{(i,j)\in W}(I1(i,j) - I2(x+i,y+j))^2 \qquad (1)$$

Where I1 is vehicle location (i, y) and I2 is position in subsequent frame (Figs. 3 and 4).

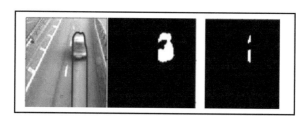

Fig. 3. (a) Vehicle detection (b) Vehicle segmentation without shadow (c) Vehicle shadow

Fig. 4. (a) Vehicle detection (b) Segmentation without shadow (c) Vehicle shadow

3.2 Inter Frame Vehicle Association

Inter-frame vehicle association is performed between the consecutive frames. On the appearance of vehicle in the scene, coordinates of the vehicle are stored. In addition, using the proposed inter-frame association algorithm, it is compared whether the vehicle is new or existing one. Algorithm counts the number of frames since the vehicle appearance. Tuned threshold for vehicle association between consecutive frames is 1000 pixel. If detected vehicle in the frame sequence is exiting one, then association between the previous and the following frames is performed.

3.3 Optical Flow Based Estimation of Motion

In optical flow based techniques, there is distinction among moving objects and the static background. And computation is based on the relative motion among the observer and the scene Direct integration of global information eliminates the need of post processing vector field to eradicate outliers. Optical flow results in less accumulate error due to less changes in the brightness and long distance coverage while accommodating the effects of noise [22]. The ultimate postulation behind the optical flow is brightness constancy [23], or some property of image, despite pixels changing position. Mathematically

$$(i,j) = I_{t+1}(i + u_{itj}, j + v_{itj}) \tag{2}$$

Where It + 1 and It illustrates the current and next frames, i and j are the horizontal and vertical pixels. Vertical and horizontal motion of pixels is represented as u_{itj} and v_{itj}. Individual tracking of pixels is possible with flow vectors of each pixel in blob. Two dimensional vectors having velocity comprises velocity and direction of motion of each pixel.

$$\int \Omega(\frac{\partial I}{\partial x}\delta x + \frac{\partial I}{\partial y}\delta y + \frac{\partial I}{\partial t} + \alpha(|\Delta\delta x|_2 + |\Delta\delta y|))d_x d_Y \tag{3}$$

Velocity of pixel (x, y) is neighborhood average of other pixels. In the computation and estimation of optical flow, noisy data should be taken into account. The proposed approach performs robust estimation of optical flow as compared to existing approaches. The coordinates of corresponding flow vectors are tracked and the generated motion vectors are stored as complex numbers. Initial threshold value is set and if the velocity of both horizontal and vertical components are less than or equal to threshold value, the coordinate value is set to zero. After applying the threshold, all horizontal and vertical components having values greater than the threshold value are added together. The average of horizontal and vertical components is calculated if and only if the sum of vertical and horizontal components is higher than the threshold value.

Density of vehicles crossing the scene over successive frames is calculated for given period of time. Individual and mean speed of vehicle in each frame is estimated in pixels per second by using the distance travelled by the vehicle.

$$S = \frac{n}{\sum_{i=1}^{n} \frac{1}{V_i}} \tag{4}$$

4 Fuzzy Logic Based Analysis

The parameters are extracted using the image processing based video analysis. These parameters are used to estimate congestion on the road. A fuzzy logic based congestion analysis algorithm is proposed for the estimation of traffic congestion. Fuzzy inference system performs the mapping of given inputs to outputs. Congestion is decided by the fuzzy inference systems.

4.1 Fuzzy Logic Base Alternate Route Planning

Based on the fuzzy logic based congestion estimation, we can suggest alternative routes to the vehicles.

The proposed framework application as shown in Fig. 5 can provide alternative route information to vehicles along all roads. Road side units can be installed along roads and based on the congestion information alternate route information can be provided all along the road. Existing alternate route planning systems provides alternate routes to vehicles only at junctions or where the sensors are installed.

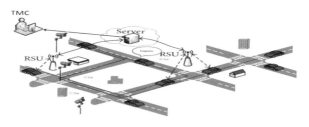

Fig. 5. Alternate route planning

5   Experimental Results

5.1   Dataset

To evaluate the proposed method, a dataset is collected from Motorway (UK) with the different traffic conditions. A camera is mounted on bridge of 20 ft high over motorway. Proposed approach and Kalman filter are employed for density estimation and results are compared to check the efficacy of both approaches.

5.2   Flow Estimation Using Fuzzy Logic

The estimated traffic flow per frame from traffic surveillance videos is shown in Fig. 6. The peaks in Fig. 6 depicts that the vehicles flow in these frames are more than the other frame sequences. Inter-frame level analysis reveals that when the congestion level increases results in decrease in vehicle mean speed. Due to congestion, users are limited to drive with slow or limited speed.

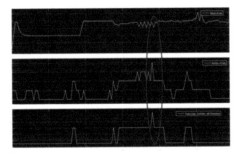

Fig. 6. Fuzzy logic based results.

Figure 7 illustrates the real time video results, which comprises of 1100 frames. As the number of vehicles increases, the mean vehicle speed tends to decrease showing the trend of congestion.

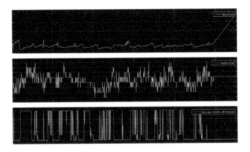

Fig. 7. Fuzzy logic based results.

5.3   Noise Removal

The proposed method is less sensitive to noise than the most recent technique [21], because this method uses sum of optical flow vectors. The temporal and spatial gradient of pixels and neighborhood average of previous iteration are used for next flow vector computation. As it can be seen from Fig. 8, motion vector computation is performed and there is a lot of noise in the relevant image.

Fig. 8. (a) Video sequence (Original) (b) Motion vectors

The averages of those pixel velocities are considered whose values are greater than certain threshold value (0.25) for accurate measurement of speed of vehicle while minimizing the error. In Fig. 9, noisy motion vectors are eradicated and only motion vectors within the bounding box of objects of interest are kept for analysis.

Fig. 9. (a) Video sequence (original), (b) Motion vectors (after noise removal)

Proposed approach removes the background noise and is more robust in accurate estimation of flow of vectors.

6   Comparison with Kalman Filter

A comparison is made between proposed approach and kalman filter for moving object detection per frame. Vehicle which is occluded is detected and tracked by proposed tracker. As proposed tracker saves the tracks of individual vehicle, if the vehicle is miss detected (in few frames), interpolation is employed for missing tracks.

Figure 10 shows the density estimation using the proposed system and Kalman tracker. In Fig. 11, few of miss detections have been observed in video frames.

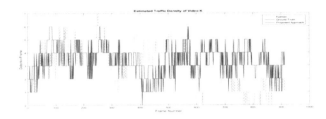

Fig. 10. Density estimation comparison with Kalman tracker of video K

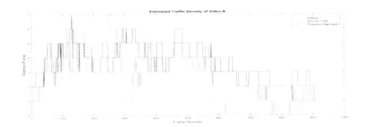

Fig. 11. Density estimation comparison with Kalman tracker of video B

It is obvious from experiments, that the proposed method yields better outcomes for vehicle detection and estimation of density with high fidelity.

The proposed technique is compared with the state of the art techniques in the literature as shown in Table 1. Comparisons are done demonstrating the uniqueness of the algorithms, set of violations and their performance with the other techniques.

Table 1. Comparison of proposed algorithm with other techniques

Ref.	Purpose	Technique	Accuracy	Year
[34]	Vehicle detection	Kalman filter	66%	1998
[35]	Vehicle detection & tracking	Kalman filter	70%	2005
[36]	Vehicle tracking	Kalman filter	80%	2008
[37]	Vehicle tracking	Kalman filter	60%	1995
[38]	Vehicle detection & tracking	Kalman filter	87%	2002
[39]	Vehicle detection & tracking	Kalman filter & GMM	97%	2015
[40]	Vehicle tracking	Kalman filter & probability product kernel	95%	2016
	Vehicle detection & tracking	**Proposed approach**	99%	2018

As shown in Fig. 12, Kalman tracker linked some detection which were noisy whereas the proposed tracker generates paths without linking noisy detections (Fig. 13).

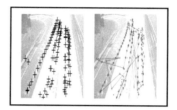

Fig. 12. (a) Vehicle detection of standard video using Kalman filter (b) Vehicle tracking

Fig. 13. (a) Vehicle detection of standard video using proposed technique (b) Vehicle tracking

Tracking results varies using Hungarian tracker as it depends on the linking distance as shown in Fig. 14.

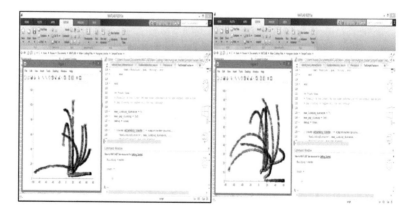

Fig. 14. Vehicle detection & tracking of standard video using Hungarian tracker

7 Conclusion

A practical surveillance video based system is proposed based on improved optical flow for density estimation. Moving object position and speed is successfully estimated in different situations i.e. occlusion, changes in moving object direction and abrupt change in object speed. In addition, multiple objects are tracked within video frame sequences; however, processing time required for individual tracking of object is minimum in comparison to existing techniques.

Algorithm efficacy is tested on several real life videos, which comprises of different resolution, environment and frame rates. Performance calculation on the basis of real input video illustrates that proposed system offer better density estimation accuracy compared to Kalman Filter. Experimental outcomes prove that the proposed system offers excellent performance accuracy regardless of frame resolution.

References

1. Ozkurt, C.: Automatic traffic density estimation and vehicle classification for traffic surveillance systems using neural networks. Math. Comput. Appl. **14**(3), 187–196 (2009)
2. Sen, R.: Accurate speed and density measurement for road traffic in India. In: DEV 2013, Bangalore, India, 11–12 January 2013
3. Lee, J.: Estimation and analysis of urban traffic flow. In: ICIP (2009)
4. Sun, D.: From Pixels to Layers: Joint Motion Estimation and Segmentation, The Chinese University of Hong Kong (2007)
5. Lipton, A.J., Fujiyoshi, H., Patil, R.S.: Moving target classification and tracking from real-time video. In: Fourth IEEE Workshop on Applications of Computer Vision, WACV 1998. Proceedings, pp. 8–14. IEEE (1998)
6. Liu, Y., Ai, H., Xu, G.: Moving object detection and tracking based on background subtraction. In: Multispectral Image Processing and Pattern Recognition, pp. 62–66. International Society for Optics and Photonics (2001)
7. Collins: Motion Detection in Static Backgrounds, SpringerBriefs in Computer Science. https://doi.org/10.1007/978-1-4471-4216-4_2,2012
8. Bas, E.: Road and Traffic Analysis From Video, August 2007
9. Zhu, Z., Xu, G.: VISATRAM: a real-time vision system for automatic traffic monitoring. Image Vis. Comput. **18**(10), 781–794 (2000)
10. Sellent, A.: Optical Flow Estimation Versus Motion Estimation, August 2012
11. Horn, B.K.P., Schunck, B.G.: Determining optical flow. Artif. Intell. **17**(1–3), 185–203 (1981)
12. Sen-Ching, S.C.: Robust Techniques for Background Subtraction in Urban Traffic Video (2007)
13. Kalman, R.E.: A new approach to linear filtering and prediction problems. Trans. ASME J. Basic Eng. **82**(Series D), 35–45 (1960)
14. Catlin. D.E.: Estimation, Control, and the Discrete Kalman Fillter. Springer, New York (1989)
15. Mandellos, N.A., Keramitsoglou, I., Kiranoudis, C.T.: A background subtraction algorithm for detecting and tracking vehicles. Expert Syst. Appl. **38**(3), 1619–1631 (2011)

16. Unzueta, L., Nieto, M., Cortés, A., et al.: Adaptive multicue background subtraction for robust vehicle counting and classification. IEEE Trans. Intell. Transp. Syst. **13**(2), 527–540 (2012)

17. Pang, C.C.C., Lam, W.W.L., Yung, N.H.C.: A method for vehicle count in the presence of multiple-vehicle occlusions in traffic images. IEEE Trans. Intell. Transp. **8**(3), 441–459 (2007)

18. Zhou, J., Gao, D., Zhang, D.: Moving vehicle detection for automatic traffic monitoring. IEEE Trans. Veh. Technol. **56**(1), 51–59 (2007)

19. Wu, B.F., Juang, J.H.: Adaptive vehicle detector approach for complex environments. IEEE Trans. Intell. Transp. Syst. **13**(2), 817–827 (2012)

20. Gangodkar, D., Kumar, P., Mittal, A.: Robust segmentation of moving vehicles under complex outdoor conditions. IEEE Trans. Intell. Transp. Syst. **13**(4), 1738–1752 (2012)

21. Chen, Z., Ellis, T., Velastin, S.A.: Vehicle detection, tracking and classification in urban traffic. In: 15th International IEEE Conference on Intelligent Transportation Systems (ITSC), Anchorage, September 2012, pp. 951–956 (2012)

22. Tsai, L.W., Hsieh, J.W., Fan, K.C.: Vehicle detection using normalized color and edge map. IEEE Trans. Image Process. **16**(3), 850–864 (2007)

23. Jia, Y., Zhang, C.: Front-view vehicle detection by Markov chain Monte Carlo method. Pattern Recognit. **42**(3), 313–321 (2009)

Features Weight Estimation Using a Genetic Algorithm for Customer Churn Prediction in the Telecom Sector

Adnan Amin[1], Babar Shah[2], Ali Abbas[3], Sajid Anwar[1],
Omar Alfandi[2], and Fernando Moreira[4,5(✉)]

[1] Institute of Management Sciences, Peshawar 25000, Pakistan
{adnan.amin, sajid.anwar}@imsciences.edu.pk
[2] College of Technological Innovation, Zayed University, Dubai, UAE
{babar.shah, Omar.alfandi}@zu.ac.ae
[3] Department of Computing, Middle East College, Muscat, Oman
aabbas@mec.edu.om
[4] REMIT, IJP, University of Portucalense, Porto, Portugal
fmoreira@upt.pt
[5] IEETA, University of Aveiro, Aveiro, Portugal

Abstract. The high dimensional dataset results in more noise, will require more computations, has huge sparsity linked with high dimensional features and has thus introduced great challenges in data analysis. To efficiently manipulate and address the impact of the challenges faced by high dimensional dataset, researchers used several features reduction methods. Feature reduction is a formidable step, when dealing with improving the accuracy and reducing the processing time, within high-dimensional data; wherein the feature set is reduced before applying data mining or statistical methods. However with attribute reduction, there is a high chance of loss of important information. In order to avoid information loss, one way is to assign weights to the attributes through domain-expert which is a subjective exercise. It is not only costly but also requires human-expert of the field. Therefore, there is a need for a technique to automatically assign more appropriate weights without involving domain expert. This paper presents a novel features weighting technique. The technique employs a genetic algorithm (GA) to automatically assign weights to the attributes based on Naïve Bayes (NB) classification. Experiments have been conducted on publically available dataset to compare the performance of the proposed approach and NB approach without the weighted features for predicting customer churn in telecommunication sector. The experimental results have demonstrated that the proposed technique outperformed through achieving an overall 89.1% accuracy, 95.65% precision which shows the effectiveness of the proposed technique.

Keywords: Naïve Bayes · Classification · Attribute weighting ·
Genetic algorithm

© Springer Nature Switzerland AG 2019
Á. Rocha et al. (Eds.): WorldCIST'19 2019, AISC 931, pp. 483–491, 2019.
https://doi.org/10.1007/978-3-030-16184-2_46

1 Introduction

High dimensional data is relevant to a large range of domains such as telecommunication, medicine, sensors & networks, and industrial applications [1], and thus its manipulation has greater associated formidable problems such as low proficiency of conventional data analysis, are very costly in term of required processing time, poses the curse of dimensionality, and data complexity [2]. Even though, with more increase speed in modern computers and cheaper parallel processing facilities, there is a need to efficiently solve the problems of high-dimensional data analysis [3–6]. For this purpose data mining and statistical methods are used to deal with multiple features set, and techniques such as feature reduction and subset selection methods can be used that improves the classification accuracy and requires low computation cost of data mining process [1]. However by overlooking the fact, that by feature reduction or subset selection from the dataset, there is a high chance of loss of important information that may be very important in future analysis.

In order to avoid data loss in the feature reduction process, researchers assign different weights to attributes with the help of domain-expert which is also costly, requires the availability of the human-expert, and is subjective exercise. Therefore, automated (self) feature weight assignment technique provides excellent motivation for the proposed study to analyze the impact of the features with different weights instead of feature reductions. Usually, weights assignment is determined by the subjective weight method in which decision makers or domain experts manually decides the weights for each attribute based on the expert's past experience and knowledge. Subjective weight assignment method is flexible but lead to much more subjective randomness and is difficult to avoid [7].

The proposed approach will take advantage of information of each attribute and accordingly determine weights for the attributes and will not depend on the subjective judgment of the domain expert to assign appropriate weights to these attributes. The weights will be assign to the attributes through GA. In this study, our focus is on another predictive study to analyze the effects of the NB classification algorithm with different weighted features using a genetic algorithm. It is observed in many cases that the performance of NB usually outperforms neural network and decision tree method [8]. The Bayes classification's strict assumption is that each instance must be independent of another. Therefore, NB assumes that the attributes are statistically independent based on the given value of the decision attribute. Although such an assumption is never inevitably insufficient and never holds in practice [9]. However, this assumption greatly affects the classification of the most real-world applications and therefore cannot be applied.

The paper is organized as follows: in Sect. 2, we explore the related studies. Section 3 introduces the methodology proposed in this work followed by results and discussion in Sect. 4. The paper is concluded in Sect. 5.

2 Related Work

In this section, we provide a concise review of the related studies.

For any business, including the telecommunication industry, its financial growth and reputation in the saturated market is greatly dependent on the efficiency of its CCP model [10]. The target domain (i.e., telecommunication companies) faces the rivalry to keep clients from access of each other. The customers are constantly searching for better service including data services, voice services, call services, and messages etc. Therefore, to retain the existing customer CCP becomes a crucial problem for the target domain. In the literature, many state-of-the-art machine learning techniques (including KNN, ANN, NB, SVM, random forest, logistic regress, and decision tree) have been applied and their performance have been investigated on CCP [11–13]. Vijaya et al. [10] focused on the unwanted features and considered that the undesired features for the CCP in telecommunication is a challenging problem. They have applied Rough Set theory (RST) for feature selection with ensemble classification. As a result, the feature set was reduced before giving to ensemble classification method (Bagging, Boosting, and Random Subspace).

For showing the predictive performance of the classification technique, many studies from the machine learning community have used feature weighting procedures especially designed for simple NB classifier. Zhang et al. [14] presented a feature weighting technique based on gain-ratio for improving the accuracy of the NB. Besides, Hall et al. [15] introduced another feature weighting method based on a decision tree which estimates the attribute's dependency on other attribute's values for standard NB. On the other hand a refined differential evolution method to assign weights to the feature set, is used by Wu and Cai [16]. Frank et al. [8] introduced a locally weighted method of NB algorithm that relaxes the independence assumption by learning local classification models during the prediction process. It enhanced the NB performance and was computationally simple as well. Zaidi et al. [17] investigated and reduced the negative conditional log-likelihood rather than to select feature weights. Lee [18] proposed another paradigm for values weights assignments of the attribute instead of assigning weights directly to each attribute. He assigned weights to each feature's values. Although weight assigning to attributes seems to be an effective procedure to increase the simple NB performance, the results obtained, however, from the current methods are not very inspiring in terms of predictive accuracy [19]. Therefore, Jiang et al. [19] only incorporated the weighted learned attributes in the classification formula of NB and not in its conditional probability estimates.

Aforementioned discussion, on all feature weighting methods, shows a somewhat good performance to improved simple NB. Consequently, the feature weighting methods using gain-ratio and decision tree are comparatively more computationally simple and therefore executes quickly. However by applying these methods directly, no such improvement in classification performance has been observed when applied to real world problems [20]. In this study, we have focused on the advantage of information of each attribute and assign weights to these attributes instead of assigning weights to values or assigning weights through domain expert. Rather we have used the genetic algorithm. The next section describe the methodology of the proposed approach, using a genetic algorithm, in detail.

3 Empirical Setup

In this section, an integrated approach is proposed comprised of weights assignments through a genetic algorithm and NB classification for dealing the high dimensional data relating to customers churn prediction in telecommunication sector.

3.1 Subject Dataset and Preprocessing

In this study the aforementioned publicly available dataset is acquired and considered for performance evaluation of the proposed approach. The subject dataset is also used in the following studies [12, 21]. There are 3333 samples and 21 attributes including the class label or decision attribute in the subject dataset. Further statistical description about subject dataset can be found from [22].

To obtain the appropriate data it is important to perform common steps of data preprocessing before developing the CCP model. In the subject dataset, around 15% of the data consists of customer churn, and nearly 85% of customers are non-churn. Therefore in order for the proposed system's consistency to increase, in this phase of the CCP model development phase we have eliminated the Phone_Number attribute from the feature set because the nature of this attribute was a unique value which was used to identify unique customers. The size of the features set is 20 including 1 decision attribute and the decision attribute consisting of two possible class label values of the churn and non-churn customers. We have transformed the class labels for simplicity such as churn to C and non-churn to NC.

3.2 Baseline Classification Algorithm

As discussed earlier in the literature review section that many studies have been proposed for fine tuning the NB algorithm and performance of the NB in binary classification. However, the improvement is little as compared to the neural network and decision tree [8]. In this study, we have used NB algorithm as baseline classification method because NB performs surprisingly well in many binary classification real-world problems for its simplicity, its better ability when the data is incomplete and computationally efficient classification technique as well as simple to implement [23]. The detail about the baseline classification algorithm is addressed in this section.

NB belongs to the Bayes family and is based on a Bayesian theorem which is a statistical classification method while more detail about NB can be found in study [24]. Let $X = \{x_1, x_2 \ldots \ldots \ldots \ldots x_n\}$ be the customer set, and Class labels C = C_1, C_2 where C1 and C2 are two class labels such as Churn and Non-Churn, respectively. The probabilities for the customers to be classified can be computed using Eq. 1.

$$P(X|C_i) = \frac{P(X|C_i).P(C_i)}{P(X)} \tag{1}$$

The highest probability value of the customers can be obtained from Eq. 1, categorizing the customers to a specific class.

3.3 Measure Index

The classifier's performance is commonly measured by using the evaluation measures (i.e., Recall, Precision, Accuracy, and F-Measure) while further detail can be obtained from [23]. The measured index can be methodically expressed as:

$$Recall = \frac{True\ Positive}{True\ Postive + False\ Negative} \tag{2}$$

$$Precision = \frac{True\ Positive}{True\ Positive + False\ Postive} \tag{3}$$

$$F_{Measure} = 2 \times \frac{precision \cdot recall}{precision + recall} \tag{4}$$

3.4 Proposed Solution

We have implemented the proposed solution in MATLAB toolkit[1] environment. The step by step procedure (i.e., Algorithm) is as follows:

Algorithm 1. Proposed solution

```
Step 1: Load original dataset
Step 2: Perform the data preprocessing (see Section 3.1)
Step 3: Select feature set from f₁ ...fₙ from step 2
Step 4: Assign initial random weights W₁ ... Wₙ where Wᵢ
        values in the range between 0 to 1
Step 5: Integrate Genetic Algorithm module where 10
        samples are selected in each iteration (genera-
        tion) and chromosomes are the set of the weights
        for every features fᵢ Such that
        S₀=W₀₁, W₀₂, W₀₃, ...Wₒₙ
        S₁=W₁₁, W₁₂, W₁₃, ...W₁ₙ
        S₂=W₂₁, W₂₂, W₂₃, ...W₂ₙ
        .. ... .. ... .. .. ...
        Sₙ=Wₚ₁, Wₚ₂, Wₚ₃, ...Wₚₙ
        Where p is population.
Step 6: For each population k=1 ... 10
        Sₖ  = Wₖ₁ f₁, Wₖ₂ f₂, Wₖ₃ f₃.. Wₖₙ fₙ
Step 7: Cross-over
        Sk=Wₖ₁f₁,..Wₖₙ/₂ fₙ/₂,W₍ₖ₊₁₎ₙ/₂₊₁ fₙ/₂₊₁ .. W₍ₖ₊₁₎ₙ fₙ
        Sk+1=W₍ₖ₊₁₎₁ f₁,..W₍ₖ₊₁₎ₙ/₂ fₙ/₂, W₍ₖ₎ₙ/₂₊₁ fₙ/₂₊₁..W₍ₖ₎ₙ fₙ
Step 8: Mutation Wₖₙ=rand()*Wₖₙ
Step 9: Objective Function, the Minimum least square er-
ror
```

---

[1] https://www.mathworks.com/products/matlab.html.

Initially, we have considered all the features with random weighted value and created 10 samples for each generation. The next step evaluates the performance of the baseline classifier on each sample (i.e., S0, S1, S2, ... S10) for each generation (i.e., number of generation is set to 100). Then we performed the crossover and mutation operations of genetic algorithm (GA) as these are two basic and depended operations of GA. Mutation allows the multiplicity by letting an offspring to change in such a way which cannot be merely identified by the inherited characteristics. Similarly, the crossover is applied to generate new solutions from the population's samples from genetic information (i.e., crossover samples 10 + 10 and selected best 10 samples) while mutation is used to create new genetic information.

For next every generation (iteration), the weights (e.g., between 1 to 10) were assigned to features and we evaluated the baseline classifier's performance. Finally, same was applied to the best results obtained from each generation. The same procedure is tested until there is no further improvement and observed result is smooth. Figure 1 depicts the performance of the baseline classifier.

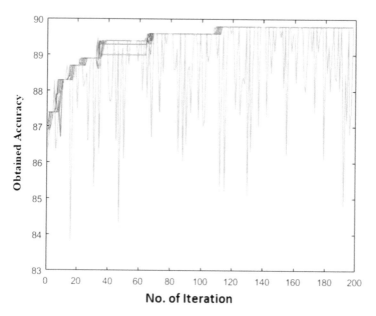

Fig. 1. Obtained accuracy on weighted features for the best 10 samples in each generation.

Models are created using the baseline classification algorithm to make CCP on the given dataset with and without weight assignments. To see the performances of the CCP models, the dataset is divided, through cross-validation technique, into training set and test sets. The training set data is used for model training or supervision purpose while the test set data evaluates the performance of models in every generation.

4   Results and Discussion

In this paper, we have studied various weights assigned to features using a genetic algorithm instead of features reduction or subjective weight assignment methods. Various weights were assigned to features and only those weights were selected which provided the best performance in CCP models based on baseline classifier (see Sect. 3.2). Additionally, we have also compared the performance of the proposed technique with and without weight assignments to the features using the same baseline classifier to evaluate our technique.

Table 1 depicts the results that were obtained from baseline classifier without weighted features which reflects the performance of the CCP models using baseline classifier in the telecommunication industry.

Table 1. Performance of the baseline classifier without weighted features

Measure index	Results
Accuracy	87.50
Precision	58.06
Recall	13.85
F-measure	23.36

The performance evaluation of our proposed technique was made through a series of experiments and the best-performed model was obtained after assigning different weight to certain features (see steps of proposed solution). Table 2 reflects the best-performing model based on baseline classifier with best weighted features.

Table 2. Performance of the baseline classifier with best-weighted features

Measure index	Results
Accuracy	**89.10**
Precision	**95.65**
Recall	**16.92**
F-measure	**28.76**

The experiments show the facts that without weights assignment, the CCP model did not perform well as compared to the results obtained from the proposed technique (by more appropriate weights assignments to the features). The proposed technique achieved 89.10% overall accuracy which is higher than the accuracy of simple Naïve Bayes technique. Similarly, the proposed technique performed well in attempting to answer what proportion of true customer churn was actually classified correctly because the precision value 95.65 of the proposed technique is much higher than without weighted features (simple Naïve Bayes). From these experimental results, it is also observed that the feature weighting approach proposed in this study rarely degraded the performance of usual Naïve Bayes based CCP model in the telecommunication sector. Furthermore, we have calculated the recall values using Eq. 2 which shows the ability of the proposed technique that it not only correctly classified the customer churn but also correctly classified churn labels that were incorrectly classified as not churn in other approaches. In order to avoid the trade-off between precision and recall, the model was evaluated for the performance through F-measure (i.e. the harmonic mean of the obtained precision and recall). The obtained values 28.76 and 23.26 for the proposed technique and standard naïve Bayes, respectively show that the proposed feature weighting technique with all aspect based on the state-of-the-art measure index performed well then the without feature weighting using baseline classifier.

5   Conclusion

In this paper, we adopted an efficient feature weighting technique using GA. The experimental results on a publically available telecommunication customer data validated the performance in terms of avant-garde evaluation measures of accuracy, f-measure, precision, and recall. The proposed feature weighting technique outperformed (accuracy 89.10, precision 95.65, recall 16.92 and f-measure 29.76) as compared to without feature weighting (simple naïve Bayes which achieved accuracy 87.5, precision 58, recall 13.83 and f-measure 23.36).

The future work may be towards exploring the performance of the proposed feature weighting technique in other classification methods and balanced dataset. Furthermore, evaluating the proposed technique on multiple datasets from other domains to generalize the results.

Acknowledgement.   This research was supported by the Research Incentive Fund (RIF) Activity code # R18051, Zayed University, Abu Dhabi, United Arab Emirates.

References

1. Sharma, N., Saroha, K.: Study of dimension reduction methodologies in data mining. In: International Conference on Computing, Communication & Automation, pp. 133–137 (2015)
2. Yang, X.-S., Lee, S., Lee, S., Theera-Umpon, N.: Information analysis of high-dimensional data and applications. Math. Probl. Eng. 1–6 (2015)
3. Amin, A., Rahim, F., Ramzan, M., Anwar, S.: Prudent based approach for customer churn prediction (2015)
4. Houaria, R., Bounceur, A., Kechadic, T., Taria, A.-K., Euler, R.: Dimensionality reduction in data mining: a Copula approach. Expert Syst. Appl. **64**, 247–260 (2016)
5. Fodor Imola, K.: A survey of dimension reduction techniques. Cent. Appl. Sci. Comput. Lawrence Livermore Natl. Lab. **9**, 1–18 (2009)
6. Martins, J., Costa, C., Oliveira, T., Gonçalves, R., Branco, F.: How smartphone advertising influences consumers' purchase intention. J. Bus. Res. **94**, 378–387 (2019)

7. Hongjiu, L., Yanrong, H.: An evaluating method with combined assigning-weight based on maximizing variance. Sci. Program. **2015**, 1–8 (2015)
8. Frank, E., Hall, M., Pfahringer, B.: Locally weighted Naive Bayes. In: Proceedings of the 19th Conference on Uncertainty in Artificial Intelligence, pp. 249–256 (2003)
9. Wang, L., Ji, P., Qi, J., Shan, S., Bi, Z., Deng, W., Zhang, N.: Feature weighted naïve Bayes algorithm for information retrieval of enterprise systems. Enterp. Inf. Syst. **8**, 107–120 (2014)
10. Vijaya, J., Sivasankar, E.: Computing efficient features using rough set theory combined with ensemble classification techniques to improve the customer churn prediction in telecommunication sector. Computing **100**, 839–860 (2018)
11. Amin, A., Faisal, R., Muhammad, R., Sajid, A.: A prudent based approach for customer churn prediction. In: 11th International Conference, BDAS 2015, Ustroń, Poland, pp. 320–332 (2015)
12. Adnan, A., Babar, S., Asad Masood, K., Thar, B., Hamood ur Rahman, D., Sajid, A.: Just-in-time customer churn prediction: with and without data transformation. In: IEEE CEC 2018, Rio de Janeiro, Brazil, pp. 1–7 (2018)
13. Amin, A., Anwar, S., Adnan, A., Nawaz, M.: Comparing oversampling techniques to handle the class imbalance problem: a customer churn prediction case study. J. IEEE Access **4**, 7940–7957 (2016)
14. Zhang, H., Shengli, S.: Learning weighted Naive Bayes with accurate ranking. In: Proceedings of the Fourth IEEE International Conference on Data Mining (ICDM 2004), pp. 4–7 (2004)
15. Hall, M.: A decision tree-based attribute weighting filter for Naive Bayes. In: Research and Development in Intelligent Systems XXIII - Proceedings of AI 2006, the 26th SGAI International Conference on Innovative Techniques and Applications of Artificial Intelligence, pp. 59–70 (2007)
16. Wu, J., Cai, Z.: Attribute weighting via differential evolution algorithm for attribute Weighted Naive Bayes (WNB). J. Comput. Inf. Syst. **5**, 1672–1679 (2011)
17. Zaidi, N.A., Cerquides, J., Carman, M.J., Webb, G.I.: Alleviating naive Bayes attribute independence assumption by attribute weighting. Mach. Learn. Res. **14**, 1947–1988 (2013)
18. Kim, K., Jun, C.-H., Lee, J.: Improved churn prediction in telecommunication industry by analyzing a large network. Expert Syst. Appl. **41**, 6575–6584 (2014)
19. Jiang, L., Li, C., Wang, S., Zhang, L.: Deep feature weighting for naive Bayes and its application to text classification. Eng. Appl. Artif. Intell. **52**, 26–39 (2016)
20. Zhang, L., Jiang, L., Li, C., Kong, G.: Two feature weighting approaches for naive Bayes text classifiers. Knowl.-Based Syst. **100**, 137–144 (2016)
21. Amin, A., Shah, B., Khattak, A.M., Baker, T., Durani, Hamood ur Rahman, D., Anwar, S.: Just-in-time customer churn prediction: with and without data transformation. J. Bus. Res. 1–5 (2018)
22. Amin, A., Anwar, S., Adnan, A., Nawaz, M., Alawfi, K., Huang, K., Hussain, A.: Customer churn prediction in telecommunication sector using rough set approach. Neurocomputing **4**, 1–18 (2016)
23. Shuo, W., Xin, Y.: Using class imbalance learning for software defect prediction. IEEE Trans. Reliab. **62**, 434–443 (2013)
24. Han, J., Kamber, M., Pei, J.: Data Mining: Concepts and Techniques, 3rd edn. Morgan Kaufmann, Burlington (2011)

Nonlinear Identification of a Robotic Arm Using Machine Learning Techniques

Darielson A. Souza[1(✉)], Laurinda L. N. Reis[1], Josias G. Batista[1],
Jonatha R. Costa[1], Antonio B. S. Junior[2], João P. B. Araújo[1],
and Arthur P. S. Braga[1]

[1] Electrical Engineering Department, Federal University of Ceará-UFC,
Campus Pici, Fortaleza, CE, Brazil
{darielson,laurinda,josiasgb,arthurp}@dee.ufc.br,
jonatha.costa@ifce.edu.br, jpbaraujo@alu.ufc.br
[2] Federal Institute of Ceará-IFCE, Maracanaú, CE, Brazil
barbosa@dee.ufc.br

Abstract. With the advancement of intelligent algorithms more and more robots perform human tasks, be they due to dangerousness or simply by reducing human costs, for that to happen requires precision. This work has the objective of making an identification of a robotic arm with three phase induction motor through machine learning techniques to obtain a better model that represents the plant. The techniques used were Artificial Neural Network (ANNs): MLP, RBF and MLP + PSO. The techniques obtained a good performance, and they were evaluated through the multi-correlation coefficient (R^2) for a comparative analysis.

Keywords: Robot arm · System identification · Machine learning · ANNs

1 Introduction

Robotics applied in the industry is an area where control, is fundamental to the success of the task to be performed, pick-and-place robots possess characteristics such as speed, torque and precision, attributes essential for its smooth operation. In industry, where jobs are repetitive, these types of machines are ideal for use in large-scale jobs. One of its characteristics is the presence of a claw at its end, and it is designed for the components that are desired to be transported taking into account the characteristics of the material as weight and shape, and it should be adjusted to handle from the rigid used metal plates in metallurgy to fragile glass ampoules very common in the pharmaceutical industry.

Machine learning techniques are increasingly present in industries where they use robots for tasks and also in applications in medical diagnostics, electronic game development, cybersecurity, among others. In the work of [1], it is presented the use of ANNs for detection of failures in electric machines. In the paper of [2] the use of neural networks for the identification of diseases in people is presented. Already [3] makes the study of a fuzzy identification of the dynamics of the system that is developed with data generated by a hydrogen fuel cell simulator.

© Springer Nature Switzerland AG 2019
Á. Rocha et al. (Eds.): WorldCIST'19 2019, AISC 931, pp. 492–501, 2019.
https://doi.org/10.1007/978-3-030-16184-2_47

The challenge of this work is to identify the faces of a robotic manipulator with 1 degree of freedom. The degree of freedom to be modeled will be the basis of the manipulator that is rotating. The system to be identified is a sound system (one input and one output system), input is current and output is speed. The methods to be used are: Multilevel Artificial Neural Network (MLP ANN), Neural Network with Radial Base Function (RBF ANN) and MLP + PSO (Particle Swarm Optimization).

This paper is structured in 5 sections relying on this, Sect. 2 presents the details of the robotic arm used to be modeled, the following section the methodology used to identify the system, in Sect. 4 present the results, and finally will be seen the conclusions about work.

2   Robotic Arm System

For the implementation of this work it was necessary to carry out the development and construction of the structure. This structure will have a robotic manipulator of five degrees of freedom. However, since this work is the beginning of the development of the manipulator developed by three-phase induction motors, the first degree of freedom will be controlled, referring to the manipulator base, so that future work can extend to the other grades. Other works have made use of this bench of tests for applications of identifications and control [4, 5].

As can be observed, the first degree of freedom has a rotational movement around the main axis of the structure, while the second and third have prismatic movements, characterizing it as cylindrical type RPP (Rotational-Prismatic-Prismatic). The last two grades relate to the claw, not yet installed, which should be connected to the third degree of freedom. Figure 1 shows the plant.

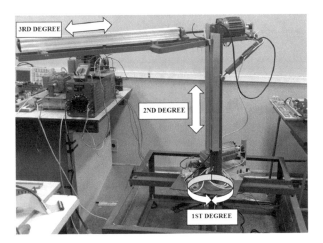

Fig. 1. Robot arm

The structure is driven by a three-phase induction motor, highlighted in Fig. 10, of squirrel cage type, with 0.5 cv of rated power, nominal voltage of 380/220 V, 4 poles and 1.18A of delta-connected nominal current. The power of the motor was chosen so that it was possible to move the structure of the manipulator. Initially an engine of 0.25 power of nominal power was installed, but this was not able to move the structure, making it necessary a motor of greater power. The transmission of movement to the first degree of freedom of the manipulator occurs through the use of belt and pulleys, as can be seen.

For the identification of the manipulator, two manipulator parameters, input current and output velocity were collected, with a sampling period of 002 s. The data collected to make the identification are shown in Fig. 2, with a size of 1001 sample.

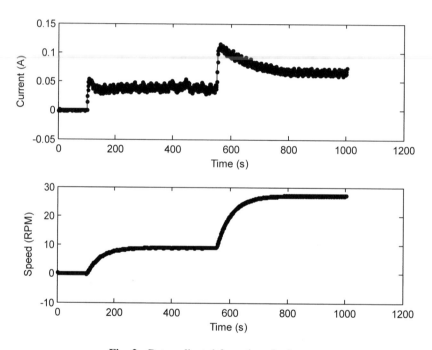

Fig. 2. Data collected from the robotic arm

In Fig. 2 the collected data are presented, in which the first image is the behavior of the current measured in Amperes which is the input of the system, the second image is the velocity of the joint one, which is measured in revolutions per minute (RPM).

3   Systems Identification

According to [6], determination of the mathematical model of a real system under study representing its characteristics adequately for a particular use. There are two steps for pre-processing the data before identification: 1 - database shuffling, 2 - separation of training labels and a part for testing. The results are measured with the coefficient of determination comparing the identification of the model with the actual system data.

In the present study we used some ANNs to make these identifications, MLP ANN, RBF ANN and MLP ANN + PSO. For both techniques, the same validation process was used, as the data was shuffled as starting point, and the base was then divided into 80% for training and 20% for validation. In both techniques the number times were the same totaling 120 iterations in each, not placed precision for the stop, thus the convergence only to stop when it reached the total times quantity. There were still 10 executions of each method, thus observing the best results of each method and describing the architecture and facings used.

3.1   MLP ANN

Artificial Neural Networks are computational techniques that present a mathematical model inspired by the neural structure of intelligent organisms and that acquire knowledge through experience. A large artificial neural network may have hundreds or thousands of processing units; already the brain of a mammal can have many billions of neurons [7].

The most common type of neural network to solve nonlinear separation problems is the MLP neural network which stands for Multiayer Perceptron. This type of ANN is a super-network because it requires a kind of output wanted to learn. The purpose of this network is to create a model that can correctly map input and output, even when the desired output is unknown. Figure 3 shows an MLP ANN.

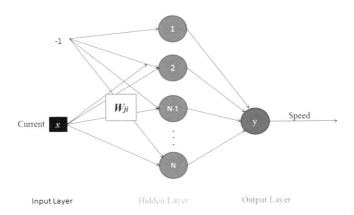

Fig. 3. MLP ANN architecture

The MLP ANN has a training called backpropagation, in which the objective is to train the weights with descending gradient method. The weights of an ANN represent the mapping obtained after training, to have a relation with the estimated outputs. That is why of all the dresses adjusted the weights is the main.

During the implementation of the MLP ANN several parameters were modified as learning steps architectures (quantities of neuroses and hidden layers) and ranges of values of weights. Figure 4 presents the steps of the MLP ANN algorithm with the backpropation training.

---

begin
Step 1: Generate W$_{aji}$ randomly the weights and bias B$_i$, where i = 1, ..., N of the N neurons
Step 2: Calculate the output of the hidden layer
Step 3: Calculate the output weights
Step 4: Error calculation
Step 5: Backpropation Training
end.

Fig. 4. MLP ANN algorithm

3.2   RBF ANN

The basic structure of the RBF neural network consists of three layers: the input layer, interfaces between the model and the medium. The hidden or hidden layer performs a non-linear transformation of the input vector into an internal vector space, typically with a larger dimension and the output layer, performs the transformation of the inner vector space into an output using a linear process [8].

The operation of the hidden layer is based on the Cover Theorem on the separability of the patterns. In this theorem, a non-linearly separable problem can be probabilistically transformed into a linearly separable problem through a nonlinear transformation that maps space to another of a larger order. As its name implies, symmetric radial base function is used as activation functions of intermediate layer neurons. The function used in this work was the Gaussian.

The output of the neurons from the intermediate layer is given by (1):

$$y = F_i(x) = \sum_{J=1}^{J} \omega_{ij} e^{-\frac{1}{\sigma^2}\|x - w_j\|^2} \tag{1}$$

Where is the center of the curve along which the values are distributed, is the input vector, is the radius or aperture of the function, is the synaptic weight between the hidden layer neuron and the output layer neuron.

The transformation from the input neurons to the neurons of the hidden layer is non-linear, and formation of this part of the network is usually performed without supervision. The training of network parameters (weight) between hidden layers and output occurs in a supervised way based on target outputs.

The adjustment of the centers was used the k-means and the choice of the spreading was by the normalization of the greater distance between the centers. The least squares were used to adjust the weights. Equation (2) was used as a function of activation of neurons of the hidden layer.

$$j_i(x) = e^{-\frac{\|x - c_i\|^2}{s^2}} \tag{2}$$

3.3 MLP ANN + PSO

Hybrid systems are becoming more widespread and have a very large advance in problem solving by improving simple systems. One of the problems of MLP ANN is usually the training of weights when it drops to the local minimum, thus making it difficult to train the network properly. One of the solutions to this problem would be to use search algorithms [9].

A metaheuristic can be defined as a heuristic method to solve generic optimization problems (usually in the area of combinatorial optimization). For the problem in question ANN will be combined with a metaheuristic named PSO which stands for Particle Swarm Optimization.

Particle swarm optimization (PSO) is a stochastic optimization technique based on population developed by Dr. Eberhart and Dr. Kennedy in 1995, inspired by the social behavior of bird flocking or fish schooling [10].

PSO shares many similarities with evolutionary computing techniques, such as Genetic Algorithms (GA). The system is initialized with a population of random solutions and optimal search updating generations. However, unlike GA, the PSO has no evolution operators, such as crossover and mutation. Figure 5 shows the pseudocode of PSO

PSO begin Algorithm
Step 1: Gbest := ∞
Step 2: **for** i at 1 a n **Make**
Step 3: **begin** x_i e v_i **randomly**
Step 4: $p_i := x_i$
Step 5: $pbest_i = f(p_i)$
Step 6: **if** $pbest_i < vbest$ **then**
Step 7: $vbest_i := pbest_i$
Step 8: $g := p_i$
Step 9: **end if**
Step 10: **end for**
Step 11: **end PSO**

Fig. 5. PSO algorithm.

3.4   Multi-correlation Coefficient (R^2)

For the evaluation of the techniques will be used a method called multi-correlation coefficient (R^2) it is used to measure the quality of the estimate, that is, the estimated in relation to the real. The assumed value varies from 0 to 1, the closer to 1 means that the estimate is in excellence. Equation (3) presents the formula for obtaining R^2.

$$R^2 = 1 - \frac{\sum_{i=1}^{n} (y(i) - \hat{y}(i))^2}{\sum_{i=1}^{n} (y(i) - \overline{y}(i))^2} \tag{3}$$

4   Results

Based on data collection from Fig. 1 the results presented good performances in this way, validating the identifications of the neural models. Figure 6 shows the identification of MLP ANN, where its error is almost null. But its computational cost was quite heavy.

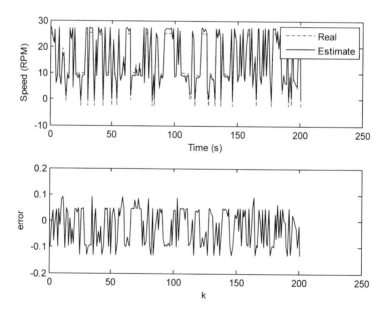

Fig. 6. MLP ANN results.

The Fig. 7 shows the results of the RBF ANN, in which it was shown to be slightly better than the MLP ANN.

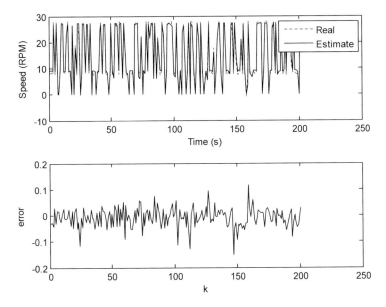

Fig. 7. RBF ANN results.

The results of the hybrid system MLP and PSO surpassed the previous identifications, besides having a better performance the architects were very reduced, making the computational cost become lighter. The Fig. 8 presents the estimation results of the MLP ANN + PSO

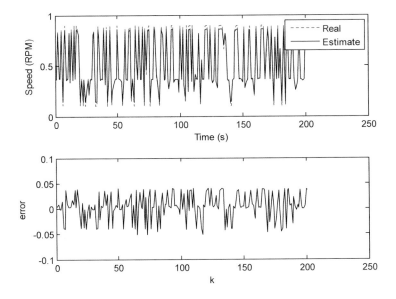

Fig. 8. MLP ANN + PSO results.

Table 1 shows all the identifications being evaluated by the multi-correlation coefficient.

Table 1. Result of the R^2

Methods	R^2
MLP ANN	0.9702
RBF ANN	0.9733
MLP ANN + PSO	0.9935

Table 2 presents the mean of the mean square error of each method proposed in the work.

Table 2. Result of the mean square error (MSE)

Methods	MSE
MLP ANN	0.0381
RBF ANN	0.0364
MLP ANN + PSO	0.0276

Table 3 shows the standard deviation of the mean quadratic error of the result of each method.

Table 3. Result of the Standard Deviation (SD)

Methods	SD
MLP ANN	0.0258
RBF ANN	0.0235
MLP ANN + PSO	0.0158

According to the results of Table 1, the hybrid method MLP ANN + PSO achieved better results than the other methods with the R^2 of **0.9935**. Table 2 shows the results of the mean square error, that the MLP ANN + PSO method obtained the lowest mse of **0.0276**. Finally, the standard deviation of the mse of all the techniques and MLP ANN + PSO was also evaluated, obtaining the best result of **0.0158**. Although all the results were good, the method that most stood out was MLP ANN + PSO.

5   Conclusion

The work presented acceptable identification results and with good performances. One of the degrees of freedom of the robotic arm was identified with computational intelligence, MLP ANN, RBF ANN and MLP ANN + PSO. The MLP ANN + PSO results were able to overcome the other methods used in the work with the lower MSE of 0.0276, smaller SD 0.0158 and greater R^2 0.9935. According to [6], the R^2 to be considered good has to be between 0.8 to 1, so all methods are evaluated as satisfactory. The use of statistical variables could evaluate the best method, this choice may be very important for other researchers to have similar problems.

As a future work proposal it is expected to identify the other joints of the manipulator at the same time. Thus the methods will have a greater challenge for the identification with the mime systems.

Acknowledgment. The authors thank to Capes for the financial support to this work.

References

1. Paya, B.A., Esat, I.I., Badi, M.N.M.: Artificial neural network based fault diagnostics of rotating machinery using wavelet transforms as a preprocessor. Mech. Syst. Signal Process. **11**, 751–765 (1997)
2. Al-Shayea, Q.K.: Artificial neural networks in medical diagnosis. Int. J. Comput. Sci. Issues (2011)
3. Bertone, A.M.A., Martins, J.B., Yamanaka, K.: Black-box fuzzy identification of a nonlinear hydrogen fuel cell model. Tendencias em Matematica Aplicada e Computacional (2017)
4. Rebouças, L.R., Filho, C.G.M., Junior, A.B.S., Reis, L.L.N., Nogueira, F.G., Neto, T.R.F., Barreto, L.H., Torrico, B.C.: Predictive control applied in 3-phase squirrel cage induction motor for zero speed. In: SBAI, 2017 (2017)
5. Rebouças, L.R., Filho, C.G.M., Junior, A.B.S., Reis, L.L.N., Neto, T.R.F., Barreto, L.H.: Identificação de um motor de indução trifásico aplicado para posicionamento utilizando modelo de Hammerstein. Revista Tecnológica de Fortaleza (2015)
6. Coelho, A.A.R., Coelho, L.S.: Identificação de sistemas Dinâmicos Lineares, Editora UFSC, 2º Edição (2015)
7. Haykin, S.: Neural Networks: Principios and Pratice (in Portuguese), Bookman, ed. (2001)
8. Buhmann, M.D.: Radial Basis Functions: Theory and Implementations. Cambridge University Press, Cambridge (2003)
9. Anders, U., Korn, O.: Model selection in neural networks. Neural Netw. **12**(2), 309–323 (1999)
10. del Valle, Y., Ganesh, K.: Particle swarm optimization: basic concepts, variants and applications in power systems. IEEE Trans. Evol. Comput. **12**, 171–195 (2008)

Study on the Variation of the A Fund of the Pension System in Chile Applying Artificial Neural Networks

Alexander Börger[✉] and Pedro Vega

Department of Industry and Business, University of Atacama, Copiapó, Chile
alexander.borger@uda.cl, pedro.vega@alumnos.uda.cl

Abstract. In Chile, all people have a legal obligation to enter and contribute to a pension system, capitalizing their savings through profits from investment funds at their reference value. The profits obtained depend on the prices of the financial instruments of the investment firms that manage these funds, and determine the amount contributors will receive at the time they collect their pensions, as well as the positive or negative variations in the value of their shares. The objective of this study is to evaluate the predictive capacity of the artificial red neural Red Ward Model by evaluating the percentage of prediction of signs and the weekly profitability the value share of Fund A of the Chilean pension system. In this research, we apply models based on neural networks to the Chilean pension system, as well as predictions of the weekly variations in the fund, to obtain an improved growth forecast. Our research shows that in five of the eight months we considered, a percentage higher than 65% was obtained, all the models had statistical significance, and most active investment strategies are superior to passive ones. Finally, we conclude that the built network has a strong predictive capacity, and that the use of artificial neural networks for the prediction of variations in financial values is a viable alternative since the results obtained are consistent with other existing methods such as the Vector Support Machine approaches.

Keywords: ANN · Prediction · Profitability

1 Introduction

Under Decree Law 3,500 of 1980, the pension system in Chile is directed by Pension Fund Administrators (PFAs) who manage the savings of contributors in five pension funds. These savings are capitalized by the profits they generate based on their quotas (The fee, or the quota value, is the unit of measure of the pension fund used by the PFAs to calculate the profitability). This is important for contributors, since the incomes they will receive when they retire will depend on the positive or negative variations in the quota value.

An Artificial Neural Network (ANN) model called "Red Ward" [1] was proposed as a means of predicting the weekly variations of national and international stock indices through the Sign Prediction Percentage (SPP) and the Directional Accuracy Test (DAT). The active strategy pursued under this model generated increased profitability, demonstrating the strength and predictive capacity of ANN itself [1].

© Springer Nature Switzerland AG 2019
Á. Rocha et al. (Eds.): WorldCIST'19 2019, AISC 931, pp. 502–512, 2019.
https://doi.org/10.1007/978-3-030-16184-2_48

To affirm that the Red Ward model has predictive capacity in its models resulting from training, the three indicators mentioned above are analyzed, where it is expected that: the SPP of our sample would be high (>65%), the profitability of the model (active strategy) would exceed the profitability of the buy & hold model (passive strategy), and the DAT would be rejected.

The article is structured as follows: In the following section, we present some antecedents about artificial neural networks. In Sect. 3, we review previous work in the field. Section 4 describes our research methodology and the ANN model, while Sect. 5 shows the results obtained in this study. Finally, Sect. 6 summarizes the conclusions obtained from the study.

2   Background

The PFAs are private companies exclusively responsible for granting and administering the benefits established under Law 3,500 [2] to each worker affiliated with an individual capitalization account. The contents of these accounts are deposited by the PFAs, according to the will of the affiliate, in 5 pension funds, denominated by the letters A, B, C, D and E.

Around 75% of Fund A is invested abroad. Of this, around 80% is destined for variable income products, mainly (75%) mutual funds, which constitute 45% of Fund A, while the 25% invested in national instruments is mainly placed in variable instruments [3].

ANN are systems that allow the establishment of a linear or non-linear relationship between inputs and outputs. Their characteristics, modeled after those of the nervous system, give them several advantages. For example, they are capable of adaptive learning, they are self-organizing, they can work in parallel in real time, and they offer fault tolerance due to redundant information coding [4].

According to Haykin [5] *"A neural network is a massively parallel distributed processor that has a natural propensity for storing experiential knowledge and making it available for use. It resembles the brain in two respects: (1) Knowledge is acquired by the network through a learning process. (2) Interneuron connection strengths known as synaptic weights are used to store the knowledge."*

3   Related Work

The literature contains various proposals for applying ANN to make predictions oriented towards decision-making in different areas, such as electricity [6–8]; meteorology [9, 10]; transport [11]; health [12]; and others. In the field of economics, proposals for prediction have been made in relation to the currency market [13], the price of gold [14], the price of oil [15], and the economic sciences [16], among others.

The Red Ward model, specifically, has been used in relation to the pension system [17] to determine how well it can predict the share value of the AFP Cuprum SA multifunds. The results showed that 60% of the models obtained a higher SPP than

65%, and the DAT showed 95% confidence with 5% statistical significance that the active profitability strategy is better than the passive strategy. It was concluded that this technique presents the most reliable way to predict the value of AFP Cuprum S.A.

4   Research Methodology

The model used for the present research is a neural network called "Red Ward," which has a multilayer network architecture (see Fig. 1), and which, in many cases, exceeds the ARIMA model [17].

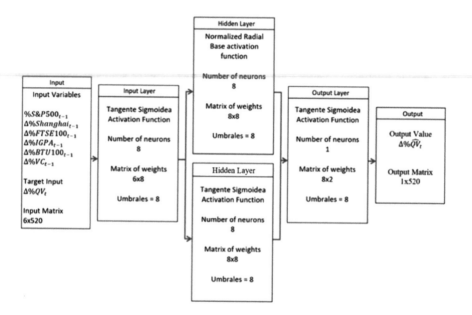

Fig. 1. Architecture of the Neural Network Model of the Red Ward type.

The first variable in the prediction of the variation in the value of Fund A is the variation in the previous period of 5 stock indices. Three of these—the S&P500 (USA), Shanghai (Asia) and FTSE100 (Europe)—are international, while the other 2—IGPA and BTU10—are national. In addition, the previous variation of the quota value $(\Delta\%QV_{t-1})$ is considered. That is, the value of the variation will be a function of the values, in the previous period, of the aforementioned variables, as shown in (1):

$$\Delta\% \widehat{QV_t} = f(\Delta\%variables_{t-1}) \qquad (1)$$

Where:

- $\Delta\%\widehat{QV_t}$: Estimated or projected percentage variation
- f : process performed by the neural network.

The previous variation in the quota value will be a function of the value of the aforementioned variables in the previous period, as shown in (2).

$$\Delta\% variables_{t-1} = \Delta\% S\&P500_{t-1}, \Delta\% Shanghai_{t-1}, \Delta\% FTSE100_{t-1}, \Delta\% IGPA_{t-1}, \Delta\% BTU100_{t-1}, \Delta\% VC_{t-1}$$

(2)

The components of the network architecture are explained below:

- Input block: A 6×520 matrix, which contains the 6 variables and the 520 weekly variations of each.
- Input layer: The first layer of the network, responsible for receiving the input matrix, has eight neurons. Its activation function is the Tangent Function Sigmoidea, which transforms the data by normalizing them in a range between -1 and 1, where the negative variations will always be negative, and likewise with the positive variations.
- Hidden layers: These 2 parallel layers have eight neurons each and have the same data from the input layer as their own input. One has the Normalized Radial Base Function (Normalized Gaussian Function), which takes the sign of the most probable value, as its activation function; the other uses the Sigmoidea Logistic Function, which looks for the most probable value between 0 and 1. The result of these two functions is to multiply the most probable sign by the most probable value, which is always positive.
- Output layer: layer responsible for generating the value that will represent the variation of the projected value, therefore, because only one value will be generated, it will have only one neuron Here, an activation function, Tangent Sigmoidea, is applied to reverse the function applied in the input layer.
- Output block: The 1×520 matrix that contains the result of the network process.

The percentage variation of the dependent variables (the value of the quota) and the independent variable (the value obtained by the network) correspond to the weekly variation between each Monday. Here, we used data from the period between 01 January 2007 and 31 December 2016. Taken together, these constitute 520 points of data were obtained for each of the variables mentioned. This variation was calculated according to the equation noted above.

$$\Delta\% x_t = \frac{x_t - x_{t-1}}{x_{t-1}}$$

(3)

The Red Ward neural network operates in two stages: the training process that uses intra-sampling data to perform the learning and adjust the parameters for the mini-mum mean squared error, which corresponds to 70% of the total data (364 intra-sampling data), and the testing and validity process that uses extra-sample data, which is the remaining 30% (156 extra-sampling data).

The research methodology is shown in Fig. 2.

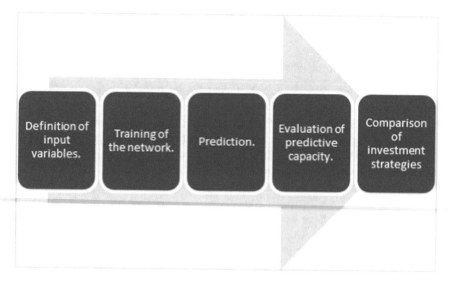

Fig. 2. Research methodology

1. Definition of input variables: The five stock indices mentioned above are considered (S&P500, Shanghai and FTSE100, IGPA and BTU10), along with the previous variation in the share value ($\Delta\%QV_{t-1}$).
2. Network training: The algorithm "Backpropagation based on Levenberg-Marquardt" [18–20] was executed eight times for the training of the NN, using 70% of the data, as indicated. The "Mean Square Error" (MSE), which compares the desired historical data (objective data or target, presented as error indicator $\Delta\%QV_t$) with the results of the NN ($\Delta\%\widehat{QV_t}$) to get parameters that minimize that error, was also executed.
3. Prediction: For the prediction of the network values, the network was executed with the input data mentioned in step 1, but only the extra-sample data was considered.
4. Evaluation of the predictive capacity: To evaluate the predictive capacity of the network, 3 indicators needed to be analyzed: SPP, DAT [21], and the comparison of the obtained profitability through a passive investment strategy and an active investment strategy.
5. Comparison of investment strategies: We verified that the ANN-Red Ward model, when applied to the prediction of the calculation of the weekly future values of fund A, allows for the improvement of profitability through an active investment strategy, as compared to a passive buy and hold strategy.

5   Results

During neural network training, 8 different models with different predictions were obtained. These were differentiated only by the configurations in the weights and threshold of each neuron.

5.1   Signal Prediction Percentage

Table 1 shows the total intra-sampling and extra-sampling SPP for each of the eight models.

As summarized in Table 1, all the models obtained a large percentage of success in their results, given that they reached high SPP values in both intra-sampling and extra-sampling processes, with a difference of 12.18% or more among them. This demonstrates that most versions of the model are stable and can accommodate new training, except for models 2, 4 and 7, which did not exceed the expected extra-sampling SPP of 65%.

Table 1.   Result of the Neural Network Model of the Red Ward type.

Models	Total SPP	Intra-sampling SPP	Extra-sampling SPP
Model 1	69,62%	69,23%	70,51%
Model 2	64,81%	65,66%	62,82%
Model 3	67,12%	67,58%	66,03%
Model 4	62,31%	62,91%	60,90%
Model 5	64,42%	64,01%	65,38%
Model 6	62,69%	61,26%	66,03%
Model 7	59,42%	59,89%	58,33%
Model 8	69,04%	69,51%	67,95%

However, the analysis should focus on the result of the extra-sampling, since it corresponds to the simulation of the model with the historical observations of the variables that are not used for training, but are used to train the built network, in which case, five of the eight models exceed 65% of SPP.

5.2   Directional Accuracy Test

Table 2 presents the eight models and the statistical value of the DAT. All the models managed to reject the null hypothesis of the DAT, which means that the results are not random, but depend on the real observations.

Table 2. DAT results.

Models	Model 1	Model 2	Model 3	Model 4	Model 5	Model 6	Model 7	Model 8
DAT	8.68	6.15	7.30	4.92	6.01	4.90	3.85	0.68

5.3   Profitability Strategy

Table 3 shows the complete result of the network, which includes the eight models, the profitability generated by both the active strategy and the passive (buy and hold) strategy, and the statistical value of the DAT. The profitability achieved through the active strategy of the whole model exceeds 600%, which is much higher than the return obtained by a passive investment strategy (68.75%).

Table 3. Result of the Neural Network Model of the Red Ward type.

Models	Total SPP	Intra-sampling SPP	Extra-sampling SPP	Active profitability strategy	DAT
Model 1	69,62%	69,23%	70,51%	625,09%	8,68
Model 2	64,81%	65,66%	62,82%	515,09%	6,15
Model 3	67,12%	67,58%	66,03%	474,52%	7,30
Model 4	62,31%	62,91%	60,90%	365,56%	4,92
Model 5	64,42%	64,01%	65,38%	383,76%	6,01
Model 6	62,69%	61,26%	66,03%	321,72%	4,90
Model 7	59,42%	59,89%	58,33%	244,25%	3,85
Model 8	69,04%	69,51%	67,95%	665,40%	0,68
Buy and hold strategy	–	–	–	68,75%	–

Table 4 shows the results of the extra sample set, where it is observed that the eight resulting models have generated returns, through an active strategy, superior to obtained through the passive (buy & hold) strategy. Model 1 obtained the highest profit (see chart in Fig. 3). This means that, if a contributor's savings are reversed in Fund A within the period where the same variations as the extra-sample observations are suffered, and no other investment in this fund is made in the future (in order to invest and maintain in the long term), the account will earn a nominal return of 26.60%. On the other hand, if the active strategy guided by Model 1 is chosen, the account will have a nominal return of 70.87%. Despite presenting a prediction percentage lower than the expected 65%, Models 2 and 4 managed to obtain a higher return than the passive strategy.

If M $ 1,000 is deposited in Fund A, under the assumption that, during a period of 156 weeks, said fund presents the same weekly nominal returns as in the extra-sampling historical observations, a profit of $1,265,960 (nominal return of 26.60%)

will be obtained with the buy & hold strategy. If, however, we opt for an active strategy guided by Model 1, which obtained the highest nominal return of 70.87%, the total profit will be $1,708,651.

According the results, the models 1, 3, 5, 6 and 8 obtained a PSS higher than the 65% proposed the profitability of the 7 models through an active investment strategy was greater than the passive investment strategy (Buy and Hold); and at the same time that all the models rejected DAT. Then, when fulfilling the aforementioned, it is determined that the models 1, 3, 5, 6 and 8 have predictive capacity to predict the sign of the nominal profitability of Fund A.

Table 4. Cost-effective extra-sampling passive and active strategy.

Models	Profitability extra-sampling active strategy	Profitability obtained with an investment of M $ 1,000
Model 1	70,87%	$ 1.708.651
Model 2	62,82%	$ 1.628.205
Model 3	43,95%	$ 1.439.492
Model 4	41,22%	$ 1.412.231
Model 5	43,52%	$ 1.435.184
Model 6	46,57%	$ 1.465.737
Model 7	24,46%	$ 1.244.647
Model 8	49,33%	$ 1.493.291
Buy and hold cost effectiveness	26,60%	$ 1.265.960

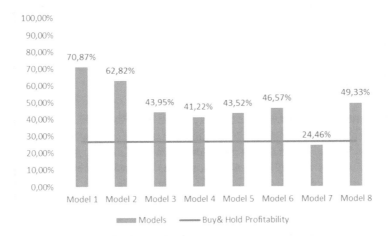

Fig. 3. Profitability strategy active v/s strategy buy and hold.

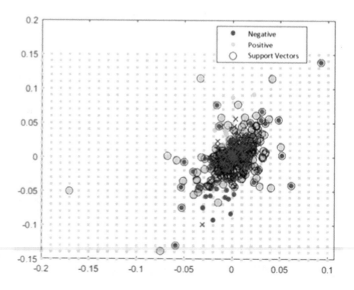

Fig. 4. Vector Support Machine approach.

In addition, the results obtained are concordant with other predictive approaches such as VSM, as can be seen in the Fig. 4.

The Graphic of Fig. 4, represents the classification of the VSM approach where the area of the blue points represent the positive variation and the area of the red points the represent the negative one. The classification is done by the inside geometrical sector of the figure (positive) and the surrounding area (negative).

By means of a VSM approach, a PPS of 63% was obtained, the DA test was refused, and the profitability obtained was a 40%, who is the higher profitability in the time interval using the extra-sampling data (used to test and validate).

6   Conclusions

This article evaluates the predictive capacity of the Red Ward NN, which can be used to anticipate changes in Fund A of the Chilean pension system. The active strategy outperformed the passive strategy with regard to returns. This shows that implementing an active investment strategy generates value for both a contributor to the AFP pension system and a private investor in Fund A, because an increase in profitability is a percentage increase in the amount of the individual capitalization account.

On the other hand, the directional accuracy test showed, with a confidence interval of 95%, that all the models rejected DAT. That is, the data are statistically significant and the results depend on the objective data or target, confirming that the prices of Fund A shares are not a product of chance, but follow a trend generated by changes in the environment that is predictable, with defined variables like stock indices.

Finally, this study indicates that the built network has a stable predictive capacity in forecasts of variations in the quota value of Fund A in five of the eight trained models. The statistical significance of the network, and the high profitability of its models as compared to that of a buy & hold strategy, was demonstrated by the SPP (65%) of many of those models. The use of artificial neural networks for the prediction of variations in financial values is therefore a viable alternative to existing methods, and the results of VSM approach are son concordant with the ANN method.

References

1. Parisi, F., Guerrero, J.L.: Modelos de redes neuronales aplicados a la predicción del tipo de cambio del dólar observado en Chile. Estudios de Administración **10**(1), 25–48 (2003)
2. Biblioteca del Congreso Nacional. https://www.leychile.cl/Navegar?idNorma=7147. Accessed 30 Nov 2018
3. Centro de Estadísticas Superintendencia de Pensiones. http://www.spensiones.cl/safpstats/stats/.sc.php?_cid=0. Accessed 30 Nov 2018
4. Villada, F., Muñoz, N., García, E.: Aplicación de las Redes Neuronales al Pronóstico de Precios en el Mercado de Valores. Información tecnológica **23**(4), 11–20 (2012)
5. Haykin, S.S.: Neural Networks and Learning Machines, 3rd edn. Pearson, Ontario (2009)
6. Velásquez, J.D., Franco, C.J., García, H.A.: Un modelo no lineal para la predicción de la demanda mensual de electricidad en Colombia. Estudios Gerenciales **25**(112), 37–54 (2009)
7. González, C.M.: Predicción de la demanda eléctrica horaria mediante redes neuronales artificiales. Rect@: Revista Electrónica de Comunicaciones y Trabajos de ASEPUMA **5**(1), 5–28 (2004)
8. Henao, J.D.V., Mejia, V.M.R., Cardona, C.J.F.: Electricity demand forecasting using a Sarimamultiplicative single neuron hybrid model. DYNA **80**(180), 4–8 (2013)
9. Camacho, J.M.G., Osornio, R.C., Bustamante, W.O., Cruz, I.L.: Predicción de la evapotranspiración de referencia mediante redes neuronales artificiales. Ingeniería hidráulica en México **23**(1), 127–138 (2008)
10. Jiménez-Carrión, M., Gutiérrez-Segura, F., Celi-Pinzón, J.: Modelado y Predicción del Fenómeno El Niño en Piura, Perú mediante Redes Neuronales Artificiales usando Matlab. Información tecnológica **29**(4), 303–316 (2018)
11. de Lara, M.A.S.: Estudio predictivo de costes y financiación del servicio de transporte urbano colectivo en las empresas españolas mediante la aplicación de redes neuronales artificiales. Doctoral dissertation, Universidad Rey Juan Carlos (2013)
12. Guerra, A., Rivas, J.: Detección de microcalcificaciones en imágenes mamográficas usando redes neuronales. Revista de la Facultad de Ingeniería Universidad Central de Venezuela **26**(3), 7–14 (2011)
13. Ayala, M.: Un modelo de predicción para el valor TRM: Un acercamiento desde las redes neuronales artificiales (No. 004995) (2008)
14. Villada, F., Muñoz, N., García-Quintero, E.: Redes neuronales artificiales aplicadas a la predicción del precio del oro. Información tecnológica **27**(5), 143–150 (2016)
15. Villada, F., Arroyave, D., Villada, M.: Pronóstico del precio del petróleo mediante redes neuronales artificiales. Información tecnológica **25**(3), 145–154 (2014)
16. Patiño, V.M.O.: Redes neuronales artificiales en las ciencias económicas (No. 009938). UN-RCE-CID (2012)

17. Parisi, A., Parisi, F., Guerrero, J.L.: Modelos predictivos de redes neuronales en índices bursátiles. El Trimestre Económico **70**(280), 721–744 (2003)
18. Hagan, M.T., Menhaj, M.B.: Training feedforward networks with the Marquardt algorithm. IEEE Trans. Neural Networks **5**(6), 989–993 (1994)
19. Hagan, M.T., Demuth, H.B., Beale, M.H., De Jesús, O.: Neural Network Design, 2nd edn. PWS Pub, Boston (1996)
20. Marquardt, D.W.: An algorithm for least-squares estimation of nonlinear parameters. J. Soc. Ind. Appl. Math. **11**(2), 431–441 (1963)
21. Pesaran, M.H., Timmermann, A.: A simple nonparametric test of predictive performance. J. Bus. Econ. Stat. **10**(4), 461–465 (1992)

Assessing the Horizontal Positional Accuracy in OpenStreetMap: A Big Data Approach

Roger Castro[✉], Alfonso Tierra, and Marco Luna

Department of Earth and Civil Sciences, University of the Army,
170550 Sangolquí, Ecuador
rrcastrol@espe.edu.com

Abstract. OpenStreetMap is perhaps the most successful project from those that produce Volunteered Geographic Information. Its characteristics are the same as the Big Data, which challenges its study. The present paper assesses the horizontal positional accuracy of OSM streets in administrative zones of the Metropolitan District of Quito, with the purpose of understanding what methods allow us to know the quality of these data. Open GIS and statistical programming software were used in combination, and then using the second one as a standalone. Considerations introduced in order to correct two inconveniences of the method are presented. The paper concludes discussing hypotheses regarding positional accuracy and identifying ways to continue the research in this area.

Keywords: OpenStreetMap · Big Geodata · Positional accuracy

1 Introduction

Technological advances during the past decade -Web 2.0 in particular- allowed the emergence of a series of collaborative-media-based projects [1, 2]. In geographic context, such trend was coined Volunteered Geographic Information (VGI) by Goodchild. OpenStreetMap (OSM) is arguably the most important and accepted VGI project all around the world [3]; since 2004, is "an international effort to create a free source of map data through volunteer effort" [1] and has become an inexpensive and timely solution to the problem of obtaining the digital map of certain place.

Nevertheless, due to the manner they are generated and shared, the quality of OSM data has been a discussion issue. Reasons adduced to cast doubt on it include: unknowing the motivations of users, lacking a reputation (in contrast to official institutions), the digital division (not everyone has access to internet), the inexperience of users, and the menace of digital terrorism [1, 4]. Responding to these doubts, virtually everything related to VGI has been researched: the quality issues [5–7], the social aspects [4, 8], and the applications [9–11]. This paper aims to contribute to these researches, but in the Ecuadorian ambit. VGI projects have peculiar geographies, which vary among times and places [3]; therefore, OSM characteristics must be addressed in every place this project has reached.

Among the aspects of geoinformation, positional accuracy has received the most attention; that is because this reflect the very essence of geographic data [12]. Furthermore, according to ISO 19113 norm, positional is the first of five geoinformation

© Springer Nature Switzerland AG 2019
Á. Rocha et al. (Eds.): WorldCIST'19 2019, AISC 931, pp. 513–523, 2019.
https://doi.org/10.1007/978-3-030-16184-2_49

quality aspects -the other four being thematic and temporal accuracy, logic coherence and completion. The goal of this work is to assess the horizontal positional accuracy of roads in OSM data, in an Ecuadorian location. A data science tool -a statistical programming environment- will be used, since a large volume of data is expected.

2   Related Work

In order to comprehend the VGI phenomenon, several authors have coined new concepts: "neogeography" (Turner, 2006; as cited by Sui [2]) refers to the inclusion of people with little in the way of scientific knowledge; "humans as sensors" [1] refers to the value added when traditional sensors are enhanced with local knowledge from those people; "wikification of GIS" [2] refers to the implications for Geographic Information Systems (GIS) due to this new participative trend; "produsers" [4] refers to the double role of users, as producers and consumers of information.

An additional condition that can be ascribed to VGI is to be Big Data. Although there is not a formal definition for this concept, it can be narrowed with a few ideas; traditionally, those are the "3 Vs" (Laney, 2001, as cited by Goodchild [13]): volume: they are very large amounts of data; velocity: they are generated by the minute and are available almost timely; and variety: they come from many different sources and have not underwent a normal quality control procedure.

Clearly, OSM reflects these three ideas: its collaborative nature excludes itself from certain applications which require generalization. In contrast, information generated by an authority -having underwent strict processes- may fit generalization. Problems arise when it is noted that -due to spatial heterogeneity- in geographic researches "generalization can become more a matter of faith than logic" [13]. In general, there are three ways for ensuring the quality of Big Geodata: firstly, benefit from the collective data generation, relying on independent observations of the same entities; secondly, create a user hierarchy, in order to moderate and correct contributions (OSM has already implemented it); and lastly, implement automated processes, which respond well to volume and velocity. At the end of the day, only the third method is robust [13]. Curating Big Data and assuring its quality area crucial issues, in a world where decision making and model generation relies more and more on it [14].

Plenty of research has been made about the possible applications for OSM data; some examples include predicting missing urban areas [9], refining population estimates [10] and estimating land use classification [11]. In the social context the contribution levels [8] and the motivations of users [4] have been discussed. Regarding the positional accuracy measurement, there are two main approaches in the literature. The first one advocates using information from inside the free dataset; for instance, source information can be used for inferring the precision [6]. The second one compares the free dataset with another one that -because of its authority- serves as reference, and then computes errors; a remarkable example can be found in Helbich et al. [5].

In this paper, an interpretation of the second approach is applied. It should be mentioned, however, that the first one has the advantage of relying on a single dataset; such approaches are usual in the state of the art [6, 7, 15]. By either extracting intrinsic information or applying computational algorithms, data highlights are obtained and quality, assessed. What makes these kind of approaches more attractive is that they are easily automated and, therefore, well-suited for addressing Big Data issues.

3 Materials and Methods

3.1 Study Zone

Study zone is the Manuela Sáenz administrative zone, which is one of eight divisions of the Metropolitan District of Quito, located at the very center of it (Fig. 1). It was chosen with the premise that, being central and urban, it is well mapped area in OSM. The official urban streets shapefile was obtained from local government website [16].

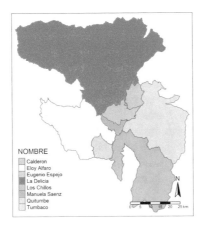

Fig. 1. Location of the Manuela Sáenz administrative zone

3.2 GIS Software Processing

OSM data is freely downloaded from its website [17] in .osm extension files, which happen to have a XML structure; in this sort of structure information is stored in tags, composed of key plus value [18]. A GIS was needed in order to transform the .osm into shapefiles; in this case, the open QGIS 3.0 was used. It was also used for spatial manipulation, in order to fitting the data for analysis; Fig. 2 summarizes this process.

3.3 Programming Software Processing

The software RStudio allowed to implement the required programming environment [19]. The R packages used were: XML and foreign, for reading the .xml and the .dbf respectively; dplyr and tidyr, for data manipulation; and ggplot2, for plotting. While linestrings (roads) were the input in QGIS (Fig. 2), points (roads intersections) were the input in RStudio, since it is easier to compute the positional accuracy from these.

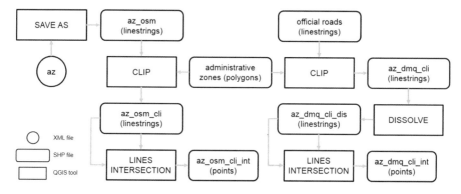

Fig. 2. Flowchart with the process made in QGIS software

It was noted that all the official road names were uppercase and without accent marks; through base R functions the OSM roads were forced into this convention. This is crucial because equivalences between both datasets are found using the pair of intersected roads names; previously, roads that lack the name and highway tags were deleted. Inconveniences derived from this method are discussed in the next section. Positional error -distance between equivalent OSM and official intersections- were computed while seeking a balance between two criteria: solve the methodology inconveniences, and maximize the number of computed errors. Afterwards, tags read directly from the XML were joined using OSM ids, with the purpose of extending the analysis. Finally, the method was applied to two additional administrative zones - Eugenio Espejo and Calderón- in order to test its performance.

3.4   Programming Software as Standalone

In the previous section it was mentioned that velocity is an important issue in Big Geodata, and automated processes are robust ways of assuring its quality. Because the GIS manipulation took rather too much time, it was replaced through the inclusion of spatial-dedicated R packages: osmdata, for downloading OSM data; sf, and raster for data handling; and tmap, for plotting maps. Particularly, sf functions replaced the QGIS tools in Fig. 2. RStudio as a standalone was also used in the additional zones.

4   Results and Discussion

Table 1 presents a summary of obtained, modified and computed entities, for each of the three studied administrative zones while applying the two software method. An early observation is that entities counts in official roads (2 and 3) are congruent with area (1) in each administrative zone (Eugenio Espejo > Calderón > Manuela Sáenz); however, entities in OSM (4 to 6) are not congruent with area (Manuela Sáenz > Calderón). This gives initial evidence for incompletion (absence of geographic entities) in the Calderón administrative zone. In order to comprehend what these numbers are trying to communicate, Table 2 presents the same data but in a comparative manner.

Table 1. Summary of entities by administrative zone

Type of entity	Manuela Sáenz	Eugenio Espejo	Calderón
Area of the zone [Ha] (1)	4844.32	11975.76	8753.92
Number of segments in official roads (clipped) (2)	4442	11406	6704
Number of intersections in official roads (3)	1962	4988	2457
Number of segments in OSM (clipped) (4)	3525	8671	3082
Number of useful segments in OSM (name, highway) (5)	2398	6066	1680
Number of intersections in OSM (6)	3095	8444	2373
Number of found errors (equivalences) (7)	941	1756	741
Number of correct errors (8)	833	1480	614
Number of segments with at least one correct error (9)	650	1190	486

Table 2. Comparative summary of entities by administrative zone

Type of entity	Manuela Sáenz	Eugenio Espejo	Calderón
Percentage of segments: OSM respect to official (4/2)	79.36	76.02	45.97
Percentage of intersections: OSM respect to official (6/3)	157.75	169.29	96.58
Percentage of found errors: intersections in OSM (7/6)	30.40	20.80	31.23
Percentage of correct errors: intersections in OSM (8/6)	26.91	17.53	25.87
Percentage of correct errors: segments in OSM (9/5)	27.11	19.62	28.93

The fact that there are more OSM than official intersections is consequence of having dissolved only the second one (Fig. 2); this is untrue in Calderón, where there are fewer OSM intersections. Furthermore, while there are always fewer segments in OSM, in Calderón the percentage is much fewer (45.97%); this contributes to the idea that incompletion is greater there (Fig. 3). Missing roads were either deleted because they were not named, or not there from the beginning; therefore, these percentages offer an initial idea of the thematic lacking in OSM, which is another quality aspect.

Fig. 3. Incompletion of OSM roads (solid lines) respect to the official ones (dotted lines)

In both tables, a "useful" OSM segment is one with both name and highway tags. What is considered a "found" or "correct" error depends on the considerations introduced in order to solve the inconveniences of this methodology. "Found" errors are simply the result of joining two tables (OSM and official roads); in the best scenario, the percentage of found errors was 31.23% of intersections, which is rather low. Regarding "correct" errors, percentages are fairly similar between intersections and roads, and the ones in Manuela Sáenz are the most similar (26.91% y 27.11%).

Unfortunately, the join method has two inconveniences; first one (Fig. 4, left) happens when a road has lanes with the same name, and extra equivalences are generated; second one happens when there are more than two roads in a crossroad (Fig. 4, right). Solution to both inconveniences consists in grouping the equivalences that have the same east and north coordinates, and pair of OSM ids, and selecting the least distance in each case. Doing so, the number of errors is reduced, but it is secured that only the "correct" ones are preserved in every location (for instance in Fig. 4, right, the arrow points a unique coordinate where there are actually three intersections).

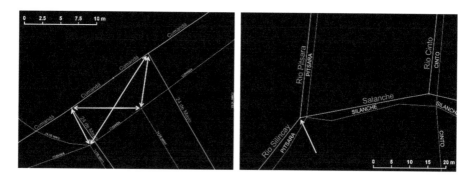

Fig. 4. Inconveniences generated by roads with many lanes (left) and crossroads with more than two roads (left). OSM roads are solid, officials are dotted, and inconveniences are arrows.

Executing the R summary function for correct errors in Manuela Sáenz reveals that 746 intersections were not grouped according to the solution; 833 is total of correct errors. The summary function also delivers error statistics, the mean (6.26 m) being greater than both median (2.66 m) and third quartile (4.20 m). This distributions, and a 406.32 m maximum value, suggests outliers. A boxplot (Fig. 5) with log scale shows a cleaner representation of error distribution and outliers (points).

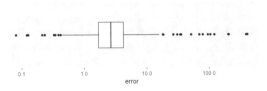

Fig. 5. Boxplot of correct errors in the Manuela Sáenz administrative zone

Through QGIS id queries, intersections related to outliers were identified, realizing (Fig. 6) that those were actually errors in the drawn geometry in OSM. Since in this context they are truly positional, they were preserved for the analysis.

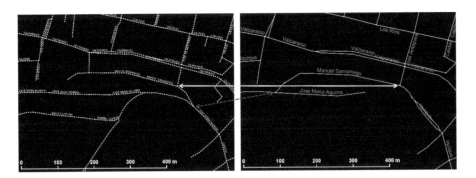

Fig. 6. Examples of correct (solid arrow) and badly drawn geometry errors (dotted arrow)

In order to conclude the analysis in Manuela Sáenz, three plots were created (Fig. 7): first, an scatter plot showing the distribution (east, north) and magnitude (size) of the errors; second, a histogram of the errors; third, a jittering plot (scatter plot with noise) showing the relation between mean error by segment and segment version, as read from the XML. Only the first plot deploys all of the errors; for the remaining two, outliers were deleted using ggplot2 boxplot interval (Q1 − 1.5 IQR: Q3 + 1.5 IQR). Jittering in the third plot helps to solve overplotting, since precision is not important here; the color scale (n) shows number of errors by segment.

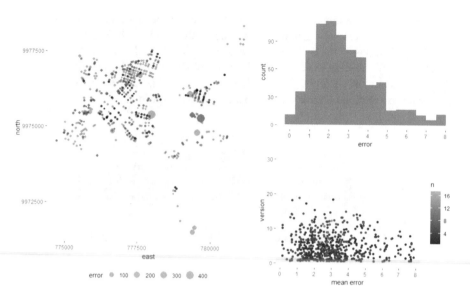

Fig. 7. Spatial distribution (left), histogram (upper right) and relation with segment version (lower right) for positional errors in the Manuela Sáenz administrative zone

Analysis as presented in Fig. 7 is repeated for the remaining two administrative zones: Eugenio Espejo and Calderón. Conducting this kind of analysis is related with the three ways for assuring the quality of geographic Big Data [13]. The spatial distribution plots may benefit those users who possess higher hierarchy in OSM, with the purpose of identifying where the badly drawn geometries related errors are. On the other hand, none of the three cases showed a relation between version and mean error in a segment. Some may argue that, the higher the version, the better the positional accuracy; however this hypothesis does not seem to stand. This is related with the quality assuring [13] in the sense that accumulating observations -even from different users- of the same entities is a dubious method for addressing the quality.

Finally, the method which is presumed to be more robust -automation- relies on scripts, such as the developed in the present study. The advantage of this methodology is that allow us, in a few seconds, to identify in which intersections of a set of roads corrections are required. Nevertheless, it was noted that it would be risky to remove the user decisions, for example, by applying directly the same algorithm in different zones. Furthermore, automating this methodology carries several difficulties; particularly, the use of two datasets and two software limits the implementation extend and speed. For this reason, implementing the RStudio software as a standalone was preferred. Even with increased computing time the overall process is faster, requires less manipulation and has just the official roads shapefile as input -since OSM data is automatically downloaded. Figure 8 presents the result of the standalone method, for the Eugenio Espejo administrative zone.

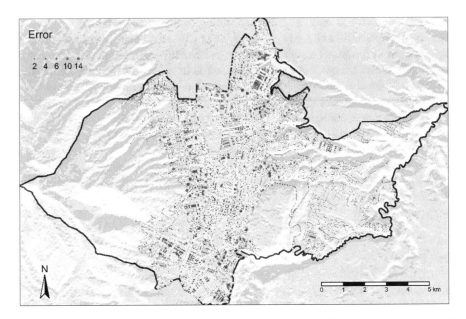

Fig. 8. Computed (red), outliers (purple) and missing (yellow) positional errors in the Eugenio Espejo administrative zone.

The abundance of yellow points in the previous figure exemplifies the great incompletion that was discussed at the beginning of this section. Regarding the errors that were actually computed, on the brighter side, the benefits of this second approach should be interpreted as a better response to the velocity aspect of Big Geodata; other state of the art methods [6, 15] also benefit from computational algorithms in order to work with VGI. Regarding the volume aspect, the method is expected to behave well but further investigation is required. However, regarding the variety aspect, this method is rather limited, mainly because it depends strictly on XML structure and R packages that may change in the future.

5   Conclusions

In the present study two types of software -GIS and programming- were used for assessing the OSM horizontal positional accuracy in three zones of the Metropolitan District of Quito. This is part of an endeavor for understanding an open dataset which has great volume, grows constantly, and comes from diverse sources; these three characteristics have granted it the condition of Big Data.

It was demonstrated that the combination of two software is effective not only for drawing conclusions, but also for identifying where corrections are needed. Moreover, the analysis was improved when only the second software was deployed. It should be noted, however, that the two inconveniences of the present method were recognized through GIS, and that dropping the GIS is only recommendable when the user is already comfortable with the OSM data structure.

Certain hypotheses may be denied according to the results. The independent and collaborative generation of many observations of the same entity is not related to any improvement in positional errors. Moreover, particularly large errors -outliers- are not restricted to specific zones, but rather related to bad drawn roads. Incompletion is also a serious problem, since a missing road produces a decrease in quality; even unnamed roads may have an impact in the positional quality aspect. It is concluded that, at least in the studied zones, more attention should be put in the thematic quality aspect.

The suggested hypotheses still suffer from the limitation of not being generalizable. Overcoming this limitation involves developing automated methods, which could be applied in a greater scale and would allow us to know the characteristics of this and other VGI projects. It is important to keep in mind that this kind of projects have particular characteristics in each place of the world; hence the need of studying them in every place they are available.

References

1. Goodchild, M.: Citizens as sensors: the world of volunteered geography. GeoJournal **69**(1), 211–221 (2007). https://doi.org/10.1007/s10708-007-9111-y
2. Sui, D.: The wikification of GIS and its consequences: or Angelina Jolie's new tattoo and the future of GIS. Comput. Environ. Urban Syst. **32**(1), 1–5 (2008). https://doi.org/10.1016/j.compenvurbsys.2007.12.001
3. Jokar, J., Zipf, A., Mooney, P., Helbich, M.: An introduction to OpenStreetMap in geographic information science: experiences, research, and applications. In: Jokar, J., et al. (eds.) OpenStreetMap in GIScience, LNGC, vol. 75, pp. 1–15. Springer, Basilea (2015). https://doi.org/10.1007/978-3-319-14280-7_1
4. Coleman, D., Georgiadou, Y., Labonte, J.: Volunteered geographic information: the nature and motivation of produsers. Int. J. Spat. Data Infrastruct. Res. **4**(1), 332–358 (2009). https://doi.org/10.2902/1725-0463.2009.04.art16
5. Helbich, M., Amelunxen, C., Neis, P., Zipf, A.: Comparative spatial analysis of positional accuracy of OpenStreetMap and proprietary geodata. In: Jekel, T., et al. (eds.) GI_Forum 2012: Geovizualisation, Society and Lerning, pp. 24–33. Herbert Wichmann Verlag, Berlín (2012)
6. Touya, G., Reimer, A.: Inferring the scale of OpenStreetMap features. In: Jokar, J., et al. (eds.) OpenStreetMap in GIScience, LNGC, vol. 75, pp. 81–99. Springer, Basilea (2015). https://doi.org/10.1007/978-3-319-14280-7_5
7. Hashemi, P., Abbaspour, R.: Assessment of logical consistency in OpenStreetMap based on the spatial similarity concept. In: Jokar, J., et al. (eds.) OpenStreetMap in GIScience, LNGC, vol. 75, pp. 19–36. Springer, Basilea (2015). https://doi.org/10.1007/978-3-319-14280-7_2
8. Neis, P., Zipf, A.: Analyzing the contributor activity of a volunteered geographic information project - the case of OpenStreetMap. Int. J. Geo Inf. **1**(1), 146–165 (2012). https://doi.org/10.3390/ijgi1020146
9. Hagenauer, J., Helbich, M.: Mining urban land-use patterns from volunteered geographic information by means of genetic algorithms and artificial neural networks. Int. J. Geogr. Inf. Sci. **26**(6), 963–982 (2012). https://doi.org/10.1080/13658816.2011.619501
10. Bakillah, M., Liang, S., Mobasheri, A., Jokar, J., Zipf, A.: Fine-resolution population mapping using OpenStreetMap points-of-interest. Int. J. Geogr. Inf. Sci. **28**(9), 1940–1963 (2014). https://doi.org/10.1080/13658816.2014.909045

11. Jokar, J., Vaz, E.: An assessment of a collaborative mapping approach for exploring land use patterns for several European metropolises. Int. J. Appl. Earth Obs. Geoinf. **35**(1), 329–337 (2015). https://doi.org/10.1016/j.jag.2014.09.009
12. Ariza, F.: Fundamentals of Geographic Information Quality Assessment. Jaén University, Jaén (2013). (in Spanish)
13. Goodchild, M.: The quality of big (geo)data. Dialogues Hum. Geogr. **3**(3), 280–284 (2013). https://doi.org/10.1177/2043820613513392
14. Miller, H., Goodchild, M.: Data-driven geography. GeoJournal **80**(1), 449–461 (2015). https://doi.org/10.1007/s10708-014-9602-6
15. Noskov, A.: Computer vision approaches for big geo-spatial data: quality assessment of raster tiled web maps for smart city solutions. In: 7th International Conference on Cartography and GIS, Sozopol (2018)
16. Open Government (in Spanish). http://gobiernoabierto.quito.gob.ec/?page_id=1122
17. OpenStreetMap. https://www.openstreetmap.org
18. OpenStreetMap Tags. https://wiki.openstreetmap.org/wiki/Tags
19. R Core Team: A Language and Environment for Statistical Computing. R Foundation for Statistical Computing, Vienna (2018)

Optimized Community Detection
in Social Networks

Lamia Berkani[1,2]($\boxtimes$), Sara Madani[2], and Soumeya Mekherbeche[2]

[1] Laboratory for Research in Artificial Intelligence (LRIA),
Bab Ezzouar, Algiers, Algeria
[2] Department of Computer Science, USTHB University,
Bab Ezzouar, Algiers, Algeria
{lberkani, smadani, smekherbeche}@usthb.dz

Abstract. We focus in this paper on the identification of groups of members who share similar tastes and preferences in the context of social networks. We address the need for community detection based on two different approaches: the first one is graph-oriented, while the second one concerns an optimized classification algorithm. The graph oriented approach is based on the calculation of the cycles and considers two variants: the modularity computation and the external links. The optimized classification approach applies the K-means algorithm as an initial classification solution, and then optimizes this classification by using two different meta-heuristics: the Tabu Search (employing local search methods to explore the solution space beyond local optimality) and the Bee Swarm Optimization algorithm (a population-based search algorithm). In order to validate our proposal, experiments were conducted on different datasets. The results obtained are promising in terms of modularity and response time.

Keywords: Social networks · Community detection · Graph cycles ·
Optimization · Classification · Tabu Search · Bee Swarm Optimization ·
Modularity

1 Introduction

With the development of Web 2.0 and social media technologies, social network sites have attracted millions of users, and many of them have integrated these sites into their daily practices. A social network can be represented by a graph consisting of a set of nodes and edges connecting these nodes. The nodes represent the users, and the edges correspond to the interactions among them. However, given the complexity of these networks, new challenges have emerged, requiring a restructuring of these networks to facilitate the efficient interaction of users. According to Bedi and Sharma (2016), the tendency of people with similar tastes and preferences to get associated in a social network leads to the formation of virtual communities/clusters. The experience and practice on social networks show that the detection of these communities can be beneficial in several ways and for many applications such as finding likeminded users for marketing and recommendations or a common research area in collaboration networks. This could strengthen links/create new links between people with similar interest profiles.

© Springer Nature Switzerland AG 2019
Á. Rocha et al. (Eds.): WorldCIST'19 2019, AISC 931, pp. 524–533, 2019.
https://doi.org/10.1007/978-3-030-16184-2_50

Community detection is one of the research fields of social network analysis. A large number of research works and algorithms have been proposed. The state of the art shows that some works are graph-oriented based on the modularity function (Newman and Girvan 2004), while other works use classification techniques. Our goal in this research is to propose a new approach taking advantage of the graph structure and using a classification technique. However, in order to optimize the detection of communities, we combine the classification algorithm with two different meta-heuristics: the Tabu Search and the Bee Swarm Optimization algorithms.

The remainder of this paper is organized as follows: Sect. 2 provides some basic definitions and concepts on community detection and presents some related work on social community detection. Section 3 presents our approach for the detection of communities in a social context. The results of the experiments are given in Sect. 4. Finally, the conclusion summarizes the most important results and describes some future perspectives.

2 Literature Review

The most commonly used definition of the term community is that of Yang et al. (2010): "a community is a group of network nodes, within which the links connecting nodes are dense but between which they are sparse". According to Fortunato (2009), communities, also called clusters or modules, as "groups of vertices that probably share common properties and/or play similar roles within the graph". Papadopoulos et al. (2012) defines communities as: "groups of vertices that are more densely connected to each other than to the rest of the network".

A panoply of community detection algorithms exists in the literature. The first idea using static networks was proposed by Newman and Girvan (2004). The proposed method is based on a modularity function, aiming to obtain the optimum partitioning of communities. In the same direction, Blondel et al. (2008) have proposed Louvain algorithm to detect communities using the greedy optimization principle to optimize the gain of modularity. According to Babers and Hassanien (2017), traditional methods such as graph partitioning used to detect community within networks by dividing it based on predefined size. Spectral clustering method is based on similarity matrix and integrating similar groups used by hierarchical clustering technique.

Recently, some works are using optimization algorithms for the community detection: Babers and Hassanien (2017) presented a cuckoo search optimization algorithm with Lévy flight for community detection in social networks. Sharma and Annappa (2016) have introduced modularity metrics and Hamiltonian function combined with meta-heuristic optimization approaches of Bat algorithm and novel Bat algorithm.

The review of related works shows the limitations of the existing graph-based methods, and the current trend to apply optimization techniques for identifying communities in social networks. In this same direction, we attempt in this work to apply other meta-heuristics and to explore other research methodologies for communities' detection in social networks.

3 Our Approach to Community Detection

In order to detect user communities we proposed the following approaches: (1) two graph-oriented methods based on cycles computing; (2) a classification based on the K-means algorithm; and (3) an optimized classification using two different meta-heuristics, the Tabu search and the Bee Swarm Optimization algorithm.

3.1 Graph-Oriented Approaches

We propose two different cycle-based approaches to avoid the problem of clique size (i.e. cliques can present a fundamental problem in considering a graph as a single community when the latter contains a large number of cliques).

Let's consider G (V, E) an undirected and non-weighted graph, V representing all the nodes of the graph G (the users) and E the set of edges (the friendships between the users). The two approaches result in another graph G', consisting of cycles (groups of users with the existing links between them).

First Approach: Cycle-Modularity. This approach consists of two steps: the calculation of cycles and then the application of an optimization using the objective function of Newman to obtain the final communities. We describe these steps as follows:

- *Step 1 - Cycles Computing:* consists of detecting all the cycles of the graph G, giving as output a new graph G 'consisting of small groups (cycles). Initially, it is a question of identifying all the existing cycles, which will represent the initial communities. Each cycle represents a group of friends where each two users of the cycle are either:
  - Friends (There is a friendship between the two users).
  - There is at least one friend in common between them.
- *Step 2 - Modularity Computing:* The process is similar to that of Newman, which consists of calculating the modularity for each pair of communities and merging the two communities with the largest gain each time. At each iteration we save the modularity that gave the best partitioning. The same process is redone until there is no more gain in modularity. Finally, at the end of this process we will have several partitions and the best partitioning will be the one with the maximum modularity.

Second Approach: External Links. This approach consists of two steps as follows:

- *Step1 - Cycle-Modularity:* The first step is similar to that of the previous approach, which aims to identify a set of cycles representing the initial communities.
- *Step 2 - Taking external links into account:* since the initial communities are formed from highly connected cycles, we have been interested in the external links between these communities to build new communities by merging the community pairs, having an external link rate higher than a given threshold.

3.2 K-Means-Based Classification Approach

The k-means classification is one of the unsupervised classifications. The principle of the algorithm is the initial definition of the number of clusters. The algorithm assigns to each cluster a vertex, extracted randomly, from all the vertices.

In order to assign each vertex to the most appropriate community, a similarity criterion is developed indicating the nearest center of a given vertex. The similarity function used is the geodesic distance µij. This function calculates the number of edges of the shortest path connecting a vertex i and another vertex j. In order to elect the new centers we calculate the average centrality of each summit with respect to its community, the summit with the lowest value of centrality of intermediacy will be elected as new center. The average centrality being the average of the distances from the summit to all the others, as indicated by the following formula:

$$C_{AVG}(V_i) = \frac{1}{n-1} \sum_{i \neq j} \mu_{i,j} \tag{1}$$

where:
Vi: is the ith vertex; n: is the number of vertices; and µij: is the geodesic distance.

3.3 Optimized Classification Approach

In this section, we present the two meta-heuristics that we have adapted to the problem of community detection.

Tabu Search-Based Classification. The Tabu search is an optimization meta-heuristic used to solve complex and/or large problems (Glover 1986). Tabu search overcomes the local optimum problem encountered in local search using a tabu list. The tabu list is represented by a hash table such that the key is the concatenation of the identifiers of the communities of the solution. The tabu list will contain all the solutions explored in the search. The implementation steps are as follows:

1. Generate an initial solution using K_means.
2. Generate the neighborhood of this solution by using local search.
3. Return the best non-tabu neighbor solution.
4. If the best neighbor does not exist then diversification.
5. Return the least worst non-tabu neighbor.
6. If the least worst non tabu neighbor does not exist then return the best solution tabu if the number of chances is greater than 0.
7. Otherwise generate a solution from k-means preserving the contents of the tabu list.
8. Repeat the process from Step 2 with the new returned neighbor.

BSO-Based Classification. Swarm optimization of bees manipulates a set of bees where each bee is a feasible solution to a given problem. In order to make the best use of meta-heuristics and to prove their effectiveness, it would be necessary to consider a codification that makes it possible to model the problem. In our case, each possible partitioning represents a solution, where each bee corresponds to a feasible partitioning. The solution can be represented by a vector containing the different existing communities. The vector indices represent the keys of each user in a hash table. Each box of the vector contains the identifier of the community to which the user belongs, where each user having as key the index of this box.

Illustrative example: Let's consider the following three communities:

- Community 1: u5, u6, u7and u10
- Community 2: u3, u2 and u11
- Community 3: u1, u8 and u9

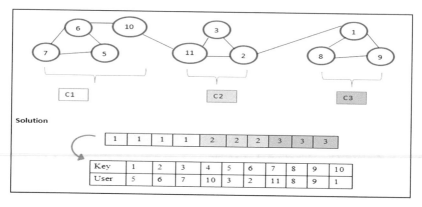

Fig. 1. Example of applying the BSO algorithm on a social network.

The general operating process of Bee Swarm Optimization is as follows:

```
Algorithm2: BSO Algorithm
Begin
Determine the search areas.
For each bee
Reach a peak of the search area
Perform a local search.
Communicate the local optimum
End.
```

- Initialization of BSO
  For the initialization of the BSO algorithm we used the K-means algorithm. For the latter, we give as input a set of vertices and we obtain at the output K communities representing an initial partition that represents an initial solution for the BSO algorithm (Fig. 1).
- Diversification Strategy
  In order to determine the different search areas, continuous overlap flip was used. The latter being part of the parameters of the BSO algorithm, allows diversification from a feasible solution. The principle is to change m values, where flip = m, to the vector representing the solution and to keep the rest of the vector values as they are, starting from the beginning while allowing the overlap.

  Illustrative example: Let's consider the flip = 3. We consider the following partitioning and we obtain the following areas:

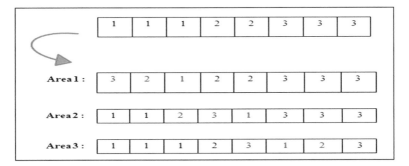

Fig. 2. Example of determining search areas using BSO algorithm.

After generating the search areas, each bee conducts a local search and communicates its local optimum to the other bees. We synthesize the communication process between bees in the following algorithm:

```
Algorithm 3: BSO Bee-Communication Algorithm
Begin
Best:= Best local optimum at time t;
Sref:= Best solution at the moment (t-1);
Δf:=Modularity (Best) - Modularity (Sref);
If Δf> 0 then
   Sref:= Best;
   Nb_chance = Max_chance;
   If Nb_chance> 0 then
      Nb_chance = Nb_chance-1;
      Sref = Best;
   Else Sref:= new solution (generated by K-means);
   End If
End If
Nb_chance:= Max_chance;
End.
```

The quality of a solution is expressed using the modularity of Newman. At the end of the process, the solution with the highest value of modularity will be considered as the final solution (Fig. 2).

4 Experiments

4.1 Datasets

To evaluate our community detection approach in a social context, we used the following well known datasets (Plantié and Crampes 2013; Wang et al. 2015): Zackary Karate Club, the Lusseau Dolphins network, a network of books on American politics, the American Football Network and the Rich Epinions Dataset (RED[1]). We randomly selected two data samples from this dataset (n = 50, m = 147 and n = 100, m = 468).

4.2   Evaluation Metrics

To evaluate our approaches, we have used the modularity of Newman (Newman and Girvan 2004), widely used in the work on community detection.

$$Q = \frac{1}{2m} * \sum_{j,k} \left(a_{V_j, V_k} - \frac{d(V_j)d(V_k)}{2m} \right) * \delta(V_j, V_k), j = 1 \ldots n; k = 1 \ldots n \qquad (2)$$

where:
  n: is the number of nodes in the graph; m: is the number of links in the graph;

  $d(vj)$: the number of neighbors of the node v_j.
  $\delta(vj, vk)$: is equal to 1 if v_j and v_k belong to the same community, and 0 otherwise.
  avj, vk: is equal to 1 if the nodes v_j and v_k are linked, and 0 otherwise.

4.3   Experimental Results

Evaluation of the Graph-Oriented Approaches. We compared the two proposed algorithms, Cycle-Modularity and External-Links, in terms of response time, number of communities and modularity (see Table 1).

Table 1. Evaluation of Cycle-Modularity and External-Links algorithms

	Cycle-Modularity			External-Links		
	Time	NbC	Q	Time	NbC	Q
Zackary Club Karate	13790	9	0.51	93	3	0.40
Lusseau Dolphin	274934	9	0.55	203	7	0.45
Politics Books	7965668	9	0.55	686	7	0.46
American Football	6144119	12	0.58	702	6	0.31
RED-50	71100	9	0.41	124	7	0.31
RED-100	4089018	14	0.28	858	13	0.16

We performed a comparison between our two algorithms with four existing approaches: EdgeBetweeness, Label propagation, Fast Greedy and Walktrap. However, we did not find evaluations done with the RED database for these same algorithms. The results obtained are illustrated in Table 2:

---

[1] https://projet.liris.cnrs.fr/red/.

Table 2. Modularity of proposed approaches vs. existing approaches

Graph-oriented algorithms	Zackary Club	Lusseau Dolphin	Politics Books	American Football	RED-50	RED-100
CycleModularity	**0.51**	**0.55**	**0.55**	0.45	**0.41**	**0.28**
External-Links	0.40	0.45	0.46	0.31	0.31	0.16
Edge Betweeness	0.40	0.52	0.51	0.59	–	–
Label propagation	0.35	0.41	0.50	**0.60**	–	–
Fast Greedy	0.38	0.49	0.50	0,01	–	–
Walktrap	0.35	0.48	0.50	0.60	–	–

We can see that the proposed Cycle-Modularity algorithm gave better performance in terms of modularity compared to the proposed External-Links algorithm, which on the other hand gave better performance in terms of response time. The Cycle-Modularity algorithm outperforms the other algorithms for the majority of datasets.

Evaluation of the K-Means-Based Classification Approach. Table 3 presents the modularity results obtained with the application of the K-means algorithm, where we varied the value of the number of clusters from 2 to 6 for each dataset.

Table 3. Modularity obtained with the K-means algorithm

Datasets	K = 2	K = 3	K = 4	K = 5	K = 6
Zackary Club Karate	0.27	0.29	0.25	0.24	0.26
Lusseau Dolphin	0.38	0.37	0.37	0.35	0.37
Politics Books	0.28	0.27	0.14	0.23	0.24
American Football	0.13	0.12	0.13	0.11	0.13
RED-50	0.17	0.14	0.20	0.17	0.18
RED-100	0.11	0.10	0.10	0.13	0.10

Evaluation of the Tabu Search-Based Approach. The Tabu Search algorithm is evaluated using two different experiments. First, we applied the Tabu Search based on the results of the Cycle-Modularity algorithm (cycles are identified as an initial solution). The results are presented in Table 4.

Table 4. Modularity of Tabu Search using the cycle identification algorithm

	10	20	50	100	300
Zackary Club Karate	0.38	0.47	0.47	0.47	0.48
Lusseau Dolphin	0.38	0.45	0.51	0.51	0.51
Politics Books	0.22	0.27	0.48	0.48	0.48
American Football	0.25	0.25	0.30	0.38	0.38
RED-50	0.27	0.327	0.32	0.32	0.34
RED-100	0.17	0.18	0.26	0.26	0.27

Then we tested the Tabu Search algorithm using the results of the K-Means algorithm as the initial solution. We set the value of the number of iterations to the best value found experimentally which is equal to 300. Then we varied the number of clusters. The obtained results are shown in Table 5.

Table 5. Modularity obtained with the Tabu Search algorithm

Datasets	K = 3	K = 4	K = 5
Zakary Karate Club	0.170	0.190	0.340
Lusseau Dolphins	0.044	0.210	0.190
Politics Books	0.290	0.100	0.150
American Football	0.096	0.110	0.170
RED-50	0.015	0.092	0.120
RED-100	0.047	0.087	0.051

Evaluation of the BSO-Based Classification Approach. We chose the combination of the parameters of the BSO algorithm that gave the best results: flip = 3, number of iterations = 100, number of chances = 4. The modularity results obtained on the different datasets are shown in Table 6.

Table 6. Modularity values obtained with BSO algorithm

Datasets	K = 2	K = 3	K = 4	K = 5	K = 6
Zakary Karate	0.37	0.37	0.28	0.11	0.22
Lusseau Dolphins	0.31	0.37	0.25	0.43	0.37
Politics Books	0.16	0.13	0.12	0.24	0.26
American Football	0.13	0.12	0.13	0.12	0.11
RED-50	0.31	0.26	0.18	0.21	0.27
RED-100	0.11	0.10	0.12	0.10	0.08

Figure 3 compares the modularity between K-means and the different optimization-based algorithms.

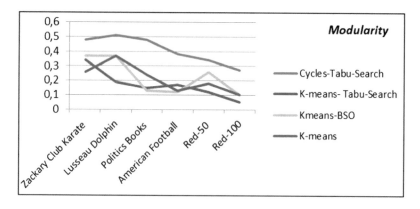

Fig. 3. Comparison between the different optimization-based algorithms.

We can notice that BSO algorithm has given a better quality of communities partitioning than K-means algorithm for the different datasets and with a variation of the number of clusters. On the other hand, the Tabu search based on cycles' computation gave better performance in terms of modularity.

5 Conclusion

We are interested in this paper to the problem of communities' detection. We have proposed two graph-oriented approaches based on cycles' identification called Cycles-Modularity and External Links. Furthermore, we have used classification techniques, applying the K-means algorithm which constitutes an initial solution. In order to optimize this classification, two meta-heuristics were used: the Tabu Search and the Bee Swarm Optimization algorithm. The results of experiments on well-known datasets show that the proposed Cycle-Modularity algorithm outperforms the other algorithms in terms of modularity and the proposed External-Links algorithm gave better performance in terms of response time. Moreover, the Tabu search based on cycles' computation has given better performance in terms of modularity. The perspectives of our work concern mainly the application of other optimization and classification algorithms.

References

Babers, R., Hassanien, A.E.: A nature-inspired metaheuristic cuckoo search algorithm for community detection in social networks. Int. J. Serv. Sci. Manage. Eng. Technol. (IJSSMET) **8**(1), 50–62 (2017)

Bedi, P., Sharma, C.: Community detection in social networks. WIREs Data Mining Knowl. Discov. **6**(3), 115–135 (2016)

Blondel, V.D., Guillaume, J.-L., Lambiotte, R., Lefebvre, E.: Fast unfolding of communities in large networks. J. Stat. Mech. Theory Exp. **2008**(10) (2008). https://doi.org/10.1088/1742-5468/2008/10/P10008

Fortunato, S.: Community detection in graphs. Phys. Rep. **486**(3–5), 103 (2009)

Glover, F.: Future paths for integer programming and links to artificial intelligence. Comput. Oper. Res. **13**(5), 533–549 (1986)

Newman, M., Girvan, M.: Finding and evaluating community structure in networks. Phys. Rev. **69**(2), 026113 (2004)

Papadopoulos, S., Kompatsiaris, Y., Vakali, A., Spyridonos, P.: Community detection in Social Media. Data Mining Knowl. Discov. **24**(3), 515–554 (2012). https://doi.org/10.1007/s10618-011-0224-z

Plantié, M., Crampes, M.: Survey on Social Community Detection. Springer Publishers. Social Media Retrieval, Springer Publishers. Computer Communications and Networks, pp. 65–85 (2013)

Sharma, J., Annappa, B.: Community detection using meta-heuristic approach: Bat algorithm variants. In: Ninth International Conference on Contemporary Computing (IC3) (2016)

Wang, C., Tang, W., Sun, B., Fang, J., Wang, Y.: Review on community detection algorithms in social networks. In: IEEE International Conference on Progress in Informatics and Computing (PIC), pp. 18–20 (2015)

Yang, B., Liu, D., Liu, J., Furht, B.: Discovering communities from Social Networks: Methodologies and Applications. Springer, Boston (2010)

Requirements for Training and Evaluation Dataset of Network and Host Intrusion Detection System

Petteri Nevavuori and Tero Kokkonen[✉]

Institute of Information Technology, JAMK University of Applied Sciences,
Jyväskylä, Finland
{petteri.nevavuori,tero.kokkonen}@jamk.fi

Abstract. In the cyber domain, situational awareness of the critical assets is extremely important. For achieving comprehensive situational awareness, accurate sensor information is required. An important branch of sensors are Intrusion Detection Systems (IDS), especially anomaly based intrusion detection systems applying artificial intelligence or machine learning for anomaly detection. This millennium has seen the transformation of industries due to the developments in data based modelling methods. The most crucial bottleneck for modelling the IDS is the absence of publicly available datasets compliant to modern equipment, system design standards and cyber threat landscape. The predominant dataset, the *KDD Cup 1999*, is still actively used in IDS modelling research despite the expressed criticism. Other, more recent datasets, tend to record data only either from the perimeters of the testbed environment's network traffic or from the effects that malware has on a single host machine. Our study focuses on forming a set of requirements for a holistic Network and Host Intrusion Detection System (NHIDS) dataset by reviewing existing and studied datasets within the field of IDS modelling. As a result, the requirements for state-of-the-art NHIDS dataset are presented to be utilised for research and development of NHIDS applying machine learning and artificial intelligence.

Keywords: Intrusion detection · Dataset · Cyber security ·
Machine learning · Artificial intelligence

1 Introduction

During the last few years, we have seen some great developments in both domains of machine learning and cyber security. Companies and researchers have investigated various ways to join these two domains with a goal of reliably detecting malicious activity in complex networked systems. The development of Intrusion Detection Systems (IDS) is one of the state-of-the-art domains where machine learning and cyber security are combined. However, research and development

© Springer Nature Switzerland AG 2019
Á. Rocha et al. (Eds.): WorldCIST'19 2019, AISC 931, pp. 534–546, 2019.
https://doi.org/10.1007/978-3-030-16184-2_51

of IDS requires generating realistic network traffic and intentionally mixing it with malicious attack traffic.

Generally, Intrusion Detection Systems are classified in two distinct ways. One way is to classify them according to the machine learning task they aim to fulfill. These tasks are *supervised signature detection* utilising labelled data and *unsupervised anomaly detection* using the normally functioning system as the baseline for comparison [7]. Another way to distinguish between the IDSs is by their operational context. Systems that focus on inter-host connection analysis are collectively called Network Intrusion Detection Systems (NIDS) while systems developed for intra-host activity analysis are called Host Intrusion Detection Systems (HIDS) [19].

Although the research and development within IDSs has been constant, the availability of required training and evaluation datasets is weak and commonly used misuse and anomaly detection datasets have fallen out-of-date. In some cases, the datasets date back to ten or even twenty years. These aged but still commonly used datasets fail to capture the complexity of modern systems and the evolved cyber threats imposed on them. Although several datasets are publicly available for research, they have been quite extensively criticised for lacking representativeness and generalisability to modern complex environments [3,9,31,32]. As stated in [25], datasets are often overly anonymised, they do not include modern threats and intrusions, have weak statistical characteristics or are not released because of the security reasons.

As these datasets are effectively just snapshots from their relevant time period, using them is bound to negatively affect the usability of threat detection models trained with these datasets. Because publicly available datasets have limitations, researchers have at times utilised existing operational systems for producing evaluation datasets. Such data, however, requires anonymisation for protecting systems and people using them, which imposes another burden on the researchers willing to make their datasets publicly available. This has led researchers to either build small-scale laboratory-like environments for producing and collecting data, or resort to disputed existing public datasets. Small-scale laboratory-like environments do not tend to mimic real world IT environments with multiple connected systems and users. Some of the datasets are also produced in environments that are outdated with regard to both the target systems and the attacks performed on them. It is also usual for the datasets to focus only either on the network or host data, which disallows the holistic modelling of the state of a system.

1.1 Motivation and Structure

Relevant and effective implementation of anomaly-based IDS models requires realistic data. There exists a risk of developing irrelevant and out-of-date models right from the beginning when using outdated and misleading data.

In this paper, we will review the prominent intrusion detection datasets to identify the necessary requirements for a modern IDS modelling dataset. The requirements are defined in terms of overall dataset composition, collected data features and the systems used to produce the data by comparing existing datasets. We differentiate between requirements for network flow data and host-based data, and review the data collection methods for each data type. We will then aggregate this information to identify the requirements for a modern NHIDS training and evaluation dataset. As will be shown in Sect. 2, there is a deficiency of datasets providing a snapshot of the environment in terms of both network traffic and host-based activities.

The paper is organised as follows. In Sect. 2 we first review the datasets selected for our review. We then compare the datasets cited and used in cyber security ID modelling research. We then define the requirements for an IDS modelling dataset capable of serving the needs of both network and host detection in Sect. 3. In Sect. 4 we discuss the environment in which the dataset could be generated. Lastly, we conclude our study in Sect. 5.

2 Dataset Review

As there already exist several cyber security method or dataset reviews providing information about distinct characteristics and features of existing public datasets, we have focused just on a concise representation of the existing datasets. The overview of datasets is divided into two subsections. In the first subsection we describe the prominently used datasets. The criterion that we used for defining a dataset as prominently used was that it should be either well-known or at least cited in several studies related to cyber security threat detection modelling. We give brief explanations about the selected datasets and refer to sources providing more in-depth dissection of their contents. However, restricting ourselves to review only datasets with a track record of utilisation means also that some of the most recently generated datasets are not included in the study. In the second subsection, we compare the features present in distinct datasets in order to see where the datasets agree and diverge across distinct features.

2.1 General Overview

First we review common datasets used with network and host IDS modelling. As the focus of our study is on building a set of dataset requirements for NHIDS modelling, we focus on datasets providing network traffic and host-based data. As these two are seldom combined, we have gathered information about datasets from both contexts.

KDD Cup 1999 [29,33] has been widely used for nearly two decades [19, 36]. The dataset was created for an IDS development competition containing aggregated and processed network and host data. *KDD Cup 1999* dataset is based on the *DARPA 1998* dataset [5]. Chattopadhyay et al. [6] stated that the *KDD Cup 1999* dataset has been predominantly used in intrusion detection studies. For these reasons, we have only included the *KDD Cup 1999* dataset and omitted the *DARPA* datasets. Although several researches indicate that the dataset of *KDD Cup 1999* is not realistic, that dataset is still used as a benchmark for new methods and results of IDS study [4].

NSL-KDD [30] builds upon the *KDD Cup 1999* dataset by addressing the problems of redundant records and imbalance of malicious samples. Because the dataset is effectively only a filtered version of its predecessor, the dataset describes an at least ten-year old IT environment with expired attacks although constructed in 2009. While the use of the dataset is trumped by the use of the *KDD Cup 1999* dataset [6], it is still present in multiple studies either as the training or evaluation dataset. In their survey of machine and deep learning method research within the context of intrusion detection, Xin et al. [36] discuss multiple studies either utilising the *NSL-KDD* only, partially or in conjunction with other datasets.

The *Sperotto* [27] dataset contains labelled network flow records collected from a honeypot installed at a public network location. The authors refrained from performing explicit attacks on the monitored host systems to have the dataset further conform to real-world conditions. After the initial collection of data, the researchers labelled the dataset utilising a correlation process based on domain knowledge. The flow-based intrusion detection method survey by Umer et al. [32] mentions several studies that use the dataset for evaluation purposes.

ISOT Botnet [24,35] is a labelled botnet network flow dataset merged from a collection of publicly available datasets during the time of the dataset's collection. The merging of the datasets was performed by replaying the network traffic from selected distinct datasets within a testbed environment. However, no details are given about the processes used to construct the malicious samples. While *ISOT* is a broader set of intrusion detection datasets, the *Botnet* dataset from 2011 has been used in several detection method studies [2,32]. The dataset contains flow data for botnets and background traffic, however, it lacks normal data [12].

CTU-13 [12] is a flow-based botnet dataset like the *ISOT Botnet*. It consists of malicious data from an infected network and legitimate background data from a university's network. The dataset consists of multiple distinct scenarios with varying statistics regarding malicious and benign traffic instead of a single collective dataset. The dataset has been used in studies focusing on botnet detection from network flow data [17,32].

ISCX IDS 2012 [25] dataset attempts to mimic real-world network events with traffic profiles constructed for distinct protocols. The dataset consists of network traffic flows collected from a simulated testbed environment. The profiles are essentially combinations of probability distributions fitted to features extracted from distinct protocols. The distributions are used for network flow data generation. This way the dataset should conform to real-world distributions of normal and background network traffic. Malicious traffic was generated by deliberate attacks performed on the testbed environment. Like *CTU-13*, the dataset contains several scenarios with distinct malicious activities. The detailed survey by Mishra et al. [19] reports dataset usage in several IDS studies.

ADFA-LD [8–10] and *ADFA-WD* [8] anomaly detection datasets are discretely host-based and built out of tokenized system call sequences. The datasets following in the footsteps of the 1998 *UNM* [13] system call dataset were

Table 1. Overview of prominently used IDS datasets.

Dataset	Year	Network data	Host data	Labelled	Acquisition[a]	Scenarios[b]	Malicious methods[c]	Avg. samples	Sample units
KDD Cup 1999	1999	Y	Y	Y	S	1	22	5,2 M	TCP packets
NSL-KDD	2009	Y	Y	Y	S	1	22	1,2 M	TCP packets
Sperotto	2009	Y	–	Y	R	1	?	14,2 M	Flows
ISOT Botnet	2010	Y	–	Y	M	1	3 (5)	1,7 M	Flows
CTU-13	2011[d]	Y	–	Y	S/R	13	7 (9)	80 M	Flows
ISCX IDS 2012	2012	Y	–	–	S	7	4	2,5 M	Flows
ADFA-LD	2013	–	Y	–	R	1	1 (6)	5,3 K	Syscalls
ADFA-WD	2014	–	Y	–	R	1	6 (12)	5,8 K	Syscalls
UNSW-NB15	2015	Y	–	Y	S	1	9	2,5 M	Flows

[a] S = Simulation, M = Merging of existing datasets, R = Real traffic/data.
[b] Split datasets (training, validation and testing) are considered a single dataset.
[c] Unbracketed numbers imply methods of conducting attacks (i.e. botnets) and bracketed numbers reported attack vectors.
[d] Associated report was first submitted in 2013, but dataset's website states 2011.

constructed from Ubuntu and Windows host systems separately and they aim to facilitate the use of sequence-based models in host intrusion detection. System call tokenization enables learning the host system's benign baseline in terms of sequences of system calls performed by benign applications. Deviation from this baseline can be considered anomalous and potentially malicious. The datasets are featured in multiple host-based IDS studies referenced in relevant reviews [1,36].

UNSW-NB15 [20,21] is a synthetic network flow dataset constructed by sampling, filtering and aggregating flow information to form two sets of artificially generated network traffic captures. The testbed environment which the dataset was constructed consisted of three virtual servers attached to a traffic generator. While the design schematics of the testbed imply incorporation of normal client-side traffic, no mentions of background traffic generation are given. The dataset was built with the intention of replacing the predominant *KDD Cup 1999* dataset and it is reported being used in studies [19].

The general overview of the selected datasets occurring prominently in studies is presented in Table 1. The table contains information about the general context of the dataset, labelling status, acquisition method, scenario count, the numbers of malicious methods and possibly reported distinct attack vectors, sample counts and sample types. There is a notable dispersion across sample counts, scenarios and the number of malicious methods found in the datasets. Only two of the datasets, the *ADFAs*, are distinctly host-based while the rest are predominantly network traffic datasets.

Table 2. Public availability and formats of IDS datasets.

Dataset	Public	Pcap	Argus	Weblog	Text	CSV	DB	Netflow
KDD Cup 1999	Y	–	–	–	Y	Y[a]	–	–
NSL-KDD	Y	–	–	–	Y	Y	–	–
Sperotto	–	–	–	–	-	–	Y	–
ISOT Botnet	Y	Y	–	–	–	–	–	–
CTU-13	Y	Y	Y	Y	–	Y	–	Y
ISCX IDS 2012	–	Y[b]	–	–	–	–	–	–
ADFA-LD	Y	–	–	–	Y	–	–	–
ADFA-WD	Y	–	–	–	Y[c]	–	–	–
UNSW-NB15	Y	Y	Y	Y	–	Y	–	–

[a] Data is comma-separated albeit text.
[b] Format inferred from dataset sizes.
[c] Executables can be handled as text.

While there is research citing that various datasets exist, the search for the availability of the datasets produced varying and surprising results. Intuitively, public availability should have a great impact on how well a dataset is received by the community of researchers. The format of source data is also an important factor. The rawer the data, the greater the effort to reproduce the ready-to-use dataset.

The availabilities and data formats of the datasets are provided in Table 2. Public availability means that the datasets are readily available for anyone to obtain without limits and the rest of the columns indicate file formats in which the dataset is provided. The most common formats are the raw network traffic captures in the *pcap* format, the text file and the comma-separated value (CSV) file.

2.2 Network Data

In this subsection, we will compare the features of the datasets. We will focus only on network traffic datasets, as only the *KDD Cup 1999* dataset contains host-based data in conjunction with similar *ADFA* datasets. The host-based datasets are discussed briefly in Subsect. 2.3. As with varying formats, there is also variance of features present across datasets. Directly comparing the features across datasets just by using the feature labels is misleading, as the naming conventions vary. The features have to be thus compared by their descriptions.

The descriptions of features for network traffic datasets are given distinctly in the following sources:

- *KDD Cup 1999* and *NSL-KDD* network traffic features are found in [5], as the latter is a subset of the former.
- *Sperotto* is described with details about its features in [26].

Table 3. Common network traffic dataset features.

Features	KDD Cup 1999	UNSW-NB15	ISOT Botnet	Sperotto	CTU-13	ISCX IDS 2012
Bytes, source to dest	Y	Y	Y	Y	Y	Y
Protocol, type	Y	Y	Y	Y	Y	Y
Bytes, dest to source	Y	Y	Y	Y	Y	–
Packets, dest	–	Y	Y	Y	Y	Y
Packets, source	–	Y	Y	Y	Y	Y
Timestamp, start	–	Y	Y	Y	Y	Y
Communication duration	Y	Y	–	Y	Y	–
Port, dest	–	Y	Y	Y	Y	–
Port, source	–	Y	Y	Y	Y	–
IP, dest	–	Y	Y	Y	Y	–
IP, source	–	Y	Y	Y	Y	–
Service accessed	Y	Y	–	–	–	Y
Protocol, state	–	Y	–	–	Y	Y

– *ISOT Botnet* is described in [24].
– *CTU-13* feature descriptions were derived from data headers.
– *ISCX IDS 2012* is described in [25].
– *UNSW-NB15* is documented and described in [34].

After collecting the features of distinct datasets, they were matched according to the provided descriptions. Some of the features were easy to match either directly with their labels or using descriptions. Others had to be looked into at a more general level e.g. how the number of packets or bytes relate to a flow or a connection. Some datasets report the counts separately for source and destination, while others only report a total count for communications. Because these describe essentially the same feature, they were handled effectively as similar. The features found in at least three of the datasets are presented in Table 3 with additional information about the dataset where the feature is present.

There were a total of 76 unique features across the datasets. Only 13 of these were common at least between any three datasets. The most commonly recorded features were the type of the communication protocol, the timestamp information and the counts for bytes and packets per communication. The source and destination address information was present four times, as was the communication duration. Some datasets incorporated records of accessed service and the state of the communication protocol. Even though some features were present just in a single dataset, multiple datasets had derived and aggregated features present complementing the features straightforwardly extractable from the raw traffic. While the complementary features were rather unique, their existence is common.

2.3 Host Data

Our selection of datasets has three either completely dedicated or network traffic complementing occurrences of host-based data. These are the *KDD Cup 1999*, the *ADFA-LD* and the *ADFA-WD*. The former of these contains the most features describing host activities performed by malicious actors. However, as discussed in [8–10], the operating system (OS), namely Solaris, from which the host activities were tracked was not selected with a focus on high deployment on markets. This poses a problem on generalisability even without discussing the agedness of the OS. Host data features are also highly processed and aggregated in terms of binary occurrences of certain indicators, which leaves a little room for more diverse modelling; detailed information of OS processes is altogether missing from the *KDD Cup 1999* dataset.

The *ADFA* datasets developed one and half decades after the *KDD Cup 1999* aim to address these two main shortcomings. The suffixes in the names point to Linux (*LD*) and Windows (*WD*) datasets. The host operating systems deployed for the corresponding datasets were Ubuntu 11.04 and Windows XP SP2, both of which were widespread in terms of installations and use commercially at the time. The approach to forming the datasets differs from the *KDD Cup 1999* and rather follows in the steps of the late 1990's *UNM* [13] dataset using raw system calls for detecting anomalous behaviour within a host system [9,36]. This in turn allows for the sequential modelling of the host's benign baseline for detecting anomalies.

3 Requirements for a NHIDS Dataset

As a result of the study, the requirements for a complete, consistent and unambiguous NHIDS dataset are presented as follows.

The Dataset Shall Include Network Traffic Data. It is safe to conclude from the recent and past reviews of intrusion detection methods that network traffic data is the predominant choice for trying to detect malicious activities aimed at IT environments [6,14,18,19].

The Dataset Shall Include Host Activity Data. However, strictly sticking to flow-based approaches tends to leave a great amount of information outside the reach of available anomaly or signature detection methods. This is because the network communications are nowadays commonly encrypted, leaving the payload of the traffic unobserved [32]. Inability to access and dissect the payloads of the traffic for detecting semantic attacks can be considered the main drawback of the flow only approach [28]. Thus, enriching the dataset with host-based data allows the use of network traffic data without limiting the detection possibilities of payload-induced manifestations in the target environment and its subsystems.

The Dataset Shall Contain Multiple Scenarios. A static dataset is also always bound to be only a snapshot in time capturing a limited representation of the observed phenomenon. In data domains, where the rate of change is slow, the effect of limited observations is somewhat negligible. In the case of IT systems in general, the rate of change is high and observable within the time frames of single years. The direction of change, however, is not only limited in the direction of time, but also easily observable across the various implementations of IT environments. This in turn creates a dilemma for intrusion detection - how to determine if an intrusion detection methodology is valid outside the dataset it was tested upon? Providing variability in the dataset itself is a way to tackle this. Like *CTU-13* and *ISCX IDS 2012* datasets, building scenarios with distinct activity profiles is a good starting point [25]. However, malicious activities are not the only form of variance introduced to the target environments. When enterprise networks are considered, it is normal that the topology is bound to change at least on the edges due to recruitments of new employees. Thus, introducing variation in the form of varying the target environment is another way to build up scenarios.

The Data Shall Be Representative of the Real-World Circumstances. Otherwise the dataset is insignificant. This requires paying attention to collecting data from background and normal network traffic [25] as well as mundane activities of host systems [8,9] regardless of whether the activities or traffic are simulated or not.

The Format of the Data Shall Enforce Usability. Considering the format in which the dataset should be provided, the data should be as readily usable as it can be from the viewpoint of the user of the data. For example, if two methods are developed independently of each other with the same raw data, the comparability of the results is left heavily dependent on the processes of aggregating and collecting the dataset. Thus, while some details are inevitably lost in the process, at least the network dataset should be in a commonly utilised row data format, for example CSV. However, mimicking the approach of Creech et al. [8], host data should consist of the appropriately tokenized sequences of system calls. While the text format is sufficient for host data, providing the data in CSV would help in removing a preprocessing step required to convert the system call tokens to lists or arrays. There should also be verbose descriptions of the features recorded within the dataset.

4   Producing NHIDS Dataset in Cyber Range

Producing an NHIDS dataset cannot be done in real production environments and a small-scale laboratory environment cannot produce a dataset with the required complexity. One possibility is to use a Cyber Range environment for dataset production.

Cyber Range functions as a research, development, training and exercise environment for the domain of cyber security. It is a closed and controlled environment providing the capability to mimic the required networks and systems for the purposes of research and development and supporting cyber security training and exercises. There can be replicated representations of the required organisation's network, systems, tools, and simulated Internet with background traffic from applications and users. In the Cyber Range environment, it is risk-free to use various attacks and intrusions with the required scenario [11,15,22].

For example, JAMK University of Applied Sciences has implemented Cyber Range called Realistic Global Cyber Environment (RGCE). RGCE mimics global Internet services with botnet-based traffic generation and attached organisation environments [15,16]. RGCE is used in scenario-based data generation for the development of anomaly based NIDS applying machine learning [23].

5 Conclusions

In this paper, we reviewed prominent intrusion detection datasets to identify the necessary requirements for producing a network and host data combining state-of-the-art intrusion detection system (NHIDS) training and evaluation dataset. We previewed the datasets in terms of overall composition of the datasets, the appearance of common features across them and the method their generation. A distinction was made between network and host datasets due to the differences in the nature of the data.

We found out that the coexistence of network and host data within a single dataset is uncommon. While flow-based network traffic datasets form the majority of datasets utilised in IDS modelling research, the encryption policies enforced nowadays render flow-based approaches unable to grapple attack vectors present in network traffic payloads. This is mitigable with host activity inclusion, which has data about the effects of payloads injected to the target environment.

While there are several network traffic features sharing common ground across multiple datasets, utilising these might not be enough. We found that, even though having unique formulations, having aggregated and derived features enriches the network flow dataset.

Utilisation of a state-of-the-art Cyber Range environment would be a prominent future research subject for creating a real-world circumstance matching IDS dataset according to found requirements. As small-scale testbed environments fail to mimic the complexity of real-world IT environments, the use of a multifaceted, holistic Cyber Range could prove out to be a prominent platform for generating a state-of-the-art NHIDS training and evaluation dataset compliant with modern equipment and design standards.

Acknowledgment. This research is partially funded by the Regional Council of Central Finland/Council of Tampere Region and European Regional Development Fund as part of the *New Business Innovations from Data-analytics* project of JAMK University of Applied Sciences Institute of Information Technology.

References

1. Abubakar, A.I., Chiroma, H., Muaz, S.A., Ila, L.B.: A review of the advances in cyber security benchmark datasets for evaluating data-driven based intrusion detection systems. Procedia Comput. Sci. **62**, 221–227 (2015). https://doi.org/10.1016/j.procs.2015.08.443

2. Alejandre, F.V., Cortés, N.C., Anaya, E.A.: Feature selection to detect botnets using machine learning algorithms. In: 2017 International Conference on Electronics, Communications and Computers, CONIELECOMP 2017, pp. 1–7. IEEE (2017). https://doi.org/10.1109/CONIELECOMP.2017.7891834

3. Aviv, A.J., Haeberlen, A.: Challenges in experimenting with botnet detection systems. In: Proceedings of the 4th Conference on Cyber Security Experimentation and Test, CSET 2011, p. 6. USENIX Association, Berkeley (2011). http://dl.acm.org/citation.cfm?id=2027999.2028005

4. Bodström, T., Hämäläinen, T.: State of the art literature review on network anomaly detection with deep learning. In: Galinina, O., Andreev, S., Balandin, S., Koucheryavy, Y. (eds.) Internet of Things, Smart Spaces, and Next Generation Networks and Systems, pp. 64–76. Springer, Cham (2018). https://doi.org/10.1007/978-3-030-01168-0_7

5. Buczak, A., Guven, E.: A survey of data mining and machine learning methods for cyber security intrusion detection. IEEE Commun. Surv. Tutor. **18**(2), 1153– (2015). https://doi.org/10.1109/COMST.2015.2494502

6. Chattopadhyay, M., Sen, R., Gupta, S.: A comprehensive review and meta-analysis on applications of machine learning techniques in intrusion detection. Australas. J. Inf. Syst. **22**, 1–27 (2018). https://doi.org/10.3127/ajis.v22i0.1667

7. Chio, C., Freeman, D.: Machine Learning and Security. O'Reilly Media Inc., Sebastopol (2018)

8. Creech, G.: Developing a high-accuracy cross platform host-based intrusion detection system capable of reliably detecting zero-day attacks. Ph.D. thesis (2013). http://handle.unsw.edu.au/1959.4/53218

9. Creech, G., Hu, J.: Generation of a new IDS test dataset: time to retire the KDD collection. In: IEEE Wireless Communications and Networking Conference, WCNC, pp. 4487–4492. IEEE (2013). https://doi.org/10.1109/WCNC.2013.6555301

10. Creech, G., Hu, J.: A semantic approach to host-based intrusion detection systems using contiguousand discontiguous system call patterns. IEEE Trans. Comput. **63**(4), 807–819 (2014). https://doi.org/10.1109/TC.2013.13

11. Ferguson, B., Tall, A., Olsen, D.: National cyber range overview. In: 2014 IEEE Military Communications Conference, pp. 123–128 (2014). https://doi.org/10.1109/MILCOM.2014.27

12. García, S., Grill, M., Stiborek, J., Zunino, A.: An empirical comparison of botnet detection methods. Comput. Secur. **45**, 100–123 (2014). https://doi.org/10.1016/j.cose.2014.05.011

13. Hofmeyr, S.A., Forrest, S., Somayaji, A.: Intrusion detection using sequences of system calls. J. Comput. Secur. **6**(3), 151–180 (1998). https://doi.org/10.3233/JCS-980109

14. Husak, M., Komarkova, J., Bou-Harb, E., Celeda, P.: Survey of attack projection, prediction, and forecasting in cyber security. IEEE Commun. Surv. Tutor. (2018). https://doi.org/10.1109/COMST.2018.2871866

15. JAMK University of Applied Sciences, Institute of Information Technology, JYV-SECTEC: RGCE Cyber Range. http://www.jyvsectec.fi/rgce/. Accessed 23 Nov 2018

16. Kokkonen, T., Puuska, S.: Blue team communication and reporting for enhancing situational awareness from white team perspective in cyber security exercises. In: Galinina, O., Andreev, S., Balandin, S., Koucheryavy, Y. (eds.) Internet of Things, Smart Spaces, and Next Generation Networks and Systems, pp. 277–288. Springer, Cham (2018). https://doi.org/10.1007/978-3-030-01168-0_26

17. Mathur, L., Raheja, M., Ahlawat, P.: Botnet detection via mining of network traffic flow. Procedia Comput. Sci. **132**, 1668–1677 (2018). https://doi.org/10.1016/j.procs.2018.05.137

18. Mishra, P., Pilli, E.S., Varadharajan, V., Tupakula, U.: Intrusion detection techniques in cloud environment: a survey. J. Netw. Comput. Appl. **77**, 18–47 (2017). https://doi.org/10.1016/j.jnca.2016.10.015

19. Mishra, P., Varadharajan, V., Tupakula, U., Pilli, E.S.: A detailed investigation and analysis of using machine learning techniques for intrusion detection. IEEE Commun. Surv. Tutor. (2018). https://doi.org/10.1109/COMST.2018.2847722

20. Moustafa, N., Slay, J.: UNSW-NB15: a comprehensive data set for network intrusion detection systems (UNSW-NB15 network data set). In: 2015 Military Communications and Information Systems Conference (MilCIS), pp. 1–6 (2015). https://doi.org/10.1109/MilCIS.2015.7348942

21. Moustafa, N., Slay, J.: The evaluation of network anomaly detection systems: statistical analysis of the UNSW-NB15 data set and the comparison with the KDD99 data set. Inf. Secur. J. **25**(1–3), 18–31 (2016). https://doi.org/10.1080/19393555.2015.1125974

22. National Institute of Standards and Technology NIST: Cyber Ranges. https://www.nist.gov/sites/default/files/documents/2018/02/13/cyber_ranges.pdf. Accessed 23 Nov 2018

23. Puuska, S., Kokkonen, T., Alatalo, J., Heilimo, E.: Anomaly-based network intrusion detection using wavelets and adversarial autoencoders. In: Lanet, J.-L., Toma, C. (eds.) Innovative Security Solutions for Information Technology and Communications, pp. 234–246. Springer International Publishing (2019). https://doi.org/10.1007/978-3-030-12942-2_18

24. Saad, S., Traore, I., Ghorbani, A., Sayed, B., Zhao, D., Lu, W., Felix, J., Hakimian, P.: Detecting P2P botnets through network behavior analysis and machine learning. In: 2011 Ninth Annual International Conference on Privacy, Security and Trust, pp. 174–180 (2011). https://doi.org/10.1109/PST.2011.5971980

25. Shiravi, A., Shiravi, H., Tavallaee, M., Ghorbani, A.A.: Toward developing a systematic approach to generate benchmark datasets for intrusion detection. Comput. Secur. **31**(3), 357–374 (2012). https://doi.org/10.1016/j.cose.2011.12.012

26. SimpleWiki: Labeled Dataset for Intrusion Detection. https://www.simpleweb.org/wiki/index.php/Labeled_Dataset_for_Intrusion_Detection.   Accessed   19 November 2018

27. Sperotto, A., Sadre, R., Van Vliet, F., Pras, A.: A labeled data set for flow-based intrusion detection. Lecture Notes in Computer Science (including subseries Lecture Notes in Artificial Intelligence and Lecture Notes in Bioinformatics), vol. 5843, pp. 39–50. Springer, Heidelberg (2009). https://doi.org/10.1007/978-3-642-04968-2_4

28. Sperotto, A., Schaffrath, G., Sadre, R., Morariu, C., Pras, A., Stiller, B.: An overview of IP flow-based intrusion detection. IEEE Commun. Surv. Tutor. **12**(3), 343–356 (2010). https://doi.org/10.1109/SURV.2010.032210.00054

29. Stolfo, S.J., Fan, W., Lee, W., Prodromidis, A., Chan, P.K.: Cost-based modeling for fraud and intrusion detection: results from the JAM project. In: Proceedings DARPA Information Survivability Conference and Exposition, DISCEX 2000, vol. 2, pp. 130–144 (2000). https://doi.org/10.1109/DISCEX.2000.821515

30. Tavallaee, M., Bagheri, E., Lu, W., Ghorbani, A.A.: A detailed analysis of the KDD CUP 99 data set. In: 2009 IEEE Symposium on Computational Intelligence for Security and Defense Applications, pp. 1–6. IEEE (2009). https://doi.org/10.1109/CISDA.2009.5356528

31. Tavallaee, M., Stakhanova, N., Ghorbani, A.A.: Toward credible evaluation of anomaly-based intrusion-detection methods. IEEE Trans. Syst. Man Cybern. Part C Appl. Rev. 40(5), 516–524 (2010). https://doi.org/10.1109/TSMCC.2010.2048428

32. Umer, M.F., Sher, M., Bi, Y.: Flow-based intrusion detection: techniques and challenges. Comput. Secur. 70, 238–254 (2017). https://doi.org/10.1016/j.cose.2017.05.009

33. KDD Cup 1999 Data. University of California, Irvine. http://kdd.ics.uci.edu/databases/kddcup99/kddcup99.html. Accessed 23 Nov 2018

34. University of New South Wales: The UNSW-NB15 Dataset Description. https://www.unsw.adfa.edu.au/unsw-canberra-cyber/cybersecurity/ADFA-NB15-Datasets/. Accessed 19 Nov 2018

35. University of Victoria, ISOT Research Lab: Datasets. https://www.uvic.ca/engineering/ece/isot/datasets/. Accessed 23 Nov 2018

36. Xin, Y., Kong, L., Liu, Z., Chen, Y., Li, Y., Zhu, H., Gao, M., Hou, H., Wang, C.: Machine learning and deep learning methods for cybersecurity. IEEE Access 6, 35365–35381 (2018). https://doi.org/10.1109/ACCESS.2018.2836950

A Fuzzy Inference System as a Tool to Measure Levels of Interaction

Ricardo Rosales[✉], Margarita Ramírez-Ramírez, Nora Osuna-Millán,
Manuel Castañón-Puga, Josue Miguel Flores-Parra, and Maria Quezada

Autonomous University of Baja California, 22390 Tijuana, Mexico
{ricardorosales,maguiram,nora.osuna,puga,mflores31,
maria.quezada}@uabc.edu.mx

Abstract. The interaction between humans and their environment is a current research area. Technology has been part of the environment with which humans have forged their progress, making this socio-technical phenomenon a field of study of interest in the field of research and technological development. In this paper, we present a FIS as a Tool to Measure Levels of Interaction during human-machine interaction. The proposal describes a system that bases its reasoning on user performance during user-exhibition interaction context in an interactive museum. Thte performance is determined by levels of interaction derived from intercativity actions and situations within the user-exhibition context. The FIS objective is to improve the user experience by managing the interaction process in user-exhibition relationships.

Keywords: Levels · Interaction · Fuzzy inference system

1 Introduction

The use of technology focused on information and communication systems is increasingly common in everyday activities. The current technological devices help users in their tasks during the day, providing an effective way of interaction that improves their experience in achieving goals. However, although interaction seems to be a fundamental aspect in the design of such technology, how can such interaction be modeled? Is it possible to incorporate the evaluation of the interaction itself as part of this technology? What impact would these improvements have?

Much suggests that the interaction is a complex phenomenon that depends largely on the subjective evaluation of the observer, so Gayesky and Williams [1] propose a theory based on levels based on some parameters they consider important. Some of the aspects to be considered are presence, interactivity, control, feedback, creativity, productivity, communication and adaptation.

By identifying the level of interaction by observing how the users experience the systems, it could help to design new forms of action that consider their

© Springer Nature Switzerland AG 2019
Á. Rocha et al. (Eds.): WorldCIST'19 2019, AISC 931, pp. 547–556, 2019.
https://doi.org/10.1007/978-3-030-16184-2_52

preferences. Evaluating the interaction requires knowing criteria based on the qualitative aspects of user behavior.

In this paper, we develop a Fuzzy Inference System (FIS) [2] as an intelligent system that implements an evaluation system that infers the level of interaction according to the Gayesky and Williams scale. A system is described that bases its reasoning on the user's performance determined by levels of interaction. Once the level of interaction is inferred, this information can be used to improve the user experience through a feedback system.

We exemplify these levels of interaction as the result of interactive actions and situations that are presented within the user-exhibition context in an interactive museum. The example system uses the inputs present in the interaction to properly configure and feedback to it. The objective is to improve the user experience by managing the interaction process in user-exhibition relationships.

The concept of interaction can be difficult to represent. However, interactions often occur as well as important aspects of behavioral science. The interaction involves at least two entities or can be a product of communication between these entities, e.g., the user and the system; both are complex, are not similar to each other, interact and communicate given the domain, actions, tasks, goals, and objectives. The interaction represents the interface. Therefore it must efficiently translate the messages among those involved helping in the fulfillment of goals of an application domain [3]. Therefore, the interaction helps to have dynamic cooperation and an end of cooperation between an entity and another entity or between several entities to achieve the objectives of the individual entities or to achieve some collective objective. The interaction often uses distributed artificial intelligence behaviors, as well as behaviors based on the multi-agent system, to propose tasks or perform tasks [4].

How can the interaction be represented? To answer this question you have to know the meaning of representation that the representation is the act or instances that represent an image or ideas of something or someone in mind, things that are not known. Otherwise, it is pure imagination or disfigurement of reality.

A description takes advantage of the conventions of a representation to describe something in particular. The representation is good because they show the restrictions of the problem, the representation of the interaction can be by mathematical Formulation [5], Analysis of the simple effects in factorial models [6, 7], Contrast of interaction [8–10], Patterns of interaction [2, 11, 12], Fuzzy inference system [2, 13, 14].

2   Related Work

Nowadays, to provide services or information to the user, it is necessary to measure interaction levels based on their performance. In general, the most common measurements of interaction levels are based on numerical and metric values, such as some clicks, feedback, logins to a system, etc. This measurement practice involves the following reflection: Does a metric allow measuring the level of interaction of a user? On the other hand, helping the user with a guide can be essential for the improvement of the interaction. Therefore, it is desirable to have interactive activities specific to the level reached by the user, helping to measure the interaction.

The measurement of the interaction takes place in the natural environment of the user often interrupted by the lack of controlled conditions and the inadequate replication of the problems caused by the user's interaction. In [15], he explains the idea that multiple measurement effects should be studied as isolated effects of a single variable.

There are investigations on the measurement of interaction levels focusing on measurable quantitative variables. For example, in [16] proposes a novel framework with characteristics to measure interactions quantitatively, quantifying the contribution made by each variable as well as the interactions involved.

Wachs et al. [17] proposes a framework of analysis to measure the human-computer interaction with four quantitative variables: ease of use, time of realization, the number of errors and the accuracy of the recognition, as well as with two variables qualitative "individual interview" to obtain feedback on experiences using a vision based on an interface compared to a conventional joystick, and the second "interview of questions", which focuses mainly on determining what features of the interface based on gestures of the hand are the most important and determine what additional features would be especially important.

The fuzzy logic [2,13,14], can help us to represent the levels of interaction of the user, just as it can help us to model the approximate fuzzy reasoning during the interaction. The relations between elements and sets follow a transition between belonging and non-belonging that is gradually represented by values of intermediate belonging between the true and the false in classical logic.

Jiang and Adeli [18] use the fuzzy logic to study investigate complexities and chaotic behavior in such systems. The fuzzy logic is used to for determining the embedding dimension a fuzzy c-means clustering approach is proposed for finding the optimum embedding dimension accurately.

Joelianto et al. [19] use an ANFIS to determine a time series estimation on Earthquake. An algorithm in the backward pass by using a mapping function maps the inputs to all corrected values obtained via error correction rules in the first layer by means of an interpolation of the inputs.

Pozna et al. [20] applied to fuzzy modelling in terms of mapping them to fuzzy inference systems. An example expressed as two applications dealing with modelling of fuzzy inference systems. A framework for the symbolic representation of data represented by the signatures and suggested a data configuration referred to as signatures offering possibilities of symbolic representation and of symbolic manipulation of fuzzy inference systems when the fuzzy modelling of MIMO systems is involved.

Nowakova et al. [21] present a novel method for fuzzy medical image retrieval (FMIR) using vector quantization (VQ) with fuzzy signatures in conjunction with fuzzy S-trees. The method is going to be added to the complex decision support system to help to determine appropriate healthcare according to the experiences of similar, previous cases.

3 Case Study

The case study is based on observation where scenes of interactive environments are analyzed, researched and modeled, to represent real examples of interaction. Interactive museums have a wide variety of interactive exhibitions and show various situations that arise due to the presence of interaction in groups of people. To understand the dynamics present in a context of interaction, the facilities of the interactive museum El Trompo, located in Tijuana, Mexico, with an annual attendance of 154,070 visitors per year is a magnificent place for the case study. The Fig. 1 shows some detail of the museum.

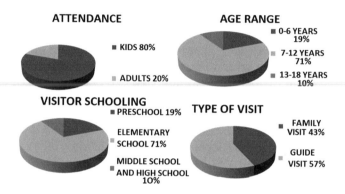

Fig. 1. Percentage of attendance, age, schooling and type of visits of users of the interactive museum El Trompo.

Because it is a museum dedicated to education, its main objective is to provide an interactive place where users can play while they learn. The museum was an ideal place to observe the behavior of users, especially of children and adolescents, being these the central objective of the research.

The behavioral patterns of the interactive museum environment were analyzed such as displacements, trajectories, place of interaction, those who are part of the interaction, interaction time, reading time of information labels, family, fatigue, social behavior, repeat visits, Information learning impact of exhibitions, attitudes, behaviors, and interests.

This information is crucial for the construction of models to help design the presentation format of the exhibition modules, signage, media design, and content information. Likewise, this information improves the measurement methods, the analysis to evaluate the behavior, performance, actions of the visitor, as well as the impact in the short, medium and long-term of the experiences of interaction, participation, patterns of social behavior and assistance.

3.1 Observation and Analysis of Users

To analyze the behavior patterns of the users, the exhibition modules (content, objective, media interface, etc.) were studied, particularly an interactive module with interactive features was chosen, which allowed obtaining the necessary data for analysis performed. The name of this display module is "move domain" The educational and interactive experience of this exhibition is that users play in four stations, where each teaches how to manipulate a means of transport (car, plane, motorcycle and hot air balloon). These show a virtual world simultaneously on all four screens, providing users with experiences of the four modes of transport as they travel through the simulated virtual world.

4   Proposed Fuzzy Inference System

Unlike other models that use different paradigms through heuristics, the research proposal adopts a theory of fuzzy sets to build the knowledge that an accredited user possesses. The FIS was built with the help of the MATLAB Diffuse Logic Toolbox [22]. The configuration of the FIS was carried out through an empirical process.

The inputs of the proposed FIS are the variables that are perceived by the display using sensors; these variables are the performance information of the user interaction which is: Presence, Interactivity, Control, Feedback, Creativity, Productivity, Communication, and Adaptation. Each of these variables was

Table 1. Inputs of the fuzzy inference system (Interaction level)

Inputs	Parameters (Values)					
	Very Bad	Bad	Regular	Good	Very Good	Excellent
Presence	0.085 3.5e−18	0.085 0.2	0.085 0.4	0.085 0.6	0.085 0.8	0.085 1
Interactivity	0.085 3.5e−18	0.085 0.2	0.085 0.4	0.085 0.6	0.085 0.8	0.085 1
Control	0.085 3.5e−18	0.085 0.2	0.085 0.4	0.085 0.6	0.085 0.8	0.085 1
Feedback	0.085 3.5e−18	0.085 0.2	0.085 0.4	0.085 0.6	0.085 0.8	0.085 1
Creativity	0.085 3.5e−18	0.085 0.2	0.085 0.4	0.085 0.6	0.085 0.8	0.085 1
Productivity	0.085 3.5e−18	0.085 0.2	0.085 0.4	0.085 0.6	0.085 0.8	0.085 1
Communication	0.085 3.5e−18	0.085 0.2	0.085 0.4	0.085 0.6	0.085 0.8	0.085 1
Adaptation	0.085 3.5e−18	0.085 0.2	0.085 0.4	0.085 0.6	0.085 0.8	0.085 1

Table 2. Outputs fuzzy inference system (Level of interaction)

Outputs	Parameters (Values)	
	Low	High
Level 0	0.05 0	0.3 1
Level 1	0.05 0	0.3 1
Level 2	0.025 0	0.3 1
Level 3	0.025 0	0.5 1
Level 4	0.025 0	0.5 1
Level 5	0.05 0	0.3 1

Table 3. Fuzzy inference rules (Interaction level)

Num.	Fuzzy inference rules
1	If (Presence is VeryBad) and (Interactivity is VeryBad) and (Control is VeryBad) and (Feedback is VeryBad) and (Creativity is VeryBad) and (Productivity is VeryBad) and (Communication is VeryBad) and (Adaptation is VeryBad) then (level0 is High)(level1 is Low)(level2 is Low) (level3 is Low)(level4 is Low)(level5 is Low)
2	If (Presence is Bad) and (Interactivity is Bad) and (Control is Bad) and (Feedback is Bad) and (Creativity is Bad) and (Productivity is Bad) and (Communication is Bad) and (Adaptation is Bad) then (level0 is Bad)(level1 is High)(level2 is Low) (level3 is Low)(level4 is Low)(level5 is Low)
3	If (Presence is Regular) and (Interactivity is Regular) and (Control is Regular) and (Feedback is Regular) and (Creativity is Regular) and (Productivity is Regular) and (Communication is Regular) and (Adaptation is Regular) then (level0 is Low)(level1 is Low)(level2 is High) (level3 is Low)(level4 is Low)(level5 is Low)
4	If (Presence is Good) and (Interactivity is Good) and (Control is Good) and (Feedback is Good) and (Creativity is Good) and (Productivity is Good) and (Communication is Good) and (Adaptation is Good) then (level0 is Low)(level1 is Low)(level2 is Low) (level3 is High)(level4 is Low)(level5 is Low)
5	If (Presence is VeryGood) and (Interactivity is VeryGood) and (Control is VeryGood) and (Feedback is VeryGood) and (Creativity is VeryGood) and (Productivity is VeryGood) and (Communication is VeryGood) and (Adaptation is VeryGood) then (level0 is Low)(level1 is Low)(level2 is Low) (level3 is Low)(level4 is High)(level5 is Low)
6	If (Presence is Excellent) and (Interactivity is Excellent) and (Control is Excellent) and (Feedback is Excellent) and (Creativity is Excellent) and (Productivity is Excellent) and (Communication is Excellent) and (Adaptation is Excellent) then (level0 is Low)(level1 is Low)(level2 is Low) (level3 is Low)(level4 is Low)(nivel5 is High)

described in the previous section. In the case of the output variables, the interaction levels discussed above were taken.

In the case of entries, it was partitioned into 6 Gaussian membership functions, since it was desired to have a partition for each existing exit in the FIS, to facilitate the construction of the FIS knowledge system. The parameters chosen for these functions were configured through the MATLAB Fuzzy Logic Toolbox.

This Toolbox has a function of adding an amount "x" of membership functions, and it distributes them evenly, in the range selected for each entry.

In the case of the outputs, it was partitioned into only two linguistic values (Low and High) for each of the outputs, as well as the entries the membership function selected was the Gaussian type. Initially, the parameters were calculated in the same way as the inputs. Only the outputs were subjected to a small process of empirical adjustment.

The parameters for the output inputs can be observed in the Tables 1 and 2.

The proposed FIS is defined by 6 rules of fuzzy inference, the Table 3 shows the base rules; this is the representation of our knowledge base.

To verify the performance of the users according to their inputs and that they have the corresponding level of interaction, the output that is the reunification of the resulting level is evaluated. In this way, users move from one level to another, that is, when their membership function value is close to the nearest integer, the corresponding value is taken, e.g., if the level is 0.2 it belongs to level 0 of interaction, but if the level is 0.9 it belongs to level 1 of interaction, therefore the system updates the knowledge base for the next interaction.

Therefore, the values obtained form the basis of the behavior and knowledge of the users in the context of interaction, on the other hand, these values can change dynamically, and the values of their membership functions can be modified, characterizing from a greater uncertainty to a lower uncertainty considering the user's performance. Once the level of interaction is known, information or services could be sent according to the level of interaction that results.

5   Experiments and Results

From a sample of 500 users, an evaluation of the interaction of each of them was made with the interactive module present in the museum. According to the parameters described above, values were generated that served as input to validate the proposed FIS.

The evaluations obtained showed a great variety of interaction ranging from very low interaction to an extremely high one. These results allowed us to obtain a universe of different user-exhibition interaction scenarios.

The Table 4 is a sample of some of the results obtained from the level of interaction evaluated by the expert. The final interaction level is calculated simply by the average of the values multiplied by the maximum number of interaction levels according to Eq. (1).

$$InteractionLevel = Average(Presence, Interactivity, Control, Feedback,$$
$$Creativity, Productivity, Comunication, Adaptation) * 5 \qquad (1)$$

To verify the percentage that was correctly classified, the results of the classification made by the evaluator were compared with the classification made by the FIS through a confusion matrix. The results of the confusion matrix are shown in the Fig. 2.

Table 4. Results obtained from the interaction level evaluated

Subject	Presence	Interactivity	Control	Feedback	Creativity	Productivity	Communication	Adaptation	Level of interaction
1	0.5	0.4	0.5	0.3	0.3	0.2	0.1	0.2	1
2	0.4	0.4	0.4	0.5	0.3	0.3	0.5	0.4	2
3	0.7	0.8	0.8	0.6	0.5	0.5	0.6	0.6	3
4	1	1	1	1	0.8	0.8	1	0.7	4
5	1	1	1	1	1	1	1	1	5
6	0.1	0.1	0.1	0	0	0	0	0	0
...	...	...	...	...	...	...	...	...	...
500	0.4	0.3	0.6	0.2	0.1	0.3	0.5	1	3

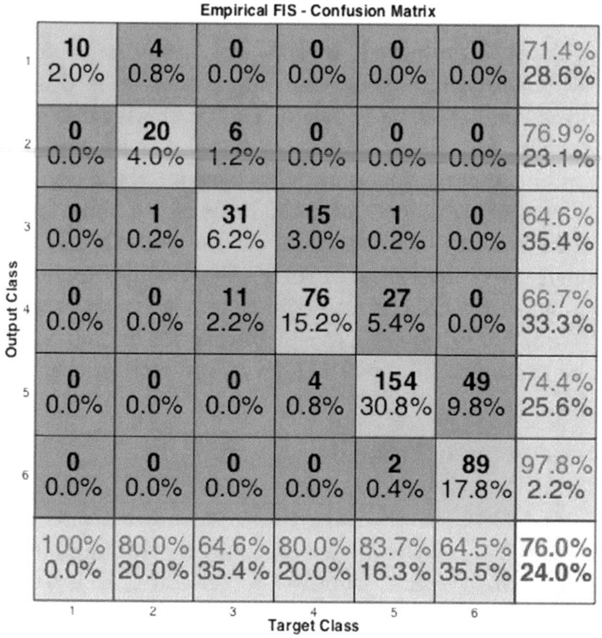

Fig. 2. Confusion matrix.

From the sample of 500 users, it was divided into two equal parts, 250 as a basis to configure the FIS and 250 to perform the tests.

The choice of the elements of the sample was made through the randperm method of MATLAB. This method allows choosing randomly the 250 elements of each sample. The configuration of the FIS was carried out with the help of the base sample, that is, this sample served as the basis for obtaining the configuration through an empirical process.

In the case of entries, it was partitioned into 6 Gaussian membership functions, since it was desired to have a partition for each existing exit in the FIS, to facilitate the construction of the FIS knowledge system. The parameters chosen for these functions were configured through the MATLAB Fuzzy Logic Toolbox.

This Toolbox has a function of adding an amount "x" of membership functions, and it distributes them evenly, in the range selected for each entry.

As mentioned in previous sections, visitors were evaluated, based on this observation, each of them was rated in each of the input variables with a score of 0 to 1, according to these ratings was classified at a level of interaction. The results of this classification made by an expert were used to compare concerning the results obtained from the FIS from a confusion matrix. The percentages obtained in each of the classifications were used to know which outputs needed an adjustment in the parameters to be able to make a classification as similar to that of the expert. The parameters of the outputs obtained are displayed in the Table 2.

6   Conclusion and Future Work

In this work, a Mamdani Type 1 Diffuse Inference System was presented that evaluates and classifies the interaction conditions in a Human-Computer Interface (HCI). This evaluation is carried out according to the levels proposed by Gayesky and Williams [1] which basically consists of a set of qualitative observations based on their experience and research.

The configuration of the system was also described and showed how it could be built using the MATLAB Fuzzy Logic Toolbox. This is a well-known tool that is commonly used in research.

An exhibition inside an interactive museum was shown as an example of application. In this case, the system observes the user evaluating the level of interaction to offer later new possibilities that help to maintain their attention. This intelligent system could help improve the user experience by keeping your attention captive.

To validate the model, a comparison was made between the response of the system and assessments made by humans. The system showed that the output of the system is similar to the evaluation done, obtaining an approximate 76% accuracy between the predicted values and the test values, being that precision is to some extent acceptable.

We will expand the sample of users to at least 1000 to have a more meaningful sample that helps refine the tool obtaining a level of interaction according to the user's interaction.

Although the validation obtained a minimum acceptable approach, as future work, we believe that the accuracy of the system can be improved if it could be generated using a data mining process based on the real data. It could also be improved by changing from a type 1 Mamdani system to type 2, or a Sugeno type to consider the uncertainty.

Acknowledgements. We thank the National Council of Science and Technology (CONACYT), the Autonomous University of Baja California and the Interactive Museum El Trompo for all the support provided during this research.

References

1. Gayesky, D., Williams, D.: Interactive video in higher education. In: Zubber-Skerrit, O. (ed.) Video in Higher Education. Kogan Page, London (1984)
2. Barros, L., Bassanezi, R.: Topicos de Logica Fuzzy e Biomatematica. Universidad de Estadual de Campinas (Unicamp), IMECC. Campinas, Brazil (2006)
3. Dix, A., Finlay, J., Abowd, G., et al.: Human-Computer Interaction, 3rd edn. Pearson and Prentice-Hall, Harlow (2004)
4. Poslad, S.: Computing: Smart Devices, Environments and Interactions. John Wiley and Sons, Chichester (2009)
5. Dodge, Y.: The Concise Encyclopedia of Statistics. Springer-Verlag, New York (2009)
6. Boik, R.J.: Interactions, partial interactions, and interaction contrasts in the analysis of variance. Psychol. Bull. **86**(5), 1084–1089 (1979)
7. Martinez, H.R.: Analysing interactions of fitted models. R J. (2013)
8. Levin, J., Marascuilo, L.: Type IV errors and games. Psychol. Bull. **80**(4), 308–309 (1973)
9. Umesh, U., Peterson, R.: Type IV error in marketing research: the investigation of ANOVA interactions. J. Acad. Mark. Sci. **24**(1), 17–26 (1996)
10. Meyer, D.: Misinterpretation of interaction effects: a reply to Rosnow and Rosenthal. Psychol. Bull. **110**(3), 571–576 (1991)
11. Tabares, M., Pineda, J., Barrera, A.: Un patron de interaccion entre diagramas de actividades uml y sistemas workflow. Revista EIA, Escuela de Ingenieria de Antioquia **10**, 105–120 (2008). ISSN 1794-1237
12. Cengarle, M., Graubmann, P., Wagner, S.: Semantics of UML 2.0 interactions with variabilities. Electron. Notes Theoret. Comput. Sci. **160**, 141–155 (2006)
13. Nguyen, N., Walker, E.: IA First Course in Fuzzy Logic, 3rd edn. Chapman and Hall/CRC, Las Cruces (2006)
14. Jafelice, R., Barros, L., Bassanezi, R.: Teoria dos Conjuntos Fuzzy com Aplicacoes. In: Sociedad Brasileira de Matematica Aplicada e Computacional (SBMAC). Notas em Matematica Aplicada, vol. 17, Sao Carlos, SP, Brazil, 66 p. (2005)
15. Pedhazur, S., Gilbert, K., Silva, R.: Multidimensional scaling of high school students perceptions of academic dishonesty. High School J. **93**(4), 156–165 (2010)
16. Dalei, W., Song, C., Haiyan, L., et al.: A theoretical framework for interaction measure and sensitivity analysis in cross-layer design. ACM Trans. Model. Comput. Simul. **21**(1), 1–26 (2010)
17. Wachs, J., Duertsock, B.: An analytical framework to measure effective human machine interaction. Adv. Hum. Factors Ergon. Healthc., 611–621 (2010)
18. Jiang, X., Adeli, H.: Fuzzy clustering approach for accurate embedding dimension identification in chaotic time series. Integr. Comput. Aided Eng. **10**(3), 287–302 (2003)
19. Endra, J., et al.: Time series estimation on earthquake events using ANFIS with mapping function. Int. J. Artif. Intell. **3**(9), 37–63 (2009)
20. Pozna, C., et al.: Signatures: definitions, operators and applications to fuzzy modeling. Fuzzy Sets Syst. **201**, 86–104 (2012)
21. Nowakova, J., et al.: Medical image retrieval using vector quantization and fuzzy S-tree. J. Med. Syst. **41**(18), 1–16 (2017)
22. Sivanandam, S., Sumathi, S., Deepa, S.: Introduction to Fuzzy Logic Using MATLAB. Springer-Verlag, Heidelberg (2007)

Rational, Emotional, and Attentional Choice Models for Recommender Systems

Ameed Almomani[1], Cristina Monreal[1], Jorge Sieira[2], Juan Graña[2], and Eduardo Sánchez[1(✉)]

[1] CITIUS, University of Santiago de Compostela, Santiago de Compostela, Spain
{ameed.aliahmad,eduardo.sanchez.vila}@usc.es, cmonreal124@gmail.com
[2] NEUROLOGYCA, Vitoria-Gasteiz, Spain
{jsieira,jgrana}@neurologyca.com

Abstract. This work analyzes the decision-making process underlying choice behavior. First, neural and gaze activity were recorded experimentally from different subjects performing a choice task in a Web Interface. Second, choice models were fitted using rational, emotional and attentional features. The model's predictions were evaluated in terms of their accuracy and rankings were made for each user. The results show that (1) the attentional models are the best in terms of its average performance across all users, but (2) each subject shows a different best model.

Keywords: Recommender systems · Decision-making · Dual Process Theory · Choice models

1 Introduction

We believe that the recommendation problem may be solved more accurately by facing it as a choice prediction task [1]. The current paradigm in the Recommender System field mainly focuses on rating prediction and item's recommendations based on the highest predicted value. However, this approach can easily assume unrealistic situations. For instance, recommendations could be based on predicted ratings on high-quality items that, considering budget limitations, could not reach the choice set of the users, and will thus never be chosen. We argue that modeling actual choice behavior is a promising approach to overcome these limitations.

In order to predict user's choice we need to understand the decision-making (DM) process of humans. The state-of-art about DM is dominated by the Dual Process Theory that considers two main types of cognitive processes [2]: the first one is fast, unconscious and effortless; the second one is slow, conscious and energy-consuming. These processes were associated to cognitive systems by psychologist Daniel Kahneman who coined the terms System I and System II [3]. System I is the fast system, which is related to intuition and driven by emotional factors, while System II is the slow one, linked to reasoning and rational outcomes. Traditionally, humans were considered rational agents and human

© Springer Nature Switzerland AG 2019
Á. Rocha et al. (Eds.): WorldCIST'19 2019, AISC 931, pp. 557–566, 2019.
https://doi.org/10.1007/978-3-030-16184-2_53

choices explained by means of modeling the rational processes of System II. The fields of Rational Choice and Utility Theory joined forces to build a conceptual model of serious economic choices [4]. The mathematical description of humans being agents looking to maximize their own utility seemed to fit well with the behavior in the economic domain and also allowed to forecast the evolution of markets. The rise of behavioral economics, however, revealed significant deviations between human behavior and the predictions of rational models [5]. A number of cognitive heuristics were identified suggesting that logical decision-making might not be as important as it was supposed. Herbert Simon described these findings as examples of "bounded rationality". In the last decade, System I models have gained major appeal in order to understand the processes behind a cognitive outcome. In this context, fields like Neuromarketing or Neuroeconomy has grown in order to apply neuroscientific methods to gather relevant data [6]. EEG monitoring, for instance, constituted a useful source to record the neural activity of the cerebral cortex. The data is then used to map neural responses onto emotional variables. Gaze monitoring and eye-traking is another popular tool to track the degree of attention. The underlying rationale is that the scanpath is driven by attentional processes, and therefore a proxy for subject's preferences. By learning the features of the items that the user's is attending, we can then predict user's choice.

This work is aimed at evaluating and comparing the performance of System I and System II models in order to predict human choices. For System I, two types of models, emotional and attentional, were built. For System II, only one type of model, rational, was estimated. In what follows, the hypothesis are presented, the choice task and the experiments performed with a neuromarketing setup are described, the models built with the recorded data are explained, and the results of the models' evaluation are shown and commented.

2   Hypotheses

On the basis of the somatic marker hypothesis, we expect the emotional models to be the best candidates to predict human choices [7]. According to it, two hypotheses were stated: (H1) System I models will show better performance than System II models, and (H2) among System I, the emotional models will overcome their attentional counterparts.

3   Methods

3.1   Experimental Setup

The experiments were carried out at CITIUS (Research Center on Information Technologies at USC) in collaboration with two Business partners: Neurologyca and Movistar. The first one is a Neuromarketing company that provided the experimental setup used to record subject's activity. The second one is the largest Telco in Spain that contributed with the stimuli presented to the subjects.

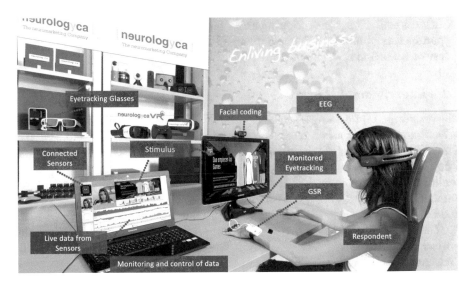

Fig. 1. The experimental setup with the set of recording devices.

The experimental setup was aimed at monitoring and recording neural as well as gaze activity in order to find the drivers of human decision-making. Figure 1 shows the setup that consisted on a set of recording devices plus a manager application that facilitates the control of the experimental procedure. The devices, described in Table 1, were placed in a room in which both temperature and illumination were controlled. With this setting, two main types of activity was recorded:

– Emotional. EEG and FACET provided emotional variables. EEG records the electrical activity generated at the cerebral cortex and the frequency bands are then processed to obtain the emotional variables. FACET works by identifying facial expressions and matching them with emotional or mood states.

Table 1. Devices and recorded variables.

Devices	Models	Variables	Variable type
EEG	Emotiv EPOC Headset	Frustration, Excitement and Engagement	Emotional
Facial Coding (FACET)	Logitech HD Pro webcam C920	Joy, Anger, Fear, Surprise, Contempt, Disgust, Sadness, Neutral, Positive and Negative	Emotional
Eye-Tracking	TOBii X2 @30Hz	TimeSpend and Fixations	Attentional
iMotions (Software)	Version 6.2		

– Attentional. The eye-tracker focused on recording the gaze activity. The parameters obtained, TimeSpend and Fixations, indicate the degree of interest or attention of each subject.

3.2 Experimental Design

The experimental task was a choice experiment in which subjects were presented a Web interface and asked to choose a movie from a set with four alternatives. The movies were labeled with three attributes (see Table 2): Genre, Novelty and Price. As Novelty and Price were fully dependent, eight possible movie profiles could be created. Each profile was presented 10 times by each subject, so a total number of 80 movies were included on each trial. This set was finally structured in 20 choice situations with 4 stimuli on each one.

The task protocol comprised the following steps: (1) Welcome, where the subject entered the room and asked to sign an informed consent; (2) Device Placement, where the recording equipment was placed on each subject and the corresponding calibration was carried out; (3) Directions, where the task was explained, and (4) Choice experiment, in which 20 choice situations with 4 movies on each set were presented trough the Web Interface (Fig. 2). The subject was asked to choose one of the four movies of the set and continue with the next choice problem.

3.3 Sample

Data were collected from 39 subjects (20 females, 19 males, age range = 18–51 years). All subjects had prior experience using Internet and Web applications and had normal or corrected-to-normal vision.

3.4 Models

Choice Models. The recommendation problem can be viewed as the problem of predicting user's choices on any particular context (see [1] for further details). Under this approach, a decision-maker c_n will choose the alternative a_i from a choice set A by estimating the probability P_{ni} of utility U_{ni} of a_i being higher than utility U_{nj} of the other alternatives a_j in A. The process is formally represented as a choice rule (CR):

$$CR(A, \geq) = \{a_i \in A \,\|\, \mathbb{P}_{ni} \geq \mathbb{P}_{nj}, \forall a_j \in A\} \tag{1}$$

Table 2. Characterization of movies: attributes and values.

Feature	Values
Genre	Action, Comedy, Science Fiction, Drama
Novelty	Release, Catalog
Price	4,99 euros (Release), 0 euros (Catalog)

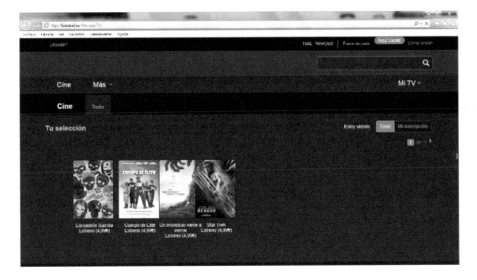

Fig. 2. Web Interface.

The most popular form of the probability is the standard logit model [8], which is obtained under the assumption that the unobserved portion of utility U_{ni} is distributed independently, identically extreme value. In this case, P_{ni} becomes:

$$\mathbb{P}_{ni} = \frac{e^{V_{ni}}}{\sum_j e^{V_{nj}}}. \tag{2}$$

This model estimates the probability on the basis of the observed utilities V_{ni} and all V_{nj}. These values are specified to be linear in the set of the features used to label the alternatives: $V_{nj} = \beta_{nj} \cdot x_j$, where x_j is the vector of observed features, and β_{nj} the vector of preferences of decision-maker c_n on the features of the alternatives a_j. The preferences β_{nj} are model coefficients that need to be estimated by fitting Eq. 2 to a dataset of choices. When a matching exist between the observed features and the preferences of the decision maker, V_{ni} rises and the probability $\mathbb{P}_{ni}$ approaches to one.

Observed Features. The data gathered in the experiment were aggregated in the Neurologyca-Movistar Dataset. Each choice situation was characterized with the choice set (4 movies), the chosen movie, and a set of features describing the process. These features included both the attributes of the movies and the subject's variables recorded during the experiment. Three different set of features were considered (Table 3): rational, emotional and attentional.

Models. According to the available features, three types of choice models were built: rational, emotional and attentional. Table 4 lists the 16 models that were fitted and evaluated with the Neurologyca-MoviStar Dataset.

Table 3. Observed features.

Models	Features
Rational	Action, Comedy, Science Fiction, Drama, Release and Catalog
Emotional	Frustration, Excitement, Engagement, Joy, Anger, Fear, Surprise, Contempt, Disgust, Sadness, Neutral, Positive and Negative
Attentional	TimeSpend and Fixations

Table 4. Choice models used in this work: Rational (R), Attentional (A), and Emotional (E).

Model	Features
R	Action, Comedy, Fiction, Premiere
A1	TimeSpend
A2	FixationTime
E1	Action, Comedy, Fiction, Premiere, Anger, Frustration, Negative, Fear
E2	Action, Comedy, Fiction, Premiere, Excitement, Joy, Engagement
E3	Action, Comedy, Fiction, Premiere, Frustration, Excitement
E4	Action, Comedy, Fiction, Premiere, Frustration, Excitement, Engagement
E5	Action, Comedy, Fiction, Premiere, Anger, Fear
E6	Action, Comedy, Fiction, Premiere, Fear, Contempt, Disgust, Sadness
E7	Action, Comedy, Fiction, Premiere, Joy, Anger, Fear, Surprise
E8	Action, Comedy, Fiction, Premiere, Frustration, Excitement, Engagement, Joy, Anger
E9	Action, Comedy, Fiction, Premiere, Joy, Anger, Fear, Surprise, Contempt, Disgust, Sadness
E10	Action, Comedy, Fiction, Premiere, Sadness, Neutral, Positive, Negative
E11	Action, Comedy, Fiction, Premiere, Frustration, Excitement, Engagement, Neutral, Positive, Negative
E12	Action, Comedy, Fiction, Premiere, Anger, Fear, Surprise, Contempt, Disgust, Sadness, Neutral, Joy, Negative
E13	Action, Comedy, Fiction, Premiere, Frustration, Excitement, Engagement, Joy, Anger, Fear, Surprise, Contempt, Disgust, Sadness, Neutral, Negative

To preserve individual preferences, we decide to build choice models for every single user. As a minimum number of observations is required to learn a model, we run the experiment with the threshold for the least number of observations is 12 (the maximum number is 20). They determined the number of total users available on the experiment is 18 users. Two problems were considered, computing the best model for each user among rational, emotional and attentional models, and the best average model for all users and ranking this model for every user.

3.5   Evaluation

We applied the Accuracy metrics to measure the performance of models. Accuracy is just the ratio between the correct and all prediction results:

$$Accuracy = \frac{T_{CorrectPrediction}}{T_{AllPrediction}} \tag{3}$$

10 trials of each training-evaluation cycle were carried out for each model. Thereafter, the average of the 10 Accuracy values was estimated.

3.6   Software

The main tool used was R, free software environment for statistical computing. Specifically, we used the mlogit package for estimation the multinomial logit model (see [9] for further details).

4   Results

4.1   Best Model for Each Subject

To explore the performance of the models we first provide some results on single subjects. Based on the Accuracy metrics, Table 5 shows the ranking of the Top-5 models for three randomly chosen subjects. The point here is to observe that each user had a different ranking and therefore shows a different best model. The decision-making process of subject $J02$ seems to follow an emotional-driven process, while subjects $J06$ and $V09$ are better explained with attentional and rational features, respectively. This pattern, the best model is different for each subject, is observed for all subjects. In order to summarize the results, a frequency chart representing the number of times each model type ranked first is shown in Fig. 3.

4.2   Best Model for All Subjects

To find the best model for all user, first, we computed the accuracy of each model for each user, next, computing the average of all results for all users. (Table 6) shows the average of the accuracy on the top five models for all users. It can be discovered the first and second are attentional models and the third one is rational. Finally, we represented the frequency of the ranking obtained by the best model across all users (Fig. 4).

5   Discussion

According to the results shown in Fig. 3 we can conclude that hypothesis H1 is confirmed. Both emotional and attentional models yield better prediction performance than the rational model. Surprisingly, on the basis of Fig. 3 and the

Table 5. Performance of choice-based models for each user. Accuracy results for top five models.

Users	Number of choices	Top five models	Accuracy
J02	12	**E8**	0.787879
		E11	0.787879
		E4	0.757576
		E2	0.666667
		E6	0.636364
J06	14	**A2**	0.681818
		A1	0.621212
		E4	0.560606
		E9	0.560606
		R	0.545455
V09	20	**R**	0.522727
		E1	0.477273
		E0	0.454545
		E5	0.431818
		E6	0.431818

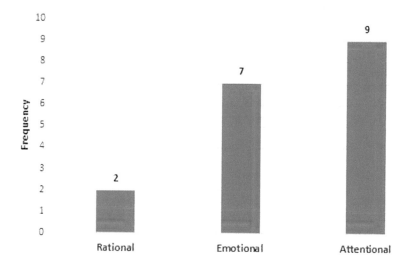

Fig. 3. Frequency of the Rational, Emotional, and Attentional models as best model.

data of Table 6, hypothesis H2 was found to be incorrect. The attentional models in general and the Fixations model in particular, they both made a better job than emotional models. These findings suggest that the factors underlying the attention process may play a key role to understand the decision-making

Table 6. Performance of choice-based models for average model for all users. Accuracy results for top five models.

Models	Average of accuracy
Fixation	0.528282828
TimeSpend	0.522306397
Rational	0.520911496
Emotional6	0.464850890
Emotional4	0.448328523

Fig. 4. Frequency of the rank of the best average model (Fixation model).

process. Further research is required to bring light into this issue and pave the way (1) to build more reliable choice models, and (2) to create a paradigm to develop novel recommender systems.

Acknowledgments. We want to acknowledge the collaboration of Movistar, the leading Spanish Telco, which provided the list of real movies that were used as stimuli during the choice experiments.

References

1. Saavedra, P., Barreiro, P., Duran, R., Crujeiras, R., Loureiro, M., Vila, E.S.: Choice-based recommender systems. In: RecTour@ RecSys, pp. 38–46 (2016)
2. Vaisey, S.: Motivation and justification: a dual-process model of culture in action. Am. J. Sociol. **114**(6), 1675–1715 (2009)
3. Kahneman, D.: Think Fast, Think Slow. Farrar, Straus and Giroux, New York (2011)

4. Scott, J.: Rational choice theory, p. 129. Understanding Contemporary Society, Theories of the Present (2000)
5. Tversky, A., Kahneman, D.: Judgment under uncertainty: heuristics and biases. Science **185**(4157), 1124–1131 (1974)
6. Pradeep, A.K., Patel, H.: The Buying Brain: Secrets for Selling to the Subconscious Mind. Wiley, Hoboken (2010)
7. Damasio, A.R.: The somatic marker hypothesis and the possible functions of the prefrontal cortex. Phil. Trans. R. Soc. Lond. B **351**(1346), 1413–1420 (1996)
8. McFadden, D.: Conditional logit analysis of qualitative choice behavior (1973)
9. Croissant, Y.: Estimation of multinomial logit models in R: the mlogit packages. R package version 02–2 (2012)

Providing Alternative Measures for Addressing Adverse Drug-Drug Interactions

António Silva[1(✉)], Tiago Oliveira[2(✉)], Ken Satoh[2(✉)], and Paulo Novais[1(✉)]

[1] Algoritmi Centre/Department of Informatics, University of Minho, Braga, Portugal
asilva@algoritmi.uminho.pt, pjon@di.uminho.pt
[2] National Institute of Informatics, Tokyo, Japan
{toliveira,ksatoh}@nii.ac.jp

Abstract. Clinical Practice Guidelines (CPGs) are documents used in daily clinical practice that provide advice on how to best diagnose and treat diseases in the form of a list of clinical recommendations. When simultaneously applying multiple CPGs to patients, this can lead to complex multiple drug regimens (polypharmacy) with the potential for harmful combinations of drugs. The need to address these adverse drug events calls forth for systems capable of not only automatically represent the common potential conflicts or interactions that can happen when merging CPGs but also systems capable of providing conflict-free alternatives. This paper presents a solution that represents CPGs as Computer-Interpretable Guidelines (CIGs) and allows the automatic identification of drug conflicts and the provision of alternative measures to resolve these conflicts.

Keywords: Computer-Intepretable Guidelines ·
Clinical decision support · Ontologies · Drug-drug interactions ·
Conflict resolution

1 Introduction

Drug-drug interactions occur when an effect of one drug alters the effect of another co-administered drug [6]. Such interactions are common in *multimorbid* patients since they suffer multiple health conditions and need the application of different disease-specific treatment plans. To help ease the burden of health care professionals, Clinical Practice Guidelines (CPGs) were developed in order to provide patient-specific advice. These documents accumulate and reflect knowledge on how to best diagnose and treat diseases in the form of a list of clinical recommendations. When treating *multimorbid* patients, health care professionals need to retrieve clinical recommendations from multiple chronic disease CPGs. The result is concurrent execution of treatment recommendations from different CPGs, which may cause conflicts. From the combination of these recommendations, several problems can happen, such includes adverse drug events, and increased treatment complexity and cost of treatment [2].

© Springer Nature Switzerland AG 2019
Á. Rocha et al. (Eds.): WorldCIST'19 2019, AISC 931, pp. 567–576, 2019.
https://doi.org/10.1007/978-3-030-16184-2_54

Several projects were developed not only to formally represent CPGs as Computer Interpretable-Guidelines (CIGs). Through the formalisation of CPGs as CIGs it is possible to develop decision support systems that offer a better possibility of affecting clinical behaviour in relation to narrative documents of the corresponding text versions. The representation of CPGs in digital format can have distinct benefits over paper-based CPGs in that they increase flexibility, minimise errors, and generalise the use of CPGs across institutions. However, few works address potential conflicts or interactions that can happen when merging CIGs.

Although some approaches offer ways to represent the conflicts and interactions between concurrent CIGs, they lack in: provision of alternative measures to resolve conflicts in treatment plans, dynamic search for solutions to conflicts outside the existing knowledge base, and provision of methods to rank and select treatment plans. The first contribution of this work is a characterisation of existing approaches to handle the combination of CIGs, especially for *multimorbid* patients. Then, it presents a system that automatically identifies recommendation interactions, conflicts, and provides alternative measures (mainly in the form of alternative drug recommendations) that resolve the identified conflicts.

The paper is organised as follows. Section 2 describes related work regarding systems for combining CIGs. Section 3, presents an architecture for combining CIGs as well as the contributions for the deployment of CIGs in Clinical Decision Support Systems (CDSSs). Section 4 describes the functionalities supporting care with a case example of how the system processes drug-drug interactions and provides alternative measures. Finally, Sect. 5 presents conclusions about the work developed so far and future work considerations.

2   Related Work

Several formalisms are proposed in the literature that are aimed to represent the conflicts and interactions among different CPGs. They provide various methods to model the conflicts of CPGs into their knowledge base. In this section, we describe different systems that automatically identify the possible interactions between concurrent CPGs for *multimorbid* patients.

Wilk et al. [13] represent CIGs as an activity graph. They use constraint logic programming and combines it with constraint satisfaction problems. By using constraint logic programming, they identify and mitigate possible adverse interactions when applying multiple guidelines on the same patient, namely identifying conflicts associated with potentially contradictory and adverse activities. They provide notification features that inform the healthcare professionals about the possible conflicts during the definition of the treatment plans. This approach provides automatic identification of conflicts, however, it depends on the availability of the information in the knowledge base about the conflicts between both CIGs in the form of constraints and of pre-existing operators to mitigate these conflicts. This requires substantial manual effort for combining CIGs. Thus, in order to provide automatic identification and resolution of conflicts, solutions need to be defined in a *medical background knowledge* as protocol-dependent rules/constraints.

Lopez et al. [8] used a rule-based methodology in order to identify and reconcile drug conflicts between recommendations of two concurrently executed CIGs. In order to provide a treatment plan without interactions, they utilise a standard terminology called ATC (Anatomical Therapeutic Chemical Classification System for drugs). The outputted treatment plan comprises a set of ATC-codes of medicines, without interactions, which should be prescribed. They manually build knowledge units for the pairwise combination of diseases in their knowledge base. These knowledge units rely on the existence of drug-drug interactions, the presence of a drug which is adverse to a specific disease (drug-disease interaction) and the absence of a necessary medicine for a combination of diseases. Although this approach can only combine CPGs pairwise, it is possible to achieve a final treatment plan for any number of CIGs by combining a pair of CIGs into a general CIG and then combining the latter with a new CIG. This approach requires significant manual effort as each combination has to be hardcoded.

OntoMorph [5] represent guidelines as a collection of ontologies. They use information such as the general domain, the mappings between CPGs and decision rules for simultaneous execution of CPGs that are provided by domain experts. Based on these ontologies, they developed a system capable of merging two concurrent CIGs into a co-morbid personalised guideline. By representing the CIGs as ontologies, it allows retrieving the clinical tasks from the CPG and converts them to computer-interpretable rules in Ontology Web Language (OWL). Using ontologies is one of the possible solutions to CPG representation. It allows the representation of declarative knowledge (medical statements and propositions) and procedural knowledge (workflow structures and actions) as rules. Ontomorph also has a merging representation ontology, which allows capturing merging criteria to achieve the combination of CIGs. By using Semantic Web Rule Language (SWRL) rules, they can identify potential conflicts during the merging process. Since all conditions need to be defined in their model during the merging process, this increases the effort to maintain the system up-to-date and reduces the possibility of sharing knowledge. In their work, some of the identified limitations were not yet entirely addressed, such as potential contradictions between rules, the scalability of the merging model to combine several CIGs, and how the ontology/rules are maintained up-to-date.

The Transition-based Medical Recommendations for Interactions (TMR4I) model is a model that automatically infers the interaction between recommendations [14] by using meta-rules for the identification and reconciliation of three categories of drug conflicts using SPARQL queries (SPARQL is a W3C-standard for semantic queries). Using meta-rules allows defining how a conflict is identified and how similar drugs without interactions and conflicts can be selected as alternatives. The categories of conflicts within CPGs are *repetition interactions, contradiction interactions* and *alternative interactions*. The model was extend in [14] with additional interaction types and several measures such as deontic strength, causation belief, and belief strength. This work provides only a representation of conflicts but does not afford reasoning or any form of decision support.

The limitations of above-mentioned approaches include: restrictions in the number of CIGs that can be combined, necessity of all solutions to be available in a knowledge base, and decidability of reasoning mechanisms. In the approaches that require hard-coded solutions, if a conflict that is not accounted for in the knowledge base appears, the reasoning component will not be able to provide a response. Also worth mentioning is that current approaches do not provide support for ranking sets of guideline recommendations that are consistent.

3   An Architecture for CIG Management with CIG Interaction Detection and Resolution

The present work not only aims to provide recommendations to support medical decision-making but also to represent automatically the conflicts and interactions that can happen when merging CIGs. In this work we focus on drug-drug interactions and propose a system capable of automatically identifying recommendation (drug) interactions using existing terminology services, namely the RxNorm API [7]. Once interactions are identified, we provide alternative measures, i.e., alternative drugs to the ones recommended that would not cause any conflict, through a mitigation function. This function calculates the solutions for the identified conflicts using different mitigation principles such as similarity between drugs or user preferences. The architecture is shown in Fig. 1. This architecture is a three-level solution that encompasses the following stages for the CIG deployment: representation of CPGs in CIGs, identification of recommendation interactions and provision of recommendation alternatives in case that some recommendations, when applied together, are adverse. The following sections explains the architecture that integrates these three levels.

3.1   Representation of CPGs in CIGs

The work described herein uses the CompGuide ontology to represent CPGs in the form of a task network. The CompGuide ontology [10] contains different types of clinical tasks such as *Question, Action, Decision, End, Plan* and provides different types of clinical constraints expressed in the form of conditions on the patient's state, such as *TriggerConditions, PreConditions* and *Outcomes*. Through the utilisation of object properties to connect instances of the subclasses of the clinical tasks, it is possible to define the relative order between tasks. In the Compguide ontology, it is possible to define sequential tasks, parallel tasks, and alternative tasks. Moreover, it provides a model of temporal representation [11] that aims to represent the temporal constraints placed on clinical tasks. This model represents temporal constructors on the execution of tasks such as *Durations, Repetitions, Periodicities, Waiting Times* and *Repetition Conditions* and temporal constraints about the state of a patient. To acquire and represent CPGs we use the CompGuide plugin which provides information

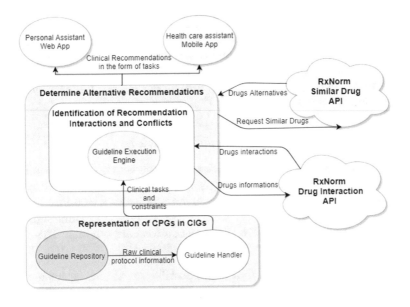

Fig. 1. Architecture of CompGuide system

step-by-step on how to fill the data for the guideline entries [4]. This plugin performs the role of managing the creation and editing of CIGs.

The output of guideline encoding is a will CIG that will be saved in the *Guideline Repository*. This component is responsible for keeping different CIGs defined according to the CompGuide ontology. The *Guideline Handler* is responsible for managing the access to CIG recommendations in the *Guideline Repository*, providing the clinical tasks and constraints placed on the tasks to the *Guideline Execution Engine*.

3.2 Identification of Recommendation Interactions

The *Guideline Handler* provides clinical task details to the *Guideline Execution Engine* in order to produce recommendations. This component provides information about temporal constraints on the execution of the clinical tasks. Using this information, the *Guideline Execution Engine* produces task enactment times, and by using RxNorm Interaction API, it determines if drug-drug interactions exist. As before-mentioned, the clinical tasks are defined, in the CompGuide ontology, by a set of subclasses. One of these subclasses is *Action*, in which it is possible to define a collection of drugs that require to be prescribed. The *Action* class has several subtypes of actions, namely exams, procedures, medication recommendations and simple recommendations [11]. The Recommendation medication provides a set of drugs that can be prescribed and is used by *Guideline Execution Engine* to determine drug-drug interactions. By using the RxNorm interaction API, it is possible to determine drug-drug interactions, without the need to manually define drug interactions in the knowledge base. Its interaction API [7] uses two data sources, ONCHigh and DrugBank and provide information such as source name, severity and description of the interaction. Thus, the *Guideline Execution Engine* processes all the clinical tasks that are being executed, retrieves all drugs and for each pair of drugs calls the RxNorm Interaction API to obtain the severity and description of the interaction.

3.3   Generating Alternative Recommendations

After processing all drug-drug interactions between concurrently executed clinical recommendations, the alternative measures are evaluated by the system. If an adverse drug event exists, the systems automatically try to find alternative recommendations to resolve the conflict. Through a mitigation function, the system determines which alternative recommendations are advised. This function encompasses a set of steps that include the following:

– **Step 1:** The system tries to find if it is possible to get alternative recommendations, i.e., alternative drugs, in the guidelines. For each specific guideline recommending a drug, the system calculates if an alternative recommendation exists within the guideline. If it is not possible to retrieve the alternative recommendation, the system moves to step 2;
– **Step 2:** Using the RxNorm API, the system tries to find conflict-free alternative drugs. The system provides these alternative drugs by determining the set of alternatives that have the high similarity score concerning the given drugs. The RxNorm API provides the similarity score. If there are no alternative drugs the system moves to step 3;
– **Step 3:** The system evaluates all possible solutions using Multiple-criteria Decision Analysis (MCDA). Since drug-drug interactions are yielding multiple solutions with conflicting objectives, it is useful to score the solutions. The patient and physician score all possible solutions, so this is a shared patient-clinician evaluation supported by the system.

In step 1, the system tries to resolve the conflicts by analysing within the guideline the different task alternatives. In this particular case, if alternative tasks (recommending alternative drugs) exist in the guideline, the system tries to retrieve them. Then, it gets all the recommended drugs of the alternative tasks and tries to find if drug-drug interactions exist in them, by calling RxNorm Interaction API for each pairwise drugs of the task. In step 2, a ranking of alternative drugs is produced based on the similarity score provided by the RxNorm API. The similarity score between drugs is a score that determines the similarity between drugs. Thus, the system calls the RxNorm API to get alternative drugs for the given conflicted drugs and calculates the highest similarity score for the alternative drugs and for each alternative with the higher score it tries to encounter conflict-free drug. If there is a conflict, the system finds the next alternative with the higher score, if there is no conflict, it stores the alternative in the knowledge base. Table 1 presents the MCDA approach for step 3.

This approach uses a value measurement model where for each criterion the patient assigns a score. The objective of this model is constructing and comparing numerical scores (overall value) to identify the degree to which one decision alternative is preferred over another. Each alternative to be scored is a combination of drugs, for instance, if there is an adverse effect between drugs A, B and C, all possible combinations are: The system automatically defines the criteria by which decision-makers should orient. Thus, when an adverse drug event occurs, and the system moves to step 3, the criteria established are: severity of disease for which drugs are advised, adverse drug-drug interactions and expected outcomes for the drug application. The criteria are defined on the basis of some types of health care decisions that are implemented in projects such as EMA's Benefit-Risk Methodology Project [12] and shared patient-clinician decision [3]. The total score for each alternative is obtained by multiplying a numerical score for each option on a given criterion by the relative weight for the criterion and later summing these weighted scores. Thus, the total score is provided by the following expression:

$$f(n) = \sum_{n=1}^{n} S^n * WeightC^n, \tag{1}$$

where n is the number of solutions to be scored, S^n a score of a specific solution and $WeightC^n$ relative weight for a specific criterion.

The relative weight for the criterion is a value defined by the healthcare professional. This value is requested before starting the evaluation of the solutions. After getting the user scores, the system determines the total score of each solution by using the aforementioned equation. Thus, the total scores of each solution are made available through the Personal Assistant Web App and Healthcare assistant Mobile App, presenting the selected solution. Moreover, after processing the constraints of clinical tasks, determining the interactions between drugs and their alternatives, the clinical recommendations are made available in before-mentioned assistants. In this assistants, it is possible to visualise the clinical recommendations that currently are being applied to the patient, in a calendar and timeline view. Thus, each clinical recommendation can have a set of drugs or alternative solutions that were previously evaluated and scored by decision makers.

Table 1. Assessment of all possible solutions. The symbol C indicates a certain criterion to be evaluated for a given solution α. S means the score of the solution.

Solutions (α)	Criterion (C)			Total Score
	C^1	...	C^n	
α^1	$S^1 C^1$	...	$S^1 C^n$	$f(1) = \sum_{n=1}^{1} S^1 * WeightC^n$, (2)
...	...	...	...	...
α^n	$S^n C^n$	...	$S^n C^n$	$f(n) = \sum_{n=1}^{n} S^n * WeightC^n$, (3)

4   Case Example

This section describes how CompGuide processes the interactions between drugs given a case test example. For this purpose, we used two CIGs based on the NCCN Clinical Practice Guideline for Prostate Cancer [9] and the IDF Clinical Practice Recommendations for managing Type 2 Diabetes [1]. These guidelines were a comprehensive case study since it was possible to test several aspects of the deployment of CIGs. However, in this section, we only address the conflicts between recommendations from many guidelines and provision of alternative recommendations using step 2 described in Sect. 3.2.

For demonstration purposes, we will consider two recommendations from the mentioned guidelines. The first one, named recommendation A belongs to the guideline for managing Type 2 Diabetes: "Apply insulin 0.2 units/kg and titrate once weekly at one unit each time during six months to achieve a target fasting blood glucose between 3.9 and 7.2 mmol/L (70 and 130 mg/dL)". The second recommendation, named recommendation B belongs to the guideline for prostate cancer: "Apply goserelin, leuprolide, histrelin 180 mg/m2 or Triptorelin 100 mg/m2 as part of Androgen Deprivation Therapy".

Recommendation A has the action apply insulin, a periodicity value of 1 with a temporal unit of week, a duration value of six, the respective temporal unit of month and medication recommendation insulin. In this case, starting on the 18th of July of 2018 the system will create one event for each week with a duration of one day, during 6 months. The expected conclusion of this task will be on the 18th of January of 2019. As for recommendation B, the action to apply goserelin, leuprolide, histrelin or triptorelin can be identified, with a duration value of 1 and temporal unit of day, starting and finishing on the 18th of July of 2018. The recommendation medications are goserelin, leuprolide, histrelin and Triptorelin. The application tries to provide alternative drugs to address the identified conflicts, by calling RxNorm API and will provide alternative medicines according to step 2, as described in Sect. 3.2. Also, in this step the system calculates a ranking of conflict-free alternative drugs, using the similarity score provided by RxNorm API. The ranking of alternative drugs is calculated by comparing similarity scores and sorting in descending order the medicines according to this similarity values. For the alternative drug (drug provided by RxNorm) with a higher score, the system determines if it is conflict-free over the prescribed drugs. If there is a conflict, the system finds the next alternative with the higher similarity score, if there is no conflict, it stores the alternative in the database and displays the alternative drug as the selected solution.

In the work described herein, we provide a system that automatically identifies conflicts and interactions between drugs for many guidelines. Comparing with the works of Jafarpour et al. [5], Wilk et al. [13] and López-Valverdú et al. [8], where conflicts are defined as constraints in the knowledge base having to be manually specified, CompGuide uses existing terminology services that aggregate different drug sources such as ONCHigh and DrugBank. Thus, through the reuse and integration of existing terminology services such as RxNorm, it is possible to identify conflicts and interactions automatically, without the need to manually define them in the knowledge base. Therefore, using existing terminology services and resorting to external knowledge sources is one of the possible solutions for the limitation mentioned above. Another solution concerns the use of meta-rules such as those used by the TM4I model. Meta-rules can be reused since they can be applied to many CIGs, and conflicts do not need to be manually identified for each guideline, because they can be automatically derived from the guideline representation. However, the bottleneck will be in converting guidelines to computer-interpretable rules. Besides, these systems do not consider aspects such as decision-making. In most cases, there are several alternatives that can lead to conflicting objectives by the decision makers. In other cases, it is necessary to decide which recommendation we want to choose, or which recommendation, in the case at hand, is less adverse. For this specific case, we provide an MCDA approach that allows to evaluate all possible solutions based on conflicting criteria.

5   Conclusions and Future Work

The application of multiple clinical protocols individually can result in complex multiple drug regimens (polypharmacy) with the potential for harmful combinations of drugs. Some of the studied approaches are unable to detect the conflicts for combinations of protocols automatically. Other approaches cannot propose alternative measures that would resolve the conflicts. Other CIG models require all the possible conflicts and their solutions to be available in a knowledge base. Moreover, they cannot lead with cases where decision makers have conflicting solutions or cannot decide on the best treatment alternatives.

As a means to solve these issues, we provide a multiple criteria decision-making approach for not only assessing the benefit-risk of applying the recommendations but also getting patient preferences on best treatment alternatives. This allows to evaluate all possible solutions and to specify different criteria to solve conflicts with medical recommendations, beyond the simple comparison of drug interactions. We also offer a system that allows to combine the knowledge of several guidelines and to identify drug interactions and conflicts among many recommendations automatically. Comparing with some of the studied systems, the CompGuide has an additional advantage since it presents a method to automatically identify drug-drug conflicts among many recommendations, without a necessity of manually define them in the knowledge base. Also, when decision makers have conflicting solutions and cannot decide on the best treatment alternatives, the CompGuide presents an approach that allows to evaluate all possible solutions and to specify different criteria to solve conflicts with medical recommendations. As future work, we intend to make a proper assessment of the fitness of the system for CIG deployment, by performing a study involving physicians interacting with the system in the clinical environment. In this ways, it is possible to analyse if the system meets the requirements of health professionals and if it is user-friendly.

Acknowledgements. This work has been supported by COMPETE: POCI-01-0145-FEDER-0070 43 and FCT – Fundação para a Ciência e Tecnologia within the Project Scope UID/CEC/ 00319/2013. The work of Tiago Oliveira was supported by JSPS KAKENHI Grant Number JP18K18115.

References

1. Aschner, P.: New IDF clinical practice recommendations for managing type 2 diabetes in primary care (2017)
2. Boyd, C.M., Darer, J., Boult, C., Fried, L.P., Boult, L., Wu, A.W.: Clinical practice guidelines and quality of care for older patients with multiple comorbid diseases: implications for pay for performance. JAMA **294**(6), 716–724 (2005)
3. Dolan, J.G., Boohaker, E., Allison, J., Imperiale, T.F.: Patients' preferences and priorities regarding colorectal cancer screening. Med. Decis. Making **33**(1), 59–70 (2013)
4. Gonçalves, F., Oliveira, T., Neves, J., Novais, P.: Compguide: acquisition and editing of computer-interpretable guidelines. In: World Conference on Information Systems and Technologies, pp. 257–266. Springer (2017)
5. Jafarpour, B., Abidi, S.S.R.: Merging disease-specific clinical guidelines to handle comorbidities in a clinical decision support setting. In: Conference on Artificial Intelligence in Medicine in Europe, pp. 28–32. Springer (2013)
6. Kennedy, C., Brewer, L., Williams, D.: Drug interactions. Medicine (United Kingdom) **44**(7), 422–426 (2016), http://dx.doi.org/10.1016/j.mpmed.2016.04.015
7. Liu, S., Ma, W., Moore, R., Ganesan, V., Nelson, S.: Rxnorm: prescription for electronic drug information exchange. IT Prof. **7**(5), 17–23 (2005)
8. López-Vallverdú, J.A., Riaño, D., Collado, A.: Rule-based combination of comorbid treatments for chronic diseases applied to hypertension, diabetes mellitus and heart failure. In: Process Support and Knowledge Representation in Health Care, pp. 30–41. Springer (2013)
9. Mohler, J.L., Lee, R.T., Antonarakis, E.S., Armstrong, A.J., D'Amico, A.V., Davis, B.J., Dorf, T., Eastham, J.A., Ellis, R., Enke, C.A., Farrington, T.A.: National Comprehensive Cancer Network - Prostate Cancer. Technical report, National Comprehensive Cancer Network (2018). http://linkinghub.elsevier.com/retrieve/pii/B9780323358682000803
10. Oliveira, T., Novais, P., Neves, J.: Representation of clinical practice guideline components in owl. In: Trends in Practical Applications of Agents and Multiagent Systems, pp. 77–85. Springer (2013)
11. Oliveira, T., Silva, A., Neves, J., Novais, P.: Decision support provided by a temporally oriented health care assistant. J. Med. Syst. **41**(1), 13 (2017)
12. Phillips, L.D., Fasolo, B., Zafiropoulos, N., Beyer, A.: Is quantitative benefit-risk modelling of drugs desirable or possible? Drug Discov. Today Technol. **8**(1), e3–e10 (2011)
13. Wilk, S., Michalowski, M., Michalowski, W., Rosu, D., Carrier, M., Kezadri-Hamiaz, M.: Comprehensive mitigation framework for concurrent application of multiple clinical practice guidelines. J. Biomed. Inform. **66**, 52–71 (2017)
14. Zamborlini, V., Da Silveira, M., Pruski, C., ten Teije, A., Geleijn, E., van der Leeden, M., Stuiver, M., van Harmelen, F.: Analyzing interactions on combining multiple clinical guidelines. Artif. Intell. Med. **81**, 78–93 (2017)

Convolutional Neural Network-Based Regression for Quantification of Brain Characteristics Using MRI

João Fernandes[1](✉) , Victor Alves[2] , Nadieh Khalili[3],
Manon J. N. L. Benders[4] , Ivana Išgum[3] , Josien Pluim[5],
and Pim Moeskops[5]

[1] Department of Informatics, School of Engineering,
University of Minho, Braga, Portugal
joaovieirafernandes@hotmail.com
[2] Algoritmi Centre, University of Minho, Braga, Portugal
valves@di.uminho.pt
[3] Images Sciences Institute, University Medical Center Utrecht,
Utrecht, The Netherlands
{n.khalili,i.isgum}@umcutrecht.nl
[4] Department of Neonatology, University Medical Center Utrecht,
Utrecht, The Netherlands
m.benders@umcutrecht.nl
[5] Eindhoven University of Technology, Eindhoven, The Netherlands
j.pluim@tue.nl, moeskops.pim@gmail.com

Abstract. Preterm birth is connected to impairments and altered brain growth. Compared to their term born peers, preterm infants have a higher risk of behavioral and cognitive problems since most part of their brain development is in extra-uterine conditions. This paper presents different deep learning approaches with the objective of quantifying the volumes of 8 brain tissues and 5 other image-based descriptors that quantify the state of brain development. Two datasets were used: one with 86 MR brain images of patients around 30 weeks PMA and the other with 153 patients around 40 weeks PMA. Two approaches were evaluated: (1) using the full image as 3D input and (2) using multiple image slices as 3D input, both achieving promising results. A second study, using a dataset of MR brain images of rats, was also performed to assess the performance of this method with other brains. A 2D approach was used to estimate the volumes of 3 rat brain tissues.

Keywords: Preterm infants · Rat brain · Brain quantification · Deep learning · Convolutional neural networks · Regression · Magnetic resonance imaging

1 Introduction

Preterm babies now have a higher survival rate due to the recent progresses in neonatal care and obstetrics. However, preterm births are associated with developmental impairments [1]. In the third trimester of pregnancy, the cerebral cortex begins to

© Springer Nature Switzerland AG 2019
Á. Rocha et al. (Eds.): WorldCIST'19 2019, AISC 931, pp. 577–586, 2019.
https://doi.org/10.1007/978-3-030-16184-2_55

develop rapidly, from a simple and smooth structure to a complex, folded and adult-like structure [1, 2]. In preterm infants this important part of brain development takes place outside the uterus, which is an important reason why they are at greater risk for problems related to cognition, socialization and behavior later in life. Under these conditions, various abnormalities can be observed, and MRI brain scans play a crucial role in this identification.

Rats are very similar to humans in terms of biological, genetic and behavioral traits [3, 4] and can therefore aid in brain imaging research.

Deep learning (DL) techniques have shown promising results in various tasks in medical image analysis [5].

Moeskops et al. [6] developed an approach for the automatic segmentation of 6–8 different tissue classes using a CNN. Based on these segmentations several descriptors that quantify the state of brain development were computed [2]. These descriptors were shown to have potential value in predicting cognitive and motor outcome [7].

In this paper we investigate the possibility of directly estimating these quantitative descriptors from the image, without explicit segmentation. Such a system could show that a CNN-based system can extract relevant information from brain MR images instead of relying on segmentations, which could potentially be useful in improving prediction systems.

Several studies have already investigated directly estimating descriptors from images. Based on a dataset with systole and diastole volumes as labels of the 2015 competition provided by Kaggle [8], Liao et al. [9] developed a fully CNN-based method segment both the cardiac MRI scan and to quantify the left ventricular volume with the aim of diagnosing heart disease. Dubost et al. [10] also performed a quantification study, but this time with enlarged perivascular spaces in the basal ganglia in the brain. De Vos et al. [11] showed that calcium scoring can be performed directly from the image, i.e. without explicit segmentation.

In this work an automatic method for the quantification of neonatal and rat brain descriptors based on MRI scans is presented using deep learning, in particular convolutional neural networks (CNNs). For the preterm infants, the descriptors are computed from images at 30 and 40 weeks of postmenstrual age (PMA). To assess how generalizable the developed techniques are, they are furthermore applied to a rat brain MRI dataset.

2 Quantification of Brain Characteristics Using Neonatal MRI

2.1 Materials

A cohort of preterm infant MRI scans, born before 28 weeks gestational age (GA) was acquired on a Phillips (Best, The Netherlands) Achieva 3T scanner. Both the consent of the parents and permission to use the clinically collected data were obtained from the University Medical Center Utrecht (UMCU) institutional review board. The infants that were considered clinically stable by the neonatologist were scanned using a MR compatible incubator (LMT Medical Systems, Lübeck, Germany), creating a dataset of 86 coronal T_2-weighted studies at an average PMA of 30.8 ± 0.8 weeks. At 40 weeks PMA, the infants were scanned using a Phillips SENSE head coil. The dataset consisted on 153 coronal T_2-weighted studies at an average PMA of 41.2 ± 0.7 weeks. Images with motion artifacts and brain abnormalities were excluded if this resulted in inaccurate automatic analysis.

The images were automatically segmented into 8 different tissues, including gray matter (GM) and unmyelinated white matter (uWM), using the system constructed in [6]. The brain masks were generated automatically based on the T2-weighted images using the Brain Extraction Tool (BET) from the FMRIB Software Library (FSL), using the same setting for all images. This work can be viewed in [12]. The other 5 descriptors, i.e. inner (IS) and outer cortical surface area (OS), gyrification index (GI), global mean curvature (MC) and median cortical thickness (MT) were quantified in [7] based on the neonatal segmentations.

2.2   Methods

Two different approaches are used: 3D using multiple image slices and 3D for full brains.

3D Approach Using Multiple Slices. This approach uses multiple slices of 3D images of the brain to estimate the same 8 volumes previously described.

Pre-processing techniques. Described below:

- Image cropping, in which images are cropped considering the corresponding masks.
- Image downscaling by 2, to complete the reduction of the image in order to tackle memory issues.
- 3D transformation of images: after cropping and downscaling, images are stacked according to a certain size of choice, creating chunks with multiple image slices of brain. Since now 3D chunks with multiple image slices are considered instead of 2D slices, the labels must be summed according to the number of slices constituting the chunks.
- Normalization of the labels, between 0 and 1, to prevent numerical instability in the labels.
- Data augmentation of the images: rotations ([−10;10] degrees), scaling (scaling factor ∈ [0,8;1]), flipping and intensity variation according to the equation: New image = image^{number}, number ∈ [0,95;1,05].

Neural Network. As Fig. 1 depicts, the CNN developed for this approach is inspired by VGG-Net [13] and adapted for 3D data. The choice for such a network is due to the fact that VGG network was a successful CNN in many deep learning studies in and out the medical field. Besides, it is also a network that can be adapted from a classification to a regression task.

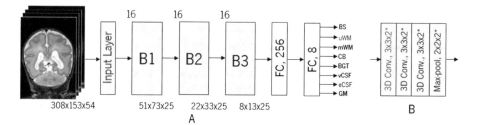

Fig. 1. Architecture of the adapted 3D neural network, based in VGG-Net: A- a general overview; B- convolutions and pooling used in each step (B1, B2 and B3). The * means that for steps B2 and B3 all 3D convolutions used are with $3 \times 3 \times 1$ kernels and max-pooling is $2 \times 2 \times 1$ due to the number of slices. The number of filters is represented on the top left of each step.

In Fig. 1, on steps B1, B2 and B3, the convolutions are performed with 16 filters each, followed by two fully connected layers, respectively with 256 and 8 nodes.

For the training task, for each minibatch, images were randomly selected, according to a certain size. In order to build and train the CNN a combination of DL libraries is used: Theano and Lasagne [14]. The training is done from scratch with a hold-out methodology.

Estimation and Model Evaluation. The estimated values are compared with the ground truth values, i.e. the descriptors computed from the segmentations. The Pearson correlation coefficient (R) and normalized root mean squared error (nRMSE) are calculated to evaluate the performance. While the Pearson correlation coefficient evaluates the magnitude of the linear relationship between the estimated and the ground truth results, the nRMSE provides a standard deviation of the estimation errors with respect to the range of volumes estimated. This last metric evaluates the error, in percentage (%), in relation to the full range of volumes of each tissue.

3D Approach Using Full Brains. For this approach, full neonatal 3D brains are used to estimate all 13 descriptors. The neural network used for this approach is the same as the described in the 3D approach using multiple slices, the only difference being to estimate 13 outputs instead of 8.

It is important to note that in the data augmentation step, scaling is removed because these 5 new descriptors change in a non-linear manner in relation to a decrease (or increase) in brain size. Augmentation is performed in terms of rotation, intensity variation and flipping, with the same values described in the previous approach.

2.3   Results and Discussion

Separate experiments are performed for each approach. For images at 30 weeks PMA, 86 studies are used: 60 for training, 13 for validation and 13 for testing. For the 153 studies made at 40 weeks PMA, 107 are used for training, 23 for validation and for 23 for testing. For each set of results, the comparison between estimated volumes and ground truth volumes are shown in graphs for each brain tissue along with the correlation (R) and nRMSE, for comparison.

3D Approach Using Multiple Slices

The hyperparameters set for the training for each dataset are given in Table 1. Figure 2 shows the estimated volume vs ground truth volume of 2 tissues (GM and uWM), using unseen images (test set) of each dataset. Table 2 shows the R and nRMSE results for each graph shown.

Table 1. Parameters used in the 3D approach using multiple image slices, using both datasets.

Parameters	30 weeks PMA	40 weeks PMA
Minibatches	24000	30000
Number of slices	6 slices (approx. 11% of the full brain)	10 slices (approx. 9% of the full brain)
Batch size	200	42
CNN	VGG-like	VGG-like
Filters per conv.	16	8
Learning rate	0,001	0,001
Optimizer	RMSProp	RMSProp
Filters' size	$3 \times 3 \times 2$ and $3 \times 3 \times 1$	$3 \times 3 \times 2$ and $3 \times 3 \times 1$
Max-pooling size	$2 \times 2 \times 2$ and $2 \times 2 \times 1$	$2 \times 2 \times 2$ and $2 \times 2 \times 1$

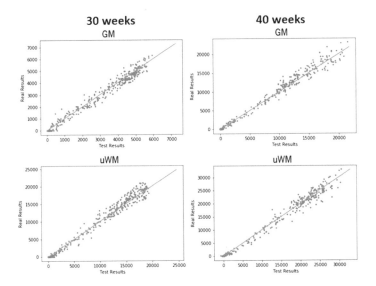

Fig. 2. Estimated volumes vs segmentation volumes of GM and uWM for the 3D approach using multiple image slices. The left and right column correspond, respectively, to the results using the 30 weeks and 40 weeks dataset. The name of the tissue is on top of its corresponding graph. Values are in mm^3.

GM and uWM have good estimation in the graphs and in terms of metrics, probably due to their larger size compared to other tissues. However, the larger the number of slices, the harder it is to make an accurate estimation. Small filter sizes in the convolutions also allow a better differentiation between examples, as they focus on smaller and local features.

3D Approach Using Full Brains
The hyperparameters fixed for training are given in Table 3. Figure 3 shows the estimated volume vs ground truth volume of 2 tissues (GM and uWM) as well as of 2 cortical descriptors. Due to the limited number of images, the original test set (red dots in the graphs) and labels are augmented for the purpose of reliable analysis of the results. The blue dots correspond to these augmented examples. Table 4 shows the metrics.

Table 2. Correlation and nRMSE for the approach, using both test sets.

Metric	30 weeks PMA		40 weeks PMA	
	GM	uWM	GM	uWM
R	0,986	0,987	0,987	0,986
nRMSE (%)	4,4	4,9	5,0	4,8

Table 3. Parameters used in the 3D approach using full 3D brains.

Parameters	30 weeks PMA
Minibatches	25000
Batch size	22
CNN	VGG-like
Filters per conv.	16
Learning rate	0,001
Optimizer	RMSProp
Filters' size	$3 \times 3 \times 3$ and $3 \times 3 \times 1$
Max-pooling size	$2 \times 2 \times 2$ and $2 \times 2 \times 1$

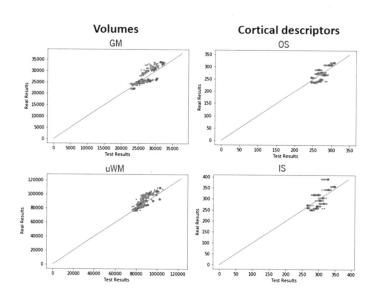

Fig. 3. Estimated volumes vs segmentation volumes of GM and uWM and cortical descriptors OS and IS, for the 3D approach using the full brain. The name of the tissue is on top of its corresponding graph. Values are in mm^3

The performance using the full images is lower than the performance using a smaller set of image slices as input. The use of downscaling leads to a deterioration of the images. This could be a factor that influences the training and consequently the estimations. The small number of diverse brains can be another issue that hampers the performance.

In addition to the two comparable methodologies previously described, a CNN inspired by ResNet-34 [15] was also tested on the 3D approach using multiple slices. As shown in Table 5, this network based on residual mapping did not lead to any improvements in either of the two metrics.

Table 4. Values of correlation and nRMSE for the 3D approach using the full brain.

Metric	30 weeks PMA			
	GM	uWM	IS	OS
R	0,812	0,836	0,704	0,766
nRMSE (%)	12,6	11,3	19,3	12,8

Table 5. Comparison of correlation and nRMSE for the 3D approach using multiple slices for the 30 weeks PMA dataset.

Metric	VGG-Net		ResNet	
	GM	uWM	GM	uWM
R	0,986	0,987	0,966	0,972
nRMSE (%)	4,4	4,9	8,3	7,8

3 Quantification of Brain Characteristics Using Rat MRI

3.1 Materials

The dataset with rats MRI scans was acquired from a part of a project called SIGMA, partially financed by Fundação para a Ciência e Tecnologia and Agence National de Recherche (ref: FCT-ANR/NEU-OSD/0258/2012). Wistar rats were the species of rodents used in this study. Using a 2×2 surface coil associated with software Paravision 6, the 139 scanning sessions were carried out on a Bruker Biospec 11.7T preclinical scanner. SE-EPI diffusion sensitive acquisitions were used, with Time to Repetition = 5 s, Time to Echo = 20 ms, in-plane resolution of 0.375×0.375 mm, slice thickness of 0.5 mm over 40 slices and a Field-of-View of 24 mm. Each acquisition acquired and averaged 10 volumes. Using WM, GM and CSF priors, SPM Segment was used with the objective of creating the ground truth for segmentation [16].

The volumes for estimation are GM, WM and CSF. The ground truth volumes were counted in voxels of GM, WM and CSF for each mouse brain 2D slice.

3.2 Methods

A 2D procedure that quantifies volumes per slice is performed. Instead of presenting the estimated volumes for each slice in the 2D approach, these quantifications are summed for the entire brain.

Pre-processing. Since there is a possibility that some studies with different parameters and scanners are acquired, an image normalization step will be performed. In this way, reducing these intensity variations in MRI scans can be an influential pre-processing step for more accurate MRI analysis [17], [18]. In the brain studies of the mouse, normalization is performed by dividing all intensity values by the maximum intensity of the studies.

Neural-Network. The 2D CNN is shown in Fig. 4 and is another adaptation of the VGG-like [13]. As shown in Fig. 4A, after the input layer 3, similar steps (B1, B2 and B3) are performed, which are detailed in Fig. 4B. All convolutions have ReLU as activation function. Finally, the network ends with 2 FC layers, with the last one outputting the 3 desired volumes.

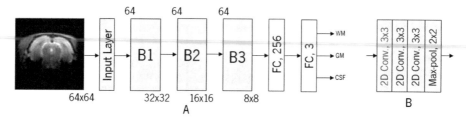

Fig. 4. Architecture of the adapted 2D neural network, adapted from the original VGG network: A-a general overview; B-convolutions and pooling used in each step (B1, B2 and B3). The number of filters associated to each step is located on the top left of that step.

3.3   Results and Discussion

For the 139 studies, 89 are used for training, 22 for validation and 28 for testing. The hyperparameters fixed for this test are shown in Table 6. Figure 5 shows estimated volume vs ground truth volume for WM and GM using unseen images (test set) by the model. The quantifications for each 2D slice are obtained through deep learning techniques and summed in the end to quantify the entire rat brain. Table 7 shows the correlation and nRMSE results for each graph shown.

Table 6. Parameters used in the 2D approach on the rat dataset.

Parameters	Rats' dataset
Minibatches	10000
Batch size	20
CNN	VGG-like
Filters per conv.	64
Learning rate	0,001
Optimizer	RMSProp
Filters' size	3×3
Max-pooling size	2×2

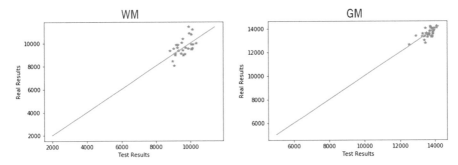

Fig. 5. Estimated volumes vs segmentation volumes for the volumes of WM and GM using the rats' dataset. The name of the tissue is on top of its corresponding graph. Values are in mm^3.

The low correlation does not overwhelm the results as it may seem due to the fact that the number of examples in the test set is low and their labels have similar values. In absolute terms, the error is similar for both tissues. However, because the range of volumes of GM is larger due to its size, the nRMSE value for this tissue is lower.

Table 7. Values of correlation and nRMSE for the 2D approach on the rat dataset.

Metric	Rats' dataset	
	WM	GM
R	0,555	0,648
nRMSE (%)	5,7	2,0

4 Conclusion

The proposed DL system for quantifying volume and cortical descriptors shows promising results in preterm neonatal brain MRI as well as in rat brain MRI. These results indicate that a DL based system can learn relevant features directly from images instead of relying on explicit segmentations, which could be of use to improve image-based prediction systems.

Acknowledgements. This work was supported by COMPETE: POCI-01-0145-FEDER-007043 and FCT – Fundação para a Ciência e Tecnologia within the Project Scope: UID/CEC/00319/2013. We gratefully acknowledge the support of the NVIDIA Corporation with their donation of a Quadro P6000 board used in this research.

References

1. Larroque, B., et al.: Special care and school difficulties in 8-year-old very preterm children: The Epipage Cohort study. PLoS One **6**(7) (2011)
2. Moeskops, P., et al.: Development of cortical morphology evaluated with longitudinal MR brain images of preterm infants. PLoS ONE **10**(7), 1–22 (2015)

3. Van Kampen, J.M., Robertson, H.A.: The BSSG rat model of Parkinson's disease: progressing towards a valid, predictive model of disease. EPMA J. **8**(3), 261–271 (2017)
4. Petrasek, T., et al.: A rat model of Alzheimer's disease based on Abeta42 and pro-oxidative substances exhibits cognitive deficit and alterations in glutamatergic and cholinergic neurotransmitter systems. Front. Aging Neurosci. **8**(83), 1–12 (2016)
5. Litjens, et al.: A survey on deep learning in medical image analysis. Med. Image Anal. **42**, 60–88 (2017)
6. Moeskops, P., Viergever, M.A., Mendrik, A.M., De Vries, L.S., Benders, M.J.N.L., Isgum, I.: Automatic segmentation of MR brain images with a convolutional neural network. IEEE Trans. Med. Imaging **35**(5), 1252–1261 (2016)
7. Moeskops, P., et al.: Prediction of cognitive and motor outcome of preterm infants based on automatic quantitative descriptors from neonatal MR brain images. Sci. Rep. **7**(2163) (2017)
8. Kaggle's competition: Second Annual Data Science Bowl. https://www.kaggle.com/c/second-annual-data-science-bowl. Accessed 26 Nov 2018
9. Liao, F., Chen, X., Hu, X., Song, S.: Estimation of the volume of the left ventricle from MRI images using deep neural networks. arXiv:1702.03833v1 (2017)
10. Dubost, F., et al.: 3D regression neural network for the quantification of enlarged perivascular spaces in brain MRI. arXiv:1802.05914v1 (2018)
11. de Vos, B.D., Lessmann, N., de Jong, P.A., Viergever, M.A., Isgum, I.: Direct coronary artery calcium scoring in low-dose chest CT using deep learning analysis. In: 103rd Annual Meeting Radiological Society of North America (2017)
12. Moeskops, P., et al.: Automatic segmentation of MR brain images of preterm infants using supervised classification. Neuroimage **118**, 628–641 (2015)
13. Simonyan, K., Zisserman, A.: Very deep convolutional networks for large-scale image recognition. arXiv:1409.1556v6 (2015)
14. Theano Development Team: Theano: A Python framework for fast computation of mathematical expressions. ArXiv e-prints, vol. abs/1605.02688 (2016)
15. He, K., Zhang, X., Ren, S., Sun, J.: Deep residual learning for Image Recognition. arXiv: 151203385v1 (2015)
16. Magalhães, R., et al.: The dynamics of stress: a longitudinal MRI study of rat brain structure and connectome. Mol. Psychiatry, 1–9 (2017)
17. Sun, X., et al.: Histogram-based normalization technique on human brain magnetic resonance images from different acquisitions. Biomed. Eng. Online **14**(1), 73 (2015)
18. Shah, M., et al.: Evaluating intensity normalization on MRIs of human brain with multiple sclerosis. Med. Image Anal. **15**(2), 267–282 (2011)

Information System for Monitoring and Assessing Stress Among Medical Students

Eliana Silva[1], Joyce Aguiar[1(✉)], Luís Paulo Reis[2],
Jorge Oliveira e Sá[1], Joaquim Gonçalves[3,4], and Victor Carvalho[4]

[1] ALGORITMI Research Centre, University of Minho, Guimarães, Portugal
jcaguiar2@gmail.com
[2] Artificial Intelligence and Computer Science Laboratory (LIACC),
Faculty of Engineering, University of Porto, Porto, Portugal
[3] 2AI/EST/IPCA – Technology School of Polytechnic
Institute of Cávado and Ave, Barcelos, Portugal
[4] LITEC – Innovation and Knowledge Engineering Lab, Porto, Portugal

Abstract. The severe or prolonged exposure to stress-inducing factors in occupational and academic settings is a growing concern. The literature describes several potentially stressful moments experienced by medical students throughout the course, affecting cognitive functioning and learning. In this paper, we introduce the EUSTRESS Solution, that aims to create an Information System to monitor and assess, continuously and in real-time, the stress levels of the individuals in order to predict chronic stress. The Information System will use a measuring instrument based on wearable devices and machine learning techniques to collect and process stress-related data from the individual without his/her explicit interaction. A big database has been built through physiological, psychological, and behavioral assessments of medical students. In this paper, we focus on heart rate and heart rate variability indices, by comparing baseline and stress condition. In order to develop a predictive model of stress, we performed different statistical tests. Preliminary results showed the neural network had the better model fit. As future work, we will integrate salivary samples and self-report questionnaires in order to develop a more complex and intelligent model.

Keywords: Stress · Heart rate variability · Wearable devices ·
Big data mining · Medical students

1 Introduction

Stress is a huge problem in today's society. Modern lifestyle exerts an enormous burden on individuals by pushing towards increasing productivity and longer working hours. This feeling of persistent underachievement, that modern working conditions exert, leads to burnout and stress-related mental disorders such as anxiety. According to the Aon EMEA Health Survey 2018, stress and mental health are the main concerns affecting the well-being of employees in Portugal [1]. It is well known that an excessive level of stress interferes with productivity and has an impact on the individual's physical and emotional health [2–4]. In fact, job stress is responsible for poor work

© Springer Nature Switzerland AG 2019
Á. Rocha et al. (Eds.): WorldCIST'19 2019, AISC 931, pp. 587–596, 2019.
https://doi.org/10.1007/978-3-030-16184-2_56

performance, high absenteeism, and several diseases, such as coronary heart disease, as well changes in lifestyle [5].

The stressful environments can be seen everywhere in our daily life, although for some occupations there is an increase in risk factors. So that, it is of the utmost importance to monitor the stress levels of these individuals on a regular basis, in order to anticipate chronic stress and thus to intervene early. Usual stress-assessment methods present some limitations: they cannot be applied continuously, in real-time, and commonly use invasive ways for evaluation, affecting the individual's routine, may skewing the results of the assessment.

In this study, we describe the EUSTRESS project that aims to create an Information System (IS), adapted to the type of stress profile of the individual, to monitor stress levels of the individual, continuously and in real time, and also predict chronic stress. The IS will use a measuring instrument based on mobile/wearable devices and machine learning techniques to collect and process stress-related data from the individual without his/her explicit interaction, and present the results in easily interpretable graphic environment. A predictive model will analyze stress recovery patterns, and a stress-control mechanism will emit alerts in case of excessive levels or cumulative effects (chronic stress).

The remaining of this paper is as follows. Section 2 presents the review of the literature in terms of theoretical and methodological rational for the study of stress among medical students. Following this background, we discuss in the Sect. 3 our research project called Eustress Solution. Section 4 presents preliminary results of our research, which will be discuss in the Sect. 5, to conclude and give some advices for future work.

2   Background

In common sense, the term *stress* usually takes a negative connotation. However, stress is a part of human life and the series of organic reactions it triggers can be quite healthy, pushing the human being to a response to a threatening situation (e.g. running away from a ferocious animal or assertively resolving a situation of conflict). Stress affects an individual's cognitive performance in a biphasic mode. Too little stress impairs adequate performance, increasing with physiological levels of acute stress and followed by a decrease in performance again with prolonged or disproportioned levels of stress.

When an individual is exposed to stressful stimuli (physical or psychological), the organism perceives it as a threatening event and mounts an adequate biological and behavioral responses. However, it is very difficult to distinguish between an optimal stress level – it is called *eustress* - and an exacerbated level – it is called *distress*. The first distinction between eustress and distress was made in the 1960s by Selye [6], to refer to a number of physiological and psychological reactions to harmful or unfavorable conditions or effects.

2.1   Methods for Stress Assessment

Among the different methods to evaluate stress in its different levels, the most common in the literature are in the psychological, physiological and behavioral domains. Usually, the psychological stress assessment is done using self-administered questionnaires, which are widely used and considered reliable, such as Stress Self-Assessment Scale, Perceived Stress Scale - PSS [7], and Stress Response Inventory – SRI, [8]. However, these questionnaires only provide information on stress levels at the time of evaluation, not covering the stressors or the evolution of stress levels [9].

Concerning the physiological data, biological markers include acute phase response hormones/mediators (cortisol, interleukins, ferritin) that are released during the stress response. For physiological assessment, there are several types of bio-signals that can be used, such as hormone levels (e.g. cortisol). The limitation of this assessment is the need for equipment and interaction with the individual, such as saliva, urine, or blood samples. Stress affects also another physiological process, in specific, the exposure to a stressor triggers the autonomic nervous system activating the sympathetic nervous system and inhibiting the parasympathetic nervous system [10, 11]. As a result, there are changes in heart rate – HR – and in heart rate variability – HRV, which can be assessed through an electrocardiogram, for example [12]. Other examples of tests that require specialized equipment are electroencephalogram, electrodermal activity, electromyogram, pupil diameter, speech analysis, and functional magnetic resonance imaging. However, again, this test does not monitor continuously, nor do they detect patterns related to the initial stages of stress. They also require moving to a place with equipment and are usually expensive. To bridge this gap, recent focus on mobile/wearable devices equipped with physiological signal sensors can be an effective way for continuous and non-invasive assessment.

Behavioral assessment, although it is fewer studied, has as its main advantage the possibility of being carried out without affecting the daily life of the individuals. As an example, we can analyze the dynamics of computer keyboard and mouse use, posture and facial expressions.

2.2   Stress Among Medical Students

Higher education is a transition period before students reach the working environment. Students are subjected to increasing periods of work with a progressive focus on autonomy and continuous assessment. The increasing workload is perceived as stressful and commonly leads to mental disorders and perception that their cognitive performance is bellowed their expected standards.

Previous works describe several potentially stressful moments experienced by medical students throughout the course. In addition to medical preparation and activity being considered as having a high potential for stress, there are other factors, such as: the student's first contact with the patient, the fact that he or she often lives alone and away from home, long hours of study, and the concerns about professional performance at the end of the course. Typically, most medical schools follow a traditional model that does not address a focus on emotions, which ultimately does not prepare the students to deal emotionally with events such as caring for patients seriously ill, with suffering and with death situations [13].

The period of university studies is commonly associated with a time of anxiety, stress, and even depression. There are several reasons, including the pressure for success in exams [13]. A study in the United States, comparing medical students and students from other courses, found there was a greater dependence or abuse of alcoholic beverages on medical students [14]. In Portugal, previous works have shown that among medical students, stress is a prevalent risk factor, affecting the decision-making process and altering the activity of brain networks [15, 16].

Academic exams are potential sources of stress for college students. Although it is a fundamental phase in the training and certification process, it is also one of the strongest stress factors due to the high-stake implications in the academic progress and self-perceived image. In particular, in medical students, there is an increase in levels of anxiety, salivary cortisol and stress in the pre-examination period compared to the post-exam period [17, 18]. Regarding stress recovery time, an empirical study with medical students [19] indicated that the fatigue levels started to decrease on the day after the examination, and most of the sample of the students needed, on average, six days to recover completely.

For these reasons, understanding the causes of the stress of medical students has also been a privileged topic in research, so that university institutions can adopt measures of prevention and stress management. In this study, we analyzed HR and HVR metrics from a sample of Portuguese medical students in two moments – without the stress induced from evaluations (baseline condition), and during university exams (stress condition). These data have been collected regarding to build a big database, in order to evaluate stress levels and predict chronic stress. Thus, the Eustress Solution has been developed in order to give a contribution in this field and help to understand this phenomenon.

3   The Eustress Solution

The research project named "EUSTRESS – Information system for the monitoring and evaluation of stress levels and prediction of chronic stress" aims to evaluate and develop an IS that monitors and evaluates in real-time the stress levels of an individual in order to develop a predictive model of stress response and chronic stress. For that, it assesses several kinds of information: salivary cortisol samples as a biological marker of stress reaction, self-report measures of experienced stress and coping strategies, physiological data evaluated through a wearable device – smartband and data from an e-assessment management system used for academic exams (MeddQuiz®).

Concerning the physiological data, it was used an application named "EUSTRESS". This application, specially developed for this project, was implemented in Android on a Samsung Galaxy 3/5 phone and received data from the wearable device – *Microsoft Smartband 2*. This Smartband evaluates skin conductance, body temperature, heart rate variability, calorie intake and expenditure, sleep patterns, and quality [20]. All of these data was sent to the mobile application via Bluetooth.

The MedQuizz is an e-assessment collaborative tool that provides several layers of information on students' performance during an exam. Basic data includes student's score, time to completion and number of correct/incorrect/avoid answers in order to understand their stress levels. Metadata obtained concerns behavioral data of the student, pertaining performance (e.g., time spent reading a question) and decision-making (e.g., number of visits to a question and of times an answer is changed). Additionally, the MedQuizz allows for the collection of direct human-machine interaction data to assess tension during an exam, such as mouse movement patterns or keystroke dynamics, and provides the possibility to test impulsivity (time lag between question) or affect environmental conditions (dim/brighten screen light) in user's experience.

Data provided from all of these sources will be integrated through statistical models in order to develop stress profiles (by baseline stress patterns and stress and reactivity patterns) to consequently predict stress states of individuals. In the architecture of this project, a set of proprietary and OpenSource technologies, linked and exchanging data, allow the collection and storage of biometric data. Figure 1 shows the architecture of this information system.

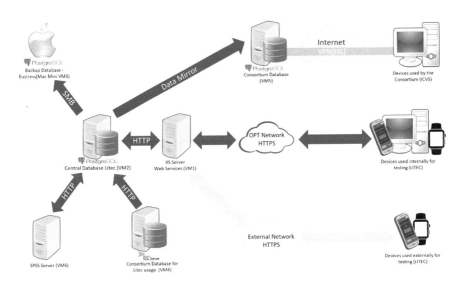

Fig. 1. Architecture of EUSTRESS Solution

The solution has three main goals. The first of them is to determine stress profiles (by baseline levels of stress and by patterns of reactivity and stress recovery) to classify individuals, making the IS adaptable to any individual. The second goal is to build a bigdata using psychological, physiological and behavioral assessments. Finally, the third goal is to interpret the stress reactivity patterns of the individual and predict stress states, cumulative effects of stress, and chronic stress.

3.1   Data Collection and Ethical Procedures

This project was reviewed and approved by the ethical committee of the Life and Health Sciences Research Institute at the University of Minho (Portugal) and the National Data Protection Commission. All participants were informed about the objectives of the study and provided written informed consent. Data were collected from September 2017 to October 2018. Participants were medical students at the Life and Health Sciences Research Institute at the University of Minho. All students received an informative email about the project and during classes the researcher explained the purpose of their participation, inviting collaboration.

Data were collected in two different conditions for each participant: the first condition was at the beginning of the academic year, a time without the stress induced from evaluations (baseline condition), and the second was during university exams (stress condition) that occurred at three different times. At baseline, participants' completed some global self-reported measures (e.g., sociodemographic questionnaire evaluating gender, age, nationality, and academic year; the PSS Portuguese version [21] Brief-COPE Portuguese version [22]) and used the Smartband during a week. In the baseline condition, physiological data were collected at 5 min each hour. At the end of each day participants answered the PSS 4 Items [21] and during three days they provided salivary cortisol (in the morning, in the afternoon, and in the evening).

In the stress condition, participants used the Smartband during multiple choice computer-based exams. Physiological data were continuously evaluated during the time of the exam. Participants provided samples of salivary cortisol before and after their exam. Salivary cortisol was collected through *Salivettes*. They also answered the questionnaires concerning stress perception (PSS and Brief-COPE) at the end of their exam. The participants did not receive any compensation for their participation.

These data (i.e., salivary cortisol sample, physiological data, data from the e-assessment management system, and self-report measures) were part of the broad project. However, in accordance with the purposes of the present study and for preliminary analyses, we only presented and analyzed HR and HRV metrics.

3.2   Data Analyses

Thirteen HR and HRV time domain indices [11] that quantify the "amount of variability in measurements of the interbeat interval" were calculated from the heartbeat interval data of Smartband sensors. Some examples of these indices were: mean heartbeat intervals (Mean RR), minimum (Min RR) and maximum values of RR (Max RR), median value of RR (Median RR), standard deviation of the RR intervals between normal beats (SDNN), root mean square differences of consecutive RR intervals (RMSSD), and percentage of consecutive RR intervals that differ by more than 50 ms (pNN50).

4   Results

Data were collected from 83 medical students who volunteered to enroll both in baseline and stress conditions. Sixty-three (76.8%) were female and 19 (23.2%) were male aged 17 to 38 years ($M = 22.13$; $SD = 5.55$). For approximately 63% of the participants, the attendance of the course did not imply the change of the residence.

Participants' academic year ranged from 1st to the 5th, with 19.5% of the participants from the 3rd year alternative program for graduate individuals. About 85% of the sample defined as the reason for the application to higher education the vocational interest.

We performed an Independent Samples t-Test in order to compare the HR and HRV variables at baseline and at exam condition. Table 1 presents the mean and standard deviation for each one. It was found significant differences between baseline and exam condition, except Diff Mean, Diff p mean and pNN50. The Mean, Min, Median, and Diff Min were significantly higher in baseline condition than in the exam condition. In contrast, SDDN, Max, Diff SDSD, and diff Max were significantly higher in the exam than in baseline condition.

Table 1. Comparing HR and HRV across two different conditions (baseline and exam)

	Baseline	Exam
Mean	.48(.06)	.45(.04)***
SDNN	.12(.02)	.14(.02)***
Min	.48(.06)	.45(.04)***
Max	1.22(.17)	1.25(.19)***
Median	.82(.12)	.72(.09)***
Diff Mean	.11(.02)	.11(.02)
SDSD	.11(.02)	.11(.01)***
Diff Min	.48(.06)	.45(.04)***
Diff Max	.57(.10)	.69(.09)***
Diff Median	.07(.03)	.07(.02)***
Diff p mean	.02(.01)	.02(.01)
RMSSD	.15(.03)	.16(.02)*
pNN50	.54(.12)	.54(.10)

$*p < .05$; $***p < .001$

Following these results, we performed different statistical tests: logistic regression, neural network, naïve Bayes, support vector machines, random forest, and k-nearest neighbor in order to predict stress based on HR and HRV variables. For each test, two models were established. In the model 1, we include all variables listed; in the model 2, we only include the significant variables. Table 2 presents the comparison between Model 1 and Model 2 for each test calculated. The neural network revealed the better results for both models. In specific, the sensitivity value was 77.8% for Model 1 and 77.21% for Model 2. The specificity values were 77.9% and 78.1%, respectively for Model 1 and Model 2.

Table 2. Comparing statistical tests to predict stress from the HR and HRV variables

		Model 1 (All variables)			Model 2 (Significant variables from the independent samples *t*-test)		
Logistic regression		Predicted			Predicted		
	Actual	0	1	Σ	0	1	Σ
	0	72.9%	27.1%	480	73.1%	26.9%	480
	1	27.7%	72.3%	480	27.7%	72.3%	480
	Σ	483	477	960	484	476	960
Neural network	Actual	0	1	Σ	0	1	Σ
	0	77.9%	22.1%	480	78.1%	21.9%	480
	1	24.8%	75.2%	480	25.8%	74.2%	480
	Σ	493	467	960	499	461	960
Naive Bayes	Actual	0	1	Σ	0	1	Σ
	0	65.0%	35.0%	480	62.7%	37.3%	480
	1	30.6%	69.4%	480	25.8%	74.2%	480
	Σ	459	501	960	425	535	960
Support vector machines	Actual	0	1	Σ	0	1	Σ
	0	43.1%	56.9%	480	47.5%	52.5%	480
	1	14.2%	85.8%	480	17.9%	82.1%	480
	Σ	275	685	960	314	646	960
Random forest	Actual	0	1	Σ	0	1	Σ
	0	75.2%	24.8%	480	74.8%	25.2%	480
	1	26.5%	73.5%	480	28.8%	71.2%	480
	Σ	488	472	960	497	463	960
k-nearest neighbors	Actual	0	1	Σ	0	1	Σ
	1	70.8%	29.2%	480	69.0%	31.0%	480
	Σ	19.8%	80.2%	480	24.4%	75.6%	480
	0	435	525	960	448	512	960

5   Conclusions and Future Work

In this paper, we discussed our ongoing research EUSTRESS and presented its main goals. Therefore, we focused on describe the architecture and the framework related with the development of this project.

Our results represent the firsts steps in order to develop the machine learning algorithms. For this, we performed different statistical tests and compare them. The neural network had the better model fit. In fact, this technique is robust enough to deal with the possibility of some possible badly classified data. Despite of we had classified the moments as without stress (baseline) and with stress (during exam), it is possible that some individuals had experienced stress in the first condition. As well, the exam situation could be not stressful for some of them. Therefore, it is possible that some of our data could not correspond to the real condition. For overpass this limitation, as next steps we will validate these results with the salivary cortisol samples and self-report questionnaires.

Future steps intend to design an intervention program where failing students are identified based on the real-time collection from exams (reactivity to stress/anxiety, biological markers, cognitive performance, and decision-making behavior) and recovered by customized coaching programs. Although the Information System is being validated in a sample of medical students, it will be applied to any individual in any occupational setting.

Acknowledgements. This work was funded by projects: the EUSTRESS (Sistema de Informação para a Monitorização e Avaliação dos Níveis do Stress e Previsão de Stress Crónico"), funded by European Regional Development Fund and by National Funds through the Portuguese Foundation for Science and Technology, within NUP = NORTE-01-0247-FEDER-017832; and the QVida+ project (Estimação Contínua de Qualidade de Vida para Auxílio Eficaz à Decisão Clínica), funded by European Structural funds (FEDER-003446), supported by Norte Portugal Regional Operational Programme (NORTE 2020), under the PORTUGAL 2020 Partnership Agreement.

References

1. AON. https://www.aon.com/getmedia/2626c505-cd46-4d01-9703-0ff5f30c118e/aon-emea-health-survey-2018-en.pdf.aspx. Accessed 5 Dec 2018
2. Stansfeld, S., Fuhrer, R., Shipley, M., Marmot, M.: Work characteristics predict psychiatric disorder: prospective results from the Whitehall study. J. Occup. Environ. Med. **56**, 302–307 (1999)
3. Hicks, T., McSherry, C.: A Guide to Managing Workplace Stress. Universal Publishers, California (2006)
4. Weiss, T.W.: Workplace stress: symptoms and solution. http://www.disabled-world.com/disability/types/psychological/workplacestress.php. Accessed 12 Dec 2018
5. Massida, D., Giorgi, I., Vidotto, G., Tringali, S., Imbriani, M., Baiardi, P., Bertolotti, G.: The Maugeri stress index - reduced form: a questionnaire for job stress assessment. Neuropsychiatr. Dis. Treat. **13**, 917–926 (2017)
6. Oksman, V., Ermes, M., Katu, T.: Eustress – findings concerning the indication and interpretation of positive stress among entrepreneurs – a case study. Bus. Manag. Rev. **7**(3), 342–347 (2016)
7. Cohen, S., Kamarck, T., Mermelstein, R.: A global measure of perceived stress. J. Health Soc. Behav. **24**(4), 358–396 (1983)
8. Bong, K., Park, J.K., Kim, C.H., Cho, S.: Development of the stress response inventory and its application in clinical practice. Psychosom. Med. **63**, 668–678 (2001)
9. Alberdi, A., Aztiria, A., Basarab, A.: Towards an automatic early stress recognition system for office environments based on multimodal measurements: a review. J. Biomed. Inf. **59**, 49–75 (2016)
10. Berntson, G., Cacioppo, J.T.: Heart rate variability – stress and psychiatric conditions. In: Dynamic Electrocardiography, pp. 56–63. Wiley, New Jersey (2007)
11. Shaffer, F., Ginsberg, J.: An overview of heart rate variability metrics and norms. Front. Public Health **5**, 1–17 (2017)

12. Taelman, J., Vandeput, S., Spaepen, A., Van Huffel, S.: Influence of mental stress on heart rate and heart rate variability. In: ECIFMBE 2008 Proceedings, vol. 22, pp. 1366–1369. Springer (2008)
13. Santos, F.S., Maia, C.R.C., Faedo, F.C., Gomes, G.P.C., Nunes, M.E., Oliveira, M.V.M.: Stress among Pre-University and Undergraduate Medical Students. Rev. Bras. Educ. Med. **41**(2), 194–200 (2017)
14. Jackson, E., Shanafelt, T., Hasan, O., Satele, D., Dyrbye, L.: Burnout and alcohol abuse/dependence among U.S. medical students. Acad. Med. https://doi.org/10.1097/acm. 0000000000001138 (2016)
15. Soares, J., Sampaio, A., Ferreira, L., Santos, N., Marques, F., Palha, J., Cerqueira, J., Sousa, N.: Stress-induced changes in human decision-making are reversible. Transl. Psychiatry **2**, e131 (2012). https://doi.org/10.1038/tp.2012.59
16. Soares, J., Sampaio, A., Marques, P., Ferreira, L., Santos, N., Marques, F., Palha, J., Cerqueira, J., Sousa, N.: Plasticity of resting state brain networks in recovery from stress. Front. Hum. Neurosci. **7**, 919–929 (2013)
17. González-Cabrera, J., Fernández-Prada, M., Iribar-Ibabe, C., Peinado, J.: M: Acute and chronic stress increase salivary cortisol: a study in the real-life setting of a national examination undertaken by medical graduates. Stress **17**(2), 149–156 (2014)
18. Kurokawa, K., Tanahashi, T., Murata, A., Akaike, Y., Katsuura, S., Nishida, K., Masuda, K., Kuwano, Y., Kawai, T., Rokutan, K.: Effects of chronic academic stress on mental state and expression of glucocorticoid receptor α and β isoforms in healthy Japanese medical students. Stress **14**(4), 431–438 (2011)
19. Blasche, G., Zilic, J., Frischenschlager, O.: Task-related increases in fatigue predict recovery time after academic stress. J. Occup. Health **58**, 89–95 (2016)
20. Nogueira, P., et al.: A review of commercial and medical-grade physiological monitoring devices for Biofeedback assisted quality of life improvement studies. J. Med. Syst. **42**(6), 275–285 (2017)
21. Trigo, M., Canudo, N., Branco, F., Silva, D.: Study of the psychometric properties of the Perceived Stress Scale (PSS) in the Portuguese population. Psychologica **53**, 353–378 (2010)
22. Ribeiro, J.L., Rodrigues, A.P.: Some questions about coping: the study of the Portuguese adaptation of the Brief COPE. Psicolog. Saúde Doenças **5**(1), 3–15 (2004)

Deep Reinforcement Learning for Personalized Recommendation of Distance Learning

Maroi Agrebi[1]([⊠]), Mondher Sendi[2], and Mourad Abed[1]

[1] Univ. Polytechnique Hauts-de-France, CNRS, UMR 8201 - LAMIH - Laboratoire d'Automatique de Mécanique et d'Informatique Industrielles et Humaines, 59313 Valenciennes, France
`{maroi.agrebi,mourad.abed}@uphf.fr`
[2] MARS Research Lab LR 17ES05, University of Sousse, Sousse, Tunisia
`mondher.sendi@isitc.u-sousse.tn`

Abstract. Nowadays, distance learning becomes more diverse and popular. Increasingly universities are currently working to offer their online courses (MOOC, SPOC, SMOC, SSOC, etc.) in the form of courses providing learners with a wide variety of choices. However, this multi-criteria choice is complex. In this paper, we propose a personalized recommendation system based on Deep Reinforcement Learning that suggests for learners a most appropriate course according to specificities of each one such as their profile, needs and competences. To validate our system, the later has been tested over a set of real students. The obtained results of our study are in favor of the robustness of our system.

Keywords: Distance learning · Deep Reinforcement Learning · Personalization · Recommendation · Learner profile

1 Introduction

In higher education, an explosion in the proliferation of distance learning courses and programs was produced by the development increasingly of new instructional technologies. These technologies are essentially the MOOC (Massive Open Online Courses) and the SPOC (Small Private Online Courses) which are asynchronous in term of time dependency as well as the SMOC (Synchronous Massive Online Course) and the SSOC (Synchronous Small Online Course) which are synchronous with regard to the time dependency as their name indicates. These four types of distance learning differ not only in the degree of time dependency but also in the student populations sizes. In fact, MOOCs and SMOCs offer unlimited (massive) number of places when SPOCs and SSOCs allow for limited (small) number of participants.

The MOOC was coined in 2008 by Dave Cormier from the University of Prince Edward Island in Canada with regard to a course called Connectivism

© Springer Nature Switzerland AG 2019
Á. Rocha et al. (Eds.): WorldCIST'19 2019, AISC 931, pp. 597–606, 2019.
https://doi.org/10.1007/978-3-030-16184-2_57

and Connective Knowledge [1]. Since then, MOOCs have rapidly expanded in the USA, Europe, Asia-Oceania, etc. (E.g. Cousera and edX in the USA, FutureLearn (UK), Iversity (Germany), MiriadaX (Spain) in Europe, KMOOC (Korea), and OpenLearning (Australia) in Asia-Oceania) [2]. From [3], it is estimated that to January 2018, more than 800 universities worldwide have launched at least one online course, with over 9400 announced MOOCs and that around 81 million learners signed up. Overall, the distribution of courses across subjects is varied. Here is the list of top five subjects: 19,9% Technologies, 18,5% Business, 10,6% Social sciences, 10% Science, 9,5% Humanities [3].

With the huge amounts of distance courses available more and more at the disposal of learners, the choice of the appropriate course between the different courses becames difficult since it is considered as multi-criteria problem [4,5]. One way to overcome this difficulty is to use personalized recommendation system based on Deep Reinforcement Learning. The later represents a solution in distance learning of ever-growing online learning resources.

In this paper, we propose a personalized recommendation system. The aim of which is to suggest to the learners the most appropriate course from available courses. The recommended course should be designed for individual needs, competences and profile [6,7], etc. of each learner. However, before doing this, we, first, present related literature in recommendation systems, in Sect. 2, in order, to identify limitations of the existing systems. Then, in Sect. 3, we present our proposed system based on Deep Reinforcement Learning. The system validation is discussed, in Sect. 4. Finally, Sect. 5 dedicates the conclusion and perspectives.

2 Related Literature

2.1 Recommender Systems

Recommender systems are personalized complex information agents [8] that provide recommendations, suggestions for items likely to be of use to a user [9]. The recommendation is based on past behavior [10], explicit and implicit preferences of concerned users, the preferences of other users, and concerned user and item attributes [11]. In fact, preferences can be selected and aggregated in such a way as to provide a reasonable prediction of the active user's preference [12].

In the literature, since the mid-1990s, recommender systems have gained much attention as an important research area. These systems have been proposed in order to recommend items primarily to the customers of online e-commerce sites [13]. Examples of such applications include recommending books, CDs and other products at Amazon.com [14], movies by MovieLens [15], and news at VERSIFI Technologies [16]. However, the beginning of recommender systems applications in the area of education it was only in the of 21th century [17]. Industry and academia have invested on proposing new approaches to recommender systems. Their main objective is to help users to deal with information overload and provide personalized recommendations, content and services to them [18].

2.2   Recommender Systems for e-Learning

In the literature, there have been a lot of recommendation systems proposed in the field of distance learning. In this section, we present the recent works in this domain.

Bousbahi and Chorfi [19] have proposed a recommender system, in order to suggest the most appropriate MOOCs (from different providers) in response to a specific request of the learner have proposed. The suggestion takes into consideration learner profile, needs and knowledge. The proposed system has been developed using the Case Based Reasoning (CBR) approach and a special retrieval information technique.

Fu et al. [20] have focused on proposing an undergraduate-oriented framework of MOOCs recommender system. The later emphasizes the effects of individual characteristics of the participants in MOOCs such as cognitive level, knowledge background, personal expectation, learning interest, learning motivation and learning style.

Drachsler et al. [21], in order to support educational stakeholders by personalising the learning process, have analysed the recommender systems in Technology-Enhanced Learning which are propose during 2000–2014. Then they have investigated and categorised the existing systems according into seven clusters by comparing to their characteristics and analysed for their contribution to the evolution of the RecSysTEL research field.

Holotescuhve [22] have developed MOOCBuddy - a MOOC recommender system as a chatbot for Facebook Messenger. The role of this educational chatbot, related to MOOCs, is to assist users to discover news about MOOCs, to find MOOCs for personal and professional development, but also teachers to integrate MOOCs in their courses. The system is based on user's social media profile and interests.

Heras et al. [23] have explored the use of argumentation-based recommendation techniques as persuasive technologies. The objective of them is to evaluate how arguments can be used as explanations to influence the behaviour of users towards the use of certain items. His proposed system has been implemented as an educational recommender system for the Federation of Learning Objects Repositories of Colombia that recommends learning objects for students taking into account students profile, preferences, and learning needs.

Tarus et al. [24] have introduced a recommendation architecture that is able to recommend interesting post messages to the learners in an e-learning online discussion forum based on a semantic content-based filtering and learners' negative ratings. Their proposed e-learning recommender system has evaluated against exiting systems that use similar filtering techniques in terms of recommendation accuracy and learners' performance.

The main limits of the approaches presented above to recommend online courses are: Firsly, the ignorance of the importance of the different attributes of the learners and the characteristics of the courses. Secondly, non-consideration of the learners feedback. To satisfy more the courses personalization recommendation, we propose in this paper an approach based on the Deep Reinforcement

Learning which allows to learn at the same time the characteristics of problem of recommendation in the platforms of e-learning and to take into account the feedbacks of the users.

3   Proposed Approach

3.1   System Architecture

Figure 1 presents our architecture for the courses recommendation system. The first step consists in constructing a model learning. To this end, we use the student profile and his learning history to identify the model parameters. Then, our system starts to recommend courses to students. During this recommendation process, the model can improve these parameters by using student's feedback. Moreover, to learn the characteristics of learning concepts, we use the Deep Reinforcement Learning. The next section presents our proposed Deep Learning model.

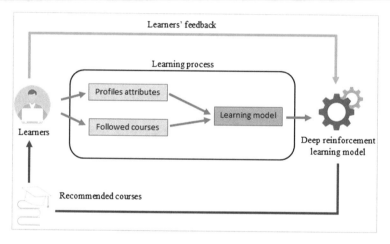

Fig. 1. Course recommendation system architecture.

3.2   Deep Learning Model

a. Problem formulation

The recommendation system proposed in this paper takes explicit information from courses and profile attributes. Assuming having $C = \{c_1, \ldots, c_k\}$ courses criteria and $F = \{f_1, \ldots, f_l\}$ courses. These laters are represented by an $l-by-k$ matrix M where each line is a vector that contains the course criteria.

The values of the vector represent the criteria importance for such course. For example, the credits of a course can be seen as a variable that can take a value between 0 and 30. We used a predefined list of criteria in this matrix.

By the same way, the learners profiles are represented by an $n - by - m$ matrix E where each line contains the learners competences $e_i = \{s_1, \ldots, s_m\}$

(eg programming, language). The learning data is constructed by making a join between the matrix E and the matrix M and we keep only the courses that are chosen by the students in a matrix A. This matrix contains the model learning data. Table 1 contains the list of criteria and profile attributes used in our model.

The problematic posed in this paper is as follow; given a list of students $E = \{e_1, \ldots, e_n\}$ represented by their competences and a set of preferences specified by each learner, we should return a list of appropriate courses for these learners.

Table 1. Criteria and attributes for personalized recommendation system.

Profile attributes	Course criteria
Language	Level
Competences	Complexity
Study level	Duration
Abandonment rate	Price
Course/Month	Credits
Knowledge background	Discipline
Learning interest	Certification
Learning motivation	Course support

b. Model architecture

Deep Learning model that we propose for the personalized recommendation of online courses is presented in Fig. 3. The first layer contains the course criteria (such as price, duration, course credits, etc.) and student profile attributes (such as language, study level, learning interest, etc.). For each given configuration in the first layer (course criteria, learner profile) there is a most appropriate course in the output layer. The layers in the middle are used for model learning. We thought to use the minimum number of layers to learn the characteristics of the courses recommendation problem.

To learn the model, we use the courses studied by the learners according to the criteria chosen in advance. Besides, we use the attributes of these learners profile in order to learn the relationships between the learners profile and the characteristics of the available courses in the system. When our model is trained, it should be able to suggest online courses for the learners based on the attributes of their profiles and their preferences.

c. Deep Reinforcement Learning

The Deep Reinforcement Learning model that we used in this paper is based on the model proposed in [25] (Fig. 2). We model the problem recommendation system with a Markov Decision Process. The Deep Reinforcement Learning uses

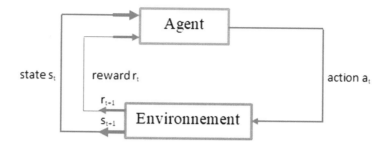

Fig. 2. Standard scenario of the Deep Reinforcement Learning.

sequence of states, actions, rewards, transition probability and discount factor (S, A, R, P, λ) defined as follows:

- S: The set of states that allows us to model the attributes of the profile and the studied course by the learners $S = \{s_1, \ldots, s_k\}$.
- A: The set of actions that consists of recommending a list of courses for each learner based on his profile $A = \{a_1, \ldots, a_m\}$.
- R: The reward is the feedback returned by the learner on the studied or recommended courses by the system $R = \{r_t, \ldots, r_{t+l}\}$.
- P: The transition probability between each state and the next state (t, t + 1) by taking into account the action performed by the system a_t.
- λ: The discount factor of the reward over time. This value is specified in advance and comprises between 0 and 1.

The standard scenario of the Deep Reinforcement Learning is presented in the Fig. 3. The agent interacts continuously with his environment. It proposes action according to its knowledges. Then, the environment returns the reward based on the recommendation and the state. More formally, given the model parameters, the objective of recommender system is to return a list of courses that allows us to maximize cumulative reward. Using learners feedback, the recommender system can learn at the same time the needs of the learner and make correspondence with existing courses in the platform. When the cumulative rewards increases, the result returned by the model gets closer to the learner profile. The cumulative reward is defined as follows:

$$r_t = \sum_{k=1} \lambda^{k-1} U^k, \tag{1}$$

where U_x^k is the reward given to course k.

In the reinforcement learning process, the Q-learning function allows us to determine what our return would be. Thus, it allows us to choose the right policy that maximizes our rewards:

$$Q(s_t, a_t) = E_{s_{t+1}} \left[r_t + \gamma \, max_{t+1} \, Q(s_{t+1}, a_{t+1})] \right], \tag{2}$$

this function represents the expected return based on state s_t and the action a_t.

Given the model parameters θ^μ, the objective function to maximize in our problem is as follows:

$$L(\theta^\mu) = E_{s_t,a_t,r_t,s_{t+1}} \left[(y_t - Q(s_t, a_t; \theta^\mu))^2 \right], \tag{3}$$

with y_t is the value of the current iteration.

To learn our model, we use the Deep Deterministic Policy Gradients [26]. The two main steps for this algorithm are the followings:

- The transition generating: In this step, the system recommends a list of courses according to the given parameters. The second state is updated according to the returned reward (feedback).
- Parameter updating: In this step the parameters of the model are updated according to the standard Deep Deterministic Policy Gradients procedure.

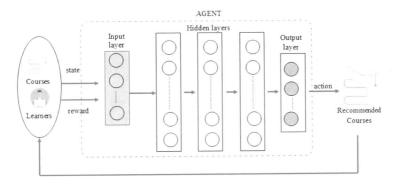

Fig. 3. Deep learning network for course recommendation.

4 Experimentation

In this section, we present the experimentation with a dataset from a real e-learning students. The aim is to evaluate the effectiveness of the proposed model. We, firstly, present the strategy used to collect the model learning data. Then, we present the analyzes that we carried out to test the performance of our model.

4.1 Experimental Settings

We evaluate our system on a dataset of October 2018 from a real e-learners. We have collected, from diverse e-learning platforms, informations about price, duration, complexity, level, etc. of 100 online courses. These courses have been evaluated by 120 learners according to the profile attributes (as shown in Table 1) of each learner. Learners have used the evaluation scale presented in Table 2 [4].

From 12000 evaluations (formation/learner), 70% of these combinations are used to learn our personalized recommendation system. The rest of combinations are reserved to test the system. Afterwards, 70 iterations have been made using

Table 2. Evaluation scale

Linguistic term	Rating
Very poor (VP)	(1)
Poor (P)	(2)
Fair (F)	(3)
Good (G)	(4)
Very good (VG)	(5)

learning model in order to find the appropriate courses. After each iteration, learners have given feedback on recommended courses. Using this feedback, the learning model have been improved.

4.2   Proposed System Evaluation

Precision and Recall are measure to evaluate the relevance of our system. Table 3 shows the precision and recall amelioration. The Precision and Recall are improved at each iteration. Taking the precision in the first iteration which is equal to 11%. This result has been improved after 70 iterations and has been reached 48%. Like the recall, the result have been improved from 20% in the first iteration to 70% after 70 iterations.

   The obtained results with our method of courses recommendation are encouraging. Nevertheless, we can not ignore certain limits like the dataset size and depth of Deep Learning model. In fact, on the one hand, the number of learners used to evaluate our system is not sufficient to confirm its relevance in such situations.

   On the other hand, the number of hidden layers in the Deep Reinforcement Learning can be increased to capture the different features of recommendation problems in the e-learning platforms. The construction of a Deep Learning model requires an analysis and correspondence between the number of inputs and the number of nodes in the hidden layers. We expect to study these limits in depth in our future work.

Table 3. Precision and recall evaluation

Measures	Iterations			
	1	20	50	70
Precision	11%	24%	46%	48%
Recall	20%	32%	66%	70%

5   Conclusion and Perspectives

Our study introduces a personalized recommendation system based on Deep Reinforcement Learning on the domain of distance learning. The aim of which is to suggest the most appropriate courses for learners based on their profile, needs, competences, etc. For this purpose, our system identifies the model parameters using learner profile and learning history. Then, it recommends for learners the courses. During the recommendation process, our system can improve the model parameters by using learners feedback.

To validate our proposed system, a dataset from a real e-learning students has been exploited. The obtained results are in favor of effectiveness of our proposition. Future researches may focus, in a first step, on the exploitation of a largest dataset from differents universities and countries in order to have diversified profiles and requests therefore on adaptation of our system. Then, in a second step, we expect to expand personalized recommendations to suggest for students assistance services such as transport, housing, catering, etc.

References

1. Kaplan, A.M., Haenlein, M.: Higher education and the digital revolution: about MOOCs, SPOCs, social media, and the Cookie Monster. Bus. Horiz. **59**(4), 441–450 (2016)
2. Kim, S.W.: MOOCs in higher education. In: Virtual Learning, InTech (2016)
3. Class Central. MOOCs by the numbers in 2017 (2018). https://www.classcentral.com/report/mooc-stats-2017/
4. Agrebi, M., Abed, M., Omri, M.N.: ELECTRE I based relevance decision-makers feedback to the location selection of distribution centers. J. Adv. Transp. **2017**, 10 (2017)
5. Agrebi, M., Abed, M., Omri, M.N.: A new multi-actor multi-attribute decision-making method to select the distribution centers' location. In: IEEE Symposium Series on Computational Intelligence (SSCI), pp. 1–7. IEEE (2016)
6. Sendi, M., Omri, M.N., Abed, M.: Discovery and tracking of temporal topics of interest based on belief-function and aging theories. J. Ambient Intell. Hum. Comput., 1–17 (2018)
7. Sendi, M., Omri, M.N., Abed, M.: Possibilistic interest discovery from uncertain information in social networks. Intell. Data Anal. **21**(6), 1425–1442 (2017)
8. Smyth, B.: Case-based recommendation. In: The Adaptive Web, pp. 342–376. Springer, Heidelberg (2007)
9. Burke, R.: Hybrid web recommender systems. In: The Adaptive Web, pp. 377–408. Springer, Heidelberg (2007)
10. Hung, L.P.: A personalized recommendation system based on product taxonomy for one-to-one marketing online. Expert Syst. Appl. **29**(2), 383–392 (2005)
11. Schein, A.I., Popescul, A., Ungar, L.H., Pennock, D.M.: CROC: a new evaluation criterion for recommender systems. Electron. Commer. Res. **5**(1), 51–74 (2005)
12. Ekstrand, M.D., Riedl, J.T., Konstan, J.A.: Collaborative filtering recommender systems. Found. Trends® Hum. Comput. Interact. **4**(2), 81–173 (2011)
13. Resnick, P., Varian, H.R.: Recommender systems. Commun. ACM **40**(3), 56–58 (1997)

14. Linden, G., Smith, B., York, J.: Amazon.com recommendations: item-to-item collaborative filtering. IEEE Internet Comput. **1**, 76–80 (2003)
15. Miller, B.N., Albert, I., Lam, S.K., Konstan, J.A., Riedl, J.: MovieLens unplugged: experiences with an occasionally connected recommender system. In: Proceedings of the 8th International Conference on Intelligent User Interfaces, pp. 263–266. ACM (2003)
16. Billsus, D., Brunk, C.A., Evans, C., Gladish, B., Pazzani, M.: Adaptive interfaces for ubiquitous web access. Commun. ACM **45**(5), 34–38 (2002)
17. Manouselis, N., Drachsler, H., Verbert, K., Duval, E.: Recommender systems for learning. Springer Science & Business Media (2012)
18. Adomavicius, G., Tuzhilin, A.: Toward the next generation of recommender systems: a survey of the state of the art and possible extensions. IEEE Trans. Knowl. Data Eng. **6**, 734–749 (2005)
19. Bousbahi, F., Chorfi, H.: MOOC-Rec: a case based recommender system for MOOCs. Procedia Soc. Behav. Sci. **195**, 1813–1822 (2015)
20. Fu, D., Liu, Q., Zhang, S., Wang, J.: The undergraduate-oriented framework of MOOCs recommender system. In: 2015 International Symposium on Educational Technology (ISET), pp. 115–119. IEEE (2015)
21. Drachsler, H., Verbert, K., Santos, O.C., Manouselis, N.: Panorama of recommender systems to support learning. In: Recommender Systems Handbook, pp. 421–451. Springer, Boston (2015)
22. Holotescu, C.: MOOCBuddy: a Chatbot for personalized learning with MOOCs. In: RoCHI, pp. 91–94 (2016)
23. Heras, S., Rodríguez, P., Palanca, J., Duque, N., Julián, V.: Using argumentation to persuade students in an educational recommender system. In: International Conference on Persuasive Technology, pp. 227–239. Springer, Cham (2017)
24. Tarus, J.K., Niu, Z., Mustafa, G.: Knowledge-based recommendation: a review of ontology-based recommender systems for e-learning. Artif. Intell. Rev. **50**(1), 21–48 (2018)
25. Zhao, X., Zhang, L., Ding, Z., Yin, D., Zhao, Y., Tang, J.: Deep reinforcement learning for list-wise recommendations. arXiv preprint arXiv:1801.00209 (2017)
26. Lillicrap, T.P., Hunt, J.J., Pritzel, A., Heess, N., Erez, T., Tassa, Y., Wierstra, D.: Continuous control with deep reinforcement learning. arXiv preprint arXiv:1509.02971 (2015)

An Approach to Assess Quality of Life Through Biometric Monitoring in Cancer Patients

Eliana Silva[1](✉), Joyce Aguiar[1], Alexandra Oliveira[2,3],
Brígida Mónica Faria[2,3], Luís Paulo Reis[2,4], Victor Carvalho[5],
Joaquim Gonçalves[5,6], and Jorge Oliveira e Sá[1]

[1] Centro ALGORITMI – University of Minho, Guimarães, Portugal
eliana.marisa@hotmail.com
[2] Artificial Intelligence and Computer Science Laboratory (LIACC),
Porto, Portugal
[3] School of Health, Polytechnic Institute of Porto, Porto, Portugal
[4] DEI - FEUP - Department of Informatics Engineering,
Faculty of Engineering of the University of Porto, Porto, Portugal
[5] LITEC – Innovation and Knowledge Engineering Lab, Porto, Portugal
[6] EST/IPCA – Technology School of Polytechnic Institute of Cávado and Ave,
Barcelos, Portugal

Abstract. Cancer is a serious disease that causes significant disability and suffering, so naturally Health Related Quality of Life (HRQoL) is a major concern of patients, families and clinicians. This paper intends to relate biometric indices, in terms of HRV metrics, with self-perceived HRQoL from patients with lymphoma. Patients ($N = 12$) answered FACT questionnaire and used a smartband that collected biometrical data in real-time along the chemotherapy treatment. Our results revealed that Physical Well-Being, Total, Lymphoma subscale and FACT-Lym Trial Outcome domains seem to have a similar pattern that HRV metrics across the treatment cycles. In specific, the FACT domains and the HRV metrics have the lowest average levels on the first cycle and seem to increase along the following cycles (3rd and 6th cycles). This approach of continuous assessment of HRQoL will enable a better accuracy and more supported clinical decision.

Keywords: Health-related quality of life · Haemato-oncological diseases · Self-reported measures · Physiological data · Wearable technology · Heart Rate Variability

1 Introduction

Cancer is a generic term for a large group of diseases that is characterized by an uncontrolled and abnormal growth of cells [1]. It can affect almost any part of the body and has many anatomic and molecular subtypes each requiring specific management strategies.

© Springer Nature Switzerland AG 2019
Á. Rocha et al. (Eds.): WorldCIST'19 2019, AISC 931, pp. 607–617, 2019.
https://doi.org/10.1007/978-3-030-16184-2_58

People living with cancer can experience significant disability and suffering, so naturally Quality of Life (QoL) is a major concern of patients, families and clinicians [2]. Moreover, due to factors such as population ageing, it is estimated that by 2030 it will be registered in Portugal 60.000 new cases of cancer per year. In fact, cancer is the second cause of death in this country [3]. Being one of the European countries with the greatest population ageing, according to the United Nations report [4], Portugal has more than one million people over 75 years old. Beyond prolonging life, it is essential also increase patient's QoL. In particular, the treatment of haemato-oncological diseases, such as lymphoma, often implies high-dose chemotherapy, which can be associated with severe symptoms and psychological distress creating difficulties in fulfilling family and social roles (e.g. be able to work or participating in daily social activities) and having a major impact in patient QoL [5]. Despite there is no consensus in the literature, QoL term is seen as a multidimensional (physical, psychological, social, and spiritual), subjective, and dynamic concept [1]. In other words, QoL is a broad concept that encompasses all aspects of human life. For that reason, a distinction regarding focus health-related quality of life (HRQoL) was introduced. This could be defined as self-perceived aspects of well-being that are related to or affected by the presence of a disease or treatment [6]. Therefore, HRQoL provides a more holistic evaluation and positive concept of health, which includes an individual's experiences, beliefs, expectations, and perceptions [7]. Being a subjective concept modulated by cultural and care patterns, the most used way to assess HRQoL is through self-perception questionnaires [2]. However, there is also an objective component of HRQoL [7], related to clinical indicators that evaluate symptoms and individual ability to do daily activities. Because of this multidimensional conceptualization, an appropriate and effective assessment of HRQoL should incorporate both objective functioning and subjective well-being.

Recent technological advances in wearable devices have created new opportunities to collect continuous in real-time, objective patient data in a non-obtrusive manner [8]. Actually, wearable biometric monitoring devices (BMDs) can be an important tool in the continuous monitoring of biometric data, which can be used to assess patient's health status, disease progression and treatment effects [9].

BMDs can be defined as a "biosensor that collects a biological recognition element (such as blood glucose or sodium levels), anatomical structure (such as tumor size, infarct size or hippocampal volume) or integrated physiological parameter (such as heart rate, blood pressure, electroencephalography, mobility, speech and sleep patterns or speed of information processing)" [9]. In particular, Heart Rate Variability (HRV) has been widely used as a diagnostic and prognostic tool [11]. In this paper, we focus on HRV time-domain indices, which quantify the amount of HRV, observed during monitoring periods [10].

Therefore, the purpose of this study was to evaluate HRQoL in lymphoma patients throughout chemotherapy treatment and relate its domains to the most common HRV time-domain metrics assessed in real-time by mobile devices – QLife+ solution.

2 State of the Art

2.1 Health Related Quality of Life in Cancer Patients

HRQoL is considered as a powerful predictor of mortality and morbidity. Specifically, HRQoL is regarding to both self - reported chronic diseases (e.g. diabetes, arthritis, hypertension, and cancer) and their risk factors (e.g. body mass index, physical inactivity, and smoking status). So, HRQoL assessment can provide insights into the relationships between HRQoL and risk factors as well as determine the burden of preventable disease, injuries and disabilities. Moreover, HRQoL is now considered an important aspect in clinical practice for patients with chronic illnesses [12].

One of the most used questionnaires to evaluate HRQoL in cancer patients is the Functional Assessment of Chronic Illness Therapy (FACT) focusing on the previous 7 days. It includes a combination of inputs from experts and patients ensuring the inclusion of clinically important issues that are also relevant to the patient [13]. It has a general core (FACT-G), which is common for all patients with cancer, it is composed by 27 general items divided into four domains, namely Physical well-being (PWB; 7-items), Social/Family well-being (SWB; 7-items), Emotional well-being (EWB; 6-items), and Functional well-being (FWB; 7-items). Adding to this general core, there are several extensions specific to other chronical illness conditions such as lymphoma (FACT-Lym; 15 items) [13]. In particular, the Trial Outcome Index score have been reported as an efficient summary index of physical/functional outcomes and it is very responsive to change of the patient [13]. "While social and emotional well-being are very important to HRQoL, they are not as likely to change as quickly or dramatically over time or in response to physical health interventions" such as chemotherapy [13]. All items are evaluated in a 5-point Likert scale (from 0 to 4).

It is important to notice that assessing HRQoL throughout questionnaires face some challenges such as: time constraints and frequency overload, the response and recall bias where the patient may not have a complete recollection of their feelings and symptoms and by the desire to enroll in a trial or receive therapy resulting in the over reporting of their health status [8]. Other important challenge associated to questionnaires is "the static nature of the assessment only captured periodically, and the patient's health status is dynamic over the course of the treatment and can change on a daily basis" [8]. Therefore, it is important to complement the questionnaire information with biometric data in an objective and non-invasive manner.

2.2 Biometrics Data – Heart Rate Variability

The autonomic nervous system (ANS), composed by the sympathetic and parasympathetic branches, acts as a control system of blood vessels, glands and muscles, including the heart [14]. The continuous non-linear modulation of the ANS, results in complex and non-linear variations in HRV [10, 15]. "HRV is an emergent property of interdependent regulatory systems which operate on different time scales so that the patient has the flexibility to adapt to the environmental and psychological challenges"

[10]. It has been recognized to be a useful non-invasive tool to predict several pathologies such as myocardial infarction, diabetic neuropathy, sudden cardiac death and ischemia, among others [16].

Typically, statistical variables are calculated over 5 min length ECGs segments) [10, 15]. Time-domain indices of HRV quantify the amount of variability in measurements of the interbeat interval, which is the time period between successive heartbeats [10]. There exist a wide variety of time domain parameters, but we will focus on one described on Table 1.

Table 1. Time domain parameters description

Parameters	Unit	Description
Mean HR	bpm	Mean heart rate
Mean RR	ms	Mean RR interval
RMSSD	ms	Root mean square of successive differences
SDNN	ms	Standard deviation of the RR interval
SDANN	ms	Standard deviation of the average RR intervals calculated over short periods
pNN50	%	Proportion of successive RR intervals greater than 50 ms

The intervals between consecutive heart beats needed to construct the time series are called RR intervals and the instantaneous Heart Rate (HR) is the number of beats per minute.

3   Methodology

3.1   QLife+ Architecture

The communication architecture implemented in the collection of biometric parameters in the QLife+ project, is based on the Microsoft Band 2 device, connected via Bluetooth protocol, to an Android smartphone, which in turn collects data from the wearable device's sensors. In this architecture, there is a set of proprietary and OpenSource technologies, linked and exchanging data, allow the collection and storage of biometric data. For attending this purpose, we need database tools, web server, mobile application - where these tools have adopted technologies such as PostgreSQL, Microsoft IIS Web Server - and an Android application. In order to provide communication between these technologies, different protocols are used, such as HTTPS in the communication between the mobile application and the web server, and an HTTP in the communication between the web server and the database PostgreSQL, given in a secure internal network.

3.2   Data Collection

Participants were recruited from one public Hospital in Portugal. Patient eligibility criteria included: histological diagnosis of lymphoma, being aged 18 years or older, confirmed to receive treatment at that Hospital, and considered by the oncologist and the researchers to be emotionally and physically capable of participating. After oncologist referral, participants were invited to participate and received an informative flyer. All participants provided their written informed consent to participate in this study.

Participants were interviewed at baseline (pretreatment or on the day of their first chemotherapy treatment), at several times during active treatment (first, third, and sixth chemotherapy cycle), and every three months at the follow-up.

At baseline, participants were asked to answer a brief sociodemographic questionnaire. They were given brief training on the BMD, specifically a smartband (Microsoft Band 2), and received written information with general instructions. Participants used the smartband during one or two weeks at pretreatment, and during two consecutive weeks at first, third, and sixth chemotherapy cycle. At the end of each week, they answered FACT-Lym. For this study, data were collected between February until October 2018.

3.3   Participants

Forty-seven patients were asked to participate in the study. Of these, 16 patients agreed to take part in the study. However, 4 participants were not included in the analyses due to the reduced biometric data. Six participants (50.0%) were female and 6 (50.0%) were male. Participants averaged 53.17 years old (SD = 16.73, range: 19–71 years). Concerning marital status, five (41.7%) participants were married, three (25.0%) were single, two (16.7%) were widowed, and two (16.7%) were separated. Four participants (33.3%) have the 4th grade, two (16.7%) have incomplete or complete 9th grade, four (33.3%) have incomplete or complete 12th grade, and two (16.7%) have higher education.

3.4   Questionnaires and Biometric Data Preprocessing

All domains of the FACT were standardized to 0–100 scale. The physiological sensor in the wearable device is a light source that calculates a RR. The signals were recorded wirelessly and features were extracted after preprocessing the heart rate series, which were created with the standard approach: considering only 5 min length.

Artifacts were removed from the signal using a "adaptive threshold for rejecting value differs from previous and following beats, and from a mobile mean more than a threshold value and removing points that are not within acceptable physiological values" [17].

Measurements described on Table 1 were calculated and it was only selected the ones that matched the dates covered by the questionnaires. All parameters in the selected range were averaged. Moreover, since the measure are calculated based on short-term (5-min) variations it may experience some correlation and so it was selected the HR, RMSSD and the SDNN to analyze.

4 Results

Considering the FACT scores, patients experienced lower scores on the Functional Well – Being domain and higher on the Social/Family Domains. Throughout the cycles, patients experienced a decrease of the Lymphoma Subscale, Physical Well-Being and Total in the first cycle regaining the QoL in the third and sixth cycles. FACT-G experienced slight decrease until the third cycle. The Emotional subscale presented an oscillatory trend and Social/Family decreased on the third cycle. The average of the HR did not appear to follow similar trend across the cycles as all the FACT domains scores. It is also important to notice the high variability of the Emotional Well-Being scores in the Third and Sixth cycle and in the Functional domain in the pre-treatment and first cycles. The Physical Well – Being has higher variability in the third cycle.

(Figure 2, Table 2).

Table 2. Mean and standard deviation per domain and per cycle of FACT score, HR, SDNN and RMSSD

Domain	Cycle	FACT Mean ± SD		HR Mean ± SD		SDNN Mean ± SD		RMSSD Mean ± SD	
Emotional well-being	PreTr	63.7	±12.43	78.6	±5.0	79.1	±19.5	81.1	±19.2
	1st	74.1	±13.28	80.1	±5.7	78.0	±14.2	78.6	±16.1
	3rd	69.7	±18.86	82.5	±6.0	80.5	±16.5	82.8	±20.3
	6th	78.4	±17.85	82.4	±5.8	83.4	±21.0	86.4	±24.5
Functional well-being	PreTr	58.7	±19.10	78.6	±5.0	79.1	±19.5	81.1	±19.2
	1st	53.6	±15.5	80.1	±5.7	78.0	±14.2	78.6	±16.1
	3rd	49.9	±13.2	82.5	±6.0	80.5	±16.5	82.8	±20.3
	6th	59.4	±12.1	82.4	±5.8	83.4	±21.0	86.4	±24.5
Lymphoma subscale	PreTr	74.0	±8.4	78.6	±5.0	79.1	±19.5	81.1	±19.2
	1st	68.1	±11.2	80.1	±5.7	78.0	±14.2	78.6	±16.1
	3rd	74.8	±16.2	82.5	±6.0	80.5	±16.5	82.8	±20.3
	6th	81.2	±12.1	82.4	±5.8	83.4	±21.0	86.4	±24.5
Physical well-being	PreTr	79.1	±12.9	78.6	±5.0	79.1	±19.5	81.1	±19.2
	1st	57.6	±15.0	80.1	±5.7	78.0	±14.2	78.6	±16.1
	3rd	72.9	±21.1	82.5	±6.0	80.5	±16.5	82.8	±20.3
	6th	79.9	±17.0	82.4	±5.8	83.4	±21.0	86.4	±24.5
Social/family well-being	PreTr	79.9	±19.1	78.6	±5.0	79.1	±19.5	81.1	±19.2
	1st	82.3	±15.5	80.1	±5.7	78.0	±14.2	78.6	±16.1
	3rd	74.4	±15.7	82.5	±6.0	80.5	±16.5	82.8	±20.3
	6th	73.1	±13.4	82.4	±5.8	83.4	±21.0	86.4	±24.5
FACT-G total score	PreTr	74.7	±9.5	78.6	±5.0	79.1	±19.5	81.1	±19.2
	1st	72.3	±10.7	80.1	±5.7	78.0	±14.2	78.6	±16.1
	3rd	70.7	±11.1	82.5	±6.0	80.5	±16.5	82.8	±20.3
	6th	78.1	±12.5	82.4	±5.8	83.4	±21.0	86.4	±24.5

(*continued*)

Table 2. (*continued*)

Domain	Cycle	FACT Mean ± SD		HR Mean ± SD		SDNN Mean ± SD		RMSSD Mean ± SD	
FACT-Lym	PreTr	71.6	±9.9	78.6	±5.0	79.1	±19.5	81.1	±19.2
trial outcome	1st	62.1	±7.8	80.1	±5.7	78.0	±14.2	78.6	±16.1
index	3rd	68.3	±14.6	82.5	±6.0	80.5	±16.5	82.8	±20.3
	6th	75.6	±10.7	82.4	±5.8	83.4	±21.0	86.4	±24.5
Total	PreTr	71.8	±10.0	78.6	±5.0	79.1	±19	81.1	±19.2
	1st	67.2	±6.2	80.1	±5.7	78.0	±14.	78.6	±16.1
	3rd	69.5	±11.5	82.5	±6.0	80.5	±16	82.8	±20.3
	6th	75.6	±9.7	82.4	±5.8	83.4	±21	86.4	±24.5

In the Physical Well-Being, Total, Lymphoma subscale and FACT-Lym Trial Outcome domains, the SDNN parameters followed the similar pattern exhibited by the domains scores across the cycles (Fig. 3, Table 2). The RMSSD parameter also exhibited similar pattern as the scores across the cycles in the FACT-Lym Trial Outcome Index, Lymphoma subscale, Physical Well-Being and Total (Fig. 4, Table 2). Biometric data exhibited high variability in the SDNN index in all chemotherapy cycles (Fig. 1).

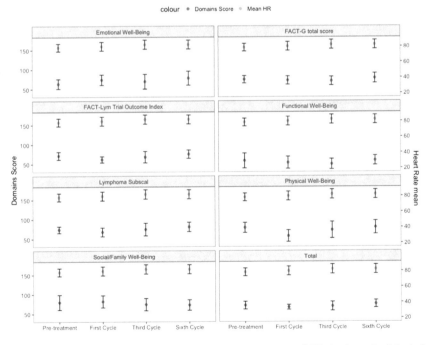

Fig. 1. Mean score of FACT (and standard deviation) and mean of HR (and standard deviation) per Domain and per cycle of chemotherapy

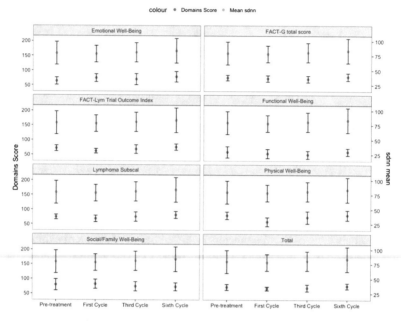

Fig. 2. Score of FACT and mean of SDNN per domain and per cycle of chemotherapy

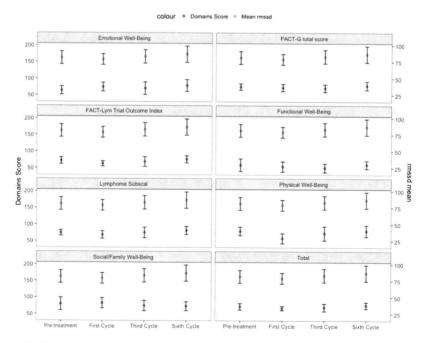

Fig. 3. Score of FACT and mean of RMSSD per domain and per cycle of chemotherapy

5 Discussion

The QLife+ Solution is an ongoing research, whose main goal is creating a new paradigm for the evaluation of HRQoL in clinical practice. The solution devised is based on an adaptive information system (IS), able to use physical and behavioral data of the patient, allowing continuous assessment of HRQoL that significantly reduce the questionnaires response time without affecting the patient's daily life. Continuous assessment of HRQoL will enable a better accuracy and more supported clinical decision [12].

In this paper, we focused to evaluate the existence of a possible relationship between self-perceived HRQoL domains and physiological data in terms of HRV metrics. Our results revealed that lymphoma patients experienced variations in HRQoL score throughout the cycles, being the Functional Well-being the domain with the lowest scores and the Social/Family with the highest. The first cycle presents as a particularly difficult moment reflected on the FACT-G, FACT-Lym Trial Outcome Index, Physical Well-Being, Lymphoma Subscale and Total.

In general, those results are in accordance with previous researches. In fact, it is common to verify a decline in Functional well-being both in lymphoma patients and survivors [18] that increases with the existence of additional chronic conditions. In comparison, Social/Family well-being has the highest values as found in a previous study [19]. Lymphoma patients commonly report a decrease in Physical Well-Being and overall HRQoL with treatment, as we have found with the lowest levels of well-being in the first cycle of chemotherapy. However, these effects may be reversed with physical activity which is recommended in cancer treatment [18, 20].

HRV seems to provide valuable information in the comprehension of pathological conditions in order to monitor the individuals' well-being [21]. In specific, higher HRV means a good physiological adaptation of the organism whereas lower HRV is a predictor of diseases or adverse events in patients with already diagnosed diseases [22]. Our results revealed that for Physical Well-Being, Total, Lymphoma subscale and FACT-Lym Trial Outcome domains, the SDNN and RMSSD parameters revealed a similar pattern as the domain scores. In specific, the FACT domains and the HRV metrics have the lowest average levels on the first cycle and seem to increase along the following cycles (3rd and 6th cycles). These results seem to suggest that individuals are more severely affected in the first cycle exhibiting lower HRV and lower self-perceived HRQoL. According to the literature, it is expected to found lower HRV in individuals during chemotherapy treatment. However, this effects "appeared to be reversible with treatment cessation" [21], which will be evaluated in our future studies, when the participants will have completed the follow-up evaluations. In addition, HRV values depend on the disease stage, being reduced in patients with the more affected disease. Accordingly, our future studies will integrate the patients' clinical data (e.g., clinical stage) to provide a more complete understanding of the HRV and HRQoL according to the disease. This study extends the current knowledge regarding the HRQoL in lymphoma patients by evaluating this construct through self-reported data and biometric information collected in real-time along the chemotherapy treatment.

Acknowledgements. This work is part of QVida+: Estimação Contínua de Qualidade de Vida para Auxílio Eficaz à Decisão Clínica, funded by European Structural funds (FEDER-003446), supported by Norte Portugal Regional Operational Programme (NORTE 2020), under the PORTUGAL 2020 Partnership Agreement.

References

1. World Health Organization. https://www.who.int/cancer/en/. Accessed 30 Nov 2018
2. Gonçalves, J., Faria, B., Reis, P., Rocha, A.: Data mining and electronic devices applied to quality of life related to health data. In: 2015 10th Iberian Conference on Information Systems and Technologies (CISTI), pp. 1–4. IEEE (2015)
3. SNS Homepage. https://www.sns.gov.pt/. Accessed 30 Nov 2018
4. United Nation. http://www.un.org/en/development/desa/population/publications/pdf/ageing/WPA2017_Highlights.pdf. Accessed 30 Nov 2018
5. Dürner, J., Reinecker, H., Csef, H.: Individual quality of life in patients with multiple myeloma. SpringerPlus **2**(397), 1–8 (2013)
6. Ebrahim, S.: Clinical and public health perspectives and applications of health-related quality-of-life measurement. Soc. Sci. Med. **41**, 1383–1394 (1995)
7. Lin, X., Lin, M., Fan, S.: Methodological issues in measuring health-related quality of life. Tzu Chi Med. J. **25**(1), 8–12 (2013)
8. Gresham, G., Schrack, J., Shinde, A.M., Hendifar, A.E., Tuli, R., Rimel, B.J., Figlin, R., Meinert, C.L., Piantadosi, S.: Wearable activity monitors in oncology trials: current use of an emerging technology. Contemp. Clin. Trials **64**, 13–21 (2018)
9. Arneric, S., Cedarbaum, J., Khozin, S., Papapetropoulos, S., Hill, D., Ropacki, M., Rhodes, J., Dacks, P.A., Hudson, L.D., Forrest Gordon, M., Kern, V.D., Romero, K., Vradenburg, G., Au, R., Karlin, D., Facheris, M.F., Fitzer-Attas, C., Vitolo, O., Wang, J., Kaye, J.A.: Biometric monitoring devices for assessing end points in clinical trials: developing an ecosystem. Nat. Rev. Drug Discov. **16**, 736 (2017)
10. Shaffer, F., Ginsberg, J.: An overview of heart rate variability metrics and norms. Front. Public Health **5**, 258 (2017)
11. Akhter, N., Gite, H., Rabbani, G., Kale, K.: Heart rate variability for biometric authentication using time-domain features. In: International Symposium on Security in Computing and Communication, pp. 168–175. Springer (2015)
12. Reis, L.P., Faria, B.M. Gonçalves, J.: QVida+: quality of life continuous estimation for clinical decision support. In: 11th Iberian Conference on Information Systems and Technologies (CISTI), pp. 1–7 (2016). IEEE (2016)
13. Webster, K., Cella, D., Yost, K.: The functional assessment of chronic illness therapy (FACIT) measurement system: properties, applications, and interpretation. Health Qual. Life Outcomes **1**, 79 (2003)
14. Kuntz, A.: The autonomic nervous system. Lea and Febiger, Reading (1953)
15. Garcia, C., Otero, A., Vila, X., Mendez, A., Rodrigues-Linares, L., Lado, M.: Getting started with RHRV, Version 2.0. (2014)
16. Kautzner, J., John Camm, A.: Clinical relevance of heart rate variability. Clin. Cardiol. **20**, 162–168 (1997)
17. Martinez, C.A.G., Quintana, A.O., Vila, X.A.: Tourine: Heart Rate Variability Analysis with the R package RHRV. Springer, Heidelberg (2017)
18. Leak, A., Mayer, D.K., Smith, S.: Quality of life domains among non-Hodgkin lymphoma survivors: an integrative literature review. Leuk. Lymphoma **52**(6), 972–985 (2011)

19. Hlubocky, F.J., Webster, K., Cashy, J., Beaumont, J., Cella, D.: The development and validation of a measure of health-related quality of life for non-Hodgkin's lymphoma: the functional assessment of cancer therapy—lymphoma (FACT-Lym). Lymphoma **54**(9), 1942–1946 (2013)
20. Courneya, K.S., Sellar, C.M., Stevinson, C., McNeely, M.L., Peddle-McIntyre, C.J., Friedenreich, C.M., Tankel, K., Basi, S., Chua, N., Mazurek, A., Reiman, T.: Randomized controlled trial of the effects of aerobic exercise on physical functioning and quality of life in lymphoma patients. J. Clin. Oncol. **27**(27), 4605–4612 (2009)
21. Kloter, E., Barrueto, K., Klein, S.D., Scholkmann, F., Wolf, U.: Heart rate variability as a prognostic factor for cancer survival–a systematic review. Front. Physiol. **9**, 623 (2018)
22. Lopes, P.F.F., de Oliveira, M.I.B., Max, S.: Aplicabilidade clínica da variabilidade da frequência cardíaca. Rev Neurocienc **21**(4), 600–603 (2013)

Big Data Analytics and Applications

Big Data Analysis in Supply Chain Management in Portuguese SMEs "Leader Excellence"

Fábio Azevedo[1] and José Luís Reis[2(✉)]

[1] ISMAI, Maia University Institute, Maia, Portugal
[2] Research Unit CEDTUR/CETRAD,
ISMAI, Maia University Institute, Maia, Portugal
jreis@ismai.pt

Abstract. With the market becoming increasingly competitive, companies are looking for ways to differentiate themselves from competitors, thereby increasing the interest of organizations in analysing large data and their potential benefits to Supply Chain Management (SCM). The objective of this study is to understand the involvement of the Portuguese SMEs "Leader Excellence" with the Big Data theme and their analysis as a function of the SCM, as well as to understand if these companies are in the same line as the companies of worldwide reference, in what concerns to the topic in question. For this study, a survey was carried and applied to 80 SMEs distinguished as "SMEs Leader Excellence" certified by IAPMEI. With this random sample, from the analysis of the results, it was possible to verify that the Portuguese SMEs are not yet at the level of the big world companies with respect to the use of Big Data in the management of the supply chain. The analysis of the study results also concluded that the greatest benefit of the use of Big Data analysis it is in operations and customer service, that the SMEs recognize the benefits of the Big Data analysis and are aware of their importance in the SCM.

Keywords: Big Data analysis · Logistics · Supply Chain Management · SMEs

1 Introduction

Market competitiveness and increasingly demanding and well-informed customers, force companies to seek new ways of distinguishing themselves from competition and providing better and better services. The maximum efficiency in Supply Chain Management (SCM) has become one of the ways in which organizations can stand out in a business world increasingly competitive [1]. With the technology increasingly developed and with companies "loaded" of data of sales, of purchases, of customers, began to appear references and publications to the analysis of this data, as well as its advantages for the SCM. In order not to lose competitiveness, companies were forced to invest in Big Data analysis software [2].

In this context, it is intended with this research to understand the involvement of SMEs Portuguese Excellence with the theme Big Data and its analysis as a function of

© Springer Nature Switzerland AG 2019
Á. Rocha et al. (Eds.): WorldCIST'19 2019, AISC 931, pp. 621–632, 2019.
https://doi.org/10.1007/978-3-030-16184-2_59

Supply Chain Management, as well as to see if these SMEs invest in associated Big Data analysis systems to SCM as the world's leading companies do.

2 Supply Chain Management

Supply Chain Management (SCM) is a term associated with logistics but is more comprehensive than logistics, going beyond the boundaries of organizations, as it involves relationships and partnerships between companies thus distinguish logistics management and SCM, since SCM involves a larger set of processes and functions [3].

Meeting the expectations of the final consumer and, if possible, overcoming it is one of SCM's main objectives, always aiming at reducing costs. It is the end consumer who has the power in the economy through their buying decisions. When a good or service is requested, the whole supply chain comes into action with the aim of providing the best possible service. While all actions in a supply chain aim to satisfy its own direct consumer, SCM's primary objective is to ensure that all actions have a full appreciation of how to satisfy the end consumer [1]. Taking these considerations into account, it is safe to say that this new organizational strategy is one way for companies to stand out from the competition, giving them a new vision of competition in the market focused on customer satisfaction and cost reduction [1, 4, 5]. However, some authors go even further arguing that SCM is the last means for companies to stand out from the competition and that competition is no longer between organizations, but rather between supply chains [6].

Pragmatically SCM is a set of approaches to effectively and efficiently interconnect the various suppliers, manufacturers and warehouses, so that the product is produced and distributed in quantity right, at the right time, in order to reduce costs and achieve the best level of service possible [7].

3 Big Data

The term Big Data has become synonymous with competitive advantage to the world's largest organizations [8]. Aware of its importance to achieve competitive advantage, well-known names such as Walmart, Amazon, Netflix, DHL among others, have invested in Big Data analysis programs in order to generate valuable knowledge for new ideas, new forms of value. The innovation, process transparency and operational optimization are some of the reasons that drive companies into this world of great data [9, 10]. In a study conducted by IBM Global Business Services, even show that operational optimization, risk management, new business creation, and improvement in human resources are the major goals of companies with the implementation of Big Data analysis systems [11].

Some authors argue that Big Data is a subject at the forefront of business, society, and the world at large, and has become, for information technology, the latest frontier of innovation [12].

3.1   Evolution of Big Data

Although it is a fairly recent issue, Big Data was first approached in 1997 by Cox and Ellsworth, NASA scientists, to refer to the challenge for computer systems to view "large" data sets [13]. With the development of technology and the broad growth of the internet, some authors have begun to focus attention on information produced by computers.

In 1998, John R. Mashey made a publication where he refers to Big Data (Big Data ... and the Next Wave of InfraStress). It is the first publication on the subject in question, however, the following year came a study of Bryson, Kenwright, Cox, Ellsworth and Haimes, named "Visually exploring gigabyte data sets in real time" with Big Data as the main theme [14]. This article has strengthened interest in the study of the subject and since then several new publications and approaches have emerged, and Douglas refers to the three dimensions of Big Data, namely volume, variety and velocity [15], and Normandeau says beyond this three dimensions is the issue of big data veracity [16].

There is no consensus regarding the definition of Big Data, however, several authors share their concepts based on the quantity, velocity and variety of data. Davenport refers to Big Data as a common concept for all data, which because of their volume can't be contained in a single repository, unstructured enough to fit an organized database of conventional way and fast enough to be stored in a static database [17]. Microsoft published in April 2013 skips some of the 3 V's - based definitions by setting Big Data as: "A term increasingly used to describe the application of true computing-the latest in artificial intelligence-to severely large and complex sets of information" [18]. In this definition, Microsoft refers to advanced technology as essential for working data, a reference that is "forgotten" by other authors and essential for the analysis of large data.

3.2   The 3 V's of Big Data

Data have always existed and have always been analyzed in a more or less complex way. The big difference between data and large data lies in three words: volume, velocity and variety. These three aspects make the great data unique and, thanks to the technological advances, the reach of anyone. Virtually everything that the individual does is recorded in some way, whether it is in shopping, car trips, phone calls or simple access to social networks [19].

If we can find correlations with great volume of data that can generate valuable insights, with the speed at which we have access to the data we can have information in real time or very close to it. This speed of data collection generates privileged information for a company, making it more agile. An example of this is transport and logistics companies such as DHL or UPS, among others, which use real-time route optimization systems to reduce delivery time and costs [20]. Variety, as the name implies, represents an abundance of data diversification.

Practically everything we do is registered in a certain way and the tendency is to increase, since the interest and bet on IoT (Internet of Things) is growing. Signals of mobile phone GPS, sensor readings, behavior in social networks and images of security cameras are just a few examples of sources of wide variety of data, whether structured or unstructured data [21].

4   Big Data in Supply Chain Management

Decision-making based on data analysis is not new in companies, in fact, had its peak of popularity in the mid-1980s [22]. However, with the market becoming increasingly dynamic and volatile, and with increasing consumer demand and the increase in electronic commerce, companies were forced to follow development by investing in Big Data analysis software to improve the decision, not only strategically but also operational [23]. With this, a new organizational paradigm was created, where less experience, expertise, and management practices are attributed, and more value to managers' ability to analyze data to generate insights that lead to competitive advantage [24]. Table 1 presents some aspects associated to the data sources that influence demand, operations, transportation and supply.

Table 1. Data sources that influence demand, operations, transportation and supply – adapted from [23]

Search	Warehouse operations	Transportation	Supply
News and Blogs	Bar code systems	Call records	Complaints data
Competitor's price	CRM data	ERP data	Delivery and collection records
Customer location	Customer questionnaires	Delivery times	Demand forecast
Purchase orders	Email registers	Equipment data	Facebook status
GPS	Intelligent transport system	Internet of things	Inventory in transit
Inventory costs	Inventory data	Local and global events	Local and global events
Loyalty programs	Machine data	Mobile location	Availability on the shelf
Source and destination data	Price range data	Product reviews	Product monitoring system
Public assessments	Public information available	Assessments and reputation (by third parties)	Raw material price volatility
Sales history	Supplier and customer capacity	Supplier financial information	Traffic density
Transport costs	Tweets	Warehouse costs	Meteorological data

The applications of insights from the Big Data analysis lie along the supply chain from purchase to the end consumer. Whether in the marketing area, with a view to better market intelligence, or in the logistics to optimize its operations, or in the sourcing itself (process to find the sources to meet the needs of the business), with the objective of segmenting suppliers and measuring the risk, all sectors gain advantages with Big Data, taking into account maximum efficiency and the best customer experience [25].

Companies working on the large data seek to measure supplier risk and the best options, predict demand for products or services, optimize inventory, observe capacity constraints, analyze workforce, optimize distribution across routes and schedules, alternative transportation, vehicle maintenance, analyze customer behavior, avoid unnecessary financial risks, among many other purposes. Among all previous applications, demand forecasting is of major interest to researchers [26, 27]. When a company can predict the exact demand, it manages the perfect management of warehouse, transport, optimizing and satisfying the needs of the final consumer, get loyalty.

5   SMEs Leader and Excellence - IAPMEI – Portugal

Portuguese SMEs have a large impact on the economy, representing 99.9% of the Portuguese business fabric [28]. For a company to get the designation micro the number of employees should be less than 10 and the turnover not exceeding two million, to get the designation small the number of employees should be less than 50 and the total turnover should be less than or equal to 10 million. An medium-sized company must have a total turnover not exceeding 50 million and the number of employees is less than 250 (Decree Law 372/2007).

The leader SME status, awarded by the Institute for Support to Small and Medium Enterprises and Innovation (IAPMEI), is a reputable brand that aims to distinguish the Portuguese SMEs superior performance and with better economic and financial indicators. Among the list of companies that have achieved SME Leader certification, the ones with the best performance obtain SME Excellence certification [29].

6   Investigation Methodology

In order to carry out the present research on the involvement of Portuguese SMEs Leader with the theme Big Data and their analysis in relation to SCM, a quantitative exploratory methodology was adopted in which the non-documentary data collection technique was used, made by inquiry, not direct, presenting clear, direct and objective questions in a written form, with the main objective of obtaining information. This technique was based on a questionnaire, where questions were presented in a closed, dichotomic and multiple response, range and evaluation or estimation, with several alternative responses, in order to answer the research questions formulated [30].

6.1   Sample Characterization

The focus of this research is Portuguese SMEs awarded the "SME 2017 Leader" status by IAPMEI, however, as in 2017 the number of companies recognized under this statute was 7 200, it was necessary to restrict the present investigation to a part of this universe. The companies that stood out from this list were chosen, with the attribution of the "Excellence SME 2017" status, also recognized by IAPMEI, in its total 1 948 companies [29]. Of these 1 948 companies, 500 were randomly selected to participate in this investigation.

The listing made available in Excel by IAPMEI, of free consultation, was used to obtain the information of the SMEs to be contacted. The 500 companies to contact were selected through the "rand between" functionality of Excel. This feature allows for a totally neutral and random draw.

6.2   Collection and Processing of Data

The questionnaire survey used is structured and divided into 18 direct and objective questions. The survey was adapted from the questionnaire used in the research "Going beyond the data: Achieving actionable insights with data and analytics" carried out by KPMG [23].

The survey structure comprises three divisions. The first division corresponds to the characterization of the company, the second is the analysis of Big Data in the Supply Chain Management of the company and the third and final division concerns the personal opinion of the respondent regarding the importance and benefits of this topic for the company. Two questions were also raised that should be in accordance with the criteria for awarding the SME Excellence Leader status, disregarding random answers and seeking the credibility of the questionnaire.

Before the survey was passed to the 500 selected companies, a "test" was made with 10 organizations in order to test the perceptibility of the same and to miss some possible error. Once everything was in place, the data collection process began. Between April 24 and May 30, 2018, the questionnaire was passed to all 500 companies by e-mail, in some cases the delivery and collection was done in person. Some companies were also contacted via telephone in order to raise awareness of the questionnaire.

7   Analysis and Discussion of Results

The results obtained through the quantitative methodology, using a questionnaire survey of 500 companies surveyed, as a data collection instrument, consists of a sample of 80 respondent SMEs, in the following points of this article, the results obtained according to the research questions are presented.

7.1   Relationship of the Size of the Company with the Use of Big Data

Considering that the first research question of this dissertation is to understand if the size of the company is related to the adoption or not adoption of Big Data analysis systems, it is necessary to know the size of the company. It was necessary to relate the answers to the three company statutes (micro, small and medium-sized companies), with answers that ask if companies use Big Data analysis software for the benefit of SCM.

After analyzing the results, it was verified that only 8% of micro-enterprises use Big Data analysis software, aimed at SCM. The percentage of small businesses is considerably higher, accounting for 25%. In the case of medium-sized companies, this figure reaches 71%. These data allow us to conclude that today the size of the company is clearly related to the adoption of Big Data analysis systems. One explanation for these results is that larger companies have more capable and more complex organizational structures and have other arguments that facilitate the adoption of these Big Data systems, exploiting the supply chain as much as possible. Nevertheless, the costs and difficulties associated with the implementation of this type of software are outlined by outsourcing. Already in 2001 a study by Thong concluded that smaller companies, to succeed in the implementation of Information Technology, should resort to outsourcing [31].

7.2   Big Data Analysis as a Differentiating Factor for the Successful SCM of SMEs

In the second research question, it is asked whether the adoption of Big Data analysis systems is a differentiating factor for the success of SMEs, taking into account the values obtained, as previously mentioned, in which only 8% and 25% of micro and small businesses, respectively, use this type of software. Meanwhile, the number of medium-sized companies reaches 71%. Since the sample refers to successful and reference companies in Portugal, distinguished as SMEs Excellence Leader, we can conclude that, despite the 71% in medium-sized companies, having Big Data analysis systems this may not be a factor differentiation for the success of Portuguese SMEs.

7.3   Relationship Between the Sector of Activity and the Territorial Scope of SMEs with the Use of Big Data

The third research question is divide in two parts; the relationship between the activity sectors and territorial scope and the use of big data, the results, as can be seen from the analysis in Fig. 1, shows that the companies in the transportation and logistics sector are the ones that most use Big Data analysis systems (53.8%). Industry SMEs have the second highest percentage with 45.8%, followed by the service area where 33.3% of the companies responded affirmatively to the issue. The retail, health and construction and furniture areas (22.2%, 12.5% and 11.1% respectively) are the ones that have the least companies making use of the Big Data analysis. It is important to note that no "other" SMEs sector has reported large data analysis to improve SCM.

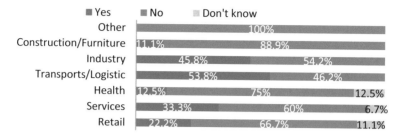

Fig. 1. Percentage distribution by activity sector of SMEs using Big Data

Regarding the second part of the third research question, the results are represented in Fig. 2. From the analysis of the data obtained, it is verified that 81.8% of SMEs with international operations have large data analysis systems.

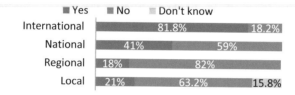

Fig. 2. Percentage of SMEs that use Big Data based on geographical distribution

Companies operating throughout the national territory, without international market, only 41% makes analysis of Big Data. Local companies have a greater percentage (21%) than regional companies (18%) in regards to the adoption of Big Data analysis software with SCM view. It should be noted that 15.8% of local businesses are unfamiliar with the issue at hand. The results presented, mainly in SMEs in the international market and national market, suggest that there is a relationship between the issue in question and the company's activity market.

According to the results obtained, it is verified that sector of activity and the territorial scope are related to the adoption of Big Data analysis systems. Regarding the sector of activity, the transport, logistics and industry sectors stand out clearly from the rest. This phenomenon is explained by the fact that some companies obtain more benefits than others with this analysis, as Swaminathan argues, even stating that in the area of logistics Big Data analysis is an essential element in competitive differentiation [32]. Of course, the construction area only accounted for 11%, since there is little data to explore compared to areas such as logistics or industry. Regarding the territorial scope, this significant difference in the results obtained is explained by the organizational structure of the company, since a company that acts internationally, needs to have more capacity to act in different markets.

7.4   The Size of the Company Influences the Choice Between Analyzing the Large Data Internally or in Outsourcing

From the answers obtained, it is verified that the outsourcing regime is clearly preferred by companies, representing 68% of all SMEs' respondents. This tendency is most likely based on the need to achieve ROI more quickly and not because of the need for major structural changes in the company, as well as not hiring data analysts [32].

In the answers obtained, no micro-enterprise has Big Data analysis controlled by the organization itself. It is perceived that it is very difficult and costly for a company with less than 10 employees to have a structure capable of supporting a department specialized in data analysis. In the responses of small companies, 83.3% use outsourcing. For medium-sized enterprises, this figure stands at 60%, certainly because they are companies of a different size and with other means.

From the analysis of the survey data we can conclude that the size of the company is clearly a preponderant factor in the choice between data analysis done by the company itself or in outsourcing. Restrictions on means, resources, knowledge in smaller companies explain this option by outsourcing. To the same conclusion reached Thong [31], in his research on resource constraints and implementation of information systems. With the presented data, we can also conclude that the bigger the company, the greater the predisposition to resort to the analysis of Big Data done internally.

7.5   The Need of Organizational Restructurings to Take Advantage of the Big Data Analysis

In order to understand if SMEs needed to make major organizational restructurings to take advantage of the Big Data analysis, the fifth question posed obtained the following results: - The training and hiring of information technicians, of analysts and data scientists, has 9.4% and 12.5% of responses by the companies addressed. This low percentage is due to the fact that most of the companies opt for outsourcing and for the possible existence of information technology departments, and 18.8% affirm that they have updated these systems.

With regard to the restructuring of decision-making processes, 18.8% of companies affirm this restructuring. The training of managers to use insights in decision making (40.6%), contracting entities external to the company (53.1%) and creation and/or improvement of data collection processes (40.6%) concentrate at the top of the actions most carried out by these SMEs. These results clearly demonstrate the companies' choice of outsourcing in the Big Data analysis.

Responding to the fifth research question, on the whole, companies do not need to make major organizational restructurings, however, organizations that choose to do a Big Data analysis internally logically need to implement more change than outsourcing firms. These findings are in line with the KPMG study [23], which states that large companies need to make restructurings to capitalize on the benefits of using Big Data.

7.6   Factors that Drives SMEs to Adopt or not to Adopt Big Data Analysis Systems

With regard to the sixth research question, which asks whether SMEs are adopting Big Data analysis systems or not, only 23% of companies replied that their main objective is innovation, with the vast majority (59%) responded to operational optimization. 10% of SMEs have put operational optimization and innovation at the same level. It should be noted that 8% answered not knowing what the main objective for the company's SCM, with the analysis of Big Data.

The responses concern about costs is clearly a priority for companies, accounting for 68.2% of the responses. Achieving best practices in general (36.4%) is the second most relevant item for these SMEs, while the management of the company's resources and the development of opportunities stand at 22.7%. With regard to achieving greater transparency in the processes, companies show less interest, totaling 9.1% of the answers.

The analysis of the findings shows what leads SMEs not to adopt this type of data processing software. The lack of knowledge on the subject in question (52.8%) is clearly the biggest barrier referenced by SMEs. The scarce information about data analysis software, as well as its advantages, may be the cause of this lack of knowledge. Data collection capacity appears to be the second largest barrier, with 37.5% of responses. Knowing which data are most relevant to collect and being able to apply the ideal solutions to interpret those same data does not seem to be the biggest obstacle for organizations, since they have only 23.6% and 22.2%, respectively, of answers.

7.7   SMEs Recognition of the Benefits of the Big Data Analysis for Their SCM

With the last investigation question, the intent is to understand what perception companies have concerning the benefits that Supply Chain Management can have with the Big Data analysis.

The results show that 53% of the SMEs surveyed are aware of the importance of the Big Data analysis in the SCM, since 53% of companies consider SCM to benefit greatly from data analysis and 33% reveal even if the company can benefit greatly. As we have seen in the previous results, a significant number of companies are not familiar with this topic, 9% of SMEs do not know if the SCM can benefit from data analysis and 5% believe that the company can benefit little.

8   Conclusions

The results presented in this study show that SMEs are confident that they can achieve more and better benefits for managing their supply chains using SCM big data analysis. Operational optimization is considered to be the main reason why companies use big data, and reducing operational costs is the main concern to justify their use. This study allowed us to conclude that the main reason why the number of SMEs performing the big data analysis is still low is due to a lack of knowledge about the topic. It has also been found that the part of the supply chain that achieves the greatest benefit is operations and customer service.

Concerning the recognition of the benefits of the Big Data analysis in the SCM, it is concluded that SMEs are aware of their importance, and most of them even have a strategy outlined for Big Data investment in the coming years.

Although the sample of respondents is not significant for the entire universe of Portuguese SMEs, it shows that companies still fall short of what big companies in the world do when it comes to the big data analysis, for the benefit of SCM. However, it was also possible to conclude that although they are not in the same line as these large companies, they are moving in that direction.

References

1. Quayle, M.: A study of supply chain management practice in UK industrial SMEs. Supply Chain. Manag.: Int. J. **8**(1), 79–86 (2003)
2. Waller, M.A., Fawcett, S.E.: Data science, predictive analytics, and big data: a revolution that will transform supply chain design and management. J. Bus. Logist. **34**(2), 77–84 (2013)
3. Cooper, M.C., Lambert, D.M., Pagh, J.D.: Supply chain management: more than a new name for logistics. Int. J. Logist. Manag. **8**(1), 1–14 (1997)
4. Stevens, G.C.: Integrating the supply chain. Int. J. Phys. Distrib. Mater. Manag. **19**(8), 3–8 (1989)
5. Castro Melo, D., Alcântara, R.L.C.: A gestão da demanda em cadeias de suprimentos: uma abordagem além da previsão de vendas. Gestão & Produção **18**(4), 809–824 (2012)
6. Christopher, M.: Logistics & Supply Chain Management, 4ª edn. Pearson, UK (2011)
7. Kaminsky, P., Simchi-Levi, D., Simchi-Levi, E.: Designing and Managing the Supply Chain: Concepts, Strategies, and Case Studies, 3ª edn. McGraw-Hill, New York (2003)
8. Manyika, J., Chui, M., Brown, B., Bughin, J., Dobbs, R., Roxburgh, C., Byers, A.: Big data: the next frontier for innovation, competition, and productivity (2011). https://www.mckinsey.com/business-functions/digital-mckinsey/our-insights/big-data-the-next-frontier-for-innovation
9. Halaweh, M., Massry, A.E.: Conceptual model for successful implementation of big data in organizations. J. Int. Technol. Inf. Manag. **24**(2), 2 (2015)
10. Loshin, D.: Big Data Analytics: From Strategic Planning To Enterprise Integration With Tools, Techniques, NoSQL, and Graph. Elsevier, Amsterdam (2013)
11. Turner, D., Schroeck, M., Shockley, R.: Analytics: the real-world use of big data in financial services. IBM Glob. Bus. Serv. **27**, 1–12 (2013)
12. Porter, M.E., Heppelmann, J.E.: How smart, connected products are transforming competition. Harv. Bus. Rev. **92**(11), 64–88 (2014)
13. Assunção, M.D., Calheiros, R.N., Bianchi, S., Netto, M.A., Buyya, R.: Big data computing and clouds: trends and future directions. J. Parallel Distrib. Comput. **79**, 3–15 (2015)
14. Bryson, S., Kenwright, D., Cox, M., Ellsworth, D., Haimes, R.: Visually exploring gigabyte data sets in real time. Commun. ACM **42**(8), 82–90 (1999)
15. Douglas, L.: 3D data management: controlling data volume, velocity and variety (2001). https://blogs.gartner.com/doug-laney/files/2012/01/ad949-3D-Data-Management-Controlling-Data-Volume-Velocity-and-Variety.pdf
16. Normandeau, K.: Beyond Volume, Variety and Velocity is the Issue of Big Data Veracity (2013). https://insidebigdata.com/2013/09/12/beyond-volume-variety-velocity-issue-big-data-veracity/

17. Davenport, T.: Big Data at Work: Dispelling the Myths, Uncovering the Opportunities. Harvard Business Review Press, Boston (2014)
18. Howie, T.: The Big Bang: How the Big Data Explosion Is Changing the World. Microsoft UK Enterprise Insights Blog (2013). http://blogs.msdn.com/b/microsoftenterpriseinsight/archive/2013/04/15/big-bang-how-the-big-data-explosion-is-changing-theworld.aspx
19. Wu, X., Zhu, X., Wu, G.Q., Ding, W.: Data mining with big data. IEEE Trans. Knowl. Data Eng. 26(1), 97–107 (2014)
20. DHL Trend Research: Logistics Trend Radar: Delivering Insight Today. Creating Value Tomorrow! (DHL Trend Research, Germany) (2016)
21. Ashton, K.: That 'Internet of Things' thing. In the real world, things matter more than ideas. RFID J. (2009). http://www.rfidjournal.com/articles/view?4986
22. Chen, H., Chiang, R.H., Storey, V.C.: Business intelligence and analytics: from big data to big impact. MIS Q. 36(4), 1165–1188 (2012)
23. KPMG Capital: Going Beyond the Data: achieving actionable insights with data and analytics (2014). https://assets.kpmg.com/content/dam/kpmg/pdf/2015/04/going-beyond-data-and-analytics-v4.pdf
24. Rowe, S., Pournader, M.: Supply Chain Big Data Series Part 1 (2017). https://assets.kpmg.com/content/dam/kpmg/au/pdf/2017/big-data-analytics-supply-chain-performance.pdf
25. Stamford, C.: Gartner Predicts Business Intelligence and Analytics Will Remain Top Focus for CIOs Through 2017 (2013). www.gartner.com/newsroom/id/2637615
26. Jun, S.P., Park, D.H., Yeom, J.: The possibility of using search traffic information to explore consumer product attitudes and forecast consumer preference. Technol. Forecast. Soc. Change 86, 237–253 (2014)
27. Sagaert, Y., Kourentzes, N., Aghezzaf, E. H., Desmet, B.: Sales forecasting with temporal big data: avoiding information overload for supply chain management. In: Informs International, Technology and Engineering conference, Hawaii, United States of America, 12–15 May (2016)
28. Instituto Nacional de Estatística: Empresas de Portugal 2016 (2018). https://www.ine.pt/xportal/xmain?xpid=INE&xpgid=ine_publicacoes&PUBLICACOESpub_boui=318224733&PUBLICACOESmodo=2
29. IAPMEI: PME Líder e PME Excelência 2017 (2018).https://www.iapmei.pt/PRODUTOS-E-SERVICOS/Qualificacao-Certificacao/PME-Lider.aspx
30. Hill, M., Hill, A.: Investigação por questionário. Edições Sílabo, Lisboa (2005)
31. Thong, J.Y.: Resource constraints and information systems implementation in Singaporean small businesses. Omega 29(2), 143–156 (2001)
32. Swaminathan, S.: The Effects of Big Data on the Logistics (2012). http://www.oracle.com/us/corporate/profit/archives/opinion/021512-sswaminathan-1523937.html

Big Data Analytics and Strategic Marketing Capabilities: Impact on Firm Performance

Research in Progress

Omar Anfer and Samuel Fasso Wamba[✉]

Toulouse Business School, Toulouse, France
o.anfer@tbs-education.org,
s.fasso-wamba@tbs-education.fr

Abstract. Big data analytics are currently considered as an important changer in the digital economy. However, the emerging literature on the topic has not yet looked at the ways to capture business value from big data analytics capabilities (BDAC). Based on resource-based view (RBV) and dynamic capabilities theory (DC) we propose a research model that assesses the impact of BDAC on firm performance (FP) in a turbulent environment (market and technological uncertainties). The study argues that this impact is mediated by the firm's adaptive marketing capabilities (AMC). To test the proposed model, mixed methods research approach (quantitative and qualitative) will be adopted. The key significance of this research study is that it seeks to explore the mechanism through which BDAC could lead to firm competitive advantage and thus enriching the literature on information technology business value.

Keywords: Big data analytics · Capabilities ·
Adaptive marketing capabilities · Strategic marketing · Business value ·
Value creation

1 Introduction

The availability of data represents new opportunities for firms to find new insights to enhance their business. Some scholars assume that big data analytics is the "next management revolution" [4]. The potential of big data technologies is limitless and can transform the way how firms do business.

Big Data is a new source of idea generation for product development, customer service, shelf location, distribution, dynamic pricing, and so on [5]. In a hyper-competitive marketplace where great ideas are easily copied, a firm must enhance its speed of idea generation (i.e., creative intensity) to achieve a sustainable competitive advantage [6]. Marketers are starting to recognize the potential power of big data as a new capital and that access to big data offers a firm new ways to differentiate its products [5]. Researches and practitioners highlight the role of big data analytics in the effectiveness of ads [7], Brand Management [8], prevision of tourist trends [9], prevision success of new movies [10], and improving the process of new product development [11]. This demonstrates the beneficial role of BDAC on marketing

© Springer Nature Switzerland AG 2019
Á. Rocha et al. (Eds.): WorldCIST'19 2019, AISC 931, pp. 633–640, 2019.
https://doi.org/10.1007/978-3-030-16184-2_60

function, and the impact of the use of BDAC on firms' marketing strategies. When firms analyze huge amount of data provided from different sources, social network, sensors, mobiles, web, and from different kind consumers', competitors' and providers' data, they may build a strong marketing strategy and enhance analytics to a strategical level. It's more true when organizations operate in an environment of uncertainties (market and technological) [3].

Indeed, the use of big data in marketing function could enhance organizational strategic marketing capabilities. The most widely accepted strategic marketing capabilities are new product development and proactive market orientation [12], those capabilities are high-order and cross-functional capabilities, which are result of other capabilities within organization. Companies that are better equipped to respond to market requirements and anticipate changing conditions are expected to enjoy long-run competitive advantage and superior profitability [13]. [14] Assume that marketing capabilities have a direct impact on firm performance; they suggest that complementary capabilities may help firms to both acquire a market-orientation and to more fully unlock their value creating potential.

For example, Netflix is using big data analytics to better understand their consumer behaviors and suggests the right content. Analyzing their data, Netflix can predict customer's needs, and with producing new content (Netflix Originals) the company is changing its business model moving from a streaming media company to movie producer company [15]. Lufthansa is using big data for renovating its business model and improving its customer's experience. "It wants to provide personalized services when they are least expected pre-travel, en-route, on arrival and post-travel" [16]. With selling ancillary services or advertising products, and helping customers use the time according to their own preferences, such as conducting their business, doing on-line shopping, booking activities at travel destinations [16] Lufthansa is more customer oriented.

Prior studies have demonstrated the impact of BDAC [17] on FP. This study follow this stream of research and argue that the combine effects of strategic marketing capabilities [18] and BDAC could lead to an improved firm performance (FP). We look at the indirect impact of BDAC on firm performance (financial, market, Overall) mediating by AMC. The research verify the moderating effect of environment, market and technological uncertainties [3] on relationships AMC/FP and BDAC/FP. More precisely, this study answer to this research question **"How big data analytics impact strategic marketing capabilities?"**. A conceptual model (Fig. 1) is proposed to demonstrate the existing relationship BDAC and AMC. This relationship has an impact on the FP.

This paper is organized as follow, first part defines BDAC, demonstrate different capabilities as major IT capabilities in the big data era. The second part, defines the strategic marketing capabilities and discuss the importance of AMC [19], the lack of dynamic capabilities and how BDAC can influence those capabilities in the marketing function. The third one describes the conceptual model and hypothesis. Forth part, describe research's methodologies. We conclude with expected results and managerial implications of the research.

2 Big Data Analytics Capabilities

As defined by [20] BDA is "a holistic approach to manage, process and analyze the 5 V data-related dimensions (Volume, Variety, Velocity, Veracity and Value) to create new insights for sustained value delivery, measuring performance and establishing competitive advantage". The 5Vs of BDA is largely explained and widely accepted as a key dimension of BDA by scholars and managers. This big data revolution [4] impacts business value by adding new BDAC [20] at a set of IT capabilities. By enhancing business value, firms can influence their performance [21]. According to IT capabilities, as defined by [21], [17] suggest three BDA capabilities categories: BDA infrastructure capability, BDA management capability, BDA personal capability. [22] identify three BDA capabilities: tangible capabilities, Human capabilities, and intangible capabilities.

Table 1. BDAC literature review.

Dimensions	Sub-Dimensions	Sources
Tangible	Data	[22]
	Technology	
	Basic resources	
Human	Managerial skills	
	Technical Skills	
Intangible	Data driven culture	
	Intensity of organizational learning	
BDAC infrastructure flexibility	Connectivity	[17, 21, 23, 25]
	Compatibility	
	Modularity	
BDAC management	Planning	
	Investment	
	Coordination	
	Control	
BDAC personal expertise	Technical knowledge	
	Technology Management capability	
	Business knowledge	
	Relational knowledge	

3 Adaptive Marketing Capabilities

Strategic Marketing Capabilities (SMC) is defined as "a firm-specific resource that is superior, rare, non- transferable, and idiosyncratic and its importance to strategic planning has been distinctive as it relates to acquiring and sustaining competitive advantages" [3]. Distinctive capabilities are important to support market position [12]. Marketing capabilities enhance the capacity of the firm not only on communication and promoting new products, but they can positively impact development new competitive products and services [10]. By mobilizing technological resources, the strategic marketing capabilities have positive impact on the product differentiation and cost advantage for better market performance [3].

When [19] diagnose the marketing capabilities gap, he identifies three types of capabilities, static, dynamic and adaptive, these capabilities create a sustained competitive advantage. Based on resources based view, [1] suggest that the resource and capabilities that give a competitive advantage is fixed by nature. [14] Suggest that market orientation liberate those capabilities to be more dynamics. Market orientation enhance interaction between market and firm Marketing capabilities to generate new market insights and reconfigure capabilities to align resources better than rivals [18]. Dynamic capabilities are defined as the capacity of the firm to rapidly adapt internal and external resources by integrating, building new and reconfiguring existing one to response to environment changes [2, 24, 26]. [19] Suggests that dynamics capabilities theory is lack by their inside-out perspective; it limits firm's ability to anticipate rapid market and become more resilient in front of volatility and complexity. To outdate this limitation, firms need adaptive capabilities to extend dynamic capabilities and enhance agility of organization [18]. Three AMC [19] was defined:

Vigilant market learning: The capacity of firm's learning from increasingly volatility and unpredictable market, it requires:

- A willingness to be immersed in the lives of current, prospective, and past customers and observe how they process data and respond to the social networking and social media space, without a preconceived point of view
- Open-minded approach to latent needs
- Ability to sense and act on weak signals from the periphery

Adaptive market experimentation: Small experiment that can generate new insights as long as that is a credible team to interpret and share the learning. It needs:

- Nurture an experimental mind-set
- Share insights and successful practices across organization
- Learning from peers and competitor's experiences

Open marketing: The capacity of the marketing function to benefit from the open network (Social media, data analytics, strategic partners …) sharing data to improve firm's resources.

Big data analytics allow firms to interact with their environment, external and internal, firms can have new insights from market, sense environment's threats and opportunities, and improve the understanding of the market. Implementing big data analytics can improve the agility, create or improve adaptive strategic capabilities.

4   Research Methodology

A mixed methodological research for developmental objective [31] will be used. To have a deep understanding of BDAC and AMC, operationalize and measure constructs, we will use a qualitative methodology which gives an exploratory way to more understand the phenomena and generate new theories. Semi-directional interviews of Managers (CMO, CIO and CEO) will be conducted to specify items which better define capabilities.

In parallel, for analyzing our model a survey will be conducted. The questionnaire was send to doctorate students and professors to validate and give us a feedback about the clarity and effectiveness of content. An online pilot survey is conducting on, http://surveymonkey.com/; More than 2500 Managers on LinkedIn was contacted to participate to the survey started from September 2018. Collecting answers for the pilot test (50 Completed responses) is in progress, the preliminary analysis to validate our constructs will be conducted at the end of this phase. At the last step a confirmatory study [32], quantitative methodology, will be conduct to collect approximatively 150 answers to be more effective.

Latent variables in our model: BDA technology capability, BDA management capability, BDA talent capability, vigilant market learning, adaptive market experimentation, open marketing, and their impact on financial and market performance, will be measured at the end of the research. To account the differences among organization, we include control variables (number of employees, activities, size, income, and localization). A seven-point Likert scale ranging from "strongly disagree" (1) to "strongly agree" (7) is used to measure constructs and for data analyzing we will apply PLS-PM for a regression modeling on XLSTAT or SmartPLS software.

5   Research Model

Based on Resources based view [1] and dynamic capabilities theory [2, 27– 30]. The proposed model describes the relationship between BDAC, AMC and FP. The study argue that there is an impact from BDAC and AMC, the model, suppose that the relationship between BDAC and FP is mediating by strategic marketing capabilities, and the environmental factors (technology and Market uncertainties) have a moderating impact on those relationships.

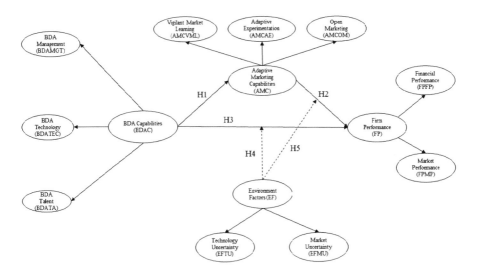

Fig. 1. Research model

We will try to define the relationship between BDAC and AMC, and verify the first hypothesis:

H1: Implementing BDAC in the marketing function has a positive impact on AMC

As a strategic Marketing capability, we will suggest that:

H2: AMC has a positive impact on firm performance

We will study the mediation role that AMC play in the indirect impact of BDAC on the firm performance (FP):

H3: AMC play a positive role for indirect mediating of BDAC on FP

Technological uncertainty is characterized by a short product life cycle and fast technological obsolescence, whereas high market uncertainty is characterized by significant changes, heterogeneity, and unpredictability of customer needs and market demands [3]. We suppose that:

H4: The impact of BDAC on FP is moderating by environment uncertainty (Technology and Market).
H5: The impact of AMC on FP is moderating by environment uncertainty (Technology and Market).

6 Discussion

This research aims to contribute to theoretical and practical finding. Theoretically, the study contributes to better accurate the AMC as strategic marketing capabilities and explain how to improve IT capabilities. Based on the dynamics capabilities theory we try to demonstrate that BDAC help firms to build marketing strategic capabilities to create value for organizations. The main goal is to seek to extend the work of [13] who define the capabilities of market-orientation and the impact of this strategic capabilities on the long term competitive advantage and with the identification of the AMC, [19] extends firm's capabilities to be more market-oriented, and try to explain how organization can combine external and internal resources to enhance AMC in the open market.

Extending literature of BDAC value creation. The model explains, like IT capabilities, that the application of BDAC is not enough to enhance firm performance. Through AMC, BDAC can contribute on a high-order marketing dynamics capability (new product development, proactive market orientation) [12] which are a cross-functional capabilities.

The practical finding, will contribute to the use of BDAC as a key IT capability that impacts the marketing function. The framework gives managers tool for evaluating their BDAC and how they can use those capabilities to improve firm performance through marketing function. IT capabilities, and specially BDAC can improve the understanding of the market, using the deluge of data, and enhance AMC to fit the need of an organization to be better market-oriented and for product success. We will contribute to the work of [17, 24, 25] to demonstrate the impact of BDAC on the firm performance by enhancing IT capabilities.

References

1. Barney, J.B., Clark, D.N.: Resource-Based Theory: Creating and Sustaining Competitive Advantage. Oxford University Press. Oxford **313**, 309–313 (2007)
2. Teece, D.: Dynamic Capabilities and Strategic Management. Knowledge and Strategy (1999)
3. Kim, N., Shin, S., Min, S.: Strategic marketing capability: mobilizing technological resources for new product advantage. J. Bus. Res. (2016)
4. McAfee, A., Brynjolfsson, E.: Big data. The management revolution. Harvard Bus. Rev. **90**(10), 61–68 (2012)
5. Erevelles, S., Fukawa, N., Swayne, L.: Big data consumer analytics and the transformation of marketing. J. Bus. Res. (2016)
6. Erevelles, S., Horton, V., Fukawa, N.: Imagination in marketing. Mark. Manag. J. **17**(2), 109–119 (2007)
7. Malthouse, E.C., Li, H.: Opportunities for and pitfalls of using big data in advertising research. J. Advert. (2017)
8. Liu, X., Burns, A.C., Hou, Y.: An investigation of brand-related user-generated content on Twitter. J. Advert. (2017)
9. Miah, S.J., Vu, H.Q., Gammack, J., McGrath, M.: A big data analytics method for tourist behaviour analysis. Inf. Manag. (2017)
10. Lash, M.T., Zhao, K.: Early predictions of movie success: the who, what, and when of profitability. J. Manag. Inf. Syst. **33**(3), 874–903 (2016)
11. Tan, K.H., Zhan, Y.: Improving new product development using big data: a case study of an electronics company (2017)
12. Barrales-Molina, V., Martínez-López, F.J., Gázquez-Abad, J.C.: Dynamic marketing capabilities: toward an integrative framework. Int. J. Manag. Rev. (2014)
13. Day, G.S.: The capabilities of market-driven organizations. J. Mark. (1994)
14. Morgan, N.A., Vorhies, D.W., Mason, C.H.: Market orientation, marketing capabilities, and firm performance. Acad. Manag. J. **51**(2), 315–334 (2008)
15. Fernández-Manzano, E.-P., Neira, E., Clares-Gavilán, J.: Data management in audiovisual business: Netflix as a case study. El Prof. la Inf. (2016)
16. Kauai, H., Herterich, M., Uebernickel, F., Brenner, W.: How Lufthansa capitalized on big data for business model renovation. MIS Q. Exec. **1615**(14), 299–320 (2017)
17. Akter, S., Wamba, S.F., Gunasekaran, A., Dubey, R., Childe, S.J.: How to improve firm performance using big data analytics capability and business strategy alignment? Int. J. Prod. Econ. (2016)
18. Hult, G.T.M., Ketchen, D.J.: Does market orientation matter? A test of the relationship between positional advantage and performance. Strateg. Manag. J. **22**(9), 899–906 (2001)
19. Day, G.S.: Closing the Marketing Capabilities Gap. J. Mark. (2011)
20. Wamba, S.F., Akter, S., Edwards, A., Chopin, G., Gnanzou, D.: How 'big data' can make big impact: findings from a systematic review and a longitudinal case study. Psychol. Lang. Commun. **16**(2), 1–33 (2015)
21. Kim, G., Shin, B., Kwon, O.: Investigating the value of Sociomaterialism in Conceptualizing IT capability of a firm. J. Manag. Inf. Syst. (2012)
22. Gupta, M., George, J.F.: Toward the development of a big data analytics capability. Inf. Manag. (2016)
23. Ji-fan Ren, S., Fosso Wamba, S., Akter, S., Dubey, R., Childe, S.J.: Modelling quality dynamics, business value and firm performance in a big data analytics environment. Int. J. Prod. Res. **55**(17), 5011–5026 (2017)

24. Fosso Wamba, S., Gunasekaran, A., Akter, S., Ji-fan Ren, S., Dubey, R., Childe, S.J.: Big data analytics and firm performance: effects of dynamic capabilities ☆. J. Bus. Res. **70**, 356–365 (2017)
25. Garmaki, M., Boughzala, I., Fosso Wamba, S.: The effect of big data analytics capability on firm performance. ICIS-RP **5**(7), 1–10 (2016)
26. Greenley, G.E., Hooley, G.J., Rudd, J.M.: Market orientation in a multiple stakeholder orientation context: implications for marketing capabilities and assets. J. Bus. Res. (2005)
27. Katkalo, V.S., Pitelis, C.N., Teecey, D.J.: Introduction: on the nature and scope of dynamic capabilities. Ind. Corp. Chang. (2010)
28. Teece, D.J.: Explicating dynamic capabilities: the nature and microfoundations of (sustainable) enterprise performance. Strateg. Manag. J. (2007)
29. Teece, D., Leih, S.: Uncertainty, innovation, and dynamic capabilities: an introduction. Calif. Manage. Rev. (2016)
30. Winter, S.G.: Understanding dynamic capabilities. Strateg. Manag. J. (2003)
31. Venkatesh, V., Brown, S.A., Bala, H.: Bridging the qualitative-quantitative divide: guidelines for conducting mixed methods research in information systems. MIS Q. (2013)
32. Teddlie, C., Tashakkori, A.: Foundations of Mixed Methods Research. Integrating Quantitative and Qualitative Approaches in the Social and Behavioral Sciences, Los Angeles Sage, vol. 20, no. 1, pp. 183–202 (2009)

Study of the Mobile Money Diffusion Mechanism in the MTN-Cameroon Mobile Network: A Model Validation

Rhode Ghislaine Nguewo Ngassam[1],
Jean Robert Kala Kamdjoug[1(✉)], Sylvain Defo Wafo[2],
and Samuel Fosso Wamba[3]

[1] Université Catholique d'Afrique Centrale, FSSG, GRIAGES,
Yaoundé, Cameroon
ghislainengassam5@gmail.com, jrkala@gmail.com
[2] MTN Cameroon, Douala, Cameroon
[3] Toulouse Business School, Université Fédérale de Toulouse Midi-Pyrénées,
20 Boulevard Lascrosses, 31068 Toulouse, France
s.fosso-wamba@tbs-education.fr

Abstract. With a high rate of subscribers to mobile and a low rate of bank account, mobile money is a clear opportunity for Cameroonians to better manage financial transactions and risks. While in certain countries it has taken less than two years to be adopted massively, its diffusion among mobile subscribers in Cameroon is still very slow, with less than 5% regular users four years after being launched.

This work explores the impact of the mobile subscriber network on mobile money and tries to identify the network diffusion mechanism of mobile money through a study of subscribers whose adoption behavior was observed during 14 weeks. From different analysis, we confirm the influence of mobile money users on new adopters. However, the decision to use mobile money is not directly related to the number of mobile money users or of potential users. This last point gives room for further research.

Keywords: Mobile money · MTN Cameroon · Diffusion of innovation ·
Social network diffusion model

1 Introduction

While developing countries still record debilitating banking rates, mobile banking is proving to be a way to emerge [1]. Banking broadly refers to the access of population to financial services. Banking rate is an indicator of development and a lever for socio-economic development [2], as it reflects the level of confidence in the economy and the degree of business open-mindedness. Indeed, studies show that greater access to the financial system can stimulate job creation, increase investment in education, and directly help poor people to manage risk and absorb financial shocks [3].

Mobile banking in Cameroon is expanding significantly at a time when the banking rate is below 20%. As a definition, mobile banking is the use of mobile banking

© Springer Nature Switzerland AG 2019
Á. Rocha et al. (Eds.): WorldCIST'19 2019, AISC 931, pp. 641–653, 2019.
https://doi.org/10.1007/978-3-030-16184-2_61

services, and more precisely the use of mobile money. Taking advantage of enabling factors such as the current exponential mobile phone penetration rate, and an impressive number of subscribers (more than 17 million for 22 million inhabitants in 2014 [4]), mobile money is steadily gaining ground in Cameroon.

However, it appears that the evolution of mobile money subscriptions in Cameroon remains quite below than that observed in countries such as Kenya, which after only two years, registered 40% [5] of mobile banking subscribers, against 20% in Cameroon after 4 years. Numerous studies have looked at Mobile Money as a technological innovation whose diffusion follows models based on the work of Everett Rogers [6], with the famous curve of adoption over time; although these studies have regularly established the important effect of the social environment on diffusion, they only give an overall measure of the phenomenon. However, as Mobile Money adopters are generally the available mobile telephone subscribers and given that this network always tends to be used in duality, a macro-explanation of the evolution of the phenomenon is limited and does not allow for developing a holistic package of effective measures on individuals.

To better understand Mobile Money's distribution mechanisms, and to cope with the challenges of accelerating its expansion, it is important to understand how mobile money is channeled across the mobile network. Otherwise it will be difficult to replicate and effectively promote this solution. In an attempt to answer the aforementioned concerns, this paper aim to explore how Mobile Money is distributed among mobile subscribers.

Based on the theory of diffusion of innovations [7], we will see whether the structure of subscribers' networks influences Mobile Money adoption, using a dynamic panel approach and the following hypothesis: **the mobile money is channeled across the network of mobile subscribers according to the linear model of thresholds of adopters' influence (threshold Model).**

The next section makes a summary description of the dynamics of mobile money in Cameroon, followed by a discussion of the main theories on in-network diffusion, the presentation of the implemented methodology, the presentation and discussion of the results obtained, the managerial and theoretical implications, research limitations, directions for future studies and, finally, the conclusion.

2   Mobile Money in Cameroon

Mobile money is a tool that allows individuals to make financial transactions using cell phone technology [5]. This term also includes all services associated with mobile money (also called electronic money) that are deemed mobile financial transactions. These include mobile banking, mobile payment and money transfer (Fig. 1).

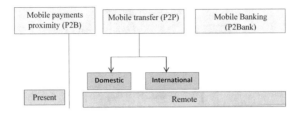

Fig. 1. Structure of financial services market. Source: [8]

2.1 Value Chain

The value chain of mobile money services can be summarized according to the main links presented in the Fig. 2 below.

Fig. 2. Mobile money value chain (adapted from [9])

Depending on the value chain, it is clear that the base of mobile money services is the network, which explains why telecommunications operators have a strategic advantage in the deployment of mobile money. Moreover, it presents the different operations made possible by mobile money.

On the one hand, there are transactions that convert conventional money into electronic money. It is at this level that banking partners intervene through several mechanisms ensuring and guaranteeing this convertibility. In such an environment, financial transactions (move money) refer to those achieved in the system and which involve account holders. These include payments and money transfers. On the other hand, there is a savings function. Once a deposit is made in an account, it can be managed like any bank savings.

2.2 Actors

Many players including the mobile operator must come into play to ensure a successful implementation of MM services. As in Fig. 3, there are five of them, all of whom can be divided into two groups:

The regulatory function actors which contains the financial regulators and mobile network regulator. The financial regulators are COBAC and BEAC, while ART and ANTIC are those of the telecommunications sector;

The operational function actors as banks, mobile money operators, and others operators in financial which are involved.

There are fifteen (15) banks operating in the Cameroonian context which are in partnership with mobile money service providers. Three mobile money service

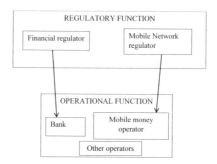

Fig. 3. Mobile money actors (Mujuru 2013).

providers are active, namely MTN (sponsored by Afriland First Bank), Orange (sponsored by BICEC) and Express Union, a microfinance institution. Other operators refer to the network and the agencies used by operators.

Mobile money really took off in Cameroon in 2016, when a new Minister of Posts and Telecommunications came in, implemented user identification measures and enabled Orange and MTN (telephony operators) to launch Orange Money and MTN Mobile Money (with secured accounts for users).

A study of the determinants of the adoption of mobile money based on the UTAUT theory, as well as a study of data mining on more than 180 variables, shows that the social influence appears as the main determinant of mobile money adoption.

3   Literature Review

3.1   Diffusion in the Networks Theories

The study conducted by Eduardo, H. D., João, P. d., and Adrian, K. C [9], which carried out a literature review on mobile money works, shows a significant shortage of studies focused on customer databases. As a result, the context of diffusion of mobile money is difficult to understand and even to reproduce elsewhere.

It is against such a backdrop that this work is also aimed at exploring theories and methods that will allow us to fill this gap.

3.1.1   Social Network Analysis

The social network analysis is an interdisciplinary methodology that was first developed by sociologists and psychologists, and then adopted in the fields of mathematics, statistics and computer science, and finally disciplines such as economics and marketing. The analysis of social networks is based on the hypothesis that relations between the units of a network are of paramount importance [10]. As part of its implementation, there are two measures: (1) the so-called local measures, which characterize a vertex or a link: degree, proximity, intermediary, etc.; and (2) global measures, which characterize the network as a whole (density, diameter, order, etc.). Thus, four types of networks can be identified from two main criteria: (1) the weighting of the links, which makes it possible, if necessary, to associate a positive quantitative value with each link existing between two vertices; (2) oriented or non-leading links in relation to oriented networks or not.

The social network analysis can be assimilated to descriptive statistics when we observe the set of aggregated measures and the development of the data collected for each summit. But it is more complex than other methodologies in that several levels of analysis are integrated.

3.1.2   Diffusion of Innovation Theory

According to Valente [7], the diffusion theory of innovation tries to explain how new ideas and practices spread among communities. Rogers [11] is of the opinion that this theory sees innovations as being communicated through certain channels over time and within a particular social system. Individuals are seen as possessing different degrees of willingness to adopt innovations, and thus it is generally observed that the portion of the population adopting an innovation is almost normally distributed over time. Indeed, Rogers [12] distinguishes four main elements in the dissemination of new ideas: (1) innovation, (2) communication channels, (3) time, and (4) the social system. Also, four essential dimensions: (1) generally, the speed of diffusion of an innovation can be defined by its relative advantage, compatibility, complexity, probability and observability; (2) Communication is the process by which participants create and share information with each other in order to achieve mutual understanding; (3) the time factor is included in the diffusion by acting on three factors: the decision, the innovation, and the rate of adoption; and (4) the fourth essential element in the diffusion of new ideas is the social system. It is a set of interdependent units that solve problems to achieve a common goal.

3.1.3   Main Approaches to Implementing the Diffusion of Innovation in a Network

The diffusion of innovation in social networks is underpinned by a combination of the theory of diffusion of innovation and the analysis of the said social networks, the objective of which is to explore their various advantages. This method focuses on the principles of dissemination of innovation, which seeks to establish the mechanisms thereof in the network and explain how innovation is spread across such an environment. For example, social media outreach attempts to explain how new ideas and practices (innovation) are intertwined within and between communities. Explaining the different features of innovation looks more complex than for a mere contagion, because there are thresholds that must be overcome in order to trigger the expansion of innovation within the network. In previous studies, these theories have been used individually, though their complementary enables a full understanding of the problem. There are two main categories of mathematical models associated with this theory: (i) cascade models inspired by probability theories; and (ii) threshold models, which are widely used in sociology.

At the initial moment ($t = 0$), the innovators are the only active summits at time t ($t > 0$), each vertex that was active during the previous period has only one chance to activate its neighbors, when they are not yet active, according to a given probability; the different activated neighbors are mutually independent. There is also two possibilities: (1) at the end of this period, the activated neighbors repeat the process of the previous step during the period t + 1. In this model we do not take into account cumulative activation attempts, but we consider the probability that an active vertex has to bring about change in its non-active neighbor. However, this training probability may differ depending on whether the non-active neighbor has previously been exposed to change as a result of an activated vertex. As before, at the initial moment (t = 0), the innovators are the only active summits at the time t (t > 0), each vertex which was active at the preceding period becomes a pressure agent on all its neighbors that are not yet active, as the different activated neighbors are mutually independent, but in terms of performance these neighbors must exceed the threshold in order to become active. At the end of this period, the activated neighbors repeat the process of the previous step during the period t + 1. Unlike the cascading model, this model takes into account the cumulative effect to cause the change on a non-active vertex of the overall active neighboring vertices.

We will use in this paper this last approach according to Everett Rogers works on the diffusion theory.

4   Methodology

Our working methodology mainly consisted in analyzing customer data from the MTN telephone network. It followed two main stages, namely the constitution of the database and the complementary verification of hypotheses.

4.1   Constitution of the Database

To build the database, we used filters at three levels: (1) Time: we constituted a dataset in three months (from 01 August to 31 October 2016), which is the period during which a telephone company can decide to deactivate a dormant subscriber. (2) The service provider: only MTN subscribers were selected based on telephone number identifiers (prefixes) allocated to that operator[1]: 680, 681, 683, 67, 650, 651, 652, 653, and 654. (3) The last level of filter was the constitution of a relational network around each user of mobile money. Therefore, we considered accounts involved in SMS or call-type transactions with mobile money subscribers registered before the end of the observation period, while ensuring that the number of transactions is greater than 2 (in order to avoid errors due to sporadic actions). However, a relationship likely to influence adoption during the observation quarter should have taken place during the three-month period. Thus, we set a static diagram of the relational network and made a dynamic view of adoption statuses from one week to another. Afterwards, we divided data into 3 categories and fourteen periods (weeks): (1) The first period was established for those accounts that were active before the 1st August 2016; (2) From the second to the fourteenth period, we had accounts that become active between the 01st August and the 31 October 2016; and (3) The last group is concerned with accounts that did not subscribe to mobile money until the end of the observation period. From the large

---

[1] Prefixes obtained following removal of the country code.

database obtained, we applied a probabilistic sampling method. We used a stratified survey to keep the specificity of each region of the country.

After the database constitution, we investigated the sensitivity of the data sample mainly with respect to size, strata structure, and processing time in order to test the chosen sampling method.

4.2   Verification of Hypotheses

General Hypothesis: Impact of the Social Network of Subscribers on Mobile Money Diffusion. We first sook to establish a relationship between old mobile money users and other phone subscribers who are potential mobile money users. The UTAUT (Unified theory of acceptance and the use of a technology), which demonstrates that an individual's social environment can significantly affect their intention to adopt a technology [13], is at the core of the above-mentioned relationship.

By transposing this principle to the adoption of mobile money, we can formulate the following hypothesis:

H1: Mobile money users significantly influence other subscribers' acceptance of mobile money in a social network. In other words, mobile subscribers tend to imitate the mobile users of their network by adopting mobile money.

In order to verify this hypothesis, a Chi-square test was performed on two variables to observe the behavior of non-users of mobile money at the beginning of the period concerned: (i) proximity to mobile money users at the beginning of the observation period or during the intermediate periods; (ii) the adoption or not of mobile money at the end of the observation period.

In our case, proximity to mobile money users will be evaluated by the existence or not of a relationship between single phone subscribers and former mobile money users by the end of the observation period. Thus these two variables are dichotomous because they take only two possible values.

The chi-square test was developed by Karl PEARSON, and is used in particular to test the independence between two variables, at least one which is qualitative. In this case, it makes it possible to compare observed numbers with theoretical numbers in case of independence.

To implement this test (using the R software), the following statistic had to be qualified:

$$\chi^2 = \frac{(f_o - f_e)^2}{f_e} \tag{1}$$

Where f_o is the observed frequency, f_e is the theoretical frequency, and χ^2 is the chi-square.

This test fits well in our case because we have two qualitative variables. The null hypothesis in this case will be: (Ho) Proximity to users is independent to adoption at the end of the observation period. When the p-value obtained is less than 5%, this hypothesis is rejected; but otherwise it is accepted. We prepared the sample for this test by building the following Table 1:

Table 1. Contingency table of the representative study population.

| | | Existence of a link with a mobile money subscriber at the beginning of the observation period or before the adoption | |
		Yes	No
Adopters of mobile money at the end of the observation period	Yes		
	No		(deducted value)

Specific Hypothesis: Diffusion Mechanisms Between Mobile Money Users and their Neighbors. While the previous hypothesis suggested a causal relationship, the aforementioned study by Valente [7] allowed for establishing the mechanism of diffusion.

Therefore, the following hypothesis is being set forth:

H2: Mobile money is diffused across the network of mobile subscribers according to the model of thresholds of adopters' influence (threshold model).

This hypothesis was verified by means of the netdiffuseR structure test, with the netDiffuseR package integrated into the R software application. This package was used to perform a descriptive network analysis and to develop a test of adequacy for the adopters threshold diffusion model. The null hypothesis in this case is:

(Ho) The observed diffusion within the network follows the process of the threshold model.

The obtained p-value had to be compared against the acceptable error threshold, which in our case was 5%. Thus, the null hypothesis is supported if the p-value is less than 5%. otherwise, it is rejected.

We prepared a sample by creating (1) the adjacency matrix, and (2) the diffnet.

5   Results and Discussion

Our sample was based on a panel of subscribers with telephone numbers. These individuals are very active in the GSM plan as they could contact in average up to fifty (50) persons during the observation period.

The graph below describes mobile money transactions (Fig. 4).

At the end of the observation period, our sample for mobile money adoption appeared as follows:

The graph on the right shows the distribution of adopters (19%) over the 13-weeks observation period and the week before the observation period (Fig. 5).

The graph below describes the distribution of new adopters by period and by proximity threshold with mobile money users (Fig. 6).

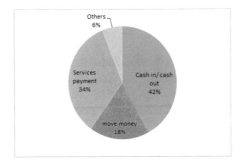

Fig. 4. Transactions structure

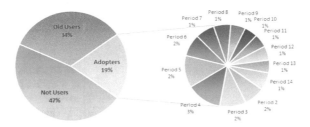

Fig. 5. Subscribers' status.

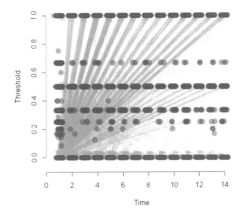

Fig. 6. Adoption time and proximity thresholds with users in the network.

In accordance with the principle of the model, the thresholds are calculated by comparing the number of neighboring users and the total number of neighbors connected to this summit.

In this graph, we have three (3) dominant thresholds and two other that are less important: (1) "0%" for the adopters who have no links with actual mobile money users; (2) "50%" for the adopters whose half of their relationships are mobile money users; and (3) "100%" for the adopters whose most of their relationships are mobile money users.

5.1 Verification of the General Hypothesis

With 346 individuals, we have the following contingency Table 2:

Table 2. Contingency table of the representative study population.

		Existence of a link with a mobile money subscriber at the beginning of the observation period or before the adoption	
		Yes	No
Adopters of mobile money at the end of the observation period	Yes	14	5
	No	47	280

The results obtained by the chi-square test in R software are:

$$\chi^2 = 39.51 \, df = 1; and \, p - value = 3.264 \times 10^{-10} \tag{2}$$

As the p-value is inferior to 5% and is almost null, we rejected the null hypothesis (no link between the adoption of mobile money and the existence of mobile interaction with previous mobile money users.

A near-to-zero p-value implies that the adoption of mobile money by potential users will be strongly influenced by their relationship with actual mobile money users.

5.2 Verification of the Specific Hypothesis

The test is implemented by calculating the statistics from the simulation of the network evolution alternatives in case certain active actors are removed from the diffusion process.

The simulation is repeated a large number of times, in our case 200 times (Table 3).

To validate the test, the p-value should be below 5%. In our case, p-value = 29%. This means that if we reject the null hypothesis of non-existence of a dependency structure, it would be with a risk of being wrong by 29%. This is why we will admit

Table 3. Result of the diffusion structure test.

Number of Simulations: 200 Number of time periods: 14	H0: E [beta(Y,G) / G] – E [beta(Y,G)] = 0 (no structure dependency) E[beta(Y,G) / G] (observed) = 0.1029 E[beta(Y,G)] (expected) = 0.1015 p-value = 0.2900

that there is no clear dependency structure governing diffusion according to the threshold model.

The obtained results is only half consistent with the theory of diffusion of innovations [7, 12] because only one of the two hypotheses was supported.

As a reminder, this study aimed to analyze the diffusion mechanism of mobile money in the MTN network. At the end of the study, we have confirmed the general hypothesis concerning the link between adoption of mobile money and influence by previous users, but the specific hypothesis (mobile money is not diffused across the network of mobile subscribers according to the model of thresholds of influence of adopters in the network) was not confirmed.

Although the general hypothesis has been maintained and confirmed by several previous studies [14], it should be noted that there are about 26% of mobile money adopters who did not have any contact with users before. As a result, other dissemination channels that could significantly impact the expansion of mobile money should be identified. They may include advertising, systematic enrollment during new registrations to the network, among others.

The specific hypothesis aimed to demonstrate whether new mobile money adoptions across a network are influenced by the contact with those already using mobile money. In other words, does the relationship with a mobile money user cause people to adopt mobile money? The results obtained may be in contrast with GSM products, where the number of contacts of a subscriber influence his/her degree to using the service [15]. Thus, results of this study and those of previous studies have established the influence of social network relations in the adoption of mobile money, but also they have shown that the linear model of threshold does not explain the adoption mechanisms within a subscribers' network. This may be explained, as indicated by [16], by that fact that a significant number of adopters in a subnet is necessary to trigger large-scale adoption. Whatever the case may be, complementary studies with different methodological tool variants are welcome, as are also other studies on different fields of study with related or quite distinct data.

6   Implications, Limits and Future Directions

As a theoretical implication, this work broadens the literature on the spread of mobile money by focusing on a study based on actual customer data, unlike previous studies that were either case studies or survey-based studies. In addition, this article highlights a quantitative approach to the evaluation of diffusion models, which is very unusual in the existing literature, with the possible explanation of the recency of evaluation tools such as the Netdiffuse R package that we used to test our hypothesis.

In practical terms, the results of this work could help decision-makers in telephone companies to better guide marketing actions to boost the expansion of mobile money; in particular, by finding incentive techniques that would encourage traditional users to join their loved ones, because this is one of the main reasons for adoption: 74% of people have adopted this technology when they contacted mobile money users.

According to Fallah and Luo work [17], which focuses on a deeper understanding of the needs of customers before action, our paper provides direction for examples of customer segments that might be of interest as part of the marketing actions to be carried out by MTN or any other company: (1) Non-adopters in contact with actual mobile money users; (2) Adopters with no contact with actual mobile money users; (3) Non-adopters with no contact with actual mobile money users; and (4) Adopters in contact with actual mobile money users.

As a limit, this work highlights a single determinant of mobile money adoption, out of many others that could impact expansion. Future studies could exploit the trail of models such as UTAUT or other information system models to find the explanation to the 26% of adopters who did not need contact with users. Another glaring limitation is the fact that the study focuses on data from a single mobile money provider while, as stated earlier in our work, there are two other companies offering the same product in the Cameroonian context. Taking into account the subscribers of these different companies would certainly make it possible to highlight more conclusive models.

Moreover, the rejection of our theoretical hypothesis suggests that another model would explain the mechanism of diffusion of mobile money in Cameroon; and therefore, it would be appropriate in the future to look at that mechanism through a model such as the other cascaded mathematical model that is being referred to hereinabove.

7 Conclusion

This study was intended to explain the spread of mobile money in the network of mobile subscribers. Our postulates were a general hypothesis and a specific hypothesis, both of which were assessed using the chi-square. Results indicated that the social link has a great impact on mobile money adoption, but other driving factors may exist. They also explained the significant share of adopters who have not had prior contact with mobile money subscribers.

However, diffusion according to the threshold model of adopter's influence was not confirmed. This could be explained by the fact that a large proportion of subscribers do not use the service for P2P transactions, but rather for service payments (bill, insurance, etc.). Nonetheless, the mixed nature of our results does not constitute a limitation, but it gives room to more in-depth research perspectives.

References

1. Mago, S., Chitokwindo, S.: The impact of mobile banking on financial inclusion in Zimbabwe: a case for Masvingo Province. Mediterr. J. Soc. Sci. **5**(9), 221 (2014)
2. GSMA: State of the industry: Mobile financial services for the unbanked, United Kingdom, London: Groupe Speciale Mobile Association (2014). http://www.gsma.com/mobileforde velopment/wp-content/uploads/2015/03/SOTIR_2014.pdf
3. Munyanyi, W.: Banking the unbanked: is financial inclusion powered by Ecocash a veracity in rural Zimbabwe. Greener J. Bank. Financ. **1**(1), 001–009 (2014)
4. ART: Observatoire annuel 2014 du marché des communications, in Agence de Regulation des Telecommunications, Yaounde (2016)
5. Jack, W., Suri, T.: Mobile money: the economics of M-PESA. National Bureau of Economic Research (2011)
6. Rogers, E.M., et al.: Complex adaptive systems and the diffusion of innovations. Innov. J. Public Sect. Innov. J. **10**(3), 1–26 (2005)
7. Valente, T.W.: Social network thresholds in the diffusion of innovations. Soc. Netw. **18**(1), 69–89 (1996)
8. GSMA: Mobile Money Transfer: Introduction to MMT, **15** (2010), Accessed January 2008
9. Diniz, E., Porto de Albuquerque, J., Cernev, A.: Mobile Money and Payment: a literature review based on academic and practitioner-oriented publications (2001–2011) (2011)
10. Coulon, F.: The use of social network analysis in innovation research: a literature review. Lund University (2005)
11. Rogers, E.: Diffusion of Innovations, 4th edn., pp. 15–23. The Free Press, New York (1995)
12. Rogers, E.M.: The Diffusion of Innovation, 5th edn. Free Press, New York (2003)
13. Venkatesh, V., et al.: User acceptance of information technology: toward a unified view. MIS Q. **27**, 425–478 (2003)
14. Zhou, T., Lu, Y., Wang, B.: Integrating TTF and UTAUT to explain mobile banking user adoption. Comput. Hum. Behav. **26**(4), 760–767 (2010)
15. Huang, B., Kechadi, M.T., Buckley, B.: Customer churn prediction in telecommunications. Expert Syst. Appl. **39**(1), 1414–1425 (2012)
16. Jackson, M.O., Yariv, L.: Diffusion on social networks. In: Economie Publique. Citeseer (2005)
17. Fallah, M., Luo, J.: Mobile Money in developing markets: What should Mobile Money providers consider when trying to drive activity from the already registered user base? (2014)

Geospatial Modeling Using LiDAR Technology

Leyre Torre-Tojal[1]([⊠]), Jose Manuel Lopez-Guede[2,4],
and Manuel Graña[3,4]

[1] Department of Mining and Metallurgical Engineering and Materials Science,
Faculty of Engineering, University of the Basque Country (UPV/EHU),
Nieves Cano 12, 01006 Vitoria-Gasteiz, Spain
leyre.torre@ehu.es
[2] Department of Systems Engineering and Automatic Control,
Faculty of Engineering, University of the Basque Country (UPV/EHU),
Nieves Cano 12, 01006 Vitoria-Gasteiz, Spain
[3] Department of Computer Science and Artificial Intelligence,
Faculty of Computer Science, University of the Basque Country (UPV/EHU),
Paseo Manuel de Lardizabal, 1, 20018 Donostia-San Sebastian, Spain
[4] Computational Intelligence Group, University of the Basque Country
(UPV/EHU), Donostia-San Sebastian, Spain

Abstract. Light Detection and Ranging (LiDAR) is a relatively new surveying technology with the ability to capture and represent a physical environment as never before. This technique has revolutionized the way the data are gathered in the topographical mapping, going from discrete data collection to a massive one. The main advantage of the technique is that it provides a direct method for 3D data collection with high accuracy. Unlike the traditional photogrammetric methods, it can directly collect accurately georeferenced sets of dense point clouds, which can be almost directly used in a variety of applications. Due to the relative novelty of this technology and the increasing interesting of its applications in different scopes, it is likely that new approaches for data processing will be developed in the near future. This paper introduces the technology and analyses the main processing stages of the LiDAR point clouds until the creation of relevant digital models.

Keywords: LiDAR · Remote sensing · DTM

1 Introduction

Due to the characteristics of the Light Detection and Ranging (LiDAR) systems, they have become a very powerful tool for gathering accurate and dense topographic data at very high speed. Laser scanning is a remote sensing tool with the ability to retrieve surface elevations at high spatial resolutions, in both rough terrain and in heavily forested areas [1].

This technology is based on a laser sensor (working in the infrared spectral region) emitting pulses that are returned to the sensor once they have impacted against any object or surface.

© Springer Nature Switzerland AG 2019
Á. Rocha et al. (Eds.): WorldCIST'19 2019, AISC 931, pp. 654–662, 2019.
https://doi.org/10.1007/978-3-030-16184-2_62

Opposite to raster three-dimensional data, that use a grid arrangement, LiDAR data are not, they are point clouds. This particular data organization has the advantage of being able to collect precisely confidence data from in a nonuniform pattern such as most LiDAR systems do [5]. LiDAR technology has reached high economic importance due to its high accuracy capability and has been recognized as one of the standard method for topographic data acquisition [6]. LiDAR mapping systems allow scientists and mapping professionals to study natural and built environments across a wide range of scales with greater accuracy, precision and flexibility than ever before.

The paper first introduces the methodology and it's principal components. Then, the main phases of the LiDAR data processing are exposed, emphasizing in classification of the point cloud and creation of digital models. Finally, the paper outlines future applications of the technology.

2   LiDAR Background

Once the round-trip time of each emitted pulse is measured and taking into account that the light speed is known, the distance to an object is easily calculable by Eq. (1):

$$R = v \cdot t/2 \tag{1}$$

where R is the slant distance or range, v is the speed of electromagnetic radiation, and t is the measured time interval. The three-dimensional coordinates (e.g., x, y, z or latitude, longitude and elevation) of the target objects are computed from (1) the time difference between the laser pulse being emitted and returned, (2) the angle at which the pulse was "fired," and (3) the absolute location of the sensor on or above the surface of the Earth [2].

Besides x, y and z coordinates for every impact, LiDAR systems provide information about the intensity of the reflected signal. LiDAR systems can be mounted in a satellite, in an aircraft or helicopter, drone, or in the case of terrestrial LiDAR, it can be static (in a tripod) or dynamic (in ground-based vehicles). Despite of the basic principle of the technology is the same, there are some differences in data capture and processing steps, applications of data, etc. For more information about the basic physical principles behind the technology, see [3].

When the sensor is mounted in a dynamic platform (aircraft, car, etc.) the system needs to integrate additional positioning systems: a Position and Orientation System (POS) and an Inertial Movement Unit (IMU) to reach an accurate positioning of the lasers and a ground segment comprised of GPS reference stations in the ground segment, as shown in Fig. 1. Major part of the last two decades has been dominated by single wavelength single pulse Linear Mode LiDAR (LML) with single or multi-pulse ability. However, recently several new types of LiDAR sensors have appeared enabling new applications in earth sciences [4].

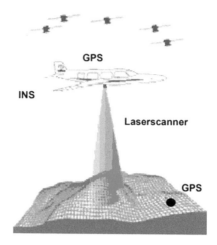

Fig. 1. Simplified conceptual diagram of an airborne LiDAR system [7].

3   LiDAR Data Processing

Both airborne and land-based LiDAR platforms can collect 3D data in large volumes at an unprecedented accuracy. The fact that the complexity of the processing of LiDAR data is relatively modest, the application of this technology to a very wide variety of scopes is rapidly increasing. In this point, a brief description of the main processes involved in LiDAR data processing pipeline are going to be explained (Fig. 2), where some of the processes can be run concurrently because they are independent.

3.1   LiDAR Data Registration and Calibration

The point clouds gathered by the LiDAR are the input for calculating geocoded data in a process known as registration. This process is mathematically described by Eq. (2):

$$\vec{G} = \vec{r}_L + \vec{s} \tag{2}$$

where G is the vector from the Earth center to the ground point, $\vec{r}_L$ is the vector from the earth center to the LiDAR's point of origin and $\vec{s}$ is the slant ranging vector, see Fig. 3.

As vector r_L is measured in WGS84 coordinate reference system, vector G should also be described in the same reference system, taking into account that vector s is measured in the coordinate system L (the origin of the LiDAR coordinates), the general approach can be expressed by Eq. (3):

$$G_{WGS84} = \underline{r}_{L_{WGS84}} + (\Lambda_0, \Phi_0)_H^{WGS84} \cdot (\omega, \kappa, \varphi)_{IMU}^H \cdot (\delta\omega, \delta\kappa, \delta\varphi)_L^{IMU} \cdot s_L(s_{add}, \gamma_M) \tag{3}$$

where the product of the matrices $()_{IMU}^H \cdot ()_L^{IMU}$ describes the orientation of coordinate system L with respect to the horizontal coordinate system H. Λ_0 and Φ_0 are the

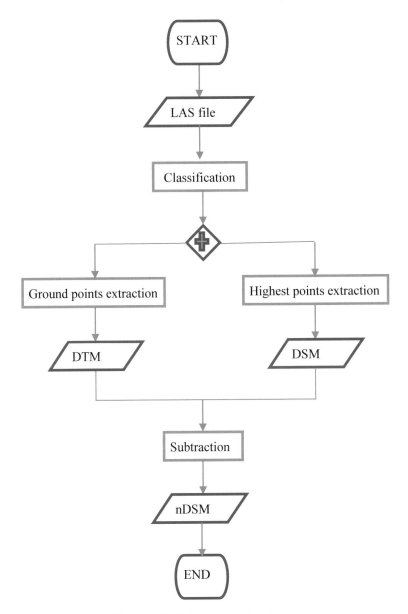

Fig. 2. LiDAR data processing pipeline

geographical longitude and latitude; ω, κ, φ the orientation angles between the IMU and the horizontal system H; γ_M is the so called wobble angle (defined as the maximum scanning angle of a flipping mirror) and s_{add} is an additive parameter of the slant range, being $\delta\omega, \delta\kappa, \delta\varphi$ (the misalignment angles in roll, pitch and heading, respectively) and s_{add}, γ_M the calibration parameters. The calibration is carried out applying a Gauβ-Markoff model.

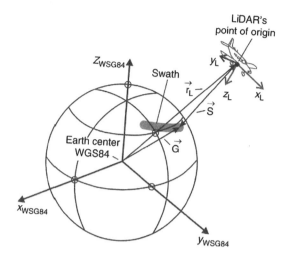

Fig. 3. Vector setup for geocoded LiDAR data [5].

3.2   LiDAR Data Filtering and Classification

After the registration and calibration of the point clouds, a classification process must be carried out, separating the point cloud in different classes such as ground, vegetation or buildings. For this goal, several filters could be applied, grouped in four main categories:

Triangulated Irregular Network (TIN)-Based Filters
These filters are based in the premise that topographical surfaces do not show big discontinuities in the relief. A first triangulation, using minimum height points would be the first reference surface to the computation of the Digital Elevation Models. One of the most used algorithm is the one developed by Axelsson [8], where the reference surface is generated from disperse minimum height points. Then, new minimum points will be searched establishing two conditions: the angle between the new triangle and the candidate point should be smaller than a certain threshold and the new point must be to a specific distance from the nearest point of the modified triangle. After every iteration, both the TIN and the threshold are recalculated. TerraSolid, which is the most advanced, powerful and used software for LiDAR data manipulation, processing and analysis, also incorporate TIN-based filters for point classification, selecting initial local low points that are confident hits on the ground for TIN creation and then TIN models would be iteratively calculated, adding more and more points. Each added point makes the model moves closer to the true ground surface [9].

Morphological Filters
These filters use mathematical morphology techniques based on the idea that great height differences between very close points suggest that, surely, the highest point does not belong to ground. Vosselman [10] defined a reference surface, whose height increases according to the planimetric distance between points, and every point below such surface is considered as ground point. Progressive morphological filters can

improve the detection of nonground object of various sizes using increasing window size for filtering points, instead of fixed window size as do the nonprogressive ones [11].

Surface-Based Filters
On the opposite to the morphological filters, in this case, it is assumed that all points belong to ground. In the first iteration, a reference surface using all points is computed, then that surface moves between the ground points and the ones which are not, obtaining, presumably, negative residuals for ground points, while nonground points should have positive or very small negative residuals. In the first iteration all the points have the same weight in order to generate the reference surface, but once the first residuals are calculated, they will determine the weight of each point for the next iteration [12]. This method was extended to filter buildings and trees in urban areas [13].

Segmentation Filters
Most of the filtering algorithms tend to perform well in flat landscapes, but a significant difference in accuracy appears where steep slopes and discontinuities occur. Discontinuities while detecting large objects is one the reasons of these discrepancies. This circumstance can be solved by the segmentation of the point clouds, analyzing segments instead of individual points when the classification is being carried out. Different homogenization criterion can be chosen to decide whatever a segment belongs to ground or not, for example, height differences between segments [14] or maximum and average gradients [15] (Fig. 4).

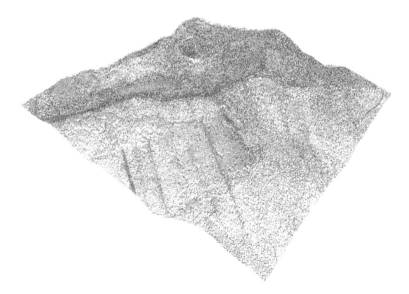

Fig. 4. LiDAR point cloud, classified following the American Society for Photogrammetry and Remote Sensing (ASPRS) standard.

3.3 Digital Terrain Model (DTM) and Digital Surface Model (DSM) Creation

Once the entire point cloud has been classified, the modelling of the ground surface is accomplished, estimating ground elevation at places where measurements are missing. This ground surface is called DTM and is defined as a continuous function which represents the 2D planimetric position (x, y) and the elevation (z), which is x, y depended (Fig. 5). For this purpose, interpolation algorithms are applied categorized by different criteria [16]:

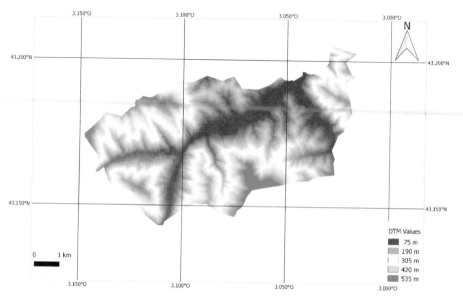

Fig. 5. DTM of the municipality of Gordexola (Biscay, Spain) in the WGS84 Coordinate Reference System using the ground point classification made with TerraScan Software.

(a) Exact or inexact algorithms: Exact algorithms produce DTM surfaces, which pass through every original data point, while approximate methods do not, since they follow the general trend in the data.

(b) Deterministic and stochastic algorithms: Deterministic methods assume that every data point has influence at an unsampled location, being this influence inversely proportional to its distance from the location. On the other hand, stochastic algorithms make use of geostatistic to estimate a surface with some level of uncertainty.

(c) Local or global algorithms: Local interpolation uses only the closest data points to estimate a value at an unsampled location. On the opposite, the global algorithms take into account the entire data set to estimate values.

Examples of commonly used interpolation algorithms in DTM generation are Inverse Distance Weighting (exact, deterministic, local), Kriging (approximate, global, stochastic) and cubic spline (exact, local, deterministic) algorithms. Several factors determine the most appropriate for each case, including the quality of data and required level of accuracy.

Unlike DTM, DSM include the elevation of the point located in the highest locations. In the urban area, this surface represents the roofs of the building, but in forested areas, the highest part of the surface is the tree crowns. The difference between the DTM and the DSM is the so called normalized DSM (Fig. 6), that is given by Eq. (4):

$$nMDS(x, y) = MDS(x, y) - MDT(x, y) \tag{4}$$

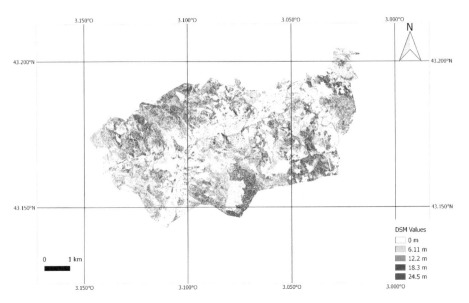

Fig. 6. Normalized DSM of the region of Gordexola (Biscay, Spain) in the WGS84 Coordinate Reference System.

4   Conclusions and Future Works

LiDAR technology has demonstrated to be a very powerful tool for surveying our environment, but as it is relatively new, novel algorithms and methodologies to improve the classification of the point clouds are being developed, being this point a key aspect of this measurement system.

Due to the promising results obtained with this technology in different applications, especially in forest research, a study case to develop new models to estimate biomass using linear and nonlinear regression techniques using exclusively low density (0.5 pulse/m^2) public LiDAR data is being developed by the authors of this paper.

Acknowledgements. The work in this paper has been partially supported by FEDER funds for the MINECO project TIN2017-85827-P, and projects KK-2018/00071 and KK-2018/00082 of the Elkartek 2018 funding program of the Basque Government.

References

1. Reutebuch, S.E., McGaughey, R.J., Andersen, H., Carson, W.W.: Accuracy of a high-resolution lidar terrain model under a conifer forest canopy. Can. J. Remote. Sens. **29**(5), 527–535 (2003)
2. National Oceanic and Atmospheric Administration (NOAA) Coastal Services Center: "Lidar 101: An Introduction to Lidar Technology, Data, and Applications." Revised. Charleston, SC: NOAA Coastal Services Center (2012)
3. Baltsavias, E.P.: Airborne laser scanning: basic relations and formulas. ISPRS J. Photogrammetry Remote Sens. **54**, 199–214 (1999)
4. Lohani, B., Ghosh, S.: Airborne LiDAR technology: a review of data collection and processing systems. Proc. Natl. Acad. Sci. India Sect. A Phys. Sci. **87**, 567–579 (2017)
5. Shan, J., Toth, C.K.: Topographic Laser Ranging and Scanning: Principles and Processing, 1st edn. Taylor and Francis Group, Boca Raton (2009)
6. Ullrich, A., Studnicka, N., Hollaus, M., Briese, C., Wagner, W., Doneus, M., Mucke, W.: Improvements in DTM generation by using full-waveform airborne laser scanning data. In: 7th Annual Conference and Exposition "Laser Scanning and Digital Aerial Photography. Today and Tomorrow", Moscow, Russia (2008)
7. Geist, T., Stötter, J.: First results on Airborne Laser Scanning Technology as a tool for the quantification of glacier mass balance. In: Proceedings of EARSel-LISSIG-Workshop Observing Our Cryosphere From Space, Bern (2002)
8. Axelsson, P.: DEM generation from laser scanner data using adaptive TIN models. Int. Arch. Photogrammetry Remote Sens. **33**, 110–117 (2000)
9. TerraSolid 2018: TerraScan User's Guide (2018)
10. Vosselman, G.: Slope based filtering of laser altimetry data. IAPRS **43** (2000)
11. Zhang, K., Lin, X., Whitman, D., Shyu, M., Yan, J., Zhang, C.: A progressive morphological filter for removing nonground measurements from airborne LiDAR data. IEEE Trans. Geosci. Remote Sens. **41**(4), 872–882 (2003)
12. Kraus, K., Pfeifer, N.: Determination of terrain models in wooded areas with airborne laser scanner data. ISPRS J. Photogrammetry Remote Sens. **53**, 193–203 (1998)
13. Pfeifer, N., Stadler, P., Briese, C.: Derivation of digital terrain models in the SCOP++ environment. In: Proceedings OEEPE Workshop on Airborne Laserscanning and Interferometric SAR for Digital Elevation Models, Stockholm, Sweden (2001)
14. Sithole, G., Vosselma, G.: Experimental comparison of filter algorithms for bare-earth extraction from airborne laser scanning point clouds. ISPRS J. Photogrammetry Remote Sens. **59**(1–2), 85–101 (2004)
15. Schiewe, J.: Ein regionen-basiertes verfahren zur extraktion der geländeoberfläche aus digitalen oberfächen-modellen. Photogrammetrie, Fernekundung, Geoinformation **2**, 81–90 (2001)
16. Maguya, A.S., Virpi, J., Tuomo, K.: Adaptive algorithm for large scale DTM interpolation from lidar data for forestry applications in steep forested terrain. ISPRS J. Photogrammetry Remote Sens. **85**, 74–83 (2013)

ND-GiST: A Novel Method
for Disk-Resident k-mer Indexing

János Márk Szalai-Gindl[1(✉)], Attila Kiss[1], Gábor Halász[1], László Dobos[2],
and István Csabai[2]

[1] Department of Information Systems, Eötvös Loránd University,
Pázmány Péter sétány 1/C, Budapest 1117, Hungary
{szalaigindl,kiss,hagtabi}@inf.elte.hu
[2] Department of Physics of Complex Systems, Eötvös Loránd University,
Pázmány Péter sétány 1/A, Budapest 1117, Hungary
{dobos,csabai}@complex.elte.hu

Abstract. Several challenges are related to metagenomics, one of which
is the data management. A related central concept is k-mer which means
a possible subsequence of length k from a DNA (sub)sequence. In this
work, the focus is on indexing k-mers and supporting box queries where a
query string of length k might have multiple allowed nucleobases per position. A novel index structure: ND-GiST is introduced which has capability to handle box queries. Comparing it with full table scan and the
traditional B-tree, the performance results of ND-GiST are encouraging.

Keywords: Genome data · Metagenomics · ND-tree · GiST ·
PostgreSQL · Box query · Indexing

1 Introduction and Related Works

Since Deoxyribonucleic acid (DNA) has been discovered, which stores genetic
information in cells of each living organism, DNA sequences have an important
role. These sequences consist of nucleobases (Adenin, Cytosin, Guanin, Thymin)
and can be expressed as a series of their initials: A, C, G, T. A genome is
the complete set of DNA sequences. A read is a generated subsequence from a
genome in DNA sequencing which is the process for the purpose of determining
the order of nucleobases in a DNA.

The main objective of metagenomics is to reconstruct genomes from a sample.
This can be done, for example, if one has a priori approximate knowledge about
which genomes can be sequenced and reads are tried to align to only these reference genomes. See [6] for comprehensive review on tools for metagenomic data
analysis. Related central concept is k-mer which means a possible subsequence of
length k from a DNA (sub)sequence. Several challenges are related to this topic,
see [8]. In the light of cited paper, it is established, on the one hand, that the
major bottleneck is data management and analysis, rather than data generation

© Springer Nature Switzerland AG 2019
Á. Rocha et al. (Eds.): WorldCIST'19 2019, AISC 931, pp. 663–672, 2019.
https://doi.org/10.1007/978-3-030-16184-2_63

because of the steep decrease in costs associated with next generation sequencing which causes data megatsunami in bioinformatics. On the other hand, the available tools require significant main memory resources which is main limiting factor for the size of metagenomic samples. For example, this is the case with Kraken which is an ultrafast tool for metagenomic sequence classification but it "runs on a computer with enough RAM to hold the entire database", according to [9], moreover this is no database management system (DBMS) whose components might be needed: multiuser environment, authorization, storing auxiliary metadata (i.e. "data about data"), etc.

Our work explicitly addresses the problem of storing data in disk-based relational DBMS because we want in principle our approach to be generally usable for any dataset regardless of its size. However, main-memory based systems is a better choice for use case scenarios of bioinformatics where data can reside in RAM (see e.g. [3,5]). In this paper, the focus is on indexing k-mers. Special access method is required to support so-called box queries where a query string of length k might have multiple allowed nucleobases per position which can represent uncertainty derived from sequencing results. ND-tree and its variant can be used for this purpose (see [2,7]). To the best of our knowledge, there has been no attempt to embed it into real DBMS. Furthermore, previous research has not examined the case how can so-called packing method be applied to insert data which are available a priori, i.e., how one can build the index in a bottom-up fashion. Our contribution is to realize this construction inside PostgreSQL with usage of Generalized Search Tree (GiST) framework[1].

This paper is organized as follows. Used ND-tree structure and the novel packing method are described in Sect. 2. In Sect. 3, we introduce novel ND-GiST index structure. In Sect. 4, it is compared with with full table scan and the traditional B-tree. Section 5 contains the summary and future work.

2   ND-tree

2.1   Structure

ND-tree can index multidimensional non-ordered discrete-valued vectors (see [7]), i.e. the domain of each dimension in the k dimensional space is a discrete set which has no natural ordering. A Non-ordered Discrete Data Space (NDDS) consists of such vectors. In this work, we utilize this tree to index k-mers of the genomic/metagenomic data thus we assume that each component of a vector corresponds to a letter in a common alphabet $\mathcal{A}$. Let $S_i \subseteq \mathcal{A}$ ($j = 0, \ldots, k-1$) denote j^{th} component set of a (discrete) rectangle S, i.e. $S = S_0 \times \cdots \times S_{k-1}$. By the area of S we mean $|S_0| \cdot |S_1| \cdots |S_{k-1}|$. A NDDS vector can be regarded as a special rectange with $|S_j| = 1$ for $j = 0, \ldots, k-1$. We say that a rectangle S overlaps with another rectangle S' if the area of intersection $S \cap S' = (S_0 \cap S_0') \times \cdots \times (S_{k-1} \cap S_{k-1}')$ is not null. A rectangle (or a vector) S is contained in another rectangle S' if $S_j \subseteq S_j'$ for $j = 0, \ldots, k-1$.

---

[1] See: https://www.postgresql.org/docs/10/gist.html.

ND-tree is a height balanced search tree having variable fanout (i.e., allowed number of children of a node) between m and M ($2 \leq m \leq \lceil M/2 \rceil$). Leaf nodes contain (ptr, key) entries where ptr is a pointer to a record in the database, and key is a search key which represents this record as a NDDS vector. Non-leaf nodes contain (ptr, key) entries where ptr is a pointer to a child node, and key is a so-called discrete minimum bounding rectangle (DMBR) of the child node. DMBR of a leaf node is the rectangle which contains all records pointed by the entries of the leaf node, and it has minimum area among all containing rectangles. By the recursive definition, DMBR of a non-leaf node is the minimum-area rectangle which contains all DMBRs of the child nodes. The tree methods to manage objects are not discussed here. For a thorough treatment we refer the reader to [7].

2.2 Packing

The question arises whether the data indexing would be optimized to perform the best possible average behavior for the box queries when static, non-time-varying data appear. In fact, there is no a priori information about biological sequence data to be stored therefore we cannot make the previous assumption. However, database space being occupied by all possible k-mers can be not too large because the NDDS can contain manageable quantities of vectors, i.e., $\mathcal{A}^k$ can be relatively small. For example, this is the case when $\mathcal{A} = \{A, C, G, T\}$ and $k = 16$ (see below) which correspond to about four billion rows. The process of indexing this amount of data is still time consuming therefore these data can be loaded into a table, be indexed only once and be referenced by other tables. In order to exploit the potential of static data, we create an ND-tree by preprocessing data with a packing method which is described in this section. It does not only enhance performance of retrieval speed, but also contributes to the acceleration of the tree creation process because complex heuristics must not be used in contrast with the proposed solutions so far.

We have been working under assumption that the box queries are uniformly distributed (thus values along dimensions are independent of each other and their marginal distributions are also uniform). Note: we investigate these box queries from slightly different perspective than the previous work [2] which was focused on the determination of the splitting dimension during insertion process because of improving node splitting heuristics as compared to the original ones of ND-tree. However, our splitting heuristics can be as simple as possible because of static data. The first goal is to minimize the expected number of nodes to be traversed in the search tree when a box query is processed. Let us denote by $S = S_0 \times \cdots \times S_{k-1}$ the key of a child node and let $W = W_0 \times \cdots \times W_{k-1}$ denote a box query. A child node is pruned away in W box query processing for which there is a jth component set such that $S_j \cap W_j = \varnothing$. Let $X_j^{S,W}$ be a random variable for all $j = 0, \ldots, k-1$ whose distribution is Bernoulli:

$$X_j^{S,W} = \begin{cases} 0 & \text{if } S_j \cap W_j \neq \varnothing \\ 1 & \text{if } S_j \cap W_j = \varnothing \end{cases} \tag{1}$$

with respective probability:

$$\mathbb{P}\left(X_j^{S,W} = 1\right) = \frac{2^{|\mathcal{A}|-|S_j|} - 1}{2^{|\mathcal{A}|} - 1} = \frac{2^{|\mathcal{A}|}}{2^{|S_j|}\left(2^{|\mathcal{A}|} - 1\right)} - \frac{1}{2^{|\mathcal{A}|} - 1} \tag{2}$$

because the favorable cases are when W_j is taken from a subset of $\mathcal{A} \setminus S_j$ (note that the empty box query is not considered). We find that the smaller the cardinality of S_j is, the more likely the W_j does not overlap with it. Let $X_j^{S,W}$ be a random vector: $X_0^{S,W}, \ldots, X_{k-1}^{S,W}$ with expected value:

$$\mathbb{E}\left[X_j^{S,W}\right] = \left(\mathbb{E}\left[X_0^{S,W}\right], \ldots, \mathbb{E}\left[X_{k-1}^{S,W}\right]\right)^T \tag{3}$$

$$= \left(\mathbb{P}\left(X_0^{S,W} = 1\right), \ldots, \mathbb{P}\left(X_{k-1}^{S,W} = 1\right)\right)^T \tag{4}$$

therefore we want to minimize the area of S because the expected number of overlapping box queries with S is also minimized in this case. The question comes up: what is the lower limit for the area of S? There is a simple answer to this question: the number of actual data elements covered by the node of S. Therefore, the less elements are covered by a given node, the more likely it is that an arbitrary box query does not overlap with them. However, there is a trade-off between area minimization and disk page utilization. (By the disk page we mean the physical storage unit, i.e., actually fixed-size data file on which records and index entries are stored.) This paper does not generally examine this issue. We want to exploit the fact that $|\mathcal{A}| = 4$ when $\mathcal{A}$ contains nucleobases in DNA, i.e. $\mathcal{A} = \{A, C, G, T\}$. Also, we choose $k = 16$ because, as will be explained later, k-mers can be binary encoded in a practical manner in this case. Furthermore, there always exists a power of two between m and M because $m \leq \lceil M/2 \rceil$. Our approach is to build an ND-tree where fixed fanout of each node is a power of two: 2^n which is between m and M. For simplicity, we assume throughout the paper that n is a divisor of $2k$. (This condition will be met with the default page size of PostgreSQL DBMS and $k = 16$.) Levels of the tree is at least $\log_{2^n} 2^{2k} = 2k/n$. One of our goals is that the number of levels is minimized. As will be seen below, it can be feasible that the keys of each node are overlap-free for the retrieval performance of the tree. Additional objectives of the construction are to minimize the standard deviation σ of $|S_0|, \ldots, |S_{k-1}|$ to make sure that there is no privileged dimension.

In order to achieve the above objectives, data are packed in the following way. Logically, the area of the key is 2^{2k} at root node. This value should be $2^{2k}/(2^n)^l = 2^{2k-ln}$ at each node of level l since the data should be evenly distributed among the nodes. (Note: $2k/n - 1$ for leaf level thus the areas of the keys are 2^n at each node of leaf level.) Therefore, there are the area of an arbitrary key $S = S_0 \times \cdots \times S_{k-1}$ is the power of two thus $|S_j| = 1, 2$ or 4 for all $j = 0, \ldots, k-1$ because of $|\mathcal{A}| = 4$. The problem in question can be interpreted as placing 1,2 or 4 into k factors of $|S_0| \cdot |S_1| \cdots |S_{k-1}|$ with a requirement that this product is 2^{2k-ln} for all $l = 0, \ldots, 2k/n - 1$. We will denote by μ the mean of $|S_0|, \ldots, |S_{k-1}|$. The following two lemmas are true. The proofs are omitted due to lack of space which are not difficult and the details are left to the reader.

Lemma 1. *Suppose* $1 < k$. *If* $|S_{j_1}| = 1$, $|S_{j_2}| = 4$ *pair exists and it is replaced by* $|S'_{j_1}| = |S'_{j_2}| = 2$ *(*$|S_j| = |S'_j|$ *for* $j \notin \{j_1, j_2\}$*), then the standard deviation* σ *decreases.*

Remark 1. This proposition yields information about how to choose cardinalities of $S_0, \ldots, S_{k-1}$: efforts should be made to choose as many $|S_j| = 2$ as possible.

Lemma 2. *Suppose* $n \nmid k$. *Furthermore, here* l *stands for the level where* $2k - ln \leq k$, *but* $2k - (l-1)n > k$. *Under the above assumptions,*

$$k - (l-1)n = ln - k \tag{5}$$

hence there are the same total number of $|S_j| = 4$ *at the level* $l - 1$ *and* $|S_j| = 1$ *at the level* l. *Furthermore, this event occurs at the level* $l = \lceil k/n \rceil$. *The proof also gives that* n *must be even.*

If $k < n \leq 2k$, then $n = 2k$ because n is a divisor of $2k$. In this degenerate case, the tree has only one level and $|S_j| = 4$ for all $j = 0, \ldots, k-1$ belong to the key of the sole node, in other words, the page of the root node contains all pointers to records. In the remainder of this section we require n to be less than or equal to k. If $2k - ln \leq k$ at a level l, then let $|S_j|$ be 2 for $2k - ln$ factors of $|S_0| \cdot |S_1| \cdots |S_{k-1}|$ and remaining $ln - k$ factors equal to 1; otherwise, let $|S_j|$ be 4 for $k - ln$ factors and be 2 for ln factors ($4^{k-ln} \cdot 2^{ln} = 2^{2k-ln}$). So far our investigations have only been on the cardinalities. Concrete values of S_j for all $j = 0, \ldots, k-1$ can be provided by the proof of the following important theorem which utilizes the above lemmas, however, it is also omitted due to lack of space. Instead, we give a description of a concise method which provides the same results (see below the Theorem 1).

Theorem 1. *There is a construction for ND-tree where the number of levels is minimized, fixed fanout of each node is* 2^n, *their keys are overlap-free and the standard deviation of the component sets of a key is minimized.*

Let us consider the serial numbers of the records:[2] $ID = 0, \ldots, 2^{2k} - 1$. A given record value will be deduced by the binary representation B of ID in the following way. We will use notation $B[t]$ for the value of B at position t. For that purpose, let

$$f : [0, k-1] \times \{\text{binary representation } B \text{ of } ID \,|\, ID = 0, \ldots, 2^{2k} - 1\} \longrightarrow \mathcal{A}$$

denote the function whose first argument is a position of a given record value and the second argument is the binary representation B of its ID:

$$f(i, B) = \begin{cases} g(B[i], B[i+k]) & \text{if } n \mid k \\ g(B[i], B[i+k+n/2]) & \text{if } n \nmid k \text{ and } i \in [0, k - n/2 - 1] \\ g(B[i], B[i+1]) & \text{if } n \nmid k \text{ and } i \in [k - n/2, k-1] \end{cases} \tag{6}$$

where $g(0,0) = A$, $g(0,1) = C$, $g(1,0) = G$, $g(1,1) = T$.

---

[2] The records are listed in the order of insertion into the tree.

3   ND-GiST

In our work, ND-tree is implemented under Generalized Search Tree (GiST) framework of PostgreSQL DBMS which is called ND-GiST. To this end, it is needed to create suitable key class (user defined data type) and key methods (user defined functions). The structure of GiST can be considered as a generalization of B-tree [1], R-tree [4], or ND-tree (see Sect. 2.1). We draw attention to the fact that *key* corresponds to the concept of predicate in the GiST terminology.

In the remainder of this paper we assume k to be 16 and data are non-time-varying. For the sake of convenience, leaf node *key* (NDDS vector) is also considered as a DMBR. The jth component set S_j of a DMBR can be binary encoded into a 4-bit length sequence $b_A^j b_C^j b_G^j b_T^j$ where b_x^j is 0 if $x \notin S_j$, and 1 otherwise. The instance of the key class for a given DMBR is the concatenation of binary representations of all S_j. E.g., let $\{A, G, T\} \times \cdots \times \{C, T\}$ be a DMBR then corresponding instance is $1011 \ldots 0101$. Note: a search key is also converted into binary format behind the scenes but it can be provided as a string in the WHERE-clause of a SQL query where IUPAC nucleic acid notation[3] can be used to improve user experience (e.g., in the above example, "$D \ldots Y$" is the given search key). Different query predicates are supported in the key class:

$$overlap(dmbr_1, dmbr_2), \ same(dmbr_1, dmbr_2), \ contains(dmbr_1, dmbr_2),$$
$$is\_contained\_by(dmbr_1, dmbr_2).$$

(These types are called "strategies" in the terminology of PostgreSQL.) We follow C-like notation for bitwise operations, i.e., & and | denote bitwise AND and OR, respectively, and the subrange of the binary representation of a DMBR between t_1 and t_2 positions will be denoted by $dmbr[t_1 : t_2]$.

The implementations of these query predicates are the followings:

overlap$(dmbr_1, dmbr_2)$: If $dmbr_1[4j : 4j + 3] \& dmbr_2[4j : 4j + 3] \neq 0$ for all $j = 0, \ldots, k - 1$, it returns true, and false otherwise.

same$(dmbr_1, dmbr_2)$: If $dmbr_1 = dmbr_2$, it returns true, and false otherwise.

contains$(dmbr_1, dmbr_2)$: If $dmbr_1[4j : 4j + 3] = (dmbr_1 \& dmbr_2)[4j : 4j + 3]$ for all $j = 0, \ldots, k - 1$, it returns true, and false otherwise.

is_contained_by$(dmbr_1, dmbr_2)$: This is similar to the previous function, except that $dmbr_1$ is swapped with $dmbr_2$.

The necessary key methods with their implementation are as follows:

consistent(E, q) : For a given entry $E = (ptr, key)$ of a tree node and a query predicate q, if key does not exactly match q in the sense below, it returns false, and true otherwise. In the former case, the subtree of the record pointed by ptr can be pruned away in the search process. If E is in leaf node, the consistency between key and q is straightforward. Now, suppose that E is in non-leaf node and q is *overlap, same, contains* or *is_contained_by* on the argument *dmbr*. Different query predicates should be examined separately:

---

[3] https://www.bioinformatics.org/sms/iupac.html.

(a) if q is *overlap* or *is_contained_by* predicate, it returns
   $overlap(key, dmbr)$,
(b) if q is *contains* or *same* predicate, it returns $contains(key, dmbr)$.

$union(P)$: For a given set P of entries $(ptr_1, key_1), \ldots, (ptr_{|P|}, key_{|P|})$, it returns
   such a key which holds for all entries included in P. In the present case,
   $key = key_1 \mid \ldots \mid key_{|P|}$.
$penalty(E_1 = (ptr_1, key_1), E_2 = (ptr_2, key_2))$: For given two entries, it returns
   a domain-specific penalty for inserting E_2 into the subtree whose root at E_1.
   In our work, it returns zero till the required number of entries (2^n) is reached.
   After that, the penalty is calculated as the absolute difference of M and the
   current number of entries. It is needed to continue loading the page which
   has begun to load because the data should be packed into pages by the order
   of insertion.
$picksplit(P)$: For given a set P of $M + 1$ entries, it splits P into two sets P_1,
   P_2 which have size at least m in principle. However, we get the entries in
   the order of insertion. Therefore, take the first 2^n entries to form P_1 and P_2
   consists of the remainders.

(Note: the compress and decompress methods were omitted because these are
not relevant here.)

4   Experimental Results

4.1   Preparation for Experiments

Experiments are performed on an instance of PostgreSQL 10.4 DBMS installed
with the default settings, running on a single virtual machine (Intel Core Processor (Skylake) 2.2 GHz (16 Cores) with 32 GB RAM) with Ubuntu 18.04.1 LTS
operating system. Our proposed approaches are implemented as an extension for
PostgreSQL.

Let us denote by dmbr_bin the PostgreSQL type of a binary encoded DMBR.
All possible k-mers are stored in table kmers(id bigint, seq dmbr_bin) in
order in which they are listed by the packing method. This table is an auxiliary table used to accelerate the access to records. Therefore the ND-GiST is
built on top of kmers on column seq. Elapsed time of this index creation is
563034245.197 ms (6 day, 12 h, 23 min and 54.245 s). For comparison, we create the table kmers_str(id bigint, seq varchar) which is similar to kmers,
except that all k-mers are represented by strings. B-tree is built on column
seq of kmers_str. Elapsed time of this index creation is 11173652.604 ms (3 h,
6 min and 13.653 s). Note: we must create this index with a special operator
class varchar_pattern_ops, which supports pattern-matching queries because
our database does not use the C locale.[4] This assistance is important because
operator LIKE is involved for box queries in this table.

---

[4] See: https://www.postgresql.org/docs/10/indexes-types.html.

4.2   Performance of ND-GiST

In this section, we report the results of basic performance tests of ND-GiST and compare them with full table scan and the traditional B-tree. The total storage needs of kmers_str (with its B-tree) is 656 GB, as opposed to 396 GB in kmers (with its ND-GiST). The simplest way to justify the usage of ND-GiST is to perform exact match queries. One million random k-mers are generated where all nucleobases have equal chance to take part in sequences. The mean elapsed time of these queries against kmers is 27.524 ms with a standard deviation of 12.854 ms, and against kmers_str, it is 24.114 ms with a standard deviation of 12.883 ms which is a little bit better. However, it should be noted that k-mers are binary encoded in kmers and a dmbr_bin-to-varchar conversion must be applied when a query is processed. It is a trade-off between response time and storage requirements.

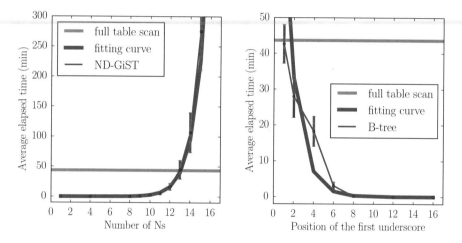

Fig. 1. Average elapsed time (minutes) of ten random queries are plotted as a function of contained N amount (left panel) and the position of the first underscore (right panel). A detailed description is in the text.

In order to highlight the potential benefits of the usage of ND-GiST, comparative studies are carried out on the box queries. For the sake of simplicity, a search key for kmers can contains $\{A, C, G, T, N\}$ characters where N symbolizes that any nucleobase is allowed in its position. It works because IUPAC nucleic acid notation is supported by dmbr_bin (see Sect. 3). However, search keys of box queries take different forms against kmers_str because box queries should be given as pattern-matching queries in this case and N should be replaced by underscore ('_') which matches any single character. Full table scan can be used as a baseline because ND-GiST and B-tree cannot be directly compared in this case. The explanation of the former is that B-tree uses the lexicographical order on seq in kmers_str, and its performance does not really depend on how many underscores are contained by a search key but on where the first underscore is in a search key. The left panel of Fig. 1 shows the average time costs of random box queries for kmers (blue curve) which contain different N amount. There are ten runs for each N amount. In the right panel of Fig. 1, the average time costs of random box queries for kmers_str are plotted as a function of the position of the first underscore (blue curve). There are ten runs for each position. Horizontal green line indicates the results of full table scan, standard deviations are indicated by red bars, respectively, on Fig. 1. As shown in left panel, ND-GiST outperforms full table scan if the number of Ns is less than 13. The exponential fitting curves, of the form $y(x) = A \cdot 2^{Bx}$, are displayed in the figure (magenta curves) with $A = 27.58$, $B = 1.26$ (left panel) and $A = 9341024.047$, $B = -1.116$ (right panel), respectively. As reflected in panels, fitting curves show the trend. In the left panel of Fig. 1, the blue curve is remarkably well fitted. In order to compare elapsed time results obtained by different trees, B-tree time costs must be averaged over all underscore positions because privileged position for the first underscore is inadmissible. This mean (cyan line) is shown with the time costs of ND-GiST (blue curves) in Fig. 2. From this it is clear that ND-GiST outperforms B-tree on average if the number of Ns is less than 12.

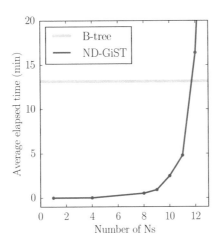

Fig. 2. Comparison of elapsed time results obtained by B-tree and ND-GiST. See the text for discussion.

5   Conclusion

In this work, a novel index structure: ND-GiST was introduced based on ND-tree and GiST. It was shown that it has capability to handle box queries. Packing method was also designed for tuning the original ND-tree. ND-GiST was compared with full table scan and the traditional B-tree. ND-GiST outperformed B-tree on average and full table scan for box queries, respectively, if the number of Ns was less than 12 and 13, respectively. The practical viability of our approach will be demonstrated in the classification of metagenomic sequences in a follow-up paper.

Acknowledgments. The project has been supported by the European Union's Horizon 2020 research and innovation program under grant agreement no. 643476 (COMPARE), by the Novo Nordisk Foundation Interdisciplinary Synergy Programme [Grant NNF15OC0016584] and by the European Union, co-financed by the European Social Fund (EFOP-3.6.3-VEKOP-16-2017-00002).

References

1. Bayer, R., McCreight, E.M.: Organization and maintenance of large ordered indexes. Acta Inform. **1**(3), 173–189 (1972). https://doi.org/10.1007/978-3-642-59412-0_15
2. Chen, C., Watve, A., Pramanik, S., Zhu, Q.: The bond-tree: an efficient indexing method for box queries in nonordered discrete data spaces. IEEE Trans. Knowl. Data Eng. **25**(11), 2629–2643 (2013). https://doi.org/10.1109/TKDE.2012.132
3. Dorok, S., Breß, S., Teubner, J., Läpple, H., Saake, G., Markl, V.: Efficiently storing and analyzing genome data in database systems. Datenbank-Spektrum **17**(2), 139–154 (2017). https://doi.org/10.1007/s13222-017-0254-9
4. Guttman, A.: R-trees: a dynamic index structure for spatial searching. SIGMOD Rec. **14**(2) (1984). https://doi.org/10.1145/602259.602266
5. Janetzki, S., Tiedemann, M.R., Balar, H.: Genome data management using RDBMSs. Technical report, Otto-von-Guericke Universität, Magdeburg, Germany (2015). https://doi.org/10.13140/RG.2.1.4047.6006
6. Oulas, A., Pavloudi, C., Polymenakou, P., Pavlopoulos, G.A., Papanikolaou, N., Kotoulas, G., Arvanitidis, C., Iliopoulos, I.: Metagenomics: tools and insights for analyzing next-generation sequencing data derived from biodiversity studies. Bioinform. Biol. Insights **9**, 75–88 (2015). https://doi.org/10.4137/BBI.S12462
7. Qian, G., Zhu, Q., Xue, Q., Pramanik, S.: The ND-tree: a dynamic indexing technique for multidimensional non-ordered discrete data spaces. In: Proceedings 2003 VLDB Conference, pp. 620–631. Elsevier (2003). https://doi.org/10.1016/B978-012722442-8/50061-6
8. Scholz, M.B., Lo, C.C., Chain, P.S.: Next generation sequencing and bioinformatic bottlenecks: the current state of metagenomic data analysis. Curr. Opin. Biotechnol. **23**(1), 9–15 (2012). https://doi.org/10.1016/j.copbio.2011.11.013
9. Wood, D.E., Salzberg, S.L.: Kraken: ultrafast metagenomic sequence classification using exact alignments. Genome Biol. **15**(3), R46 (2014). https://doi.org/10.1186/gb-2014-15-3-r46

Human-Computer Interaction

Sustainable Physical Structure Design of AR Solution for Golf Applications

Egils Ginters[✉]

Department of Modelling and Simulation, Riga Technical University,
Daugavgrivas Street 2-434, Riga 1048, Latvia
egils.ginters@rtu.lv

Abstract. One of the applications of virtual and augmented reality (VR/AR) is the social sphere, such as communication among individuals and also sports activities. It is important to select a sustainable technical solution when designing the AR application. At present, AR smart glasses, which are offered by several manufacturers on the market, can be considered the most suitable. The author's article discusses possible options for assessing the sustainability of the proposed hardware, based on the methods of system dynamics simulation.

Keywords: Augmented reality · Systems sustainability · Simulation

1 Introduction

Virtual and augmented reality (VR/AR) technologies are introduced in various sectors of the economy as well as in social life. One of the applications is the golf game, which is a democratic and widespread sport and recreation. Mostly these technologies are used in the analysis of the game or in preparation for the beginning of the game [1]. Players have the opportunity to explore the playground, hole placement, make terrain analysis, and receive meteorological data. After the game, it is possible to analyze the gathered data about the shots. Both before and after the game, different situations simulations and scenario studies can be performed [2]. However, the emergence of AR smart glasses brought new opportunities because of the compact and mobile equipment that is connected to the smart phone. In turn, the smart phone provides connectivity to the Internet environment, which allows to transfer the most resource-consuming data processing to the Cloud. The aim of the authors is to design an AR solution for putting training in golf game.

An important issue is the right choice of hardware. An AR solution must be sustainable and useful not just for golf. It needs to be adaptable and transformable for realization of other communication tasks. An AR device should be applicable every day, and its use should not be cumbersome. In turn, the manufacturer must provide an easily accessible service. The price of the equipment must also be in line with the purchasing capacity of potential consumers.

AR's solution for golf is a classic sociotechnical system whose architecture is determined by the requirements of the logical and physical structures, that is, the functionality determines the selection of hardware.

© Springer Nature Switzerland AG 2019
Á. Rocha et al. (Eds.): WorldCIST'19 2019, AISC 931, pp. 675–685, 2019.
https://doi.org/10.1007/978-3-030-16184-2_64

High quality VR/AR technologies are functional but usually incompatible, complicated and relatively expensive. It is therefore very important to predict their sustainability. Existing VR/AR technology assessment methods mainly determine acceptance, i.e. whether the audience is ready to use this technology today. Different methods are used to evaluate VR/AR technologies, such as expert opinions and comparison of technical parameters [3, 4], platform benchmarking [5], case studies comparison [6], human factors evaluation [7–9], heuristic assessment [10] etc. However, these methods do not assess the sustainability of the VR/AR platform that the designer will use in the application. In the following sections of the article we will look at a complex approach, linking the conceptual modeling of AR requirements with sustainability assessment.

2   Conceptual Requirements to Sustainable AR Solution for Golf Training Development

The development of sustainable AR solution aimed to communication and collaboration enhancement depends of requirements to logical and then to physical structures. Major part of logical structure involves requirements of functionality. The reference model of typical AR solution [11] determines the set of features for two interoperable subsystems of visual recognition and voice processing. Apart of specific features the requirements are as follows.

For identification, recognition, visualization and tracking:

-eyesight responsive visualization; -picture digital conversion; -determination of distance; -zooming; -recording; -scene and separate objects recognition; -multi-objects spatial identification (face, marker, landscape, virtual object, text (printed and handwritten)); -multi-objects recognition; -recognition improvement by learning; -multi-objects spatial tracking.

For voice processing and translation:

-voice/audio digital conversion; -volume tuning; -recording; -language recognition; -translation; -voice/audio reproduction; -language recognition and translation improvement by learning.

The above functionality requirements create the conditions for the physical structure or AR glasses selection:

-weight no more than 100 g; -battery operation time without recharge no less than 4 h; -everyday style sunglasses or compatible; -built-in video camera at least 720 pix and 120 angle view or more; -display ration 16:9; -built-in microphone and headphone; -Bluetooth (Wi-Fi is optional); -built-in RAM at least 1 GB; -additional SD memory; -wireless access to inertial measurement (IMU) and GPS units; -wireless access to and compatibility with data processing equipment.

However, important is sustainability. How sustainable is provided AR solution? Therefore, the set of necessary conditions consists of three components (1):

$$AR\, selection = \langle Functionality, Technical\, specs, Sustainability \rangle \qquad (1)$$

Technical requirements are determined by the functionality and are analyzed below.

3 Technical Requirements Justification

Our task is to choose an appropriate AR glasses solution for a simplified conceptual model that meets the previously defined functional requirements. AR glasses must provide the golfer with putting skills improvement. An earlier study by the authors [11] identified several models of AR glasses that could be successfully applied to improve communication skills. The research also can be used as the basement for current task. The models were analyzed respecting the set of technical requirements $\langle TR_i \rangle$, such as weight, operating system, Wi-Fi interface, resolution, RAM, additional memory, CPU, Bluetooth, battery capacity and price. The following models were analyzed: Recon Jet, Telepathy Walker and Jumper, Vue, Moverio BT-200, Jins Meme ES, ODG R9, SED-E1, Google Glass and Vuzix. The authors did not analyze the items that are focused mainly on VR functions, such as Oculus Rift VR, Samsung Odyssey, and others. Of these models, only HoloLens was reviewed. Further the technical requirements are discussed:

{*TR*1}: *Weight of glasses cannot be more than 100* g.

The AR equipment must not interfere with its user. A certain weight limit is not absolute and may be slightly higher, but it is understood that the glasses should not cause discomfort. In addition, the design must ensure the stability of the spectacles during rapid movement.

{*TR*2}: *Battery operation time without recharge no less than 4* h.

The duration of the golf game, including the duration of the workout before the game, can be reached in about 3–4 h. If it is not possible to provide adequate battery capacity, then it must be possible to easily replace the power supplies by pausing the game for a short time. It should be remembered that the battery life is not infinite and provides for a limited charging number. In addition, the operating time will decrease according to the life of the battery. Similarly, the battery operation time will be shorter under lower temperature conditions.

{*TR*3}: *Everyday style sunglasses or compatible.*

We have already mentioned that AR glasses should not cause discomfort. In this case, cause discomfort to the community. They cannot be striking, attract attention and disturb other players. Specific solutions may pose a public risk of identity and data security threats. This means that glasses should not look like specific technical equipment. Ideally, they would look like ordinary glasses or sunglasses.

{*TR*4}: *Built-in video camera at least 720 pix and 120 angle view or more.*

Since one of the eyepieces serves to visualize the eyesight, it must first be recorded. When leaving aside the specific applications where high resolution is required, such as medical procedures, design, etc., then use an ordinary enough with 720 pix. Sufficient peripheral vision is required. The golf game and everyday life are not so dangerous that a 180° angle of sight would be required, but the play area needs to be reviewed. Eyesight should include both a golf ball and a target hole, so visibility at around an angle of 120° would be desirable.

{*TR*5}: *Display ration 16:9.*

Horizontal and vertical ratio correspond to the usual practice of modern video and provide good scene transparency. In the meantime, the built-in display should be inconspicuous and transparent when not on, and also provide good energy efficiency performance.

{*TR6*}: *Built-in microphone and headphone.*

In the future, the AR solution can be used to collaborate with members of the group by exchanging voice information. So the both a microphone and at least one headphone are necessary. This will allow to receive instructions and recommendations from the coach, as well as information from other gaming partners, possibly through phone conversations. However, AR glasses are not intended to be used as a music player, so special technical requirements for this aim are not required. Constructive, of course, should not be overwhelming and worsen the visual appearance of the spectacles.

{*TR7*}: *Bluetooth (Wi-Fi is optional).*

Understandably, the AR-based glass computing resource is limited because it requires paying for battery consumption. Thus, the basic processing of data must take place on an additional device that is traditionally a smart phone of the user, or even cloud resources can be accessed through an external communication environment. Therefore, the minimum requirement is the availability of Bluetooth, which could also provide data from IMU/GPS devices. Of course, it would be even better if the AR glasses had a Wi-Fi connection to ensure higher data transfer rates. Unless this add-on essentially affects the glasses' constructive and damaging visual image, Wi-Fi is desirable. The above-mentioned communication capabilities could provide the opportunity for the player to see the other player's field of vision in his glasses or on the smart phone screen and give him the recommendations. Therefore, the functionality requirements include the demand for responsive visualization.

{*TR8*}: *Built-in RAM at least 1* GB.

The technical specification does not have special requirements for the CPU, since it is clear that AR glasses will be able to provide only a limited set of image processing operations. Because the CPU will work faster, the lower will be its energy efficiency, or CPU price will increase so much that purchase for the everyday user will be economically unjustified and the potential business audience will decrease. So it does not make sense for a large amount of RAM, which should only cache the most important data and store the screen.

{*TR9*}: *Additional SD memory.*

Currently, the sizes and price of SD memory cards allow them to be placed in very compact devices. In this case, several GB-based extra memory cards would be usable. SD card can be used to deploy both geospatial map, video, additional service software and other necessary information.

{*TR10*}: *Wireless access to inertial measurement (IMU) and GPS units.*

This feature will be provided via Bluetooth and/or Wi-Fi, as well as smart phone resources. The use of a built-in GPS device would be unimportant as signal reception requires a significant battery consumption.

{*TR11*}: *Wireless access to and compatibility with data processing equipment.*

We have already noted that AR glasses mainly serve only for recording and visualization of information, but data processing operations have to take place on external devices. Traditionally it will be a smart phone. Here, the requirement for AR-glasses compatibility with a smart phone appears, which states that both operating systems must be the same. Unless something extraordinary happens, the ball is on the

Android side. There are AR-glasses solutions that support iOS and MS Windows, but these are just some of the solutions that will not provide enough coverage for the product. Emulators can also be used, but will have to be paid with slowdown in data processing and inefficient computing power consumption that will affect battery life. However, it is good that such emulators exist because they provide the ability to design and validate AR glasses in a wider group of designers through a variety of software designing tools and services.

An analysis of the current models of AR glasses [11] determine the choice between two manufacturers, Google and Vuzix.

4   AR Glasses Comparison

4.1   Google Glass Development

Google Glass [12] is a monocular glasses form AR device, actually wearable computer that allows voice commands to be used instead of touchscreen. The first version of Glass Explorer was released in 2014 (see Fig. 1). The headset received a great deal of criticism and legislative action due to privacy and safety concerns. The reason was the opportunity offered, to record the image and use the capabilities of Google and the Internet to recognize the object and receive up-to-date information about it. Even more it was possible in real time. Due to public protests, Google stopped the sale of Glass, which was renewed in 2017 when the new Google Glass Enterprise Edition was announced.

The new Google Glass [13] version in 2018 is no significantly different from the Explorer. Additional storage capacity is increased to 32 GB, but RAM up to 2 GB. Intel Atom CPU performance is higher, but the battery capacity is 780 mAh, which could provide 8 h of operation, of course, depending on the operating conditions. Enterprise edition has GPS/GLONASS support and barometer. Wi-Fi Dual-band 802.11n/ac has a higher performance as the previous one. Google Glass like majority of AR devices is supported by the Android operating system. It has a 640×360 Himax HX7309 LCoS display, an 8 megapixel camera, capable of 720p video recording, a hinge sensor, ambient light sensing and proximity sensor, 3 axis (gyroscope, accelerometer, compass) and a mike with headphone. Users can control key functions via audio commands, without selecting buttons manually. The Google Glass weighs at least 43 g.

Since all these features consume battery power, no specific operating time savings compared to the Explorer is expected. In general, the Google Glass battery is a critical resource. The battery is built-in and cannot be exchanged without special skills. And, of course, you must first find where you can buy it. If the device is not running for a long time, then Google Glass become difficult to be resuscitated, and the user can drop 1800 USD paid for this device in the basket. Google's marketing policy is rather obscure, as if Google Glass can be purchased, however, the user is not entitled to full control because of the legal inability to dispose of the glasses.

Concerned about privacy violations against which the society objected, it has not been eliminated, otherwise the purpose of using the AR device will be lost. The visual design is still well noticed. As noted by "... *the Google Glass is now back with a focus on utility, not fashion* ..." [14]. To reassure the society, Google Glass places more emphasis on industrial application.

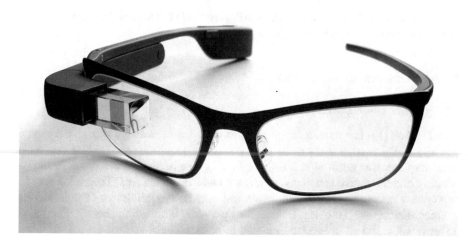

Fig. 1. Google Glass AR unit [12].

In 2018 Google Cloud announced its cooperation with the Israeli company Palatine [15]. The award-winning company specializes in artificial intelligence (AI)-based software for optimizing manufacturing processes. Factory operators and managers can now communicate continuously with Plataine's Digital Assistant to receive real time, AI-based alerts and optimized recommendations based on the current production progress.

One of the partners is Boeing [14], where Google Glass is used to improve efficiency during the hard wire-framing process. The device is used in DHL warehouses during the picking process showing the workers the item's location rather than using voice commands. DHL expects time reduction by 25%, and plans to implement the technology in 2000 warehouses across the globe. Google Glass is also used in General Motors, Picavi and Volkswagen in the picking and packing processes. The hands-free factor means increased safety for workers in hazardous environments and allows them to be always connected to the network.

Google Glass [12] is also used in other areas where there is no significant privacy violation. Augmedix app allows physicians to live-stream the patient visit possibly saving them up to 15 h a week and improving record quality. The video stream is passed to remote secure rooms where the doctor-patient interaction is transcribed. Glass are used also during a live surgical procedures. Several groups began developing Google Glass based technologies to help children with autism learn about emotion and facial expressions. Voice of America Television in 2014 explored the technology's potential uses in journalism. Google Glass in Nanjing 2014 Youth Olympic Games were used by number of athletes to record their exercises.

Nowadays hardware functionality and applications are determined by the software. Google Glass uses many applications, such as Google Now, Google Maps, Google+, and Gmail. Many developers and companies have built applications for Glass, including news apps, facial recognition, photo processing, translation, and sharing to social networks. Because Google Glass is a Google product, application development and verification capabilities are almost endless, and the potential audience is limited only by the price of an AR device.

The sustainability of the Google Glass product is not detrimental to Google's existence. Parent company Alphabet's [16] market capitalization in June 2018 was valued at 793 billion U.S. dollars. In 2016, Google Sites revenue amounted to 77.8 billion U.S. dollars.

4.2 Vuzix Smart Glasses

Founded in 1997, Vuzix [17] is one of a leading suppliers of AR Smart-Glasses technologies and products. The Vuzix products offer users a portable high-quality viewing experience and provide solutions for mobility and augmented reality.

Vuzix M100 Smart Glasses (2014) is an Android-based wearable computer enhanced with a wearable monocular display and onboard processor, recording features and wireless connectivity capabilities designed for a wide variety of enterprise applications. Ergonomic and rugged the M100 is currently in large scale productive use in fields such as telemedicine, remote assistance and warehousing. Hands free access to data, direct and remote video capabilities, direct onboard processing of video capture for lag free AR and more. Visually the glasses are similar to Google Glass Enterprise model. Price is 1079 euro.

Vuzix M300 Smart Glasses industry adapted model have been designed from the very beginning with enterprise applications in mind. The unit has higher battery capacity than M100. Price is 1699 euro. The model demonstrates succession and continuity of development of Vuzix AR glasses models.

Vuzix Blade AR smart glasses (see Fig. 2) [17] provides a wearable smart display with a waveguide optics and Cobra II display engine. Vuzix Blade AR smart glasses allows users to interact with smartphone hands-free. Finally fashion meets technology in the wearable display arena. Vuzix launched the Vuzix Blade Edge Developer kit for 1099 euro. The Vuzix Blade® Smart Glasses received 'Best in Show Overall' Auggie Award at AWE Europe 2018 held in Munich, Germany October 18–19.

Vuzix AR has many and versatile applications. Vuzix smart glasses improves the process of manual order picking, incoming/outgoing goods, sorting and packing of goods, as well as inventory and deficiencies in warehouses.

Vuzix enables onsite clinicians to connect with remote providers and partners anywhere enabling streaming quality video/audio through smart glasses and tablets. Vuzix smart glasses empowers field technicians to diagnose and resolve mechanical problems with online data and manuals, as well as remote assistance from experts. Training and courses can be conducted directly to a single location or multiple sites with Vuzix glasses. Vuzix is used also in golf applications. GENiUS Ball [18] is a smart ball with embedded chip-set technology that captures data including golf ball velocity, spin rate, spin angle, trajectory, GPS location, distance to the tee, and more. With the use of a Bluetooth connection, the Genius Ball pairs with a user's mobile phone.

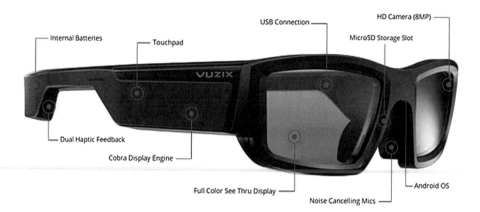

Fig. 2. Vuzix Blade AR smart glasses [17].

Golfers can play their courses using AR glasses and see a visual trace of golf ball's flight path combined with display readout of their performance data.

4.3   Google Glass vs. Vuzix Blade Sustainability Assessment

Eight years ago, the team of authors started to develop a technology sustainability assessment methodology. One of the most significant problems was the assessment of public acceptance of the technology. The UTAUT methodology was first used, but it was based on a potential audience survey that was too time consuming and costly. In the following solutions, the UTAUT approach was replaced by Rogers diffusion theory, but the sustainability assessment model was simplified. A decision was taken to use Skypes's reference curve for sustainability index assessment to ensure its comprehensibility and perceptibility. To describe the overall Integrated Acceptance and Sustainability Assessment Model (IASAM) [19] the following expression is used (2):

$$Sus_T^i(t) = Sus_T^i(t - dt) + \left(Accept._T^i + Manag._T^i + Quality_T^i + Domain_T^i \right) * dt \quad (2)$$

In this expression $Sus_T^i(t)$ is sustainability index curve for the i-technology. Each change determines by the set of factors: $Acceptance_T^i$ flow (12 factors), $Mangement_T^i$ (22 factors), $Quality_T^i$ (18 factors) un $Domain_T^i$ (7 factors), but the changes are modeled in the system dynamics simulation environment *InsightMaker*. All the parameters are assessed using 7-points Likert psychometric scale from strongly disagree until agree. IASAM index has breakdown in four groups with 0.25 points deviation. First group is problematic, but in upper group technology is reasonable for investments.

In order to properly choose the Google Glass or Vuzix as basic solution for AR application the sustainability simulation was done, based on both market information and practical experience (see Fig. 3). The quantitative values of the evaluation parameters are, of course, subjective, and therefore the overall assessment is subjective. For all streams the weight equals 25%. The Google Glass technology's IASAM sustainability index is 0.78 *skypes*, while the Vuzix Blade received 0.91 *skypes*.

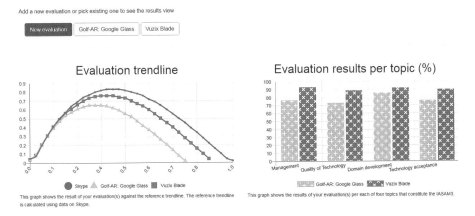

Fig. 3. Google Glass vs. Vuzix Blade sustainability comparison.

Reasons are to be found in the technologies reviews and practical experiments already provided. Google Glass is still not eliminates the problems related with the first project failure. This had to be foreseen in the development of a business plan. Service and object ownership issues raise doubts about the applications development. An important issue affecting the audience is the relatively high price of the glasses. In addition, the Vuzix AR product nomenclature and applications are more diverse, and AR is the core business of the company. Even if Vuzix will lose its independence, then the competitor will not be interested in canceling the range of successful products. Of course, Google, as a manufacturer of a variety of software applications, is far ahead of Vuzix, but at the moment they could be the only benefits of Google Glass vs. Vuzix Blade.

5 Conclusion

Despite the rapid development of VR/AR technologies, they are still poorly standardized, difficult compatible and therefore expensive. It limits the spread of these technologies. In turn, the development of good and useful applications is financially roomy. Therefore, when choosing the hardware, it is important to make the right choice. One of the methodologies is the Integrated Acceptance and Sustainability

Assessment Model (IASAM), which is based on system dynamics simulation. When evaluating two possible AR hardware solutions Google Glass and Vuzix Blade, it turns out that despite the wide-ranging capabilities of Google's software offer, Vuzix Blade is currently on the rise.

In this article, authors' approach to developing the physical structure of an AR application is demonstrated by choosing smart glasses for improving golf game. Golf is a democratic kind of sport and at the same time a hobby with a growing number of people in the world. However, golf is a relatively complicated and technical challenge that requires long-term training to achieve a good result. The AR solution offered by the authors can reduce the duration of training and significantly improve the quality of the game. However, the appropriate AR equipment is not cheap, so there is a need for a sustainable application that could be upgraded later without changing hardware.

The purpose of the article is to introduce virtual application developers with the augmented reality conceptual requirements model and digital technology sustainability assessment methodology (IASAM). Using conceptual and sustainability modelling at the same time will avoid possible errors and significant losses that may occur when choosing the wrong digital platform for the development base. Of course, this is only the opinion of the authors, so the development of a particular AR application is one of the validation paths of the proposed approach.

The proposed approach will be useful for VR/AR application developers, allowing them to avoid short-term solutions that could compromise the sustainability of applications.

The author's further work will be related to the development and implementation of a fully-fledged and widespread AR solution for putting improvement in golf game. The validated approach will be used to design usable AR application for senior communication skills renewal and improvement.

Acknowledgements. The article publication is initiated by FP7 FLAG-ERA project FuturICT 2.0 (2017–2020) "Large scale experiments and simulations for the second generation of FuturICT".

References

1. Winters, N.: The future of golf: simulators, V.R., & Robot Caddies. http://www.fairfieldhillsgolfcourse.com/the-future-of-golf/. Accessed 18 Oct 2018
2. O'Brien, M.: GolfAR™ releases PuttAR™: a golf simulator for smartphones and tablets. http://www.insidevtknowledgeworks.com/2015/11/golfar-releases-puttar-a-golf-simulator-for-smartphones-and-tablets.html. Accessed 14 Nov 2018
3. Gearbrain: 11 Questions about virtual reality and augmented reality headsets to ask before you buy one. https://www.gearbrain.com/11-questions-virtual-reality-headsets-1635372294.html. Accessed 2 Jan 2019
4. Kaiser, R., Schatsky, D.: For more companies, new ways of seeing. https://www2.deloitte.com/insights/us/en/focus/signals-for-strategists/augmented-and-virtual-reality-enterprise-applications.html. Accessed 2 Jan 2019
5. Benchmarks. https://benchmarks.ul.com/vrmark. Accessed 2 Jan 2019

6. Jayaram, S., Vance, J.M., Gandh, R., Jayaram, U., Srinivasan, H.: Assessment of VR technology and its applications to engineering problems. J. Comput. Inf. Sci. Eng. **1**, 72–83 (2001)
7. LaValle, S.M.: Virtual Reality. Chap. 12. Evaluating VR Systems and Experiences. Cambridge University Press, Cambridge (2017)
8. Mariani, C., Ponsa, P.: Improving the design of virtual reality devices applying an ergonomics guideline. In: Rebelo, F., Soares, M. (eds.) Advances in Ergonomics in Design, Advances in Intelligent Systems and Computing, vol. 588, pp. 16–25 (2018)
9. Stone, R.J.: Human Factors Guidelines for Interactive 3D and Games-Based Training Systems Design. Human Factors Integration Defence Technology Centre, p. 86 (2008)
10. Sutcliffe, A., Gault, B.: Heuristic evaluation of virtual reality applications. Interact. Comput. **16**, 831–849 (2004)
11. Ginters, E., Puspurs, M., Griscenko, I., Dumburs, D.: Conceptual model of augmented reality use for improving the perception skills of seniors. In: Bruzzone, A.G, Ginters, E., González Mendívil, E., Gutierrez, J.M., Longo, F. (eds.) Proceedings of the International Conference of the Virtual and Augmented Reality in Education (VARE 2018), Rende (CS), pp. 128–139 (2018). ISBN 978-88-85741-20-1
12. Google Glass. https://en.wikipedia.org/wiki/Google_Glass. Accessed 14 Nov 2018
13. Glass Enterprise Edition: VR/AR headsets. https://www.aniwaa.com/product/vr-ar/google-glass-enterprise-edition/. Accessed 14 Nov 2018
14. Klara, L.: Why Google Glass Enterprise is a game changer for warehouses. https://www.wonolo.com/blog/google-glass-enterprise-game-changer-warehouses/. Accessed 14 Nov 2018
15. Von Kentzinsky, A.M.: Hey, Google: Google Glass is back and packed with AI. https://www.xovi.com/2018/08/google-glass-is-back-and-packed-with-ai/. Accessed 14 Nov 2018
16. Statista: The statistics portal. Google - Statistics & Facts. https://www.statista.com/topics/1001/google/. Accessed 14 Nov 2018
17. Vuzix. https://www.vuzix.com/. Accessed 26 Oct 2018
18. ProShopOnly: The Genius ball. https://proshoponly.com/genius-ball/. Accessed 26 Oct 2018
19. Ginters, E., Aizstrauta, D.: Technologies sustainability modeling. In: Rocha, A., Adeli, H., Reis, L.P., Costanzo, S. (eds.) Trends and Advances in Information Systems and Technologies, vol. 2. Advances in Intelligent Systems and Computing, vol. 746, pp. 659–669. Springer, Cham (2018). ISSN 2194-5357

The Effect of Multisensory Stimuli on Path Selection in Virtual Reality Environments

Guilherme Gonçalves[1](✉), Miguel Melo[2], José Martins[1,2,3],
José Vasconcelos-Raposo[1,2], and Maximino Bessa[1,2]

[1] University of Trás-os-Montes and Alto Douro, Vila Real, Portugal
guilhermeg@utad.pt
[2] INESC TEC, Porto, Portugal
[3] EsACT, Instituto Politécnico de Bragança, Bragança, Portugal

Abstract. Virtual Reality (VR) has as a key feature, the users' interaction with a virtual environment. Depending on the purpose of a given
VR application, it can be essential to use multisensory stimulus without
biasing users towards specific actions or decisions in the virtual environment (VE). The goal of the present work is to study if the choice of paths
can be influenced by the addition of multisensory stimulus when navigating in a VE using an immersive setup. The awareness of having to take
such decisions was also considered. For the purpose, we used a VR game-
like application contemplating three levels. Each level was symmetrical
and had two possible paths to move to the next level (left or right). For
each level, there was a multisensory stimulus on the right path (from a
subject orientation): wind, vibration, scent respectively. The sample of
the study consisted of 50 participants, and the results showed that none
of the multisensory stimuli had a significant impact users' decision. The
users' awareness of having to decide also did not affect their path. We
conclude that multisensory stimuli can be used to raise the credibility of
the virtual environments without compromising the users' decisions.

Keywords: Virtual Reality · Multisensory · Credibility ·
Decisions Awareness

1 Introduction

The concept of Virtual Reality (VR) has been around for many years. However,
only recently immersive VR equipment became accessible to the general public.
Since then, immersive VR evolved at a fast pace giving the opportunity for users
to become the actors in such virtual environments (VE) [1]. By actors, we refer
to the possibility of interaction with the VEs by using the same physical movements we would use in the real world. Such possibilities open new ways to use
VR in application fields like education [2] or training [3,4]. In both areas, VR

© Springer Nature Switzerland AG 2019
Á. Rocha et al. (Eds.): WorldCIST'19 2019, AISC 931, pp. 686–695, 2019.
https://doi.org/10.1007/978-3-030-16184-2_65

can be used to test or train concepts that require users to make the right decisions. It allows simulations of situations that could be difficult to mimic in real life regarding logistics, costs or safety. For example, in surgery, students could practice procedures repeatedly in a safe environment without compromising the health of real patients.

1.1 Related Work

The feeling of "being there" while in a VR application is known among the literature as Presence [5]. If a user feels high levels of presence, their behaviour in the VE should be more like the behaviour in a real analogous situation [6,7]. This fact suggests that higher levels of presence should be required to efficiently use VR applications to train or educate users to real-world situations. To achieve this feeling of presence, the VE should be credible [8]. Studies conducted proving multisensory increases presence [9], being that multisensory stimuli can play an important role.

Works suggest multisensory stimuli are essential for users to make choices or take decisions in certain situations [10]. Barbosa et al. [3] conducted an experiment where users use the haptic feedback of temperature to perform a firefighter procedure correctly. Regarding scents, those associated with dangerous situations can, for example, raise the level of awareness [11] and possibly influence how a user makes his decision. The correct use of multisensory cues can also manipulate the users' decisions without them knowing - e.g. using proper stimuli can increase how much users spend time and money in markets just by manipulating sounds, smells, colours and touch [12]. This fact suggests that these stimuli, even if we don't realise, can influence our decisions in real life.

From state of the art, we know that users' decisions can depend on the multisensory stimuli they receive. But in these cases, the multisensory stimuli were critical for the task at hand. For instance, users would have to be aware and constantly monitor the stimuli to decide based on it. To the best of our knowledge, no works could be found in the literature that studies how multisensory stimuli influences the path of users in immersive VR applications.

Our work investigates wherever the use of multisensory stimuli delivered in a subtle manner (not critical for the task), can influence users' decisions in VR. We will test each stimulus individually (wind, vibration, and smell) and analyse which path users take (left or right). Each stimulus will have its source on only one of the two paths. Another factor that we included in this study was if the knowledge of the fact that there is a decision to be made affects the decision itself. We theorise that if the player was not advised about the fact that he must decide between paths, he could opt randomly for the first path that he/she sees. Such could imply that users could not feel the stimuli coming from one of the sides if they went directly to the path without a multisensory stimulus associated. If the users know beforehand that there are two paths and they must choose one of them, we expect them first to analyse both paths and pay extra attention to possible differences between them. We then propose two hypotheses:

- **H1**: The multisensory stimuli will influence the user's path.
- **H2**: The awareness of having to decide between paths will influence how they do it.

To test these hypotheses, we investigate in this work whether subtle multisensory cues and awareness about decision-making in VEs can influence the user's path when compared to the typical audiovisual VE setup.

2   Methods and Materials

An experimental cross-sectional study, of comparative nature following a between-group design was conducted to investigate the knowledge about the decision-making process and the impact of multisensory feedback in the user's path in VR applications as described below.

2.1   Sample

We used a non-probabilistic sampling technique, namely convenience sampling (sample taken from a group of people easy to reach). In this case, we've taken our sample from university students. The sample consisted of 50 participants (37 males and 13 females) aged between 17 and 44 years old ($M = 24.88$) and most were students. The participants were randomly divided into two main groups. In group A ($N = 25$), participants did not know that they could make decisions in the VE. A researcher told participants in group B (N = 25) that they could make decisions in the VE. Each group was subdivided into 2 groups, resulting in a total of 4 groups. These subgroups consisted of 2 different conditions: one where participants experienced multisensory stimuli ($N = 12$) and one where experienced only audiovisual stimuli ($N = 13$).

2.2   Materials

For this experiment, the research team has developed an immersive VR application using Unity® 2017, named "Illusions" that allowed to have full control over the study variables. The VR application is game-like and depicts a dark environment with a suspense component and composed by three levels. The levels are designed (Fig. 1) in such a way that the player must choose one of two possible paths to move through the next level.

   In every level, the player must find a key to open a door in each and proceed to the next one (Left image on Fig. 2). The door to be unlocked is always situated in the middle of the paths. So, when opening the door, the player is always in the middle of the paths. In each level, both paths are symmetrical and look the same to avoid a possible bias towards a side [13] (Right image on Fig. 2). However, there is always a multisensory stimulus coming from the right path relative to the participant's orientation that can be felt in the middle. In the first level, there was a slight breeze coming from the right side. In the second level, there

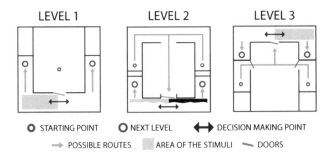

Fig. 1. Illusions level design

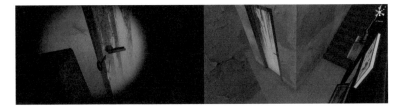

Fig. 2. Left: Player opening door after finding the hidden key. Right: Level 1 showing the left path (symmetrical to the right path)

was haptic feedback (vibration) from the ground beneath him that intensifies if he goes through the right side and that stops if he goes left. In the third and last level, the player could sense a burning scent coming from the right side of the level. However, there were problems with the users recognising the stimulus direction or even feeling the scent in the middle of the paths in the development stage of the game. To solve this, we implemented a game objective to fetch an object in the far end of both paths and go back to the middle so the subject could have a higher possibility to have sensed the scent in one of the sides while fetching the object. Based (or not) in this information the player could then choose the path to finish the game.

The levels were designed in a way that allows the player to walk physically through all the game in an area of 4 by 4 m.

The experiments were run using a desktop computer with the following specifications: CPU Intel® Core™ i7-5820K @ 3.30 GHz, 32 GB RAM, Geforce GTX 1080Ti. The visual stimulus, as well as the interaction, was ensured via the HTC VIVE system. Headphones with active noise cancelling were used to deliver audio and to isolate the participant from outside noises. A ground board delivered the force feedback stimuli in the ground with a mounted transducer. The wind stimulus was delivered using compressed air through a hose that was synced with the VE. For delivering olfactory stimuli, the SensoryCo SmX-4D smell machine was used. All the stimuli were synced with the VE.

2.3   Variables

In this study, it was considered two independent variables: Multisensory Stimulus and Decision Knowledge. Regarding the Multisensory Stimulus, there were two levels: without multisensory stimulus (only visual and audio) and with the multisensory stimulus (visual and audio plus wind, force feedback, or smell). Regarding the Decision Knowledge, there were also two levels: without knowledge (participants were not informed that they would have to choose one from two possible routes) and with knowledge (participants were informed that they would have to pick one from two possible paths).

The dependent variable considered is the side (left or right) which the player decides to proceed to the next level.

2.4   Procedure

All the experiments were taken in an experimental room where the research team had full control over the ambient variables. When receiving the participants, they were briefed about the study - the purpose of the study was not disclosed to avoid bias. To formalise their agreement in participating in the study, they were asked to fill a consent. A generic sociodemographic questionnaire was also filled to be possible to characterise the sample (age and gender) - no sensitive personal data was collected as well as all the data was anonymous. As the experimental setup included a ground board that required participants to step on it (Fig. 3), participants were told to be careful with the ground obstacles as they could physically exist. This ground board was synced between the VE and reality, so the player could see this ground board in the game and step on it physically like it was real. To be sure players could sense the scent, we asked them if they had any limitations in breathing through the nose or detecting scents. If the response was positive, and to not compromise the experiment, players would perform conditions that did not consider multisensory stimuli.

Fig. 3. Participant interacting with the ground board

Every participant began the experiment in the centre of the room, which corresponded to the centre of the game level, and in the same orientation.

Throughout the whole experiment, the participants were always accompanied by a researcher to provide support if needed and to help the participants avoid the HMD cables. In each level of the game, players had to choose one of two possible routes to move through the next level. The VR application automatically logged the side the player selected.

The experiment ended when the player managed to reach the end of the game. The mean time that took for the participant to finish the game was around 14 min with the longest taking 22 min and the fastest 10 min to complete. After finishing the virtual experiment, a debriefing was conducted to collect data about the experiment (assess if participants had any factor in mind when deciding to go through one side or the other).

2.5   Statistical Procedures

Due to our dependent variable being dichotomic (left or right), we conducted a Pearson Chi-Square test to analyse if multisensory stimuli (namely wind, vibration, and scent) individually could influence participants to take one side more than the other. First, we performed the Pearson Chi-Square test between the groups A (W/ Multisensory) and B (W/ Multisensory) as well as between the groups A (W/O Multisensory) and B (W/O Multisensory) to verify if the variable Decision Knowledge affects how participants choose their path. If results reveal that this independent variable does not have any impact in the dependent variable, we will group the samples that experimented with multisensory stimuli across groups and compare it against the grouped sample that performed without multisensory stimuli.

3   Results

For ease of presentation, results are divided into two subcategories, corresponding to the two independent variables of the study: Decision Awareness and Multisensory Stimuli.

3.1   Decision Awareness

To verify if Decisions Awareness could influence the participants' path (H2), a Pearson Chi-Square test was performed for every level of the game to compare the side taken by the subjects between the multisensory condition (group A and B) and non-multisensory condition (group A and B).

Analysing the **multisensory** conditions, groups A (No knowledge) ($N = 12$) and B (With knowledge) ($N = 12$), results reveal that:

- **For the first level (wind)** no differences were found in the dependent variable Side ($\chi^2(1) = 0.000$, $p = 1.000$). Both group A and B (W/ Multisensory), 50% of the participants ($N = 6$) went right and 50% went left ($N = 6$).

- **The second level (vibration)** also had no differences ($\chi^2(1) = 0.000$ $p = 1.000$). Similar to the first level, in both groups A and B (W/ multisensory) 50% of the participants ($N = 6$) chose the right path and 50% ($N = 6$) the left one.
- **The third level (scent)** also revealed no differences between the conditions ($\chi^2(1) = 1.510$, $p = 0.219$). In this level, group A had 66.7% of participants ($N = 8$) opting to go left, and 33.3% ($N = 4$) to go right. Regarding group B, 41.7% of the participants ($N = 5$) went left and 58.3% ($N = 7$) went right.

Analysing the **non-multisensory** condition, groups A ($N = 13$) and B ($N = 13$), results revealed that:

- **For the first level (wind)** no differences were found between groups ($\chi^2(1) = 0.000$, $p = 1.000$). In both group A and B (W/O Multisensory) 69.2% of the participant ($N = 9$) chose to go left and 30.8% ($N = 4$) to go right.
- **The second level (vibration)** revealed the same results, no differences between groups ($\chi^2 = 0.000$, $p = 1.000$). In both group A and B (W/O Multisensory) 46.2% of the participants ($N = 6$) chose to go left and 53.8% ($N = 7$) to go right.
- **The third level (scent)** also revealed no differences ($\chi^2(1) = 2.476$, $p = 0.116$). In group A (W/O Multisensory) 30.8% of the participants ($N = 4$) went through the left path and 69.2% through the right path ($N = 9$). In group B (W/O Multisensory), 61.5% chose the left path ($N = 8$) and 38.5% the right one ($N = 5$).

3.2   Multisensory Stimuli

As no differences were found regarding the previous knowledge about having to decide one path to proceed to the next level, the samples were grouped by the two levels of the independent variable multisensory stimuli. A Pearson Chi-Square test between the conditions that involved multisensory ($N = 24$) and no multisensory ($N = 26$) was applied to verify if it can affect how subjects chose their path (H1) (Table 1). The results revealed that:

- **For the first level** wind had no impact in the variable side ($\chi^2(1) = 1.923$, $p = 0.166$). In the non-multisensory condition (Group A + B), 69.2% ($N = 18$) chose the left side and 30.8% ($N = 8$) chose the right one. Regarding the multisensory condition (Group A + B), 50% ($N = 12$) went right and 50% ($N = 12$) went left.
- **For the second level** vibration also had no impact in the users decisions ($\chi^2 = 0.075$, $p = 0.786$). In the non-multisensory condition (Group A + B), 46.2% ($N = 12$) chose the left side and 53.8% ($N = 14$) chose the right one. Regarding the multisensory condition (Group A + B), 50% ($N = 12$) went right and 50% ($N = 12$) went left.

Table 1. Percentage of participants that opted to go through the left or right paths based on Multisensory Stimuli

	W/ Multisensory		W/O Multisensory	
	Left	Right	Left	Right
Level 1	50	50	69.2	30.8
Level 2	50	50	53.8	50
Level 3	54.2	45.8	54.2	58.3
Mean	55.57	44.43	47.23	52.77

- **The third level** revealed that scent also did not affect the dependent variable ($\chi^2(1) = 0.321$, $p = 0.571$). In the non-multisensory condition (Group A + B), 46.2% ($N = 12$) chose the left side and 53.8% ($N = 14$) chose the right one. Regarding the multisensory condition (Group A + B), 54.2% ($N = 13$) went right and 45.8% ($N = 11$) went left.

4   Discussion

In this work, we studied how multisensory stimuli and previous knowledge about having to take a decision affects decision-making in VR. He hypothesised that multisensory stimuli, delivered in a subtle manner, (not critical for the task or objective) could influence participants path (H1). We also hypothesised that the awareness of the existence of two paths and having to decide between them could also influence the user's path (H2). Results demonstrated that none of the stimuli managed to influence the participant's path. We attribute these results to the fact that the stimulus, individually, was not important for the task or objective. For example, Spencer [10] says that the sense of smell has a vital role in both diagnosis and surgery. Thus, the sense of smell in this situation is important for the task, so users will pay more attention and make decisions according to the feedback of this stimuli. Another example is Barbosa et al. [3] work. It consisted in using temperature stimuli to perform a firefighter procedure. Here the user would also be aware of the presence of that stimuli and constantly seek temperate changes to perform the procedure correctly.

In this works the stimuli were important, and users would have to pay attention to it to perform the task and take the correct decisions. None of the stimuli we used (wind, vibration, smell) represented any strong importance in completing the game or any of the tasks of it. In other words, the game levels could be completed without the multisensory stimuli as easily as with them. From a gameplay perspective, these extra stimuli only helped to create an immersive environment. There was no challenge or objective associated with the multisensory stimuli. An example of a task which would require the user to pay attention to the stimuli and thus change how he/she acts could be: "The key that allows the player to unlock the doors and move through the next levels are always in the source of the multisensory stimuli". We also justify these results to the fact that some players did not explore both ways before going to the next level possibly not noticing that there was a difference which could influence the path.

Regarding the awareness of having to decide which path to take, results demonstrated no impact across conditions. From the debriefing session, we learned that none of the users told they went through one way because any multisensory stimuli attracted them, being the decisions random.

The results suggest that the use of multisensory stimuli in a subtle manner does not influence the user's path. This could prove useful, for example, if developers want to increase the credibility of a VE but they don't want it to influence the users' path. They could then add multisensory stimuli to archive that purpose - for example, a firefighter simulation where the researchers want the subjects to feel immersed and with high levels of presence. They could add multisensory stimuli to raise the credibility of the VE, but they don't want it to distract, overwhelm or influence the subjects out of their training steps.

5   Conclusion

The main objective of this study was to investigate multisensory stimulation and its impact regarding the user's path. The results suggested that multisensory stimuli did not have any effect on the subjects' path - also, the awareness of having to make a decision regarding the path they wanted to proceed revealed no differences when compared with the participants that were not aware of that. The study had some limitations. Different people can act differently to some of the stimulus (ex. the same scent can attract or repeal different persons) which could to a certain point influence the results). The last level had an extra objective that consisted in fetching an object in the far end of each path because it was impossible to determinate the direction and sometimes even sense the smell in the middle of the paths. This extra objective forced the player to experience both paths. This did not happen in the previous levels, and it could have influenced to some degree the results in that level. Also, the stimuli were always present to the right side of the subject in the moment of deciding which path to take which could have provoked some laterality bias.

The present work opens doors for further studies regarding the investigation whether the importance of the stimuli for the task at hand will influence the decisions of the users. Other stimuli, such as temperature, could also be incorporated into new studies. Also, the combination of multiple stimuli at the same time (ex. wind and temperature, or scent and wind) can be studied to verify its impact on the subject's path and decisions. The intensity of the stimuli could also be analyzed to understand from which threshold they can start to influence the user's paths and counterbalancing the stimulus positions to avoid a possible laterality bias. Also, the importance of the stimuli for the objectives should also be studied to investigate the points where subjects start to use the stimuli conscientiously to decide.

Acknowledgement. This work is financed by the ERDF – European Regional Development Fund through the Operational Programme for Competitiveness and Internationalisation - COMPETE 2020 Programme and by National Funds through the Portuguese funding agency, FCT - Fundação para a Ciência e a Tecnologia within project POCI-01-0145-FEDER-031309 entitled "PromoTourVR - Promoting Tourism Destinations with Multisensory Immersive Media.

References

1. Fuchs, P., Moreau, G., Guitton, P.: Virtual Reality: Concepts and Technologies, 1st edn. CRC Press, London (2011)
2. Freina, L., Ott, M.: A literature review on immersive virtual reality in education: state of the art and perspectives. In: The International Scientific Conference eLearning and Software for Education, vol. 1, p. 133 (2015)
3. Barbosa, L., Monteiro, P., Pinto, M., Coelho, H., Melo, M., Bessa, M.: Multisensory virtual environment for firefighter training simulation: study of the impact of haptic feedback on task execution. In: 2017 24th Encontro Português de Computação Gráfica e Interação
4. Aïm, F., Lonjon, G., Hannouche, D., Nizard, R.: Effectiveness of virtual reality training in orthopaedic surgery. Arthrosc. J. Arthroscopic Relat. Surg. **32**(1), 224–232 (2016)
5. Slater, M.: A note on presence terminology. Presence Connect **3** (2003)
6. Slater, M., Brogni, A., Steed, A.: Physiological responses to breaks in presence: a pilot study. In: Presence 2003: The 6th Annual International Workshop on Presence, vol. 157 (2003)
7. Bessa, M., Melo, M., de Sousa, A.A., Vasconcelos-Raposo, J.: The effects of body position on Reflexive Motor Acts and the sense of presence in virtual environments. Comput. Graph. **71**, 35–41 (2018)
8. Bouvier, P.: The five pillars of presence: guidelines to reach presence. In: Spagnolli, A., Gamberini, L. (eds.) Proceedings of Presence, pp. 246–249 (2008)
9. Fröhlich, J., Wachsmuth, I.: The visual, the auditory and the haptic – a user study on combining modalities in virtual worlds. In: Virtual Augmented and Mixed Reality. Designing and Developing Augmented and Virtual Environments, pp. 159–168. Springer. Heidelberg (2013)
10. Spencer, B.S.: Incorporating the sense of smell into patient and haptic surgical simulators. IEEE Trans. Inf. Technol. Biomed. **10**(1), 168–173 (2006)
11. Köster, E.P.: The specific characteristics of the sense of smell. In: Rouby, C., Schaal, B., Dubois, D., Gervais, R., Holley, A. (eds.) Olfaction, Taste, and Cognition, pp. 27–44. Cambridge University Press (2002)
12. Soars, B.: Driving sales through shoppers' sense of sound, sight, smell and touch. Int. J. Retail Distrib. Manag. **37**(3), 286–298 (2009)
13. Vilar, E., Rebelo, F., Noriega, P., Duarte, E., Mayhorn, C.B.: Effects of competing environmental variables and signage on route-choices in simulated everyday and emergency wayfinding situations. Ergonomics **57**(4), 511–524 (2014)

The Impact of Gender, Avatar and Height in Distance Perception in Virtual Environments

Hugo Coelho[1]([⊠]), Miguel Melo[2], Frederico Branco[1,2],
José Vasconcelos-Raposo[1,2], and Maximino Bessa[1,2]

[1] University of Trás-os-Montes and Alto Douro, Vila Real, Portugal
hcoelho@utad.pt
[2] INESC TEC, Porto, Portugal

Abstract. Virtual Reality is becoming more popular over the years because it allows the user to be the main actor in another environment and interact with it in real time. New interaction methods are being studied, like tangible interfaces, but there is little work done related to small distances when grabbing objects through a virtual environment. This study is important because, in our perspective, interaction in virtual reality will be at arms reach, meaning that the user will interact within very close distances (under 1 m). In this paper, the research team further evaluate distance perception using gender, the presence of avatar and height (fixed or personalised). The sample consisted of 64 participants (32 females and 32 males) evenly distributed between all four conditions (8 males and 8 females for each condition). Results revealed that gender does have an impact on small distance estimation; height does not have an impact on distance estimation; and avatar does make a difference when trying to grab a real object through the virtual environment.

Keywords: Virtual Reality · Distance perception · Gender · Height · Avatar

1 Introduction

Virtual Reality (VR) allows the user to play the main role in a virtual environment (VE) and develop the feeling of being there as they interact with the it [1]. Thus, interactions are an important characteristic of VR, and companies are working to implement a truly intuitive and natural interaction [2]. All interaction methodologies can be used to interact with the VE, the only problems are recognising the right movement, the gesture of the action the user wants to perform and syncing them with the VE. There are ways to overcome those problems like tracking real objects or even the user and replicating everything in the VE. One way to do it is by using a tracking system like OptiTrack [3], but even if the object is well placed in the VE, there is no guarantee that the user will reach it when trying to grab the object. [2] said that, of all interaction techniques, direct

© Springer Nature Switzerland AG 2019
Á. Rocha et al. (Eds.): WorldCIST'19 2019, AISC 931, pp. 696–705, 2019.
https://doi.org/10.1007/978-3-030-16184-2_66

picking and direct manipulation are the most natural methods for the user to interact with the VE. Since direct picking and direct manipulation are easier for the user, it is important to understand how people perceive small distances. With this, one can establish how to include objects naturally inside the VE and generate knowledge about small distances.

Since direct picking or manipulation is the most natural methods to interact with the VE, our work aims to study if the presence of an avatar (hand representation present or not) and height (personalised or fixed height) changes the way people perceive virtual objects within small distances. The research team also considered gender as the independent variable in the data was collected by the research team. The remainder of this paper is structured as follows: the second section will present a literature review of distance perception and some concepts. The third section is where the methodology and procedure will be presented. The results are shown in the fourth section, following by the discussion in the fifth section and conclusions on the sixth section.

2   State-of-the-Art

Distance perception has been studied using different methods and different variables such as real environment vs VE, depth cues, the presence of avatar, and height. For instance, [4] studied the comparative perception between the real world, a semi-spherical projection and an immersive large screen. Authors concluded that in the real environment people could be precise when estimating distance but in the VE they tend to underestimate distances by 50%. Moreover, [5] performed a study where they shifted the horizon upwards by 11.5°. The shift performed in his experiment did not change the way people perceived distances, and they also concluded that for distances higher than 2.5 m people tend to overestimate while when presented with distances less than 1.5 m people tend to underestimate them.

Studies on how the perception of distances changes from the real world to the VE and from interior to exterior in real environments was made by [6]. They concluded that the distance estimation is different comparing the real-world observations and VE observations, but the results obtained from the interior experience are not statistically different from the results obtained in the exterior environment.

There is no consensus about the presence of avatar and distance estimation. Some authors state that the presence of avatar increases the accuracy in distance estimation [7], others say that even a dislocated and/or static avatar increases the accuracy [8] and others state that only a well tracked-avatar leads to an increase of accuracy [9]. However, there are also authors stating that the presence of avatar does not change the way people perceive distances [10].

Regarding the height in the VE, [11] studies if decreasing or increasing the virtual height changes distance perception. The results showed that decreasing the virtual height by 50 cm had no impact on distance estimation being the explanation that people sometimes have to perform tasks being crouched so they are used to being lower than their actual height. On the other hand, increasing the virtual height by 50 cm decreases the accuracy of distance estimation by 20%.

Studies are showing that females and males see the VE differently; for example, females have a natural experience when presented with the computer-generated environment while males have a more natural experience when presented with the captured content environment [12]. While there are differences in natural experience, studies on distance estimation show that there is no statistical difference when comparing females with males [13].

Besides all the variables described above, there is also work that studied technological variables like Field of View and Head Mount Display (HMD) comparison. These studies are essential since its needed such technologies to see the VE, which influences the user's perception of the VE. [14] studied if the Field of View of the HMD could change the way people perceive distances. They reported that the Field of View does not have an impact on distance estimation. Regarding the state-of-the-art HMD comparison, [15] studied if different HMD could lead to a different distance perception. Their results showed that there is no statistical difference in distance estimation between them.

Distance perception is widely evaluated by using the Blind Walk methodology [16]. In this methodology, the user walks blindfolded to a particular object and stops where he thinks the object is. This approach has some problems since people are hesitant to walk blindfolded, but an acoustic signal or a researcher walking alongside the participant seems to solve that issue. Perceptual correspondence is another technique that can be used to measure distance perception; it uses a reference distance to compare other distances [17]. Verbal reports can also be used, and they measure the distance between the participant and the target object [16]. Time Imagined Walking is a methodology where the user is standing still, blindfolded and uses a chronometer to indicate the time to reach the target object. Using the mean velocity of the participant one can calculate the distance that the participant thought the target object was at [16]. These distance evaluation methodologies can be used to measure two types of distances: Egocentric and Exocentric. Egocentric distance is measured from the feet of the participants to the target object; and Exocentric distance is the distance between two target objects placed in an environment [8]. Usually, both of them are measured using absolute distances, meaning that the participant has to report absolute values about the distance he thinks the target object is at [8].

3   Methodology

The adopted methodology consists of a quasi-experimental design (the researcher controls the assignment of the condition the user will perform), cross-sectional study (at a specific point of time) with a quantitative focus. The sampling technique used was the non-probabilistic convenience sampling technique.

3.1 Sample

In this study, the sample consisted of 64 participants (32 males and 32 females) equally distributed in four groups. Their age ranged between 18 and 44 years old ($M = 22.22$ and $SD = 4.199$) with all participants reporting normal to corrected-to-normal vision. Also, the samples had heights between 152 and 195 cm ($M = 169.98$ and $SD = 11.077$).

(a)   (b)

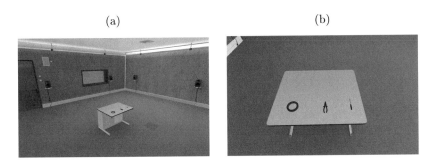

Fig. 1. Virtual representation of the VE. (a) Virtual replica of the experimental room; (b) Representation of the virtual objects.

3.2 Materials

To perform the actions of touching a real object throughout an HMD, one created a VE which is a replica of the real room where the experiments took place, Fig. 1a. Also, all three virtual objects and the two avatar hands were replicated to be used in the VE, Fig. 1b. The equipment used to deliver the visual stimulus was the Oculus Rift DK2 HMD. The tracking of the user's head movement and the positions of the plier, tape, and screwdriver was achieved by using the Optitrack's Motive unified motion capture software platform. The application ran on a desktop computer with an Intel i7-5820K, three NVIDIA GTX 980 GPU and 32 GB of RAM. The Optitrack system was mounted in a room (7 by 7 m) and consisted of eight cameras, two cameras on each wall, allowing to track objects in an area of 4 by 4 m. To track each object, three or more Optitrack markers per object were used so that the Optitrack software could get the position and rotation of the object, allowing us to reproduce it in the VE.

3.3 Variables

In this study, the research team defined three independent variables: Gender (female and male), Height (Fixed or personalised) and Avatar (not present vs present). As for dependent variables, the research team only have the reports saying if the participant touched the real object or not.

3.4   Procedure

Before the user enters the room, the researcher calibrates the room using the OptiTrack system, ensuring that the VE is aligned with the real room. As the participants entered the room, they were asked to sign a consent form and a sociodemographic questionnaire. Next, each participant put the HMD. In the VE, they had the same height as their real one measured using the OptiTrack system. Regardless, the researcher asked the participants if the height felt natural. The researcher changed the height until the participant felt comfortable (increasing or decreasing the virtual height accordingly). After adjusting the virtual height, the researcher chose an object randomly and asked the participant to touch the virtual object and stop where they think the virtual object was. This task was performed one time for each object. When the user's hand stopped, the researcher took notes whether the hand touched the chosen object or not.

3.5   Statistical Procedures

Since all the independent variables were dichotomous, the research team did the statistical procedure called Chi-square Test for Association. In this test, all the independent variables have to be nominal or ordinal variables, and one has to have the independence of observations, meaning that there is no relationship between observations. Another requirement is that all expected cell counts have to have a minimum of 5 observations, and if the number is above that, one report the Pearson's chi-square; otherwise, one report the Fisher's exact test.

4   Results

Since the research team had three independent variables (female or male, fixed or personalised height and with or without avatar), the first analysis performed was the comparison between gender for each group. A chi-square test for association was conducted between gender and the frequency with which participants placed the hand on top of the objects correctly.

4.1   Gender Reports

For all the groups, all expected cell frequencies were greater than five. There was no statistically significant association between gender and the frequency with which participants placed the hand on top of the objects correctly for without an avatar and fixed height condition, $\chi^2(1) = 1.343$, $p = 0.247$. The same happened for the condition without an avatar and personalised height where we got $\chi^2(1) = 0.000$, $p = 1.000$.

As for the group with avatar and fixed height, there was a statistically significant association between gender and the frequency with which participants placed the hand on top of the objects correctly for this group, $\chi^2(1) = 10.243$,

$p = 0.001$. There was a moderate association between gender and the frequency with which participants placed the hand on top of the objects correctly, $\Phi = 0.462$, $p = 0.001$.

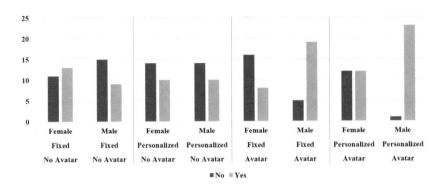

Fig. 2. Gender comparison for all conditions.

For the group with avatar and personalised height, there was a statistically significant association between gender and the frequency with which participants placed the hand on top of the objects correctly for this group, $\chi^2(1) = 12.765$, $p < 0.001$. There was a strong association between gender and the frequency with which participants placed the hand on top of the objects correctly, $\Phi = 0.516$, $p < 0.001$. The frequencies are shown in Fig. 2.

4.2   Height Reports

After analysing gender variable for each group and realising that gender does make a difference on touching objects through the virtual environment, one analysed the height and avatar variables taking into account the previously reported differences between gender.

For height, a chi-square test for association was conducted between height and the frequency with which participants placed the hand on top of the objects correctly. For the two groups (without avatar group; and with avatar and female participants group), all expected cell frequencies were greater than five. There was no statistically significant association between height and the frequency with which participants placed the hand on top of the objects correctly for the group without an avatar, $\chi^2(1) = 0.169$, $p = 0.681$. One got the same result for the group with avatar and female participants, $\chi^2(1) = 1.371$, $p = 0.242$.

As for the group with avatar and male participants, not all expected cell frequencies were greater than five. Therefore a Fisher's Exact test was conducted between the height and the frequency with which participants placed the hand on top of the objects correctly. There was no statistically significant association between height and the frequency with which participants placed the hand on top of the objects correctly, $p = 0.188$. The frequencies can be seen in Fig. 3.

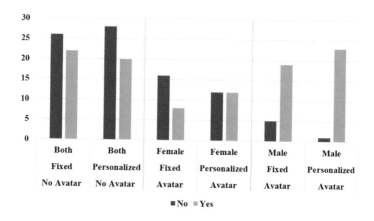

Fig. 3. Height comparison taking into account the gender differences

4.3 Avatar Reports

A chi-square test for association was conducted between avatar and the frequency with which participants placed the hand on top of the objects correctly. All expected cell frequencies were greater than five for the comparisons between: without an avatar and fixed height versus female participants with avatar and fixed height; without an avatar and fixed height versus male participants with avatar and fixed height; and without an avatar and personalised height versus female participants with avatar and personalised height. There was no statistically significant association between avatar and the frequency with which participants placed the hand on top of the objects correctly when comparing the groups without an avatar and fixed height versus female participants with avatar and fixed height, $\chi^2(1) = 1.029$, $p = 0.310$.

On the other hand, there was a statistically significant association between avatar and the frequency with which participants placed the hand on top of the objects correctly when comparing the groups without an avatar and fixed height versus male participants with avatar and fixed height, $\chi^2(1) = 7.251$, $p = 0.007$. There was a moderate association between avatar and the frequency with which participants placed the hand on top of the objects correctly, $\Phi = 0.317$, $p = 0.007$. The frequencies can be seen in Fig. 4.

There was no statistically significant association between avatar and the frequency with which participants placed the hand on top of the objects correctly when comparing the groups without an avatar and personalised height versus female participants with avatar and personalized height, $\chi^2(1) = 0.450$, $p = 0.502$.

Comparing participants without an avatar and personalised height vs male participants with avatar and personalised height, not all expected cell frequencies were greater than five. Therefore a Fisher's Exact test was conducted between avatar and the frequency with which participants placed the hand on top of the objects correctly. There was a statistically significant association between avatar

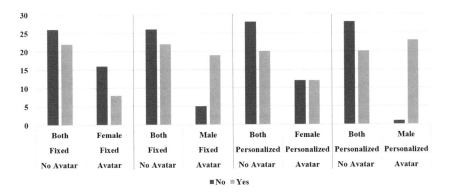

Fig. 4. Avatar comparison taking into account the gender differences.

and the frequency with which participants placed the hand on top of the objects correctly, $p < 0.001$. The frequencies are shown in Fig. 4. Also, when participants could adjust their height to fell more natural, they tended to increase it, G2 (M $= 0.047$ and SD $= 0.085$) and G4 (M $= 0.022$ and SD $= 0.058$).

5   Discussion

Having as motivation the little literature regarding the perception of small distances, the research team also studies how gender, height and avatar influences distance perception regarding those distances. Results of the Pearson's chi-square test for association found differences between gender for all the conditions with avatar present but no statistically significant differences in the absence of avatar. Results show that males have a higher rate than females when grabbing a real object through VE. One explanation for females underestimating distances can be because of the avatar's hand was not close enough, regarding representation, to their hand, causing perception changes when trying to grab the objects.

Regarding height, the Pearson's chi-square test for association shows that there is no statistically different accuracy between fixed and personalised height. An explanation would be that the changes in height made by the participants who had this possibility, was so insignificant that they made no difference when grabbing the real objects (changes in height in meters M $= 0.022$ and SD $= 0.058$).

As for the avatar variable, results of the Pearson's chi-square test for association show that there are statistically different results between both genders of the condition "fixed height with no avatar" and "males with fixed height and avatar". There were also statistically different results between both gender of the condition "personalized height with no avatar" and "males with personalized height and avatar". As for females, the presence of avatar did not make any difference. In the condition with an avatar, people tend to correct the hand position accordingly to the virtual representation, allowing males to increase their accuracy; but because females had perception changes, the presence of an avatar did not contribute to a better understanding of the distances.

6 Conclusion

This work the impact of gender, avatar and height in distance perception of small distances in the VE. Results show that gender does make a difference in distance perception in some cases (any condition where an avatar was present). Moreover, results also evidentiate that height (Fixed or Personalized) did not make any statistical difference when grabbing objects through a VE, and the avatar (present or not present) made a difference in carrying out the task given.

This work is not free of limitations, one of the limitations is that one only has reports whether they grabbed the real object or not and not a quantitative value. We also need to evaluate further why the gender had differences in distance estimation. As for future work, one intends to evaluate small distance estimation with quantitative reports increasing the knowledge in this field. The research team also want to implement different hands for each gender and compare the new results with the results obtained in this work.

Acknowledgments. This work is financed by the ERDF - European Regional Development Fund through the Operational Programme for Competitiveness and Internationalisation - COMPETE 2020 Programme and by National Funds through the Portuguese funding agency, FCT - Fundação para a Ciência e a Tecnologia within project POCI-01-0145-FEDER-028618 entitled PERFECT - Perceptual Equivalence in virtual Reality For authEntiC Training.

References

1. Fuchs, P., Moreau, G., Guitton, P.: Virtual Reality: Concepts and Technologies, 1st edn. CRC Press, Boca Raton (2011)
2. Koutek, M.: Scientific Visualization in Virtual Reality: Interaction Techniques and Application Development. Delft University of Technology, mathesis, January 2003
3. OptiTrack: OptiTrack - Motion Capture (2018). http://optitrack.com/
4. Piryankova, I.V., de la Rosa, S., Kloos, U., Bülthoff, H.H., Mohler, B.J.: Egocentric distance perception in large screen immersive displays. Displays **34**(2), 153–164 (2013)
5. Williams, B., Rasor, T., Narasimham, G.: Distance perception in virtual environments. In: Proceedings of the 6th Symposium on Applied Perception in Graphics and Visualization - APGV 2009, Chania, Crete, Greece, pp. 7–10. ACM Press (2009)
6. Bodenheimer, B., Meng, J., Wu, H., Narasimham, G., Rump, B., McNamara, T.P., Carr, T.H., Rieser, J.J.: Distance estimation in virtual and real environments using bisection. In: Proceedings of the 4th Symposium on Applied Perception in Graphics and Visualization - APGV 2007, Tubingen, Germany, pp. 35–40. ACM Press (2007)
7. Creem-Regehr, S.H., Kunz, B.R.: Perception and action. Wiley Interdisc. Rev. Cognitive Sci. **1**(6), 800–810 (2010)
8. Mohler, B.J., Creem-Regehr, B.J., Thompson, W.B., Bülthoff, H.H.: The effect of viewing a self-avatar on distance judgments in an HMD-Based Virtual Environment. Presence Teleoperators Virtual Environ. **19**(3), 230–242 (2010)

 9. Ries, B., Interrante, V., Kaeding, M., Phillips, L.: Analyzing the effect of a virtual avatar's geometric and motion fidelity on ego-centric spatial perception in immersive virtual environments. In: Proceedings of the 16th ACM Symposium on Virtual Reality Software and Technology - VRST 2009, Kyoto, Japan, pp. 59–66. ACM Press (2009)
10. McManus, E.A., Bodenheimer, B., Streuber, S., de la Rosa, S., Bülthoff, H.H., Mohler, B.J.: The influence of avatar (self and character) animations on distance estimation, object interaction and locomotion in immersive virtual environments. In: Proceedings of the ACM SIGGRAPH Symposium on Applied Perception in Graphics and Visualization - APGV 2011, Toulouse, France, pp. 37–44. ACM Press (2011)
11. Leyrer, M., Linkenauger, S.A., Bülthoff, H.H., Kloos, U., Mohler, B.: The influence of eye height and avatars on egocentric distance estimates in immersive virtual environments. In: Proceedings of the ACM SIGGRAPH Symposium on Applied Perception in Graphics and Visualization - APGV 2011, Toulouse, France, pp. 67–74. ACM Press, August 2011
12. Coluccia, E., Louse, G.: Gender differences in spatial orientation: a review, vol. 24, no. 3, pp. 329–340. http://www.sciencedirect.com/science/article/pii/S0272494404000477
13. Naceri, A., Chellali, R.: The effect of isolated disparity on depth perception in real and virtual environments. In: 2012 IEEE Virtual Reality (VR). IEEE, March 2012
14. Creem-Regehr, S.H., Willemsen, P., Gooch, A.A., Thompson, W.B.: The influence of restricted viewing conditions on egocentric distance perception: implications for real and virtual indoor environments. Perception **34**(2), 191–204 (2005)
15. Peer, A., Ponto, K.: Preliminary exploration: perceived egocentric distance measures in room-scale spaces using consumer-grade head mounted displays. In: IEEE Virtual Reality (VR), Los Angeles, CA, USA, pp. 275–276. IEEE, March 2017
16. Renner, R.S., Velichkovsky, B.M., Helmert, J.R.: The perception of egocentric distances in virtual environments - a review. ACM Comput. Surv. **46**(2), 1–40 (2013)
17. Li, Z., Phillips, J., Durgin, F.H.: The underestimation of egocentric distance: evidence from frontal matching tasks. Attention, Percept. Psychophysics **73**(7), 2205–2217 (2011)

Usability Evaluation of a Virtual Assistive Companion

Ana Luísa Jegundo[1]($\boxtimes$), Carina Dantas[1], João Quintas[2], João Dutra[1],
Ana Leonor Almeida[1], Hilma Caravau[3], Ana Filipa Rosa[3],
Ana Isabel Martins[3], Alexandra Queirós[3],
and Nelson Pacheco Rocha[3]

[1] Cáritas Diocesana de Coimbra, Coimbra, Portugal
{analuisajegundo, carinadantas, joaodutra,
anaalmeida}@caritascoimbra.pt
[2] Instituto Pedro Nunes, Coimbra, Portugal
jquintas@ipn.pt
[3] Universidade de Aveiro, Aveiro, Portugal
{hilmacaravau, filiparosa, anaisabelmartins,
alexandra, npr}@ua.pt

Abstract. As a result of the great technological advances in the last decades, new solutions are emerging to avoid social isolation and to delay the institutionalization of older adults, as is the case of virtual assistants. This paper presents the usability evaluation of a Virtual Assistive Companion (VAC), the CaMeLi. The usability evaluation was based on a multi-method approach that comprises self-reported usability, usability reported by an evaluator and critical incidents registration. The usability tests were performed with 46 participants with an average age of 64 years. The results showed a good usability and satisfaction level, although usability reported by evaluators and critical incidents registration suggest that some functionalities need to be improved in order to facilitate the VAC interaction and understandability.

Keywords: Usability evaluation · Usability testing ·
Virtual Assistive Companion · Avatar

1 Introduction

Home care emerges as a potentially cost-effective solution to the challenges that health and social care systems need to face due to actual demographic trends. Most of the older adults desire to stay at home as long as possible, with independence and autonomy, and integrated in the community rather than become institutionalized [1, 2]. However, the changes in mental and physical abilities, resulting from the normal aging process, make this preference a challenge to be addressed.

To satisfy this preference, avoiding or delaying the institutionalization and preventing social isolation, alternative solutions have been given increased attention. Integrated home care services are supposed to be one of the possible solutions available to promote aging in place. Concerning this, new intelligent technologies to support

© Springer Nature Switzerland AG 2019
Á. Rocha et al. (Eds.): WorldCIST'19 2019, AISC 931, pp. 706–715, 2019.
https://doi.org/10.1007/978-3-030-16184-2_67

older adults remaining longer independent and active at home are under development. Although these solutions cannot and should not replace the human role, they represent an additional support to the human caregiving [3]. One example of these solutions are personal assistants that provide practical tools to help older adults to maximize independence and autonomy at home. Personal assistants include virtual assistant applications capable of receiving human voice communications from a user, and, after processing that information, able to select an appropriate responsive action to the query or instruction received [4].

In the scope of the CaMeLi European Project [5], funded under the AAL Joint Program, a Virtual Assistive Companion (VAC), the CaMeLi, was developed [6]. This system works through an avatar with a human-like figure with which older adults can speak and interact by asking to perform some actions, such as consulting the news [6]. The online version can be found at https://cameli-gui.las.ipn.pt/.

The usability of products and services is a key issue during the development of technological solutions. There are several methods and instruments used to ensure the system's usability, concerning the stages of design and development, and the type of required data (data from users or usability experts) [7].

This paper presents the results of the usability evaluation of the CaMeLi based on a multi-method approach that comprises self-reported usability, usability reported by evaluators and critical incidents registration. Additionally, the paper presents the state of the art related to VAC developments, followed by the methods used to evaluate usability, the respective results, discussion and some concluding remarks.

2   State of the Art

2.1   Virtual Assistive Companions

A VAC is a computer-animated character exhibiting a certain level of intelligence and autonomy as well as social skills to simulate human face-to-face conversation and abilities to sense and respond to user affect. Several VACs have been successfully developed for various target applications to monitor, encourage, and assist older adults.

Different researchers have explored VACs that interact with users over multiple conversations, ranging from a handful of interactions to hundreds of interactions spanning long-term periods [8–10]. Most of the developed VACs are designed for controlled environments and have rarely made the step out of the laboratory as autonomous applications in real-world settings. To achieve useful and successful virtual agents, that maintain their users engaged in beneficial long-term relationships, it is necessary to integrate these systems seamlessly in real-world environments and make them capable of interacting with humans autonomously, in an intuitive, natural and trouble-free way in everyday situations [11, 12].

VACs are typically represented in the form of human or animal bodies that should be lifelike and believable in the way they behave. They simulate human-like properties in face-to-face conversation, including abilities to recognize and respond to verbal and non-verbal inputs, generate verbal and non-verbal outputs (e.g. mouth, eye and head movements, hand gestures, facial expressions or body posture) and can deal with conversational functions such as turn taking, feedback and repair mechanisms [13].

Studies suggest that VACs dealing with emotion and affect are particularly capable of capturing the user's attention, engaging them in active tasks and entertaining them [15], leading to the development of affinity relationships with human partners [16].

2.2   Examples of State-of-the-Art Virtual Assistive Companions

A variety of VACs has been developed aiming to provide social support to isolated older adults [18] and to address daily needs for an autonomous living [14].

In an exploratory pilot [21], a VAC designed to provide longitudinal social support to isolated older adults using empathetic feedback was placed in the homes of 14 older adults for a week. Results demonstrated significant reductions in loneliness based on self-reported mood.

Many examples of VACs have been developed within the scope of Active and Assisted Living (AAL) applications (besides CaMeLi, which is the object of the study reported by this paper).

Greta[1] is an embodied conversational agent that uses a virtual three-dimensional model of a female character. Greta uses verbal and nonverbal communication to interact with the users and can be used with different external text-to-speech software. Facial animation parameters are required to animate faces of different sizes and proportions.

Virtask[2] has developed Anne, a VAC that provides organizational and individual users with novel opportunities in regard to the execution of their tasks. Currently, several companies and organizations are successfully using Anne, for example, this agent was adopted in DALIA and MyLifeMyWay AAL projects, where it is being employed as a virtual caregiver of elderly and disabled persons, acting as an assistant for daily life activities at home. Older adults can use speech to interact with an avatar, which will answer based on data collected through a set of sensors deployed in the household. Moreover, informal careers have access to the same avatar, which can tell them what they should do in different situations or just to talk with the person cared for.

V2me is a virtual coach to combine virtual and real-life social networks to prevent and overcome loneliness of older adults. V2me supports active aging by improving integration into society through the provision of advanced social connectedness and social network services and activities. Professional caregivers and elderly family can monitor user's activity with the system by a specific web interface.

3   Methods

3.1   Study Design

The aim of the observational study reported by this paper was to assess the usability of CaMeLi, a VAC developed in a research and development project [5]. The observational study took place between September and October 2018. The participants were

---

[1] https://perso.telecom-paristech.fr/~pelachau/Greta/.

[2] http://www.virtask.nl/wordpress/en/.

selected by convenience between users of a non-profit social organization, *Cáritas Diocesana de Coimbra*.

Previously to the evaluation sessions, the participants received an information sheet with the study objectives, duration and methods. Elements of the research team explained that participants could request additional information about the study at any moment and abandon the study at any time without any explanation or personal prejudice. All data collection was anonymized, and the involved participants received all the information regarding the study and their participation before completing the informed consent. Additionally, a written informed consent was obtained prior to data collection.

The inclusion criteria defined were age over 18 years, ability to read, understand the study, accept voluntarily to participate in the study, and willingness to sign the informed consent. The exclusion criteria considered were limitations in terms of health conditions that, according to the principal researcher, could compromise the subject's ability to integrate the study.

Each usability evaluation session followed three stages: (i) Pre-test: the participants filled a sociodemographic questionnaire for sample characterization, with demographic information and one question about the technological literacy; (ii) Test: the participants interacted with the CaMeLi following the tasks described in a session script. Simultaneously, an observer took notes about the performance of the participants and critical incidents registration; (iii) Post-test: the participants filled the self-reported usability questionnaire: Usefulness, Satisfaction, and Ease of use (USE) [22] and evaluators filled the International Classification of Functioning Disability and Health based Usability Scale (ICF-US) according to their opinion about the user interaction with CaMeLi.

3.2   Assessment Instruments

USE is an auto-reported questionnaire to identify issues that may influence and determine frequency of use and user satisfaction. It contains 30 items to assess four dimensions of usability: ease of use, ease of learning, usefulness and satisfaction. It allows the perception of the users, as it is a self-reported questionnaire [22]. Each item is scored in a seven-point Likert scale, ranging from 1 (strongly disagree) to 7 (strongly agree). The questionnaire analysis is based on the average of the user's answers, and the average of each dimension can be analyzed in a disaggregated way [22].

The ICF-US is a generic usability assessment scale that can be used for self-reported usability, as wells as usability evaluation based on the opinion of evaluators. This tool is in line with the International Classification of Functioning Disability and Health conceptual model and consist of two subscales: (i) ICF-US I, that allows a comprehensive usability assessment and (ii) ICF-US II, that allows the classification of application components as barriers and facilitators, identifying the application strengths and weaknesses [23, 24]. This subscale is adaptable in order to address each component of the application being assessed. To fulfill the two subscales evaluators must consider the performance of the participants during the evaluation. Both subscales score all items from −3 (less positive) to 3 (most positive) values. The final score of the ICF-US I is calculated by the sum of the scores of all items. A value above 10 points indicates a good usability and less than 10 is considered as a prototype with opportunities to improve. Concerning ICF-US II, when an item is classified as a barrier, it is possible to identify the feature that is influencing that classification, in order to report the features that can be improved. On the other hand, the features identified as facilitators can be pointed as good practices for future developments [23, 24].

3.3 Critical Incident Register

During the test, while participants were executing the tasks, an observer recorded the critical incidents, to systematically identify behaviors that contribute to the participant's success or failure in specific tasks. In this record some details were considered such as easy/difficult interaction with the application or learning ability of the sequence needed to complete a task.

4 Results

4.1 Assessment Instruments

The total sample consisted of 46 participants, 34 female (73,91%) and 12 males (26,09%) and had an average age of 63,63 years (SD = 20,48). More than half of the sample have internet at home (60,87%), almost half use computer and tablet (45,65%) and only 28,26% use smartphone.

The average total score of the USE for all participants was 5,06 out of a maximum of 7 (SD = 0,31), which indicates that CaMeLi has a good degree of usability and satisfaction. The results for the four dimensions associated with USE, listed in descending order of average scores, were: Satisfaction (mean value = 5,28; SD = 0,36), Usefulness (mean value = 5,12; SD = 0,12), Ease of learning (mean value = 5,02; SD = 0,13) and Ease of use (mean value = 4,89; SD = 0,33).

Items with higher scores are related to the satisfaction and usefulness of the CaMeLi, namely: "I would recommend it to a friend" (mean value = 5,70; SD = 1,33); "It is fun to use" (mean value = 5,65; SD = 1,22) and "It is useful" (mean value = 5,46; SD = 1,13). The items that obtained lower scores are related with the ease of use dimension: "I can use it successfully every time" (mean value = 4,41; SD = 1,67); "I can use it without written instructions" (mean value = 4,59; SD = 1,71) and "It is user friendly" (mean value = 4,65; SD = 1,65).

Regarding usability evaluation depending on the evaluator's opinion about user's performance, the results of ICF-US I show that CaMeLi was a facilitator for 35 participants and a barrier for 11 participants. In a range from −30 to 30, the mean score for all participants was 15,9 (SD = 16,6), indicating that in general the application was a facilitator. The participant with the highest score had 30, and the participant with the lower score had −25.

Considering the average of the different items of ICF-US I, none of the items had scores lower than zero, which means that, in average, barriers were not identified (see Fig. 1). In turn, the facilitators, in order of importance, were: "The degree of satisfaction with the use" (mean value = 2,0; SD = 1,5); "The similarity of the way it works on different tasks (e.g. to confirm an action is always equal)" (mean value = 1,8; SD = 1,5); and with mean values of 1,6 points were the items "The ease of use" (SD = 1,9), "The ease of learning" (SD = 1,9) and "Overall, I consider that the application was" (SD = 1,7). Concerning the items identified as facilitators, it seems that most participants were satisfied with the easiness and learnability of CaMeLi.

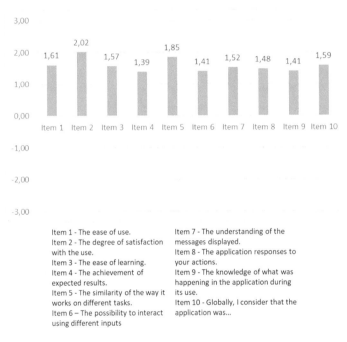

Item 1 - The ease of use.
Item 2 - The degree of satisfaction with the use.
Item 3 - The ease of learning.
Item 4 - The achievement of expected results.
Item 5 - The similarity of the way it works on different tasks.
Item 6 – The possibility to interact using different inputs

Item 7 - The understanding of the messages displayed.
Item 8 - The application responses to your actions.
Item 9 - The knowledge of what was happening in the application during its use.
Item 10 - Globally, I consider that the application was...

Fig. 1. ICF-US I results.

During the observational study, the ICF-US II subscale was applied to 13 participants and the detailed results are presented on Fig. 2.

The results of the parts associated with ICF-US II were analyzed according to the mean values of the items (range −3 to 3):

- Application Components (3 items) - average of the responses = −0,90 (SD = 0,62);
- Detailed Usability (12 items) - average of the responses = −0,70 (SD = 1,19);
- Overall Assessment (1 item) - average of the responses = −1,23 (SD = 0,93).

The results of the application of the ICF-US II indicate that the studied parts of CaMeLi could be improved, particularly because the lowest result was the ones referring to the overall assessment. The identified barriers, in order of severity, were: "The voice interaction" (mean value = −2,46; SD = 0,78); "The layout change (hide the avatar)" (mean value = −1,69; SD = 1,32); "The avatar was" (mean value = −1,54; SD = 1,13); "The icons representativeness" (mean value = −1,54; SD = 1,27) and "The understanding of the application operating mode" (mean value = −1,46; SD = 1,05).

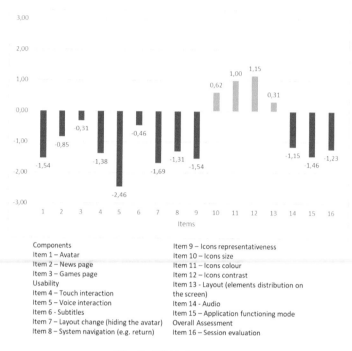

Fig. 2. ICF-US II results.

Thereby, it seems that participants had difficulty in tasks that encompass the voice interaction and hiding the avatar, as well as to perform tasks that involve the buttons identification due to the low representativeness of the icons. Despite this, the items with the higher results identified as facilitators were all related with icons specifications, namely: "The contrast of the icons" (mean value = 1,15; SD = 1,77); "The color of the icons" (mean value = 1,00; SD = 1,63) and "The size of the icons" (mean value = 0,62; SD = 1,56).

4.2 Critical Incident Register

Concerning the critical incidents resulting from the observation process, the participants reported difficulties in obtaining CaMeLi response to voice commands and, in several cases, the task had to be carried by the evaluators or through touch control. Also, almost all the participants reported difficulties identifying the icon correspondent of each button.

Hiding the avatar and turning it visible again, as well as turning off the microphone, identifying the return and locating it on the screen were other tasks that raised problems to the participants.

Some participants suggested that these actions could be completed through voice commands. When the avatar is hidden the layout of the elements on the screen change, spreading on the screen, and in these situations some participants revealed problems executing tasks. Some participants had difficulty interacting with the computer's touchscreen due to their lack of manual dexterity, and also revealed problems browsing the system pages using the scroll function to select element/buttons on the screen. Some of them, unintentionally, touched two points of the screen simultaneously and was notorious the lack of understanding of what was happening in the system. These results are consistent with the outcomes of the application of the USE and ICF-US II scale.

The existence of links for external sites, such as advertising or news sites, had led to confusion among some participants when they inadvertently clicked on these options, without realizing what was happening.

5 Discussion

Most of the developed VACs were designed for controlled environments and, in this sense, CaMeLi states the difference, being developed for real-world environment. CaMeLi has the potential to help older adults to actively engage in daily tasks and entertaining themselves, stimulating also their socialization.

The results collected during the usability evaluation, made it clear that participants considered CaMeLi a useful tool that may represent an added value for their daily lives, both in institutional or home context. Participants enjoyed the available functionalities and activities, were satisfied with the application and referred that they would recommend it to friends.

Considering the results of the application of the USE questionnaire, the participants reported a good level of usability and acceptance, specifically, in the dimensions of satisfaction, usefulness and ease of learning, which have the best results. These results were confirmed by the outcomes of the application of ICF-US I, with all the items identified as facilitators.

The positive results obtained by the application of USE and ICF-US may be a good predictor of a frequent and independent use of the CaMeLi in the future.

Despite the good results of the participant's perspective, the usability evaluation was also important to capture some details that, if improved, would favor the interaction with CaMeLi.

The item with the lowest score, in USE questionnaire, was related to the ability to use the system successfully, which indicates that participants had some difficulties in tasks completion. These problems were also identified in the ICF-US II and of critical incidents register that also showed voice interaction problems for a significant number of participants.

These results clearly showed that improvements should be implemented. To address these issues, the authors suggested the following: voice interaction should be improved (e.g. capture distant sounds or capture sounds at low volume); other actions trough voice commands should be enable (e.g. hide the avatar); some of the features/buttons should be more intuitive and evident on the screen (e.g. return or hide/turn visible the avatar); text labels should be included on buttons; and links for external websites should be disabled. The authors further suggest the redesign of the icons and the improvement of touchscreen interaction (e.g. the sensitive touch area should be enlarged).

The CaMeLi research team should continue working on defining new functionalities and menus that older adults consider relevant to include in the system.

6 Conclusions

This paper reports the usability evaluation of the CaMeLi, using a multi-method approach for evaluating the respective main functions. The results showed good usability in terms of the user's perspective and a high level of satisfaction. However, the evaluator's assessment and the critical incidents registration identified several usability problems that need to be considered in more detail.

The usability of VACs such CaMeLi is crucial, as this type of technologies can have an important role in promoting aging in place with independence, autonomy and access to services that help older adults to stay connected, informed and socially active. In this respect, the multi-method that was applied for the CaMeLi usability evaluation prove to be efficient, which means that it can be used to evaluate the usability of other VACs.

Acknowledgements. This work was partially supported by the GrowMeUp project, funded by the European Commission within the H2020-PHC-2014, (Grant Agreement: 643647) and by the DAPAS project co-funded by the European AAL Joint Programme (Active and Assisted Living — ICT for ageing well, Call 2017) and by FCT (Portuguese Funding Authority) - Fundação para a Ciência e Tecnologia, I.P. (AAL/0005/2017).

References

1. Wiles, J.L., Leibing, A., Guberman, N., Reeve, J., Allen, R.E.S.: The meaning of 'Aging in Place' to older people. Gerontologist **52**(3), 357–366 (2012)
2. Bedaf, S., et al.: Which activities threaten independent living of elderly when becoming problematic: inspiration for meaningful service robot functionality. Disabil. Rehabil. Assist. Technol. **9**(6), 445–452 (2014)
3. Tsiourti, C., Joly, E., Wings, C., Ben Moussa, M., Wac, K.: Virtual assistive companions for older adults: qualitative field study and design implications. In: Proceedings of the 8th International Conference on Pervasive Computing Technologies for Healthcare, pp. 57–64 (2014)
4. Cooper, R.S., McElroy, J.F., Rolandi, W., Sanders, D., Ulmer, R.M., Peebles, E.: Personal virtual assistant, 21 March 2008
5. Tsiourti, C., et al.: A virtual assistive companion for older adults: design implications for a real-world application, pp. 1014–1033. Springer, Cham (2018)
6. Cáritas Diocesana de Coimbra: International Project Cameli (2017). https://www.caritascoimbra.pt/en/project/cameli/
7. Martins, A.I., Queirós, A., Silva, A.G., Rocha, N.P.: Usability evaluation of ambient assisted living systems using a multi-method approach. In: Proceedings of the 7th International Conference on Software Development and Technologies for Enhancing Accessibility and Fighting Info-exclusion - DSAI 2016, pp. 261–268 (2016)
8. Bickmore, T., Gruber, A., Picard, R.: Establishing the computer-patient working alliance in automated health behavior change interventions. Patient Educ. Couns. **59**(1), 21–30 (2005)

9. Schulman and Daniel: Embodied agents for long-term interaction, Northeastern University (2013)
10. Bickmore, T.W.: Relational agents: effecting change through human-computer relationships, Massachusetts Institute of Technology (2003)
11. Quintas, J., Menezes, P., Dias, J.: Information model and architecture specification for context awareness interaction decision support in cyber-physical human-machine systems. IEEE Trans. Hum. Mach. Syst. 47(3), 323–331 (2017)
12. Quintas, J., Martins, G.S., Santos, L., Menezes, P., Dias, J.: Toward a context-aware human-robot interaction framework based on cognitive development. IEEE Trans. Syst. Man, Cybern. Syst., 1–11 (2018)
13. Cassell, J.: Embodied conversational interface agents. Commun. ACM 43(4), 70–78 (2000)
14. Kramer, M., Yaghoubzadeh, R., Kopp, S., Pitsch, K.: A conversational virtual human as autonomous assistant for elderly and cognitively impaired users? Social acceptability and design considerations, vol. 220. Ges. für Informatik (2013)
15. Klein, J., Moon, Y., Picard, R.W.: This computer responds to user frustration. Interact. Comput. 14(2), 119–140 (2002)
16. Nijholt, A.: Disappearing computers, social actors and embodied agents. In: Proceedings. 2003 International Conference on Cyberworlds, pp. 128–134 (2003)
17. Kelley, J.F.: An iterative design methodology for user-friendly natural language office information applications. ACM Trans. Inf. Syst. 2(1), 26–41 (1984)
18. Vardoulakis, L.P., Ring, L., Barry, B., Sidner, C.L., Bickmore, T.: Designing relational agents as long term social companions for older adults, pp. 289–302. Springer, Heidelberg (2012)
19. Bickmore, T.W., et al.: A randomized controlled trial of an automated exercise coach for older adults. J. Am. Geriatr. Soc. 61(10), 1676–1683 (2013)
20. Kasap, Z., Magnenat-Thalmann, N.: Building long-term relationships with virtual and robotic characters: the role of remembering. Vis. Comput. 28(1), 87–97 (2012)
21. Ring, L., Shi, L., Totzke, K., Bickmore, T.: Social support agents for older adults: longitudinal affective computing in the home. J. Multimodal User Interfaces 9(1), 79–88 (2015)
22. Lund, A.M.: Measuring Usability with the USE Questionnaire. Usability Interface 8(2), 3–6 (2001)
23. Martins, A.I., Rosa, A.F., Queirós, A., Silva, A., Rocha, N.P.: Definition and validation of the ICF - usability scale. Procedia Comput. Sci. 67 (2015)
24. Martins, A.I., Queirós, A., Silva, A.G., Rocha, N.P.: ICF based Usability Scale: evaluating usability according to the evaluators' perspective about the users' performance. In: Proceedings of the 7th International Conference on Software Development and Technologies for Enhancing Accessibility and Fighting Info-exclusion - DSAI 2016, pp. 378–383 (2016)

Internet Access in Brazilian Households: Evaluating the Effect of an Economic Recession

Florângela Cunha Coelho[1,2,4($\boxtimes$)] (ID), Thiago Christiano Silva[2,3($\boxtimes$)] (ID), and Philipp Ehrl[2($\boxtimes$)] (ID)

[1] Defensoria Pública da União, SAUN Q5, Asa Norte, Brasília, DF, Brazil
florangela.coelho@dpu.def.br
[2] Universidade Católica de Brasília,
SGAN 916 Módulo B Avenida W5, Brasília, DF, Brazil
{thiago.christiano,philipp.ehrl}@ucb.br
[3] Universidade de São Paulo,
Av. Bandeirantes, 3.900, Monte Alegre, Ribeirão Preto, SP, Brazil
[4] Universidade de Brasília, Campus Universitário Darcy Ribeiro,
Brasília, DF, Brazil

Abstract. The present article analyzes how the Brazilian recession in 2014 affected internet access at home using annual data between 2012 and 2017 from a large scale, representative household survey by the Regional Center for Studies for the Development of the Information Society (Cetic). Pooled OLS and Probit regression models show that the Brazilian recession had a substantial negative impact on Internet access. We find that, on average, the demand for internet access at home decreased 8% due to the deteriorated economic situation. The data also reveal that regions in which internet access was more widespread suffered a higher percentage decline. Households with lower levels of income and schooling also experience a stronger negative impact on internet access rates due to the crisis.

Keywords: Internet access at home · ICT · Recession · Probit

1 Introduction

Information and communication technologies (ICT) are important for the development of the economy because they are present in the main productive sectors. The ICT industry is known for its constant innovation and for delivering significant spillovers to other industries. According to the World Bank [18], research has shown that ICT investment is associated with economic benefits such as increased productivity, lower costs, new economic opportunities, job creation, innovation and increased trade. ICTs also help deliver better health and education services and strengthen social cohesion.

Access to ICTs has grown tremendously since the late 1990s, driven mainly by wireless technologies, the liberalization of telecommunications markets and lower utilization costs, according to the World Bank [18]. The internet today is an essential tool for businesses and private households. Table 1 shows the performance of the ICT sector in terms of different access rates in the world and in Brazil. By 2016 there were

© Springer Nature Switzerland AG 2019
Á. Rocha et al. (Eds.): WorldCIST'19 2019, AISC 931, pp. 716–725, 2019.
https://doi.org/10.1007/978-3-030-16184-2_68

more than 3.5 billion people worldwide were using the Internet, of which 2.5 billion were from developing countries, according to the World Bank [18]. At least by 2016, Brazil shows above average performance in the five selected ICT access indicators. More than half of all Brazilian households were using internet access at home and reported that they possess a personal computer. In 2010, only 27% of households had the internet access at home. Between 2010 and 2016 fixed broadband subscribers went from 7.2 to 13.0 per 100 inhabitants, while fixed and mobile telephone subscriptions show considerably lower growth rates.

Table 1. Evolution of ICT access in Brazil and in the world by type of access, 2010 and 2016.

Period	2010	2016	2010	2016
Region	World		Brazil	
Fixed-telephone subscriptions (per 100 people)	17.8	13.6	21.6	20.4
Mobile-cellular telephone subscriptions (per 100 people)	76.5	101.6	100.9	118.9
Fixed-broadband subscriptions (per 100 people)	7.9	12.5	7.2	13.0
Households with a computer (%)	36.0	46.6	34.9	51.0
Households with Internet access at home (%)	30.1	51.5	27.1	52.4

Source: World Bank (2018)

Despite the apparent positive development of ICT usage rates in Brazil, the aggregate data is silent regarding a possible non-linear evolution over time and whether there are considerable regional heterogeneities. For several reasons, we suspect that these two aspects are interesting and underexplored research topics. First, Brazil is known for its pronounced and persistent regional economic inequality, particularly between the richer South and Southeast and the lagging regions in the North and Northeast [3]. Despite the convergence in social inequality [1], the persistent income divergence may be associated with inequalities in terms of ICT access. Second, the real growth rate of Brazil's GDP clearly indicates that the economy entered into recession during second half of the 2010 s. GDP started to decrease since 2014, showing declines of more than 4% during 2015 and 2016. Despite the stabilization of the GDP growth rate, the crisis is all but over in 2017. Table 2 also shows the associated steady increase in unemployment rates reaching almost 12% in 2016 and 2017. According to the National Development Bank BNDES [4], household consumption also fell by 4.0% in 2015 and 2016.

Motivated by this context, the purpose of the present paper to analyze the evolution of the internet access rate (as one specific example ICT) in Brazil during the period from 2012 to 2017 with a special focus on the effects of the economic crisis. The use of microdata allows us to explicitly analyze how education and income of Brazilian households affect the probability of having internet access and whether these socioeconomic characteristics interact with the effects of the economic crisis.

Table 2. Evolution of the per capita GDP growth rate and average unemployment rate in Brazil, 2010–2017

Year	2010	2011	2012	2013	2014	2015	2016	2017
GDP per capita growth (%)	6.45	2.97	0.97	2.08	−0.36	−4.34	−4.40	0.20
Unemployment rate (%)	6.70	6.00	5.50	5.40	4.80	8.50	11.50	11.73

Source: IBGE (2018) [9].

In addition to the present introduction, this paper is divided into four more sections. In the second section, we briefly review related literature. Section three presents the data and econometric methodology. The fourth section contains our results and the fifth section offers some brief conclusions and possibilities for further research.

2 Related Literature

According to Prieger [14] and Wirthmann [17], the generalized growth of ICTs helps to create incentives to increase the participation of individuals, which in turn presupposes developing skills to deal with innovations in the digital world, which is especially important for developing countries, such as Brazil. Carvalho et al. [5] document the expansion of internet access, cable TV, their increased service quality and the implied effects on the growth rate of municipalities.

Notable previous studies that analyze the determinants of access to ICTs, considering the socioeconomic characteristics of families such as Nishijima et al. [11], Demoussis and Giannakopoulos [8], Vicente and López [15] or Macedo [10] demonstrate that access inequalities are strongly related to income and education, amongst others. Particularly close to our analysis is the paper by Nishijima et al. [11] on the digital divide in Brazil. In contrast to our approach, the authors used socioeconomic data from the Household Sample Survey (PNAD) for the pre-crisis years between 2005 and 2013, using quarterly data, and estimated the highly similar logistic model instead of probit models to evaluate the disaggregation among different population groups.

Out of the Brazilian context, Dasgupta et al. [7], Pâmtea and Martems [13], Vicente and Lopez [15] and Zhang [16] showed that the role of socioeconomic, cultural and geographic characteristics can be defining with regard to opportunities for the ICT access as well as for its use among individuals in different countries.

3 Methodology

This section discusses the data base and the econometric specifications of our analyses.

3.1 Data

The data stem from the TIC research, which is an annual survey realized by the Research Center of Studies for the Development of the Information Society (Cetic)

[12], acting under the auspices of UNESCO. The Cetic is the executive arm of the Internet Management Committee in Brazil (CGI), which is responsible for the coordination and integration of internet services in the country and consists of members of government, business sector and the academic community. We obtained the microdata from 2012 to 2017 in electronic format for research purposes. The main purpose of the TIC survey is to map the access to internet and ICT infrastructure in urban and rural households in Brazil and the ways in which these technologies are used by individuals 10 years or older. In addition to the Brazilian Federal capitals, the survey covers municipalities that are part of the government program "Digital Cities".[1]

We exploit the some basic household characteristics such as family income and education level, the region of residence and whether the family had internet and cable TV access at home. Following the definition of the Economic Cycle Dating Committee (CODACE) [6], we define the crisis variable as an indicator for 2014 and the following years with negative or zero GDP growth rates, see Table 2 above. Table 3 shows the descriptive statistics of the variables used in the study.

Table 3. Description of statistics, 2012–2017

Statistic	N	Mean	St. Dev.	Min	Pctl (25)	Pctl (75)	Max
Cable TV?	114,671	0.307	0.461	0	0	1	1
Internet access at home?	114,671	0.519	0.500	0	0	1	1
Household income	103,885	2.490	1.302	1	1	3	6
Schooling	114,450	2.298	0.878	1	2	3	4
Recession	114,671	0.783	0.412	0	1	1	1

3.2 Econometric Specification

We first test whether the Brazilian recession had a significant effect on Internet access of households across different regions in Brazil. We follow the Brazilian Institute of Geography and Statistics (IBGE) and divide Brazil in five disjoint regions: Midwest, North, Northeast, South, and Southeast. The Southern regions are the most developed, while the Northern regions are the least. We use the following econometric model:

$$y_{hrt} = \alpha_r + \beta\,\text{Recession}_t + \gamma\text{Time}_t + \epsilon_{hrt},$$

in which h, r and t index household, region and time, respectively. The dependent variable y_{hrt} is a dummy indicator whether a household h in region r had internet access at home in year t. When $y_{hrt} = 1$, then household r in region r at time t had Internet access at home, and 0, otherwise. The term α_i represents region fixed effects that

---

[1] This program aims to promote digital inclusion in municipalities with a focus on improving the quality of services and public management, through the installation of networks, public Internet access points, management systems in the public area and training.

capture time-invariant idiosyncrasies of Brazilian regions. Such term is important because Brazilian regions significantly differ from each other. Time$_t$ is a linear time trend that captures how Internet access in Brazil evolves over time. Recession$_t$ is our recession dummy, which equals one if $t \geq 2014$, and zero otherwise. Our coefficient of interest is β, which we except to be negative, i.e., the recession in Brazil impacted negatively the probability of having Internet access at home.

We also test whether the recession had different impacts over distinct Brazilian regions. For that, we use the following augmented model:

$$y_{hrt} = \alpha_r + \alpha_t + \sum_r \beta_r \text{Recession}_t \times \text{Region}_r + \epsilon_{hrt},$$

i.e., we interact the recession indicator with the five different Brazilian region dummies. While the region fixed effects absorb the average time-invariant idiosyncratic factor of each region from 2012 to 2017, the interaction term absorbs only the differential of Internet access due to the Brazilian recession since 2014. Note that we also substituted the linear time trend for time fixed effects α_t to capture time-varying but homogeneous factors across different Brazilian regions.

Third, we include explicit determinants of Internet access and verify whether they have changed due to the Brazilian recession using interaction terms with the recession time dummy. We use the following econometric model:

$$y_{hrt} = \alpha_r + \alpha_t + \kappa \text{Determinants}_{ht} + \sum_r \beta_r \text{Recession}_t \times \text{Determinants}_{ht} + \epsilon_{hrt}$$

where the determinants of interest are either:

- Household income[2]: we expect that households with low income will be affected more severely by the recession than those with high income;
- Schooling[3]: We expect that households with low levels of education will be affected more severely by the recession than those with a high level of education;
- Cable TV[4]: We expect that households with cable TV access are more likely to have internet access and that the internet access rates of households with cable TV are less sensitive to the economic recession.

For robustness, we also re-run every the econometric specifications using a non-linear estimation model with a Probit link function. Such robustness is important to confirm the validity of our estimates, because the dependent variable is binary.

---

[2] We discretize the variable in sextiles: up to 1 minimum wage (MW) [1]; between 1 and 2 MWs [2]; between 2 and 3 MWs [3]; between 3 and 5 MWs [4]; between 5 and 10 MWs [5]; more than 10 MW [6].

[3] We discretize the variable in quartiles: illiteracy/early childhood education [1]; elementary school [2]; high school [3]; higher education [4].

[4] We use a dummy variable: 1 when there is cable TV and 0, otherwise.

4 Results

The OLS and Probit model regression results are presented in Table 4, which provides evidence that the Brazilian recession of 2014–2017 had a negative impact on Internet access. Considering the variable recession, the share of households with internet access decreased by 8.2% points. Independent from this crisis effect, we observe a linear trend that increases the share of households by 0.5% points each year meeting federal government's ambitions of increasing the portion of the population that can benefit from broadband Internet access, see Carvalho et al. [5]. When we analyze the interaction between the recession indicator and five Brazilian regions, we observe that the South shows a fall of 11.4% points in Internet access at home compared to the North (the omitted reference region).

This reduction is the largest among the regions of the country. In the Probit model in column 2, the estimated coefficient is equal to −0.21, which corresponds to a marginal effect of 0.08% points, i.e., the negative effect of the economic crisis the probability of having internet access at home is almost equal to the previous OLS estimation. The same applies to the model with interaction terms between the recession and regions.

Table 4. Estimates of OLS and Probit regression for Internet access at home, 2012–2017

	Dependent variable			
	Internet access at home?			
	OLS	Probit	OLS	Probit
	(1)	(2)	(3)	(4)
Recession	−0.082***	−0.210***		
	(0.006)	(0.014)		
Year	0.005***	0.012***		
	(0.001)	(0.004)		
Recession × Northeast			−0.092***	−0.161***
			(0.012)	(0.017)
Recession × Southeast			−0.061***	−0.190***
			(0.012)	(0.015)
Recession × South			−0.114***	−0.332***
			(0.014)	(0.023)
Recession × Midwest			−0.101***	−0.264***
			(0.016)	(0.032)
Constant	−9.481***	−24.718***		−0.283***
	(2.955)	(7.511)		(0.024)
Region FE?	Yes	Yes	Yes	Yes
Time FE?	No	No	Yes	Yes
Observations	114,671	114,671	114,671	114,671
R^2	0.026		0.066	
Log likelihood		−79,206.600		−77,825.230
Akaike Inf. Crit.		158,419.200		155,670.500

Note: *$p < 0.1$; **$p < 0.05$; ***$p < 0.01$.

Table 5. OLS Estimates of the determinants of internet access and recession, 2012–2017

	Internet access at home?		
	(1)	(2)	(3)
Household income2	0.169***		
	(0.009)		
Household income3	0.376***		
	(0.010)		
Household income4	0.492***		
	(0.011)		
Household income5	0.559***		
	(0.013)		
Household income6	0.627***		
	(0.016)		
Recession × Household income1	−0.049***		
	(0.015)		
Recession × Household income2	−0.053***		
	(0.013)		
Recession × Household income3	−0.068***		
	(0.014)		
Recession × Household income4	−0.031**		
	(0.015)		
Recession × Household income6	−0.011		
	(0.021)		
Schooling2		0.200***	
		(0.013)	
Schooling3		0.408***	
		(0.013)	
Schooling4		0.586***	
		(0.014)	
Recession × Schooling1		−0.049***	
		(0.015)	
Recession × Schooling2		−0.024**	
		(0.010)	
Recession × Schooling3		0.018*	
		(0.010)	
Cable TV			0.548***
			(0.007)
Recession × Cable TV			−0.154***
			(0.008)
Region FE?	Yes	Yes	Yes
Time FE?	Yes	Yes	Yes
Observations	103,885	114,450	114,671
R^2	0.211	0.205	0.200

Notes: *p < 0.1; **p < 0.05; ***p < 0.01

For the determinants of internet access, the OLS regression results are presented in Table 5. Given the resemblance between the previous OLS and the Probit estimates, we continue only with the OLS model to save space and for ease of interpretation. The interaction terms between income and the recession indicator reveal that families with a monthly income of between 2 and 3 minimum wages[5] in relation to households with income between 5 and 10 minimum wages (the reference group), suffered from a decrease in their access to the internet at home equal to 6.8%. In fact, all income levels below the reference group have lower average internet access rates, but the lower middle income group was relatively hit the hardest. Note also that, independent of the economic recession, the average probability of having internet at home increases linearly with income, as expected. A similar linear increase is observed for the education levels in column (2).

Those families with the lowest level of schooling experience a relative drop in internet access equal to 4.9%, compared to households with the highest education level, we observed a drop in internet access of 4.9%. The recession had no significant impact on families with at least higher school education which can be explained by the fact that income tends to increase with the level of schooling.

These observations are in line with Nishijima et al. [11], who observe that income is one of the main determinants of ICT use capacity. Finally, column (3) indicates that more than 50% of all families who have a cable TV access in their household also have internet access.

This positive relation may be stronger than in other countries due to the low number of suppliers in both markets and their promotion offers for both TV and broadband internet access. In the wake of the economic crisis, however, this share decreased by 15%. Nevertheless a cable TV subscription is a significant predictor for having internet access.

5   Conclusion

Using data of the TIC Domicile Research of the Regional Center for Studies on the Development of the Information Society, we analyze if the recent economic recession in Brazil affected the access to the internet in households across the country. We also investigate how certain socioeconomic determinants of Brazilian households affected the access to the internet at home and how these determinants interacted with the crisis.

We find that the Brazilian recession of 2014–2017 had a substantial negative impact on Internet access rates. Our results reinforce the importance of socioeconomic characteristics, such as education and income, in the determination of internet access as previously noted by Nishijima et al. [11], Demoussis and Giannakopoulos [8], Dasgupta et al. [7]. Higher unemployment and lower GDP thus do not only directly harm households but they also reduce their capability of sustaining their access to ICT. Furthermore, we observed that regions in which internet access was more widespread suffered a higher relative reduction of access rates.

---

[5] The minimum wage in Brazil is currently around 250 US$.

In general, the results suggest that the increase in the penetration of the broadband internet access service may be related to the increase in the population's purchasing power through a rise in per capita income or a decrease in prices, which reinforces the importance of public policies to stimulate the expansion of the broadband service in the country. To make the economy grow, it is necessary to invest in the State and finance the economic activity, generating quality employment that will positively affect the development of the ICT sector.

References

1. Almeida, R., Moreira, T., Ehrl, P.: Social and economic convergence, Fiscal Federalism Conference Version. (2017) http://mesp.unb.br/images/eventofederalismo/trabalhos/Social-and-economic-convergence.pdf. Accessed 21 Nov 2018
2. Arelano, M.: Panel Data Econometrics. Oxford University Press, Oxford (2003)
3. Azzoni, C.R., Menezes-Filho, N., Menezes, T., Silveria-Neto, R.: Geography and income convergence among Brazilian States. Inter-American Development Bank. Latin American Research Network. Research network Working paper #R-395, 30p. (2000)
4. BNDES - Banco Nacional de Desenvolvimento Social: O crescimento da economia brasileira 2018-2023 (2018). https://web.bndes.gov.br/bib/jspui/bitstream/1408/14760/1/Perspectivas%202018-2023_P.pdf. Accessed 11 Nov 2018
5. Carvalho, A.Y., Mendonça, J.M., Silva, J.J.: Avaliando o efeito dos investimentos em telecomunicações sobre o PIB. Texto para Discussão nº 2336. Instituto de Pesquisa Econômica Aplicada – IPEA (2017)
6. CODACE – Comitê de Datação de Ciclos Econômicos: Comunicado de início da recessão. https://portalibre.fgv.br/data/files/12/17/48/F4/978FE410F9AC5BD45C28C7A8/Comite%20de%20Datacao%20de%20Ciclos%20Economicos%20-%20Comunicado%20de%204_8_2015.pdf
7. Dasgupta, S., Lall, S., Wheeler, D.: Policy reform, economic growth, and the digital divide – an econometric analysis. World Bank, Washington, D.C (2011)
8. Demoussis, M., Giannakopoulos, N.: Facets of the digital divide in Europe: determination and extent of internet use. Econ. Innov. New Technol. 15(3), 235–246 (2006)
9. IBGE – INSTITUTO BRASILEIRO DE GEOGRAFIA E ESTATÍTICA: SIDRA. https://sidra.ibge.gov.br/tabela/6601. Accessed 11 Nov 2018
10. Macedo, H.R.: Análise dos possíveis determinantes de acesso à internet em banda larga nos municípios brasileiros. Texto Paraense Discussão, 1503 Brasília, IPEA (2010)
11. Nishijma, M., Ivanauskasb, T.M., Sarti, F.M.: Evolution and determinants of digital divide in Brazil (2005–2013). Telecommunications Policy (2016). http://dx.doi.org/10.1016/j.telpol.2016.10.004
12. Núcleo de Informação e Coordenação do Ponto BR - NIC.br: Pesquisa sobre o uso das tecnologias de informação e comunicação: Pesquisa TIC Domicílio, ano 2012,2013,2014,2015,2016,2017
13. Pantea, S., Martens, B.: Has the digital divide been reversed? Evidence from five EU countries. Digital Economy Working Paper 2013/06. JRC Technical (2013)
14. Prieger, J.E.: The broadband digital divide and the economic benefits of mobile broadband for rural areas. Telecommun. Policy 37(2013), 483–502 (2013)
15. Vicente, M.R., López, A.J.: Assessing the regional digital divide across the European Union-27. Telecommun. Policy 35, 220–237 (2011)

16. Zhang, X.: Income disparity and digital divide: the internet consumption model and cross-country empirical research. Telecommun. Policy **37**, 515–529 (2013)
17. Wirthmann, A.: The European survey on the use of information and communication technologies in households and by individuals, 30. Centre for Social Innovation, Vienna (2012)
18. World Bank: The Little Data Book on Information and Communication Technology (2018). https://www.itu.int/en/ITU-D/Statistics/Documents/publications/ldb/LDB_ICT_2018.pdf. Accessed 11 Nov 2018

The Impact of the Digital Economy on the Skill Set of High Potentials

Mariana Pinho Leite[1], Tamara Mihajlovski[1], Lars Heppner[1],
Frederico Branco[2], and Manuel Au-Yong-Oliveira[1,3(✉)]

[1] Department of Economics, Management, Industrial Engineering and Tourism,
University of Aveiro, 3810-193 Aveiro, Portugal
{marianal, tamara.mih, lars.heppner, mao}@ua.pt
[2] INESC TEC, University of Trás-os-Montes e Alto Douro, Vila Real, Portugal
fbranco@utad.pt
[3] GOVCOPP, Aveiro, Portugal

Abstract. Hiring outstanding employees is the goal of every company. Digitalization impacted the way companies work and the environment they are surrounded by. Considering this change, employees are facing new challenges for which a concrete skill set is needed. By conducting qualitative interviews, a distinct skill set required by companies was identified. The outcome is that soft skills moved to the forefront, playing a major role when coping with the digital future. In a further step, recommendations for talent management regarding recruitment and development of high potentials are given. With the implementation of technology and the adaption to the human resources perspective, companies will be able to master digital transformation.

Keywords: High potentials · Digital economy · New way of working ·
Future of employment · Digital transformation · Soft skills ·
Talent management · War for talent

1 Introduction

With a fast-changing market where technological innovation has proven to be one of the main keys to the growth of a business [1], said innovation has also become a major competitive advantage. There are many ways in which technology helps the value creation of a company. This assignment focuses on the impact digitalization has on the management of Human Resources, specifically on high potential (HIPO) employees.

Human Resource Management (HRM) has become a contemporary topic in the business world, as its importance has been increasing alongside digitalization. Employers are learning the impact of having employees that go beyond their expectations in their company. Nowadays, employers are looking for people who can do more than their job. The "expectations of customers have gone up drastically in terms of the services they expect" [2], and so have the job requirements of young employees. Recruitment has become a hunt for the best of the best, commonly referred as "diamonds in the rough". "High potentials, then, are those core employees whose skills are high in value and in uniqueness from the point of view of their particular employers [3]." It has

© Springer Nature Switzerland AG 2019
Á. Rocha et al. (Eds.): WorldCIST'19 2019, AISC 931, pp. 726–736, 2019.
https://doi.org/10.1007/978-3-030-16184-2_69

been proven by a SHL study that high potentials are 91% more valuable than non-HIPOs [4]. Thus, two questions remain: how do we spot high potentials and how do we become one? Above-average aptitude and employee engagement [3] have been two trends that differentiate high potentials from most employees, including high performing employees. Studies concluded that although "high performance may be a precondition to being identified as a high potential, learning agility is an overriding criterion for separating high potentials from non-high potentials" [3].

This study addresses the digital transformation in the skill set required by employers. The authors conducted interviews with experts on the area, in order to reach viable conclusions on the path the job market is taking in this digital era. Additionally, a clear and distinctive analysis was developed of high potentials and how they are shaped in the work environment established. Identifying and managing high potential employees is a key to success, not only for the company, but also for the employee himself/herself, as is discussed by this study.

2   Literature Review

Digitalization and automation precisely are the reasons why there is a decrease of need for low-skilled workers. Consequently, as the jobs that have repetitive tasks are being automated, there is a higher demand for the highly skilled workers and more interactive jobs. Hence, no human input is required anymore [5].

It is debatable that unemployment will rise due to automation. However, in the modern digital landscape, jobs will require different skills. Therefore, the human workforce will have to develop new skills and/or change their areas of work. This, however, does not mean that only the people whose jobs were automated have to develop new skills, but also workers who will obtain their jobs will have to adapt to new technologies and required skill sets. Furthermore, soft skills and skills which cannot be easily automated will be more valuable [6].

Due to digitalization, globalization is possible, and it is one of the main trends nowadays. This also has an impact on the job market and skill sets of the workforce. With Labor being mobile, it is possible to find skilled employees from all around the globe. This leads us to the next topic, which explores how companies are more and more competitive to attract the best employees - high potentials [7].

A company's most valuable resource is the employee [8]. Thus, employing the right people for the job is crucial for the success of any company. War for talent makes it hard for companies to find a skilled worker. Hence, nowadays there is a huge rivalry between companies to attract the most talented employee [7]. Companies are investing in the personal development of their employees. In the last 20 years, research and studies about high potential employees were conducted with the goal of reaching a better and faster identification of those employees and making strategies on how to maximize their potential. It could be argued that high potential employees bring along not only the already existing expertise, but also the motivation to learn and develop further. Thus, additional training and development of high potential employees tends to show greater results in their work and progress, compared to results after training low-potential employees [9].

To start the analysis of possible high potential employees, a company must first have a clear picture of what makes a certain employee a high potential and what value the employee brings to the company [10]. Although the definitions of a high potential employee may differ from one company to another, there are some characteristics that could be taken as a rule when companies are defining them:

- Ability: High potential employees are usually the ones who deliver everything on time, whose work is always excellent and far above average or expected [11];
- Aspiration or strong ambition for achievement: High potentials are ambitious, and they like to have a possibility to grow within the company [10];
- Engagement: High potentials see themselves growing concurrently with the company [11];
- Learning-oriented: Usually, while for other employees a difficult situation may be a drawback or a problem, for a high potential every challenge is an opportunity to learn and acquire new skills [10];
- High adaptability to new situations and surroundings: As the business world is constantly changing due to technological innovations, it is very important to be up to date and cope well with new situations and technologies [12];
- Always seeking for new opportunities: High potentials are likely to take on more responsibilities and work than it is strictly stated in their job description, and they will take risks, since each risk can also be an opportunity as well [10].

Up until now, research has been conducted regarding identification of high potentials in a company. As for the impact of digitalization on the future job market, studies have been focused on jobs in general, and those that will be lost and created in the future in particular.

Nevertheless, no research has been conducted regarding the impact of digitalization on high potentials and how their skill set will change.

3   Methodology

To answer the research question stated in the introduction, two methods were used: a literature review and interviews. This order was chosen and performed beforehand to identify gaps in the research about high potentials, while obtaining knowledge about trends, which allowed the interviewers to ask precise and distinct questions. The subsequent conducted interviews give proof and show examples of how the praxis deals with high potentials. A qualitative and interpretative research methodology addresses the explanatory approach of providing evidence for the coherence of digitalization and a transformed skill set of high potentials [13, p. 1 ff.].

A first literature screening revealed the penetrative impact of digitalization on companies. The third technological wave with its radical transformation of several jobs revealed the importance to analyze the skill set that the best of the best employees have to offer, to generate value in the future world of business [14, p. 8 f.].

Publications of consulting companies conducted and interpreted questionnaires of market analysis; articles of subject-related journals show secondary sources to describe and understand the characteristics of the transformed digital economy and society. The term 'high potential' is sufficiently and pertinently described. Analysis of the impact of digitalization on the future job market and way of work is a more debatable and vaguer topic. Nevertheless, by interpreting this information and comparing it with the impact of digitalization, a picture can be sketched. Connecting these main research topics to each other is the transfer task that has been undertaken.

The analyzed literary sources were mainly published by digital competitive countries such as Germany and the USA [15]. Additionally, the two mentioned countries have a similar definition of the term "high potential" as being a young employee, and support a masculine culture where the focus is on the best performers of the country [16].

In interviews as a primary source, the authors were given the chance to precisely tackle the research question and to obtain a practical perspective on what engages companies.

Three main channels were identified to acquire interviewees. Firstly, the personal network of the authors, secondly, the universities network of start-ups, and lastly, externals. The personal network was the most valuable source, since none of the people from the universities and the externals agreed on doing an interview. It came down to merging the authors' networks together, presenting a large group of potential interviewees.

Advantage was taken of the authors' multinationalism, which offered the opportunity to interview people from Germany, Austria and Portugal. The total number of eight interviews was split up among four interviewees categorized as managers, two interviewees categorized as experts and two as human resource officers. The interviewees' companies' background was versatile, covering information technology (IT) consulting, IT technology, IT services, technology, and accounting companies, as well as the public administration sector. The companies in focus are in the technology sector, which is an industry mostly affected by digitalization, including how digital is put to work and "how intensively [...] employees use digital tools in every aspect of [...] daily activities" [17].

To ensure comparability of the answers, a semi-structured interview type was chosen. The questions asked were mainly open to ensure collecting own opinions and experiences [18]. The interviews conducted on phones were given a duration of 30 min, which were only in minor circumstance exceeded by a few minutes. Due to language barriers, all authors contributed equally to performing the interviews.

Due to well-structured interview questions, the authors were able to get a deep insight into the culture and thoughts of the companies that contributed highly to the outcome of this research paper. Since the interviews were held in different languages, the answers given had to be translated into English, a process which was based on a logical approach and it was ensured that no information was lost. Finally, the data was evaluated by putting the answers into a logical context and quantifying the relevance of each response.

4 High Potentials: The Digital Redeemers for a Company's Future

The following abstract explains the exceptional value that high potentials generate for a company. Thereby, this topic will be divided into three parts: describing the impact of digitalization, analyzing the modern high potentials' characteristics and interpreting the data to introduce a set of actions. In the first subsection, the core changes of companies themselves and their environment based on digitalization are going to be defined. This fundamental understanding is used to analyze which skill set has to be brought by future employees to cope with prospective challenges. The final part interprets the company's perspective on how to acquire valuable human resources and how to ensure that the full potential will be revealed [19].

4.1 The Impact of Digitalization on Companies

Bloomberg stated the process of digitizing information equals digitalizing processes towards digitally transformed businesses and strategies [20]. Facing this trend, external influences were identified:

- *A disruptive environment:* Nowadays prosperous business models are eliminated by newly founded start-ups. The current situation has no guarantees that something that works at the current time will work the same way in the future. The future is described as blurred and unpredictable [21].
- *Exploding markets:* New markets occur and rise quickly. Competition in those markets is high. Projects are proposed and described openly, thereby many companies will have the chance to participate in the bidding process to win the deal.
- *Wide distribution of information:* News can be spread around the world. This also includes information about achievement and failure of one's own company or competitors'. A wide range of social networks are accessible through freely available applications.
- *Fast development:* Technology develops quickly. An example is Moore's law, which describes the doubling of processor chip performance every two years [22]. With this rapid improvement, completely new services can be provided and distributed promptly. Besides, this means that the state of affairs, including the knowledge alters very fast and needs to be refreshed permanently.

Embossed by an environment transforming at a dramatical pace, companies adapt internally to remain competitive:

- *Way of working:* Computers, mobile phones, projectors, smart boards and webcams are digital devices that belong in every office space and are used on a daily basis. Moreover, the complete design of office buildings is rearranged to have open areas for socializing and the overall well-being of the employees, but also to improve productivity and enable creative thinking.
- *Communication:* Communication takes place through different channels like video, voice or text messages. Nevertheless, the burgeoned trend towards written communication has its limitations. The liability for binding predication is reduced

because appointments can be rescheduled easily. A written word can be held against someone, which is not suitable for some critical topics, and a long time is spent just to find the correct language. Communication is ubiquitous and causes stress. As stated by one of the interviewees, "it has never been communicated that much but so little at the same time". This means that often a pertinent level is not achieved, and a flood of information is shared without an objective.

- *Organizational structure:* Complex tasks and connectivity require autonomous working. Top-down hierarchies with several management levels are transformed into lean models. This also allows a free composition of teams, where skills can be instated wherever they are needed the most.
- *Team composition:* Accepting diversity was for many companies the first step. In the following period it will be the goal of many companies to seek diversity [23, p. 12]. Global projects postulate team compositions with members from different nationalities and cultures. Gender differences as well as sexuality will not play a role in the structure of a team in the future. Digitalization, with its communication tools as well as its mainly cognitive workload allows to form virtual teams where members work together even though they are located in completely different parts of the world.
- *Processes:* Whereas many projects in the past where approached with a traditional waterfall model, nowadays' agile project management methods are implemented to tackle in the best way possible the previously mentioned open task offers with an unpredictable outcome. Apart from that, automation takes over routine jobs and established processes will be revised.

The listed internal changes also affect the skills required by employees, which must agree with and support the digitally transformed businesses.

4.2   Skills of High Potentials in Digital Businesses

Historically, technical skills, also known as hard skills, were the only skills necessary for career employment; but today's workplace is showing that technical skills are not enough to keep individuals employed when organizations are right-sizing and cutting positions [24]. Companies are looking for people that offer more than just hard skills.

As employers are progressively looking for employees who are mature and socially well-adjusted, they rate soft skills as number one in importance for entry-level success on the job [24]. The authors extracted from the interviews seven competences and/or characteristics that high potentials acquired, whose value is recognized by the employers, organized by importance:

1. *Self-management and Intrinsic Motivation*
2. *Learning Agility*
3. *People Skills*
4. *Innovation and Creativity*
5. *Education*
6. *Commitment and Stability*
7. *Analysis and Problem-Solving*

High potentials display, according to our interviewees, intrinsic motivation, and that distinguishes them from the common employee. One cannot simply teach someone how to get motivated to work. Managers and employers can boost the employee's motivation. Nevertheless, only the employee himself can change his own mind. In other words, the tasks performed in the workspace to achieve a certain objective arise from within. This intrinsic motivation can also lead to passion for the job they have, which makes positive changes in the company's environment, as well as creates an impact in the company's success. Along with intrinsic motivation, high potentials have shown high levels of self-responsibility and self-organization. Those that can manage themselves are seen as fit to manage their work successfully and generate better outcomes in comparison to others. Self-management is a technique that translates into what a manager needs to know. That includes the ability to prioritize tasks, making sure there is time to perform each task and plan what to do beforehand.

Secondly, the ability to learn and adapt to new situations is also very valuable to employers. This soft skill is at the core of maintenance techniques performed by the talent management working group. The company normally offers job positions in other areas to test out their adaptability, included in, for example, BOSCH's retention plan of employees, according to Ana Pereira, HR Business Partner of BOSCH Portugal. Despite it being a skill that has been talked about only recently, companies in the past also implemented it, in the sense that employees, if they desired to have a better salary or get a promotion, they would need to be able to adapt to their new job position. In the fast-changing market nowadays, this transition of jobs has to be made quickly since every setback can be costly to the company.

In third place came people skills. Globalization, a consequence of digitalization, has had a huge impact on the interpersonal relationships of employees and their dynamics. The interviewees agreed that social competences play a huge role in the company's success. People skills promote a positive attitude, effective communication, respectful interaction, and the ability to remain composed in difficult situations [24].

Being able to work in a team with people who have different backgrounds is a reality in most companies in this day and age. High potentials display an intercultural awareness and are able to deal with conflict and different opinions without severing their relationships with others.

This is followed by innovation and creativity. High potentials are known for having lateral thinking. They are people who think out-of-the-box and come up with creative solutions for problems that occur within the company. As game-changing innovations lead the market world, companies are always looking for people who will come up with the next success. This is important, especially because the pace of change has exponentially speeded up in this last decade, with digitalization as its motor engine.

According to the interviewees, high potentials are literate and most have a degree in higher education. Additionally, Meik Bödeker, director of the ADV-Institute in the field of consulting, stated that these employees display a "wide education" with knowledge in different areas that may not be related to the job position. This knowledge could be earned by participating in, for example, workshops or training sessions, according to Michael Köster – member of the management board in enmore AG, in the field of IT-Consulting. The interviewees also mentioned the ability to speak foreign languages, as well as digital literacy in this component. In Sect. 4.3, this topic is further discussed and explained.

At the end of the list, there are "Commitment and Stability" and "Analysis and Problem-Solving". The first one has to do with the employees' commitment to the company and their availability. High potentials tend to prioritize the company and their work, making themselves available to solve problems and obstacles even in off-hours. This also includes being able to travel to a different site if the company sees the need to do so. As for the latter soft skill, interviewees expect high potentials to have a critical sense as well as an ability to identify the problem, create solutions and decide critically which one to implement. This competence has been deemed as one of the core soft skills from high potentials directly linked with "Innovation and Creativity".

The authors would like to mention how leadership has not been included in this list. The leadership skill deemed as the number one soft skill to include on the CV [25] is not one of the main characteristics of high potentials. In fact, according to Ana Pereira, high potentials are not necessarily good leaders. When retaining high-potentials, it is not mandatory to give them positions of power, if they do not display any leadership skills. It can even be considered as a bad human resource management decision.

4.3   Acquire Value Adding Employees

Interviewees mentioned that active recruiting - meaning that the first approach is done by the company and not by the employee - became more significant than previously. Considering the disruptive environment (Sect. 4.1) and a mentality of the "survival of the fittest", a search for talents or also called "war for talents" is of high importance [26].

Being able to acquire and develop high potentials requires an understanding of their generation, the children of the millennial, an overlapping of the end of generation y and the beginning of generation z. Two major platforms were identified for recruiting purposes:

- *University:* Since it was identified that high potentials generally went through tertiary education, relations to and cooperation with universities were mentioned as a valuable source. The biggest disadvantage is the range of the network. It only includes students of the field the company engages with in a particular university. However, it is easier to stand out because competition is not so high.
- *Internet:* As digital natives, a big part of the life of high potentials takes place in the world of the internet. To recruit high potentials, companies have to enter this world [26, 27]. This includes publicity on social media networks as well as business networks. The internet offers the ability to access a wide network with billions of users. However, to address the target group is much harder. Additionally, the competition is high, and it is hard to stand out by having a lot of offers and content producers [28, 29].

Simply using a channel is not enough. It is also necessary to trigger the desires of high potentials. High salaries and promotion prospects as an incentive are not of high relevance anymore. For children of the millennial, it is more important who you are, instead of what you possess. What has to be offered is an experience and a chance for self-development (referred to in Sect. 4.2), which can be achieved by providing interesting projects and foreign deployments. They have to provide free space for self-realization. Work takes place in line with leisure. This can be achieved by "mobile working", therefore it becomes irrelevant where, when and how the work is accomplished. In recruiting processes, it is often mentioned that the job candidates' mindset must match with the company's mindset. The same accounts for the perspective of high potentials who are identified as being in a stronger position than companies in the current job market. Hence, high potentials will look for companies that openly present their mindset. In this connection, it is important to present it in an authentic way. New channels such as video streaming or events or even an evening chat can be used to deepen the relationship between the potential employees and the companies. Finally, as a global generation, millennials know only a world without borders. They "don't think twice about travelling and are connected to people from every continent" [26]. This approachability has to be lived in companies as well. Lean hierarchies are a goal but the behaviour of employees and managers between each other is of much more importance.

High potentials, as rough diamonds, will need a specific treatment to reveal full capacity. The following guidelines encourage high potentials to perform well for a company:

- *Adequate leadership:* Adequate leadership in this case implies a coaching role taken on by the supervisor. This includes paying attention to the needs of the employee, enabling professional development through training sessions and giving constant feedback [30, p. 17].
- *Freedom:* The tasks offered to high potentials should have enough freedom for them to make their own decisions. This goes along with taking responsibility for the outcome [30, p. 19].
- *Teamwork:* A fast integration into a team with strong members empowers high potentials.

5 Conclusions

Digital transformation is more than just using new technology and applications. It starts with a computer on a desk and lasts until the mindset of every individual employee, and not every company, is aware of this insight. Technology is being developed but the employees are not taken aboard on the journey to a digital economy.

The result of this economic landscape are stoked fears, especially for older generations. These fears include not being able to keep up with the digital development and furthermore being replaced by a digital native. Thus, instead of having two generations benefiting and empowering each other, businesses have two generations minimizing and working against each other.

Quoting Dr. Gunther Singer, CEO of Life and Career Design, in his interview with the authors: "Jobs that are not endangered are the ones that are not trivial [...], that require soft skills [...]." After careful analysis of the data collected, the authors daringly assume that high potentials with their valuable skill set and their learning agility will become the most searched for employees in this new era.

For future research, the focus can be shifted from the approach of looking at a generation of new employees with high potentials, towards former generations that are salaried and identify how they deal with digital transformation. The combination of the two researches can be used to develop a guideline on how to merge older generations of best performers together with younger generations of high potentials to ensure all-over value creation throughout the impracticable journey of transformation.

Acknowledgements. The authors would like to thank the following people who shared their time and knowledge for this study: Harald Heppner; Klaus Chatzidimpas; Meik Bödeker; Michael Köster; Siegmar Moltzahn; Ana Pereira; Inês Ribeiro; and Günther Singer. With the interviewees, it was possible to gather empirical data and overcome any gaps in the existing literature for this research paper.

References

1. Björkdahl, J.: Technology cross-fertilization and the business model: the case of integrating ICTs in mechanical engineering products, Research Policy (2009)
2. Grover, R.: The business benefits of digitization (and why your business needs to invest now!) (2017). https://www.linkedin.com/pulse/business-benefits-digitization-why-your-needs-invest-now-rohan-grover. Accessed 06 Nov 2018
3. Dries, N., Vantilborgh, T., Pepermans, R.: The role of learning agility and career variety in the identification and development of high potential employees. Personnel Review (2012). Accessed 06 Nov 2018
4. SHL: High-potential: realize the promise, reduce the risk. https://www.shl.com/en/solutions/identify-develop-leaders/high-potential/. Accessed 06 Nov 2018
5. Dachs, B.: The Impact of New Technologies on the Labour Market and the Social Economy. European Parliamentary Research Service (2018)
6. Manyika, J., Lund, S., Chui, M., Bughin, J., Woetzel, J., Batra, P., Sanghvi, S.: Jobs Lost, Jobs Gained: Workforce Transformation in a Time of Automation. McKinsey Global Institute (2017)
7. Zinser, J.: How to hire top talent in a competitive job market (2018). https://www.business.com/articles/hiring-top-talent/. Accessed 06 Nov 2018
8. Singh, S.: Employees are company's most valuable ASSETS (2016). https://www.linkedin.com/pulse/employees-companys-most-valuable-assets-samsher-singh. Accessed 06 Nov 2018
9. Chamorro-Premuzic, T., Seymour, A., Kaiser, R.: What science says about identifying high-potential employees (2018). https://hbr.org/2017/10/what-science-says-about-identifying-high-potential-employees. Accessed 06 Nov 2018
10. Ready, D., Conger, J., Hill, L.: Are you a high potential? (2010). https://hbr.org/2010/06/are-you-a-high-potential. Accessed 05 Nov 2018
11. Gartner: Create Agile HIPO Strategies (2018). https://www.gartner.com/en/human-resources/insights/leadership-management/high-potentials. Accessed 05 Nov 2018
12. Reynolds, J.: 20 characteristics of high-potential employees (2017). https://www.tinypulse.com/blog/20-characteristics-of-high-potential-employees. Accessed 06 Nov 2018
13. Mack, N., Woodsong, C., MacQueen, K., Guest, G., Namey, E.: Qualitative Research Methods: A Data Collector's Field Guide. Family Health International (fhi360) (2015)
14. Katz, R.: Social and Economic Impact of Digital Transformation on the Economy. International Telecommunication Union (ITU) (2017)

15. Chakravorti, B., Bhalla, A., Chaturvedi, R.: 60 countries' digital competitiveness, indexed (2017). https://hbr.org/2017/07/60-countries-digital-competitiveness-indexed. Accessed 01 Nov 2018

16. Hofstede insights: Country comparison. https://www.hofstede-insights.com/country-comparison/germany,the-usa/. Accessed 01 Nov 2018

17. Gandhi, P., Khanna, S., Ramaswamy, S.: Which industries are the most digital (and why)? (2016). https://hbr.org/2016/04/a-chart-that-shows-which-industries-are-the-most-digital-and-why. Accessed 02 Nov 2018

18. Jamshed, S.: Qualitative research method-interviewing and observation (2014). https://www.ncbi.nlm.nih.gov/pmc/articles/PMC4194943/. Accessed 02 Nov 2018

19. Wolcott, H.: Transforming Qualitative Data: Description, Analysis, and Interpretation. Sage Publications Inc., Thousand Oaks (1994)

20. Bloomberg, J.: Digitization, digitalization, and digital transformation: confuse them at your peril (2018). https://www.forbes.com/sites/jasonbloomberg/2018/04/29/digitization-digitalization-and-digital-transformation-confuse-them-at-your-peril/#320112252f2c. Accessed 05 Nov 2018

21. Hill, J.: Leading through digital disruption (2017). https://www.gartner.com/imagesrv/books/digital-disruption/pdf/digital_disruption_ebook.pdf. Accessed 05 Nov 2018

22. The Economist: After Moore's law (2016). https://www.economist.com/technology-quarterly/2016-03-12/after-moores-law. Accessed 05 Nov 2018

23. Wippmann, P.: New Work Trendbook: Die 15 wichtigsten Trends zur Arbeitswelt der Zukunft. Xing, Hamburg (2018)

24. Robles, M.: Executive Perceptions of the Top 10 Soft Skills Needed in Today's Workplace. Business Communication Quarterly (2012)

25. Leighton, M.: 4 soft skills LinkedIn says are most likely to get you hired in 2018 - and the online courses to get them (2018). https://www.businessinsider.com/best-resume-soft-skills-employers-look-for-jobs-2018-4. Accessed 06 Nov 2018

26. Burfeind, S., Generation Z.: #brightprospects (2018). https://international.brandeins.de/generation-z-brightprospects. Accessed 06 Nov 2018

27. Au-Yong-Oliveira, M., Gonçalves, R., Martins, J., Branco, F.: The social impact of technology on millennials and consequences for higher education and leadership. Telematics Inform. **35**, 954–963 (2018)

28. Faria, A., Almeida, A., Martins, C., Gonçalves, R., Martins, J., Branco, F.: A global perspective on an emotional learning model proposal. Telematics Inform. **34**, 824–837 (2017)

29. Martins, J., Gonçalves, R., Branco, F., Peixoto, C.: Social networks sites adoption for education: a global perspective on the phenomenon through a literature review. In: 2015 10th Iberian Conference on Information Systems and Technologies (CISTI), pp. 1–7. IEEE (2015)

30. Hübbe, E.: Trust in Talent: Warum mitarbeiterorientierte Unternehmen erfolgreicher sind. Kienbaum Consultants International GmbH, Köln (2018)

Designing a Dual-Layer 3D User Interface for a 3D Desktop in Virtual Reality

Hind Kharoub[✉], Mohammed Lataifeh[✉], and Naveed Ahmed[✉]

Department of Computer Science, University of Sharjah, Sharjah 27272, UAE
{hkharoub,mlataifeh,nahmed}@sharjah.ac.ae

Abstract. We present a new dual-layered 3D user interface for a 3D desktop environment in Virtual Reality (VR). VR technology has proven its potential in supporting higher levels of engagements and intuitive interactions between the human and the machine. Natural and intuitive interaction is still considered as one of the most important user interaction issues in VR. Our novel dual-layered 3D user interface allows the user to interact with multiple screens portrayed within a curved 360-degree effective field of view available for the user. Downward gaze allows the user to raise the interaction layer that facilitates several traditional desktop tasks. We perform a quantitative and qualitative user study to validate the usability and user experience of the new user interface. We show that our 3D user interface is well suited for a VR 3D desktop environment.

Keywords: Virtual Reality · 3DUI · 3D desktop · User interface ·
3D user interface

1   Introduction

The field of VR is an area of interest for both research and development for the past five decades. With the advancement of technology, specifically miniaturization of computing hardware and improved display technology, the field is finally going mainstream [1]. Indeed, the advancement of consumer grade VR devices, e.g. Oculus Rift, HTC ViVe, Gear VR etc., has opened new ways for people to interact with technology. In general, the user-interaction paradigm for VR is an extension of how people interact with the computing systems using a graphical user interface (GUI). The concept of GUI revolves around the concept of direct manipulations (DM), where the user uses some input device to directly manipulate the content on a two-dimensional (2D) display, and receives an immediate feedback of its action, e.g. point and click, dragging etc.

The VR on the other hand is a different medium, where the content is not only displayed in a true three-dimensional (3D) environment, but also because of the VR device, the user loses the ability to directly observe the input devices, e.g. mouse and the keyboard. Thus, the current user interface paradigm that relies on DM is not well suited for 3D VR environments. Therefore, for a VR environment, it is necessary to explore different types of user interface paradigms that are more suited for natural user interaction (NUI) without means of a physical input device. A NUI allows people to interact with the technology without the need of intermedial devices for the user interaction; rather the user interaction takes place directly using the hand gestures or the

© Springer Nature Switzerland AG 2019
Á. Rocha et al. (Eds.): WorldCIST'19 2019, AISC 931, pp. 737–747, 2019.
https://doi.org/10.1007/978-3-030-16184-2_70

body movement. Thus, the understanding of how a NUI can be more effectively used to communicate with computers within an engaging 3D user interface becomes more important.

Moreover, the new wave of input systems in video game consoles (such as Nintendo Wii, Xbox Kinect, and Leap Motion) is leading the new generation of Human Computer Interaction (HCI) systems to focus on creating interfaces that are more intuitive and user-friendly. While the gaming industry is currently leading the way using the afore-mentioned consoles, it will not be long before users are able to control simple and advanced VR and computer systems using body gestures that feel intuitive. A 3D User Interface (3D UI) design is now becoming a critical area for developers, students, and researchers to understand. The use of VR visualization and natural interactive gestures can increase the user engagement and consequently improve the user experience while seeking information [2].

3D VR application should be useful because it supports immersion, uses natural skills, and provides immediacy of visualization. But many current VR applications either, support only simple interaction, or, have serious usability problems [1]. Therefore, 3D UI design is an area ripe for further work, the development of 3D user interfaces is one of the most exciting areas of research in human–computer interaction (HCI) today, providing the next frontier of innovation in the field [1].

One of the main limitations of the immersive VR is that the user cannot see the input device during the user interaction. This is in complete contrast with the current systems, e.g. desktop or mobile, where there is no limitation on the visualization of the input device. The interaction with a 2D desktop is the most common user interaction paradigm prevalent for the last 30 years. It is imperative for such a system to exist in a VR environment, if the user has to spend significant amount of time in VR as their primary computing device.

We therefore propose a new dual-layered 3D user interface that allows the user to perform common desktop tasks in a VR environment. The proposed 3D user interface allows the user to visualize the content on a 360-degree curved field of view, in the form of multiple curved screens. The gaze mechanism is used to switch between the screen layer to the interaction layer. The interaction layer allows the user to perform simple tasks related to a desktop environment. We implemented three user interaction modes to test the new 3D user interface and validates its usability and user experience both qualitatively and quantitatively.

Our main contributions for this work are:

1. Designed a novel dual-layered 3D user interface allows the user to interact with multiple screens portrayed within a curved 360-degree effective field of view available for the user.
2. Evaluated the 3D user interface using three different interaction modes.
3. Perform a user study to quantitatively and qualitatively validate the usability and user experience of the new 3D user interface.

In the following sections, we will first review the related work. In the following section, the 3D user interface design is presented. Finally, the user study is discussed, followed by the conclusions.

2 Related Work

There are already some applications of 3D user interfaces used by real people in the real world (e.g., walkthroughs, psychiatric treatment, entertainment, and training). Most of these applications, however, contain 3D interaction that is not very complex. More complex 3D interfaces (e.g., immersive design, education, complex scientific visualizations) are difficult to design and evaluate, leading to a lack of usability. Better technology is not the only answer—for example, 30 years of virtual environment (VE) technology research have not ensured that today's VEs are usable [1]. Thus, a more thorough treatment of this subject is needed [1]. 3D UI design is an area ripe for further work, the development of 3D user interfaces is one of the most exciting areas of research in human–computer interaction (HCI) today, providing the next frontier of innovation in the field [1].

Within a 3D human-computer interaction mode, user's tasks are carried out in a 3D spatial context. 3D interaction involves either 3D input devices (such as hand gesture or physical walking) or 2D input devices (such as mouse) mapped into 3D virtual location. 3D User Interfaces as the intermediary between the human and the machine, become the medium where different user's tasks are performed directly within the 3D spatial context. In carrying out different tasks, users' commands, questions, intents and goals are communicated to the system, which in turn must provide feedback, with information about the status and flow for the mentioned [1]. While 3D interaction allows user to accomplish a task within 3D UI, immersive VR Systems combine interactive 3D graphics, 3D visual display devices, and 3D input devices (especially position trackers) to create the illusion that the user is inside a virtual world [1], see Figs. 1 and 2.

As more tasks are enabled within the 3D virtual environment (3D VE), 3D Interaction Techniques, defined as different ways that the user can interact with such environments, are matured. The quality of these techniques greatly influences the entire 3D UI. These qualities can be grouped into three different categories: Navigation, Selection and manipulation, and System control [1]. The tasks can be performed using a physical device, e.g. a mouse, or a gesture-based user interface that does not require any physical device.

A gesture is simply defined as a physical movement of a human body or some of its parts like hands, arms, and face with the intention to relay information or meaning. Hence, the recognition of these gestures entails not only the tracking of human movement, but also the interpretation of that movement as semantically meaningful commands. In 2009, Murthy and Jadon [3] recognized three groups of gestures based on Cadoz's work: "semiotic gestures are those used to communicate meaningful information, ergotic are those used to manipulate the physical world and create artefacts and epistemic are those used to learn from the environment through tactile or haptic exploration". Saffer [4] has identified the following advantages of gestural interfaces against traditional interfaces, which are (1) more natural interactions; (2) less cumbersome or visible hardware; (3) more flexibility; (4) more nuance; and (5) more fun. Gesture recognition can be used in many areas. Some of the application domains that employ gesture interaction are: VR, robotics and telepresence, desktop and tablet PC applications, games, and sign language.

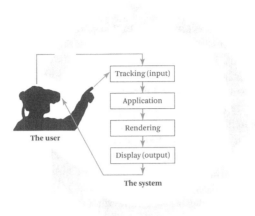

Fig. 1. VR system process.

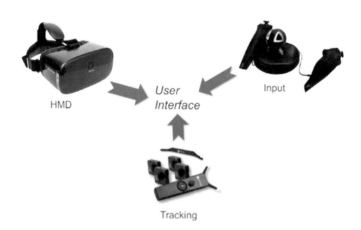

Fig. 2. Typical VR system components.

Humans rely on eye gaze and hand manipulations extensively in their everyday activities. Most often, users gaze at an object to perceive it and then use their hands to manipulate it. Chatterjee et al. [5] presented a set of interaction techniques combining gaze and free-space gesture to support common digital activities. They used a taxonomic breakdown to guide their development, organized along two basic interactive dimensions. They found that technique was robust to sensing inaccuracies in commodity hardware, suggesting their approach could be used on low cost, off the-shelf components. They compare their method against five baseline techniques; results suggest the technique could enhance systems using gaze or gesture alone.

For the usability study, Cabral et al. [6] discussed multiple issues related to the use of gestures as input modes in VR environments. The use and usability of hand gestures for tasks such as making telephone calls, operating the television, and executing mathematical calculations has been studied by Bhuiyan and Picking [7].

The type of spatial tracker used in a 3D UI would necessarily influence its usability as different trackers may demand different UI designs. UI display characteristics can also affect 3D interaction performance. For instance, when Head Mounted Displays (HMDs) are used in VR, users are unable to see their own hands or other parts of their bodies. This temporary visual imparity may affect the motivation of some users in exploring the built environment, if physical movement is required. Furthermore, HMDs vary widely in field of view (FOV). Several recent studies [8–10] have evaluated the accuracy obtained from the leap motion and Kinect devices based on different experiments setup.

In this work, we evaluate our 3D user interface with both device-based and gesture-based user interactions. User can use a 3D pointer to interact with the desktop environment and can also use hand gestures that we implement using a Leap Motion controller. In the following section, we will describe the novel 3D user interface.

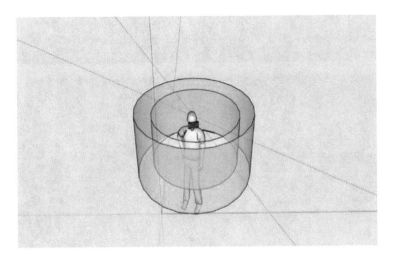

Fig. 3. Our dual-layer UI consisting of two main layers with a curved FOV.

3 3D User Interface Design

Since the immersive Virtual Reality has different medium, where the content is displayed in 3D environment. Consequently, we need to design a new 3D user interface to suit our environment that is different than the traditional 2D desktop, where all the files and tasks are carried out in 2D space and on the same layer. Our 3D user interface is a dual-layer user interface consisting of two main layers with curved screens, Fig. 3.

The upper layer is used for displaying the opened files. It allows the user to interact with multiple screens portrayed within a curved 360-degree effective field of view available for the user. While the lower layer contains the traditional desktop tasks e.g. delete, duplicate etc. Downward gaze allows the user to raise the interaction (lower) layer that facilitates several traditional desktop tasks, Fig. 4. Based on the three different interaction modes, we have three user interfaces, point-and-click user interface, controller-based direct manipulation user interface, and gesture-based interface. A visualization of the three user interfaces design is shown in Figs. 4, 5 and 6 respectively.

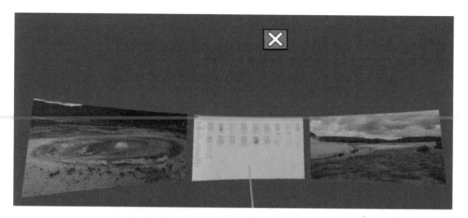

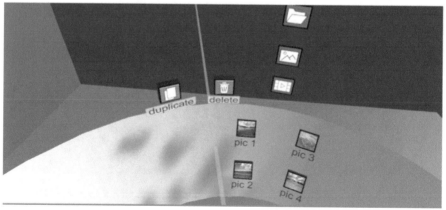

Fig. 4. Two layers with the pointer-based interface.

We use the gaze-based selection mechanism in our 3D user interface for the context-aware to interchange between the two layers, and to perform certain tasks. Applying the context-aware using the gaze tracking will activate the intended layer/object. For example, looking down makes the lower layer active, while looking up makes the upper layer active. In addition to that, the gaze is used to choose an object/component within a curved screen at the upper layer to make it active and then the user can perform the task on that object. We used object to refer to different types of

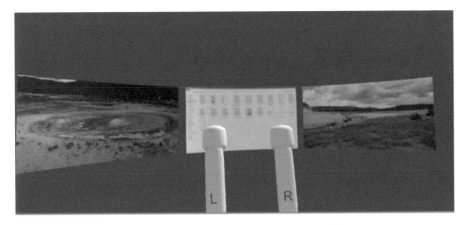

Fig. 5. Controller-based direct manipulation.

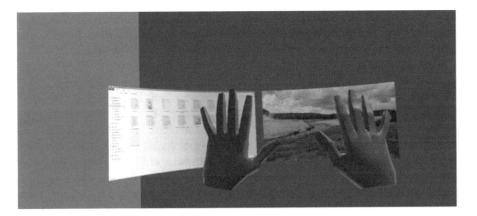

Fig. 6. Gesture-based user interaction.

files available in the 3D user interface, e.g. images, videos and folders. Specifically, for the videos files we used the gaze to play the video and if the user moves his gaze away from the video, then the video will be paused. In the design of the user interface we took into consideration the design principles such as: affordance, feedback, consistency, visibility, usability.

Several tasks can be performed by the users through our 3D user interface for the immersive VR desktop. As we mentioned earlier the context-aware of an object by gaze tracking is used to indicate which specific object is active, and then the specific task is performed on the object.

The manipulation and system control tasks/actions available in our 3D user interface are as follows:

1. Open: allows the user to open an object from the menu.
2. Close: allows the user to close an object displayed in the selected space.

3. Scroll: allows for the movement (scrolling) of the layer in the x-axis of 3D space.
4. Scale: allows for zooming in or zooming out on all objects displayed in the selected space.
5. Duplicate: allows the user to make a copy of an object on the lower layer.
6. Delete: remove the object from the lower layer.

We use HTC ViVe as our immersive VR display. Leap Motion was used to implement the gesture-based interface. The pointer and controller-based interfaces were implemented using the HTC ViVe controllers. In the following section we will discuss the user study to evaluate the usability and user experience.

4 User Study and Analysis

To evaluate the viability of our 3D user interface we performed a usability study to analyze the three different interaction modes for an immersive VR 3D desktop environment. We studied the usability and the functionality of each interaction modes. We performed a controlled experiment, using within-subjects experimental designs. In within subjects design each participant tested all the user interfaces under each condition, simple and compound tasks in both standing and seating positions.

The experiment started with a short introduction to the immersive VR 3D desktop environment, our goals, our 3D user interface, the hardware components (HMD basically) used, the three different interaction modes, and a reference to the questionnaire to be completed at the end of the experiment. Users were asked to sign a usability test consent form. After the initial presentation, we explained the three interfaces for the user, then the user was trained to practice all the tasks available in our interfaces. The interfaces were explained in the following order for all the participants: point-and-click user interface, controller-based direct manipulation user interface, and finally gesture-based user interface. Then the participant was asked to perform a case scenario (sequence of tasks) by the researcher for each interface in random order. The same case scenario is applied to all interfaces using two set-ups, standing VR experience and seated VR experience. During the testing session, the observer/researcher was monitoring the user performance and taking down relevant information e.g. completion time. Finally, the participant was asked to complete an online comprehensive questionnaire about each tested user interface and the overall experience.

Appropriate qualitative and quantitative analysis is used to analyze the data and provide a comprehensive usability study in terms of the following standard usability issues: precision, efficiency, effectiveness, ease-of-use, fun-to-use, fatigue, naturalness, and overall satisfaction.

The user study was performed on 21 participants (11 females, and 10 male). They were aged from 18 to 55 (the majority between 25 and 34). Each user was given 7 tasks to complete on the user interface. We calculated the time taken by each user to complete the tasks in both standing and seated positions using each interaction mode. The time taken to complete the tasks by each participant for each user interaction mode can be seen in Figs. 7, 8 and 9. We also collected the feedback from the users in the form a questionnaire and exit interview that was used for further analysis.

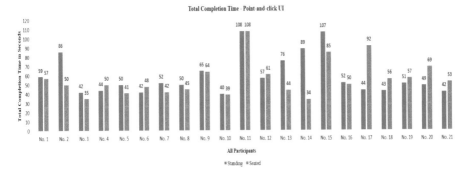

Fig. 7. Time completion, pointer-based interaction.

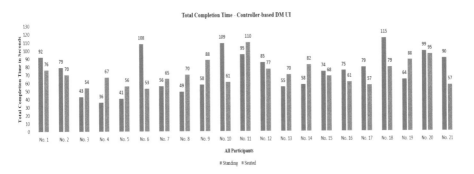

Fig. 8. Time completion, controller-based direct manipulation.

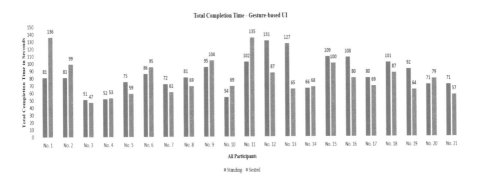

Fig. 9. Time completion, gesture-based user interaction.

As can be seen in Figs. 7, 8 and 9, the maximum time for completing the tasks using the interaction modes varies according to the users. In general, the maximum time ranges from 108 s (pointer-based) to 136 s (gesture-based). Similarly, the minimum time ranges from 34 s (pointer-based) to 47 s (gesture-based). This shows that the pointer-based interface is the preferable mode of interaction for the user even in a VR environment because of their familiarity with the interface.

Similarly, our qualitative analysis shows most of the users having a positive experience with the 3D user interface and the overall pleasing and productive experience. The usability study validates that all the usability goals are met with more than 80% approval of the 3D user interface design.

5   Conclusions

In this paper, we presented a new dual-layered 3D user interface for a 3D desktop environment in Virtual Reality. The user interface allows the user to visualize the desktop screens on a curved 360-degree curved layer. A downward gaze raises the secondary layer that allows the user to perform common desktop tasks. We evaluated the user interface using three different interaction modes. The pointer-based point and click interface along with a controller-based direct manipulation interface use the physical device. Whereas, we also evaluated the user interface using a gesture-based controller free user interaction mode. We performed a comprehensive qualitative and quantitative user study to validate both the usability and user experience of the 3D user interface. Our study shows that the pointer-based interface is fastest in terms of time taken to complete the tasks, and overall more than 80% of the users are satisfied with the interface in terms of both usability and the user experience. In future, we would like to extend the system and analyze it with an extended degree of control over the user interaction modes.

References

1. LaViola Jr., J., Kruijff, E., McMahan, R.P., Bowman, D., Poupyrev, I.: 3D User Interfaces: Theory and Practice, 2nd edn. Addison-Wesley Professional, Boston (2017)
2. Re, G.M., Bordegoni, M.A.: Natural user interface for navigating in organized 3D virtual contents. In: International Conference on Virtual, Augmented and Mixed Reality. Springer International Publishing (2014)
3. Murthy, G.R.S., Jadon, R.S.: A review of vision based hand gestures recognition. Int. J. Inform. Technol. Knowl. Manage. 2(2), 405–410 (2009)
4. Saffer, D.: Designing Gestural Interfaces: Touchscreens and Interactive Devices. O'Reilly Media, Inc., Sebastopol (2008). ISBN 978-0-596-51839-4
5. Chatterjee, I., Xiao, R., Harrison, C.: Gaze+gesture: expressive, precise and targeted free-space interactions. In: Proceedings of the 2015 ACM on International Conference on Multimodal Interaction, pp. 131–138. ACM. November 2015

6. Cabral, M.C., Morimoto, C.H., Zuffo, M.K.: On the usability of gesture interfaces in virtual reality environments. In: Proceedings of the 2005 Latin American Conference on Human-Computer Interaction. ACM (2005)
7. Bhuiyan, M., Picking, R.: A gesture controlled user interface for inclusive design and evaluative study of its usability. J. Softw. Eng. Appl. **4**(09), 513 (2011)
8. Marin, G., Dominio, F., Zanuttigh, P.: Hand gesture recognition with leap motion and kinect devices. In: 2014 IEEE International Conference on Image Processing (ICIP). IEEE (2014)
9. Potter, L.E., Araullo, J., Carter, L.: The leap motion controller: a view on sign language. In: Proceedings of the 25th Australian Computer-Human Interaction Conference: Augmentation, Application, Innovation, Collaboration. ACM (2013)
10. Weichert, F., Bachmann, D., Rudak, B., Fisseler, D.: Analysis of the accuracy and robustness of the leap motion controller. Sensors **13**(5), 6380–6393 (2013)

The Control of a Vehicular Automata Through Brain Waves. A Case Study

Christian Ubilluz[1](✉), Ramiro Delgado[1](✉), Priscila Rodríguez[1](✉),
and Roberto Lopez[2](✉)

[1] University of the Armed Forces ESPE, Sangolquí 171103, Ecuador
{cmubilluz, rndelgado, pprodriguez2}@espe.edu.ec
[2] Regional Autonomous University of the Andes, Ambato, Ecuador
caposgrado@uniandes.edu.ec

Abstract. Science and technology are constantly changing; this has boosted a better quality of life, brain signals are among the most significant advances in this area allowing us to develop a-computer human interaction in the basis of using these signals for several applications. This research propose the management and control of a vehicular actuator through brain signals that are obtained with a non-invasive device placed in the cerebral cortex called Emotiv EPOC, this commercial electroencephalogram (EEG) allows us to interpret the signals via a Brain Computer Interface (BCI). The results granted to control the actuator, and, within the future it can be integrated to new applications such as wheelchairs for people with physical disabilities, giving them greater autonomy.

Keywords: Brain signals · Electroencephalogram (EEG) · BCI ·
Emotiv EPOC · Bluetooth

1 Introduction

The behavior of human beings as well as their bodies has always been studied; it can be inferred that the brain is one of the most important organs due to its function as a central control unit of the living [1].

The development of equipment and systems that acquire, process and analyze the signals emitted by the brain, are extremely important due to they allow the development of applications that help people with motor disability, that is, the control of this type of equipment with just thoughts.

The purpose of these systems is to improve the quality of life of this population sector, increasing the level of independence when performing tasks such as controlling a computer cursor, operate a wheelchair and even control a car [2].

The collection of brain signals was made through a commercial device called Emotiv EPOC based on a training method in which a visual stimulus is presented, and the reaction is recorded. The system learns to interpret the signals emitted by the brain and executes controlled actions in the prototype vehicle by using Bluetooth. With the results we can have control of UAVs or any other electronic controller.

© Springer Nature Switzerland AG 2019
Á. Rocha et al. (Eds.): WorldCIST'19 2019, AISC 931, pp. 748–754, 2019.
https://doi.org/10.1007/978-3-030-16184-2_71

2   Electroencephalogram EEG

EEG or electroencephalogram is the process used to study the functions of the central nervous system specifically the activity of the cortex of the brain. Consists essentially in recording through special electrodes electric flows formed within brain cells and are the basis of the nervous system [3].

3   Brain Waves

All brain waves have amplitude and frequency, which are measured in cycles per second and in microvolts (waves voltage). It has been found that there is a relationship among frequency and brain waves voltage with the different states of consciousness as shown in Table 1.

Table 1. Brainwave type [3].

Wave type	Voltage	Frequency	State of mind
Delta	10–50 microvolts	0.5–4 Hz	Hypnotic, Meditation
theta	10–100 microvolts	4–7.5 Hz	Vigil, Fullness
Alpha	100–150 microvolts	8–13 Hz	Relaxation, Tranquility
Beta	150–200 microvolts	14–26 Hz	Maximum alert, Vigilant
gamma	+200 microvolts	+30 Hz	State of stress and confusion

4   Emotiv EPOC

Emotiv EPOC equipment (Fig. 1), is a detection system that captures and amplifies electrical neuro brain waves generated by different mental "actions". This device is able to obtain signals of 14 channels: AF3, F7, F3, FC5, T7, P7, O1, O2, P8, T8, FC6, F4, F8, AF4, on the basis of the international 10–20 system, which is an approved method that describes the location of scalp electrodes for recording in the EEG. This device also operates with a filter for frequencies from 0.2 to 45 Hz, which can takes up to 128 samples per second on each channel [4].

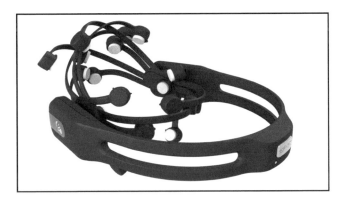

Fig. 1. Emotiv Epoc [4].

This device includes 16 sensors that are composed of a layer of felt, which are based on a gold plated contact. These sensors are mounted on plastic so that the assembly can be made easily. To establish a good conduction between the sensor and the scalp it is necessary to perform a process of hydration, which is the saturation of the sensor with saline solution in order to leave the felt completely soaked.

5   System Description

The system is described by a block diagram shown in Fig. 2, it begins with the EEG obtained brain activity conversion and transmitted to BCI within the movements: ahead, behind and stop, using an EEG Emotiv EPOC, the computer with the BCI in this case, Emotiv Xavier SDK, allows to determine the inputs signals: Expressiv (facial expression), Affectiv (emotional state) and Cognitiv (user itention for a particular movement). Afterwards the software Mind Your OSC, receives and sends data in packets OSC (Open Sound Control) completing the signal processing.

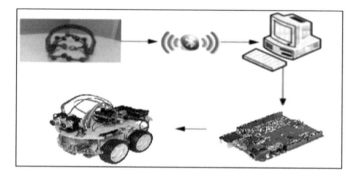

Fig. 2. Integration and communication schema.

Finally, the Processing software (Java environment) collects data from the electronic device (PLC) and carriers this data into an Arduino microcontroller; which management the actions of the controller.

To develop the experiment 60 students from different levels, age and gender were randomly chosen, all of them are coursing an IT career in an institution of higher education. With whom mental training and the automata execution effectiveness were evaluated, according to the brain triggered orders.

6   Testing Analysis

Tests were performed for each individual under apparent calm situations in order to get better brain waves, a total of five sessions were held. After analyzing the obtained data, the sixty students were categorized under three classifications.

Category 1 (Regular), whose average results are shown in Table 2, represented 20% of all individuals; they showed a low concentration capacity, however, they showed continuous improvement during the execution of a brain training activity under a state of medium relaxation. Figure 3 presents the virtual training.

Table 2. Category 1 (Regular).

Session	Ahead	Behind	Stop
S1	2	1	3
S2	2	2	2
S3	3	2	2
S4	5	3	4
S5	5	4	4

Fig. 3. Virtual training 1.

Category 2 (Restless) represented the 30% of individuals analyzed, on average it was obtained the results shown in Table 3, indicating that they were not relaxed and focused which influenced the brain training process. Figure 4 presents the virtual training.

Table 3. Category 2 (Restless).

Session	Ahead	Behind	Stop
S1	1	2	2
S2	1	3	2
S3	3	2	1
S4	2	2	3
S5	4	2	4

Fig. 4. Virtual training 2.

Category 3 (Optimal) represented the 50% of the individuals tested, the results obtained on average are described within Table 4, it showed subjects totally relaxed revealing a better management and control of their brainwaves. Figure 5 shows the virtual training.

Table 4. Category 3 (Optimum).

Session	Ahead	Behind	Stop
S1	3	2	4
S2	4	4	2
S3	4	4	4
S4	4	3	3
S5	4	4	4

Fig. 5. Virtual training 3.

Once executed the training of brainwaves, 50% of the students placed under the third category were selected in order to conduct and evaluate the movements of the automata. Data from their brain waves were sent to Mind Your OSC application as shown in Fig. 6(a), later, this information was processed by the Processing software, Fig. 6(b), which gives us a Java interface that transmit it to an Arduino microcontroller; After completing the transmission, mental commands are sent via bluethooth and the automata performs the desired movements.

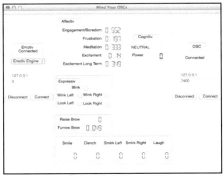
(a). Mind Your OSC.

(b). Processing.

Fig. 6. **(a).** Mind Your OSC. **(b).** Processing.

The chosen group performed three sessions in order to determine: the average percentage of precision, the average test duration time as shown in Table 5, determining that the concentration level is essential for a best brainwaves reception, which are processed, analyzed and subsequently employed in the automata motion.

Table 5. Accuracy and average time.

Subject 3	Session 1	Session 2	Session 3
Precision	93	91	96
Weather	37	34	31

7   Conclusions and Future Work

After analyzing the case study, it can be concluded:

People who showed greater concentration and relaxation were able to have a better control the automata, corroborating the results of similar researches.

BCI integration with Java environment allowed getting the data from brain signals and transmits them into the Arduino for performing the movements of the Automata.

The use of the software provided by the EEG equipment for the virtual training was the basis of the experiment success.

As future projects should be considered the following:

1. The connection to an automata by the use of Internet of Things (IoT), generating data that will be stored in the cloud and establishing connections with other automatas, medical equipment or smart homes.
2. Development of a training system that supports open and configurable multiple EEG equipment.

References

1. Chavez, V., Torres, D., Herrera, J., Hernandez, A.: Acquisition and analysis of electroencephalographic signals using the Emotiv EPOC + device. J. Technol. Innov. **3** (7), 1007–1118 (2016)
2. Chavez, V., Ramos, E., Dominguez, A.: Acquisition and analysis of electroencephalographic signals. Congress University, pp. 320–325 (2014)
3. Emotiv EPOC.: Brain computer interface & scientific contextual EEG. https://www.emotiv. com/files/Emotiv-EPOC-Product-Sheet-2014.pdf. Accessed 20 Nov 2018
4. Rodriguez, B.: Acquisition, processing and classification of EEG signals for design imagination motion-based BCI systems, VI Days introduction to research UPCT, 10–12 pp. ISSN 1888-8356 (2013)
5. Torres, F., Sanchez, C.: Acquisition and analysis of brain signals using the MindWave device. Cuenca University, Faculty of Electrical Engineering (2014)
6. Pinillos, E.: Proposal for the design of a computerized electroencephalography and self-diagnosis based on pattern recognition. Guatemala (2003). http://biblioteca.usac.edu.gt/tesis/ 08/08_0200_CS.pdf. Accessed 21 Nov 2018
7. Adams, K., Hunt, M., Moore, M.: The aware system: an augmentative communication interface prototyping. Paper presented at the Proceedings of the Rehabilitation Engineering Society of North America (RESNA) (2003)
8. Eid, M., Fernandez, A.: Readgogo! towards real-time notification on readers' state of attention. In: XXIV International Conference on Information, Communication and Automation Technologies (ICAT), Sarajevo, pp. 1–6. https://doi.org/10.1109/icat.2013. 6684047, (2013)
9. Gutierrez, J.: Systems of brain-computer interface: a tool to support the rehabilitation of patients with motor disabilities. Research on Disability, pp. 62–69 (2013)
10. Emotiv System.: Brain Activity Map. https://www.emotiv.com/product/epoc-brain-activity-map/. Accessed 01 Nov 2018
11. Gordo, M.: BCI interface high performance based on the detection and processing of brain activity. BCI-DEPRACAP. University of Granada (2009)
12. Ubilluz, C.: Brain waves processing, analysis and acquisition to diagnose stress level in the work environment. Trends Adv. Inform. Syst. Technol. **2**, 859–866 (2017)

Usability Study of a Kinect-Based Rehabilitation Tool for the Upper Limbs

Gabriel Fuertes Muñoz[1]([⊠]), Jesús Gallardo Casero[2]([⊠]),
and Ramón A. Mollineda Cárdenas[3]([⊠])

[1] Edison Desarrollos, S.L., 44002 Teruel, Spain
gabriel.fuertes@gmail.com
[2] E.U. Politécnica, Universidad de Zaragoza, 44003 Teruel, Spain
jesus.gallardo@unizar.es
[3] Institute of New Imaging Technologies, Universitat Jaume I, Castellón, Spain
ramon.mollineda@lsi.uji.es

Abstract. The execution of exercises during the rehabilitation process is very important to obtain the best result possible in an injury, but the patient often does not known the way to do correctly these exercises and it is unrealistic to have the physiotherapist controlling the execution. With the aim of solving this problem, we have developed KineActiv®, with the use of the Microsoft Kinect v2, as a tool to help the rehab process in a gamification environment, and to study the behavior of the users with the interface and the games. Each exercise is gamified, with different games for different movements. In this paper, we expose the description of the project and the results of the usability study with people with different injuries, doing the same exercises. These results show that the user experience in a rehab process, using new tendencies, with the Kinect camera, under a friendly design and funny games, is positive for the patients.

Keywords: Rehabilitation · Kinect · Virtual reality · Augmented reality · Usability study

1 Introduction

Prescribed, exercises regulated and controlled, help people with motor difficulties to recover mobility in the affected limbs and, therefore, quality of life. A frequent practice is to have specialized health personal in the rehab sessions (i.e. physiotherapists), who supervises the quantity and quality of the exercises. However, it is an expensive solution and could involve the displacement of patients with reduced mobility.

An alternative solution is usually to perform the exercises prescribed independently at home, without supervision of a professional. However, a study indicates that only the 31% of the people that suffer a mobility disorder execute the exercises correctly [1]. In addition, loss of patient motivation is frequent.

The development and the reduction in the price of sensor technologies have allowed to imagine and develop systems capable of automatically controlling the frequency, duration and correction of the exercises [2–4]. In addition, the use of

© Springer Nature Switzerland AG 2019
Á. Rocha et al. (Eds.): WorldCIST'19 2019, AISC 931, pp. 755–763, 2019.
https://doi.org/10.1007/978-3-030-16184-2_72

multimodal and interactive user interfaces, together with gaming technologies, can help to keep the user motivated and engaged during the execution of the exercises.

During recent years, gamification has become one of the most popular methods of enriching information technologies in several domains. Popular business analysts have made promising predictions about the penetration of gamification; however, it has also been estimated that most gamification efforts will fail due to poor understanding of how to design gamification solutions.

In this paper, we show the system KineActiv®[1], which has been implemented using a special camera, Microsoft Kinect v2. With the potential of Kinect, we have developed a system that allows the physiotherapist to prescribe exercises to their patients with the goal of improving the rehabilitation process between sessions of physiotherapy.

This paper is structured as follows. Section 2 includes some relevant works that have been taken into account in our approach. Section 3 presents the tool, along with a description of its operation and its design. In Sect. 4 we introduce a usability study that we have performed with real patients using the tool. Lastly, Sect. 5 is about conclusions and future work.

2   Related Work

Research in the fields of computational vision, virtual reality and augmented reality has led to the development of applications with the potential to have an impact on the lives of people with disabilities or injuries. Examples can be found in the use of computers for the development of vision based human-machine interface [5, 6].

The Kinect sensor shows robustness against partial occlusions and Chang [3] used Kinect to facilitate the necessary task patterns of people with cognitive impairment. The performance of Kinect as a tool to evaluate kinetic variables, compared to current standard methods, is a subject of great interest [7]. Although several studies have been published in this area [8], the general trend has been to compare motion-capture systems (MOCAP) based on scalar summary measures with those based on metric measurements. Other studies have shown that Kinect is a sufficiently accurate and sensitive sensor to measure gross movements, which makes it suitable for stroke rehabilitation systems [9] or to measure movement symptoms in people with Parkinson's disease [10]. Additionally, Chang [11] performed another study about the behavior of patients with cerebral palsy with the use of Kinect, without games and doing only 3 movements, with good results. An example of integration of Kinect, gaming and virtual reality can be found in the treatment of pathologies in brain injuries, which show that they are useful in rehabilitation with such patients [12]. Pastor et al. [13] ratified these results with their study of an upper limb rehabilitation system using Kinect and computer games, but, the same as the aforementioned, they did not perform an exhaustive real-time control of the movements.

---

[1] http://www.kineactiv.com.

Rand [2] studied Rehabilitation potential with the PlayStation 2 video game console and camera. It was used as a tool during the rehabilitation of stroke patients and with other neurological disorders and, in parallel, Broëren [14] investigated the effects of virtual reality technology for stroke rehabilitation. All of them provided evidence that rehabilitation systems through virtual reality media can increase cognitive abilities and demonstrated that virtual reality allows quantitative analysis of 3D tasks in real time.

Several studies have shown that offering virtual rehabilitation exercises similar to games, motivates patients to perform rehabilitation exercises and also to increase their adherence to treatment, entailing a greater efficiency of the rehabilitation process [15–17].

3   Tool Description

In this section, we are going to explain the features of KineActiv, a tool developed to be used by physiotherapists as a help in their work methodology when treating patients that need rehabilitation in upper limbs. We are going to emphasize the user interface aspects, which are the main focus of this work. The main different of KineActiv regarding similar systems is that we evaluate the movements of the patients in real time, showing warning messages to the user with the goal of making a perfect movement in all exercises.

The most important content of a visual design can be used for effective summarization or to facilitate retrieval from database but, in our case, the priority is to focus on the importance of design. Our interest is to develop not only a useful interface showing information, but also a gamified interface that involves the user and makes him perform the exercise prescribed by the physiotherapist in a perfect way. The correct execution of the movement is very important, and this is the reason why we correct the user every time that he makes an error, in a real-time basis.

We have used two differenced parts to design the interface. On the one hand, we show the information about the exercises that the patient must do in the rehab process. This interface uses only two or three basic colors (white, green and grey) and divide the screen into four columns.

The information will always be in the first and second columns, and the description of the execution of the exercise will be in the last two or three columns, depending on the screen in which we are in each moment, as we can see in Figs. 1 and 2.

On the other hand, we have determined the elements, components and key facts to design interactive interfaces for virtual reality environments, focusing on the design and development of virtual worlds for rehabilitation applications. Our effort focuses on the usability study of virtual reality applications interfaces developed for non-touchable natural environments.

To do this, we have developed different gamified approaches to guide the patient while performing the exercises prescribed by the physiotherapist, with the objective that the patient feel well, safe and comfortable during the experience, doing at the same time the effort to get an objective established by the physiotherapist. Each game is different depending on the type of the exercise, the movement, and the part of the body to rehabilitate. For instance, in Fig. 3 we can see a game which consists in keeping the arm in a 90° position holding a weight during 35 s. The gamification in this activity consists in cooking a chicken in a grill.

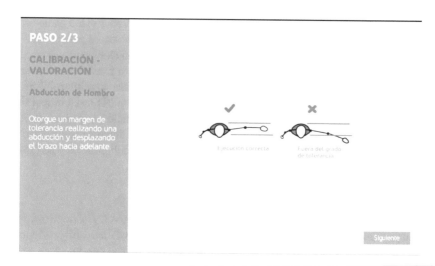

Fig. 1. Tolerance calibration

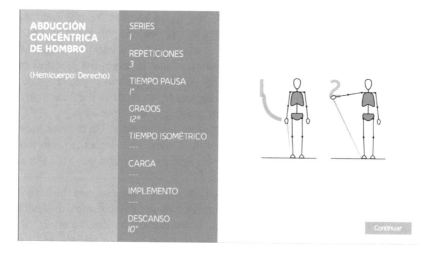

Fig. 2. Exercise explanation

In Fig. 4 we can see another activity. In this case, the patient has to kill the alien in the screen. This alien will be located in a given position that will be established by the physiotherapist, who will determine the degrees of movement that the patient must perform. The patient will do several repetitions of the movement. If the patient fulfills the goal, the alien will be destructed; otherwise the alien will destroy the city.

After every activity is completed, the patient will see a screen with the score obtained. This score is based in a 3-star system, in order to keep the patient motivated for his next session. Depending on the percentage of correct executions, the system shows one, two or three stars.

Fig. 3. Game execution 1

Fig. 4. Game execution 2

4   Usability Study

In order to have an initial validation of the tool and to test its suitability for being used in real situations, we have performed a study with real patients. The patients were 10 adults with ages between 38 and 83 years, with an average value of 54,8. Six participants were men, and the remaining four were women. All of them suffered with upper limb disorders due to diseases such as tendinopathy, luxation or pinching. All of them used the tool during three weeks, twice each week, for a total of six uses of the tool. Each session implied playing the three games included in the tool.

In this paper, we are going to conduct a usability study performed with the enrolled patients by means of a questionnaire. The empirical results of the performance achieved by the patients during the study are currently being analyzed and will be addressed in a different paper when the analysis gets complete.

In order to design the questionnaire to test usability, we analyzed questionnaires used for similar purposes in the field of rehabilitation and virtual health [18, 19]. Finally, we decided to add the items present in two questionnaires that we understand that match the goals that we want to achieve with this study. Firstly, we adopted the ten items defined in the SUS questionnaire [20], that have been used for several purposes when evaluating usability, for example, in the usability assessment of a rehabilitation system for the upper limbs [21]. But, in order to obtain more information during our study, we added also the six items used in [22] for evaluating *flow* in a virtual reality rehabilitation system, adopted from [23]. This way, we have been able to test both usability in our system and *flow* when playing with the games included in it. All 16 items were put together in one questionnaire were every item had to be ranked with a value from 1 to 5, were 1 was the lowest value and 5 was the highest.

Next, we are going to perform a descriptive analysis of the results obtained after the patients fulfilled the questionnaire. Thus, Table 1 shows the average values and standard deviations of the values obtained. Items 1 to 10 are the ones from the SUS questionnaire, and items 11 to 16 are the *flow* items.

Based on these results, we started the analysis by calculating the SUS score of the system. The SUS score has a range from 0 to 100, and is calculated from the results obtained when applying the SUS questionnaire, which in our case is made up by items 1 to 10. After calculating the score, the global result was 84,5. We understand that this value validate the suitability of KineActiv as a usable system. We have not performed an exhaustive analysis of individual items as the SUS questionnaire authors state that its results may not be relevant.

The second part of the questionnaire, as we have already stated, is about *flow* when using the system. We have taken this into account as the gamification approach in the tool should imply a full immersion in the system in order to achieve the goals that have been set out. The *flow* items (11 to 16) evaluate three different factors. The first one is attentional focus (items 11 and 12). In this case, the values obtained are not the ones desired, which suggests that the game may not maintain the attention in a strong manner. The second factor is intrinsic interest or pleasure (items 13 and 14). In this case, the values received by the items are good (1,6 in a question where boredom was valued and 4,4 in the opposite one, where they were asked by fun). Thus, the system has been considered to be enjoyable. Lastly, control was also evaluated (items 15 and 16). Here, values are also good ones, so it shows up that the use of KineActiv did not cause any negative feelings.

This way, we have performed this validation of KineActiv with in general satisfactory results. The system has proved to be usable and the flow experience when using it seems to be the proper one, in general terms. Thus, we are encouraged to continue using the tool and to perform the analysis of the empirical results, which, as we have already mentioned, will be our next step.

Table 1. Results of the questionnaire

Question	Average value	Standard deviation
1. I think I would like to use KineActiv frequently	4,7	0,48
2. I think that KineActiv is unnecessarily complex	1,4	0,52
3. I think that KineActiv is easy to use	4,5	0,53
4. I think that I would need help to use KineActiv	2,3	1,06
5. I think that the various functions in KineActiv are well integrated	4,3	0,67
6. I think there is too much inconsistency in KineActiv	1,4	0,52
7. I imagine that most people would learn to use KineActiv very quickly	4,6	0,52
8. I found KineActiv very cumbersome to use	1,5	0,53
9. I felt very confident using KineActiv	4,4	0,52
10. I would have needed to learn a lot of things before using KineActiv	2,1	0,88
11. I thought about other things when using KineActiv	2,5	0,85
12. I was aware of distractions when using KineActiv	3,0	0,47
13. Using KineActiv was boring for me	1,6	0,52
14. KineActiv was fun for me to use	4,4	0,52
15. I felt that I had the control over my rehabilitation process with KineActiv	3,9	0,74
16. I was frustrated with what I was doing when using KineActiv	1,4	0,52

5   Conclusions and Future Work

In this paper, we have introduced KineActiv, a Kinect-based rehabilitation tool with gamification elements oriented towards the rehabilitation of the upper limbs. We have performed a user study with the tool, and we have shown the results of a usability analysis with the participants in the study. The results have shown that the usability of the system is good, and that users get engaged when using the tool, although the attentional focus evaluation has not been well valued by the participants in the study.

As future work, we are going to carry out a complete analysis of the data of use collected during the empirical study, in order to perform an exhaustive evaluation of the system regarding its rehabilitation goals. With the results of that analysis, together with the information obtained in the usability study that we have explained in this paper, we will have information enough to perform modifications in the system in order to solve the problems that may have arisen.

Acknowledgements. This work has been partially supported by grants P1-1B2015-74 from Univ. Jaume I.

References

1. Shaughnessy, M., Resnick, B., Macko, R.: Testing a model of post-stroke exercise behavior. Rehabil. Nurs. **31**, 15–21 (2006)
2. Rand, D., Kizony, R., Weiss, P.L.: Virtual reality rehabilitation for all: Vivid GX versus Sony PlayStation II eye toy. In: Proceedings of the Fifth International Conference on Disability, Virtual Reality and Associated Technologies, pp. 87–94 (2004)
3. Chang, Y.-J., Chou, L.-D., Wang, F.T.-Y., Chen, S.-F.: A kinect-based vocational task prompting system for individuals with cognitive impairments. Pers. Ubiquit. Comput. **17**, 351 (2011)
4. Saposnik, G., Teasell, R., Mamdani, M., Hall, J., McIlroy, W., Cheung, D., Thorpe, K.E., Cohen, L.G., Bayley, M.E.: Effectiveness of virtual reality using Wii gaming technology in stroke rehabilitation: a pilot randomized clinical trial and proof of principle. Stroke **41**(7), 1477–1484 (2010)
5. Betke, M., Gips, J.: The camera mouse: visual tracking of bodyfeatures to provide computer access for people with severedisabilities. IEEE Trans. Neural Syst. Rehabil. Eng. **10**(1), 1–10 (2002)
6. Varona, J., Manresa-Yee, C., Perales, F.J.: Hands-freevision-based interface for computer accessibility. J. Netw. Comput. Appl. **31**(4), 357–374 (2008)
7. Napoli, A., Glass, S., Ward, C., Tucker, C., Obeid, I.: Performance analysis of a generalized motion capture system using microsoft kinect 2.0. Biomed. Signal Process. Control **38**, 265–280 (2017)
8. Dehbandi, B., et al.: Using data from the Microsoft Kinect 2 to quantify upper limb behavior: a feasibility study. IEEE J. Biomed. Health Inform. **21**(5), 1386–1392 (2017)
9. Webster, D., Celik, O.: Experimental evaluation of Microsoft Kinect's accuracy and capture rate for stroke rehabilitation applications. In: 2014 IEEE Haptics Symposium (HAPTICS), Houston, TX, pp. 455–460 (2014)
10. Galna, B., Barry, G., Jackson, D., Mhiripiri, D., Olivier, P., Rochester, L.: Accuracy of the Microsoft Kinect sensor for measuring movement in people with Parkinson's disease. Gait Posture **39**(4), 1062–1068 (2014)
11. Chang, Y.-J., Han, W.-Y., Tsai, Y.-C.: A Kinect-based upper limb rehabilitation system to assist people with cerebral palsy. Res. Dev. Disabil. **34**(11), 3654–3659 (2013)
12. Cabrera, R., Molina, A., Gómez, I., García-Heras, J.: Kinect as an access device for people with cerebral palsy: a preliminary study. Int. J. Hum. Comput. Stud. **108**, 62–69 (2017)
13. Pastor, Is., Hayes, H.A., Bamberg, S.J.M.: A feasibility study of an upper limb rehabilitation system using kinect and computer games. In: 2012 Annual International Conference of the IEEE Engineering in Medicine and Biology Society (EMBC). IEEE (2012)
14. Broeren, J., Bjorkdahl, A., Claesson, L., Goude, D., Lundgren-Nilsson, A., Samuelsson, H., Blomstrand, C., Sunnerhagen, K.S., Rydmark, M.: Virtual rehabilitation after stroke. In: Proceedings of the Medical Informatics Europe, pp. 77–82 (2008)
15. Lange, B.S., Flynn, S.M., Proffit, R., Chang, C.Y., Rizzo, A.A.: Development of an interactive game-based rehabilitation tool for dynamic balance training. Top. Stroke Rehabil. **17**(5), 345–352 (2010)
16. Flynn, S.M., Lange, B.S.: Games for the rehabilitation: the voice of the players. In: International Conference on Disability, Virtual Reality & Associated Technologies (ICDVRAT 2010), pp. 185–194 (2010)
17. Lozano-Quilis, J.A., Montesa, J., Juan, M.C., Alcañiz, M., Rey, B., Gil-Gómez, J.A., Martínez, J.M., Gaggioli, A., Morganti, F.: VR-Mirror: a virtual reality system for mental practice in post-stroke rehabilitation. In: Smart Graphics 2005, pp. 241–245 (2005)

18. Parmanto, B., Lewis Jr., A.N., Graham, K.M., Bertolet, M.H.: Development of the telehealth usability questionnaire (TUQ). Int. J. Telerehabilitation **8**(1), 3 (2016)
19. Gil-Gómez, J.A., Gil-Gómez, H., Lozano-Quilis, J.A., Manzano-Hernández, P., Albiol-Pérez, S., Aula-Valero, C.: SEQ: suitability evaluation questionnaire for virtual rehabilitation systems. Application in a virtual rehabilitation system for balance rehabilitation. In: Proceedings of the 7th International Conference on Pervasive Computing Technologies for Healthcare, pp. 335–338 (2013)
20. Brooke, J.: SUS-A quick and dirty usability scale. In: Usability Evaluation in Industry, vol. 189, no. 194, pp. 4–7 (1996)
21. Pei, Y.-C., Chen, J.-L., Wong, A.M.K., Tseng, K.C.: An evaluation of the design and usability of a novel robotic bilateral arm rehabilitation device for patients with stroke. Front. Neurorobot. **11**, 36 (2017)
22. Shin, J.H., Ryu, H., Jang, S.H.: A task-specific interactive game-based virtual reality rehabilitation system for patients with stroke: a usability test and two clinical experiments. J. Neuroengineering Rehabil. **11**(1), 32 (2014)
23. Park, J., Parsons, D., Ryu, H.: To flow and not to freeze: applying flow experience to mobile learning. IEEE Trans. Learn. Tech. **3**(1), 56–67 (2010)

Leadership and Technology:
Concepts and Questions

Ana Marisa Machado[(⊠)] and Catarina Brandão

Faculdade de Psicologia e de Ciências da Educação, Universidade do Porto,
Porto, Portugal
anamarisamachado96@gmail.com, catarina@fpce.up.pt

Abstract. In recent decades there has been a transformation based on economic, social, cultural and political integration between countries, guided and
fueled by the increasingly frenetic scientific and technical progress. Globalization triggers a new social and work paradigm in which it is possible to communicate and interact at any time, from anywhere, face-to-face or virtually,
making the organizational configurations Volatile, Uncertain, Complex and
Ambiguous. That is, VUCA. In this context, leadership processes are increasingly mediated by technology, giving space to the concepts of e-leadership, e-
leader and virtual team, whose conception gravitates around technology as a
facilitator of communication, interaction, development and task sharing. This
reality will become even more prevalent in the future. Hence, developing
research that explores the dynamics underlying the interdependent relationship
between these elements is fundamental. This paper explores the literature
regarding the role of technology in leadership. It seeks to contribute to the
theoretical clarity regarding this topic and identify relevant questions to the
discussion of what is and will be, one of the biggest debates of the 21st century.
E-leadership is the process by which a leader guides and motivates followers
through technology. This requires leaders to be communication-oriented, supportive, promote confidence building and to be capable of creating a stable
environment. Gaps in literature show that there is a lot more to know about this
topic, and that it is urgent to address it so organizations stay competitive.

Keywords: VUCA world · Leadership · Technology · Virtual teams ·
Virtual leadership · E-leadership

1   Introduction

In recent decades, scientific and technical progress has boosted globalization and the
growth of the world economy, sustaining new inventions and introducing changes in all
areas of human life. This transformational process has reached an unprecedented pace,
making it increasingly difficult to keep up with the rapid technologic, economic and
social developments propelled by the frenzy of the digital revolution and the expansion
of global markets. The speed of these changes makes the organizational context
unpredictable or, using a concept introduced by the US Army after the end of the Cold
War - VUCA (Volatile, Uncertain, Complex, Ambiguous) [1].

© Springer Nature Switzerland AG 2019
Á. Rocha et al. (Eds.): WorldCIST'19 2019, AISC 931, pp. 764–773, 2019.
https://doi.org/10.1007/978-3-030-16184-2_73

The acronym VUCA is used to describe various organizational configurations characterized by their volatility (nature, speed, volume, magnitude and dynamics of change), uncertainty (lack of predictability), complexity (the set of conditions involving the organization and its members) and ambiguity (the bustle of reality and the perceived meanings of conditions) [1]. These definitions encompass security, economy, market and workforce conditions around the world, which have rendered existing models obsolete in dealing with the inherent contextual challenges.

Globalization has sped up and expanded the flow of information, services, goods and people across borders, creating a growing labor demand for flexible employees. At the same time, there has been an extensive proliferation of Advanced Information Technology in organizational systems. That is, tools, techniques and knowledge that facilitate the functions of digitization, planning, decision, dissemination and control of information. Thus, regardless of industry and function, the phenomena of globalization, technological change, and rapid evolution present workers with new challenges, namely to those who have leadership roles. These workers have to be educated and prepared to effectively manage and respond to situations in the VUCA world where communication through technology media is indispensable.

This paper focuses on leadership processes mediated by technology. We start by focusing leadership and technology, and then introduce the concept of e-leadership as a consequence of the emerging use of technology in work environments. We next highlight some of the questions that remain unanswered regarding e-leadership and its impact on organizations and individuals, whether these be leaders or followers.

2 Leadership

Leadership is a process of influence focused on the empowerment, development and motivation of others, maximizing their effort and commitment to achieve specific goals. It is a mean that fosters the incorporation of the organization's mission, values and vision into the worker's daily tasks [2].

Literature stresses that leaders in the 21st century must consider critical issues that will guide strategic plans, resource allocation and organizational development needs, adapting their dynamics to the changing characteristics of the context. The agility of the leader, or the ability to continuously adjust and adapt to the strategic direction, is increasingly considered the vital factor of success in organizations [2–4].

Building from this, literature on leadership focus the need for a new vision, higher energy, positive approach and orientation to act, which will allow leaders to meet the dynamic and often conflicting demands of the environment. These ideas seek to avoid some of the assumptions and limitations underlying the more traditional perspectives on leadership, and although they focus on different aspects, they are similar in many others [5]. In this sense, literature points that effective leaders are able to identify change-oriented tasks, relationships and behaviors appropriate to the current situation, using them to modify objectives, strategies and work processes and to facilitate adaptation to the external environment. Empowerment is considered fundamental in this process, involving the independence of followers, as well as shared responsibility and decision-making.

Theories such as authentic leadership and ethical leadership emphasize the leader's trust, cooperation, and moral values as determinants of the reciprocal influence necessary for effective leadership. On the other hand, charismatic leadership and transformational leadership highlight the subordinate's personal identification with the leader as a potential source of influence (that is, reference power) [5]. Vielmetter and Sell [6] introduce the altrocentric leader neologism to characterize the leader who, as opposed to the alpha leader, focuses primarily on others and their needs, who understands that leadership is relational and that the difference between leaders and followers is contextual. This means that individuals may develop leadership processes in specific relationships, while in others they assume the role of followers. In each role there are values and strengths that must be acknowledged and understood. The altrocentric leader is empathic, intellectually curious, ethical, and genuinely concerned with diversity. This leader faces diverse challenges related to communication, coordination and dissemination of knowledge. These challenges highlight the difficulties associated with followers and leaders' self-management failures and the possibility that, because individuals belong to multiple groups with unclear barriers, followers will develop problems of identity and inability to establish connections and commitments with the work group, the organization and its mission, values and objectives.

3   Leadership and Technology

Nowadays, there is a growing consensus regarding the changes that technology has introduced in the way we live, work, communicate and organize our activities. The increasing flexibility of organizations, the fading of their boundaries and the change of the content of work mean that organizations need structures different from traditional hierarchical systems. It is created, to a great extent, a need for temporary systems whose elements (people and technology) are assembled and disassembled according to the changing needs associated with specific projects mediated by technology. Thus, the interdependent dynamics between leadership and technology is characterized by cognitive elements, transmitted through tools such as e-mail, knowledge management and executive information systems, among others, in which multiple sources take part, both organizational and inter-organizational [3, 7].

Information technologies have introduced new ways of working and creating value, allowing to increase the physical distance between the individuals involved in systems' development, those who use them and those whose lives are potentially affected by them. Thus, empowering communication and work in geographically dispersed teams, socially or temporarily, as well as with different technological qualifications and different life times. Organizations are able to hire the most qualified individuals for a given project or function, regardless of location, to respond more quickly to increased competition, and to provide greater flexibility to employees working at home or whilst traveling [8].

The constant updating that characterizes information technologies causes them to have a high disruptive potential in the work context, causing difficulties of concentration and focus on the task. At the same time, the decrease in communication and face-to-face coexistence negatively affects the affective and social dimension of labor relations, turning employees more reserved and self-centered [9]. Hence, it is urgent to focus relationships based on technology-mediated interactions as well as the role of different styles of leadership and communication in trust, openness, feedback and acceptance in virtual work groups [7, 9]. This requires acknowledging that the impacts of technology differ according to the nature of the existing relations, so it is fundamental to approach them in a contextualized way [9].

4   E-leadership

Given the advantages and disadvantages of technology in the workplace, and as virtual and flexible work options continue to evolve, more employers try to formalize their virtual work policies and gain a better understanding of how to manage virtual workforces [8]. This has led to the emergence of concepts such as e-leadership, digital leader, or e-leader and digital or virtual team.

E-leadership refers to the process by which a person guides and motivates the behavior on a technology-mediated environment [4]. Thus, an e-leader is the individual who is capable of influencing others through technological means and originate a change of goals, feelings, thoughts, behaviors and/or performance in individuals, groups or organizations, to achieve a specific goal [4]. This way, a virtual organization is composed by a group of people working together from different geographic and technology dependent domains to communicate and achieve organizational goals [4]. In this way, the gravitational focus concentrates on the technological side that allows, as a facilitator, communication, interaction, development and sharing of tasks. This turns a fundamental question: which are the factors that contribute to the promotion of virtual teams and in what kind of situations do these teams succeed?

There are five distinct theoretical streams that provide helpful insights about leadership and, inherently, e-leadership, namely: (1) contingency theories emphasizing situational variables which moderate the effects of e-leadership practices; (2) comparative research on roles and behaviors crucial to distinctive e-leadership positions; (3) research on external threats and opportunities that call for e-leadership; (4) studies on conditions that hamper the success of e-leadership; and (5) research on characteristics and competencies that facilitate e-leadership [10].

The first stream is based on the idea that leaders are most effective when they adjust their behavior to the situational factors that characterize the situation. Thereby, they reflect on follower characteristics and event aspects such as means of communications, time and purpose and act in accordance. In technological mediated environments, this involves thinking about the tool to communicate the message and weighing different forms of clearly and constructively build the message, among various others concerns that module the e-leader behavior for which the environment of action is virtual.

The second stream states that different hierarchical levels of leaders have a necessity for different approaches and behaviors, once the objectives and structure also differ. Thereby, by comparing different leadership roles and their particularities, it is possible to formulate models that show what is most effective in each level, allowing the development of adjusted strategies to promote performance.

Thirdly, there has been a growing interest in investigating the ever-changing organizational environment and the needs that emerge from its configurations. Technological pressures are, clearly, one of the main focus of this stream, not only by its benefits but also for the challenges that it implicates.

Literature on the fourth stream is scarce, however, there are some studies that point antecedents, moderators and/or mediators of e-leadership, such as leader's personality traits. The last stream focus on skills that e-leaders and followers have that can facilitate or enhance the success of e-leadership, such as clear communication, openness and technological knowledge.

Theoretically, it is assumed that digital leaders must be flexible and adaptable, possess broad intellectual curiosity and desire for new knowledge; be willing to see and create value in a clear different perspective and be comfortable with uncertainty. So, like all leaders, they must have true passion for what they do; they must look for solutions and challenges, and they must be willing to learn constantly, and to maintain an egalitarian and results-oriented approach [11]. An effective e-leader must be willing to let others take the lead when needed, be a supportive communicator through technology mediated means to facilitate and guide communication among team members, create clear structures, promote clarity of roles, and to improve socio-emotional relationships with limited access to face-to-face meetings [12]. If it is true that individuals in leadership roles face several challenges, one can anticipate the additional ones that technology may present to those in leadership positions.

E-leadership focuses on the management of information flow, even if incorporating in an organic way, basic characteristics such as vision, orientation, motivation, inspiration and confidence building. In this sense, the role of future leaders is and will be, fundamentally, to balance the emphasis on change, providing (a sense of) stability and continuity, as well as establishing and maintaining collective identities in the absence of traditional boundaries of identity [2]. Virtual leadership reflects a new work environment, in which human interactions are mediated by information and communication technology, and where leaders can lead complete projects at distance [7]. Similar to face-to-face leadership, virtual leadership, or e-leadership, can be conveyed through traits (who it is), behaviors (what is done), cognitions (what and how it is thought), and affection (what it is felt) associated with leaders and followers [3].

Consistent with the literature on the importance of social-emotional support in successful team development and social presence as a promoter of group cohesion, effective e-leaders are classified by their followers as very present in the virtual community. Personal involvement in online settings is essential since this is a context where nonverbal clues, inclusion, and perceived social support in interactions can be lost. In this sense, leaders face great pressures to be authentic and transparent in their interactions [3].

5   Related Work

Romand and colleagues [13] developed and tested a model that focuses on e-leadership as an online communication skill, as well as the challenges and opportunities this environment implies. Communication, independently of the form or context, is fundamental to disseminate information and orientate behaviors. Thus, it's equally or more important in virtual environments, given the lack of face-to-face contact. The authors identified six virtual competencies (e-competencies) central to successful e-leadership, namely: e-communication skills, e-social skills, e-change management skills, e-tech savvy, e-team skills, and e-trustworthiness [13]. The Six E-Competency (SEC) Model, as they named it, was tested through factor analysis of 560 valid responses to a self-administered survey. Results of the study point to a multidimensional, integrated and comprehensive understanding and conceptualization of e-leadership, given that clear interdependent associations were identified. Moreover, results show that the lack of one of the competences may not have negative short-term consequences, but it will reduce the e-leadership effectiveness on the long-term. Finally, Romand and colleagues [13] state the need of an operationalizable e-leadership definition to allow testable propositions.

Building on previous works, Liu and colleagues [14] studied e-leadership using a similar methodology, providing an e-leadership model based on individual level communication. Assuming that e-leadership requires a special set of skills and resources, they empirically tested propositions about personal traits and skills, intention to use and the use of information and communications technology (ICTs) at the individual level for leading (ECAM model) [14]. The overall model showed a good fit regarding e-leaders technological adoption. Moreover, energy, responsibility and analytical skills stood out as particularly significant.

Darics [15] also addresses e-leadership on a communication perspective, specifically in text-only channels, through Grounded Practical Theory and Interactional Linguistics. The author highlights the importance of nonverbal audio and visual cues in leadership contexts, presenting an inventory of strategies used to replace this in text-based digital interactions, ranging from orthography to typography and discourse mechanisms. Results from the analysis of conversations in a virtual team provided evidence that the nonverbal audio and visual cues can, in fact, be recreated in writing. These findings prove the need of e-leadership theories that take in account the specificity of the environment in which it's developed, reflecting the difference between face-to-face and virtual interaction [15].

Maduka, Edwards, Greenwood, Osborne and Babatunde [16] conducted an exploratory case study on the broad competencies required for e-leadership and its effectiveness. They identified constant feedback, clear directions and reliability, personal characteristics and trust and task clarity as some of the fundamental elements for an ideal e-leader. The study also reflects on transformational leadership as a theoretical framework that not only suits their findings but can also support selection processes of leaders of virtual teams. Purvanova and Bono [17] examined transformational leadership in face-to-face and virtual teams. Their laboratory experiment revealed that leaders were more effective in virtual environments, when they increased their transformational leadership. Thus, the effect of leadership style was stronger in this context/type of teams.

6   What We Still Need to Know Regarding Leadership Process Through Technology

Despite the increased interest in this subject, there are several gaps in research as to how technology impacts leadership and how the context of technological communication affects leaders' and the success of virtual teams [8]. We are particularly blind as to if processes mediated by technology activate leaders' and subordinates' system for the defensive self. We next highlight some of the areas that are still to research regarding e-leadership.

6.1   Theoretical Framework

Research focusing e-leadership has stressed the importance of understanding leadership in virtual teams, starting with the definition of the concept of e-leadership itself. Virtuality needs to be approached from an unanimous point of view to be possible to measure it consistently [18] and construct a cohesive body of knowledge about specific configurations of teams and, inherently, e-leadership. It is imperative to develop a theoretical framework that consistently explores the functions of e-leadership, especially from a multilevel perspective [19].

If there is a concern regarding the disadvantages of traditional leadership approaches in virtual environments, there is also a necessity for methods that enable to identify leaders and leadership behaviors which appropriately deal with the team context and means of action [18]. Working with longitudinal methodologies may also allow to identify antecedents and mediating and moderating mechanisms that impact on e-leadership, reducing ambiguity [18]. It is also relevant to understand if organizations where e-leadership processes are more present face specific challenges regarding workers' engagement and organizational commitment. The case study design presents itself as particularly relevant to do this.

6.2   E-leadership as a Social Influence Process

From a social point of view, understanding the ways by which virtual followers develop interpersonal perceptions may lead to diverse opportunities regarding social constructions in virtual environments. The way by which these perceptions impact on feelings of trust, support and shared sense of purpose also remains undefined. This approach could contribute to the framing of a social virtual environment that offers a base to optimize relationships, in synchronous and asynchronous forms of communication [20].

6.3   Role of the E-leader

Supervisors are the subordinates' expected caregivers at the organization [21]. Caregiving is a psychological act involving conscious and unconscious phenomena. It implies the transference of emotions through exchanging resources, times, information, counselling or services [22]. Early work on leadership and individual roles in the virtual environment suggest that teams continue to benefit from the presence of the leader as a caregiver, whose main contribution to the team is to support regular, detailed and rapid communication as well as the identification of associated relationships and responsibilities to individual roles. However, our understanding of this role is still deeply limited [12].

Given that the (conscious or unconscious) experience of a threat to one's well-being activates our careseeking system for self-defense, it is expected that difficulties associated with e-leadership might create in the subordinates the need for receiving caregiving. It is important that leaders are able to acknowledge this as well as their own dilemmas and difficulties regarding the use of technology to lead and interact with subordinates and other organizational members. Leaders need to present themselves as being emotionally and physically available to subordinates [21]. In a technology mediated work environment this may become a challenge.

6.4   Individual Level of Analysis

Addressing interculturality, personality characteristics and types of communication in this type of working process would also be fundamental to a contextualized, adjusted and adequate comprehension of individuality. Investigating the individual level of e-leadership brings comprehensiveness to theoretical models, promoting the applicability (and usefulness) of these models to organizations.

7   Final Remarks

The importance of leaders-members relationships for the well-being in an organizational context, turns fundamental to understand how leaders and subordinates experience e-leadership processes and the impact of these experiences, namely on the retention of talent in organizations. With the new opportunities brought by the digital era, the pressure for leadership to incorporate technology will continue. The consequences of this are substantial and to date very little studied. The use of technology to engage followers in accessing and integrating information changes the fundamental paradigm of leadership, moving from a leader-centered process to a more follower-centered and technology-driven mediator. Team members now have access to information previously known only to leaders, which puts pressure on the latter to be prepared to justify their decisions quickly or to develop inclusive systems that ensure members' participation [23].

Leaders are now directing parts of projects, or even entire projects, exclusively through communication technologies [23]. Virtual teams are often formed to overcome geographical or temporal separations and are, by definition, composed of members who are never or rarely physically present. These teams require new ways of working across borders, through systems, processes, technology and people. Despite the generalized increase of this type of work, there is still relatively little investment in research that focuses on the experiences of leadership mediated by technology, although it is accepted that technology affects the ability of e-leaders to transmit their social presence and non-verbal cues such as facial expressions, voice inflections and gestures.

Given this, e-leaders face diverse challenges related to the dynamics underlying trust, communication, distance and time and diversity [8].

Future digital leaders need to develop critical skills to improve the effectiveness of virtual work, starting by understanding their own capabilities and difficulties. It takes continuous reflection and self-evaluation to develop the ability to critically analyze the processes by which improvement is possible. E-leaders are constantly bombarded with instant messages, e-mails and videoconferences, facing demands from various sources and multiple, ever growing, challenges. In spite of that, they are capable of dealing successfully with VUCA contexts and also decoding them, providing a shared framework to their followers. Moreover, they develop a digital skillset [24] that adaptively responds to the gap between theory and practice in relation to technological tools. Thus, e-leaders depend on their ability to effectively select and operate those tools, for both organizational and personal objectives [24].

We have briefly examined the field of e-leadership, regarding its current concepts and shifting paradigms, as well as some emerging research questions. There are several challenges to address regarding the theoretical foundations of its understanding but is crystal clear that practice won't slow down. If researchers want to catch up with the field, it's crucial to start unravelling the layers that involve e-leadership - at an individual level, focusing followers and leaders and at an organizational level, focusing policies and implemented strategies - and addressing the pressing issues it conveys. Pairing qualitative methods like focus group and interviews with quantitative methods represents an opportunity to capture the complexity of this social phenomena by data triangulation.

Furthermore, an e-leadership approach that only applies principles of traditional conceptions and theories of leadership to virtual environments cannot be enough to address the challenges ahead. It should be taken as a starting point to an exploratory perspective much deeper and specific. E-leadership represents not only a new type of managerial work but also a new context [25] in which leadership should be fully explored as if it was a completely new thing.

References

1. Bennett, N., Lemoine, J.: What a difference a word makes: understanding threats to performance in a VUCA world. Bus. Horiz. **57**(3), 311–317 (2014)
2. Hartog, D., Koopman, P.: Leadership in organizations. In: Anderson, N., Ones, D., Sinangil, H., Viswesvaran, C. (eds.) Handbook of Industrial, Work and Organizational Psychology. Organizational Psychology, vol. 2, pp. 166–187. SAGE Publications, London (2001)
3. Avolio, J., Sosik, J., Kahai, S., Baker, B.: E-leadership: re-examining transformations in leadership source and transmission. Leadersh. Q. **25**(1), 105–131 (2014)
4. Chamakiotis, P., Panteli, N.: E-leadership styles for global virtual teams. In: Leadership in the Digital Enterprise: Issues and Challenges, pp. 143–162. IGI Global, Hershey (2011)
5. Yukl, G.: Leadership in organizations, 7th edn. Prentice Hall, Englewood Cliffs (2010)
6. Vielmetter, G., Sell, Y.: Leadership 2030: The Six Megatrends You Need to Understand to Lead Your Company into the Future. Hay Group Holdings, Inc., Filadélfia (2014)
7. Avolio, J., Kahai, S., Dodge, E.: E-leadership: implications for theory, research, and practice. Leadersh. Q. **11**(4), 615–668 (2001)

8. Lilian, C.: Virtual teams: opportunities and challenges for e-leaders. Procedia Soc. Behav. Sci. **110**, 1251–1261 (2014)
9. Haythornthwaite, C.: Social networks and internet connectivity effects. Inf. Community Soc. **8**(2), 125–147 (2005)
10. Yukl, G., Mahsud, R.: Why flexible and adaptive leadership is essential. Consult. Psychol. J. Pract. Res. **62**(2), 81–93 (2010)
11. Wilson, J.: Leadership in the digital age. In: Encyclopedia of Leadership, vol. 4, pp. 858–861. SAGE Publications, London (2004)
12. Powell, A., Piccoli, G., Ives, B.: Virtual teams: a review of current literature and directions for future research. ACM SIGMIS Database **35**(1), 6–36 (2004)
13. Roman, A., Van Wart, M., Wang, X., Liu, C., Kim, S., McCarthy, A.: Defining e-leadership as competence in ICT-mediated communications: an exploratory assessment. Public Adm. Rev., 1–14 (2018)
14. Liu, C., Ready, D., Roman, A., Van Wart, M., Wang, X., McCarthy, A., Kim, S.: E-leadership: an empirical study of organizational leaders' virtual communication adoption. Leadersh. Organ. Dev. J. **39**(7), 826–843 (2018)
15. Darics, E.: E-leadership or "How to be boss in instant messaging?": The role of nonverbal communication. Int. J. Bus. Commun., 1–27 (2017)
16. Maduka, N., Edwards, H., Greenwood, D., Osborne, A., Babatunde, S.: Analysis of competencies for effective virtual team leadership in building successful organisations. Benchmarking **25**(2), 696–712 (2018)
17. Purvanova, R., Bono, J.: Transformational leadership in context: face-to-face and virtual teams. Leadersh. Q. **20**(3), 343–357 (2009)
18. Hoch, E., Kozlowski, W.: Leading virtual teams: hierarchical leadership, structural supports, and shared team leadership. J. Appl. Psychol. **99**(3), 390–403 (2014)
19. Liao, C.: Leadership in virtual teams: a multilevel perspective. Hum. Resour. Manag. Rev. **27**(4), 648–659 (2017)
20. Avolio, J., Kahai, S., Dumdum, R., Sivasubramaniam, N.: Virtual teams: implications for e-leadership and team development. In: Applied in Psychology. How People Evaluate Others in Organizations, pp. 337–358. Lawrence Erlbaum Associates Publishers, Mahwah (2001)
21. Brandão, C.: The good parent metaphor: contributions to understand leadership processes. In: Ilieva, S., Markovic, M.R., Yankulova, Y. (eds.) Book of Papers of the International Scientific Conference "Leadership and Organization Development", Kitten, Bulgaria, pp. 24–28 (2016)
22. Kahn, W.A.: Caring for the Caregivers: patterns of organizational caregiving. Adm. Sci. Q. **38**(4), 539–563 (1993)
23. Avolio, J., Kahai, S.: Adding the "e" to e-leadership: how it may impact your leadership. Organ. Dyn. **31**, 325–338 (2003)
24. Wart, M., Roman, A., Wang, X., Liu, C.: Integrating ICT adoption issues into (e-)leadership theory. Telemat. Inform. **34**, 527–537 (2017)
25. Savolainen, T.: Trust-building in e-Leadership: a case study of leaders' challenges and skills in technology-mediated interaction. J. Glob. Bus. Issues **8**(2), 45–56 (2014)

Keystroke and Pointing Time Estimation for Touchscreen-Based Mobile Devices: Case Study Children with ASD

Angeles Quezada[1]($\boxtimes$), Margarita Ramirez Ramírez[1],
Sergio Octavio Vázquez[1], Ricardo Rosales[1], Samantha Jiménez[1],
Maricela Sevilla[1], and Roberto Muñoz[2]

[1] Facultad de Contaduría y Administración, Universidad Autónoma de Baja
California, Calzada Universidad 14418, Parque Industrial Internacional Tijuana,
22390 Tijuana, BC, Mexico
{maria.quezada,maguiram,sergio.vazquez,
ricardorosales,samantha.jimenez,
mary_sevilla}@uabc.edu.mx
[2] School of Informatics Engineering, Universidad de Valparaíso,
Valparaíso, Chile
roberto.munoz@uv.cl

Abstract. Nowadays, children with autism spectrum disorders (ASD) show great interest and ease in the use of technology such as tablets and smartphones. There is much research that has been done and is focused on helping users with this type of disorder. However, the challenge is the creation of applications that adapt to their physical, cognitive and motor skills of this type of users. This article focuses on identifying the distance of drag that users with ASD can perform with less complexity, as well as identifying the time that the user with ASD needs to complete the task. The results show that the higher the drag distance, the more complicated this type of user will be. With this result, we can conclude that when to develop mobile applications to support the teaching of this type of users should be considered a smaller drag size and an image size greater than 63 pixels.

Keywords: Usability · Keystroke · Pointing · ASD spectrum disorders

1 Introduction

Autism Spectrum Disorder (ASD) is a pervasive neurodevelopmental disorder characterized by impairments in social communication and restricted, repetitive patterns of behavior, interests or activities. On the other hand, recent studies indicate that people with ASD may present deficiencies in motor skills and cognitive limitations [1].

The rapid increase of mobile games in the market is being driven by a powerful operation of mobile phones and tablets, which are accelerating the rapid growth and use of mobile devices. Although the number of developments aimed at people with ASD has increased in recent years, this has not occurred with studies that seek to generate knowledge on how to make solutions for possible motor difficulties that people with ASD may have.

The original version of this chapter was revised: An author has been included in this chapter. The correction to this chapter is available at https://doi.org/10.1007/978-3-030-16184-2_89

© Springer Nature Switzerland AG 2019
Á. Rocha et al. (Eds.): WorldCIST'19 2019, AISC 931, pp. 774–784, 2019.
https://doi.org/10.1007/978-3-030-16184-2_74

For this reason, it is essential to develop technology that adapts to the motor and cognitive abilities of users with ASD.

Many improvements have been proposed to the Keystroke-Level Model ((KLM-GOMS [2] and FLM [3]).) in order to evaluate different techniques. However, there exists little research on improving techniques of user behavior in users with ASD; in particular, those that seek to estimate the time used to achieve common interactions when using touchscreen devices. The objective of the current study is to identify the time in which users with ASD can execute the operator Keystroke (K) and Pointing (P) using mobile applications.

The rest of the paper is organized as follows. In Sect. 2, we analyze related work. Section 3 describes the experimental design. In Sect. 4, we present the obtained results, then Sect. 5 presents the discussion, and finally, Sect. 6 presents conclusions and future work.

2   Related Work

With the rising popularity of mobile devices, the KLM-GOMS model has recently been revised to evaluate interactions based on touch-screen devices [4, 5]. The model KLM-GOMS determined 5 operators: Drawing (D), Keystroke (K), Mental Act (M), Pointing (P) and Homing (H).

Similarly, in [3] he proposed a modified version of the KLM-GOMS model called FLM (fingerstroke level model). The purpose of this study was to define the time it takes to perform the operators of mobile devices with direct movements of the fingers (Drag (D), Point (P), Move (F) and (Touch (T)), it was only applied to typical adults.

Also in [3], the Fingerstroke Level Model (FLM) was proposed, and a game was analyzed using the FLM operators. The empirical study confirmed the effectiveness and efficiency of FLM, and suggested how HCI methods can improve the design of the user interface of mobile games, but the experiment was only applied to adults with experience in video games.

In [7] a study is shown that evaluated the operators of drag, zoom, and movement for mobile devices. The research consisted of comparing efficiency and user satisfaction during navigation with 2D documents on mobile screens. Although the results obtained were positive, the experiment was only applied to users with a typical psychological development.

In the other hand [8] examines the size of the objective and the distance between each one with smartphones. The results of this study show that the larger the size of the lens (image), the easier it will be to use touch technology for this type of user.

In the same way, in [9] analyzed the interaction of autistic users with mobile devices. This study took into account 6 operators: M, K, G, I (Initial Act), T (Tapping), and S (Swipe). The results suggested that users with level 1 ASD are more likely to perform operations such as K, G, I, and T than users with level 2 ASD.

In this article, we evaluate the K (Keystroke) and P (Pointing) operators using one prototype and one different application. In subsequent works other operators were evaluated, in this article our objective is to evaluate P and K to generate reference values and to be able to propose a KLM model for users with autism.

3   Experimental Design

3.1   Method

The purpose of this experiment is to determine the time of keypress that user with ASD can perform with the least amount of effort. It is for this reason that two different image sizes were evaluated to define the appropriate size for this type of users, as shown in Fig. 1.

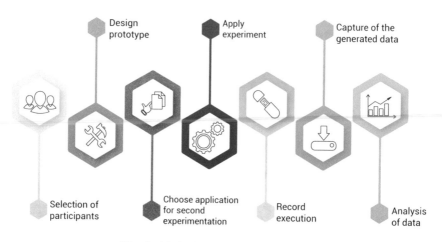

Fig. 1. Methodology of experimentation

3.1.1   Participants

The experiment was carried out with 14 users diagnosed with ASD in a special education school. A group of ASD psychologists assessed and approved the methodology used in this experiment. The users were diagnosed by specialist psychologists and each user was associated with a level of ASD according to the DSM-5 [11]. Users with ASD are between 5 and 11 years old.

3.1.2   Apparatus

For this study, a prototype of an application for mobile devices was designed for the Android platform. This prototype reflects a common interaction widely used in tablets and mobile devices: press and point a screen element.

The objective of the prototype is to present to the users different scenarios of KeyPress, Ponting and automatically collect the time necessary to carry out these interactions. The scenarios presented by the application vary the size of the item that users must press as shown in the Fig. 2.

The developed prototype consisted of pressing the objective that is the image of an orange star in size of 21, 63 and 86 pixels, the task of the user was to press the star, the task was completed when the user pressed the 5 images. This task was repeated once for each image size.

Fig. 2. Characteristics of prototype for Keypress.

The next task was to point to three different image sizes 21, 63 and 86 pixels and with a distance of 95, 324 and 553 pixels as shown in the Fig. 3.

Fig. 3. Characteristics of prototype for Pointing

Kids Animals Jigsaw Puzzles

It is an application to assemble puzzles developed for children offered by App Family Kids - Games for boys and girls. Each relaxing puzzle presents a beautiful different scene drawn by a professional cartoon artist, and a unique reward when the puzzle is completed. The scenes include things like cute animals, dragons or dinosaurs, and the rewards can be balloons, fruits, snowflakes or many more.

For this experiment, a Samsung Galaxy Tab 4 tablet was used with specifications that included a 7-inch resolution screen and 1280 × 800 pixels in the Android operating system. Figure 4 shows the characteristics of this application Kids Animals Jigsaw Puzzles, such as image size and drag size.

To measure the interaction time of the users, a video camera was used to record the interaction in the video, and then we used the ELAN is a professional tool for the creation of complex annotations on video and audio resources to measure the time of the video.

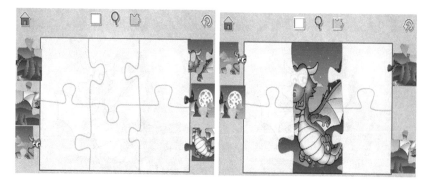

Fig. 4. Interface of app Kids Animals Jigsaw Puzzles

3.1.3   Procedure

The experiment took place in the place where users attend classes, a quiet room was chosen without distractors in which users could interact with the tablet. During the experiment the support staff (psychologists) were explained what the procedure consisted of and the use of the application. The support staff was responsible for carrying out the experiment, which consisted of helping the subjects to do each use case explained in previous sections. Before conducting the experiment, the parents of the subjects signed a letter of consent for the video recording of the subjects, only their hands were recorded, and only while the tablet was being used.

Participants used the index finger of their dominant hand to perform each of the tasks set. While the experiment was running, the participants were asked to execute the task of the first drag as quickly and accurately as possible. When the users started to interact with the applications, these interactions were recorded to measure the time later. All the tasks were repeated at least 3 times, and for the measurement they were only used from the second interaction, since the first one was considered as training.

4   Results

In this section we present the times that users need to perform each task for each size of the image and the different distances.

4.1   Pointing Task

Group of ASD Level 1. For the pointing (P) task with an image size of 31 pixels and a distance of 95 pixels, the results show that the maximum time of use of a user of level 1 was 1.30 s and the minimum of 1.10 s and a median of 1.20. In the case of the image size of 63 pixels with the same distance the results show that the maximum time was 1.63 s and the minimum of 1.37 s and a median of 1.51. For the image size of 86 pixels distance the results show that the maximum time was 0.78 s and the minimum of .45 s and a median of 0.61 as shown in Fig. 5.

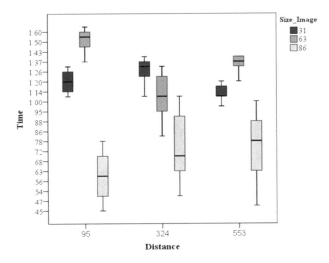

Fig. 5. Task pointing ASD level 1.

In the same task with an image size of 31 pixels and a distance of 324 pixels, the results show that the maximum time of use of a user of level 1 was 1.40 s and the minimum of 1.10 s and a median of 1.28. In the case of the image size of 63 pixels with the same distance the results show that the maximum time was 1.30 s and the minimum of 0.79 s and a median of 1.08. For the image size of 86 pixels distance the results show that the maximum time was 1.10 s and the minimum of 0.49 s and a median of 0.75 as shown in Fig. 5.

Finally the task P with an image size of 31 pixels and a distance of 553 pixels, the results show that the maximum time of use of a user of level 1 was 1.20 s and the minimum of 0.99 s and a median of 1.10. In the case of the image size of 63 pixels with the same distance the results show that the maximum time was 1.40 s and the minimum of 1.20 s and a median of 1.34. For the image size of 86 pixels distance the results show that the maximum time was 1.0 s and the minimum of 0.46 s and a median of 0.75 as shown in Fig. 5.

Group of ASD Level 2. In the case of users with level 2 ASD For the pointing (P) task with an image size of 31 pixels and a distance of 95 pixels, the results show that the maximum time of use of a user of level 1 was 1.49 s and the minimum of 1.11 s and a median of 1.35. In the case of the image size of 63 pixels with the same distance the results show that the maximum time was 1.74 s and the minimum of 1.42 s and a median of 1.61. For the image size of 86 pixels distance the results show that the maximum time was 1.40 s and the minimum of 1.20 s and a median of 1.28 as shown in Fig. 6.

In the same task with an image size of 31 pixels and a distance of 324 pixels, the results show that the maximum time of use of a user of level 1 was 1.77 s and the minimum of 1.35 s and a median of 1.56. In the case of the image size of 63 pixels with

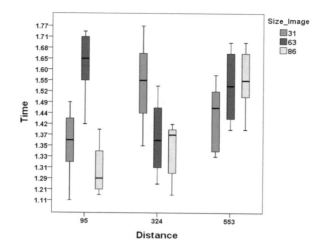

Fig. 6. Task pointing ASD level 2.

the same distance the results show that the maximum time was 1.53 s and the minimum of 1.22 s and a median of 1.38. For the image size of 86 pixels distance the results show that the maximum time was 1.42 s and the minimum of 1.20 s and a median of 1.34 as shown in Fig. 6.

Finally the task P with an image size of 31 pixels and a distance of 553 pixels, the results show that the maximum time of use of a user of level 1 was 1.56 s and the minimum of 1.33 s and a median of 1.45. In the case of the image size of 63 pixels with the same distance the results show that the maximum time was 1.70 s and the minimum of 1.40 s and a median of 1.55. For the image size of 86 pixels distance the results show that the maximum time was 1.70 s and the minimum of 1.40 s and a median of 1.56 as shown in Fig. 6.

4.2   Keystroke Task

For the Keystroke (K) task, users with ASD level 1 completed the task in a shorter time than those of level 2. For the task K with the image size of 31 pixels the Maximum time was 0.83 s and the minimum 0.33. In the case of the same task but with an image size of 63 pixels the time maximum was 0.70 and the minimum as 0.40. And for the same task but with an image size of 86 pixels the maximum time was 0.40 and the minimum time was 0.60, as shown in the Fig. 7.

In the execution of the same task Keystroke the users with ASD of level 2 for this task with an image size of 31 pixels the maximum time was 0.90 and the minimum of 0.40. In the case of the execution of the same task but with an image size of 63 pixels, the maximum time was 0.80 and the minimum time was 0.50. Similarly in the execution of the task but with the image size of 86 pixels the maximum time was 0.74 and the minimum was 0.60, as shown in the Fig. 8.

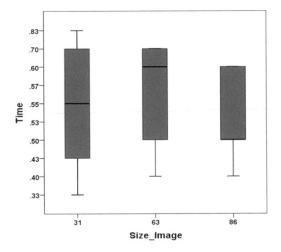

Fig. 7. Task Keystroke ASD level 1.

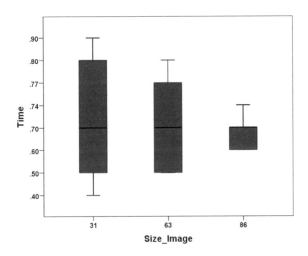

Fig. 8. Task Keystroke ASD level 2.

In the case of the interaction with the application Kids Animals Jigsaw Puzzles, users with ASD level 1 executed the task K in a maximum of 0.87 s and in a minimum time of 0.45 s. For users with level 2 ASD the maximum time to execute the same task was 1.02 s and a minimum of 0.68 s, as can be seen there is a slight variation between both groups, as shows in that Fig. 9.

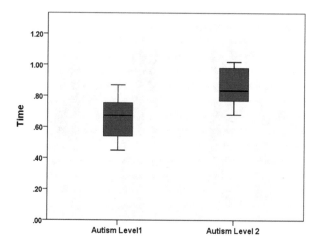

Fig. 9. Task Keystroke for Kids Animals Jigsaw Puzzles ASD level 1 and 2.

5 Discussion

The results show that there is a variation in the execution of the task Pointing (P) users with ASD level 1 executed the task 32% slower with the size of the image of 31 pixels and with a distance of 95 pixels, the same is presented with the image of 86 pixels needed 59% more time to execute it. For the distance of 324 pixels users needed 21% more time with the image of 31 pixels and the image of 63 pixels, for the case of the image of 86 pixels needed 32% more time. It can also be observed that for the distance of 553 pixels with an image size of 31 pixels the time required was 24% more than with the image of 63 pixels, it can also be seen that for the image size of 86 pixels it was necessary 32% more time.

In the case of users with level 2 ASD in the execution of the same task P, with the image size of 31 pixels and a distance of 95 pixels they made the task 26% slower than with the image of 63 pixels with the same distance. In the case of the distance of 324 pixels with an image size of 31 pixels, they executed the task 18% slower than with the image of 63 pixels and 86 pixels.

For task K, users with level 1 ASD were 20% faster with the image size of 63 pixels compared to the image of 31 pixels, the same happens with the image size 86 pixels, the task was performed 17% quicker. In the case of users with level 2 ASD they executed the task 47% faster with the image size of 63 pixels with respect to 31 pixels, in the case of the 86 pixels image the percentage was 14% difference.

We can see that the bigger the image, the better the user interacts with the interface. It was also observed that the applications should be as simple as possible to achieve the child's attention. This indicates that developers should take into account the motor skills of users as are users with ASD.

6   Conclusions

In this article an experimental design was presented to evaluate the time required for the execution of two tasks Pointing (P) and Keystroke (K), in this experiment a prototype with different image sizes was used to determine which is the size with the that users with ASD can perform the task more easily. In the first task P the results show that the bigger the image and the smaller the distance the execution, I need less time for both groups of users

In this experiment, the time required for each group to perform the operations proposed by KLM and FLM, which are variants of the original model proposed by GOMS, was evaluated.

The results obtained in this study allow us to establish that users with level 1 ASD performed tasks in less time compared to users with level ASD 2, due to the cognitive and motor deficits that each user level has, despite the fact that have the same task, the difference is identified for each of the trawls with the distances used in each of the applications.

As future work, experiments will be carried out with typical users and with users who have some motor disability, this is to be able to counteract the results obtained with this experimentation and be able to compare the results between each group of users, this is with the purpose of being able to develop applications that they adapt to the different motor and cognitive abilities of each user.

References

1. Fitzpatrick, P., et al.: Evaluating the importance of social motor synchronization and motor skill for understanding autism. Autism Res. **10**(10), 1687–1699 (2017)
2. Card, S., Moran, T., Newell, A.: The Psychology of Human-Computer Interaction. L. Erlbaum Associates Inc., Hillsdale (1983)
3. Lee, A., Song, K., Ryu, H.B., Kim, J., Kwon, G.: Fingerstroke time estimates for touchscreen-based mobile gaming interaction. Hum. Mov. Sci. **44**, 211–224 (2015)
4. Jung, K., Jang, J.: Development of a two-step touch method for website navigation. Appl. Ergon. **48**, 148–153 (2015)
5. El Batran, K., Dunlop, M.D.: Enhancing KLM (keystroke-level model) to fit touch screen mobile devices. In: Proceedings of the 16th International Conference on Human-Computer Interact with Mobile Devices & Services, MobileHCI 2014, pp. 283–286 (2014)
6. Bi, X., Li, Y., Zhai, S.: FFitts law. In: Proceedings of the SIGCHI Conference on Human Factors in Computing Systems, CHI 2013, p. 1363 (2013)
7. Spindler, M., Schuessler, M., Martsch, M., Dachselt, R.: Pinch-drag-flick vs. spatial input: rethinking zoom & pan on mobile displays. In: Proceedings of the SIGCHI Conference on Human Factors in Computing Systems, vol. V, pp. 1113–1122 (2014)
8. Leitão, R., Silva, P.: Target and spacing sizes for smartphone user interfaces for older adults: design patterns based on an evaluation with users. In: Conference on Pattern Languages of Programs, vol. 202915, pp. 19–21 (2012)
9. Quezada, A., Juárez-Ramírez, R., Jiménez, S., Noriega, A.R., Inzunza, S., Garza, A.A.: Usability operations on touch mobile devices for users with autism. J. Med. Syst. **41**(11), 184 (2017)

10. Froehlich, J., Wobbrock, J.O., Kane, S.K.: Barrier pointing - using physical edges to assist target acquisition on mobile device touch screens. In: Proceedings of the 9th International Conference on Computers and Accessibility, pp. 19–26 (2007)
11. APA: American Psychiatric Association (2013). http://www.apa.org/pi/disability/resources/publications/newsletter/2016/09/autism-spectrum-disorder.aspx

The Use of Virtual Cues in Acquired Brain Injury Rehabilitation. Meaningful Evidence

Sergio Albiol-Pérez[1]([✉]), Alvaro-Felipe Bacca-Maya[2],
Erika-Jissel Gutierrez-Beltran[3], Sonsoles Valdivia-Salas[1],
Ricardo Jariod-Gaudes[4], Sandra Cano[2],
and Nancy Jacho-Guanoluisa[5]

[1] Aragón Health Research Institute (IIS Aragón),
Universidad de Zaragoza, Zaragoza, Spain
{salbiol,sonsoval}@unizar.es
[2] LIDIS Group, University of San Buenaventura, Cali, Colombia
iam@pipebacca.com, sandra.cano@gmail.com
[3] Pontificia Universidad Javeriana, Cali, Colombia
erika.giselle.gb@hotmail.com
[4] Aragón Health Research Institute (IIS Aragón),
H.U. Miguel Servet, Zaragoza, Spain
rjariod@salud.aragon.es
[5] Universidad de Las Fuerzas Armadas ESPE Sangolquí, Sangolquí, Ecuador
npjacho@espe.edu.ec

Abstract. Virtual cues are a complement for improving quality of life in Acquired Brain Injury (ABI) patients. In the last few years, the high incidence and high costs of this pathology have produced new technological methodologies that helps alleviate motor and cognitive disorders. The use of low-cost and portable systems together with Virtual Environments allow the creation of novel solutions that can serve as a complement in traditional rehabilitation sessions. Medio-lateral weight transferences are a characteristic movement in the therapeutic sessions of these patients. For this reason, we have created a novel Virtual Rehabilitation system that is composed of virtual cues that satisfy the suggestions of clinical specialists. In this regard, the Neuro Virtual (NeuroVirt) system is a satisfactory and useful tool that is capable of improving in kinematic parameters such as reaction and completion time in ABI patients. In this study, we have tested the NeuroVirt system to obtain positive outcomes in kinematic parameters thanks to the use of Virtual Cues using a reliable questionnaire that focuses on Virtual Rehabilitation. For this purpose, we analyzed our novel system with two different populations, university students and older people. The outcomes have revealed that the NeuroVirt system is an acceptable tool to be validated in the near future with ABI patients.

Keywords: Virtual Rehabilitation · Virtual cues · Acquired Brain Injury · Usability · Kinematic parameters

© Springer Nature Switzerland AG 2019
Á. Rocha et al. (Eds.): WorldCIST'19 2019, AISC 931, pp. 785–794, 2019.
https://doi.org/10.1007/978-3-030-16184-2_75

1 Introduction

Acquired Brain Injury (ABI) is a neurological impairment produced in the brain that causes a deterioration of functional activity and reduces quality of life [1]. Patients who suffer this disease follow certain criteria. These criteria are the following: (1) ABI produces partial or total brain damage; (2) this impairment is generated in early stages from a few seconds to a few days; (3) ABI produces typical impairments that are associated to this pathology; (4) ABI reduces Activities of Daily Living (ADL) [2]; and (5) there are non-neurodegenerative and congenital disorders.

The etiology of ABI is composed of the following most common disorders: (1) stroke [3]; (2) traumatic brain injury (TBI) [4]; and (3) anoxic or hypoxic encephalopathy [5]. There are other types of anomalies related to this pathology that are produced by viral infections. These include the following: (1) herpes simplex encephalitis (HSE) [6]; (2) vasculitis [7]; and (3) acute disseminated encephalomyelitis [8].

Stroke can be described as a neurological disorder that manifests itself throughout a specific period of time (from 24 h to one week after different cognitive alterations) [9]. Traumatic brain injury is a disorder that is caused by external events, sudden movements, and impacts by accidents or shock waves in the brain [10]. Hypoxic encephalopathy is a disruption in the brain that is caused by oxygen deprivation and it is produced mainly during birth [11]. Herpes simplex encephalitis is caused by a viral infection in the central nervous system [6]. Vasculitis is a set of clinical syndromes that is produced by an inflammation of the blood vessel walls [12]. Acute disseminated encephalomyelitis is an inflammatory syndrome in the central nervous system that is caused by a viral infection, which affects the brain and the spinal cord [13].

Different sub-types of TBI are recognized [14]. These sub-types are the following: (1) epidural hematomas; (2) contusion and parenchymal hematomas; (3) diffuse axonal injury (DAI); (4) subdural hematoma (SDH); (5) subarachnoid and intraventricular hemorrhage (SAH/IVH); and (6) diffuse brain swelling.

The symptomatology in ABI produces negative motor, cognitive, and social effects; ABI is the main cause of death and incapacity worldwide. In the USA, the highest rates of TBI are especially occur in people of around 65 years of age [15] from a total population of approximately 1.7 million who suffer this disease. With a national (USA) incidence of 2,232.2 per 100,000 (people of 75 years old or more), 1,591.5 per 100,000 population (people ranging from 0 to 4 years old), and 1,080.7 per 100,000 population (subjects ranging from 15 to 24 years old), we can highlight the high rates in older people [16]. These numbers show the prevalence in the USA, which ranges from 3.32 to 5.3 million people [17]. In developed countries, the incidence rates of TBI fluctuate from 228 per 100,000 in Europe to 415 per 100,000 in the Australasian region [18].

The direct cost of TBI in the USA was around $13.1 billion US dollars in 2013, with a total cost ranging from $63.4 to $79.1 billion US Dollars for the same year [19]. The direct cost of stroke patients was $33.0 billion US Dollars in 2013, with indirect costs of $27.3 billion US Dollars in stroke patients, and a total cost ranging from $36.6 billion US Dollars to $72.7 billion US Dollars in 2013 [20]. These high rates indicate that it is necessary to alleviate the main motor and cognitive disorders of this pathology.

Typical sensorimotor motor disorders such as spasticity, coordination, or sensitivity in ABI patients produce a worsening of their quality of life. Cognitive alterations that ABI patients suffer from are focused on attention, speed of processing, memory, and executive functioning [21].

Traditional motor rehabilitation is based on different techniques that are used to obtain improvements in upper/lower extremities, postural control, gait retraining, etc. Upper extremity techniques include the following: (1) constraint-induced movement therapy (CIMT); (2) repetitive and intensive upper-limb tasks; (3) bilateral arm-training; (4) mirror therapy; (5) music intervention; (6) biofeedback functional electrical stimulation, etc. Lower extremity procedures include the following: (1) stretching; (2) heat on the affected limb; (3) pressure on the affected area, etc. Postural control is focused on static medio-lateral and antero-posterior weight transferences in the standing or sitting position. Gait retraining is based on dynamic exercises with partial suspension of ABI patients.

Visual and auditory cues in traditional rehabilitation are useful for alleviating motor disorders in ABI patients. Thanks to the use of verbal instructions during intervention periods, it is possible to obtain precision in gross and fine movements.

This paper is structured as follows: Sect. 2 explains similar studies that are related to our purpose. Section 3 presents the materials and methods that were used with the participants during the intervention period of the study and also presents the NeuroVirt system. Section 4 presents the kinematic and usability outcomes of the study. Finally, Sect. 5 presents a brief discussion of our outcomes and conclusions.

2   Related Work

The use of groundbreaking and customizable technology together with Virtual Environments (VE) is a complement to traditional rehabilitation [22]. In the last few years, Virtual Reality (VR) has been used as an innovative technique to alleviate motor disorders in multiple pathologies such as: Parkinson's disease, multiple sclerosis, ABI, etc.

VR systems can be used as a training program thanks to specific simulated and customizable VE, without any type of physical limitation or discomfort in patients with neurological disorders. VR provides an accurate measurement of repetitive patient movements, showing an enrichment of sessions, in a fun, playful, and entertaining way.

Visual and auditory cues together with VE have been tested mainly in people with Parkinson's disease to alleviate freezing of gait and balance [23], but there are few studies that focus on the use of virtual cues in ABI patients. Walker et al. [24] tested a VR system with visual and auditory cues in ABI patients. The outcomes revealed improvements in walking speed at intervention periods by using treadmill-training sessions. Faria et al. [25] analyzed a method of fading cues together with visual cues in a virtual city that had virtual cognitive aspects (memory, attention, executive functions, and visuo-spatial tasks) and actions based on ADL in nine stroke patients. The outcomes show evident improvements in cognitive functionality and overall recovery in the stroke group that participated by using their VR systems.

The analysis of kinematic parameters to obtain outcomes that are related to improvements in motor disorders allows us to study and correct mistakes during in therapeutic sessions. Reaction and completion time together with VE provide a valid assessment and training instrument for ABI patients [26].

At the present time, there are no studies that analyze the relationship between visual cues and kinematic parameters (reaction and completion time) in people without neurological disorders.

The purpose of our study is to test and analyze kinematic parameters together with visual cues in specific VE by using a safe, comfortable, entertaining, and low-cost technological system. Our tool, the Neuro Virtual System (NeuroVirt), has accomplished all of the suggestions and requirements of clinical specialists regarding motor disorders in ABI patients. Based on these recommendations we analyzed the usability and satisfaction of the NeuroVirt system using a reliable questionnaire that focuses on VR systems: The User Satisfaction Evaluation Questionnaire (USEQ) [26].

3   Methods

3.1   Participants

The participants in this study were university students without any motor and cognitive disorders and older people. We made two groups comprised of different populations. The control group was composed of 11 university students (1 man and 10 women) with a mean age of 21(2.82). The experimental group was comprised of 9 elderly subjects (4 men and 5 women) with a mean age of 55(4.09). Table 1 describes the characteristics of the participants.

Table 1. Demographic characteristics of the participants.

	Control group (n = 11)	Experimental group (n = 9)	p
Age	21(2.82)	55(4.09)	>0.05 (NS)
Gender	Female 90.9% Male 9.09%	Female 44.44% Male 55.55%	>0.05 (NS)

3.2   Instrumentation

The hardware components of our technological system are the following: (1) a laptop; (2) a high resolution monitor; (3) a low-cost optical tracking device (Orbbec Astra®).

The NeuroVirt system was designed and programmed thanks to the assistance of a multidisciplinary team composed of a psychologist, engineers, clinical specialists, and graphic designers. Our objective was mainly to reinforce the motivation of the participants in the use of visual cues.

The Virtual environment consists of a track that shows wide curves to perform medio-lateral weight-transferences on the left and on the right. Visual cues are represented by curves so that to the participants can perform motor exercises correctly.

This technological system fulfills all of the needs that must be validated in the short-term in acquired brain-injury (ABI) patients in order to improve postural control and balance disorders.

Key aspects, such as usability and satisfaction items were designed in order to obtain a quality system in the intervention and follow-up period. The NeuroVirt system was designed after multiple meetings with a multidisciplinary team in order to obtain a comfortable, safe, enjoyable, and playful environment. Thanks to these meetings, it was possible to obtain information such as: (1) specific positive auditory cues; (2) visual and auditory feedback, such as applause and a finish line at the end of the track; (3) custom curves based on the active session and degree of motor disorder of the ABI patient, etc.

The participants tried to perform movements along the track, tilting their bodies while sitting in a rest position according to those movements performed in traditional ABI rehabilitation patients. This was possible thanks to the use of wide curves inside the track.

3.3   Intervention

The present study was made at the facilities of a university in a small city. The participants were university students and older adults who trained with NeuroVirt for one session. The time that participants played with the technological system was around 10 min each session.

Thanks to the use of visual cues (custom curves in the track), the participants performed medio-lateral weight-transferences on the left and on the right.

In this period of time, our system stored moments of time in order to calculate the kinematic parameters of the participants. At the end of the session, the subjects filled

Fig. 1. Participant by using the NeuroVirt system.

out the User Satisfaction Evaluation Questionnaire (USEQ) [27] so that we could analyze the satisfaction and usability of our system. Figure 1 shows a participant interacting with NeuroVirt.

Throughout the session, the multidisciplinary team observed and analyzed a series of comments for future improvements. These comments were the following: (1) enrichments of visual cues; and (2) visual and auditory feedback when subjects get a hit with a virtual object.

The motivation of the participants increased throughout the session. Thanks to similar movements that a cyclist carries out on a track, the participants could perform medio-lateral weight-transferences throughout the session more precisely. This positive aspect was intimately related to the accuracy of medio-lateral weight-transferences and to kinematic parameters.

4   Results

A first analysis based on kinematic parameters was performed with the participants in our study. We tested our technological system and we stored two parameters, response time and completion time. Our study was tested in a public university under specialized researchers in the field.

4.1   The Kinematics Outcomes

The primary outcome measures related to kinematic parameters (response and completion time) are shown in Table 2.

Table 2. Kinematic parameters.

	Control group		Experimental group	
p-value: Response time > 0.05; Completion time > 0.05				
	Response time	Completion time	Response time	Completion time
Mean	1.52	2.52	1.87	3.18
Standard deviation	0.64	0.64	0.98	1.42

The outcomes revealed non-significant difference between university students (the control group) and older adults (the experimental group).

Figures 2 and 3 show the reaction time in both groups. As can be observed, the outcomes for reaction times are lower in the control group with respect to the experimental group.

Figures 4 and 5 present the completion time in our study. It can be observed that the completion time of the university students (the control group) is lower than the outcomes obtained by older adults (the experimental group).

Fig. 2. Control group (university students).   **Fig. 3.** Experimental group (older adults)

Fig. 4. Control group (university students).   **Fig. 5.** Experimental group (older adults)

4.2 USEQ Outcomes

The USEQ questionnaire is composed of specific items related to enjoyment (Q1), success (Q2), control of the system (Q3), information provided by the system (Q4), feeling of discomfort (Q5), and usefulness for rehabilitation processes (Q6). The participants answered the first five questions. We considered the sixth question of the USEQ (Do you think that this system will be helpful for your rehabilitation?) to be irrelevant to our study because the subjects had no motor or cognitive disorders. Table 3 shows the outcomes for both groups.

Table 3. Score percentage on questions of the USEQ.

Control group	Mean	Std. dev
Q1	13	28.18
Q2	4.3	4.33
Q3	17.9	6.8
Q4	32.1	18.98
Q5	32.7	20.07
Experimental group		
Q1	18.5	40.16
Q2	1.9	4.54
Q3	7.4	11.48
Q4	7.4	9.07
Q5	64.8	36.8

5 Discussion and Conclusions

We have tested the NeuroVirt tool, which is a novel technological system that focuses on motor improvements in ABI patients. For this purpose, we validated our system with subjects who did not have any motor or cognitive disorders. The use of visual cues inside a customizable virtual environment allowed the participants to anticipate the completion of medio-lateral weight transferences.

The outcomes revealed non-significant differences between the control group and the experimental group in reaction time and completion time. However, there is clear evidence of more effective reactions in the control group. We think that this is due to the cognitive processing that all subjects perform in order to achieve the objectives of the Virtual Environment, thanks to the use of visual cues represented by curves on the track.

Once the results of applying the USEQ test were obtained, we consider that NeuroVirt fulfills features such as satisfaction and usability and merits other studies focus on ABI patients in the near future. Based on their responses, there is clear evidence that Questions 4 and 5 have good outcomes (Control Group 31.1% in Q4 and 32.7% in Q5; Experimental Group 66.7% in Q5). We consider that this is because the participants felt comfortable using NeuroVirt, which provides clear and simple information in the Virtual Environment. On the other hand, in the experimental Group, there is a significant outcome for Question 1. This result indicates that older adults enjoyed the NeuroVirt system, making us think that, in the near future, our technological tool will satisfy requirements for validation in ABI patients with relevant outcomes.

These outcomes have encouraged us to improve our technological system by adding new functionalities that focus on the motor and cognitive disabilities of ABI patients.

As future work, we are going to design a protocol for ABI patients in order to test the NeuroVirt system. In that study, we will establish inclusion/exclusion criteria, intervention periods, and specific clinical tests to validate outcomes in patients with this pathology. The clinical test will focus on improvements in postural control and balance. These tests will include the following: the Trunk Control Test (TCT), the Berg scale, and the Standing balance test.

Acknowledgments. The authors would like to acknowledge Iván Verde for designing and programming the Virtual Environment. This contribution was funded by the Gobierno de Aragón, Departamento de Industria e Innovación, and Fondo Social Europeo "Construyendo Europa desde Aragón" and by grants from the Instituto de Salud Carlos III (FIS. PI17/00465) from the Spanish Government and European Regional Development Fund, "A way to build Europe".

References

1. Castellanos-Pinedo, F., Cid-Gala, M., Duque, P., Ramirez-Moreno, J.M., Zurdo-Hernández, J.M.: Acquired brain injury: a proposal for its definition, diagnostic criteria and classification. Rev. Neurol. **54**(6), 357–366 (2012)
2. Fallahpour, M., Kottorp, A., Nygård, L., Lund, M.L.: Participation after acquired brain injury: associations with everyday technology and activities in daily life. Scand. J. Occup. Ther. **22**(5), 366–376 (2015)

3. Guzik, A., Bushnell, C.: Stroke epidemiology and risk factor management. Continuum (Minneap Minn). **23**(1, Cerebrovascular Disease), 15–39 (2017)

4. Galgano, M., Toshkezi, G., Qiu, X., Russell, T., Chin, L., Zhao, L.R.: Traumatic brain injury: current treatment strategies and future endeavors. Cell Transplant. **26**(7), 1118–1130 (2017)

5. Ferris, L.M., Engelke, C.: Anoxic brain injury secondary to metabolic encephalopathy. Optom. Vis. Sci. **93**(10), 1319–1327 (2016)

6. Safain, M.G., Roguski, M., Kryzanski, J.T., Weller, S.J.: A review of the combined medical and surgical management in patients with herpes simplex encephalitis. Clin. Neurol. Neurosurg. **128**, 10–16 (2015)

7. Lakdawala, N., Fedeles, F.: Vasculitis: kids are not just little people. Clin. Dermatol. **35**(6), 530–540 (2017)

8. Gray, M.P., Gorelick, M.H.: Acute disseminated encephalomyelitis. Pediatr. Emerg. Care **32**(6), 395–400 (2016)

9. Coupland, A.P., Thapar, A., Qureshi, M.I., Jenkins, H., Davies, A.H.: The definition of stroke. J. R. Soc. Med. **110**(1), 9–12 (2017)

10. Pervez, M., Kitagawa, R.S., Chang, T.R.: Definition of traumatic brain injury, neurosurgery, trauma orthopedics, neuroimaging, psychology, and psychiatry in mild traumatic brain injury. Neuroimaging Clin. N. Am. **28**(1), 1–13 (2018)

11. Gopagondanahalli, K.R., Li, J., Fahey, M.C., Hunt, R.W., Jenkin, G., Miller, S.L., et al.: Preterm hypoxic-ischemic encephalopathy. Front. Pediatr. **4**, 114 (2016)

12. Prete, M., Indiveri, F., Perosa, F.: Vasculitides: proposal for an integrated nomenclature. Autoimmun. Rev. **15**(2), 167–173 (2016)

13. Tenembaum, S., Chitnis, T., Ness, J., Hahn, J.S.: International pediatric MS study group acute disseminated encephalomyelitis. Neurology **68**(16 Suppl. 2), S23–S36 (2007)

14. Saatman, K.E., Duhaime, A.C., Bullock, R., Maas, A.I., Valadka, A., Manley, G.T.: Classification of traumatic brain injury for targeted therapies. J. Neurotrauma **25**(7), 719–738 (2010). Workshop Scientific Team and Advisory Panel Members

15. Chan, V., Zagorski, B., Parsons, D., Colantonio, A.: Older adults with acquired brain injury: a population based study. BMC Geriatr. **13**, 97 (2013)

16. Taylor, C.A., Bell, J.M., Breiding, M.J., Xu, L.: Traumatic brain injury-related emergency department visits, hospitalizations, and deaths - United States, 2007 and 2013. MMWR Surveill. Summ. **66**(9), 1–16 (2017)

17. Ma, V.Y., Chan, L., Carruthers, K.J.: Incidence, prevalence, costs, and impact on disability of common conditions requiring rehabilitation in the United States: stroke, spinal cord injury, traumatic brain injury, multiple sclerosis, osteoarthritis, rheumatoid arthritis, limb loss, and back pain. Arch. Phys. Med. Rehabil. **95**(5), 986–995 (2014)

18. Nguyen, R., Fiest, K.M., McChesney, J., Kwon, C.S., Jette, N., Frolkis, A.D., et al.: The international incidence of traumatic brain injury: a systematic review and meta-analysis. Can. J. Neurol. Sci. **43**(6), 774–785 (2016)

19. Coronado, V.G., Xu, L., Basavaraju, S.V., McGuire, L.C., Wald, M.M., Faul, M.D., et al.: Centers for disease control and prevention (CDC). Surveillance for traumatic brain injury-related deaths–United States, 1997–2007. MMWR Surveill Summ. 6, **60**(5), 1–32 (2011)

20. Wang, Z.M., Law, J.H., King, N.K., Rajeswaran, D.K., Soh, S., Rao, J.P., et al.: Treatment of severe, disabling spasticity with continuous intrathecal baclofen therapy following acquired brain injury: the experience of a tertiary institution in Singapore. Singapore Med. J. **57**(1), 8–12 (2016)

21. Virk, S., Williams, T., Brunsdon, R., Suh, F., Morrow, A.: Cognitive remediation of attention deficits following acquired brain injury: a systematic review and meta-analysis. NeuroRehabilitation **36**(3), 367–377 (2015)

22. Albiol-Pérez, S., Gil-Gómez, J.A., Llorens, R., Alcañiz, M., Font, C.C.: The role of virtual motor rehabilitation: a quantitative analysis between acute and chronic patients with acquired brain injury. IEEE J. Biomed. Health Inf. 18(1), 391–398 (2014)
23. Park, H.S., Yoon, J.W., Kim, J., Iseki, K., Hallett, M.: Development of a VR-based treadmill control interface for gait assessment of patients with Parkinson's disease. In: IEEE International Conference on Rehabilitation Robotics, Rehab Week Zurich, Switzerland (2011)
24. Walker, M.L., Ringleb, S.I., Maihafer, G.C., Walker, R., Crouch, J.R., Van Lunen, B., et al.: Virtual reality-enhanced partial body weight-supported treadmill training poststroke: feasibility and effectiveness in 6 subjects. Arch. Phys. Med. Rehabil. 91(1), 115–122 (2010)
25. Faria, A.L., Andrade, A., Soares, L., I Badia, S.B.: Benefits of virtual reality based cognitive rehabilitation through simulated activities of daily living: a randomized controlled trial with stroke patients. J. Neuroeng. Rehabil. 13(1) 96 (2016)
26. Fong, K.N., Chow, K.Y., Chan, B.C., Lam, K.C., Lee, J.C., Li, T.H., et al.: Usability of a virtual reality environment simulating an automated teller machine for assessing and training persons with acquired brain injury. J. Neuroeng. Rehabil. 30(7), 19 (2010)
27. Gil-Gómez, J.A., Manzano-Hernández, P., Albiol-Pérez, S., Aula-Valero, C., Gil-Gómez, H., Lozano-Quilis, J.A.: USEQ: a short questionnaire for satisfaction evaluation of virtual rehabilitation systems. Sensors 17(7), 1589 (2017)

Problematic Attachment to Social Media: Lived Experience and Emotions

Majid Altuwairiqi[1(✉)], Theodoros Kostoulas[1], Georgina Powell[2], and Raian Ali[1]

[1] Bournemouth University, Poole, UK
{maltuwairiqi,tkostoulas,rali}@bournemouth.ac.uk
[2] Phone Life Balance Ltd – Creators of the SPACE App, Guildford, UK
georgie@space-app.com

Abstract. People's relationship with social media and their contacts on them can be problematic. People may engage in social media in a compulsive and hasty style to increase their popularity, reputation and enhance their self-esteem. However, this problematic attachment to social media may result in side effects on people's well-being. Therefore, people may need assistance to reform their relationship with social media in a way that it maintains different aspects of their online interaction, such as empathy with others and maintaining their popularity and relatedness. In order to provide the tools and methods to support people in reforming their relationship with social media, towards a healthier usage style, we need to understand the experience of people who suffer a problematic relationship with them. Most studies on the topic are based on methods which would lack *ecological validity*, e.g. using surveys and interviews, and do not capture or imitate such a digital experience as lived. In an attempt to better explore how people experience problematic attachment and relationship with social media, and their associated emotions, we conducted a multistage qualitative method study including a *diary study* to gather lived experience. We aim to inform both users and designers towards a managed and tool-supported reform of their problematic relationship with social media and, ultimately, having a healthier online interaction.

Keywords: Social media dependency · Digital addiction · Digital wellbeing

1 Introduction

Social media became prominent in our daily lives, determining in a major way how an individual shares information and exchanges knowledge. The use of social media has created a new set of cyber social norms around expectancy and responsiveness as well as group membership and relatedness. While many benefits can be identified regarding these norms in human communication and socialness, negative experiences emerging from them have become evident.

Recent studies have demonstrated similarities between certain symptoms of using social media and symptoms of behavioural addiction [1, 2]. Such symptoms include those associated with (a) *withdrawal*, e.g. feeling anxiety when unable to connect as

© Springer Nature Switzerland AG 2019
Á. Rocha et al. (Eds.): WorldCIST'19 2019, AISC 931, pp. 795–805, 2019.
https://doi.org/10.1007/978-3-030-16184-2_76

desired, (b) *tolerance*, e.g. increasing presence, interaction and accounts, (c) *relapse*, e.g. after attempting to minimise or adjust the current style of usage, (d) *conflict*, e.g. using social media despite having other priorities as well as (e) *mood modification*, e.g. feeling better when receiving posts, likes and comments. Moreover, research has indicated that excessive and obsessive usage of technology is associated with undesirable life experiences characterised by measures such as reduced creativity, increased anxiety as well as neglecting reality [3].

Despite the increasing awareness of the negative effects of a problematic usage style of social media, certain individuals seem to have a strong feeling and intimate engagement with them and tend to ignore the risks associated with such digital experience. Research has shown that when some people disconnect from social media or are asked to spend less time and interaction than desired, they may become anxious despite the lack of clear purpose of that online presence [3]. The overwhelming use of social media combined with the peer pressure to be online can lead one to lose track of time spent online and of the interactions made with social media platforms. This has prompted research on tools to aid people in self-regulating their digital media usage [4, 5].

Despite being a medium for an emotional and behavioural problem, social media can also contribute to the solution. It can host persuasive techniques to encourage healthy usage [6]. Unlike other problematic mediums for addiction, such as alcohol, technology can provide the means for monitoring its usage by users and, hence, report back to them to be more informed as well as conscious of the usage [7]. It is possible to have software tools to predict whether someone uses a phone or social media in an anxious and uncontrolled way. We advocate that, similarly to online gambling; some users may need assistance so that they stay in control of their addictive digital experience, e.g. self-exclusion as well as lock-out schemes. For example, users can authorise software to alert them when their usage indicates risk, and then send them messages about ways to combat it, e.g. through goal setting techniques [5].

Most research about social media addiction, online identity and online attachment has utilised offline data collection methods introducing limitations about recall bias and ecological validity. For example, questionnaires were used in [8, 9], online surveys method were used [10, 11] and focus groups were used in [12]. In order to know how to design software-assistance to combat problematic social media attachment, an in-depth understanding of the problematic attachment itself in a naturalist, or close to the naturalistic setting, is needed.

In this paper, we aim to gain insights into the real-world experience of people who have a problematic attachment to social media. We adopt a multi-stage qualitative research method employing diary studies designed as a primary data collection technique to elevate ecological validity. We report on the findings regarding the negative and positive emotional states experienced by the participants. The paper is aimed at informing better use of social media that preserve well-being and to help research-informed development of software tools to help that requirement.

2   Research Method

Our research method is qualitative and follows an exploratory approach. We conducted multi-phase studies consisting of an exploratory phase (described in Sect. 2.1) and a confirmatory and refinement phase (described in Sect. 2.2). The data analysis and the framework used as theoretical underpinning are described in Sect. 2.3.

2.1   Exploratory Phase

The objective of this first phase was to explore people's problematic attachment to social media. We used a qualitative method founded on focus groups and a diary study where the diary study was the core method in this stage. Focus groups were first used to gather initial insights that were then elaborated via a diary study allowing the capture and refinement of users' problematic attachment on a daily basis.

A set of 18 participants were recruited via convenience sampling. Participants aged between 18 and 50, with an equal number of males and females were recruited based on the following criteria: being an adult social media user and self-declaring to have a problematic attachment to social media, e.g. excessive and obsessive usage and constant preoccupation about online presence and content. A pre-selection questionnaire was used to assure the existence of the problematic attachment with social media. This questionnaire was adapted from the Generalised Problematic Internet Use Scale [13]. We modified the phrasing to fit the usage of "social media".

We conducted two focus group sessions with the 18 participants, with each session consisting of nine participants. The sessions aimed to familiarise the participants with the objective of the study and to get insights into their problematic attachment to social media. At the end of each session, the participants were explained a practical example to the Evernote application[1]. The application was employed in the next part of this phase, the diary study. The application allowed users to take notes, pictures, and voice and share it with the research team on a daily basis. We also used this application to send forms for completion as well as reminders.

The diary study was conducted with the same 18 participants and lasted two weeks. Participants completed the task which was focused on elaborating their online experience and online behaviour with an emphasis on social media features known to promote attachment such as *profiling* features and online *presence*. The participants provided their notes three times a day, i.e. morning, afternoon and night, via the Evernote application. Reminders were sent on a daily basis through the application. After 14 days of daily diaries, we conducted interviews with the same participants for clarifying and gaining further insight into their diaries entries.

2.2   Refinement and Confirmation Phase

The goal of this phase was to confirm the results of the first phase on the mapping between the emotions accompanying the problematic attachment to social media and

---

[1] https://evernote.com/.

the usage pattern. To achieve this, a card sorting technique within two focus group sessions was carried out with 14 participants; six of them participated in the first phase and eight new participants so that we had a balanced sample and avoid analysis bias. The six participants also served our member checking validation technique [14]. The remaining eight participants were recruited via convenience sampling. They were recruited on the same criteria as Phase 1. A similar pre-selection survey was also utilised for self-assessment in order to check the suitability of the participants. During the session, the participants were required to provide clarifications to questions pertinent to the rationale of their sorting choices.

2.3   Data Analysis

We conducted a thematic analysis for the data collected through the interviews and diary study. The conceptual framework of the analysis was the one proposed in [15]. Positive and negative emotions formed the main themes of the first iteration analysis process. We utilised the Parrott's framework [16] to differentiate between primary, secondary and tertiary emotions. We only use the primary emotions in this work. This is mainly because the participants may not be fully aware of the subtle differences in the secondary and tertiary level, e.g. between worry and anxiety. Induction and analysis around those levels would require a larger scale study. The findings were further validated through the card-sorting within the second phase of the study.

3   Emotions vs Problematic Attachment to Social Media

Social media can have a significant influence on human emotions such as joy, happiness, anger, sadness, fear as well as surprise. Social media users can take advantage of various features such as posting videos and pictures, commenting on them, posting events, searching for new friends, expressing themselves and sharing daily activities with others. Research has found that young adults used social media to communicate with friends and family [17]. In addition, other research has reported that college students spend most of their time reading about the activities of their friends rather than adding content to their profiles [18]. Others explained that seeking information and social interaction was a drive to using social media [19]. Accordingly, people can experience *emotions* through online interaction, especially in the case of problematic attachment to social media. Indeed, their problematic attachment to social media may reinforce and provide strength to the emotions evoked by interaction. The following sections will explain the states and examples of these emotions. Specifically, we present the usage experiences and negative emotions in Sect. 3.1 and the usage experiences and positive emotions in Sect. 3.2.

3.1   Usage Experience vs Negative Emotions

Depending on the nature of the interaction on social media, a person may experience either positive or negative emotions. For example, problematic social media attachment may lead to experiencing negative emotions such as sadness, anger and fear. Negative emotions accompany and contribute to the individuals' problematic online attachment to social media. In Fig. 1, we present frequent usage experiences encountered by people with problematic attachment and correlated emotions. It is important to note that the relation between the experiences and emotions are complex. In this sense, user experience can trigger the negative emotions and vice versa.

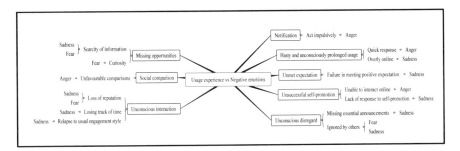

Fig. 1. Problematic attachment to social media: usage experience vs negative emotions

Notifications. People receive notifications online to inform them about update information from others. There are different styles of receiving notifications, such as status, vibration, sound, email, message or a set of these to notify the people that they have received a notification. However, notifications seem to trigger people with problematic attachment to *act impulsively and irresistibly*. Reacting uncontrollably due to notifications can evoke negative emotions; *"I dislike notifications. They are annoying [Parrot's primary: Anger], and it interrupts things"*. Even when people are actively trying to reform how they behave online, notifications can trigger negative emotions such as sadness. For instance, even after a conscious decision has been made to deactivate an online account, users may still receive a notification email suggesting that they reactivate their account; *"I have tried to delete my Facebook account, but I failed to do so. The Facebook system kept notifying me via email what my friends had posted; I felt bad because notifications triggered me to reactivate my Facebook page [Anger]"*.

Hasty and Unconsciously Prolonged Usage. Hasty usage means that people are acting quickly or unconsciously on social media. People with a problematic online attachment may find themselves interacting online in a hasty style, which may have a negative impact on their work productivity, their concentration and their emotions. The hasty usage triggers negative emotions such as anger and sadness when people engage impulsively with social media in the state of *quick response behaviour* or *being online for extended periods without thinking of it*; *"I felt angry because I always respond quickly to activities and communications on social media without any self-control [Anger]"*. Hasty and unconsciously prolonged usage can be to maintain self-presentation and self-concept in front of others through posting too much content; *"I felt sad because I realised I was hasty in posting and commenting on social media [Sadness]"*.

Unconscious Disregard. The notion "disregard" in the context of this paper refers to the state when people are unable or reject to notice, pay attention to or acknowledge someone or something on social media. People with a problematic attachment to social media may ignore others when they are technically available online but unconsciously overlooking messages received from others or neglecting to interact with them in an organised style. For them, social media seem to be an avoidance and escapism tool while having a disciplined interaction is not the main purpose. Social media can be a medium for attachment bonds and communication between people, but those who avoid responding are likely to become more detached from others and may become socially isolated, exacerbating their situation. Therefore, ignoring interaction may lead people to experience negative emotions such as sadness due to *missing essential announcements*; *"I remember that I was moving between groups in social media unconsciously. One day later, I realised that I had ignored important messages which made me sad [Sadness]"*. People may also feel negative emotions such as fear and sadness when *others ignore their presence* on social media or ignore their participation; *"Suddenly people on my Facebook site ignored all my posts. I feel sad and afraid that I did something wrong [Sadness]"*.

Unsuccessful Self-promotion. Self-promotion is an attempt to introduce oneself to others in an ideal or better presented image. People with problematic attachment to social media seem to make an effort to maintain their self-promotion through online interaction, e.g., through posting, commenting and giving likes to others. Self-promoters tend to believe that others will respond to their promotion activities in a positive way. Therefore, when there is *no response to their promotion*, they feel negative emotions; *"For the past two months I have shared my achievements and activities with my friends in social networking groups, but I have not found enough interaction to feel happy, and that made me feel sad [Sadness]"*. In addition, the negative feelings may affect their self-esteem or may lead to competition and excessive interaction. Conversely, if an individual is *unable to interact online* and feels incapable of maintaining their self-promotion, they could experience negative emotions such as anger.

Missing Opportunities. Missing opportunities means the desire to stay constantly connected with social media network activities out of fear of losing benefits. Thus, people with problematic attachment to social media continually wonder if something may be happening online. They need to feel a sense of relatedness and remaining connected with others via social media. The exploratory study has revealed two facets associated with missing opportunity in relation to problematic online attachment.

The first facet concerns the inability to access news or *scarcity of information* that can diminish the degree of social interaction. For instance, people who engage online may be afraid of missing news or updates posted by others which may expire within a short time. This situation could result in negative emotions; *"I feel sad because I lost a live story on Instagram from my favourite comedian [Sadness]"*.

The second facet is *curiosity*. People in social media are interested in what their friends are doing and what latest updates they make on their online profiles to retain a sense of belonging. Their preoccupation with what others do causes curiosity about others' updates which results in excessive use of social media. This can trigger negative feelings; *"I frequently use Facebook because I worry about missing group posts [Fear]"*.

Unmet Expectations. Expectations refer to what people wait to occur; a belief that is concentrated in the future which may or may not be realistic. An expectation could also be about the behaviour of others. People with problematic online attachment tend to be overly reliant on social media to initiate interaction and gain acceptance from others, and consequently establish expectations amongst each other. Therefore, expectations, when unmet, in relation to problematic attachment may facilitate negative emotions. Expectation could be dual sided when people expect others to like their content and when this does not happen, they may feel that the *failure in meeting the expectation of others*; *"I posted pictures on my Facebook account. These pictures were fantastic in my opinion, but I did not receive many likes for them which made me feel disappointed* [Sadness]".

Unconscious Interaction. Unconscious interaction means that people engage in social media interaction without conscious control or awareness. People with problematic attachment to social media typically lack concentration or self-awareness during online interaction, for example, scrolling and navigating between pages in their social media to retain a sense of interaction but without processing the content. Thus, they will repeatedly miss important information that has been posted and this has a *negative impact on their reputation* among friends and family, thereby resulting in negative emotions; *"I lost a vital event which made me feel sad [Sadness]. This was because I seemed to be quite unconscious of my presence on WhatsApp. My friend texted me many times, but I did not realise that"*.

In addition, when people experience sadness in relation to problematic attachment to their social media engagement style, they are likely to try to reform that style. Acting unconsciously can be a sign of *relapse to their usual engagement style* as it indicates a loss of control and deviation from the planned behaviour. Such relapse typically has an adverse effect on their self-esteem and triggers negative emotions; *"I decided to use social media just once a day, but last night unconsciously I logged onto my account many times and felt regret about that [Sadness]"*. Moreover, the unconscious use of social media can result in users *losing the sense of time*. When too much time is devoted to engaging with others online, this can cause sadness; *"I always spend around 4 h per day on social media without feeling it. I feel guilty about the time I waste [Sadness]"*.

Social Comparison. Social comparison means that people compare themselves to others online in order to meet the need for self-evaluation. The profiles that people create for themselves on social media platforms provide details of their work, personality, thoughts and experiences. Users are free to amend or comment on their profiles to convey their opinions and emotions and satisfy their need to belong. In problematic attachment to social media, profiles are heavily used for social comparison. Thus, people with problematic social media attachment are typically keen to compare themselves with the profiles of others. Those people, who are involved in this type of behaviour, typically use social media to an excessive degree in an attempt to present themselves in the best possible way. Online profiles disclose information about the identities of users and their activities. This disclosure results in comparisons being made between people in terms of their social characteristics, online image, reputation, belongingness, and how frequently people interact online. *Unfavourable comparisons* may cause people to experience negative emotions; *"One of my friends is popular on Facebook, and I feel jealous when I see his number of friends growing every day; especially from our group. I ask myself what's wrong with my profile [Anger]"*.

3.2   Usage Experience vs Positive Emotions

Social media users can post comments, share activities, upload photos and videos. Like others, people with problematic online attachment to social media often consider social media to be a form of entertainment, but they tend to overly rely on it for self-promotion and increasing popularity and relatedness. While this could result in excessive use of social media and facilitate negative emotions, people may also experience positive emotions as a result of that problematic attachment making it also difficult to replace with other means. These positive emotions are similar to those experienced in everyday life such as love, joy and surprise. In Fig. 2, we present the positive emotional states associated with usage experiences typically found in people problematic attachment to social media.

Fig. 2. Problematic attachment to social media: usage experience vs positive emotions

Popularity Increase. Popularity is the status of being admired, being liked and supported by others. Social media platforms provide users with an opportunity to build social capital and become popular by presenting their profiles in an attractive form to others. However, the desire to belong and be accepted by others can become obsessive and be seen as a sign of problematic attachment to social media. This desire motivates them to utilise social media excessively in an attempt to maintain their self-presentation. As an individual's group of *friends expands online*, this can result in positive emotions; *"I feel satisfied because I have become popular and an influencer on social media. This requires a high level of online activities to maintain my social standing [Joy]"*. Social media also provide a platform for individuals to express cultural and personal characteristics. Indeed, there have been instances of *divers cultural activities* making certain individuals popular. Consequently, these people are likely to experience positive emotions such as arousal *[Love]*.

Mutuality of Interaction. Mutuality of interaction is a situation in which two individuals' or groups of people are exchanging messages and interacting reciprocally and keeping their engagement active. Social media platforms provide a means for people to engage with each other mutually. Indeed, features such as likes, tags and comments actively encourage people to interact with each other. For people with a problematic attachment to social media, mutual interaction enables them to contribute to what is posted on social media platforms mutually, which seems to be critical, in their case, to boost their self-esteem and positive emotions. For example, if an individual commented on a friend's profile with a *positive comment* and the friend responds similarly, this provides them with a great sense of satisfaction. Thus, mutuality of reactions on social media could cause users to experience joy, thereby promoting their self-esteem and self-evaluation; *"I feel enjoyment because I made an effort to like my friends' posts, and then I received many likes on my own recent post on Facebook [Joy]"*.

Constant Connectedness. Social media has provided new means of social communication. Users can choose to engage one-to-one or in a group of friends or relatives. They can communicate and feel connected through posts, texts, audio, photos or video. Also, social media enables people to post and access information in real-time, regardless of the place and time of the day. Because of such connectivity and *real-time accessibility*, the distance appears less. However, people with a problematic attachment to social media typically have a high need to have a *sense of belonging* and connectedness with others. Such possibility to stay connected leads to excessive use but at the same time motivated by some positive emotions such as joy; *"social media helps me to communicate with my family at any time. I feel happy about that [Joy]"*.

Passing Time. People find engaging on social media to be an enjoyable way of passing the time, and in this sense, it can be regarded as a form of entertainment. Social media offers a means of communication and knowledge and information exchange, but it is equally valid to describe it as a source of entertainment. In addition to posting, sharing and commenting via social media, it can also be used for playing games. About problematic attachment, *entertainment* of this sort evokes positive emotions; *"I rely on social media when I have a long time to wait or when I have nothing to do. It is the only way to pass the time and make me feel pleasure [Joy]"*. Social media also provides the means to make new friends, *re-establish friendships*, or to build communities of people with similar interests. Problematic attachment is characterised by relying on that to feel positive emotion such as surprise, thereby enhancing satisfaction; *"I was exploring my friend's contact list, and I found my close friends from primary school, I felt surprised, and I sent her a friend request" [Surprise]*.

4 Conclusion and Future Work

In this paper, we aimed at investigating the emotions accompanying the usage experience of people with a problematic attachment to social media. Such understanding will facilitate building tools to enable self-regulated and a managed reform of that relation. Our work aimed at capturing the user experience as lived through employing a diary study approach. The findings covered a range of user experiences and accompanying emotional states. We note that we may still be missing additional experiences and emotions due to the difficulty for the participants to recognise what makes part of their problematic attachment despite their declaration of having it. Hence, we would still think of other methods and data sources perhaps based on objective measures and more in real-time. For example, smart watches and wrestle sensors could be thought to collect biological, physical, behavioural or environmental data which may be correlated with actions on social media for a better understanding of the user experience and emotions. This is part of our future work, with the view of understanding the different manifestations of the emotions involved in our experience with social media.

Acknowledgement. This work has been partially supported by the EROGamb project funded by GambleAware, UK, and by European H2020-MSCA-RISE-2017 project, under grant agreement No. 778228 (IDEAL-CITIES).

References

1. Griffiths, M.D., Kuss, D.J., Demetrovics, Z.: Social networking addiction: an overview of preliminary findings. In: Behavioral addictions, pp. 199–141. Elsevier (2014)
2. Enrique, E.: Addiction to new technologies and to online social networking in young people: a new challenge. Adicciones **22**, 91–95 (2010)
3. Andreassen, C.S.: Online social network site addiction: a comprehensive review. Curr. Addict. Rep. **2**, 175–184 (2015)
4. Ali, R., Jiang, N., Phalp, K., Muir, S., McAlaney, J.: The emerging requirement for digital addiction labels. In: International Working Conference on Requirements Engineering: Foundation for Software Quality, pp. 198–213. Springer (2015)
5. Alutaybi, A., McAlaney, J., Stefanidis, A., Phalp, K., Ali, R.: Designing social networks to combat fear of missing out. In: Proceedings of British HCI, p. 1 (2018)
6. Alrobai, A., McAlaney, J., Dogan, H., Phalp, K., Ali, R.: Exploring the requirements and design of persuasive intervention technology to combat digital addiction. In: Human-Centered and Error-Resilient Systems Development, pp. 130–150. Springer (2016)
7. Alrobai, A., Phalp, K., Ali, R.: Digital addiction: a requirements engineering perspective. In: International working conference on requirements engineering: Foundation for software quality (REFSQ), pp. 112–118. Springer (2014)
8. Barke, A., Nyenhuis, N., Kroner-Herwig, B.: The German version of the internet addiction test: a validation study. Cyberpsychol. Behav. Soc. Netw. **15**, 534–542 (2012)
9. Monacis, L., de Palo, V., Griffiths, M.D., Sinatra, M.: Exploring individual differences in online addictions: the role of identity and attachment. Int. J. Mental Health Addict. **15**, 853–868 (2017)
10. Lin, J.-H.: Need for relatedness: a self-determination approach to examining attachment styles, Facebook use, and psychological well-being. Asian J. Commun. **26**, 153–173 (2016)
11. Oldmeadow, J.A., Quinn, S., Kowert, R.: Attachment style, social skills, and Facebook use amongst adults. Comput. Hum. Behav. **29**, 1142–1149 (2013)
12. Balakrishnan, V., Shamim, A.: Malaysian Facebookers: motives and addictive behaviours unraveled. Comput. Hum. Behav. **29**, 1342–1349 (2013)
13. Caplan, S.E.: Problematic Internet use and psychosocial well-being: development of a theory-based cognitive–behavioral measurement instrument. Comput. Hum. Behav. **18**, 553–575 (2002)
14. Birt, L., Scott, S., Cavers, D., Campbell, C., Walter, F.: Member checking: a tool to enhance trustworthiness or merely a nod to validation? Qual. Health Res. **26**, 1802–1811 (2016)
15. Braun, V., Clarke, V.: Using thematic analysis in psychology. Qual. Res. Psychol. **3**, 77–101 (2006)
16. Parrott, W.G.: Emotions in Social Psychology: Essential Readings. Psychology Press, Chicago (2001)

17. Subrahmanyam, K., Reich, S.M., Waechter, N., Espinoza, G.: Online and offline social networks: use of social networking sites by emerging adults. J. Appl. Dev. Psychol. **29**, 420–433 (2008)
18. Pempek, T.A., Yermolayeva, Y.A., Calvert, S.L.: College students' social networking experiences on Facebook. J. App. Dev. Psychol. **30**, 227–238 (2009)
19. Johnson, P.R., Yang, S.: Uses and gratifications of Twitter: an examination of user motives and satisfaction of Twitter use. In: Communication Technology Division of the Annual Convention of the Association for Education in Journalism and Mass Communication in Boston, MA (2009)

Gender Differences in Attitudes Towards Prevention and Intervention Messages for Digital Addiction

John McAlaney$^{(\boxtimes)}$, Emily Arden Close, and Raian Ali

Bournemouth University, Poole, UK
{jmcalaney, eardenclose, rali}@bournemouth.ac.uk

Abstract. It has been suggested that excessive use of the internet and digital devices can lead to digital addiction (DA). In contrast to other industries such as the alcohol industry there appears to be very little expectation on the software industry to position itself as a primary actor in the development of DA; even though software providers have unique capabilities to engage with users in real time and in a personalised way through multi-modal interactive and intelligent prevention and intervention messages. One aspect of personalisation that has been demonstrated to be of importance in relation to DA and other compulsive behaviours is gender. This study consisted of a series of initial exploratory interviews followed by a survey of 150 respondents and several same-sex focus groups, the latter of which was recruited from a university student sample. Thematic and quantitative analyses were then conducted on the data gathered. Overall participants welcomed the idea of DA prevention and intervention messages, although they also demonstrated a clear preference for any DA prevention and intervention messages system to be adaptive and context aware. Some gender differences were evident, such as in terms for acceptance of messages generated by friends or the preference for graphical messages. The results of this study suggest that DA prevention and intervention messages may be useful to and welcomed by individuals who use digital technologies excessively. However, these users also appear to have expectations of what a successful DA prevention and intervention messages system should be able to achieve. Further research is needed on this emergent topic.

Keywords: Digital addiction · Prevention · Intervention · User modelling · Personalization · Gender

1 Introduction

Digital connectivity is a characteristic of a modern society, but its excessive and obsessive use has led to concerns about what has been termed digital addiction [1]. The concept of digital addiction (DA) and whether this qualifies as a mental disorder is a somewhat controversial one, with it listed as an area that requires further attention in the most recent version of the Diagnostic and Statistical Manual of the American Psychiatric Association [2]. Nevertheless, there is evidence to indicate that excessive internet use can become a serious issue for some individuals. It has been suggested that

© Springer Nature Switzerland AG 2019
Á. Rocha et al. (Eds.): WorldCIST'19 2019, AISC 931, pp. 806–818, 2019.
https://doi.org/10.1007/978-3-030-16184-2_77

the consequences of DA can include poor academic performance, reduced social and recreational activities, relationships breakups, low involvement in real-life communities, poor parenting, depression and lack of sleep [3, 4]. Estimates of DA vary according by country and according to the definition of DA being used, but it has been suggested 6%–15% of the general population meet the requirement of DA [5]. This figure rises to 13–18% among university students, who have been identified as a high-risk group for DA [5].

There is also a paucity of research that positions software and its developers as primary actors in the development of DA; with a few notable exceptions [6]. This contrasts with other behavioural domains such as alcohol or gambling. Industry bodies in these sectors engage in corporate social responsibility to address the potentially harmful or risky aspects of the products they are promoting, although it must be acknowledged that there is on-going debate about the underlying motivations of industry bodies in this work and how well they engage with it [7]. This corporate social responsibility can include activities such as financially supporting the provision of sources of further information or help for problematic use. An example of this would be the link to Drinkaware (www.drinkaware.co.uk) website that is included in many UK alcohol industry posters and television adverts.

In comparison to traditional prevention and intervention messages, such as that for alcohol and tobacco products, messages delivered via digital media is distinct in its ability to be issued in an interactive multimodal and intelligent style [6]. Unlike those traditional mediums for addiction, software can monitor the usage pattern of a user and decide upon whether there are symptoms of a problematic usage. This can occur in real time and whilst the behaviour is happening. For example, when a user is taking pictures and sharing these via their social networks in a hasty style accompanied using negative and emotive language. Similarly, software could be used to determine if an individual is compulsively checking for messages and social network site notifications. When a problematic usage is detected, software can generate a warning message and deliver it to the user using a suitable presentation and delivery style, e.g. through text messages, animations, avatars, progress bars, timers, a change in the colour scheme or a buzz indicating the status of the usage. This interactivity and intelligence in digital media give us space for customization and personalization of the DA messages.

When considering prevention and intervention strategies for any behaviour it is important to understand socio-demographic factors which may be relevant to the application and efficacy of such strategies. Gender is one such factor and is also something which can be relatively easily used as the basis for personalisation and customisation. Research into child and adolescent populations suggests that there are gender differences in the pathways for the development of DA, with boys being more likely to experience DA through online video gaming than girls, as well as being more likely to use the internet excessively to attempt to develop new friendships [8]. Gender has also been found to be an important factor in other behavioural domains that may relate to DA. Research into gambling for example has found that men tend to behave in riskier ways then women and exhibit higher levels of social anxiety [9]. Similarly, the importance of identifying gender specific factors that relate to prevention and relapse have been noted with regards to other addictive behaviours [10]. There remains though a lack of understanding on the role of gender in the prevention or intervention of DA.

The current study aimed to explore the issues around gender and prevention and intervention messages for DA to determine when or if gender specific approaches to prevention and intervention should be considered for this behaviour. This was done as part of a larger study into cultural differences in attitudes towards DA prevention and intervention messages.

2   Method and Research Settings

This was a mixed methods study, composed of an initial exploratory stage based on interviews and then followed by a survey and several focus groups. Both the survey and the focus groups were conducted simultaneously using the model of triangulation described by Mayring [11]. The initial exploratory stage consisted of interviews with 11 participants, five males and six females, aged between 19 and 35 years old and holding differing views on the usefulness of prevention and intervention messages for DA. Four of them were professionals and seven were students.

The purpose of the survey was to expand upon the results obtained through the analysis of the interviews. The survey was disseminated through mailing lists to students at a university, professional mailing lists and the social media and mailing lists of the authors. Screening items were used to ensure that the respondents who participated in the survey could be considered to demonstrate a problematic usage of digital media. These items were based on the first three items of the CAGE measure [12], which was developed as a tool to quickly screen for alcohol use disorders. The four items of the original measure ask individuals if they have ever felt they should (i) cut down on their drinking, (ii) been annoyed by people criticising their drinking, (iii) felt guilty about their drinking and (iv) needed to drink in the morning to alleviate a hang-over (eye-opener). For the purposes of this study 'drinking' was replaced by 'use of social networks or gaming' for the first three items. The fourth item of the original CAGE measure was not adapted for use as this was not relevant to the topic of DA. A total of 151 surveys were returned, of which 146 were fully completed.

Two same sex focus groups were conducted in parallel to the survey, one including female participants and one including male participants. These participants were drawn from the student population of a university and were aged between 18 to 22, with a range of academic disciplines represented. Informed consent was obtained from all individual participants included in each part of the study.

3   Results

The initial results on prevention and intervention messages, relating to the interviews and of the survey (with 72 completed responses), were presented at REFSQ 2015 conference [6]. The focus in that work was on the descriptive analysis and trends in prevention and intervention messages. It also presented a set of software engineering challenges for messages design. In this paper the focus is on the attitude and behaviour of digital technology users towards prevention and intervention messages and the gender-effect on that. Since the interview analysis was to provide a baseline for the survey and the focus groups, we will only report on those later studies in this paper.

3.1 Study One: Thematic Analysis of Focus Group Discussions

Each focus group was audio recorded and then transcribed by a research assistant. Thematic analysis was then applied using the method outlined by Braun and Clarke [13]. Several themes emerged from the thematic analysis, as discussed below.

Contextual Awareness. Participants in both groups commented that they would want any prevention and intervention messages software to demonstrate an awareness of the context of their online activity. For example, participants stated that they would not want messages to appear in the middle of a pivotal task, particularly during online gaming. 'when it interrupts you at a crucial point. Where it full stops you and then it just makes you want to hate the idea more often even though it is a great concept' (male participant). Certain types of digital addiction were identified as being more amenable to notification systems, such as online shopping. 'every time you made a purchase on this site and count it up and say, "Look you've spent this much, within this many days" and showed you it, and you'd be like "Wow, okay I've spent way too much money"' (male participant). Overall participants wanted a message and notification system that understands when it is appropriate to deliver messages and when it is not, or to draw upon a phrase used by one of the female participants a system which is 'intuitive'.

Control. A strong theme throughout the discussion in both groups was the desire to control the appearance and operations of messages. It was felt that this would reduce some of the defensiveness that may be felt from receiving notifications 'If the person can choose what it says then that might be more effective too because they know what's most likely gonna be like "Ah I like that"' (female participant). Participants also expressed the wish to be able to choose which software platforms they would receive notifications from. 'so like for me if Facebook notified me I'd be like "Okay go away I don't have a problem with you" if Tumblr notified me I'd be like "Oh yeah okay"' (female participant). This was echoed by male participants, who commented that they perceived different platforms to have different requirements of what could be defined as addictive use. '…in comparison with console gaming umm because in console gaming you're, you're not kind of as, like immersed and it doesn't have like, also it doesn't have the stigma of a WoW gamer or a League gamer who stays up to like 6 the next day with their guild' (male participant).

Social Comparisons. Participants were asked for their views on messages which would show them how their usage compares to their peers, or the norms of people of a similar age or profession as themselves. An issue raised by participants was that allowing people to compare their internet usage to their peers may result in individuals attempting to compete with one another to be the most 'addicted'. …if we're dealing with students that can be a dangerous way of having pop ups because they might make a game out of it like who can play for the longest' (female student). 'It's naturally like a competition, be like "Oh right, I'm gonna, yeah I'm gonna play it more, as long because oh he didn't play it as long"' (male participant). This reflects research into other risky behaviours, which has found for example that within American college student populations some groups such as Greek fraternities base their identity on the fact that that they perceive themselves to be the heaviest drinkers on campus [14]. Male participants also commented that people may use social comparisons to justify their own use, on the basis that others are more addicted than themselves. 'You'll probably think "Ah there's someone worse"' (male participant).

Message Source. A related theme to social comparison was where the DA message would originate from. Both male and female participants expressed a preference for warnings and messages generated by software, although it appeared to be the case that this was because it was the least objectionable option to them. Participants stated that they would prefer not to receive messages from official sources such as the National Health Service (NHS), which they felt could come across as commercial adverts and would therefore be ignored. Female participants spoke about messages being sent from friends and family, in what they referred to as online 'interventions'. Female participants commented that this could be effective but also lead to resentment, although having the source of messages be anonymous was seen as a solution. '...because if I had it from friends or family yes it may affect me more but even if it was anonymous I'd be like as much as I appreciate their care it's not their business it's my time, I don't want you to bitch at me about this because you can't say it to my face' (female participant). The use of peer groups as a persuasive tool to combat DA has been explored in related work by the authors [15].

Message Content. Female respondents agreed that warning messages based on risks to their public profile (e.g. social media posts that may be seen by prospective employers). 'if they like you they'll Google your name and the stuff comes up on Google and they see loads of pictures of you being out drunk every night they're not gonna wanna hire you' (female participant). This appeared to be less of an issue for male participants, who commented that a system such as this could become intrusive and repetitive. There was also a gender difference in preference over the use of figures and graphs, with female participants stating they would like this whereas male participants would not. 'you don't want going "Oh here's, you've been doing this, here's an amazing pie chart!" because you probably honestly just wouldn't care' (male participant). Similarly, male participants expressed a concern that notifications that came across as too medical could result in a sense that interventions were being held, and also that the act of focussing on digital addiction may lead individuals to self-diagnose. 'that's why you should never, it's like the same thing with never like Google your symptoms if you're ill because it normally comes up with "You have cancer". And then people naturally go "Oh no, I have cancer" kind of thing' (male participant).

Message Style. A preference was expressed for informal, positive wording, as opposed to negative content. 'it's like give me a friendly one like "Just so you know you've been playing for 3 h would you like a break?" and then you could select yes or no. So something that gives you an option to choose and isn't mean to you because then you get defensive' (female participant). Male respondents also commented that they would prefer messages phrased as suggestions rather than instructions. 'I think instead of it saying like "Take a break" it should be like "You've played for this amount of time. Why don't you take a break?" Because it just says like "Take a break" it's like "Nah"' (male participant). This theme is consistent with research into gambling, which suggests that personalised behavioural feedback may be an effective technique in reducing problem gambling [16].

3.2 Study Two: Quantitative Analysis of Survey

A total of 151 participants responded to the survey, of whom 146 completed it. Of this 68 (44%) identified as male and 80 (52%) identified as female, with the remainder declining to identify a gender. Most participants (57%) were aged between 18–25, with the remainder (27%) being aged 26–34 or 35 and above (16%). There were no statistically significant gender differences across age groups.

Perceptions of Social Networking and Gaming Addictive Use. Most participants (60%) reported that they felt they should cut down on their use of social network sites or gaming. Females (n = 55, 68.8%) were found to be significantly more likely than males (n = 36, 52.9%) to report this ($\chi^2 = 3.88$, df = 1, p < 0.05).

Gender Differences in Attitudes to Messages, and Types of Messages Considered Important. Gender differences in acceptance and type of messages, delivery style of warning messages, preferences for theme of content and preferences for choice of warning messages are presented in Table 1 (see Appendix). Messages considered important by 40% or more of participants are reported in Table 2 (see Appendix). Findings from both tables are discussed below.

    Acceptance of warning messages and type. Overall the majority (79%) of respondents reported that they were certain or somewhat likely to accept the use of warning messages. There were no significant gender differences in acceptance of warning messages (male: 51 (75%), female 66 (82.5%)). Time spent online (77%) and number of times a social networking site/gaming platform were opened or visited (51%) were the most accepted types of warning message. Females were significantly more likely to agree that warning messages based on consequences/damage to their public profile ($\chi^2 = 8.9$, df = 1, p < 0.01) would be useful.

    Delivery style of warning messages. Time based progress bars (e.g. timers of usage so far) were reported to be the most preferred delivery style for warning messages (58%), followed by pop-up notifications (48%) and dynamic colourisation of interfaces (45%).

    Preferences for theme of content. Most respondents (60%) reported that they would prefer warning message content that is positive and supportive and correspondingly reported that they would not prefer overly-negative content, with female respondents significantly more likely to report this ($\chi^2 = 4.14$, df = 1, p < 0.05). Approximately one third (29%) of respondents reported that they would prefer socially generated content (e.g. messages designed by other users, friends) Males were significantly more likely to state this than females ($\chi^2 = 4.36$, df = 1, p < 0.05). Similarly, 28% of respondents stated that they would prefer precautionary content (e.g. warnings of potential addiction), with females more likely to state this than males ($\chi^2 = 5.3$, df = 1, p < 0.05).

Preferences for choice of warning messages. The majority of respondents stated that they would like to control the frequency of receiving messages (60%), and the format in which the messages is presented, e.g. graph, email or sound etc. (50%). There were no significant gender effects for preference of choice in warning messages.

Thoughts on uses of software messages and consideration of software. The extent to which individuals agreed with thoughts on use of software messages and consideration of software was assessed on a 5-point Likert scale from 1 (strongly agree) to 5 (strongly disagree). This information is presented in Table 3 (see Appendix) and discussed below.

Thoughts on uses of software messages. Most respondents (65%) agreed with the statement that software developers are often unaware or uninterested in the addictive nature of software. In addition, most respondents (81%) were also in agreement that knowing how a message was generated would increase their acceptance of the message. There were no significant gender effects on thoughts on uses of software messages.

Thoughts on consideration of software. Approximately a third of respondents strongly agreed that messages should take into account the type of device being used (82%) and the stage of addiction of the user (86%). Females were significantly more likely ($U = 1988$, $p = .015$) to report that they would wish warning message software to take into account the timing of use (e.g. peak use at weekends but lower use at other times).

4   Discussion

As determined by the survey conducted in the second stage of the study most respondents agreed that they sometimes felt they should cut down on their gaming or use of social network sites, with female participants significantly more likely to agree to this. Participants also agreed that there could be benefits of messages or warning messages, but as evident from the focus group discussions had several expectations about how this system should operate. Participants clearly preferred an intelligent and adaptive system that would understand the nature of their usage and only deliver messages at appropriate times in a non-intrusive style, although it was appreciated that there was a need for messages to capture attention against a background of competing stimuli. This conflict between how participants would prefer the prevention and intervention messages system to work and their understanding of how it would need to work in order to be effective was also evident with regards to control of the system. Participants expressed a strong desire to have control over any warning system they used, whilst acknowledging that any system that was too easy for the user to change could lead to it being ineffective.

As demonstrated in both the focus groups and the survey participants demonstrated a strong preference for message content that was not overly negative or directive. This is in keeping with research into prevention and intervention in other behavioural domains, which has noted that people will tend to react negatively if they feel are being pushed into an action [17]. This is known as reactance and has been documented in a number of studies [18, 19]. Linked to this is the concept of the boomerang effect, in which an attempt to change a behaviour or attitude results in the individual engaging with their original behaviour or attitudes even more strongly, in an act of defiance against the attempt to change them [20].

Female participants reported in both the focus group and the survey that they would like messages which warn them of damage to their public profile due to inappropriate or hasty posting of messages on sites such as Facebook. This is consistent with research into impression management, that is how we attempt to control how we are perceived by others, which suggests that men engage more in self-enhancement and self-promotion than women do [21]. Overall it has been found that men tend to use what could be termed more aggressive impression management techniques, whereas women are more likely to use cautious and considered approaches [22]. This may in part explain why female participants in the current study appeared to be more receptive to messages warning of damage to their public profile, as they give more consideration to how a negative image could impact on their lives. This greater consideration by women may also explain the divide between male and female participants on the use of figures and graphs in warning messages, with female participants stating they would find this helpful whilst male participants commented that they would not.

The survey found that males were significantly more likely than females to agree that they would like message content that had been generated by friends, whereas in the focus groups female participants raised concerns that this would lead to resentment and tension between social network members. This is somewhat in contrast to previous research which suggests that females are more likely to rely on their social networks for support when faced with a challenge [23]. It may reflect the aforementioned results that female participants appeared to be more concerned with their public profile, and therefore perhaps more reluctant to be seen to have an issue. In addition, research suggests that women are more likely to develop digital addiction through social networking sites; whereas men are more likely to develop digital addiction through online gaming [24]. It may be that women are therefore less likely to wish to seek or receive support via a social networking site if the social networking site is itself the primary cause of their excessive internet use.

There were several limitations to the study. The focus groups participants consisted entirely of university students who, whilst as noted being a high-risk group for DA, are not representative of the general population. The survey consisted mainly of participants who are living in the UK, and therefore again may not be representative of the general population. Nevertheless, as exploratory research into an emergent area the study identified a number of gender related factors to be considered when designing messages for DA.

5   Conclusion and Future Work

This paper explored the difference and similarities in male and female perceptions and preferences of digital addiction prevention and intervention messages. Whilst some gender effects were evident there was overall a lack of differences in the views towards digital addiction prevention and intervention messages between male and female participants. Our future work will mainly take an experimental approach to tackle the nuances in the differences in accepting messages and their fit to different variables including gender, personality traits, usage style and contextual factors of the usage. While we do not expect a generalization, the understanding of the most common cases would help the design of default options when giving users the ability to customize prevention and intervention messages and thus save time and effort and maximize usability and potentially acceptance of such messages.

Acknowledgement. This research has been supported by the EROGamb project funded jointly by GambleAware and Bournemouth University and a European H2020-MSCA-RISE-2017 project, under grant agreement No. 778228 (IDEAL-CITIES).

Appendix

Table 1. Gender differences in attitudes to content, delivery style, theme and preferences of digital addiction messages

Information relating to the message	Agree (n%)		Comparison (χ^2)
	Male (n = 68)	Female (n = 80)	
Which of the following information would you like to see in the message?			
Time spent online	56 (82.3%)	60 (75%)	1.17
Times checked/visited software	36 (52.9%)	50 (50%)	0.13
Features heavily used	9 (13.2%)	12 (15%)	0.094
Potential risks	15 (22.1%)	23 (28.75%)	0.86
Usage *bill*	31 (45.6%)	34 (42.5%)	0.14
Consequences/damage on your public profile	14 (20.6%)	35 (43.8%)	8.90**
Consequences on your online relationship with others	14 (20.6%)	17 (21.3%)	0.01
Consequences on online contacts	8 (11.8%)	13 (16.3%)	0.61
Consequences on real social life	23 (33.8%)	31 (38.8%)	0.39
Effects on physiological and mental health	34 (50%)	38 (47.5%)	0.09
The ease and speed of information spread once shared	15 (22.1%)	23 (28.8%)	0.86
Suggestions/advice on how to regulate the usage style	22 (32.4%)	32 (40%)	0.93
Suggestion/advice on potentially interesting real life activities based on your usage	20 (29.4%)	26 (32.5%)	0.16
Factual and proved statements about the benefits of regulating usage style	21 (30.9%)	26 (32.5%)	0.04

(*continued*)

Table 1. (*continued*)

Information relating to the message	Agree (n%) Male (n = 68)	Female (n = 80)	Comparison (χ^2)
Delivery style of warning messages			
Pop-up notifications	32 (47.1%)	40 (50%)	0.13
Offline notifications	9 (13.2%)	21 (26.3%)	3.85*
Hardware-based interactions	14 (20.6%)	14 (17.5%)	0.23
Personalised metaphors	14 (20.6%)	30 (37.5%)	5.03*
Analogy to traditional addiction	7 (10.3%)	11 (13.8%)	0.41
Time-based progress status	40 (58.8%)	47 (58.8%)	0
Sounds	8 (11.8%)	16 (20%)	1.84
Dynamic colouring of interfaces	29 (42.6%)	39 (48.8%)	0.55
Preferences for theme of content			
Non-repetitive content	34 (50%)	39 (48.8%)	0.02
Socially generated labels	26 (38.2%)	18 (22.5%)	4.36*
Supportive content	37 (54.4%)	52 (65%)	1.72
Precautionary content	13 (19.1%)	29 (36.3%)	5.31*
Non-overly negative content	26 (38.2%)	44 (55%)	4.14*
Preferences for choice of warning messages			
Control type of information the label could contain	27 (39.7%)	34 (42.5%)	0.12
Control and specify time(s) the label will be delivered to me	26 (38.2%)	33 (41.3%)	0.14
Control the frequency of sending labels to me	45 (66.2%)	46 (57.5%)	1.17
Control how the label will be presented to me	33 (48.5%)	42 (52.5%)	0.23
Control actions that trigger a label	31 (45.6%)	27 (33.8%)	2.16
Specify accepted sources of the label	24 (35.3%)	20 (25%)	1.86
Control strategy through which the labelling is decided	15 (22.1%)	23 (28.8%)	0.86
Software to be autonomous in forming and delivering labels	22 (32.4%)	22 (27.5%)	0.41

* $p < .05$, ** $p < .01$

Table 2. Aspects of messages considered important by over 40% of participants

Aspect of messages	Percentage considering it important
Which of the following information would you like to see in the label?	
Time spent online	79%
Times checked/visited software	52%
Effects on physiological and mental health	49%
Usage *bill*	44%
Delivery style of warning messages	
Time-based progress status	59%
Pop-up notifications	49%
Dynamic colouring of interfaces	46%
Preferences for theme of content	
Supportive content	62%
Non-repetitive content	50%
Non-overly negative content	48%
Preferences for choice of warning messages	
Control the frequency of sending labels to me	62%
Control how the label will be presented to me	51%
Control type of information the label could contain	42%
Control and specify time(s) the label will be delivered to me	41%
Control actions that trigger a label	39%

Table 3. Extent to which individuals agree with thoughts on use of software messages and what the software should consider when developing messages (means are presented with standard deviations in brackets)

Statement	Male	Female	Difference (t-test)
Thoughts on uses of software messages			
Software needs to inspire my trust before I accept its offer of labelling	1.88 (0.90)	1.80 (0.73)	0.57
Labelling may lead to less natural use of the software and make me lose closeness with it	2.97 (1.00)	2.78 (0.83)	1.27
Software can only have approximation and estimation about my usage	2.63 (0.87)	2.66 (0.95)	−0.20
Knowing how the label was generated and why will increase my acceptance	2.04 (0.95)	1.97 (0.88)	0.47
I need to be able to know how my usage data and reactions to labels are used	1.99 (0.88)	2.13 (0.85)	−1.01
I feel software developers/industries are often unaware of, or uninterested in, the addictive nature of their software and its consequences	2.60 (1.34)	2.26 (1.27)	0.13

(*continued*)

Table 3. (*continued*)

Statement	Male	Female	Difference (t-test)
When measuring and judging my usage, forming a message and delivering it, software should consider			
Progress or stage of addictive usage	1.99 (0.86)	1.89 (0.65)	0.72
Type of devices being used	2.03 (0.97)	1.87 (0.70)	1.15
Time space	2.18 (0.94)	1.84 (0.90)	2.20*
Social context	2.40 (1.13)	2.07 (0.88)	2.00*
Personal profile	2.55 (1.08)	2.21 (1.08)	1.90

* $p < .05$

References

1. Moreno, M.A., et al.: Problematic internet use among US youth: a systematic review. Arch. Pediatr. Adolesc. Med. **165**(9), 797–805 (2011)
2. American Psychiatric Association: Diagnostic and Statistical Manual of Mental Disorders. Washington, DC (2013)
3. Kuss, D.J., Griffiths, M.D.: Online social networking and addiction-a review of the psychological literature. Int. J. Environ. Res. Public Health **8**(9), 3528–3552 (2011)
4. Echeburua, E., de Corral, P.: Addiction to new technologies and to online social networking in young people: a new challenge. Adicciones **22**(2), 91–95 (2010)
5. Young, K.S., Yue, X.D., Ying, L.: Prevalence estimates and etiologic models of internet addiction. In: Young, K.S., Nabuco de Abreu, C. (eds.) Internet Addiction: A Handbook and Guide to Evaluation and Treatment. Wiley, Hoboken (2011)
6. Ali, R., et al.: The emerging requirement for digital addiction messages. In: The 20th International Working Conference on Requirements Engineering: Foundation for Software Quality (REFSQ 2015). Essen, Germany (2015)
7. Lyness, S.M., McCambridge, J.: The alcohol industry, charities and policy influence in the UK. Eur. J. Public Health **24**(4), 557–561 (2014)
8. Lee, Y.H., Ko, C.H., Chou, C.: Re-visiting internet addiction among Taiwanese students: a cross-sectional comparison of students' expectations, online gaming, and online social interaction. J. Abnorm. Child Psychol. **43**(3), 589–599 (2015)
9. Wong, G., et al.: Examining gender differences for gambling engagement and gambling problems among emerging adults. J. Gambl. Stud. **29**(2), 171–189 (2013)
10. Hoving, E.F., Mudde, A.N., de Vries, H.: Predictors of smoking relapse in a sample of Dutch adult smokers; the roles of gender and action plans. Addict. Behav. **31**(7), 1177–1189 (2006)
11. Mayring, P., On generalization in qualitatively oriented research. Forum: Qual. Soc. Res. **8**(3) (2007)
12. Ewing, J.A.: Detecting alcoholism. the CAGE questionnaire. JAMA **252**(14), 1905–1907 (1984)
13. Braun, V., Clarke, V.: Using thematic analysis in psychology. Qual. Res. Psychol. **3**, 77–101 (2006)
14. Carter, C.A., Kahnweiler, W.M.: The efficacy of the social norms approach to substance abuse prevention applied to fraternity men. J. Am. Coll. Health **49**, 66–71 (2000)

15. Alrobai, A., et al.: Online peer groups as a persuasive tool to combat digital addiction. In: Persuasive Technology, pp. 288–300. Springer, Cham (2016)

16. Auer, M.M., Griffiths, M.D.: Personalized feedback in the promotion of responsible gambling: A brief overview. Responsible Gambl. Rev. **1**(1), 27–36 (2014)

17. Rhodewalt, F., Davison, J.: Reactance and the coronary-prone behavior pattern - the role of self-attribution in responses to reduced behavioral freedom. J. Pers. Soc. Psychol. **44**(1), 220–228 (1983)

18. Rhodewalt, F., Strube, M.J.: A self-attribution-reactance model of recovery from injury in type-a individuals. J. Appl. Soc. Psychol. **15**(4), 330–344 (1985)

19. Tormala, Z.L., Petty, R.E.: What doesn't kill me makes me stronger: the effects of resisting persuasion on attitude certainty. J. Pers. Soc. Psychol. **83**(6), 1298–1313 (2002)

20. Lee, S.-J., et al.: Development of a self-presentation tactics scale. Personality Individ. Differ. **26**(4), 701–722 (1999)

21. Guadagno, R.E., Cialdini, R.B.: Gender differences in impression management in organizations: a qualitative review. Sex Roles **56**(7–8), 483–494 (2007)

22. Reevy, G., Maslach, C.: Use of social support: Gender and personality differences. Sex Roles **44**(7–8), 437–459 (2001)

23. Durkee, T., et al.: Prevalence of pathological internet use among adolescents in Europe: demographic and social factors. Addiction **107**(12), 2210–2222 (2012)

Evaluating of Mobile Applications and the Mental Activation of the Older Adult

Maricela Sevilla[1]([⊠]), Ángeles Quezada[1], Consuelo Salgado[1],
Ricardo Rosales[1], Nora Osuna[1], and Arnulfo Alanis[2]

[1] Facultad de Contaduría y Administración,
Universidad Autónoma de Baja California, Calzada Universidad 14418,
Parque Industrial Internacional Tijuana, 22390 Tijuana, B.C., Mexico
{mary_sevilla, maria.quezada, csalgado, ricardorosales,
nora.osuna}@uabc.edu.mx
[2] Department and Computer Systems,
Calzada Technological s/n Unit Tomas Aquino, Tijuana, Mexico
alanis@tectijuana.edu.mx

Abstract. This article describes how ICTs have changed the ways of socializing among people. Especially the elderly. Thus, an experiment was presented that allowed evaluating 25 older adults in courses using ICT. A qualitative methodology is applied to analyze in particular the situation experienced by older adults in relation to the use of technologies, their experiences at the end of the computer course and how it influences them in their daily lives. Results are obtained derived from taking the technology courses, as well as answering a questionnaire before and after the course. The result shows that the technology allows any adult is able to use a computer, smartphone, tablet among other devices as long as you have patience and do not suffer from any health impediment that prohibits it.

Keywords: Older adults · Aging · Tablet · Computers · Technology

1 Introduction

In traditional societies the elders had a prominent and leading role in the orientation of their respective societies. They were respected, revered, and obeyed in their role as counselors and guides to the community; they were considered repositories of the wisdom gained and accumulated throughout their life. In [1], mentions in his article that this situation changed radically. In today's technologically consumed societies, particularly in underdeveloped countries, the vast majority of the elderly are victims of helplessness. From the moment they cease to be part of the productive apparatus or have an active professional life, it seems that they cease to be part of society.

Because physical abilities change with age, it's important that mobile app developers design smartphone apps with impending health challenges in mind.

As more elderly community members rely on smartphones for communication, information gathering, and entertainment, and as more adults begin using mobile apps as they age, the question app developers need to consider is, "What can be done to make interactions with mobile apps easier for older users?"

© Springer Nature Switzerland AG 2019
Á. Rocha et al. (Eds.): WorldCIST'19 2019, AISC 931, pp. 819–828, 2019.
https://doi.org/10.1007/978-3-030-16184-2_78

Mobile app developers should consider psychological and physical characteristics of elderly people when they're designing mobile apps. Some easy factors to consider in the design process include: contrast, labels, formatting, navigation, and cues [2].

2 Related Work

The forms of socialization have changed in an accelerated way, in part, due to the advancement of Information and Communication Technologies (ICT). Years ago, people only communicated through phones, letters or face to face. Today, communicating in this way seems inconceivable and all this thanks to the technologies that we have at our disposal that give us opportunities to be able to be connected with family or friends thousands of kilometers no matter where in the world we are. Through platforms or social networks, one can have contact with other people, known or unknown, for work reasons or simply looking for friendship [3].

The use of Information and Communication Technologies (ICT) is undoubtedly the key component that best symbolizes the change of time we are living. One of the challenges facing the information society is, in addition to improving the technology needed to get access to high quality content and information resources, to ensure that this access reaches a majority and on equal terms to the entire citizen.

It is a verifiable fact that ICT have, to a large extent, changed the way people relate to each other, it is also evidence that these devices are not available to the entire population, there is the so-called digital divide, not only for economic reasons, but, in our concrete case, for reasons related to age and the limited training that some older people have in this area. In this sense, and with the exception of this limitation, never before has information and knowledge been available to any individual, having certain powers to search for information and the necessary technological resources, every person can self-fashion. This is the characteristic of today's society; In the past, information was only available to a few.

Older people, like any other group, have different limitations, both physical and training level for the use of ICT, but it is more the interest shown by many of them to be up to date in this information society, by what we can say, that there are more possibilities than the limitations that have [4].

Older people have surpassed the most common ages for training and work and as a result, their learning opportunities are limited and their needs for use are limited, as opposed to young people. Many older people have not been born in the new technological era and therefore they need to learn these new innovative tools and their coping strategies, which in turn translates into fears and insecurities; On the one hand, they have to start thinking about issues that were previously unknown and, on the other, they perceive a lack of capacity in learning these tools.

There are numerous studies that have focused on the relevant role that these new technologies have in the new technological society. Many of the Research has focused on knowing the role that different variables, well related to the context, with the technologies or with the beneficiaries of them, have on the access and uses that can be made of them [5, 12].

The treatment of ICT and older adults is generally part of the lines that deal with inequalities or gaps in the information society: all authors start from the recognition of the disadvantage of this group in relation to digital technologies. The consideration of the elderly group within the "digital excluded" or as "late adopters" and the recognition of the importance of technologies in different areas, provides a framework to base the convenience in the adoption of the same or raise it in terms of social inclusion: "ICT become for the elderly as an opportunity to remain integrated into society" [6]. In that sense, digital inclusion can be registered as a transversal dimension for social integration according to the principles in favor of older persons already enunciated in 1991 by the United Nations: Independence, Participation, Care, Self-Realization and Dignity [7].

The consideration of the benefits that ICTs can provide for the quality of life of the elderly is present in the authors who highlight them as an opportunity. Most of the studies in relation to these issues specifically refer to the Internet, however today ICT devices are almost inseparable from the Internet, so in this case it has been decided to assimilate them.

At the present time, around 10% of the world's population are aged 60 years and over, but this is set to increase to about 20% by 2050. Globally, the 60–79 and 80 plus age groups are growing the fastest. While the population worldwide is growing at around 1% per annum, the number of people aged over 80 is growing at 4% per annum and by 2050 it is expected that people aged over 60 will outnumber children aged 14 and under. In much of the world, the number of children has already peaked and is now declining. Moreover, the phenomenon of population aging is not limited to developed countries.

The aging of global populations presents many challenges, not least how services can be improved in order to enhance the health and quality of life of older people in the context of limited financial resources both at the individual and governmental levels. In view of this, there has been growing recognition of the potential for ICTs to improve services and enhance the well-being and social participation of older people [8].

Technology now supports or streamlines many day-to-day activities. This continued technological development is occurring alongside the aging of global populations, creating opportunities for technology to assist older people in everyday tasks and activities, such as financial planning and connecting with friends and family. New technology also has the potential to provide timely interventions to assist older adults in keeping healthy and independent for longer. Older adults are slower to adopt new technologies than younger adults, but will do so if those technologies appear to have value, for example in maintaining their quality of life. To make technology more age-friendly, it is important to understand the advantages and disadvantages that older adults perceive in using it. We therefore explored older adults' familiarity with and barriers to using technology [9].

Mobile technological devices such as tablet computers (commonly referred to as tablets), a type of portable computer that has a touchscreen, are becoming increasingly popular. The number of adults aged 65–74 years using tablets to go online more than trebled in recent years in the UK, going from 5% in 2012 to 17% in 2013. However, this percentage remains low compared with younger age groups (e.g., 37% of adults aged 25–34 years used tablets to go online in the last 3 months [9].

The use of mobile devices reaches all age brackets equally and the attraction for these devices is the same for someone over 65 years old as for a millennial. In fact, according to the annual report of the Society of Information in Spain, published by the Foundation Telefonica, Internet users over 65 have significantly increased the use of the Internet: 14.3% use instant messaging weekly, a 52.5% of them buy online and 27.2% see or listen to multimedia content through mobile [10].

The new technologies have seen difficulties to implant in the third age, since it is a public that has not grown with them and does not know how it works; however, it is increasingly consolidated in the population over 60 years old. This digital divide has been reduced in recent years [11].

3   Experimental Design

3.1   Method

3.1.1   Experiment Sample

For this study, a group of 25 seniors was taken at their convenience, who have been part of the course: full adults learning new technologies, taught in the FCA UABC Tijuana. In addition, 2 geriatric doctors were interviewed to obtain their medical point of view about the benefit of the use of technologies by the elderly.

3.1.2   Instruments

Older adults were given a survey to know about their knowledge about the technologies, also what sports and leisure activities they like.

For the geriatric doctors an interview was used where the purpose was to obtain information about the needs of the elderly, and the opinion they have about the use of ICT by that sector of the population.

3.1.3   Procedure

The research methodology in this study is applied, qualitatively, through the case study modality. The interest is focused on the investigation of a phenomenon, population or general condition. The study does not focus on a specific case, but on a certain set of cases for the analysis of social reality, and represents the most relevant and natural form of research oriented from a qualitative perspective. This study focuses on a particular situation, which is the situation experienced by seniors in the elderly regarding the use of technologies, their experiences at the end of the computer course and how it influences the same in their daily lives.

The sources of data collection will be the testimonies of the elderly participants and assistants to the course taught at the Faculty of Accounting and Administration, UABC, Tijuana, Mexico. The interview and survey were used as data collection techniques (Fig. 1).

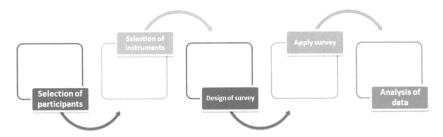

Fig. 1. Experimentation methodology

4 Results

The 92% of the older adults surveyed use mobile devices, and 67% express that it is not complicated for them to use them, one of the reasons is that they have difficulty reading or writing on a keyboard of their phone or tablet as indicated by the 42%, as shown in Fig. 2.

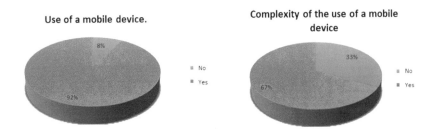

Fig. 2. Result of the use of mobile devices

Most of them (83%), express not having experience when installing mobile applications, however if they use them, and prefer the applications referring to social and health networks, however, 21% show interest in entertainment applications. What they expect from applications is that they are free, that allows them to learn new things, that is easy to use, that provides a benefit and that allows them to be in contact with family and friends, as shown in the Fig. 3.

¿Have you ever downloaded an application??

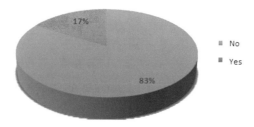

Fig. 3. Result of the experience of mobile devices

In order to consider mobile applications that focus on physical activities, some questions were asked that allowed to know the following: 64% of the respondents do not perform physical activity, and 36% that if they do, they prefer to walk, swim and do yoga, as shown in Fig. 4.

¿Do you do any physical activity?

Fig. 4. Results of physical activities.

The 80% show interest in having an application that allows them to perform aerobics or yoga, as shown in Fig. 5.

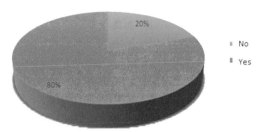

Fig. 5. Results interest in having an application

In another section that deals with what seniors prefer in terms of games or recreational activities, what stands out is games of mental agility, casino, letters, educational and creativity and drawing.

When questioning about the readings that they prefer, there is no answer that stands out with great percentage difference, there are several preferences, among them are: personal improvement, science and technology, novels, mystery, adventure and romance.

When interviewing the geriatric doctors, it was possible to complement the information obtained by the elderly, first, we must understand that a geriatric doctor is a doctor specialized in caring for the elderly and fulfills the mission to detect early and diagnose diseases, as well as to prevent possible sequelae. Its main function and responsibility is to educate family members and caregivers of the preventive or curative areas of old age, as well as physical and cognitive rehabilitation. They also contribute to the improvement of the health of the patients by educating the patient and the primary caregiver about the disease and/or present diseases and how to prevent secondary complications. It is common and well known that, when entering this state of deterioration, the older adult is more prone to suffer from certain diseases or that those already suffered complicate a little more Diseases such as mixed dementia, Alzheimer's dementia and other dementias, diabetes, hypertension, depression, dyslipidemias, sleep disorders, anemia, osteoarthritis, malnutrition, falls syndrome, some of them susceptible to improvement. All these in addition to damaging the health of the person damage their quality of life, causing the adult to become isolated or really feel like a burden to others.

According to the information gathered from the interviews applied to the geriatric doctors on the use of applications by older adults. The biggest challenges that these doctors face every day are to take care of subsequent or first-time patients who lack care on the part of their relatives, with advanced and poorly cared for diseases or who come for a second opinion about a diagnosis, especially in the cases that in public institutions tell them that there is nothing to do. Many times the patient of subsequent visit is willing to go to the consultation to talk about their ills, their mental deficiencies.

Those who expect to see them solve their problems or diseases that have been dragging on for some time.

That is why it is considered vital to acquire a lifestyle where physical activity and a good diet is essential, is the basis for all pathology and staying active brings benefits both for your health, to control anxiety, to rest in a way more appropriate, and control their comorbidities, as well as interact with other people, various health institutions manage or have groups of activities for seniors.

Regarding the use of technology by these, there is a mistaken belief that they do not have enough physical and mental capacity to incorporate into their daily routine, due to certain limitations, or that they do not use it because they are not interested, but it's not like that, it's often due to the fact that they do not know how to manipulate it and they do not have someone who has the patience to teach them. In addition to this the fact that the device does not have favorable characteristics for the recurrent use of an elderly person suffering from a bone disease that allows him to dominate or perform different movements or the simple deterioration of sight over the years, as well as The loss of memory can make them forget the procedure of searching for something on the web or how to make a call from the cell phone. Leaving behind the benefits that the good use of this could provide. In addition to making them feel that they belong or that they are part of the new society, it serves as a good distractor, thus contributing to their emotional health by keeping them occupied in things of interest to them, be it entertainment, information sources, or helping them acquire new ones knowledge.

Technology is a tool that everyone uses and now everything revolves around it, and if they do not learn to use it, they will be misplaced. Any adult is able to use a computer, Smartphone, tablets among other devices as long as they are patient and do not suffer from any health impediment that prohibits it.

However, the excessive use of these devices in addition to causing damage to the health of the elderly, damaging their vision, decreasing the quality of sleep contributing to the development or increase of anxiety, be more irritable, and aggravate depression. Over information and the simple way to maintain contact with other people through this medium could cause an addiction by unleashing the aforementioned. Most devices and applications are created without taking into account the needs of this sector of the market and are often somewhat complex, or impossible to use taking into account that certain elderly people have little or no education. That is why it is important to implement simple applications that allow you to stay physically, mentally and socially active. Some of the applications that doctors recommend and that could benefit from preventing cognitive impairment are those of messaging, memory games, alphabet soup, news, and electronic books, since they test memory, mental agility and concentration, keeping the mind active and stimulated, helping to have a better retention of information in the short term. Regarding the applications that encourage physical activity, it is essential that they be simple in such a way that allows them to adapt it to their daily activities and that motivates them to perform more physical activities taking into account their capacity, visual and joint limitations of each patient. Following the instructions on physical exercises by means of an application and not of a particular instructor would not cause conflict as long as they contain clear images and illustrations or descriptive videos.

5   Conclusions

It is important to consider the complexity involved in assisting older adults in their different personal and, in some cases, professional activities. There are a large number of non-deterministic variables that can influence the appropriate assistance for them.

As a future work, a modeling based on agents will be carried out that allows to have an abstract representation of the different scenarios presented in real life, as well as to simulate some hypothetical cases this in order to evaluate all the variables involved in the interaction.

The model will allow the intelligent and adaptive application to be developed, having an impact on a better sensitivity to the context, anticipating the emerging changes. The application will be adapted according to the behavior and characteristics of the user derived from the HCI. There will be different interaction scenarios that allow different work dynamics to be carried out based on predefined themes or objectives to be met. These objectives and themes will take an evolutionary sequence, allowing the user to progress in a greater degree of content and adaptability.

The applications to develop must contemplate adaptive behavior skills such as conceptual, social and practical. The adaptive difficulties of limitation, social intelligence and practice must be considered. The application must allow the user to develop the practical intelligence that allows the self-sustaining capacity to be independently promoted, even considering the deficiency of physical aptitudes.

The application should promote social intelligence by understanding social and behavioral expectations of others, achieving a better adaptation of behaviors in social situations. The application will be a guide for users helping to develop these skills, without age being an obstacle.

Additionally, the use of this application will allow achieving a better adaptation to the changes that occur in aging, the adaptive content of the application will allow to develop coping strategies to old age.

Finally, all the direct and indirect variables must be considered, in order to feedback and update the application with an amplified universe. Likewise, it is necessary that the case study with a more significant sample, a minimum sample of 500 to 1000 individuals in different interaction scenarios. This will help to understand in detail the context of study, as well as make the decision to make the application more specific, in order to give follow up to health problems, and exercise the mind, daily activities.

On the other hand, we have to involve groups of expert nurses, doctors, psychologists, psychiatrists, anthropologists, and pedagogues that help us to better understand the behavior and activities of this elderly. In order to make the relevant options demanded by those involved in different scenarios under different circumstances.

References

1. Gómez V.R.: Tercera Edad y TICs. Una sociedad positiva. http://www.usuaria.org.ar/noticias/tercera-edad-y-tics-una-sociedad-positiva.html. Accessed December 2016
2. Tatsiana, L.: Designing Apps for elderly smartphone users. services and solutions (2017). https://clutch.co/app-development/resources/designing-apps-for-elderly-smartphone-users

3. García, E.G., Heredia, N.M.: Facultad de Ciencias de la Educación Universidad de Granada. Personas mayores y TIC: oportunidades para estar conectados. Revista de Educación Social (2017). http://www.eduso.net/res/24/articulo/personas-mayores-y-tic-oportunidades-para-est ar-conectados
4. Casamayou, A., González, M.J.M.: Personas mayores y tecnologías digitales: desafíos de un binomio. Psicología, conocimiento y sociedad (2017). http://www.scielo.edu.uy/pdf/pcs/ v7n2/1688-7026-pcs-7-02-00152.pdf
5. Porras Moral, A.J.: TIC, alfabetización digital y envejecimiento satisfactorio: un estudio longitudinal. Universidad de Córdoba (2017). https://helvia.uco.es/bitstream/handle/10396/ 15902/TFM_Antonio_Jes%C3%BAs_Porras_Moral.pdf?sequence=1&isAllowed=y
6. Agudo Prado, S., Fombona Cadavieco, J., y Pascual Sevillano, M.: Ventajas de la incorporación de las TIC en el envejecimiento. Revista Latinoamericana De Tecnología Educativa - RELATEC, **12**(2), 131–142 (2013). http://relatec.unex.es/article/view/1169
7. Organización de las Naciones Unidas: Principios de las Naciones Unidas en favor de las personas de edad. Resolución 46/91 adoptada el 16/12/91 (1991). http://www.acnur.es/PDF/ 1640_20120508172005.pdf
8. Sixsmith, A., Mihailidis, A., Simeonov, D.: Aging and Technology: Taking the Research into the Real World. Oxford Academic (2017). https://academic.oup.com/ppar/article/27/2/ 74/3883637
9. Vaportzis, E., Clausen, M.G., Gow, A.J.: Older adults perceptions of technology and barriers to interacting with tablet computers: a focus group study. Frontiers in Psychology (2017). https://www.ncbi.nlm.nih.gov/pmc/articles/PMC5649151/
10. Corcobado, M.Á.: 'Apps' para personas mayores y para quienes les cuidan. El país (2018). https://elpais.com/tecnologia/2018/05/30/actualidad/1527674268_707515.html
11. González Oñate, C., y Fanjul Peyró, C.: Aplicaciones móviles para personas mayores: un estudio sobre su estrategia actual Universidad Oviedo, Aula abierta (2018). https://Dialnet-AplicacionesMovilesParaPersonasMayores-6292839.pdf
12. Vázquez, M.Y.G., Sexto, C.F., Rocha, Á., Aguilera, A.: Mobile phones and psychosocial therapies with vulnerable people: a first state of the art. J. Med. Syst. **40**(6), 157 (2016)

SmartWalk Mobile – A Context-Aware m-Health App for Promoting Physical Activity Among the Elderly

David Bastos[1(✉)], José Ribeiro[1], Fernando Silva[1], Mário Rodrigues[2],
Rita Santos[3], Ciro Martins[2], Nelson Rocha[4], and António Pereira[1,5]

[1] School of Technology and Management,
Computer Science and Communications Research Centre,
Polytechnic Institute of Leiria, Leiria, Portugal
david.bastos19@gmail.com
[2] ESTGA & IEETA, University of Aveiro, Aveiro, Portugal
[3] ESTGA & DigiMedia, University of Aveiro, Águeda, Portugal
[4] DCM & IEETA, University of Aveiro, Aveiro, Portugal
[5] INOV INESC INOVAÇÃO Institute of New Technologies - Leiria Office,
2411-901 Leiria, Portugal

Abstract. The population of the world is getting increasingly older, and with that new challenges appear. It is thus necessary to promote healthy aging, where people maintain their independence and autonomy for longer. The SmartWalk project's aim is to create a physical activity monitoring system for smart cities for motivating the elderly to a more physical active lifestyle, that is monitored and adjusted by health care professionals for best results. This paper describes the various components that compose the project, focusing on the mobile application and how it works.

Keywords: Elderly care · Healthy lifestyle · Context awareness · m-Health

1 Introduction

In the past decades, in almost every country there has been an increase on the proportion of elderly people that make up a nation's population [1]. One of the reasons for this is the significant way life expectancy has increased during the twentieth century, particularly from the 1950s onward, with studies predicting it to continue to increase, reaching the ninety year mark by 2030 [2].

This means that the older population will be required to remain in the workforce for longer and to be more independent when they leave, because there will be less younger people available to care for them. The problem is that increased life expectancy does not directly translate into ability to function at the same level that a person had when it was younger. While some deterioration of

© Springer Nature Switzerland AG 2019

Á. Rocha et al. (Eds.): WorldCIST'19 2019, AISC 931, pp. 829–838, 2019.
https://doi.org/10.1007/978-3-030-16184-2_79

faculties, both physical and mental, is unavoidable, steps can be taken to soften that deterioration.

The SmartWalk project's main goal is precisely that of encouraging the elderly to perform more physical activities. For that purpose, it makes use of a mobile application that provides a selection of routes for the user to walk on. During a walk various types of data are collected, by the smartphone and the sensors connected to it, which are, at the conclusion of the walk, forwarded to a server where the data is stored and made available for consultation by the user's health professional. The health professional, using another component of the system, the SmartWalk Web platform, can then make use of that data to suggest changes that the user should implement to promote better health, and he can then send a message with those suggestions immediately to the user's application.

The application developed is an m-Health context aware application – but what does that mean and why is it important? M-Health is an abbreviation for mobile health and, as the name suggests pertains to the use of mobile devices (smartphones, tablets, wearables, ...) in assisting the practice of medicine, which is helped by the ubiquity of such devices, even in nations that lack more traditional computational devices. Context awareness means that the application can sense changes in its environment and react appropriately without user intervention. For our applications this means that the application can, for example, know if the user is in movement or not and record that information accordingly without bothering the user by asking.

The next Section briefly reviews the current state of the art and how it influenced this work. Afterwards, Sect. 3 overviews the SmartWalk platform and what it does. Section 4 delves into the mobile application itself, how it works, and the design rationales that supported its construction. Section 5 discusses the current state of the project and future plans, and Sect. 6 concludes the article with some final remarks.

2   Background and Related Work

Due to smartphones possessing mobile computing capabilities, a vast array of sensors, and also to the relative ease with which other types of sensors can be connected to them wirelessly, smartphones are already an useful tool for healthcare – but what makes them really stand out is that due to their increasing omnipresence in our society and our tendency to keep our smartphone with us at all times, it provides the perfect platform to monitor the health of its user without the need for intrusive devices that might inconvenience the user on its day to day life.

This makes it far easier to gather more and better data from the user; however that is only true if the user already possesses a smartphone and is used to using it. The elderly in 2018, not having grown with these technologies, and not using them as routinely as the younger generations, are thus not as familiar with them, making it often necessary for someone else, possibly a familiar or a healthcare professional, to convince them of their usefulness [3] as well as explain how different it is from the technologies they are used to.

One way is to show the vast availability of health and fitness applications which are used by everyday people for a wide range of activities, who use these apps to maintain themselves motivated to accomplish their goals, which can range from fitness, to weight goals, nutrition, sleep and more [4].

Many methods are used to maintain the motivation of the user, from creating a game out of exercising, that maintains the person engaged in a story and distracted from the actual exercise (Zombies, Run! [5]), to creating a virtual coach, that provides the user with feedback and encouragement (Runtastic [6], MapMyFitness [7]) or apps that share your results and achievements on social media, creating friendly competition with other people, or making it so that the user has more difficulty abandoning their goals because other people are watching (Fitocracy [8], Map My Run [9]).

These applications are geared towards the general populace, and while the elderly population can use them and reap their benefits, due to their age and generally poorer health they have different needs that need to be satisfied.

For those another type of app is needed, apps such as Medicine Alert [10] that helps the user to better manage their medication, such as when to take it or when to refill a prescription. Or the Blood Pressure Companion [11] that allows the user to enter blood pressure measurements and easily visualize and analyse blood pressure history, or the Dip.io application [12] that uses a smartphone camera and a special kit, and allows a user to perform an urinalysis in the comfort of their own home, with the results being sent to a cloud server to be classified, and is as of October 2018 being put in a test trial as to ascertain its effectiveness.

While these types of applications are helpful in their own right, they are created for generic use by any elder, and do not take into account, or only on a limited level, a user's personal physical and mental idiosyncrasies, as well as any health conditions the user might have, and are thus not optimized for a specific user. A healthcare professional would be able to optimize the process, but only by offering advice to the user whenever they met for a consultation and he could monitor them.

It is, however, possible to use a mobile device itself to monitor the user in real time, without user and healthcare professional ever meeting, and in fact systems like these have already been implemented. Systems like the Ubiquitous Mobile Health Monitoring System for Elderly (UMHMSE), a system that uses a smartphone as the central node of a WWBAN that collects health data from the user, and then sends this data to a central server, that autonomously processes the data and allows the family or the healthcare professional of the user to visualize this information through a website, as well as warn them when an emergency occurs [13]. Or the system [14] where users with cardiac conditions wear sensors that communicate the user's vital statistics to a smartphone that sends it to a server, where the user's doctor can see the information via a web interface. The mUAHealth [15] is a similar project with the difference that the smartphone does not just work as a transmitter, but also possesses an interface through which the user can view is own collected vital data.

These types of apps also exist adapted for the use of elderly people, such as the Enhanced Complete Ambient Assisted Living Experiment (eCAALYX) [16] project which besides having the functionalities of the previously described apps, also possesses the ability to autonomously detect some anomalies, such as tachycardia, and a user interface redesigned to be simpler and easier to use as well as using more adequate hardware, with a smartphone without physical buttons and large screens. The PhonAge project takes this another step by allowing the user or one of his caretakers to modify the interface, such as text or icon size, to be better adapted to the user [17].

The SmartWalk project takes inspiration from these approaches so as to give the user control of its health regimen, while still allowing for remote supervision and assistance by an experienced health professional. The following Sections will explain how the ideas and principles gleaned from these projects were incorporated into our own.

3   SmartWalk – The Platform

The SmartWalk project [18,19] works on the client-server model as shown in Fig. 1. There is one server that stores all the data collected and provides the services needed by the clients. There are two types of clients: one is the website client which is used by healthcare professionals to create a fitness regimen by scheduling walks on various routes, and visualise their patients' information, such as what routes they have been walking or not walking and any health conditions that the patient has on file, as well as see an historical graph on the pain levels of the various body parts of the patient. This information is used by the healthcare professionals to make well informed changes to their patients' fitness regimen.

A healthcare professional can create map routes (using, for example, the My Maps service from Google) and then upload the generated KML files (a KML file uses the XML format to store geographic data that can then be visualized on online maps) to the server so they can be downloaded by the app as a new route for the user to experience, all from any device capable of accessing the internet. The data collected on the server is stored in a SQL database which conforms to the Fast Healthcare Interoperability Resources (FHIR) health data standard. The server provides several web services coded in PHP.

The other type of client is the mobile application, which primarily collects data about the user and then sends it to the server. It will be shown precisely what it does and how it does it in the next Section.

4   SmartWalk – The App

The mobile application, created for Android systems, is the most direct connection that the end user has with the whole project, and is the one with which it interacts. This section explains, on a screen by screen basis, the design reasoning and coding processes used to create the app.

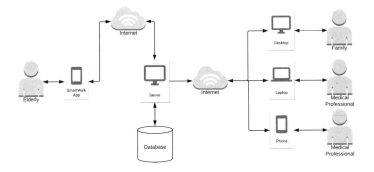

Fig. 1. SmartWalk architecture

4.1   First Steps

The first interaction the user has with the application is with a few onboarding screens as shown in Fig. 2 that give him a short overview of the app and its purpose so that all users go into the app with at least an idea of what to expect and what are some of the options available. This onboarding only appears the first time the application is used, since it is only to be used as a short introduction and not as a manual for the app.

Fig. 2. Onboarding   **Fig. 3.** Route screen

When the user enters the main screen of the application he has to log-in into his Google account. While this may be an annoying step, it is absolutely required for the application to function correctly, as the application makes use of several

Google services, in particular Google Fit to track and collect the data generated by the user. This step may be skipped and the user can continue to explore the application, but he will eventually reach a point where he must log-in to use the application's features, at which point the application will once again ask the user to log in.

Permissions are asked of the user only when they become necessary and with a clear explanation of why they are necessary, so that they are transparent to the user and he can make informed decisions and never have any reason to doubt the application's intentions.

4.2   Starting the App

The main screen is a simple screen that shows the SmartWalk logo and can be used to access the rest of the application. The menu is hidden and can be easily accessed by pressing the menu button at top left of the screen. On the first usage, instructions appear on screen to explain to the user how to access the menu, and can be dismissed by pressing anywhere on the screen, but will continue to appear in subsequent interactions of this screen until the user demonstrates that he has understood them, namely by pressing the menu button, to make the menu appear.

It is also possible for the user to make the menu appear by swiping right on the screen. From the menu the user can access the various other screens of the application and its features.

4.3   Going for a Walk

Of all these screens the one that will be used the most, will certainly be the Routes one. In this screen a user can choose one route from the many made available to him by is healthcare professional. Each route is presented with a name, a route length and an image of the route itself for easy identification.

Choosing one of the routes will send the user to a new screen show on Fig. 3 with the route chosen presented on a Google Maps fragment. In this map the path that the user actually walks it is also drawn, so that the user can see his progress, and as he walks the map centers itself on the user's position. While the user must follow the path, he is allowed a bit of leeway, that is by default set at twenty five metres but which the user can adjust to his/her preferences, from the path in either direction. However if he strays too far a short audio alarm will play every five meters he walks, until he returns to the path; this allows a user to stop for whatever reason without being disturbed by the alarm.

This screen also displays a timer to show how long the user's been walking, the distance travelled, the user's heart rate and blood oxygen level and a button that allows the user to start and stop the walk. If the heart rate goes over or under a limit that is defined by the user's healthcare professional, then an alarm will start and play continuously until the heart rate goes back to normal levels. When the user touches the button to stop the walk he is sent to a new screen where he will be able to send feedback about the walk he was just in.

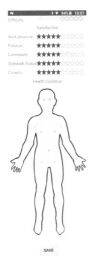

Fig. 4. Feedback screen   **Fig. 5.** Heart rate screen

4.4 Your Opinion Matters

The feedback that can be sent ranges from how difficulty the user had walking the route, to the conditions of the route itself – pollution, number of people around, if it was well illuminated or not, etc. –, to the pain levels of the user on various body parts as shown in Fig. 4. For this purpose rating bars are used, where the user can slide is finger left or right over the bar to decrease or increase the value respectively. The bar changes color from green, on the lower values, to yellow and then red as the value increases, this serves as a visual indicator for the user. For the pain values of the user, it was decided, as can be seen in Fig. 4, to use a picture of the human body with the different body parts well defined that the user can press. When the user is finished he will be returned do the main screen, where they can start another walk, exit the application or access one of the other features of the app.

4.5 Data Gathering

While the application is capable of gathering data automatically, if for any reason it is not possible to gather that data or the user's healthcare professional asks for it, the user can enter various type of data manually. The user can manually enter heart rates measured externally to the app, as can be seen in Fig. 5, and how much time he has spent walking or not. He can also indicate what medication he has been taking, as well as provide a justification of why he failed to complete a scheduled activity.

This application makes use of the Google Fit API[1] to collect the various types of health data necessary, and was chosen due to its ease of implementation and the ease of connecting sensors using the GATT protocol for BLE. It also uses Google's Awareness API[2] to register if the user is walking or not, and was chosen because of its very efficient management of the device resources while still providing accurate data.

5   Discussion

The mobile application exists inside the context of the larger SmartWalk project and its development is framed and influenced by it. The application cannot do its work by itself and must work with the web platform, which provides it with essential services. This begs the question of how the communication between the two occurs.

The project was designed to make use of the infrastructure provided by a Smart City so that the collected data, that is important and confidential, can be safely and securely transmitted. But all this is important, because it is created with the purpose of aiding people, namely: the elderly user of the application, which will gain an improvement to its health; the health professional, which will gain a better understanding of its patient routines and thus can better care for them; and the city itself, which can use the data collected to better plan their infrastructure by, for example, better knowing where people concentrate or if the public spaces in the city are adequate or not for its citizens needs [19].

To test how this whole system works the project will be deployed in a real life environment. The location chosen was the city of Águeda[3] in Portugal, because it is: an internationally recognized good example of a Smart City; and is part of a region of the country where the elderly population is increasing [18]. There is no clear definition of what a Smart City is, but the universal agreement is that Information and Communication Technologies (ICT) are the driver and enabler of city smartness. Based on this researchers found a minimum set of features that a city needs to fulfill to claim to be smart. It is broadly accepted that for a city to be considered smart it must: (1) have a smart agenda; (2) make available open data about itself; (3) provide services or software applications that can be used from a mobile device; and (4) have some sort of smart infrastructure. In the European Union the smartness of a city can be measured according to six dimensions: Smart Economy; Smart Environment; Smart Governance; Smart Living; Smart Mobility; and Smart People [18].

Smart Living which refers to the quality of life in the city is where the Smart-Walk project fits in, and that is why it is important that the city chosen for the test has an increasing elderly population, because it gives a wide selection of people with different living environments and conditions, and how those people change when using SmartWalk. That is why the study personas[4] and scenarios

---

[1] Data gathering platform for fitness and wellness data.

[2] Allows an application to react intelligently to the user's situation.

[3] A municipality in the centre of Portugal with around 50 thousand inhabitants.

[4] An archetype of a potential user.

were developed based on census data that characterize the population of Águeda. Using them the functional requirements were identified, and using UML methodology the actors and use cases for the project created, forming the basis for how this project and the application in particular were created [19].

6 Conclusions

The SmartWalk project's purpose is to provide a way for elderly people to be able to maintain a healthy lifestyle that is supervised by health professionals while minimizing the time, resources and manpower needed. It does this through the use of an app where the elderly user can walk prescribed routes and give feedback, and a health professional can supervise the user and make changes to the prescribed regimen to better optimize it.

The application collects several different types of data from the user, making use in particular of the Google Fit API to collect fitness and wellness data.

In the near future we will put the project into actual live trials with a group of elderly people. Since one of the features of the application is to connect with external sensors through BLE, one of the future steps for the project is the creation of a pulse oximeter that can be used with the application. While it will be a simpler device than the more advanced smart bands or smart watches available, it will also be significantly cheaper, meaning that a broader range of people can have access to those functionalities.

To conclude, it is important to note that while this project is a step in the right direction, it is but a small fraction of the enormous potential that information and mobile technologies can have on this field.

Acknowledgements. Research funded by CENTRO-01-0145-FEDER-024293. Project Smart-Walk: Smart Cities for Active Seniors held by ESTGA/ESSUA/ISCA - University of Aveiro, by ESTG - Polytechnic Institute of Leiria, by ESE - Polytechnic Institute of Coimbra, by Águeda Municipality, and by Globaltronic S.A.

This work is financed by national funds through the FCT - Foundation for Science and Technology, I.P., in the scope of the project UID/CEC/04524/2019.

References

1. United Nations. World Population Ageing 2017. Department of Economic and Social Affairs, Population Division, (ST/ESA/SER.A/408) (2017)
2. Kontis, V., Bennett, J.E., Mathers, C.D., Li, G., Foreman, K., Ezzati, M.: Future life expectancy in 35 industrialised countries: projections with a Bayesian model ensemble. Lancet **389**(10076), 1323–1335 (2017)
3. Boontarig, W., Chutimaskul, W., Chongsuphajaisiddhi, V., Papasratorn, B.: Factors influencing the Thai elderly intention to use smartphone for e-health services. In: 2012 IEEE Symposium on Humanities, Science and Engineering Research, pp. 479–483, June 2012
4. Higgins, J.P.: Smartphone applications for patients' health and fitness. Am. J. Med. **129**(1), 11–19 (2016)

5. Play Store. Zombies, run! (free), October 2018. https://play.google.com/store/apps/details?id=com.sixtostart.zombiesrunclient

6. Play Store. Runtastic running app & mile tracker, October 2018. https://play.google.com/store/apps/details?id=com.runtastic.android

7. Play Store. Map my fitness workout trainer, October 2018. https://play.google.com/store/apps/details?id=com.mapmyfitness.android2

8. Play Store. Fitocracy workout fitness log, October 2018. https://play.google.com/store/apps/details?id=com.fitocracy.app

9. Play Store. Run with map my run, October 2018. https://play.google.com/store/apps/details?id=com.mapmyrun.android2

10. Play Store. Medicine alert, October 2018. https://play.google.com/store/apps/details?id=com.fuzzide.medicine

11. Play Store. Blood pressure companion, October 2018. https://play.google.com/store/apps/details?id=de.medando.bloodpressurecompanion

12. Healthy.io Site. Healthy.io, October 2018. https://healthy.io/product/

13. Bourouis, A., Feham, M., Bouchachia, A.: Ubiquitous mobile health monitoring system for elderly (UMHMSE). CoRR, abs/1107.3695 (2011)

14. Kakria, P., Tripathi, N., Kitipawong, P.: A real-time health monitoring system for remote cardiac patients using smartphone and wearable sensors. Int. J. Telemed. Appl. **2015**, 11 (2015)

15. Milosevic, M., Shrove, M.T., Jovanov, E.: Applications of smartphones for ubiquitous health monitoring and wellbeing management. J. Inf. Technol. Appl. **1**, 7–15 (2011)

16. Boulos, M.N.K., Wheeler, S., Tavares, C., Jones, R.: How smartphones are changing the face of mobile and participatory healthcare: an overview, with example from ecaalyx. Biomed. Eng. OnLine **10**(1), 24 (2011)

17. Arab, F., Malik, Y., Abdulrazak, B.: Evaluation of phonage: an adapted smartphone interface for elderly people. In: Kotzé, P., Marsden, G., Lindgaard, G., Wesson, J., Winckler, M. (eds.) Human-Computer Interaction – INTERACT 2013, pp. 547–554. Springer, Heidelberg (2013)

18. Rodrigues, M., Santos, R., Queirós, A., Silva, A.G., Amaral, J., Gonçalves, L.J., Pereira, A., da Rocha, N.P.: Meet smartwalk, smart cities for active seniors. In: TISHW 2018 - The 2nd International Conference on Technology and Innovation in Sports, Health and Wellbeing, June 2018

19. Queirós, A., Silva, A.G., Simões, P., Santos, C., Martins, C., da Rocha, N.P., Rodrigues, M.: Smartwalk: personas and scenarios definition and functional requirements. In: TISHW 2018 - The 2nd International Conference on Technology and Innovation in Sports, Health and Wellbeing, June 2018

A Usability Analysis of a Serious Game for Teaching Stock Market Concepts in Secondary Schools

B. Amaro[1], E. Mira[1], L. Dominguez[1,2], and J. P. D'Amato[1,2(✉)]

[1] Univ. Nac. Del Centro, UNICEN, Tandil, Argentina
juan.damato@gmail.com
[2] National Scientific and Technical Research Council, CONICET,
Buenos Aires, Argentina

Abstract. Games have been a useful tool in education since earliest times. More recently, computer video games have become widely used, particularly in secondary education, to impart important knowledge in an attractive way. Academics have proposed a number of approaches, using games-based learning, to impart theoretical and applied knowledge, especially in different disciplines. Our research is concerned with the design of an innovative educational game platform focused on the development of business and management skills. To reach this purpose, the idea is to study how well designed is the user-interface and how students interact with the applications, in order to be more effective. In this case, we propose a new in-depth usability analysis pipeline, combining a well-documented methodology based on navigation metrics and surveys using free WEB libraries. Feedback coming from this group are necessary to improve the software, and to provide structured empirical evidence for supporting our approach.

Keywords: Usability metrics · Serious-game · WEB development

1 Introduction

According to ISO/DIS 9241-11 [1], usability could be defined as "The extent to which a product can be used by specific users, to achieve specific goals with effectiveness, effectiveness and satisfaction in a context of specific use". Within software standards, [2] defines usability as an "attribute of quality and is the ability of a product to be easily used". From the point of view of Nielsen [3], a pioneer in the diffusion of usability, it is defined in terms of five attributes: learning, efficiency, memorization, error prevention and subjective satisfaction. In the Nielsen model, usability is "Part of the utility of the system, which is part of the practical acceptability and, finally, part of the acceptability of the system". In this context, the usability of education apps is taking an increasingly important role, especially those developed with WEB technologies [5]. As the target audience is formed mostly by young people and children, several concerns should be fulfilled in order to capture their attention. In these areas, the use of video games has been increasingly adopted, mainly due to the number of interactions and situations that can be recreated. These games demand simplicity and attraction, with ease of reading,

© Springer Nature Switzerland AG 2019
Á. Rocha et al. (Eds.): WorldCIST'19 2019, AISC 931, pp. 839–849, 2019.
https://doi.org/10.1007/978-3-030-16184-2_80

loading speed and even having a design adaptable to different devices at the same time that they are used for teaching, that is why they are called "serious games", concept also related to simulators.

The stock market behavior has been learned in secondary schools in Argentina for several years. The way with which it was played was using forms and spreadsheets, and all the work to collect student's data was manual. Even though there are many free-simulators to accomplish a similar purpose [4], they only let one player each time and they are extremely full of information. Also, with these kind of simulators, the teacher could not conduct its class, and evaluate other skills as data analysis, research and team work. In these situation, the school authorities ask this group to developed an own Web game, with three main requirements: that could be customized for different age groups, that students could play in groups and that were very easy to use. With this mechanism, students were expected to learn concepts as: currency management, buy and sell operations, money management, charts readability, price evolution, among others. In [13], authors show how important is usability in education software, but they do not have a structured method for measuring it, as they evaluate third-parties' applications.

In this paper, we present the development and the usability analysis of a serious game used in teaching stock market behavior concepts in a secondary school. To reach the "ease of use requirement" a deep usability analysis was carried out during the development, using logging tools as well as the use of surveys. During this process, important elements were detected and corrected that makes the practical and efficient use of these kind of game and that could be used for improving the APP.

The work is distributed as follows. In the following section we detail antecedents related to the theme of usability. In Sect. 3, we present the structure of the game developed and the context of application. In Sect. 4, we introduce the methodologies followed for the usability analysis. Finally, we present the most relevant results and conclusions of the process followed.

2   State of the Art

The evaluation of *usability* is a process to produce a measure of *ease of use*. In the evaluation, there is an object that is being evaluated and a process through which one or more attributes are judged or given a value, as determines Obeso in [6]. Authors in [7] say that a key factor for measuring usability is the identification of the *implicit* and *explicit* characteristics of the user interface of a computer application, since the success or failure of a product of this type, is highly linked to the satisfaction and comfort perceived by the end user.

2.1   Usability Measurement

The evaluation of usability is one of the most important tasks that must be undertaken when developing a user interface. Poor interfaces can drive away potential clients or in the educational environment lead to failure of an apprentice. This evaluation, as explained in [6, 8] is an empirical study with real users of the proposed system, with the purpose of providing feedback in software development during the life cycle of iterative development. The usability evaluation allows the validation of all requirements, to make it as useful as possible and thus increase product quality and user satisfaction. On the other hand, the users and tasks should be studied in depth. Since a user-based evaluation can study only a subset of all the possible tasks that a system can support, the evaluation should take the most representative tasks, chosen for their frequency or criticality. The users' preferences and experience are also critical. It is essential that it can be evaluated by a group of users and not by the developers themselves.

In resume, there are different methods to establish if usability is "good enough" that can be used during product development. In Table 1, a resume of most known usability evaluation methods is presented.

Table 1. Usability methods and their purposes.

Method	Purpose	Detail
Inspection	Set of methods based on having evaluators that examine the principles related to usability based on the experience and knowledge	**Features:** check if features of a product meet the requirements **Consistency:** identify the degree of consistency between components and design **Standard:** an expert verifies that UI is in accordance with the standards
Heuristic [3]	It allows to test interfaces in a fast and economic way. Very used in the software industry	Each heuristic has one or more of the following elements: • Compliance questions • Evidence of conformity • Motivation • Evaluators: (3 or 5)
Cognitive walkthrough [14]	Evaluates the learning facility through prototypes in the initial stages of development	It is formed by three components: • The problem-solving • The learning • The execution one
Groups inquiries	They allow you to discover and learn to generate design ideas	**Oriented groups:** the moderator is fundamental and its procedure is decisive for the success of the session **Discussion groups:** a moderator leads and proposes the topics to be discussed
Individual inquiries	Formulation of effective questions through different methods	• Interactive surveys • Questionaries' • Interviews • Participatory inquiry
Empirical methods	Users run a prototype in operation, at the design stage or with the system in use, and evaluate it	**Test** a more or less interface to verify if the usability goals have been achieved **Formal evaluation** of a system that is still being designed

From this study, it could be highlighted that the different evaluation methods have strengths and weaknesses and they are focused on evaluating specific aspects of usability. In a whole educational application, it is desirable to combine them in an evaluation to complement each other, as proposed in [9]. In the present work, the selected methods are: *Inspection of verification guides, Individual inquiry* and *Empirical methods*.

2.2 Features to Be Evaluated

The evaluation criteria considered are based on the ISO 914 [10] and ISO 9126 [2] standards, and the features to evaluate from a WEB system are:

- **Operability:** assess whether the user can operate and navigate the site.
- **Attractiveness:** evaluate the aesthetic and visual aspect of the pages.
- **Satisfaction:** subjective evaluation of the comfort of use, familiarity, etc.
- **Learning:** measure the time it takes for users to learn how to use site-specific functions, the ease with which they do so, and the effectiveness of documentation.
- **Communication:** the possibilities that the site offers to the user.

These five criteria determine high-level usability characteristics cannot be measured directly, so they need to be decomposed into metrics and attributes that can. The metric is a function of two types of arguments: *attribution* and *measure* of the attribute.

$$\text{Metric} = f(\text{attribute } j, \text{measure } m), \ i = 1, \dots, m; \ j = 1, \dots, n$$

Where m is the numeric value of an attribute i according to a defined table. The attribute must be identified and measured qualitatively or preferably quantitatively. A list of potential measurable attributes for WEB applications is presented in Table 2.

Table 2. A list of potential attributes and metrics.

Metrics	Attributes
Effectiveness	• Completeness of the task • Length of the sequence
Easiness of learning	• Predictive, • Simple, • Familiar, • Coherent
Help	• Easy to read, Useful • Sensitive context * Documentation
Ease of use	• Ease of sending and receiving • Identifiable options • Selection of parameters • Simple and clear language • Fixed located information
Navigation	• Scrolling between site pages • Friendly navigation • Links and labeling
Extensions	• In-line interface functions • Clear explanation of entry and exit • Easy to understand the sequence • Simple and short language
Error messages	• Self-exploding error messages • Minimize recovery times • Facilitates correction • Detection and warning of input errors
Interface Esthetics	• Personalization • Aesthetically pleasing introduction • Consistent presentation • Combination of sufficient texts and graphics
Trustability	• Complete objectives comfortably and safely • Confidence
Acceptability	• Actual information • Functions

3   Serious Game Principles and Architecture

In order to use the stock market game, the teacher has to be able to organize content according on the level of the course. In general, the teachers consider two students groups based on age, from 12 to 14 years old and the older ones. This criterion is due to the fact that, initially (entering the first year), students are induced in the most important concepts of stock market behavior and stock trading. To achieve the maturation of these concepts, they must be given time to feel comfortable with the domain (first 2 years). Once that is achieved, they begin to add greater difficulties, since they already have tools with which to face them (more than 14 years). The difficulty is defined by amount of companies to operate, the quantity and types of commodities and the amount of different currencies (Dolar, Euro, Peso).

The students who wanted to play, are registered online, on the school's website. The registration could be individual or by group. Once the registration was made, each player was assigned, an initial amount of money. In the case of a group, the initial capital was defined proportional to the number of students in it.

3.1   Development Considerations

As a first step, the requirements were captured, using an initial document provided by the school, and then complementing it by specifying new requirements through the research of similar applications available on the Web, such as La Bolsa Virtual [4]. These tools were evaluated with the teacher, in order to select the main functionality. As additional information, interviews with the appropriate personnel were used, always taking into account the different types of users that would use the game (Fig. 1).

Fig. 1. (left) An existent WEB game (right) the simplistic proposed game.

Once the main functionalities were defined, the process of prototyping the graphical interface of the system began, which was iterated and incremental. First, working on the main page of the game. A demo was made to the users, and based on the received feedback, adjustments were made until reaching a final version of the prototype. In the same way, the rest of the pages were designed, completing all the pages involved in the WEB application. A distributed WEB architecture was proposed and implemented using *PHP* in the backend and *Angular* in the front-end. Stock market data was taken from a third-party open API. Every five minutes, data is collected and stored in a local database, that is later accessed from the application in a local network. This approach was necessary as Internet bandwidth was low.

UI Principles. Thinking that the main users will be high school students, with an age of between 12 and 17 years, some principles of interface design are followed:

- More images than text
- Simplicity
- Few options
- Short clear messages
- Smartphones ready

Age Groups Organization. The groups have assigned the following groups:

- Initial Category (From 12 to 14 years old):
  - Commodities: corn, sunflower, wheat and soybeans.
  - National stock: Bco. French, Mirgor, Molinos Río de La Plata, YPF.
- Advanced Category (above 15 years old):
  - Commodities: corn, sunflower, wheat and soybeans.
  - National stocks: Arcor, Molinos Río de La Plata, YPF.
  - Currency: dollar, peso.
  - Foreign Stocks: Crocs, Motorola, Netflix and Walt Disney (Fig. 2).

Fig. 2. A screenshot of how stock options are presented.

3.2   Tools for Measuring Usability

As described in Sect. 2.1, three mechanisms were combined. First, for the WEB application used data collection, 2 (two) free Web tools, a survey and usability guides were used. Web tools helps to obtain user behavior data automatically, such us clicks, navigations between pages, scrolls and time, among others. The WEB tools used were Heap Analytics and Hotjar. Combining both libraries, several quantitative metrics were obtained.

Secondly, to determine the user's perception of the application, it was decided to conduct a brief survey of the students who had access to the game. Mainly the idea was to ask questions related to specific functionalities, in order to determine the difficulty of using them, and in which cases using them was complicated or confusing. And finally, the usability guides were used based on the data collected by the aforementioned methods. It was decided to apply several methods, with the aim that, being complementary to each other, they allow covering the greatest number of ways in which data collection can be performed and in turn obtain different types of information as a result.

Heap Analytics [11]: This is a WEB library offers several features among which are:

- *Data Capture:* automatically captures every user interaction; clicks, touches, gestures, forms presentations, page visits, etc.
- *Users management*: Allows you to see each of the users, together with each action they have taken in real time.
- Funnels: it allows to identify in what place or time the user leaves the application, to find out what is the reason for which users are lost.

Hotjar [12]: This is WEB library offers these features:

- Activity recordings: with this functionality it is possible to see a recording of each user in his session.
- Analysis of WEB forms: performs an analysis of the use of web forms, so it is possible to detect which are not completed or which takes more time to finalize them.
- Heat maps: can be observed in a visual way where users make more clicks and where they move, presented through colors.

4   Results

The present development was used in a secondary school in the city of Tres Arroyos, Argentina in two courses, one for students of 13 years and another for students of 16 years. Each course has about 25 students. This system was used for three (3) months, in TICs teaching courses, using surveys, analysis tools and usability measurement. After the data was collected from the different methods, we proceeded to process data in order to establish the level of usability of the application. It was obtained that students make about 12 operations per day; which is a high valued.

In the following section the usability methodology is applied, using the features list presented in Sect. 2.2 and a final combination is proposed.

4.1   Users Evaluation

Based on the parameters established in Sect. 2.1 taking into account that the usability requirements vary depending on the audience level (children, youth, adult and elderly) it is necessary to establish a relative preference or weight to each of the parameters that make up the requirements tree. This weight will determine the importance of each parameter in meeting the usability requirements in the established hierarchy.

Table 3. Criteria's and their according weight.

Criteria	Weight
Learning	0,20
Aesthetics	0,25
Satisfaction	0,20
Communication	0,10
Operations	0,25

All assigned weights were decided by the authors of the present work, as presented in Table 3. Since it was a game with educational purposes, it was considered that the criteria of learning, attractiveness, satisfaction and operability should have similar importance.

Then, the assignment of importance was made for each attribute to be evaluated, and it is proposed to establish the measure of compliance of each one. For this, all the data obtained from the surveys and Web tools are considered. From this, the degree of compliance will be determined by assigning a value to it, these will be determined from 0 when it is **Unfulfilled**, to 1 **Completely fulfilled.**

Once the scale of values was established, the degree of compliance for each attribute was determined. Then, the calculations will be made to determine the overall usability achieved in the application, respect to data generated by the users, all normalized. To obtain the value of each criterion, it was combined with the weight of each metrics.

- Learning **(effectiveness, ease, help, documentation)** $= 0{,}20 \times (0{,}15 \times 0{,}65 + 0{,}4 \times 0{,}8125 + 0{,}25 \times 0 + 0{,}2 \times 0) = \mathbf{0{,}083}$
- **Aesthetics** (attractiveness, customization) $= 0{,}25 \times (0{,}80 \times 0{,}8625 + 1 \times 0{,}2) = \mathbf{0{,}2225}$
- **Satisfaction** (trustability, success information) $= 0{,}20 \times (0{,}625 \times 0{,}65 + 0{,}7 \times 0{,}35) = \mathbf{0{,}13025}$
- **Communication** (Message format, clarity) $= 0{,}35 \times 0{,}5 + 0{,}15 \times 0{,}5 + 0{,}25 \times 1 + 0{,}25 \times 0{,}75 = 0{,}6875 = \mathbf{0{,}06875}$
- **Operations** (Ease of use, navigation, error tolerance, accessibility) $= 0{,}25 \times (0{,}6625 \times 0{,}30 + 1 \times 0{,}25 + 0{,}4625 \times 0{,}20 + 0{,}10 \times 0{,}8 + 0{,}15 \times 0{,}75) = \mathbf{0{,}1865}$

All these metrics are combined to generate the **usability users measure**.

User   usability   measure $= 0{,}083 + 0{,}2225 + 0{,}13025 + 0{,}06875 + 0{,}1865 = \mathbf{0{,}691}$.

4.2 Experts Evaluation

As a second step, we proceeded to obtain the measurement of each of sections of the WEB App by Experts. Evaluating the different modules, each one was exploded in individual items considering the proposed in Table 2, and grouped in the same way as user (*Learning, Aesthetics, Satisfaction, Communication, Operations*) but now including *Management* (that was an item describing how easy is to administrate the site). Each item has an individual weighted and checked if it was fulfilled and then, for each category, they were sum up.

Table 4. Criteria's weights for experts

Criteria	Weight
Management	0,10
Satisfaction	0,15
Learning	0,25
Aesthetics	0,20
Operations	0,20
Communication	0,10

The weights were a bit different, having more importance in learning, as shown in Table 4. Three experts, a technical director, a software designer and a professor, applied the method.

The expert's usability measure was the following.

Expert usability measure = 0,1 × 0,67 + **0,15** × 0,75 + **0,25** × 0,837 + **0,20** × 0,89 + **0,20** × 0,64 + **0,10** × 1 = 0,067 + 0,1125 + 0,20925 + 0,178 + 0,128 + 0,10 = **0,7947**.

4.3 Composed Usability and Findings

After obtaining both usability values, from experts and user, the combined usability was obtained. It was defined as an average of both values, that is, it was considered that the measurement made by the experts and that made by the users have the same weight within the final usability.

Composed usability $= 0,50 \times 0,7947 + 0,50 \times 0,691 = \mathbf{0,7428}$

In general, a good web Site should meet a (normalized) 65% of evaluation, as propose [14], so the results were, by far, satisfying. At the same time, with this methodology, several issues were detected and corrected. For instance, one of the more important ones, considering to work with child, were notifications. The web APP lacked of confirmation messages, that is why the "Learning and communication" has poor values.

In order to increase these criteria, a newer version was deployed, including several notifications and increasing readability. Some sections in the page that were never visited were removed. This simplified the navigation in smart-phones, and the pages loaded faster.

5 Conclusions

In this work, the implementation and usability evaluation of a stock market game, using WEB and multi-device technologies was described. Since such a game would be used in a secondary school, special attention was given to generating attractive and understandable user interfaces. It was found that the use of technology in the course had a positive impact, having a lot of child playing during the lapse. This was to be expected, since the change that was made in the game mode was radical. On the one hand, because the interaction with a Web application is more motivating, intuitive and simple. In addition, as the game could be played in different platforms, and even out of the schools, make the participation of students more frequent.

At the same time, thanks to the methodology, it was possible to detect various errors/misunderstands in the User Interface, which first reduced the usability of the site. These findings were very important, especially when designing and developing kids-oriented applications. It is very important to have guides and tools that improved the user's experience when using the site. Several of the errors could have been avoided if the usability evaluation had been applied from the beginning, following guidelines or established standards.

With the analysis of the data obtained from users and experts, it was possible to determine that the usability achieved in the application developed and analyzed is at a good level. Although the value exceeds the acceptable level, it is important to continue working to increase it. There many extra data that could be included in future works.

Acknowledgments. The project has been partially financed by the Escuela Agropecuaria de Tres Arroyos (EATA) and National Scientific and Technical Research Council (CONICET).

References

1. ISO CD 9241-11: Guidelines for specifying and measuring usability (1993)
2. ISO 9126: Software product evaluation - quality characteristics and guidelines for their use (1991)
3. Nielsen, J.: Browser and Gui Chrome (2012). http://useit.com/alertbox/ui-chrome.html
4. La Bolsa Virtual. http://www.labolsavirtual.com/. Accessed 31 Dec 2018
5. Madden, M.: Internet penetration and impact. Report Internet Evolution, Pew Internet and American Life Project (2006)
6. Obeso, M.E.: Metodología de Medición y Evaluación de la Usabilidad en Sitios Web Educativos. Universidad de Oviedo: Facultad de Informática (2005)
7. What & Why of Usability. https://www.usability.gov/what-and-why/index.html. Accessed 31 Dec 2018
8. Covella, G.J.: Medición y Evaluación de Calidad en Uso de Aplicaciones Web, Grupo de Inv. y Desarrollo en Ingeniería de Software, UNLPam (2005)
9. Sweetser, P., Wyeth, P.: GameFlow: a model for evaluating player enjoyment in games (2005)
10. ISO 9241-11: Ergonomic requirements for office work with visual display terminals (VDTs) - Part 11: Guidance on usability (1998). www.iso.org/standard/16883.html

11. Heap Analytics: Plataforma de registro de actividad del usuario. www.heapanalytics.com. Accessed 31 Dec 2018
12. Hotjar: Plataforma de registro de actividad del usuario. www.hotjar.com. Accessed 31 Dec 2018
13. Crescenzi-Lanna, L., Grané-Oró, M.: Análisis del diseño interactivo de las mejores apps educativas para niños de cero a ocho años. Revista Comunicar (46) (2016)
14. Neil, T.: Mobile Design Pattern Gallery. O'Riley Media, Sebastopol (2012)

Ethics, Computers and Security

Macro and Micro Level Classification of Social Media Private Data

Paul Manuel[(✉)]

Department of Information Science, Kuwait University, Kuwait City, Kuwait
pauldmanuel@gmail.com

Abstract. In the modern world, social media private data evolves as a great asset to business and governments. While social media private data is a boon to the business, it is also causing concern to privacy regulators. We classify the social media data as the business and governments require. The social media private data is classified into two layers: macro level and micro level. The macro level classification is Static Private Data and Dynamic Private Data. The micro level classification includes four types: Personal Identity Data (Static), Relational Identity Data (Static), Personal Identity Data (Dynamic), and Relational Identity Data (Dynamic). Two software metrics "complexity" and "relevancy" are considered. Based on the macro and micro level classification, we measure the complexity and relevancy of social media private data from the perspectives of business and police communities. By conducting extensive experimental research, we study the relationship between different types of social media private data and different communities by the means of the two-metrics relevancy and complexity and justify the necessity of macro and micro level classification. The outcome of the experimental survey is interesting. Police officers are more interested in static private data than dynamic private data. Business managers are more interested in dynamic private data than static private data. While the police are interested in static private data, the business communities are interested in dynamic private data.

Keywords: Social media private data · Classification · Static private data · Dynamic Private Data · Personal Identity Data · Relational Identity Data

1 Introduction

The impact of social media was realized by the whole world when the jasmine revolution spread across the entire middle east countries. The famous quote was "Mohamed Bouazizi set fire on himself. Social Media spread the fire across the continent" [7]. Social media has revolutionized the social life across the globe. In the modern world, data privacy of social media has evolved as a great asset to every sector of the societies. There are two major clients who very much depend on the private data of social media users. One is business communities and the other one is police communities. The private data gathered from social media is huge and is classified under bigdata [3, 19]. A survey [3] shows that not all the social media data is relevant and useful. Different types of communities are interested in different types of private data. This is addressed

© Springer Nature Switzerland AG 2019
Á. Rocha et al. (Eds.): WorldCIST'19 2019, AISC 931, pp. 853–866, 2019.
https://doi.org/10.1007/978-3-030-16184-2_81

in the literature of social media intelligence [16, 18]. So, it is important to classify the private data of social media into different types based on the requirement of different social communities. This is the main objective of this research.

In this paper, we classify the social media private data into two levels: macro level and micro level. The macro level classification is Static Private Data and Dynamic Private Data. The micro level classification includes four types: Personal Identity Data (Static), Relational Identity Data (Static), Personal Identity Data (Dynamic), and Relational Identity Data (Dynamic). We consider two software metrics such as relevancy and complexity. Our research focuses on some basic questions such as "Which type of private data is important? Who is interested? To what extent are they interested?". As a part of the research, four types of social media private data were collected from pizza consumers and social activists. Two users, who are pizza restaurant managers and police officers, evaluated and measured the relevancy and complexity of the collected data. The outcome is interesting. Police officers are more interested in static private data than dynamic private data. Business managers are more interested in dynamic private data than static private data. While the police are interested in static private data, the business communities are interested in dynamic private data. By conducting extensive experimental research, we justify the necessity of the macro and micro level classification of social media private data by showing that different communities are interested in different types of social media private data.

2   Review of Literatures

In the analysis of social media private data, there are two stages:

1. Classification of social media private data - classifying the data the way the end users want
2. Application of social media intelligence system - collecting those types of data and delivering it to the end users.

There are several kinds of social media private data classifications in the literatures. Lange [11] classifies social media data as publicly private and privately public. Aggarwal [1] divides social media data into content data and linkage data. Manovich [14] partitions the social media data into deep data and surface data. Flitter [5] breaks the private data into 5 segments: Performance data, Conversation data, Industry or topic data, Traffic and link data and Conversion or results data. The data classification has been studied by Carnegie Melon University in the context of information security based on its level of sensitivity and the impact to the University [15]. A popular classification of social media private data is due to Schneier [23]. Schneier organizes the social media data as Service Data, Disclosed Data, Entrusted data, Incidental data, Behavioral data and Derived Data etc. It is further refined by Richthammer et al. [21]. Richthammer et al. [21] have given a good literature survey on the classification of social media data. Smith [24] organizes the social media data into 10 types which are Facebook's interest/social graph, Google+'s relevance graph, LinkedIn's talent graph, Twitter's news graph etc. Patel [20] arranges social media content into six categories:

Infographics, Interactive content, Content that evokes strong positive emotion, Content with images, List posts and newsworthy content.

The social media intelligence system [16, 18] identifies, collects and filters the required data from the huge volumes of social media contents. Mainly, it reads billions of social conversations to gather the required data and information by intrusive and non-intrusive channels. Several software tools available in the market [3, 12] can effectively listen in real-time to what is said across a range of sites including LinkedIn, Facebook, Google+, Twitter, Pinterest, blogs, and YouTube. It was originally introduced by Omand et al. [18] to assist only British security community. Later it was widely adopted by business communities. Business organizations apply social media intelligence system to collect customer data from social media and to use the data in their strategic plan and decision-making process. It is considered as very valuable information system for business communities because it brings down not only the data relevant to their organization but also the data relevant to their competitors. Like artificial intelligence system, it is a concept and it is not a product. It is used under different names such as social media analytics, social media monitoring system, social intelligence platform etc. Social media intelligence system uses *Learning Algorithm* (Machine learning algorithms) [3, 10, 12, 13] that extracts meaningful data from social media. As a search algorithm makes the difference between Google and other search engines, social media intelligence system makes the difference between Facebook and other social media sites. Facebook Inc. reached $27 billion in revenue in 2016. While Facebook Inc. posted more than $1 billion in quarterly net income of 2016, Twitter made significant loss. Even though Facebook and Twitter have almost the number of users, how Facebook is making profit and how Twitter is making loss in the same financial year [17]. As a search algorithm differentiates Google from its competitors, a social media intelligence system differentiates Facebook from its competitors.

3   Classification of Social Media Data

The social media private data is divided into two layers: macro level and micro level. See Fig. 1.

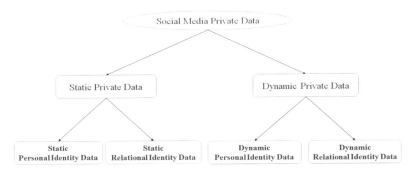

Fig. 1. Classification of social media private data

The macro level classification of social media private data is:

1. Static Private Data (remains stationary or fixed)
2. Dynamic Private Data (subject to change)

There are several definitions in the literatures. The simple one is due to Techopedia. According to Techopedia, static data is the data that does not change after being recorded. It is a fixed data set. Dynamic data may change after it is recorded and should be continually updated. Static private data usually contains past and completed activities. On other hand, dynamic private data contains future activities and the data subject to changes. Table 1 illustrates the classification by examples.

Table 1. Static vs dynamic

Examples of Static Private Data	Examples of Dynamic Private Data
The date of birth of Peter is 9 February 1957	Peter is 15 years old
Peter got his first mobile as a birthday gift	Peter is planning to buy a new iPhone
Peter was born in New York	Peter is planning to apply for home loan
Peter is a Math teacher	Peter is going to apply for Retirement Insurance Scheme
Peter got driving license at the age of 20	Peter is thinking to buy a car

Social Media Private Data [26] is further classified as follows:

1. Personal Identity Data
2. Relational Identity Data

Personal identity is about you and your family. Personal Identity Data consists of your personal data [8]. It also covers data related to your family including the parents and children. *Relational identity* is about you and your social media neighbors. Anyone who is in your social network group is your social media neighbor. In addition, friends of your friends who are in the same social network group are also your social media neighbors. Relational Identity Data consists of data that is related to you and your social media neighbors. Relational Identity differentiates you from others and consists of data how you are distinguished from others. Personal Identity Data is standalone data only about you and your identity. On other hand, Relational Identity Data is the data that contrasts and compares you with your social media neighbors.

The micro level classification of the social media private data is as follows:

1. Personal Identity Data (Static)
2. Relational Identity Data (Static)
3. Personal Identity Data (Dynamic)
4. Relational Identity Data (Dynamic)

Tables 2 and 3 illustrates the classification by examples.

Table 2. Personal Identity Data (Static) vs Relational Identity Data (Static)

Examples of Personal Identity Data (Static)	Examples of Relational Identity Data (Static)
Peter is a civil engineer	Among all his friends, Peter is only civil engineer and others are mechanical and electrical engineers
The mother tongue of Peter is French	All Peter's friends speak French as mother tongue
Peter was a Junior Tennis champion in 2016	None of Peter's friends won any championship at junior level
Peter's native town is Blacktown and Peter graduated BS from MIT in 2015	Peter is the only one from Blacktown who graduated from MIT
Peter has blonde hair by birth	Peter and 7 students in his class have blonde hair and other have black hair

Table 3. Personal Identity Data (Dynamic) vs Relational Identity Data (Dynamic)

Examples of Personal Identity Data (Dynamic)	Examples of Relational Identity Data (Dynamic)
Peter has iPhone 7	Peter and his three colleagues use iPhone. All others in his workplace use Samsung
Peter likes Pizza	Most of the employees in his workplace eat pizza as lunch
Peter does not have any pension plan	Peter's friends are considering AIG super fund pension plan
Peter is thinking to buy Toyota Camry	Every employee in the company except Peter has a car and only two employees have Toyota Camry
Peter's ambition is to pursue PhD in Stanford	Only a few employees in his workplace have graduate degrees

4   Some Basic Facts of Macro Level Classification

In this section, we analyze some basic facts about the macro level classification Static Private Data and Dynamic Private Data. Interestingly, the Static Private Data and Dynamic Private Data are collected from different sources [12].

The Static Private Data are collected from

1. Profile,
2. Tags
3. Private settings in social media.
4. Personal websites
5. CV published in sites such as LinkedIn, BranchOut, Meetup, VisualCV etc.

The Dynamic Private Data are collected from social communication and personal conversation in social media:

1. Chatting, SMS, MMS, emails
2. Pictures, Audio, and Video which are exchanged in social media
3. Likes/Dislikes/Retweets
4. Comments, Posts, blogs, forum conversation in social media

There are two main communities which are interested in social media data.

1. Government Intelligence Department (Police, Defense) [2, 6]: The Intelligence Departments require social media data
   a. To track down anti-social elements and law offenders.
   b. To monitor the activities of social groups who seek to make changes in the society.
   c. To read the mind of the people who are not happy with the existing system.
   d. To manage crisis situations from the routine (e.g., traffic, weather crises) to the critical (e.g., earthquakes, floods) [9].
2. Business Communities (Apple, Pepsi etc.) [4, 22, 25]: Business communities require these data
   a. To promote sales and marketing,
   b. To monitor customer sentiments
   c. To spy the competitors
   d. To apply the data in the decision-making process
   e. To prepare the strategic plan that includes innovation, new features, new products, future expansion etc.

5 Collecting Social Media Data

As it is mentioned in the previous section, there are two communities who are primarily interested in social media private data. One is business communities and the other is police departments. This research involves two steps: (1) Collection of private data from social media users and (2) Assessment of the collected data by police officers and business managers.

5.1 Business Managers

Fast foods such as pizza and burgers are becoming very popular in Chennai. There are more than 150 pizza restaurants in Chennai, India. Fast food mainly pizza is very popular among the teenagers and college students. This is the reason that the project focusses on pizza restaurants. Geographically Chennai is a vast place. It is difficult to reach different managers in different locations. More than 45 assistant managers and managers of pizza restaurants including Pizza Hut and Domino Pizza in Chennai agreed to cooperate in the research.

5.2   Police Officers

As usual, police officers are difficult to approach everywhere. However, the author managed to convince 27 police officers in Chennai to participate in the research. Most of them are inspectors and some of them are deputy superintendent of police who is higher level in the rank than inspector. During the period of data collection, there was a social protest in Chennai by public (mainly by students) on the issue of "Jallikattu" (Fig. 2).

Fig. 2.   Courtesy to https://en.wikipedia.org/wiki/2017_pro-jallikattu_protests.

Jallikattu is a traditional bull-taming sport in Tamil Nadu, India and it is a Tamil Nadu version of Madrid bullfighting which is a traditional spectacle of Latin American countries. The Jallikattu was banned by the Supreme Court of India and the ban created an uproar among the people. A small group of youths organized the protest in Chennai Marina beach largely through social media. An amazing number of people especially youths and college students participated in the two-week protest (8 January 2017–23 January 2017). This was called the 2017 pro-Jallikattu protest (https://en.wikipedia.org/wiki/2017_pro-jallikattu_protests). The Chennai city was in disarray and the protest posed a big challenge to police to bring law & order under control. It was a burning law & order issue during the period of our research. Thus, it was decided to focus the research on this topic.

5.3   Social Media Intelligence System

Social media intelligence system [3, 12] was used to collect social media data mainly from students. Two kinds of data are gathered: (1) data related to pizza and (2) data related to Jallikattu. Social media intelligence system gathers data based on your input. The quality of the output from the social media intelligence system depends on how the input is well-phrased and well-structured. The social media data was collected from more than 300 students in Chennai, India with their consents.

The social media intelligence system can do three jobs: (1) collecting private data from social media users, (2) filtering the required data from the collected data and (3) assessing the filtered data. The collected data is normally huge and only small percentage of information will be useful to end users (pizza mangers and police officers). In our research, we restrict the use of social media intelligence system only to the collection of data. The second and third jobs (filtering the collected data and assessing the filtered data) are carried out by pizza mangers and police officers. The data collected by social media intelligence system were handed over to pizza mangers and police officers for the second and third job.

6 Processing Social Media Private Data

The collected data by social intelligence system that is related to pizza was filtered and processed by managers of pizza restaurants. The data related to Jallikattu was filtered and processed by Chennai police officers. The data was filtered and processed by 27 police officers and 45 managers of pizza restaurants in Chennai. Two software metrics "Relevancy" and "Complexity" are considered to process the filtered data.

Relevancy – Relevancy means the importance and usefulness of the social media private data to police and business communities. Among different macro and micro types of social data, which type of social media data is more relevant to the police and business communities?

Complexity – The collection of private data from social media is not a challenge. But, the process of filtering the collected data and then assessing the filtered data poses a real challenge. Complexity is the measure of difficulty in the filtration and assessment of relevant data from the collected data. Extracting the right data from the right people is a difficult job. The metric "complexity" measures the difficulty in the extraction of relevant data from the macro and micro types of social media data. Among different macro and micro level types of social media private data, which type of private data is more difficult to extract from the social media?

6.1 Macro Level Classification of Social Media Data

Experiments were carried out for the justification of the macro and micro level classification of social media private data. Tables 4 and 5 represent the social media data related to metric "relevancy" that are assessed by police officers against the rows of macro level types of classification.

$$R_S = \frac{S_L \times 0.3 + S_M \times 0.6 + S_H \times 0.9}{S_L + S_M + S_H} \times 100$$

Table 4. Macro-level classification: assessment form for police officers to assess relevancy

Name of the Police officer:	Rank of the Police Officer:		
Mark ✓ in the appropriate box (Only one ✓ is allowed in one row)			
Type of private data	The level at which this type of private data is relevant to me		
	Low (30%)	Medium (60%)	High (90%)
Static Private data			
Dynamic Private data			

Table 5. Macro-level classification: consolidated table to compute relevancy to police officers

Type of private data	The level at which this type of private data is relevant to me			Relevancy
	Low (30%)	Medium (60%)	High (90%)	
Static Private data (S)	S_L	S_M	S_H	R_S
Dynamic Private data (S)	D_L	D_M	D_H	D_Y

Here is the description of the above formula:

1. S_L denotes the number of police officers who rate that the importance of Static Private Data to them is "LOW".
2. S_M denotes the number of police officers who rate that the importance of Static Private Data to them is "MEDIUM".
3. S_H denotes the number of police officers who rate that the importance of Static Private Data to them is "HIGH".
4. R_S denotes the relevancy factor.

6.2   Micro Level Classification of Social Media Data

Tables 6 and 7 represents the social media private data related to the metric "relevancy" that are assessed by police officers against the rows of micro level types of classification.

Table 6. Assessment by for police officers to assess relevancy

Name of the Police officer:	Rank of the Police Officer:		
Mark ✓ in the appropriate box (Only one ✓ is allowed in one row)			
Type of private data	The level at which this type of private data is relevant to me		
	Low (30%)	Medium (60%)	High (90%)
Personal Identity Data (Static)			
Relational Identity Data (Static)			
Personal Identity Data (Dynamic)			
Relational Identity Data (Dynamic)			

Table 7. Consolidated table to compute relevancy to police officers

Type of private data	The level at which this type of private data is relevant to me			Relevancy
	Low (30%)	Medium (60%)	High (90%)	
Personal Identity Data (Static) (X)	X_L	X_M	X_H	R_X
Relational Identity Data (Static) (Y)	Y_L	Y_M	Y_H	R_Y
Personal Identity Data (Dynamic) (U)	U_L	U_M	U_H	U_S
Relational Identity Data (Dynamic) (V)	V_L	V_M	V_H	V_T

$$R_X = \frac{X_L \times 0.3 + X_M \times 0.6 + X_H \times 0.9}{X_L + X_M + X_H} \times 100$$

Here is the description of the above formula:

1. X_L denotes the number of police officers who rate that the importance of Personal Identity Data (Static) to them is "LOW".
2. X_M denotes the number of police officers who rate that the importance of Personal Identity Data (Static) to them is "MEDIUM".
3. X_H denotes the number of police officers who rate that the importance of Personal Identity Data (Static) to them is "HIGH".
4. R_X denotes the relevancy factor.

In the same way, R_Y, R_U, and R_V are defined. Also, the metric "complexity" is modeled in a similar way.

7   Analysis of Processed Data

This section discusses the results derived from the experiments. The charts are provided in Figs. 3 and 4. Also, the feedbacks and comments of police and business communities are provided.

7.1   Complexity

The metric "complexity" means the level of difficulty in the extraction of relevant data from social media private data where the level is rated as Low, Medium and High. The social media private data collected by social media intelligence system was handed over to police officers and pizza restaurant managers. Police officers and pizza restaurant managers studied the collected data and filled-up the forms given in Tables 4 and 6. Tables 5 and 7 were generated by the author from Tables 4 and 6.

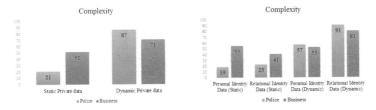

Fig. 3. Charts related to the metric "complexity"

7.1.1 Assessment of "Complexity" by Business Managers

Business managers commented on dynamic private data "It is very difficult to extract dynamic private data from social media because the quantity of dynamic private data from social conversation is massive. Even though there are automated tools, still it is time consuming. A small business cannot afford too many business hours in the extraction process". They commented on static private data "It is also difficult to collect and filter static private data from social media because significant amount of static private data from social media are fake. Filtering the genuine data from fake data is complex and cumbersome".

7.1.2 Assessment of "Complexity" by Police Officers

Police officers commented on dynamic private data "It is a hectic task to extract dynamic private data from social media because important conversations among the law-offenders are cryptic. Some intelligent law-offenders use secret metaphors". However, police officers commented on static private data "It is relatively easy to collect static private data. Criminals' fake data can be easily identified".

7.2 Relevancy

The metric "relevancy" means the level of importance of social media private data to the business and police communities where the level is rated as Low, Medium and High.

7.2.1 Assessment of "Relevancy" by Business Managers

Business managers concluded that they were more interested in dynamic private data than static private data. Business managers commented on dynamic private data "Dynamic private data is very useful and important. It helps us identify social customers (influencers) and silent customers. Social customers make loud and influential voices by means of online social media conversations. Social customers make significant influence on the success and failure of the business. In the same way, dynamic private data helps us segregate satisfied and dissatisfied customers. The complaints of the customer are mainly collected from dynamic private data. Star ratings (5 stars) are common nowadays and they are a part of dynamic private data. The dynamic private data is very helpful in monitoring social customers and dissatisfied customers. The third reason is trend analysis. Trend analysis is an integral part of business process and is used to organize the strategic plan of the business. Again, the required data and information related to trend analysis are available only from dynamic private data".

Business managers commented on static private data "Personal Identity Data (static) is not very useful because significant amount of static private data from social media are fake. Relational Identity Data (static) is also not very useful because most of the social media neighbors live outside Chennai and they are not our customers".

7.2.2 Assessment of "Relevancy" by Police Officers

Police officers concluded that they were more interested in static private data than dynamic private data. Police officers commented on dynamic private data "Proactive measures are key to thwart organized crimes, coordinated agitations and social protests. The data related to proactive security measures are mainly derived from static private data. Since the anti-social elements are aware of the internet surveillance, they do not transmit critical information in plain text. That is why dynamic private data extracted from social media is not useful to that extent. However, common people still exchange key information in plain text. In Jallikattu incident, the organizers of Jallikattu protests used cryptic texts and the participants used plain text".

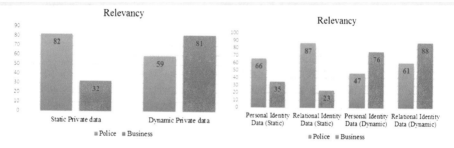

Fig. 4. Charts related to the metric "relevancy"

They commented on static private data "The static private data is found more useful than the dynamic private data because demographic segmentation is an important factor. The criminal networks are mostly formed by demographic segmentation such as religion, caste, language, race etc. Another important information is the length and depth of a criminal social network. Information such as 'A person from Chennai has a social media neighbor in Karachi' is a key information to us. These kinds of information are extracted mostly from static private data. Police officers judge a person by his friends. So, relational identity data (static) is more important than personal identity data (static). Moreover, only static private data provides past and completed activities of a person. Police decides to track a person by his past activities".

8 Conclusion

Police officers concluded that they were more interested in static private data than dynamic private data. Business managers concluded that they were more interested in dynamic private data than static private data. While the police are interested in static private data, the business communities are interested in dynamic private data. The reason is that past data (static private data) is more important to police and future data (dynamic private data) is more important to business. The experiments justify the need of micro and macro level classification of social media private data because different users require different types of social media data.

The data classification has been studied by Carnegie Melon University in the context of information security based on its level of sensitivity and the impact to the university [15]. This paper considers two metrics "complexity" and "relevancy". It is good to extend this research by including additional metrics such as "sensitivity" and "impact".

It is interesting to study the relationship between Schneier classification such as service data, disclosed data, entrusted data, incidental data, behavioral data and derived data etc. and micro level classification such as Personal Identity Data (Static), Relational Identity Data (Static), Personal Identity Data (Dynamic), and Relational Identity Data (Dynamic).

Acknowledgement. This work was supported and funded by Kuwait University, Research Project No. (QI 02/17).

References

1. Aggarwal, C.: An introduction to social network data analytics. In: Social Network Data Analytics, pp. 1–15 (2011)
2. Barnes, S.: A privacy paradox: social networking in the United States. First Monday **11**(9), 25–30 (2006)
3. Batrinca, B., Treleaven, P.: Social media analytics: a survey of techniques, tools and platforms. AI Soc.: Knowl. Cult. Commun. **30**(1), 89–116 (2015)
4. Betancourt, L.: How Companies Are Using Your Social Media Data. Mashable, Australia (2010)
5. Flitter, J.: Manuscript, 5 types of social data (2012)
6. Gallagher, R.: Software that tracks people on social media created by defense firm. World news, The Guardian (2013)
7. Ghonim, W.: Revolution 2.0: The Power of People is Stronger than the People in Power. Mariner Books, Wilmington (2013)
8. Huang, Q., Yang, Y., Fu, J.: PRECISE: identity-based private data sharing with conditional proxy re-encryption. Future Gener. Comput. Syst. **27** (2017)
9. Kavanaugh, A., Fox, E., Sheetz, S., Yang, S., Li, L., Shoemaker, D., Natsev, A., Xie, L.: Social media use by government: from the routine to the critical. Gov. Inf. Q. **29**(4), 480–491 (2012)
10. Korolova, A.: Protecting privacy when mining and sharing user data. Ph.D. thesis, Stanford University, USA (2012)
11. Lange, P.: Publicly private and privately public: social networking on YouTube. J. Comput.-Med. Commun. **13**(1), 361–380 (2007)
12. Leskovec, J.: Analytics and predictive models for social media. In: International World Wide Web Conference in Hyderabad, India (2011)

13. Lindamood, J., Heatherly, R., Kantarcioglu, M.: Inferring private information using social network data. In: Proceedings of the 18th International Conference on World Wide Web, Geneva (2009)

14. Manovich, L.: Trending: the promises and the challenges of big social data. Debates in the Digital Humanities, University of Minnesota Press (2012)

15. Markiewicz, D.: Guidelines for Data Classification. Information Security Office, Computing Services, Carnegie Mellon University (2017)

16. Moe, W., Schweidel, D.: Social Media Intelligence, 1st edn. Cambridge University Press, Cambridge (2014)

17. Molla, R.: Facebook has made more than $20 billion in profit since going public—Twitter has lost $2 billion. Recode (2017)

18. Omand, D., Bartlett, B., Miller, C.: Introducing social media intelligence (SOCMINT). J. Intell. Natl. Secur. 27(6), 801–823 (2012)

19. Bello-Orgaz, G., Jung, J., Camacho, D.: Social big data: recent achievements and new challenges. Inf. Fusion 28, 45–59 (2016)

20. Patel, N.: The 6 Types of Social Media Content That Will Give You the Greatest Value. CoSchedule, 74 (2015)

21. Richthammer, C., Netter, M., Riesner, M., Sänger, J., Pernul, G.: Taxonomy of social network data types. EURASIP J. Inf. Secur. 11 (2014)

22. Roberts, J.: How Companies Use Your Social Media Data When Taking Your Call. Fortune Tech, 1 (2017)

23. Schneier, B.: A taxonomy of social networking data. IEEE Secur. Priv. 8, 80–88 (2010)

24. Smith, C.: Social big data: the different types of user data collected by each major social network. Tech Insider, Business Insider, Australia (2014)

25. Thompson, C.: What companies are doing with your intimate social data. CNBC, USA (2013)

26. Vallor, S.: Social Networking and Ethics. The Stanford Encyclopedia of Philosophy (Winter 2016 Edition), Stanford University, USA (2016)

Implementation of Web Browser Extension for Mitigating CSRF Attack

Saoudi Lalia[(⊠)] and Kaddour Moustafa

Computer Science Department, Mohamed Boudiaf University, M'sila, Algeria
Saoudi_l@yahoo.fr, Mustafa.rtic@gmail.com

Abstract. CSRF is one of the most serious cyber-attacks and has been recognized among the major threats and among the top ten worst vulnerabilities of web applications. CSRF attack occurs when the attacker takes the advantages of implicit authentication mechanisms of HTTP protocol and cached credentials in the browser to execute a sensitive action on a target website behalf of an authenticated user without his knowledge. In this paper, we present a CSRF protection mechanism that can be added to Google Chrome browser as an extension. Our tool "CSRF Detector" is purely implemented on the client-side to defeat the attacker attempt to perform CSRF attacks by analyzing web requests and web pages content to detect all the basic and advanced CSRF attacks. Our evaluation result shows that CSRF Detector extension successfully detects all the generated attacks and it has the ability to protect users and web applications against CSRF attacks with no false positive.

Keywords: CSRF Detector · CSRF attack · Google chrome extension · Client-side attack detection

1 Introduction

Nowadays, web applications are being developed in several fields and manage user's personal, confidential, and financial data, causing the security of such applications an important question, however, significant number of attacks actively exploit user session information to perform unauthorized actions (such as perform authenticated bank transactions, or Change a user's password.) These attacks are commonly known as Cross Site Request Forgery (CSRF).

CSRF is one of the most dangerous attacks and has been recognized among the major threats to web applications and among the top ten worst vulnerabilities for web applications. Nowadays, many countermeasures and a set of techniques are available to defend against common CSRF attacks; however attackers still find vulnerabilities in websites and exploit them to carry out their attacks against servers and clients. With regard to CSRF protections, the existing countermeasures cannot defend against all scenarios encountered in the real world. Recently, many security assessment and bug bounty considering CSRF vulnerability have been reported. 133 of these suffered from at least one vulnerability enabling CSRF. The experiment also tested 132 additional web sites (from the Alexa global top 1500) and discovered that 95 of them were

© Springer Nature Switzerland AG 2019
Á. Rocha et al. (Eds.): WorldCIST'19 2019, AISC 931, pp. 867–880, 2019.
https://doi.org/10.1007/978-3-030-16184-2_82

vulnerable to CSRF (i.e. 72%). these findings include serious vulnerabilities among the web sites of Microsoft, Google, eBay... etc. [1].

To protect users and servers application from CSRF attack and overcome the attack, a set of techniques are proposed. The existing countermeasures can be classified into two main categories:

1. Server-side: The web application can use three common mechanisms known to-day on the web to defend against cross-site request forgery attacks: validating a secret token, verifying the Same Origin with Standard Headers, and including additional headers with XMLHttpRequest. Unfortunately, not any of the proposed mechanisms is fully capable of carrying out this task.
2. Client-Side: This category let researchers and developers to implement new mitigation techniques to combat CSRF attacks. The client-side protection techniques are divided into two categories:

- Browser Extensions/add-ons: It refers to mitigation techniques that are based upon the browser extensions and used by the users.

The majority of these techniques have a small amount of attack vectors, with a high false positive.

- Proxy level framework: This category represents client-side techniques that are implemented on proxy level.

This technique requires a proxy installed on local host or on intranet, which is not the case for all users.

In order to combat CSRF attacks, we proposed a novel technique for protecting users from basic and advanced CSRF attacks at client-side. Our proposed tool 'CSRF Detector' is a web browser extension implemented for Google Chrome.

Our work makes the following contributions:

- Our proposed tool 'CSRF Detector' detects CSRF attacks with a hybrid mechanism based on the notion of content checking and request analysis.
- CSRF detector can detect reflected XSS attack for the purpose of preventing attackers from using such vulnerability to carried out the CSRF attacks.
- CSRF Detector also has the ability to detect cross and sub-domain CSRF attack.
- CSRF Detector adopts Google Safe Browsing service to identify unsafe websites and notify users to protect themselves from harm websites.
- CSRF Detector also serves as a tool for web developers to detect CSRF vulnerability. It allows developers to quickly discover forms that are likely unprotected against CSRF attacks.

The paper is organized as follows: In Sect. 2, we give a brief presentation of CSRF attack mechanism, in Sects. 3 and 4 we briefly present the various CSRF attack types and different defense approaches. Section 5 presents our proposed CSRF detector extension approach, Sect. 6 consists of extension architecture and Sect. 7 presents its working mechanism, the experimentation and results are explained in Sect. 8. Finally, a conclusion is drawn in Sect. 9.

2   Understanding the CSRF Attack

A CSRF attack involves three parties, a victim, a trusted website where the victim is currently authenticated and an attacker [2]. CSRF is an attack that consists in tricking the victim to send an authenticated HTTP request to a target Web server, e.g., via visiting a malicious web page that creates such a request in the background. In cases that the user is currently in an authenticated state with the targeted Web site, the browser will automatically attach the session cookies to such requests, making the server believe that this request is a legitimate one and, thus, may cause state-changing actions on the server-side.

3   CSRF Types

The launching of CSRF attack may be carried out in different steps depending on the type of CSRF attack. CSRF can mainly be classified depending on the way in which the attacker makes the victim send the forged HTTP request (stored and reflected CSRF) and also depending on the state of CSRF victims with the trusted web application (pre-authenticated and post authenticated CSRF).

3.1   Stored and Reflected CSRF

A stored CSRF attack is one where the attacker can use the application itself to provide the victim the exploit link [3] or other content which directs the victim's browser back into the application and causes attacker controlled actions to be executed as the victim.

In Reflected CSRF attack, the attacker performs the attack outside the system application, by exposing the victim to exploit link [4] or content which is malicious. This can be done using fishing techniques.

3.2   Pre- and Post-authentication CSRF

From 2001 (first reporting of CSRF) to 2008, it was widely considered that only the state changing actions that can be caused by authenticated users need to be protected from CSRF attacks. This is mainly due to the assumption that only authenticated users can execute actions having high-impacts (e.g. transferring money from one account to another). However, in 2008, Login CSRF attack was introduced [5]. Unlike previously classic CSRF attacks, the victims of Login CSRF are note authenticated users. Depending on this, CSRF attacks can be classified into two categories: CSRF attacks that do not require the victim to have an authenticated session with the vulnerable website (refer to as pre-authentication CSRF attacks) and the type of CSRF attacks that require the victim to have an authenticated session (refer to as post-authentication CSRF attacks).

4 CSRF Protection Techniques

To protect users and servers application from CSRF attack, a set of techniques that do not require user interaction are available. The existing countermeasures can be classified into two main categories:

4.1 Server-Side Protection Techniques

The equations are an exception to the prescribed specifications of this template. You will need to determine whether or not your equation should be typed using either the Times New Roman or the Symbol font (please no other font). To create multileveled equations, it may be necessary to treat the equation as a graphic and insert it into the text after your paper is styled.

The web application can use three common mechanisms known today on the web to defend against cross-site request forgery attacks:

4.1.1 Secret Validation Token

Secret Validation Token is a well-known server-side protection scheme against CSRF attacks. This scheme sends additional information with each HTTP request to determine whether the request came from an authorized user. To apply token, web applications must first create a "pre-session", and then proceed forward a real session after successful authentication. Validation token should be hard to guess for an attacker who does not already have access to the user's account. If a request is missing a token or the token does not match the expected value, the web application should reject the request and prompt the user [6]. One disadvantage of using this scheme is that, occasionally, some users disclose the contents of web pages they view to third parties, for example via email. If the page contains the user's Validation Token, any-one who views the contents of the page can impersonate the user to the web application until the session expires [7].

4.1.2 Verifying Same Origin with Standard Headers

There are two steps to this check:

- Determining the origin of a request is coming from (source origin).
- Determining the target of a request is going to (target origin).

Both of these steps rely on examining an HTTP request header value. Although it is usually trivial to spoof any header from a browser using JavaScript, it is generally impossible to do so in the victim's browser during a CSRF attack, except via an XSS vulnerability in the site being attacked with CSRF. More importantly for this recommended Same Origin check, a number of HTTP request headers can't be set by JavaScript because they are on the 'forbidden' headers list. Only the browsers themselves can set values for these headers, making them more trustworthy because not even an XSS vulnerability can be used to modify them [8]. To identify the source origin, web application server can use one of these two standard headers:

- Origin Header.
- Referer Header.

4.1.3   Custom Request Headers

Custom HTTP headers can be used to prevent CSRF attack because the browser prevents sites from sending custom HTTP headers to another site but allows sites to send custom HTTP headers to them using XMLHttpRequest [5]. XMLHttpRequest's popularity has increased recently with more sites implementing AJAX interfaces. Sites can defend against CSRF by setting a custom header via XMLHttpRequest and validating that, the header is present before processing state modifying requests. Although effective, this defense requires sites to make all-state modifying requests via XMLHttpRequest, a requirement that prevents many natural site designs [5].

4.2   CSRF Specific Defense

There are numerous ways that web servers can specifically defend against CSRF. The flowing three mechanisms are the most recommended by OWASP as a second check and an additional precaution.

4.2.1   Double Submit Cookie

Storing the CSRF token in session is problematic, an alternative way is the use of a double submit cookie. In this case, the CSRF token is submitted twice. Once in the cookie and simultaneously within the actual request. As the cookie is protected by the Same-Origin Policy only same-domain scripts can access the cookie and thus write or receive the respective value. Hence, if an identical value is stored within the cookie and the form, the server side can be sure that the request was conducted by a same-domain resource. As a result, cross-domain resources are not able to create a valid request and therefore CSRF attacks are rendered void [9].

4.2.2   Encrypted Token Pattern

The Encrypted Token Pattern leverages an encryption, rather than comparison, method of Token-validation. After successful authentication, the server generates a unique Token comprised of the user's ID, a timestamp value, and a nonce, using a unique key available only on the server. This Token is returned to the client and embedded in a hidden field. Subsequent AJAX requests include this Token in the request-header, in a similar manner to the Double-Submit pattern. Non-AJAX form-based requests will implicitly persist the Token in its hidden field. On receipt of this request, the server reads and decrypts the Token value with the same key used to create the Token. Once decrypted, the UserId and timestamp contained within the token are validated to ensure validity; the UserId is compared against the currently logged in user, and the timestamp is compared against the current time [8].

4.3   Client-Side Protection Techniques

In this category we will present techniques and extensions implemented on web browser to defeat CSRF attack.

4.3.1   The Same Origin Policy (SOP)

SOP is the most widespread countermeasure. This policy limits access to DOM properties and methods to scripts from the same 'origin', where the origin is usually defined as the triple <domain name, protocol, TCP port> . Unfortunately, the protection offered by the SOP is insufficient. Although the SOP prevents the requesting script from accessing the cookies or DOM properties of a page from another origin, it does not prevent an attacker from making requests to other origins. The attacker can still trigger new requests and use cached credentials, even though the SOP prevents the attacker from processing responses sent back from the server [10].

4.3.2   Client-Side Proxy and Web Extensions

On top of SOP, a client-side protection technique can be used to protect users from CSRF attacks by monitoring outgoing requests and incoming responses. This countermeasure can be implemented as a browser proxy or extension to web browsers.

(a) Requestrodeo

Johns and Winter proposed RequestRodeo as a client-side protection proxy against CSRF. This technique offers protection via detecting cross-domain requests and then removal of cookie values from these requests (stripping of implicit authentication). A request is authenticated when it satisfies the Same Origin Policy (SOP) and it initiated as a result of an interaction with the currently viewed browser tab [11].

Limits: RequestRodeo is limited to only certain HTTP requests and not for HTTPS requests, so it does not scale well to web 2.0 applications. RequestRodeo fails to detect all JavaScript dynamic links in the responses, since this dynamic content has come after passing through the proxy. Also, RequestRodeo does not differentiate between malicious and genuine cross-origin requests, so it provides very poor protection against CSRF [12].

(b) CSFire

CsFire is an integrated extension of Mozilla browser to mitigate CSRF attacks, it extends the work of Maes and al [13]. CsFire strips cookies and HTTP authorization headers from a cross-origin request. The advantage of stripping cookies and HTTP authorization headers is that there are no side-effects for cross-origin requests that do not require credentials in the first place. Additionally, CsFire supports users for creating custom policy rules, which use user supplied whitelist and blacklist to certain traffic patterns.

Limits. The Limits of CsFire is it protects users against one specific type of CSRF attack which is Reflected CSRF.

(c) NoScript ABE

Application Boundary Enforcer (ABE), restricts an application within its origin, which effectively strips credentials from cross-origin requests unless specified otherwise. The default ABE policy only prevents CSRF attacks from the internet to an intranet page. The user can add specific policies, such as a CsFire-alike stripping policy, or a site specific blacklist or white list. If configured with, ABE successfully blocks the CSRF attack scenarios, but disables the payment and central authentication scenario [13].

5   CSRF Detector Extension

In our work, we propose a novel technique to protect users from basic and advanced CSRF attacks at client-side. Our CSRF detector has the following characteristics:

- Our proposed extension 'CSRF Detector' detects CSRF attacks with a hybrid mechanism based on the notion of content checking and request analysis. This approach does not rely on server side program states and it does not require storing URLs or tokens to be matched at a later stage for attack detection. Our CSRF detector analysis the HTML code before loading each page and detects potential CSRF attack, and it intercepts every HTTP request and decides whether it should be allowed or not. The suspicious request event will strip from the authentication credential and session information.
- In order to combat Image based CSRF Attacks we propose a mechanism to identify of such suspicious image elements by obtaining all IMG elements from the response page and matches the image URL linked with the image file formats (JPG, Gif … etc.).
- CSRF detector can detect XSS attack in web site for the purpose of preventing attackers from using such vulnerability to carried out the CSRF attacks.
- Our proposed solution could defend against login CSRF attack because all processes of checking are done before the login process.
- CSRF Detector has the ability to detect both types of CSRF attack (reflected and stored).
- CSRF Detector has the ability to detect cross and sub-domain CSRF attack.
- CSRF Detector adopts Google Safe Browsing service to identify unsafe websites and notify users to protect themselves from harm websites (malware, phishing …).
- Finally, CSRF Detector also serves as a tool for web developers to detect CSRF vulnerability. It allows developers to quickly discover forms that are likely unprotected against CSRF attacks.

6   CSRF Detector Architecture

The Fig. 1 shows the flowchart of our CSRF Detector extension for Google chrome browser, to monitor any interaction between web sites and web client. CSRF Detector analysis web requests and web pages content to detect all basic and advanced CSRF attacks. CSRF detector also adopts Google Safe Browsing service to check in really big global database offered by Google that contains a reported malicious website (Malware, fishing, social engineering). CSRF Detector has three modules:

6.1   Request Analysis Module

CSRF Detector intercepts the outgoing request sent by the web browser and check for:

a. Suspicious Cross-domain requests.
b. Suspicious sub-domain requests.

c. Suspicious reflected XSS (Cross-site scripting) requests.
d. Check URL in Google Safe Browsing.

For the first and the second type of requests (Cross domain requests and Suspicious sub-domain requests), CSRF Detector intercepts the outgoing requests sent by the web browser and determines the target and the origin domain. If the origin domain does not match the destination of the request, the request is considered suspicious. Suspicious requests will be stripped of authentication credentials in the form of cookies or HTTP authorization headers. There are two steps to this check:

1. Determining the origin of the request is coming from (Origin domain).
2. Determining the target of request is going to (Destination domain).

For the third type of requests (reflected XSS requests), our CSRF Detector intercepts all requests and looks for cross-site scripting signatures within parameter values. To look for signatures in parameters values, our CSRF detector must parse the URL correctly and retrieve the value part, then search for the signature on the value while overcoming encoding issues.

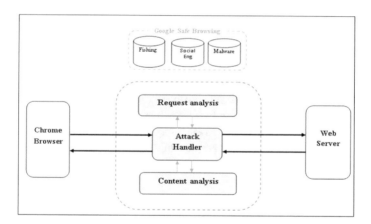

Fig. 1. CSRF detector architecture.

6.2 Content Analysis Module

Before and after the requested web page is loaded, CSRF Detector will check for:

1. Invisible iframes.
2. Suspicious image elements.
3. Unprotected forms.

After webpage is loaded, users can checks for any CSRF vulnerabilities. The module will check for HTML forms that are likely unprotected against CSRF attacks by examining forms on a webpage and matching the form elements with user-specified CSRF Token names and formats.

6.3   Attack Handler Module

This module sanitizes all suspicious requests, and generates an alert message to the user about all different types of CSRF attack. The module will strip the suspicious request from authentication credential and session information, or any suspicious XSS code.

7   CSRF Detector Mechanism

In this section we will explain how our CSRF Detector works in its different stages:

7.1   Request Analysis

Figure 2 illustrates how CSRF Detector intercepts every HTTP request sent by the browser and decides whether it should be allowed, blocked or stripped from authentication credential and session information.

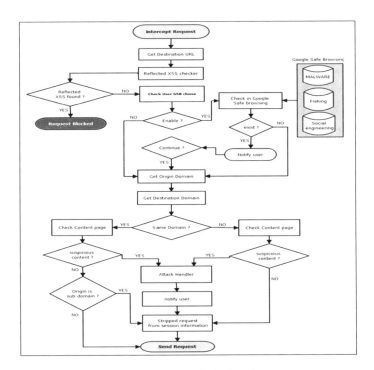

Fig. 2. Request analysis flowchart.

First of all, CSRF Detector will get the requested URL and checks for reflected XSS (Cross-site scripting attack) using reflected XSS checker. The purpose of the malicious script is to steal user session token and then send it to the attacker. For this reason our detector module matches any XSS signature (for example: <script>,"%3Cscript%3E" …) in the requested URL, blocks any suspicious request and notify the user.

In the second phase, the requested URL is checked in Google safe browsing databases if the user was enabling this service. Enabling Google Safe Browsing will offer users more protections. After that, CSRF Detector will check for cross-domain requests and requests event that originating from subdomain to parent domain, this two events are considered suspicious and will be stripped from session information.

7.2   Content Analysis

Figure 3 shows the flowchart of web page response analysis, and how it works when the response webpage is loaded. This module will check for invisible iframes and suspicious image elements. The module will extract all iframes exist in a webpage, checks CSS style and the attributes of HTML element iframe ('width', 'height', and 'opacity').

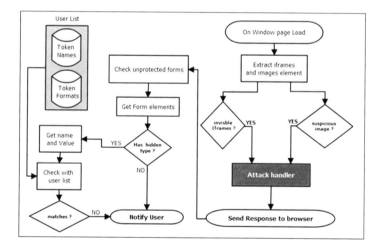

Fig. 3. Flowchart of content analysis.

We integrate in this module the CSRF vulnerability checker to notify users if any unprotected forms are found, and help developers to check their web sites for CSRF vulnerabilities.

8 Experimentation and Discussion

In this section, we show the performance of our CSRF Detector extension, for that we use two web sites families:

8.1 Testing with White List (Legitimates) Web Sites

In order to ensure the efficiency of our solution, we conducted some tests to evaluate the efficiency and the capability of our CSRF Detector against basic and advanced CSRF attacks. To make sure that its results match with what is predetermined, we developed a small web application which implements cookies as session management mechanism and simulates the CSRF attacks with it. We also tested our solution with Real-life vulnerable and white list websites.

8.2 Testing with Basic CSRF Attacks

We developed a small web application E-Money to simulate the basic CSRF attacks and to evaluate the ability of CSRF Detector to protect vulnerable applications.

Test scenario

Case1: Without CSRF Detector

The victim login to E-money transfer website. After that, he clicks on a malicious link sent by an attacker or malicious link stored in attacker's website. Because there is no protection against CSRF attack in this website, the attackers can easily trick user through the malicious link and send money to his account without knowledge of the victim.

Case 2: With CSRF Detector Extension

We enable CSRF detector. Then, we repeat the previous attacks. Our tool detected and rejected all CSRF attacks attempted from different sources correctly, which means that the false negative rate is zero.

8.3 Testing with Advanced CSRF Attacks

We conducted some tests to evaluate the efficiency and the capability of our tool against advanced CSRF attacks including both vulnerabilities (XSS and exploiting sub-domains). At first, we developed a webpage that contains malicious XSS. After the request is triggered from this form, CSRF Detector shows his ability to block such malicious request and alert user about it. We also test CSRF Detector with request that originating from sub-domains to parent domain. The Fig. 4 below shows that CSRF detector was successfully stripped cookies from such request.

Fig. 4. Sub domain detection

8.4   Testing with White List Websites

We conducted an experiment using a white list to check if our proposed tool will generate any warning on possible CSRF attack and identify its false positive rate. The experimental considering 300 websites belonging to different rank ranges (computer, science, News... etc.) of the Alexa global top 1500. The results of this experimental shown that we got no false alert which means no false positive over 300 websites visited.

8.5   Testing with Real-Life Vulnerable Websites

To test our solution with real-life websites, we used at first our extension CSRF detector to discover unprotected forms on random 100 Algerian websites from Alexa top websites in order to find vulnerable websites. We found some famous Algerian web-sites are vulnerable to CSRF attacks, including a serious vulnerability that can change credentials information such as email. Then, we used Burp 'CSRF PoC generator' [14] which is a tool used to generate a proof-of-concept (PoC) cross-site request forgery (CSRF) attack for a given request. The result of this experiment shows that CSRF Detector is successfully detects and prevents every attack generated by Burp CSRF PoC.

8.6   Performance

We visited several websites from white list sites with and without using CSRF Detector extension, and then we have used Google chrome Analyzing Performance to measure webpage loading time. Figure 5 shows line chart of the average delay time of loading webpage (in seconds) measured by webpage size (bytes).

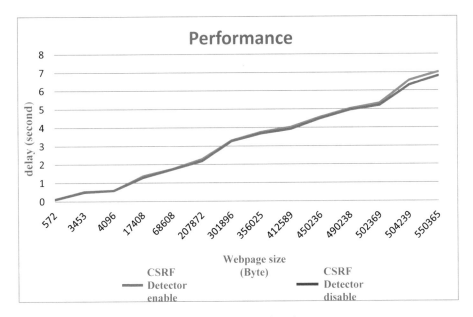

Fig. 5. Performance line chart.

The line chart shown in the Fig. 5 represents the performance of our tool. The blue line shows loading delay time when CSRF Detector is disabled and the red one when CSRF detector is enabled.

We can say that the rise of the average loading time is simultaneously, and the highest difference time between them is 0.21 s. We have also notice that in some cases when Webpage triggered a big number of cross-domains requests, the performance will be higher than normal cases, because of our tool will strip the no importance of additional information send with those requests in form of cookies.

Finally, the result of our performance testing shows that CSRF Detector effect on browser performance is negligible.

9   Conclusion

Undoubtedly, CSRF attacks present a dangerous attack on the web. As the use of web application increasing, the need for more efficient CSRF detection system becomes critical.

For this reason we developed a Google Chrome extension "CSRF Detector" to defeat the attacker attempt to perform CSRF attacks by analyzing web requests and web pages content to detect all basic and advanced CSRF attacks.

The extension was tested by determining how well it picks up a large variety of advanced CSRF attacks.

The obtained experimental results demonstrate that our CSRF detector extension successfully detects all the generated attacks with no false positive, this proves the effectiveness of our extension to detect all basic and advanced CSRF attacks and protect users from credentials theft without effect on browser performance.

References

1. Sudhodanan, A., Carbone, R., Compagna, L., et al.: Large-scale analysis & detection of authentication cross-site request forgeries. In: IEEE European Symposium on Security and Privacy, EuroS&P 2017, Conference Paper, Paris, France, pp. 350–365, April 2017
2. Siddiqui, M.S., Verma, D.: Cross site request forgery: a common web application weakness. In: 2011 IEEE 3rd International Conference on Communication Software and Networks, pp. 538–543 (2011)
3. Ramarao, R., Radhesh, M., Alwyn, R.P.: Preventing Image based Cross Site Request Forgery Attacks, July 2018. https://isea.nitk.ac.in/rod/csrf/PreventImageCSRF/icscf09PreventImageCSRF.pdf
4. Vernotte, A.: A pattern-driven and model-based vulnerability testing for Web applications. Ph.D. thesis. Franche-Comté university (2015)
5. Barth, A., Jackson, C., Mitchell, J.C.: Robust defenses for cross site request forgery. In: Proceedings of ACM Conference on Computer and Communications Security (CCS), October 2008
6. Xing, X., Shakshuki, E., Benoit, D., Sheltami, T.: Security analysis and authentication improvement for ieee 802.11 i specification. In: Global Telecommunications Conference, IEEE GLOBECOM 2008. IEEE (2008)
7. Zeller, W., Felten, E.W.: Cross-site request forgeries: exploitation and prevention. The New York Times, pp. 1–13 (2008)
8. OWASP. https://www.owasp.org/index.php/CrossSite_Request_Forgery_%28CSRF%29_Prevention_Cheat_Sheet. Accessed 15 Mar 2018
9. Lekies, S., Tighzert, W., Johns, M.: Towards stateless, client-side driven Cross-Site Request Forgery protection for Web applications. In: Sicherheit, pp. 111–121 (2012)
10. Maes, W., Heyman, T., Desmet, L., Joosen, W.: Browser protection against cross-site request forgery. In: Proceedings of the First ACM Workshop on Secure Execution of Untrusted Code, Chicago, Illinois, USA, 9 November 2009, pp. 3–10 (2009)
11. Johns, M., Winter, J.: RequestRodeo: client side protection against session riding. In: The OWASP Europe Conference, Leuven, Belgium, May 2006
12. Telikicherla, K.C., Choppella, V., Bezawada, B.: CORP: a browser policy to mitigate web infiltration attacks. In: Information Systems Security, pp. 277–297. Springer, Heidelberg (2014)
13. De Ryck, P., Desmet, L., Joosen, W., Piessens, F.: Automatic and precise client-side protection against CSRF attacks. In: Computer Security-ESORICS 2011, pp. 100–116. Springer, Heidelberg (2011)
14. Portswigger. https://portswigger.net/burp/help/suite_functions_csrfpoc. Accessed 01 May 2018

What Does the GDPR Mean for IoToys Data Security and Privacy?

Esperança Amengual[✉], Antoni Bibiloni,
Miquel Mascaró, and Pere Palmer-Rodríguez

Department of Mathematics and Computer Science,
University of the Balearic Islands,
Ctra. de Valldemossa, Km. 7.5, 07122 Palma de Mallorca, Spain
{eamengual, toni.bibiloni, mascport, pere.palmer}@uib.es

Abstract. At present, Internet of Things (IoT) is a relatively new and promising field where objects can be connected to the Internet and collaborate among them in an intelligent way. The technological potential of IoT demands improvements regarding personal data security and privacy. When connected objects are toys, that is in the Internet of Toys (IoToys) field, security and privacy aspects are even more important since the special nature of final users, in this case children. This paper presents a preliminary study of the European Union General Data Protection Regulation (GDPR) applied to IoToys in particular. The obtained results show the need for greater coordinated efforts among professionals, toy manufacturers and software developers to manage shared personal information with the free and full consent of the final users. The goal is to ensure that IoToys applications are reliable and secure in terms of the processing of personal data.

Keywords: Internet of Things (IoT) · Internet of Toys (IoToys) ·
Data security · Data privacy · General Data Protection Regulation (GDPR)

1 Introduction

The growth of Internet connected devices has led a significant increment in new technologies ubiquity, now available at any time in everyday objects. Internet connected devices, known as "Internet of Things" (IoT), are rapidly reaching our homes with familiar objects such as watches, toothbrushes or coffee machines, turning our houses into smart houses [1, 2].

Among these new everyday connected objects, toys have also found a place raising a new concept: "Internet of Toys" (IoToys) [3]. As a consequence, as well as in many other areas, new technologies are prompting a transformation in the toy industry. Prestigious toy manufacturers such as LEGO [4], Hasbro [5] and Mattel [6] have already started working with connected toys and are considering the possibility of exploring the mobile apps market [7].

A connected toy is able to connect to a Web service and to interact with a mobile device or smart TV. As highlighted by the Family Online Safety Institute (FOSI) [8], connected toys have been designed to connect to remote severs that can collect data;

© Springer Nature Switzerland AG 2019
Á. Rocha et al. (Eds.): WorldCIST'19 2019, AISC 931, pp. 881–889, 2019.
https://doi.org/10.1007/978-3-030-16184-2_83

and this ability is in fact what make them powerful and potentially dangerous in terms of data security. Moreover, another aspect to consider is the capability of sharing data, a new feature that faces a new challenge regarding data privacy and confidentiality.

According with the technical report produced by the Joint Research Centre (JRC) of the European Commission [3], there are different views from media, the toy industry, and potential consumers, around the risks and benefits of Internet connected toys. Among the benefits of IoToys, the before mentioned report highlights: the educational dimension, encouragement over passive screen time, social play with toy and with other children, the possibility of continually to be updated with new content, and even diagnostic possibilities (e.g. identify learning difficulties or medical problems). However, in contrast with these benefits, we also have to consider a wide list of potential risks: biographical data security, device security (a toy could be hacked and used as a surveillance device or to behave badly, geolocational tracking of children), overuse and balance in life (sleep, physical activity, socializing), lack of real authentic play, lack of parent child interaction or health implications from electromagnetic radiation (EMR).

The above-mentioned risks need to be analyzed from very different perspectives and the fact is that to date there is little regulation. Moreover, as pointed out in [9], in the particular case of connected toys, these risks imply hidden dangers because technology is so well integrated in the toy that it becomes that technology goes unnoticed, by contrast with standard smartphones which already relatively small children use.

In this article we focus on security and data privacy from the point of view of the EU General Data Protection Regulation (GDPR) that came into force on May 25, 2018 [10]. The GDPR main goal is to regulate the processing by an individual, a company or an organization of personal data relating to individuals in the EU.

This article is structured as follows. In Sect. 2 security and data privacy concerns in IoToys are introduced. Section 3 analyses the GDPR from the perspective of IoToys and discuss some proposals for the application of the legislation to the particular case of IoToys. Finally, Sect. 4 concludes this study and proposes some future directions.

2   Security and Data Privacy Concerns in Internet of Toys

As happened with Internet evolution, throughout technological progress in the IoT field, there have been some situations in which several security breaches have been detected. These situations allowed access to confidential data, or gave control to devices in an inappropriate context. This fact has caused IoT to be seen as an unreliable technology. As an example, in 2014 only 18% of British people and 22% of US people thought that the benefits of smart devices outweigh any privacy concerns about the personal information they may collect [11], and a year later the figures were 14% and 20% respectively [12, 13]. That means that people do not really trust IoT, and the tendency does not seem to improve with time.

The reality is that there will not be a majority acceptance of IoT (and in particular IoToys) as long as people do not trust this technology. This concept is not new; it was already identified in the first stages of IoT [14]. At the technological level proposals to address this lack of trust are aimed at ensuring security and privacy of data. Although the number of proposals to deal with this issue is large [1, 2, 15, 16], it is still an open issue.

According to [20–23] security and privacy have become an important concern for IoT technology. Devices are connected to Internet and Internet is an unsecured environment. Objects in IoT communicate with each other; hence, there is a possibility that privacy and security can be hindered. The same argument can be considered for the particular case of IoToys.

The lack of security goes beyond purely technical issues. Security does not only require that the devices are technically safe against third party actions, but, in addition, it is necessary to have guarantees that the actions carried out by the companies that provide services, directly or indirectly, are ethically acceptable.

As pointed out in [7, 8, 24] we have to distinguish between "smart toy" and "connected toy". Smart toys include electronic devices (microphone, camera, sensors, accelerometers, gyroscope, compass …) that allow them to interact with users and adapt to their actions. These toys do not necessarily have Internet connection. On the contrary, connected toys have Internet connection and are able to request data to a service and share data with other toys or with home connected devices. This is a new way of playing named "connected play" that involves a continuous flow of data between different domains: online and offline, digital and non-digital, material and immaterial, but also public and private, global and local [24, 25]. Data collected while users are playing is known as "play data" and can be used by players to improve their skills or even by developers to update and evolve the game. A step further would be sharing data with other children or parents, among schools or healthcare services and, unavoidably, with the toy provider. It is necessary to take into account that the data collected can be many more than those for which explicit consent has been given (time of use, actions taken, interactions, and many, many more).

It is important to note that stored data must be carefully managed and treated as confidential. Usually toy companies have their own privacy policies where they detail how data is collected, used and managed. In the same way, games that can be installed and executed in a toy should also have their own privacy policies.

Another aspect to consider is how data is protected once registered, from the Internet connection, the data transmission protocol, to the server storage possibly in the cloud. More specifically, it is imperative to prevent unauthorized access.

2.1   Related Work

According to the GDPR [10], personal data is any information that relates to an identified or identifiable living individual. Different pieces of information, which collected together can lead to the identification of a particular person, also constitute personal data. The Regulation protects personal data regardless of the technology used for processing that data. It also doesn't matter how the data is stored; in all cases, personal data is subject to the protection requirements set out in the GDPR.

The Regulation also applies to the processing of personal data which includes the collection, recording, organization, structuring, storage, adaptation or alteration, retrieval, consultation, use, disclosure by transmission, dissemination or otherwise making available, alignment or combination, restriction, erasure or destruction of personal data.

Until now, a limited number of works have taken into account the requirements of the GDPR from the IoToys perspective. In [26] their authors highlight the importance of the GDPR implementation for toy companies in the scope of privacy protection for children. They conclude that a good starting point for business is to assess their current practices, identify gaps, and use that information to design a compliance plan. According to [27], the entry into force of the GDPR represents the official acknowledgement of a need to protect citizens that has made a change in how we manage privacy in the digital world and in IoToys in particular.

The same reasoning is used in [28] to raise awareness about the new regulatory framework and especially for IoToys which is considered a main area with challenges to address regarding the processing of personal data regarding children.

3   Impact of GDPR in IoToys

After a detailed analysis of the GDPR for the particular case of IoToys in this section the results are exposed. In order to facilitate the discussion, the results have been organized into four categories: (1) Sensitive data; (2) Informed consent; (3) Data treatment; and (4) Flows of personal data. Table 1 gives a summary of the analysis.

3.1   Sensitive Data

The Regulation "accepts a margin of manoeuvre for Member States to specify its rules, including for the processing of special categories of personal data ('sensitive data')". Therefore, the issue is if data processed by IoToys applications should be categorized as sensitive data.

According to the Regulation's text, "children merit specific protection with regard to their personal data, as they may be less aware of the risks, consequences and safeguards concerned and their rights in relation to the processing of personal data". This seems to be exactly the case for IoToys. Hence, personal data of children that could be registered when playing with a connected toy needs to be properly protected. This is a question that needs to be always considered as a mandatory restriction in IoToys application development.

3.2   Consent of the Data

In accordance with the Regulation, "consent should be given by a clear affirmative act establishing a freely given, specific, informed and unambiguous indication of the data subject's agreement to the processing of personal data, such as by a written statement, including by electronic means, or an oral statement".

More specifically, article 8 of the Regulation details the conditions applicable to child's consent in relation to information society services that could be applicable for IoToys: "the processing of the personal data of a child shall be lawful where the child is at least 16 years old. Where the child is below the age of 16 years, such processing shall be lawful only if and to the extent that consent is given or authorized by the holder of parental responsibility over the child".

Moreover "the principles of fair and transparent processing require that the data subject be informed of the existence of the processing operation and its purposes. The controller should provide the data subject with any further information necessary to ensure fair and transparent processing taking into account the specific circumstances and context in which the personal data are processed. Furthermore, the data subject should be informed of the existence of profiling and the consequences of such profiling". All this information should be included not only in toy specifications but also in game instructions. A problem that may occur at this point is lack of parental understanding regarding potential risks which makes them not to give enough importance to toy or game specifications. This requires defining some mechanisms to raise awareness to those in charge of children. On the other side, the GDRP ensures that toy companies and software developers will also inform properly about personal data processing.

3.3   Data Treatment

As exposed in the previous section, consent of data is mandatory according to the new regulation, however this is still not enough to ensure appropriate and legal use of personal data. More specifically "personal data should be adequate, relevant and limited to what is necessary for the purposes for which they are processed. This requires, in particular, ensuring that the period for which the personal data are stored is limited to a strict minimum". IoToys applications should define time limits for data storage as nonfunctional requirements and delete data when necessary.

Furthermore, "modalities should be provided for facilitating the exercise of the data subject's rights, including mechanisms to request and, if applicable, obtain, free of charge, in particular, access to and rectification or erasure of personal data and the exercise of the right to object". At this point we have to consider the possibility of including additional functionality for connected toys. In other words, IoToys applications should offer not only game functions but also a different category of functions for parents to be able to access and manage toy registered data. This is a solution that would also ensure the requirement of "the data subject to have the right of access to personal data which have been collected concerning him or her, and to exercise that right easily and at reasonable intervals, in order to be aware of, and verify, the lawfulness of the processing".

Furthermore, there is a special consideration for personal data in IoToys. The Regulation establishes that "data subject should have the right to have personal data concerning him or her rectified and a 'right to be forgotten' where the retention of such data infringes the Regulation. That right is relevant in particular where the data subject has given his or her consent as a child and is not fully aware of the risks involved by the processing, and later wants to remove such personal data, especially on the Internet". Again a possible solution could be that IoToys applications provide this service as functionality.

3.4   Flows of Personal Data

The GDPR provides guidance in order to share personal data securely among countries even outside the European Union. These flows of personal data "are necessary for the expansion of international trade and international cooperation. The increase in such flows has raised new challenges and concerns with regard to the protection of personal data". As pointed out in the Regulation the Commission may decide if "a third country, a territory or specified sector within a third country, or an international organization,

Table 1. GDPR requirements for IoToys

Category	GDPR reference	GDPR requirement	IoToys proposal
Sensitive data	Article 6. Lawfulness of processing Article 35. General principles for transfers of personal data Article 57. Tasks	Children personal data shall receive specific protection	A mandatory restriction in IoToys applications
Informed consent	Article 8. Conditions applicable to child's consent in relation to information society services Article 9. Processing of special categories of personal data Article 13. Information to be provided where personal data are collected from the data subject Article 20. Right to data portability Article 22. Automated individual decision-making, including profiling	Clear and freely given consent	Include information in both toy specifications and game instructions Define new mechanisms to raise awareness to those in charge of children Ensure that toy companies and software developers inform properly about personal data processing
Data treatment	Article 13. Information to be provided where personal data are collected from the data subject Article 15. Right of access by the data subject Article 16. Right to rectification Article 17. Right to erasure ('right to be forgotten') Article 18. Right to restriction of processing Article 19. Notification obligation regarding rectification or erasure of personal data or restriction of processing	Adequate, relevant and limited personal data. Right to be forgotten	IoToys applications should define time limits for data storage. IoToys applications should offer functions for parents to be able to access and manage toy registered data
Flows of personal data	Article 51. Supervisory authority Article 58. Powers	International flows of personal data	Control data destinations. Code of conduct and cooperation between developers, toy industry and certification authorities

offers an adequate level of data protection, thus providing legal certainty and uniformity throughout the Union as regards the third country or international organization which is considered to provide such level of protection". This consideration implies to control data destinations from the connected toy to the rest of the world. Unfortunately, this seems to be an open question today given the complexity of the issue. In connection with this, article 40 refers to codes of conduct for the purpose of specifying the application of the Regulation in particular situations such as the transfer of personal data to third countries or international organizations among others. In any case, successful application of these requirements will depend largely on the good will and cooperation between toy companies, software developers and certification authorities.

4 Conclusions and Future Directions

IoT is a relatively new field in which efforts are clearly being made to move forward personal data security and privacy management. When connected things are toys, that is, in the particular case of IoToys, data privacy and security require special treatment as sensitive data.

This article presents a preliminary study of the data protection regulation currently applicable throughout the European Union. The requirements set in this regulation have been analysed in order to propose particular measures that could be useful for the development of IoToys applications.

The results have been grouped into four categories that cover data sensitivity, consent of parents, data treatment and international data sharing. Children personal data protection is mandatory in IoToys applications that would include information about security and privacy concerns in toys specifications and game instructions. Toy companies and also software developers should inform properly about personal data processing and the limits for data storage. Applications would also offer functions to access and manage toy registered data.

There is still a lot of work to do to ensure the compliance with the current legislation in order to offer secure and reliable IoToys applications. In this sense, the recently updated legislation is a first step to raise awareness on the importance of paying attention to personal data privacy and security. In our opinion the application of a code of conduct among toy companies and software professionals to make the compliance with legislation to become a reality is essential. Finally, we are convinced that the solution lies in parents reaching knowledge about potential risks and acting as mediators between experts and children.

References

1. Li, S., Da Xu, L., Zhao, S.: The internet of things: a survey. Inf. Syst. Front. **17**(2), 243–259 (2015). https://doi.org/10.1007/s10796-014-9492-7
2. Atzori, L., Iera, A., Morabito, G.: Understanding the internet of things: definition, potentials, and societal role of a fast evolving paradigm. Ad Hoc Netw. **56**, 122–140 (2017). https://doi.org/10.1016/j.adhoc.2016.12.004

3. Chaudron, S., Di Gioia, R., Gemo, M., Holloway, D., Marsh, J., Mascheroni, G., Peter, J., Yamada-Rice, D.: Kaleidoscope on the Internet of Toys – Safety, Security, Privacy and Societal Insights. EUR 28397 EN (2017). https://doi.org/10.2788/05383
4. LEGO Homepage. https://www.lego.com. Accessed 30 Oct 2018
5. Hasbro Homepage. https://shop.hasbro.com. Accessed 30 Oct 2018
6. MATTEL Homepage. https://www.mattel.cl/. Accessed 30 Oct 2018
7. Maynard, N.: The Internet of Toys. Juniper Research (2017). http://www.juniperresearch.com. Accessed 30 Oct 2018
8. Future of Privacy Forum: Family Online Safety Institute (FOSI). Kids & the connected home: privacy in the age of connected dolls, talking dinosaurs and battling robots. FOSI (2016)
9. Ihamäki, P., Heljakka, K.: The internet of toys, connectedness and character-based play in early education, vol. 1 (2018). https://doi.org/10.1007/978-3-030-02686-8_80
10. EU Commission website. https://ec.europa.eu/commission/priorities/justice-and-fundamental-rights/data-protection/2018-reform-eu-data-protection-rules_en. Accessed 30 Oct 2018
11. https://www.ipsos.com/ipsos-mori/en-uk/truste-internet-things-privacy-index. Accessed 11 Sept 2018
12. https://www.trustarc.com/press/41-percent-british-consumers-now-own-a-smart-device/. Accessed 11 Sept 2018
13. https://www.trustarc.com/press/35-of-americans-now-own-at-least-one-smart-device-other-than-a-phone/. Accessed 11 Sept 2018
14. Atzori, L., Iera, A., Morabito, G.: The internet of things: a survey. Comput. Netw. **54**(15), 2787–2805 (2010). https://doi.org/10.1016/j.comnet.2010.05.010
15. Yan, Z., Zhang, P., Vasilakos, A.V.: A survey on trust management for internet of things. J. Netw. Comput. Appl. **42**, 120–134 (2014). https://doi.org/10.1016/j.jnca.2014.01.014
16. Vashi, S., Ram, J., Modi, J., Verma, S., Prakash, C.: Internet of things (IoT): a vision, architectural elements, and security issues. In: 2017 International Conference on I-SMAC (IoT in Social, Mobile, Analytics and Cloud) (I-SMAC), Palladam, Tamilnadu, India, pp. 492–496. IEEE (2017). https://doi.org/10.1109/I-SMAC.2017.8058399
17. Sicari, S., Rizzardi, A., Grieco, L.A., Coen-Porisini, A.: Security, privacy and trust in internet of things: the road ahead. Comput. Netw. **76**, 146–164 (2015). https://doi.org/10.1016/j.comnet.2014.11.008
18. Lee, I., Lee, K.: The internet of things (IoT): applications, investments, and challenges for enterprises. Bus. Horiz. **58**(4), 431–440 (2015). https://doi.org/10.1016/j.bushor.2015.03.008. Accessed 12 Nov 2018
19. Stojkoska, B.L.R., Trivodaliev, K.V.: A review of internet of things for smart home: challenges and solutions. J. Clean. Prod. **140**, 1454–1464 (2017). https://doi.org/10.1016/j.jclepro.2016.10.006
20. Leloglu, E.: A review of security concerns in internet of things. J. Comput. Commun. **5**, 121–136 (2017)
21. European Commission: IoT Privacy, Data Protection, Information Security. http://ec.europa.eu/information_society/newsroom/cf/dae/document.cfm?doc_id=1753. Accessed 12 Nov 2018
22. Allhoff, F., Henschke, A.: The internet of things: foundational ethical issues. Int. Things **1–2**, 55–66 (2018). https://doi.org/10.1016/j.iot.2018.08.005

23. Di Martino, B., Rak, M., Ficco, M., Esposito, A., Maisto, S.A., Nacchia, S.: Internet of things reference architectures, security and interoperability: a survey. Int. Things **1–2**, 99–112 (2018). https://doi.org/10.1016/j.iot.2018.08.008
24. Mascheroni, G., Holloway, D. (eds.): The Internet of Toys: A Report on Media and Social Discourses Around Young Children and IoToys. DigiLitEY (2017)
25. Marsh, J.: The internet of toys: a posthuman and multimodal analysis of connected play. Teach. Coll. Rec **119**(15), 1–32 (2017)
26. Millar, S.A., Marshall, T.P., Cardon, N.A.: The Toy Association White Paper on Privacy & Data Security: New Possibilities and Perils. 3rd Edn, 9 March 2017. https://www.toyassociation.org/App_Themes/toyassociation_resp/downloads/research/whitepapers/white-paper-data-privacy.pdf. Accessed 8 Jan 2019
27. https://vaikai.com/new-blog/2018/8/20/the-privacy-of-play
28. https://www.i-scoop.eu/internet-of-things-guide/iot-regulation/

Factors Influencing Adoption of Information Security in Information Systems Projects

Landry Tafokeng Talla$^{(\boxtimes)}$ and Jean Robert Kala Kamdjoug

GRIAGES, Catholic University of Central Africa, 11628 Yaoundé, Cameroon
tafokeng@gmail.com, jrkala@gmail.com

Abstract. This article deals with information security in information systems projects, which has become vital in a context of cybercrimes and economic spying proliferation. Based on the theoretical research model derived from UTAUT 2 and TAM models, this study seeks to determine the factors that influence the intention to adopt and use information security in Information Systems (IS) projects. Both qualitative and quantitative approaches were employed. Upon data analysis on the SmartPLS 3.2.7 software, the results suggested that: (i) perceived usefulness, habit, and facilitating conditions have a positive influence on the intention to use information security in IS projects; (ii) habit has a positive influence on the use of information security in IS projects. This study surely contributes to enriching the literature in the field of information security where the level of scientific research production remains low.

Keywords: Information security · Technology adoption · Use of technology · UTAUT 2

1 Introduction

On October 21st, 2016, the following were unavailable four hours: Amazon, eBay, Spotify, Airbnb, Netflix, PayPal, Twitter; online games from PlayStation and Xbox; Media, CNN, The New York Times, The Boston Globe, The Financial Times, The Guardian. Millions of users of these services could not access their servers, and the reason was obvious: a large-scale computer attack [3] had been perpetrated.

In their daily activities, organizations use the Internet to collect personal data from employees, customers or third parties, and to build their own database. Such data constitute an asset for the organization [4]. Organizations need to compile have more or less sensitive information about themselves and their counterparts, about people, and even governments. This rise in digital technology has led to the emergence of new forms of criminality ranging from frequent actions of cyber-vandalism or cyber-crime, to hidden action modes of cyber-war or economic spying that are much more difficult to characterize or to recognize. Traditional attacker groups have been replaced by criminal organizations that are able to launch industrial attacks [5].

Information plays a vital role for organizations nowadays. Mismanagement of the latter can be fraught with consequences for the organization and the people involved if it falls into bad hands. Therefore, the question of securing information systems is a real issue for companies that want to remain competitive.

© Springer Nature Switzerland AG 2019
Á. Rocha et al. (Eds.): WorldCIST'19 2019, AISC 931, pp. 890–899, 2019.
https://doi.org/10.1007/978-3-030-16184-2_84

In view of the above, we want to answer the following question: what factors influence the intention to use and adoption of information security in IS[1] projects?

2 Literature Review

2.1 Importance of Projects Within Organizations

To achieve a project, the various stakeholders need to consider a number of prerequisites, including a precise context, the adequate deadline and sufficient defined means. But the management factor is always key to all projects. Nowadays, companies are increasingly using more sophisticated management techniques and tools to work out complex tasks. This implies mastering not only project management techniques and tools, but also for several companies, managing a portfolio of projects [6].

According to AFNOR, the company merges with the project. The degree of 'project-enterprise' relationship varies from one organization to another. If this relationship is strong, failure to efficiently implement a project can compromise the future of the company. In some cases, the project can bring together several companies, with different levels of involvement. This relationship is considered weak when the failure of a project does not endanger the organization [7].

Projects are important for organizations not only because they have a limited duration and budget, but also because the organization merge with the project, so much so that the former lives on the latter. For any project, it is possible to measure whether the objectives have been achieved or not [8].

2.2 Importance of Security in the Success of an Organization

Both for personal and professional purposes, we regularly use applications on smartphones, tablets, connected devices or computers. The use of these devices leads us to regularly exchange data that can become sensitive in bad hands. In the digital world, data has become a valuable commodity for any business strategy. However, if such data are accessible to companies, they are also accessible to hackers. On mobile devices, accounts are almost always connected; some applications require, in order to function properly, access to your calls and messages, and must be allowed to broadcast messages on your behalf. Others will request to access your mobile phone number oreven your bank account information in order to update your profile. An infected device can then become a real connected spy object [9].

Price Waterhouse Coopers reveals that there were 4165 cyber-attacks against companies in France in 2016 and that the losses incurred were estimated at 1.5 million euros [10]. In the United States of America, according to the FBI[2], losses due to cyber criminality exceed 170 billion dollars (US) per year. In China, at the estimated amount of 153 billion yuan was published for the year 2010 [11].

---

[1] Information Systems

[2] Federal Bureau of Investigation

Beyond the price value, cybercrime directly harms the image of the company. Cyberattacks can undermine the sustainability of the business. For example, 93% of US companies that lost their data for ten (10) days went bankrupt the year after the loss. A total of 60% of companies that suffer from a simple loss of data cease their activity within six (06) months [12].

3   Research Methodology

3.1   Research Model and Hypothesis

Research Model

Based on the literature, we opted for a modified research model derived from UTAUT 2 (Unified Theory of Acceptance and Use of Technology) and TAM (Technology Acceptance Model). The following independent variables were used: perceived usefulness (PU), price value (PV), effort expectancy (EE), habit (HT), facilitating conditions (FC) and intention to use (IU). We studied the influence of these variables to explain the intention to use (IU) and the use (USE) of information security in IS projects (Fig. 1).

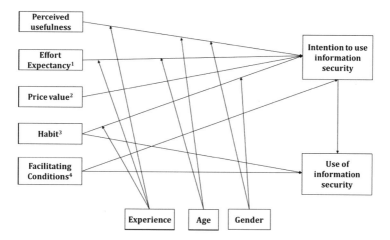

Fig. 1. Theoretical research model (Adapted from Venkatesh, Thong et al. [13], Venkatesh, Thong et al. [14])

Notes

1. Moderated by gender.
2. Moderated by age, experience and gender.
3. Age-moderating effect on intention to use, and moderating effect of age, experience and gender on the use of information security in IS projects.
4. Moderating effect of age, experience and gender on intention to use and on the use of information security in IS projects.

3.2 Data Collection

For this research, we adopted a mixed approach (both qualitative and quantitative), the advantage of which is complementarity. A mixed method can help to develop rich insights into various phenomena of interest that cannot be fully understood using only a quantitative or a qualitative method [15].

Based on the research model we developed, we designed a questionnaire divided into eight (08) sections and made up of forty-two (42) questions. Sixteen (16) questions were concerned with the identity of the respondents and the twenty-six (26) others used a Likert scale ranging from 1 (Strongly disagree) to 7 (Strongly Agree). For qualitative data, we developed an interview guide that was initially made up of seven (07) questions and eventually enlarged to nine (09) after conducting three (03) interviews.

3.3 Data Description

Out of the 214 answers that were received in terms of the quantitative research, only 169 were used in this study, as 45 answers were rejected because the respondents did not have the required profile or because the questionnaire was not fully filled out. In terms of gender distribution, 35 answers came from women (20.7%) while134 were sent by men (79.3%). With regard to their age, the respondents were distributed as follows: only one respondent in the age range 0–20 years (0.6%); 68 respondents (40.2%) aged between 21 and 30; 62 respondents aged between 31 and 40 years (36.7%); 25 respondents in the age range 41–50 years (14.8%); and13 respondents aged 51 or more (7.7%).

The sample population of this study was made of 110 respondents from the private sector (65.1%), 42 respondents from public entities (24.9%), 10 respondents from parastatals (5.9%), 04 NGO[3] employees (2.4%), and 03 respondents from other sectors (1.8%).

Regarding the ranking of the respondents in their respective organizations, 36.1% of respondents (61 persons) were senior executives; 34.3% of them were managers (that is, 58 in number), and there were 43 supervisors (25.4%) while 07 respondents (4.1%) could not be clearly ranked.

For the collection of qualitative data, we targeted solely the respondents who had already worked on at least three IS projects. A Chief Information Officer from a parastatal, a Chief Executive Officer from a private company and three managers from the public sector (working in the IT department of their organization) were interviewed.

3.4 Model's Reliability

Model's validation is a prerequisite for any PLS[4] regression analysis. Therefore, there is a need for convergent and divergent validation.

Convergent validity occurs when each manifest variable correlates strongly with its latent variable. This means that each manifest variable must have a significant t-value

---

[3] Non-Governmental Organizations
[4] Partial Least Square

on its latent variable. In general, the t-value should be at least 1.65 for 90% significance, 1.96 for 95% significance, and 2.57 for 99% significance. Discriminant validity is demonstrated when each variable has a low correlation with the other variables except for the one with which it is theoretically associated [16, 17].

The t-values for the indicators of variables PU, EE, HT, FC, IU and USE are greater than 1.65. For the variable PV, the indicator PV1 has a t-value equal to 1.567. Similarly, the AVEs[5] of PU, EE, HT, FC, IU and USE are greater than 0.5. For the variable PV, it is 0.379. Also, when comparing correlations existing between different latent variables, the t-values for each of the latent variables are greater than the correlations between these latent variables and the other latent variables, except for the variable PV.

Overall, the results obtained provide some satisfaction considering the reliability and validity of our research model's measuring instrument.

4   Results and Discussions

4.1   Results

The significance of a relation can be studied by the t-value or the p-value. For a relation to be significant, its t-value must be at least 1.65, which corresponds to a p-value less than or equal to 0.1 for a significance of at least 90% [18].

The table below summarizes the results obtained using the bootstrapping method.

As shown in Table 1, four (04) of the eight (08) hypotheses were accepted. The intention to use security is generally explained by security's perceived usefulness, habit, and the prevailing facilitating conditions. Here, it is explained only by habit.

Table 1. Hypothesis verification synthesis

H[a]	Relation	Regression	t-value	Significance level	Hypothesis accepted
H1	PU→IU	0,187	2,473	95%	Yes
H2	EE→IU	0,141	1,088	NS[b]	No
H3	PV →IU	0,103	1,442	NS	No
H4	HT→IU	0,163	1,697	90 %	Yes
H5	FC→IU	0,256	2,175	95%	Yes
H6	HT→USE	0,425	4,741	99%	Yes
H7	FC→USE	0,132	1,223	NS	No
H8	IU→USE	0,062	0,556	NS	No

a   Hypothesis
b   Not Significant

---

[5] Average Variance Extracted

4.2   Findings

Analysis of the results obtained through the SmartPLS 3.2.7 software suggests the following observations: (i) variables such as perceived usefulness, effort expectancy, price value, habit and facilitating conditions explain 41.6% of variance in the intention to use information security in IS projects. With a regression equal to 0.256, we found that the variable "facilitating conditions" contributes the most to achieving the intention to use information security, followed by perceived usefulness (0.187) and habit (0,163); (ii) habit, facilitating conditions, and the intention to use contribute by 31.0% to variance in use of information security in IS projects. Since these values are greater than 25%, we can conclude that the contribution of these variables is significant for our model [18]. With a regression coefficient equal to 0.425, habit is the one that contributes to achieving a variable use of information security.

The analysis of the moderating effect of gender shows significant differences between men and women. For women, the intention to use information security is not explained by the selected variables of this study. Among men, the variable intention to use information security is well explained by perceived usefulness, habit, and facilitating conditions. For both men and women, the use of information security in IS projects is explained by habit.

The data collected did not enable us to make a good analysis of the moderating effect of age. There were less than 30 respondents belonging to the age ranges 0–20, 41–50, and 51 years old or more, which was insufficient to apply the bootstrapping PLS method. However, we merged the 41–50-year-old interval with the interval of respondents who were over 50 years old. Thus, we studied the age-moderating effect for respondents who were between 21 and 30, 31 and 40, and over 40 years old. The intention to use information security is not explained by our constructs and indicators for the respondents belonging to the age bracket 21–40, by perceived usefulness for the respondents aged between 31 and 40 years old, and by the variable "facilitating conditions" for those who are older than 40. We also note that in different cases, the use of information security is explained exclusively by habit for respondents under the age of 40. For responders who are more than 40 years old, it is not explained by the variables that we selected for the study.

The moderation of experience shows significant differences between those with less than 06 years of experience, those with between 6 and 10 years of experience and those with more than 10 years of experience. For respondents with less than 6 years of experience, the intention to use information security in IS projects is mostly explained by perceived usefulness. For the most experienced, it is explained by facilitating conditions. As for the use of security, it is well explained by habit. For the more experienced, it is not explained at all by the variables selected for this study.

We noticed that the collected qualitative data showed almost similar results with those from quantitative data, except for the "price value" construct. For IT professionals, the cost of implementing information security has no influence. When the project clearly presents the risks, the top management will not see the cost as a problem.

4.3   Managerial Implications

Following the results of this study, the following recommendations may be made to organizations for an increased consideration of information security in IS projects: (i) it is critical to develop a security charter, an information system security policy and the risk map for information security in IS projects; (ii) the function of Chief Information Security Officer (CISO) is important for organizations of considerable size and its usefulness may depend on the habits and ages of employees. This function could facilitate the adoption and use of information security for older employees and those without any training in the field of information security; (iii) younger employees are more likely to use and implement information security concepts because of their habit of using technology. Any employing company should therefore make sure its youngest employees are trained information security, for them to take on IS projects with some automatic reflexes.

4.4   Contributions of the Study

The objective of this study was to help grasp the determinants of the intention to adopt and use information security in IS projects. The UTAUT 2 model adopted enabled us to obtain interesting results, which informed that the intention to use information security in projects is generally explained by three factors: perceived usefulness, habit, and facilitating conditions. The use of security measures is explained by habit. Our results are aligned with those of numerous other research studies using the UTAUT 2 model or the TAM model.

This work is a clear contribution to research for at least two reasons: (i) it fills an obvious gap that the paucity of studies on the subject had not permitted so far. Kotulic and Clark [1] (in: "Why there aren't more information security research studies?") argue that it is not suitable to carry out large-scale studies on information security because organizations fear to disclose security and IS information to strangers; and (ii) this study opens up new avenues for future research while contributing to refining the UTAUT 2 model, notably by taking into account other variables that can better explain and help understand the factors that influence the intention to adopt and use information security in IS projects.

4.5   Limitations and Future Research

Limits

The various limitations of this study may reside in the following: (i) the defined sample only took into account the professional field of potential respondents. Data collection was not done in a specific geographic area; (ii) this study did not take into account all the basic constructs of the UTAUT 2 model, as variables such as hedonic motivation, performance expectancy and social influence were not considered. As the result, the selected variables may not have adequately explained all aspects of the model. The intention to use information security is explained at 41.6%. The use of information security is explained at 31% only. Although these values are acceptable according to Hair Jr, Hult et al. [18], they are still low; (iii) some questions were not well formulated and therefore were not well understood by the respondents; some questions about the use of technology seemed to be too technical; (iv) the data on the moderators, age and experience should have been captured better than by predefined intervals; (v) a good apprehension for the variable "experience" could not be established; (vi) the data collected was not always sufficient for different groups to do a better analysis of moderators effects; (vii) the administration of the questionnaire using the Internet does not guarantee a good quality of answers insofar as the respondent was not always able to get clarifications concerning the questions not understood. In addition, the sensitive nature of information does not encourage employees or managers to pay attention to the questionnaire; (viii) most of people solicited for this study were those whose core business is information systems or information technology.

Recommendations for Future Research
We have a few recommendations for future research: (i) future research should consider other UTAUT 2 model constructs such as hedonic motivation, social influence, and performance expectancy. The constructs and questions could result from a prior qualitative study that will allow for variables that better explain the research model; (ii) it would be interesting for future research to have a representative sample of a specific geographical location; (iii) future work could be done with a more representative sample considering responses from respondents working in organizations whose core business is not just information systems or IT. It should also be representative for the variables: gender, experience and age. is the aim will be to have a good study of the moderating effects; (iv) a better wording of the questions is necessary so that they are better understood by the respondents; (v) due to the sensitive nature of information, a face-to-face administration of the questionnaire is preferable; (vi) for data such as age and experience, using predefined intervals or very wide intervals should be avoided; (vii) it would be more interesting to capture, for the experience variable, the number of projects on which a respondent worked rather than the number of years of experience.

5 Conclusion

Information security research is one of the most intrusive types of organization research, and there is undoubtedly a general mistrust of any "outsider" attempting to gain data about the actions of the security practitioner community [1].

The purpose of this research paper was to understand the factors that influence the intention to adopt and use information security in IS projects. We used the UTAUT 2 model by Venkatesh and Bala [19], Venkatesh, Thong et al. [13] to which we integrated the "perceived usefulness" variable of the TAM model. We got 166 valid answers from professionals working in the field of information systems. We also conducted five interviews with IT project managers. From the results we got through the PLS method, the following conclusions are being made: (i) the intention to use information security in IS projects is explained by perceived usefulness, habit, and facilitating conditions; (ii) the use of information security is explained by habit. Some of these results are similar to those obtained by Venkatesh, Thong et al. [14]. The variables used to conduct the study reveal that the intention to use information security is explained at 41.6%. The use of information security is therefore explained at 31%. These results show that there are other variables that may better explain the intention to adopt and use information security.

This study surely contributes to enriching the literature in the field of information security where the level of scientific research production remains low. Other researches by Kotulic and Clark [1] or Dagorn and Poussing [2] show the difficulties faced by researchers in this area. This study provides science with a theoretical research model that can be refined to better explain dependent variables in subsequent research works.

While enriching knowledge in the sensitive area of information systems, this study provides decision-makers with a body of useful insights for properly adopting and using information security to protect their organizations from various risks related to the use of information systems.

References

1. Kotulic, A.G., Clark, J.G.: Why there aren't more information security research studies. Inf. Manag. **41**(5), 597–607 (2004)
2. Dagorn, N., Poussing, N.: Engagement et pratiques des organisations en matière de gouvernance de la sécurité de l'information. Systèmes d'information Manag. **17**(1), 113–143 (2012)
3. Raufer, X.: De la cyber-jungle au cybermonde. Sécurité Globale **8**(4), 5–10 (2016)
4. Le Cœur, J.: Sécuriser les données personnelles de son entreprise. I2D Inf. données Doc. **53**(1), 25–26 (2016)
5. Wolf, P.: Les menaces numériques aujourd'hui. Sécurité et stratégie **3**(1), 44–46 (2010)
6. Christian, M., Dennis, K., Jochen, W.: Is it worth the effort? A decision model to evaluate resource interactions in IS project portfolios. Paderborn University, Faculty of Business Administration and Economics (2015)
7. Aïm, R.: Les fondamentaux de la gestion de projet. http://groupe.afnor.org/pdf/fondamentaux-gestion-projet.pdf. Accessed 27 May 2017
8. Coupe, C., Masanovic, S., Rodet, P.: L'humain au secours d'un management violent: le mode projet. http://www.finyear.com/L-humain-au-secours-d-un-management-violent-le-mode-projet_a11046.html. Accessed 28 June 2017
9. Chen, A.: Gérer les données personnelles fournies aux applications. I2D Inf. données Doc. **54**(3), 47–48 (2017)
10. Coopers, P.W.: La cybersécurité en France vue par les entreprises: 47% de cyberattaques en moins et des pertes financières divisées par deux en (2016). http://www.pwc.fr/fr/espace-presse/communiques-de-presse/2016/octobre/cybersecurite-france-gsiss.html. Accessed 29 March 2017
11. Yuan, T., Chen, P.: Data mining applications in E-government information security. Procedia Eng. **29**, 235–240 (2012)
12. Dajoux, J.: La perte de données, une cause importante de faillite des entreprises. https://www.informanews.net/perte-de-donnees-cause-faillite/. Accessed 31 Oct 2018
13. Venkatesh, V., et al.: Extending the two-stage information systems continuance model: Incorporating UTAUT predictors and the role of context. Inf. Syst. J. **21**(6), 527–555 (2011)
14. Venkatesh, V., Thong, J.Y.L., Xu, X.: Consumer acceptance and use of information technology: extending the unified theory of acceptance and use of technology. Manag. Inf. Syst. Q. **36**(1), 157–178 (2012)

15. Venkatesh, V., Brown, S.A., Bala, H.: Bridging the qualitative-quantitative divide: Guidelines for conducting mixed methods research in information systems. MIS Q. **37**(1), 21–54 (2013)
16. Gefen, D., Straub, D.: A practical guide to factorial validity using PLS-graph: tutorial and annotated example. Commun. Assoc. Inf. Syst. **16**(1), 5 (2005)
17. Kunstelj, M., Jukić, T., Vintar, M.: Comment tirer pleinement parti des résultats des enquêtes auprès des utilisateurs du gouvernement électronique: le cas de la Slovénie. Revue Internationale des Sci. Adm. **75**(1), 129–166 (2009)
18. Hair Jr., J.F., et al.: A Primer on Partial Least Squares Structural Equation Modeling (PLS-SEM). Sage Publications, Thousand Oaks (2016)
19. Venkatesh, V., Bala, H.: Technology acceptance model 3 and a research agenda on interventions. Dec. Sci. **39**(2), 273–315 (2008)

Privacy Preserving kNN Spatial Query with Voronoi Neighbors

Eva Habeeb$^{(\boxtimes)}$, Ibrahim Kamel, and Zaher Al Aghbari

University of Sharjah, Sharjah, UAE
{habeeb,kamel,zaher}@sharjah.ac.ae

Abstract. With the increased demand for outsourcing databases, there is a demand to enable secure and efficient communications. The concern regarding outsourcing data is mainly providing confidentiality and integrity to the data. This paper proposes a novel solution to answering kNN queries at the cloud server over encrypted data. Data owners transform their data from a native domain to a new domain to assist in nearest neighbors' classification. The transformation is achieved by Voronoi diagram, which transforms the data space into numerous small regions, simplifying the nearest neighbor search. However, because the regions that make up a Voronoi diagram are irregularly shaped, the search through the network becomes hard to accomplish.

Thus, the solution includes a Grid-based indexing approach for the Voronoi diagram to expedite the kNN search. Additionally, a strong encryption algorithm, like AES, is used to encrypt the data objects being sent from the data owner to the cloud. An authorized user sends encrypted kNN queries to the cloud where the query is processed over encrypted data. The cloud service provider utilizes the proposed indexing scheme to identify a superset of the nearest neighboring objects to be sent back to the user. The user possessing a copy of the encryption key decrypts the superset of k nearest neighbors and filters the exact k objects.

Keywords: kNN query · Database outsourcing · Spatial data · Voronoi partitioning

1 Introduction

With numerous applications available over the internet along with mobile location-based services, there is a demand to enable secure and efficient communication. This in turn has increased the amount of spatial data that is required to be managed efficiently. Spatial data used by location-based services is composed of location information (*i.e.*, longitude and latitude) as well as other descriptive components that necessitate huge storage capacity. On daily basis, location-based service users aim to issue spatial queries anonymously with fast and efficient response. Additionally, owners of the spatial data aim to keep the data from being disclosed by the service provider to maintain data confidentiality. Therefore, in our approach the goal is to achieve a balance between efficient query processing and obscuring the data at the service provider.

© Springer Nature Switzerland AG 2019
Á. Rocha et al. (Eds.): WorldCIST'19 2019, AISC 931, pp. 900–910, 2019.
https://doi.org/10.1007/978-3-030-16184-2_85

There are several requirements that must be satisfied when outsourcing spatial databases. Firstly, it is important to keep the service provider and any untrusted party (i.e., malicious attacker) from having access to the database content. The clear solution would be to have the data owners encrypt the databases and send the encrypted databases to the service provider without revealing the encryption keys. However, when an authorized user queries the cloud service provider, the user will retrieve the entire encrypted database then decrypt it to process the query and find the required result. This approach provides complete security of the database but does not provide real-time results and the communication cost will be high.

Another requirement is developing an efficient query processing algorithm that can be executed on encrypted data at the service provider. Therefore, allowing the communication cost to be reduced since all query processing would be done at the service provider, while only the results will be returned to the authorized user rather than the entire database. Nowadays, one of the most challenging issues is secure computation on the cloud assisted platform [12].

With the excessive use of mobile devices and navigational systems with GPS, location-based services (LBSs) have become widely popular. Several prior works [5, 6] provide security by adding a middleware or tamper-proof device at the CSP to ensure security. However, with numerous users, it is not practical to have an individual device for every AU at the CSP. To overcome this, Yiu *et al.* [3] present a cryptographic-based transformation scheme (CRT) for two-dimensional data to enhance the security of spatial data. Yiu *et al.* [3] also present three different spatial transformation methods that are based on partitioning and redistributing the locations in the space.

The cloud architecture model used in this paper consists of three main entities (see Fig. 1), which are the Data Owner (DO), the Cloud Service Provider (CSP), and the Authorized User (AU). The DO owns the spatial database and is responsible for encrypting it before sending it to the CSP. The AU sends queries to the CSP, which processes the query on encrypted data and returns the results back to the AU. The AU has the encryption key used by the DO and uses the key to decrypt the results returned from the CSP.

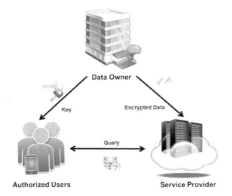

Fig. 1. Cloud architecture model

The contributions made by this paper are as follows:

- Developing a novel algorithm for processing kNN queries over spatial data at the cloud service provider.
- Developing a novel algorithm for detecting intersections between grid cells and Voronoi regions.
- Answering kNN queries on encrypted data.
- Presenting a Grid-based index for expediting the kNN search.

The remainder of the paper is organized as follows. Next section presents some existing work in the literature. Section 3 outlines the proposed scheme for answering nearest neighbor queries and describes the Grid-based index for retrieving kNN. Experiments are conducted on a real dataset [8] to evaluate the performance of the proposed approach, and results are presented in Sect. 4. Lastly, the conclusion and discussion are presented in Sect. 5.

2   Related Work

Prior works like [13] provide data privacy by hiding the real location of the query points using location anonymity. Other works like [14] propose using Asymmetric scalar-product-preserving encryption (ASPE) for securing query processing on encrypted data on the cloud service provider. However, in this approach the user is assumed to be trusted, which is unrealistic in real applications. Therefore, optimized ASPE is proposed by [15, 16], where the user is not trusted and doesn't know the encryption key, and only the data owner knows the encryption key.

Hore et al. [17] propose an algorithm for securing outsourced data by applying bucketization to compute a secure indexing tag of the data. This approach prevents the server from knowing the exact values of the data but allows checking whether a record satisfies a query predicate. Multidimensional range queries are answered in an approximate manner, in which the returned result will contain many false positives.

Our previous work [9] uses Hilbert space-filling to transform the space by mapping each spatial point in the multidimensional space to one-dimensional space. The resulting Hilbert cell values are encrypted using order-preserving encryption, while the data points are encrypted using AES. Both the transformation key and the encryption key are transmitted by the DO to the AUs over SSL without any tamper-resistant devices. The user issues spatial kNN queries to the service provider based on the Hilbert values and then uses the encryption key to decrypt the query response returned.

Similarly, Hong et al. [18] deal with kNN queries on encrypted outsourced spatial data. The scheme of encryption used is an asymmetric scalar-product-preserving encryption scheme, where diverse encryption keys are used to encrypt the data points as well as the query points. The cloud service provider determines the distance relation between the query points and the data points. Additionally, an SS-tree that supports efficient kNN is built.

A spatial transformation scheme is proposed by Hossain et al. [7] offers data security by applying a shear transformation and rotation transformation. However, these techniques preserve the coordinates of the original spatial points and assuming that an attacker can gain background knowledge of the original points and coordinates of these points in the transformed space, information about close by data points can be exposed.

Hu *et al.* [10] proposed the use of Voronoi diagram to derive neighborhood information regarding spatial data. The scheme developed is called VN-Auth, that allows a user to verify the correctness of the computed results. VN-Auth handles more than one spatial query types including, *k*NN, range queries, reverse *k*NN, spatial skyline, and aggregate nearest neighbor. The scheme is evaluated using real-world dataset with mobile devices as the query clients.

Efficient query processing is a key requirement of LBSs for the user and therefore database outsourcing techniques have to achieve a low communication cost. This requires schemes that encode the spatial data and query regions. Most of the existing techniques do not utilize the computational power of the CSP and thus, we plan to overcome this shortcoming by performing efficient *k*NN queries on the encrypted data at the server.

3 Scheme for Securely Finding *K*NN Objects

The suggested scheme of computing the *k* nearest neighboring objects over outsourced databases ensures the confidentiality of the data. Data owners (DO) outsource their encrypted data to Cloud Service Providers (CSP) to be stored. Authorized Users (AU) send encrypted query requests to the CSP to be processed. Therefore, the main contribution of this paper is allowing the CSP to process *k*NN queries over encrypted data.

Voronoi diagram used in this scheme, transforms the objects from a native space to a new space, which makes the neighborhood information and the search for nearest neighbors easier. However, due to the fact that a Voronoi diagram partitions space in an irregularly-shaped manner, we construct a Grid-based index to make the search for the nearest neighboring objects easier. The index is used by the CSP to return a superset of *k*NN to the AU, which in turn decrypts the returned set and extracts the exact *k* nearest objects by distance comparisons.

3.1 Voronoi Diagram

A Voronoi diagram consists of partitioned polygons, where each polygon is centered by a data point, these data points are called the generators. The region around each generator consists of all possible data points that are closer to the generator than any other point, meaning any data point found inside the polygon will have the generator data point as its nearest neighbor. Voronoi regions are bounded by edges called Voronoi edges, that are locations that belong to more than one Voronoi region. Any Voronoi regions that share the same edges are considered Voronoi neighbors.

3.2   Space Partitioning for *K* Nearest Neighbor

The first step of the proposed approach is the construction of a Voronoi diagram to organize the spatial data points into smaller networks. The Voronoi diagram partitions the large network of data points into smaller networks called Voronoi regions or Voronoi polygons.

After the constructing the Voronoi diagram, a Grid-based indexing algorithm is created, in which a grid overlays the entire Voronoi diagram. The size of the grid depends on the data points found in the dataset, to ensure that the grid covers the entire Voronoi diagram. The grid cells created are numbered starting from 1, making the process of searching for the nearest neighbors easier. Grid-based indexing allows the CSP to identify where to begin the search. It also focuses the kNN search on a specific grid cell containing part of the Voronoi diagram, rather than the entire Voronoi diagram.

The next step is to compute the Voronoi neighbors table, which is done by determining which Voronoi regions share an edge hence making them Voronoi neighbors. Another table is constructed, the Cell-Voronoi Mapping table, which corresponds to intersections between grid cells and Voronoi regions. The intersections were computed using a polygon-polygon intersection function, where each grid cell is considered as a polygon and each Voronoi region is also considered as a polygon, allowing us to go through the entire grid and compute the Voronoi regions that intersect with each grid cell.

Finally, the resulting data is encrypted using AES encryption at the DO and sent to the CSP. The encryption key is transmitted from the DO to the AU over SSL without the use of tamper-resistant devices.

3.3   AES Encryption

To protect the data sent from the DO to the CSP from any third-party interventions, the Voronoi neighbors and the Cell-Voronoi Mappings are encrypted using Advanced Encrypted Standard (AES). AES encryption is a block cipher that uses symmetric key encryption, the same key is used for both encryption and decryption. This algorithm is fast and secure, where it has not been proven that the encryption can be broken yet by any of the well-known cryptoanalysis attacks. Since the CSP does not have the encryption key and does not have the ability to decrypt, the data is kept confidential from the CSP, as it is not trusted. Query processing done at the CSP on encrypted data is possible, because AES allows conducting equality comparisons on the encrypted data directly. Lastly, the DO provides the AU with the encryption key to issue and encrypt queries to the CSP, and to decrypt query response from the CSP to acquire the actual results.

3.4   *K*NN Query Processing

The proposed approach utilizes the flexibility in computing distances provided by Voronoi diagrams to identify the nearest neighbors to a query point. The AU issues a query with the required *k* neighbors to be returned by the CSP. The query is encrypted using the encryption key obtained from the DO and is sent to the CSP. At the CSP where the encrypted data is stored, processing of the query is done efficiently according to Algorithm 1.

Given a query point, the algorithm locates the grid cell that contains the query point Q. Next, the initial nearest neighbors of the query point are computed, which correspond to an entry in the Cell-Voronoi Mapping table (*lines 1–2*). As mentioned before, Cell-Voronoi Mapping maps each grid cell to the Voronoi regions that are intersecting the grid cells, making these points the nearest points to any point that falls into the specific grid cell. In the case of k equals one, the candidate set is returned as variable *"Results"*. Variable *"Results"* represents the nearest neighbors to the query point as extracted from the Cell-Voronoi Mapping table for the specific cell ID. *Lines 3–9* explain the case where k is greater than one, the algorithm checks if the value of the initial nearest neighbors is greater than k. However, if it is not, the Voronoi neighbors of all the initial nearest points according to the Voronoi neighbors table are added to the candidate set *"Results"* until the values returned are greater or equal to k. The final values contained in candidate set *"Results"* (superset) are encrypted and transmitted to the AU (*lines 10–11*). In *lines 12–16*, the AU receives the superset of kNN, decrypts the result, and filters the exact k nearest neighbor by distance comparison between the query point and the returned data points.

Algorithm 1. Spatial *k*NN Query

Input: Q: Query Point, k: Value of k
Output: *k*NNResult: k Spatial Data Points
1) $QueyCell$ = GridSearch (Q), Grid cell containing Q
2) $Result$ = CellVoronoiMapping($QueryCell$)
3) **if** $(k > 1)$
4)   **while** $(k > \text{length}(Result))$
5)   **for all** $(i \in Result)$
6) $Result = Result \cup VnNeighbors(i)$
7)   **end for**
8)   **end while**
9) **end if**
10) Encrypt ($Result$)
11) Return Result to AU
12) **for all** $(R \in Result)$
13)   ResD= Decrypt R
14)   DistNN = Compute distance between Q and ResD
15) **end for**
16) *k*NNResult = Find*k*NN(DistNN)

---

4   Experimental Results

This section evaluates the performance of the proposed approach. The main goal is to investigate the effect of the parameter k of the kNN query on the redundancy in the candidate results sent to the AU. The experiments are performed on real-world spatial data obtained from [8], in which the dataset contains 18,263 points describing the road network of City of San Joaquin County (TG). We normalize the domain of the dataset to the domain [0,1]. Experiments were performed on an Intel Core i7-4710HQ CPU @ 2.50 GHz with 8 GB RAM running the 64-bit Windows operating system implemented using MATLAB. For conducting the experiments, a query request for k nearest neighbors is generated randomly. The response for the query will be generated according to the Cell-Voronoi Mapping table and the Voronoi Neighbors table. The query response represents the k nearest points to the query.

If the value of k is equal to one, the result consists of the data points of the Voronoi regions that intersect with the grid cell that the query point falls in. However, if the value of k is greater than one, the result consists of the data points of the Voronoi regions that intersect with the grid cell that the query point falls in and their Voronoi neighbors until the returned result satisfies the value of k. Note that the data points that fall in the neighbor of the neighbor can be closer than the data that falls in the neighboring cell. Therefore, our approach will result in return of a superset of the answer points.

4.1   Candidate Set Redundancy

A randomly generated query point is used to test the functionality of the algorithm, while the value of k is varied. The same query point is used in all the following parts of this experiment. Figure 2 plots the dependent variable (redundancy) at the y-axis and the independent variable (k-nearest neighbor) at the x-axis. Figure 2 shows the relationship between the values of k and the redundancy in the returned results. When the value of k is one, it can be seen in the figure that there are nine redundant result points, meaning that there are ten points returned as the superset of nearest neighboring points. As the value of k increases to two, the number of the points returned increases to 68, thus the redundant points are 66 which is the highest number of redundancies. As the value of k increases, the number of returned points stays the same 68 until k reaches 68. Therefore, the redundancy decreases until it reaches zero at the value of 68.

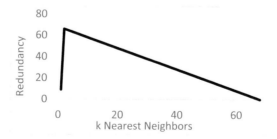

Fig. 2. Relationship between redundancy and k- value

As k increases beyond 68, the Voronoi neighbors of the previous 68 points will be retuned as the kNN query result. Thus, the new resulting superset consists of 200 points, with the highest redundancy value 131 at k equaling to 69. This number of resulting points stays constant and the redundancy decreases as the value of k increases until k reaches 200, where the redundancy will become zero. Figure 3 shows cycles of this pattern of decrease and increase in redundancy, where the redundancy reaches the maximum value of redundant points then decreases linearly until there is no redundancy in the k nearest neighbor points returned. The percentage of redundancy in the superset turns out to be 33.3%. Therefore, the number of false positive results is low.

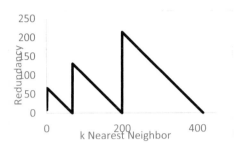

Fig. 3. Pattern for 3 cycles redundancy vs. *k*-value

4.2 Pattern of Redundancy

The pattern for each of the cycles begins with an initial redundancy value, which is the maximum redundancy of points for that cycle. Notice also that an increase in the width of each cycle until zero is reached is clear in the patterns. Figure 4(a) plots the dependent variable (redundancy) at the y-axis and the independent variable (*k*-nearest neighbor) at the x-axis. Figure 4(a) shows the peak redundancy values reached for each cycle, in which the redundancy increases as the value of k increase. As shown in Fig. 3, the peak of the redundancy follows the sawtooth pattern.

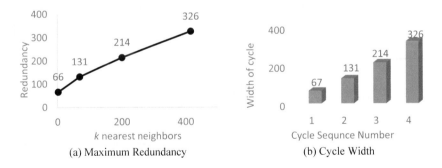

(a) Maximum Redundancy

(b) Cycle Width

Fig. 4. Redundancy trend vs. cycle width

While Fig. 4(b) plots the width of the cycles as the y-axis and the cycle sequence number as the x-axis. It is shown that the width of the cycle also increases as the value of k increase. We conclude that there is a relation between the peak values and the width of each cycles. For each cycle, the peak value is the same as the width of that cycle, because as the peak value of the cycle is reached, the redundancy is reduced by 1 for every k increase until it reaches zero. An example would be cycle number three where the peak value is 214, this cycle will go on for 214 times (width of the cycle) until the redundancy becomes zero and then begins cycle number four where the neighbors of the data points in cycles number three are taken into consideration.

4.3   Redundancy and Grid Size

There is a relation between redundancy in the candidate set and the size of the grid cells that form the grid used for indexing. Figure 5 plots the Redundancy at the y-axis and the size of the grid cell at the x-axis. Note that the relationship is linear, for a smaller size of the grid redundancy is low. As the size of the grid cell increases, the redundancy increases as well. The reason behind this trend is the fact that with smaller dimensions, the area where the search for the k nearest neighbors will be decreased. Which in turn will result in less points in the candidate set. Therefore, it can be concluded that to provide better results in term of redundant data, it is recommended to implement grid cells with small dimensions.

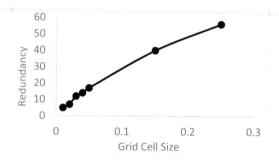

Fig. 5. Redundancy vs. grid cell size

5   Conclusion

Database outsourcing is a popular paradigm in cloud computing where storage and computing resources are virtualized at the cloud, while providing data to Authorized users. In this work, the main goal is to achieve data confidentiality and efficient query processing at the cloud. The ability of Voronoi diagrams to transform objects to a different space is utilized in this approach to assist in computing the k nearest neighbors to a query object. Additionally, Grid-based indexing method has been developed to simplify the process of searching for neighboring objects.

The proposed system returns a superset that contains the kNN, but the number of false positive objects is low. The experiment shows that the false points start high at a specific k and quickly fall to zero as k increases. We notice that redundancy increases with the value of k. The experiment shows that the average redundancy is below 35%. Thus, we conclude that our method not only protects the data but also enables the CSP to process kNN queries over encrypted data and retrieve the superset of resulting objects to be returned to the user. However, the response time for the Grid-based indexing is not fast, but due to the use of spatial data which updates rarely, the response time is considered acceptable. Additionally, the Grid-based indexing will be done only once in advance (not in real-time), as the spatial data is static. For future work, we plan to improve the response time for Grid-based indexing.

References

1. Khoshgozaran, A., Shahabi, C.: Blind evaluation of nearest neighbor queries using space transformation to preserve location privacy. In: Advances in Spatial and Temporal Data Bases. Springer, pp. 239–257 (2007)
2. Ku, W.-S., Hu, L., Shahabi, C., Wang, H.: A query integrity assurance scheme for accessing outsourced spatial databases. Geoinformatica **17**(1), 97–124 (2013)
3. Yiu, M.L., Ghinita, G., Jensen, C.S., Kalnis, P.: Enabling search services on outsourced private spatial data. VLDB J. **19**(3), 363–384 (2010)
4. Kim, H.-I., Hong, S.-T., Chang, J.-W.: Hilbert-curve based cryptographic transformation scheme for protecting data privacy on outsourced private spatial data. In: 2014 International Conference on Big Data and Smart Computing (BIGCOMP), pp. 77–82. IEEE (2014)
5. Hacigumus, H., Iyer, B., Mehrotra, S.: Providing database as a service. In: Proceedings of 18th International Conference on Data Engineering, 2002, pp. 29–38 (2002). IEEE
6. Damiani, E., Vimercati, S., Jajodia, S., Paraboschi, S., Samarati, P.: Balancing confidentiality and efficiency in untrusted relational DBMSs. In: Proceedings of the 10th ACM Conference on Computer and Communications Security, pp. 93–102. ACM (2003)
7. Hossain, A.A., Lee, S.-J., Huh, E.-N.: Shear-based spatial transformation to protect proximity attack in outsourced database. In: 2013 12th IEEE International Conference on Trust, Security and Privacy in Computing and Communications (TrustCom), pp. 1633–1638. IEEE (2013)
8. Real spatial datasets. http://www.cs.fsu.edu/lifeifei/SpatialDataset.htm
9. Talha, A.M., Kamel, I., Aghbari, Z.A.: Secure kNN queries over outsourced spatial data for location-based services. In: 2016 12th International Conference on Innovations in Information Technology (IIT), Al-Ain, pp. 1–4 (2016)
10. Hu, L., Ku, W., Bakiras, S., Shahabi, C.: Spatial query integrity with voronoi neighbors. IEEE Trans. Knowl. Data Eng. **25**(4), 863–876 (2013)
11. Kolahdouzan, M., Shahabi, C.: Voronoi-based k nearest neighbor search for spatial network databases. In: Proceedings of VLDB, vol. 30, pp. 840–851 (2004). https://doi.org/10.1016/b978-012088469-8.50074-7
12. Arora, D., Kumar, U.: Implications of privacy preserving k-means clustering over outsourced data on cloud platform. J. Theor. Appl. Inf. Technol. **96**(12) (2018)
13. Chow, C.-Y., Mokbel, M.F., Aref, W.G.: Casper*: query processing for location services without compromising privacy. ACM Trans. Database Syst. (TODS) **34**(4) (2009). Article24
14. Wong, W.K., Cheung, D.W., Kao, B., Mamoulis, N.: Secure kNN computation on encrypted databases. In: Proceedings of the International Conference on Management of Data and 28th Symposium on Principles of Database Systems (SIGMOD-PODS 2009), Providence, RI, USA, pp. 139–152, July 2009
15. Zhu, Y., Xu, R., Takagi, T.: Secure k-NN computation on encrypted cloud data without sharing key with query users. In: Proceedings of the 2013 1st International Workshop on Security in Cloud Computing, Cloud Computing 2013, China, pp. 55–60, May 2013

16. Zhu, Y., Huang, Z., Takagi, T.: Secure and controllable k-NN query over encrypted cloud data with key confidentiality. J. Parallel Distrib. Comput. **89**, 1–12 (2016)
17. Hore, B., Mehrotra, S., Canim, M., Kantarcioglu, M.: Secure multidimensional range queries over outsourced data. VLDB J. **21**(3), 333–358 (2011)
18. Hong, J., Wen, T., Guo, Q., Ye, Z.: Secure kNN Computation and Integrity Assurance of Data Outsourcing in the Cloud. Math. Probl. Eng. **2017**, 1–15 (2017)

Analysis of Authentication Failures
in the Enterprise

Richard Posso[✉] and Santiago Criollo-C

Universidad de las Américas, Quito EC170125, Ecuador
{richard.posso,luis.criollo}@udla.edu.ec

Abstract. Information security within enterprises has endured the systems and services in order to protect all the information generated. Among the tools currently adopted worldwide to comply with information security is the implementation of AAA systems, which are used to authenticate users, grant access to the network and account the events generated in the network, particularly for authentication success and failures. The latter is the interest for this journal. Hence, a research is developed in an enterprise to reveal the reasons for failures, solutions and feasibility of application which empowers the company to improve authentication systems and, consequently, information security.

Keywords: AAA · Authentication · Enterprises · Radius · Security

1 Introduction

The last decade has witnessed an exponential growth of digital information within enterprises. According to IDC, the tendency of digital info will grow from 130 exabytes in 2005 to 40.000 exabytes in 2020, similar to 40 trillion gigabytes Unfortunately, there is no enough attention to the security for these digital information systems [1, 2]. Several methods for information security risk assessment have been developed and are available for the enterprise [3].

Indeed, one of the basic recommendation for security is to implement Authentication, Authorization and Accounting (AAA) system within the network, this concept and how it works is briefly explained in the first part of the paper. The second part describes the status of the enterprise in order to understand the authentication scheme implemented. Next section seeks to discover the main reasons about authentication failures within the enterprise. The data sample was examined from a data network in Ecuador with about 3.100 internal employees and about 1,600.000 taxpayers who use the network [4].

These reasons are sorted and investigated according to the percentage of occurrence. Moreover, a research for the causes and solutions are presented and contrasted with the status of the enterprise to determine whether they are feasible to apply or not. Finally, conclusions and recommendations are stated about the research and to improve the AAA system within the data network.

© Springer Nature Switzerland AG 2019
Á. Rocha et al. (Eds.): WorldCIST'19 2019, AISC 931, pp. 911–920, 2019.
https://doi.org/10.1007/978-3-030-16184-2_86

2 Authentication, Authorization and Accounting (AAA) System

The triple-A or AAA is a term refered to authentication, authorization, and accounting processes in order to get effective network management and security [5]. Consider AAA as the security framework to empower business success. The truth is that the concept of AAA is not new; for instance, in 2001 the United States published a patent about the AAA mechanism and how it works with different protocols and clients. In 2007 the IETF published a RFC-4962 "Guidance for AAA key management". Another example is the IEEE which issued a framework for the WiMAX with AAA schemes [6]. The authentication process is also explained and simplified by Cisco. Briefly, the process for user login consists of three main steps:

(1) Incoming access request to switch
(2) The device sends the authentication request to the AAA servers configured sequentially; if the first server fails, the request is sent to the second and so on until any of the AAA server in the group responds the authentication request. If none of the AAA servers responds, then the request are sent to the next servers group. When all the servers groups failed, the local database is used for authentication instead.
(3) When the AAA request reaches an active server and if the request is successfully authenticated with RADIUS protocol, the user is granted access to the network. Otherwise, when authentication fails the user is either isolated of the network or granted access with restrictions.

3 Status of the Enterprise

Due to the security role that authentication brings about for the enterprise, the AAA process was implemented in an Ecuadorian enterprise with about 3.100 internal employees [4] and about 1,600.000 taxpayers [7] who visit the 65 branches along the country. The research approach or methodology was based on the qualitative analysis of authentication logs from these users collected in the infrastructure; however, for a better understanding about the current authentication failures, it is required to briefly

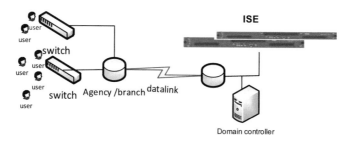

Fig. 1 Network authentication summary

describe the status of the company which includes the authentication structure, the identity stores configured along with the switchs configurations applied, and the active directory policies (Fig. 1).

3.1 Authentication Scheme

A star topology was implemented in the enterprise to connect all the branches along the country; the AAA implementation follows the same structure; hence, the switches implemented in each branch sends the authentication request to the authentication server in the data center (middle of topology). The active directory server also resides in the data center.

3.2 Identity Stores

The authentication server allows configuring the following identity stores: internal, external such as active directory, OTP or Deny. The identity stores already configured in the enterprise are only two: the internal and the active directory.

The latter is mostly used to authenticate users and computers from the enterprise domain. The internal identity store is used for equipment that does not have a unique username such as printers, alarms, biometrics, and cameras.

3.3 Active Directory Policies

In the active directory is possible to create several policies in order to safeguard the enterprise scheme and comply with corporate guidelines [8]. Among the policies configured in the active directory, the password policy requires users to change password each 90 days.

Moreover, some examinations were performed in the active directory to identify users with special characters such as "ñ", "ü", and "ä". Remarkably, none of the users have these special characters; however, the search results show that active directory does not differentiate between special characters are taken as a normal character. In other words, the search for "ñ" is the same as "n"; search for "ü" equals to "u", and "ä" equals to "a".

3.4 Switch Configuration

There exist several recommendations to configure radius in the network device [8]. In fact, the configuration implemented in the enterprise was based on these recommendations. As a result, a summary of the script implemented for radius has the following commands:

```
aaa new-model
aaa authentication dot1x default group radius
aaa authorization config-commands
aaa authorization network default group radius
aaa accounting network default start-stop group radius
dot1x system-auth-control
radius-server host X.X.X.X key SHAREDSECRET
radius-server vsa send accounting.
```

4 Reasons for Radius Authentication Failures

With an appropriate understanding of network topology, the processes of authentication, the identity stores configured and some of the security policies implemented in the active directory, it is time to analyze the causes for failures. The data authentication samples were gathered from the AAA server implemented in the institution. The data was obtained from 26th march until 24th April 2018.

Table 1. Number of authentication requests

Day	Passed	Failed	Total	Failed (%)
26/3/2018	9175	47	9222	0,510%
27/3/2018	12403	70	12473	0,561%
28/3/2018	10250	65	10315	0,630%
29/3/2018	9499	62	9561	0,648%
30/3/2018	14288	103	14391	0,716%
31/3/2018	15937	116	16053	0,723%
1/4/2018	10049	74	10123	0,731%
2/4/2018	13449	230	13679	1,681%
3/4/2018	10195	191	10386	1,839%
4/4/2018	54439	1034	55473	1,864%
5/4/2018	57869	1109	58978	1,880%
6/4/2018	56378	1084	57462	1,886%
7/4/2018	55622	1103	56725	1,944%
8/4/2018	56832	1129	57961	1,948%
9/4/2018	56833	1141	57974	1,968%
10/4/2018	52980	1067	54047	1,974%
11/4/2018	54059	1111	55170	2,014%
12/4/2018	55482	1208	56690	2,131%
13/4/2018	46630	1017	47647	2,134%
14/4/2018	51206	1135	52341	2,168%
15/4/2018	56866	1263	58129	2,173%
16/4/2018	51344	1143	52487	2,178%
17/4/2018	51512	1156	52668	2,195%
18/4/2018	49323	1129	50452	2,238%
19/4/2018	50643	1162	51805	2,243%
20/4/2018	52254	1248	53502	2,333%
21/4/2018	49710	1198	50908	2,353%
22/4/2018	48126	1207	49333	2,447%
23/4/2018	48229	1244	49473	2,515%
24/4/2018	17691	674	18365	3,670%

*Colored rows are the percentages above 2%

According with the data in this period, the percentages of authentication failures are below the 4%. Nevertheless, it is important to determine the main reasons that cause failures. The Table 1 shows the date of the statistics, the number of succeeded and failed authentication, the total number of requests and the percentage of failures per day.

The rows highlighted indicate the dates where the faults exceeded 2,00%. It can be inferred that the failures occurred mostly from April 11 to 24. To identify the main reasons for the failures, an additional analysis was implemented. Table 2 summarizes the findings about authentication failures within the same time interval.

Table 2. List of authentication failures

Item	Failure reason	Percentage
1	Rejected per authorization profile	76,945%
2	Subject not found in the applicable identity store(s)	16,489%
3	User has expired	2,141%
4	User authentication against Active Directory failed since user has entered the wrong password	1,994%
5	No response received from Network Access Device after sending a Dynamic Authorization request	0,498%
6	Received EAP packet from the middle of conversation that contains a session on this PSN that does not exist	0,387%
7	Account is not yet active.	0,322%
8	RADIUS packet contains invalid state attribute	0,257%
9	No response received from Dynamic Authorization Client in ISE	0,249%
10	Unexpectedly received empty TLS message; possible rejection by client	0,208%
11	PEAP failed SSL/TLS handshake because the client rejected the ISE local-certificate	0,204%
12	Wrong password or invalid shared secret	0,077%
13	Username attribute is not present in the authentication request	0,065%
14	Supplicant sent unexpected unencrypted TLS handshake message instead of TLS application data in PEAP protocol	0,041%
15	User authentication against Active Directory failed since user is required to change his password	0,041%
16	Invalid or unexpected EAP payload received	0,024%
17	Unexpectedly received empty TLS message; possible rejection by client	0,016%
18	Change password against Active Directory failed since user has a non-compliant password	0,008%
19	User change password against Active Directory failed	0,008%
20	Expected TLS acknowledge for TLS fragment but received another message	0,008%
21	PEAP handshake failed	0,004%
22	Supplicant sent unmatched EAP Response packet identifier	0,004%
23	Cryptographic processing of received buffer failed	0,004%
24	Client sent Result TLV indicating failure	0,004%

The rows highlighted in yellow indicate the dates where failures overpassed the 2.00%. It can be inferred that failures mostly occurred from the 11th to the 24th of April. To identify the main failures reasons, further analysis was deployed. The next table summarizes the findings about authentication failures within the same time interval.

The Table 2 shows up to 24 reason for authentication failures; nevertheless, this paper will examine the main reasons by applying Pareto rule (80/20). On this context, the first 5 reasons constitute the 20% of the total number of failures, which is approximately the 98.067% of failures. In brief, the 5 main reasons are:

1. Rejected per authorization profile
2. Subject not found in the applicable identity store(s)
3. User has expired
4. "User authentication against Active Directory failed since user has entered the wrong password"
5. "No response received from Network Access Device after sending a Dynamic Authorization request"

5   Analysis for Each of the Authentication Failures

Further details about the reasons for failures and suggested solutions are described for each of the items listed in the previous part.

5.1   Rejected Per Authorization Profile

A cause to have this error is a bug in the authentication server. This bug applies to the authentication server firmware version 1.3.

According to Cisco [9] the reason to have this failure is the existence of a default authorization rule to "deny access", instead of specify an authorization rule for the session. The solution proposed within the community is to check the appropriate results [10], or create specific rule for the session and avoid the default deny.

5.2   Subject not Found in the Applicable Identity Store(S)

According to server fault community, this error is obtained because some users have umlaut in their names [11]; hence the authentication server cannot read the special character. To solve this issue, it is recommended to change the umlaut character for a simple vowel. For instance ä → a, ö → o, ü → u. Thus a correct and consistent spelling between the active directory and authentication server is implemented and minimize the errors.

The most possible cause to have this error is that the user or device was not found in the configured identity store [9]. To check this issue is required to verify whether user exists or not in the identity store.

5.3   User Has Expired

According to Cisco, this message generally appears when the user account in the identity store has expired or has been removed. Hence, to solve the issue, it is required to check the identity store or renew the user account. Nevertheless, if the user account has expired or was deleted and is no longer valid, it is recommended to investigate the reasons for the access attempts.

5.4   User Authentication Against Active Directory Failed Since User Has Entered the Wrong Password

The description about the issue is clear enough to understand the problem, indeed, it is recommended to check if password was typed correctly [10]. Nevertheless, the investigation also reveals that the problem is caused by the shared secret between authentication server and the switch where the user is connected. The recommendation is to check the shared secret between both elements. Another reason is that the NETBIOS name format includes a dot (.); for instance, laptop.name.enterprise\name-lastname. The latter is solved by changing the NETBIOS name structure within the enterprise to this: laptopnameenterprise\namelastname.

5.5   No Response Received from Network Access Device After Sending a Dynamic Authorization Request

This issue appears when no response was received from the switch after sending a dynamic authorization request. The recommended action is to check the connectivity between authentication server and switch, also be sure that authentication server is defined as dynamic authorization client on the switch and that change of authorization (CoA) feature is supported on this device [12].

6   Solutions Feasibility for the Enterprise

Though solutions are suggested to handle each failure within the network in the last section, an analysis has to be carried out to determine their feasibility. Thus, a contrast between the current status of the network and the suggested solution is developed.

6.1   Rejected Per Authorization Profile Rejected Per Authorization Profile

According to the research, a cause to have this error is a bug in the authentication server, particularly in the firmware version 1.3. Notwithstanding, the version of the installed authentication server is firmware 2.1. The next proposed solution is to create more authorization profile to segment the hosts according to their group. Indeed, more authentication groups were created apart of "deny access".

6.2   Subject not Found in the Applicable Identity Store(s)

This failure is caused by special characters such as umlaut. Remarkably, the enterprise active directory does not have a user with umlaut in its name. Moreover, there is an internal policy in which the users are created with the initial letter of names and surnames and the ingress date. For instance, the authentication for the user: José José Dominguez Banderas hired the 1st may 2018 will be jjdb010518. On the other hand, when a user does not exist in the identity store, a detailed analysis should be deployed, but this is topic is out of context.

6.3   User Has Expired

Results points out that when this message appears the user account has been removed. Nevertheless, if the user account has expired or was deleted and is no longer valid, it is recommended to investigate the reasons for the access attempts. Unfortunately, this analysis should be covered in other journal.

6.4   User Authentication Against Active Directory Failed Since User Has Entered the Wrong Password

The enterprise implemented a policy that compels users to change password each 90 days. The time lapse differs between users because if a user has not changed the password in 90 days, it is compulsory to change the password immediately; however, if user changed his password in the 80th day, it will be granted another 90 days. Of course, each user is responsible for his account and penalties applies when user access was misused.

6.5   No Response Received from Network Access Device After Sending a Dynamic Authorization Request

Monitor and connectivity tests were performed between the authentication server and the switch. The change of authorization (CoA) or RFC 3576 is enabled in the WLC and the Cisco Prime within the enterprise to support the dynamic authorization client.

An important fact in the enterprise about the dynamic authorization is that switches are not configured with this feature. Furthermore, the configuration needs to be tested and might be part of complementary journal.

7   Conclusions

The highest percentage of authentication failures in the time frame was 3,76%. Though this percentage seems acceptable, a further analysis was deployed to discover the main failure reasons and to improve the authentication service as well as the security in the network.

The authentication failures within the interval increased on April 11th, which is the day after monthly tax declaration starts. However, further analysis must be performed to identify the causes for the increment.

Curiously, up to 24 reasons for authentication failures were disclosed in this research. All of them were sorted by percentage of occurrence and the 20% of failures were deeply examined. Top 5 reasons are:

1. Rejected per authorization profile
2. Subject not found in the applicable identity store(s)
3. User has expired
4. User authentication against Active Directory failed since user has entered the wrong password
5. No response received from Network Access Device after sending a Dynamic Authorization request

The recommended solutions discovered in the research were analyzed to determine whether they are or are not applicable in the enterprise. For instance, the solution for "Rejected per authorization profile" proposes to check the firmware version of the authentication server and create more authentication groups to grant access according to the user role and avoid the deny access profile. The current firmware version is higher than recommended 1.3 and more profiles were created to grant access instead of simply deny access.

The second failure is closely related with special characters in the username. Particularly with the umlaut. A deep examination was performed in the active directory to determine users with special characters. Oddly, none of the users had umlaut on the name. For users with special characters was implemented a policy to avoid inconsistencies between active directory and authentication server. The policy creates users with the initial letter of names, surnames and the ingress date. For instance, the authentication user for José José Dominguez Banderas hired the 1st may 2018 will be **jjdb010518**.

Another failure is about the absence of user account. For this case, supplementary analysis needs to be implemented to determine the reasons for logging from a user that no longer exists. The fourth common error for authentication is the mistyped password. Though the solution suggests to type the password correctly, the enterprise implemented a policy that compels users to change password each 90 days instead. This policy does not guarantee the correct typing of password; but it enforces the security in the network.

The next issue is about the lack of response of a dynamic authorization request which generally occurs for communication issues between the authentication server and switches, or misconfiguration to support dynamic authorization and change of authorization (CoA) on the switch and the wireless Lan Controller (WLC). The output from the examination turned out that no communication issues exist between authentication and switches, nor misconfiguration is applied in the WLC; however, the commands to support dynamic authorization are not configured in the switches. This implementation is out of the scope of this research cause the enterprise needs to evaluate the convenience of the dynamic authorization.

8 Recommendations

Extend the time interval of the authentication sample. It will empower the enterprise to determine which days of the week or the month the authentication failures increase and determine its relationships with more is related to the tax monthly declaration, which starts the 10th of each month.

By increasing the time frame data will help the enterprise to identify the major causes as long as the endpoints, or users that present more authentication failures. Hence, the enterprise will enforce security policies and minimize risks.

This analysis does not include external parameters that might affect authentication, such as failure of external authentication database. This paper examines radius authentication failures. It does not examine the failures about tacacs. Thus, the recommendation is to determine why tacacs authentication fails and which users have these issues in the network.

References

1. John, G., David, R.: The digital-universe in 2020, vol. 2007, pp. 1–16 (2012)
2. Hao, T.: The information security analysis of digital library. In: ICICTA 2015, pp. 983–984 (2016)
3. Wangen, G.: Information security risk assessment: a method comparison. Computer **50**(4), 52–61 (2017)
4. Servicio de rentas internas (SRI): Distributivo de personal - Servicio de Rentas Internas del Ecuador (2014). http://www.sri.gob.ec/web/guest/distributivo-de-personal2. Accessed 28 Nov 2018
5. Al, S.: The triple-a approach to enterprise IT security | Enterprise Security | E-Commerce Times (2013). https://www.ecommercetimes.com/story/76987.html. Accessed 28 Nov 2018
6. Sasan, A., Bin, L., Pin-Han, H., Shervin, E.: Authentication authorization and accounting (AAA) schemes in WiMAX, pp. 210–215
7. Lucía, L.: Explicativo sobre el cumplimiento del proceso de la reinversión de utilidades como requisito previo para la disminución en el pago del impuesto a la renta (2017)
8. Microsoft: Account policies (2009). https://docs.microsoft.com/en-us/previous-versions/tn-archive/dd277398(v%3Dtechnet.10). Accessed 28 Nov 2018
9. Cisco Community (a): Cisco ISE Machine failed machine authentication (2014). https://community.cisco.com/t5/policy-and-access/cisco-ise-machine-failed-machine-authentica-tion/td-p/2437971. Accessed 29 Nov 2018
10. Cisco community (b): Configuring AAA, pp. 1–18
11. Serverfaul Community: Cisco ISE Wlan user authentication fails for users with umlaut (2018). https://serverfault.com/questions/462565/cisco-ise-wlan-user-authentication-fails-for-users-with-umlaut. Accessed 29 Nov 2018
12. Cisco community (c): CoA is not working using Cisco ISE 1.1 (2012). https://community.cisco.com/t5/policy-and-access/coa-is-not-working-using-cisco-ise-1-1/td-p/2105702. Accessed 29 Nov 2018

Digital Addiction: Negative Life Experiences and Potential for Technology-Assisted Solutions

Sainabou Cham[1](✉), Abdullah Algashami[1], Manal Aldhayan[1],
John McAlaney[1], Keith Phalp[1], Mohamed Basel Almourad[2],
and Raian Ali[1]

[1] Bournemouth University, Poole, UK
{scham, aalghashami, maldhayan, jmcalaney,
kphalp, rali}@bournemouth.ac.uk
[2] Zayed University, Dubai, UAE
basel.almourad@zu.ac.ae

Abstract. There is a growing acceptance of the association between obsessive, compulsive and excessive usage of digital media, e.g., games and social networks, and users' wellbeing, whether personal, economic or social. While specific causal relations between such Digital Addiction (DA) and the negative life experience can be debated, we argue in this paper that, nevertheless, technology can play a role in preventing or raising awareness of its pathological or problematic usage styles, e.g. through monitoring usage and enabling interactive awareness messages. We perform a literature review, with the primary aim of gathering the range negative life experiences associated with DA. We then conduct two focus groups to help gather users' perception of the key findings from the literature. Finally, we perform a qualitative analysis of experts and practitioners' interviews and comments from a user survey on DA warning labels. As a result, we develop eight families of the negative life experiences associated with DA, examine the role of software in facilitating the reduction of such negative experiences, and consider the challenges that may be encountered in the process.

Keywords: Digital Addiction · Digital well-being · Responsible technology

1 Introduction

As digital media has become an integrated part of our daily lives, people are spending a considerable amount of time using it for various purposes including social networking and gaming. The *overdependence* on such media could be, in some cases, attributed to emotional escapism such as escaping from stress, depression and other real-life problems. The term Digital Addiction (DA) can be described as a high degree of behavioural dependence on software products [1]. Griffiths [2], described six symptoms or characteristics for behavioural addiction, including DA, namely; *salience*: when the use of digital media becomes a vital activity for a user; *mood modification*: when used as a coping strategy for users; *tolerance*: increasing the digital usage over time to

© Springer Nature Switzerland AG 2019
Á. Rocha et al. (Eds.): WorldCIST'19 2019, AISC 931, pp. 921–931, 2019.
https://doi.org/10.1007/978-3-030-16184-2_87

achieve the same effect; *withdrawal symptoms*: behaving unpleasantly when unable to access or interact with digital media as wished; *conflict*: inter or intrapersonal issues caused by digital usage; and *relapse*: quickly falling back to a user's old digital usage habits after a period of abstinence.

DA is becoming a recognised problem globally. For example, 40% of adults in the UK look at their phone within five minutes of waking up, and 37% of adults check their phones five minutes before turning lights out [3]. Also, the UK tops the list of the prevalence of internet addiction among university students at 18.3% followed by Poland at 16.2% and Taiwan at 15.1% [4]. While in Muller et al. [5], a 2.1% internet addiction prevalence rate was found in the German general population. We note here the different, perhaps subjective, metrics used in these tests.

Studies linked DA to various negative life experiences including lower grade point among students [6], marital discord, social isolation, reduced work performance and job loss, [7] and parent-child relationship issues [8]. The last decade has witnessed a considerable increase in the treatment of behavioural addictions including DA and most of which shared approaches similar to the treatment of substance-related addictions including counselling, pharmacotherapy, self-help therapy [9], cognitive-behavioural therapy, psychotherapy [10, 11] and motivational interview [12].

Despite the argument that technology can be designed to sense and react to addictive usage style [13], there is still limited research on how this can be realised. Classic software engineering would fall short here given the special mission of software of changing a user's behaviour. Hence, software requirements here are behavioural and potentially in conflict with the current status, mental and psychological, of their users. Current approaches for digital behavioural change, such as Apple iOS Screen Time (https://support.apple.com/en-gb/HT208982) and Google Digital Wellbeing (https://wellbeing.google), focus on conscious interaction with technology and help users to avoid seeing repeated content; setting limits, e.g. in terms of time spent and break times; avoid distraction, e.g. muting notification; and heath, e.g. advising to take a break after long hours of watching videos. These tools are focusing on the interaction between the users and the device, and they may be seen to a large extent a usage optimisation tools. Behavioural change would require a much more in-depth consideration of both the content of the interaction, not only the amount and frequency, as well as the profound reasons why the person may become over-dependent. This will be vital for correct personalisation and customisation of the tools and their suggestions.

The work in [13, 14] investigated software-based interventions as countermeasures for DA. The work argued the capability of the software to raise awareness and apply a range of persuasive techniques, interactively and in real-time to keep users in control. Still, the content of messages, as well as the various modalities of intervention, are to explore and concretise. For any intervention to be effective it is necessary to first identify the precise nature of the negative harms associated with a problematic behaviour. **In this paper**, we review the literature and conduct qualitative studies and present eight families of negative life experiences associated with DA and then explore what role software could play to help raise awareness and support user to regulate their behaviour in the online space and reduce the prevalence to such negative experiences.

2   Research Method

We performed a literature review on DA as an umbrella term for a range of terms such as internet addiction, online addiction, problematic internet usage, online gaming addiction, etc. Besides searching these relevant avenues, we applied online search using Google Scholar and main digital libraries such as ACM, IEEE Xplore and DBLP. As search criteria we used combinations of keywords related to digital media such as 'social media', 'games', 'gaming' 'internet use', 'smartphone', 'social networks' as well as addiction-related terms such as 'addictive', 'excessive', 'compulsive', 'addiction', 'problematic', 'pathological'. We used a snowballing approach to reviewing relevant references from reviewed studies so that we expand our search results [15]. The study was not meant to be a systematic literature review but rather an elicitation of the primary negative life experiences linked to the concept of DA.

After identifying the negative life experiences from the literature, we worked on classifying and categorising them. In some other cases, the findings from the follow-up focus groups informed the process detailing the experiences and introducing additional elements. We come up with eight families of negative life experiences and listed related elements under each family. The findings are presented in Sect. 3.

The second stage of the study involved two focus group sessions. The focus groups aimed to elaborate and explain our results from the literature review based on the participants experience with DA. The participants (i) were familiar with the research area, and (ii) had prior experience working in the field of digital, internet or behavioural addiction and (iii) were frequent users of social media such as Facebook, Twitter, and LinkedIn so could also comment on the software design aspect. The focus of these sessions was to have participants discuss the various findings from the literature mainly to explain them and help their organisation into categories so that we reduce redundancies and flatness of the literature review outcome. For example, while we initially considered *skipping meals* and *forgetting meals* to be similar, the focus groups highlighted a subtle difference between them, which is around *intentionality*.

The final stage of the study consisted of two activities with the aim of investigating the role of software in combatting DA and reducing negative life experiences. First, we analysed interview data collected from ten experts and practitioners' in the area of wellbeing, addiction recovery, social and cyber psychology and human factors in computing. The primary goal of the interviews was to explore the perception of online labelling and warning messages issued by software to reduce or warn against DA with a focus on social networks and games. Also, interviewing experts in one of the main addiction centres in the UK enabled us to develop an understanding of their practices and how we may use it in developing interactive software interventions. In the second part, we built on a survey initially conducted in [13] and extended it in terms of number of the participants and the comments received. The original survey was itself to validate the findings of previous interviews with people self-declaring to have DA, i.e., part of a mixed method approach. In this paper, we only analyse the survey comments on the proposed software intervention techniques including the progress bar, a timer relating to digital media usage and content of feedback relating to negative life experiences and the effect DA may have on the users and their significant others.

The results of the analysis of the interviews as well as the qualitative part of the survey are presented in Sect. 4 and meant to provide insights on how future software tools can help in combatting DA as a professional responsibility requirement.

3   Digital Addiction and Associated Negative Life Experiences

Existing research associated DA with a wide range of negative life experiences including lowering self-esteem, preoccupation, irregular sleeping patterns, reduced face-to-face communication, invasion of privacy of others and erratic dietary behaviour. Figure 1 presents eight families of such negative life experiences and the elements listed under each family. Main elements from each family are written in *italic* text and underlined. Due to the space limitation in this paper and given that most of the negative effects are self-explanatory; we will only elaborate on the primary elements from each category and insert a few comments from the users' study.

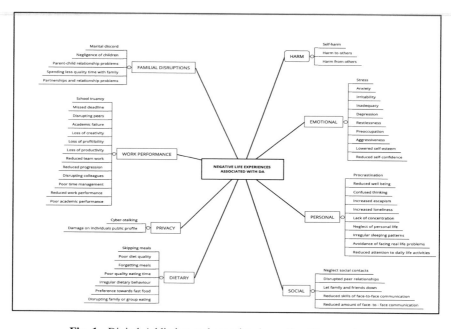

Fig. 1. Digital Addiction and associated negative life experiences

3.1   Emotional Problems

Emotional problems refer to problems that affect the psychological well-being of users. Interaction activities of some contacts could elevate *depression* especially for those with low self-esteem; "*seeing online contacts posting perfect pictures whether holiday pictures or pictures of a social function could facilitate depression for some users*". *Irritability* is another related emotional problem [16], i.e. when a user feels annoyed or impatient when they are unable to engage in an online interaction. Furthermore, users with DA could overly feel *Inadequate* which may lead to reducing self-worth; "*you feel incompetent and inferior when you see posting of an event on social media from your contacts which you were not invited to*".

3.2   Familial Disruptions

Familiar disruptions resulting from DA can be linked to the disclosure of sensitive personal information online, cyber-stalking and online surveillance of loved ones. For users with DA, this can happen unconsciously and irresistibly. It could lead to family breakdown and partnerships and relationship problems. *Partnerships and Relationship problems* can be affected when one chooses to spend time online and neglect their partners which may lead to relationship conflict, separation, and divorce in some cases [17]. *Parent-child relationship depth and strength* can be affected when "*the parent-child communication and engagement in other family activities are overpowered by their engagement in the virtual world, e.g. social networks*". Online immersion could distract and lead people to spend *less quality time with family*; "*being on the phone when other members of the family are together catching up after school, work or watching the TV together as a family*".

3.3   Personal Problems

The continuity and the real-time nature of online interaction, the attractiveness of games and social media features and the unlimited access to these features could lead to neglect of other important aspects of a person's life. For example, excessive usage reduced attention to daily life activities, e.g. not cleaning the house or doing the needed shopping, and not taking proper care of loved ones. The access to a variety of content online may *confuse* the way people think; "*disorientation, confusion and anxiety, tunnelling when shopping online and unable to control how one thinks as a result of prolonged usage sessions*". The *lack of concentration* while driving resulting from social media usages and notification could cause accidents and annoy other drivers. Another experience is *procrastination*, e.g. putting off personal chores that need to be completed or delaying the start of a task for other social online activity [1].

3.4   Social Problems

Social problems may be related to situations when users feel the need to be online at all times and therefore neglect their social contacts such as peers, colleagues, family and friends in the physical world. The negligence of such connections could affect the *real-life relationships and their face-to-face* engagement of users with DA. For example, when a user chooses to spend all their time using online software, their family and friends may feel let down especially during special occasions where their presence is highly expected such as the case when gamers may be "*turning down an invitation from family or friends or not showing up to social events such as birthday parties, weddings ceremonies and other special celebrations*". Also, spending too much time online may lead to a *loss or reduction of social skills* required in real-life communication such as confidence and concentration levels needed to engage in a meaningful conversation. As a counter-argument, some users argued that real-life communications are both slow and more inhibitive, hence the difficulty.

3.5   Work Performance Problems

The urge for students or employees to be excessively online for non-academic or non-work-related purposes could *impair their academic and work performances* [5]. It is also about the preoccupation, e.g. an employee who posts on social media and becomes worried of the reactions of other contacts could have a reduced attention span on work duties which may affect the *profitability* of an organisation. A user commented that uncontrolled and hasty *"employees' online software activities might lead to disciplinary, suspension and even job loss in extreme cases leading to a reduction in the workforce"*. A reduction in work performance could *reduce one's work-profile* which could reduce the chances of employment *progression* and this could in turn *lower other job opportunities*. Also, DA could be contagious where staff with DA may be putting pressure on colleagues to respond on their online post, and this could *disrupt* colleagues from their routine daily work. Effects on *academic performance* can happen when students cannot complete of focus on a task such as coursework due to their immersive digital media usage. Students multitasking, e.g. joining a conversation on social media while working on an assignment, could affect their performance level causing decreased productivity [18]. *Loss of creativity* could be attributed to *"the variety of content that users can access and the number of activities that can be performed while online"*. The ease of finding and sharing content on social media, rather than creating, can be argued to reduce the creativity level.

3.6   Invasion of Privacy of Others

Online software applications provide the opportunity for users to disclose personal information. This information can be assessed and used by others in a negative way such as the aforementioned *cyber-stalking* [19]. The sharing of sensitive information in a hasty and less thoughtful manner, which is often the case for users with DA, could lead to *damage on individuals' public profile*. For example, sharing pictures and tagging contacts in a post that can be accessed and used by employers and others for judging professional ability could affect one's reputation.

3.7   Harm

Harm could be personal or financial, e.g. gaming addicts buying items and neglecting health and hygiene. Users can suffer a physical injury which is developed over a period, e.g. headaches, pain on the wrist, text-neck or poor vision. The skills users learned and developed as a result of watching or being exposed to a violent game could cause *harm to others* *"if a user practices a fighting scene from an online game on other people"* and when users steal money from their partners to finance their gaming or online gambling addiction. The *financial harm* could lead to issues such as family breakdown due to a divorce or separation resulting from job loss, and this may lead to debts, impairment of assets or even personal bankruptcy. Such harms are more noticeable in the case of online gambling in particular.

3.8   Dietary-Related Problems

Dietary problems refer to those factors that could affect users' diet quality. For example, playing games or using social networks continuously for hours and *forgetting to eat or drink*. The real-time nature of the interaction, continuity of software usage, variety of information content could entice users into spending considerable amount of time using software; this could lead to not having meals or not preparing good quality meals leading to *missing or deliberately skipping meals*, and *poor diet quality*, e.g. the tendency toward eating fast food or junk food in the case of gamers.

4   Software-Assisted Digital Addiction Prevention and Awareness

To help people combat the potential negative life experiences presented in Sect. 3, we argue the need for the tech industry to play a key role and provide tools that may enhance users' online well-being and help prevent unhealthy usage behaviours. To support our argument, we analysed the interviews and survey comments described in Sect. 2 through the six phases of thematic analysis as proposed in [20]. Below, we present and discuss the mechanisms identified through the data analysis of the interviews data and users' comments in the survey. The design of these studies was based on a set of arguments we made about the role of software in monitoring and solving the problem and its unique features of interactivity, reactiveness and real-time [13].

4.1   Software as a Tool to Disseminate Educational Material

Research evidence has linked excessive digital media usage to an increased level of psychological arousal, which often leads to a lack of sleep, forgetting to eat or drink, psychological harms and reduction in physical activity [21]. There is a growing interest in the possible use of addiction-awareness software solutions to help maintain the mental and social-wellbeing of users [22]. Our analysis demonstrates the need for software to provide educational materials relating to the obsessive and excessive users' online presence and the effects this may have on their psychological and mental health status and also on other aspects of their lives; *"Providing psychological and mental health information could be good and may help some users consider their usage behaviour before it gets serious and turns into something addictive"*.

The participants also emphasised the need for such information to be tailored and personalised based on their needs and preferences. One expert stated that *"education materials should be personalised to the needs of users and only displayed to them at the right time"*. Based on our analysis and our practical expertise in the field, we conclude that such personalisation might be based on four main factors, (i) usage behaviour, (ii) recognition of DA and potential risks of DA, (iii) knowledge of psychological wellbeing and DA, and (iv) stage of change and treatment levels to help planning and relapse prevention. AI techniques could be used here to help predict the future usage behaviours, risks associated with the usage and to deliver appropriate materials at the right point, to reduce its effect early on [23]. Another factor to consider is the nature of the materials, i.e., should they be based on real-life user stories, scientific facts or usage history and context? The challenge here would be on the way to design software to deliver such information in a timely and acceptable manner, specifically: (i) how to monitor users' digital behaviour and intentionality, (ii) how to ascertain users' preferences on the nature of the materials, their presentation and delivery modes given the nature of people with problematic behaviour and their characteristics, e.g. denial and trivialisation of the problem, and (iii) when to design the educational material in a progressive and phased way in response to the change of behaviour. An approach would be to ask input from the users but this may also be challenging given the characteristics of people with problematic behaviour, e.g. subjectivity and ambivalence.

4.2   Software-Assisted Goal Setting

Users reported that they would like the software to provide the opportunity to set usage and style of living goals. The real-time nature, the traceability and tractability of software usage would facilitate goal setting, and the provision of goal performance feedback and this may enable users to make informed decisions about their usage behaviour [14]. Based on our interviews and survey comments, it is evident that users would like some degree of control when setting goals; "*giving users some control would lead to some sense of involvement which can help encourage people and improve their overall acceptance of the goals*". Others commented that giving control shall be minimal to avoid bias and "*flight-into-health*". An assessment questionnaire could be used to determine users' preferences on the level of involvement such as (i) Passive or consultative, (ii) Representative, (iii) Participative or decision-maker [24]. Another option may be to solicit the expertise of a mediator, e.g. a therapist who is an expert in the area, who can assess a user's emotional state and issues and help them to choose the right goal settings and provide data-driven feedback with the help of data analytics tools.

When setting goals, the proximity of the goals needs to be considered, i.e., how far into the future the goals are set [25]. Setting proximal goals may help reduce loss of goal interest, boost motivation and confidence in goal attainment; "*Encourage users to set proximal goals. Users only have main goals and not sub-goals. Having little sub-goals motivates people to keep doing what they are doing and the program can help them set these sub-goals and reward or remind them when they attain the sub goals*". The software interface to set up the goals should be concise; it should only contain a few questions or options which are easy to interpret "*fast and no-fuss to set up, i.e. clean interface*". Software can monitor goal achievement in real time and this is a bonus in comparison to human-based counselling.

4.3   Software Transparency

Software provides new opportunities for timely, interactive and personalised transparency with users. For example, it could demonstrate to users how the real-time monitoring and traceability of their usage are performed and how goal performance feedback information was determined. A practitioner commented; "*focus on a message that is very clear and genuine for all users. It needs to be transparent with the users on how such information was derived*". Providing this information at the right time may improve users' belief and trust in the software. Because users exhibit different skills and self-esteem levels, transparency concerning the delivering and sharing of performance feedback collectively should be considered by all parties involved [26, 27] in case group therapy approach or online peer support groups are applied [28]. Delivering such transparency information without consideration of these factors may lower some users' self-esteem and reduce their motivation, commitment towards the goals and, in the worst cases, may lead them to abandon their goals.

4.4   Data Sharing for Responsible Digital Media Usage

The question that needs to be addressed here is around the party who should be responsible for the implementation of DA awareness and prevention mechanisms. Should it be an in-house job for software companies, i.e., where technology companies think about how their software might affect users during system design and development, and provide tools and support that would help combat any negative effect resulting from its usage, or should the software companies delegate this responsibility to third parties and enable them to do that? In addition, the users should be assured that all ethical considerations have been considered and that the companies will only use data to raise awareness and to alleviate any negative effects that might be linked to addictive software usage. It has been also argued that software companies may choose to democratise the process by enabling users to allow access to their data by their own third party services. With the General Data Protection Regulation (GDPR) in Europe, such data *portability* is a right of citizens where the real-time and automation aspects are still not mandatory for software companies.

5   Conclusion and Future Work

We explored the negative life experiences associated with DA. We then classified and categorised these effects into eight families of negative life experiences. Having discussed these effects, we then advocated the need for tech companies, producing such potentially problematic products, to employ mechanisms that would help users to be aware of their usage patterns, and maintain a healthier online behaviour. As an initial step, from our analysis, we identified three ways that software could adapt to help reduce the negative effects associated with DA, by disseminating user behaviours to them, allowing goal setting and ensuring transparency. We also discussed who should take responsibility for such mechanisms. Our future work will expand on these mechanisms and discuss the major software design considerations that would enhance their successful design and implementation. We will also elaborate further on the challenges discussed and propose ways of countering them to ensure a better implementation.

Acknowledgement. This work has been partially supported by the EROGamb project funded jointly by GambleAware and Bournemouth University, SSCoDA project funded by Zayed University and by European H2020-MSCA-RISE-2017 project, under grant agreement No. 778228 (IDEAL-CITIES).

References

1. Kirschner, P.A., Karpinski, A.C.: Facebook® and academic performance. Comput. Hum. Behav. **26**(6), 1237–1245 (2010)
2. Griffiths, M.D.: A "components" model of addiction within a biopsychosocial framework. J. Subst. Use **10**, 191–197 (2005)
3. Ofcom: Making Communication Work for Everyone. https://www.ofcom.org.uk/about-ofcom/latest/features-and-news/decade-of-digital-dependency. Accessed 23 Nov 2018
4. Kuss, D.J., Griffiths, M.D., Binder, J.F.: Internet addiction in students: prevalence and risk factors. Comput. Hum. Behav. **29**, 959–966 (2013)
5. Muller, K.W., Glaesmer, H., Brähler, E., Wolfling, K., Beutel, M.: Prevalence of internet addiction in the general population: results from a german population-based survey. Behav. Inf. Technol. **37**(7), 1–10 (2014)
6. Young, K.S., Case, C.J.: Internet abuse in the workplace: new trends in risk management. CyberPsychol. Behav. **7**(1), 105–111 (2004)
7. Young, K.S.: Internet addiction: symptoms, evaluation, and treatment. Innov. Clin. Pract.: Sour. Book **17**(17), 351–352 (1999)
8. Grüsser, S.M., Thalemann, R., Griffiths, M.D.: Excessive computer game playing: evidence for addiction and aggression? Cyberpsychol. Behav. **10**(2), 290–292 (2007)
9. Griffiths, M.D.: Classification and treatment of behavioural addictions. Nurs. Pract. **82**, 44–46 (2015)
10. Young, K.S., de Abreu, C.N.: Internet Addiction: A Handbook and Guide to Evaluation and Treatment. Wiley, New Jersey (2011)
11. Khazaal, Y., Xirossavidou, C., Khan, R., Edel, R., Zebouni, F., Zullino, D.: Cognitive-behavioral treatments for "internet addiction". Open Addict. J. **5**(1), 30–35 (2012)
12. Corrigan, P.W., McCracken, S.G., Holmes, E.P.: Motivational interviews as goal assessment for persons with psychiatric disability. Commun. Ment. Health J. **37**(2), 113–122 (2001)
13. Ali, R., Jiang, N., Phalp, K., Muir, S., McAlaney, J.: The emerging requirement for digital addiction labels. REFSQ **9013**, 198–213 (2015)
14. Alrobai, A., McAlaney, J., Dogan, H., Phalp, K., Ali, R.: Exploring the requirements and design of persuasive intervention technology to combat digital addiction. In: HCI, vol. 9856, LNCS, pp. 130–150 (2016)
15. Jalali, S., Wohlin, C.: Systematic literature studies: database searches vs. backward snowballing. In: Proceedings of the ACM-IEEE International Symposium on Empirical Software Engineering and Measurement, pp. 29–38. ACM, New York (2013)
16. Abel, J.P., Buff, C.L., Burr, S.A.: Social media and the fear of missing out: scale development and assessment. J. Bus. Econ. Res. **14**(1), 33 (2016)

17. Beutel, M.E., Braehler, E., Glaesmer, H., Kuss, D.J., Woelfling, K., Mueller, K.W.: Regular and problematic leisure-time internet use in the community: results from a German population-based survey. Cyberpsychol. Behav. Soc. Netw. **14**, 291–296 (2011)
18. Li, W., O'Brien, J.E., Snyder, S.M., Howard, M.O.: Characteristics of internet addiction/pathological internet use in U.S. university students: a qualitative-method investigation. PLoS ONE **10**(2), e0117372 (2015)
19. Kuss, D.J., Griffiths, M.D.: Online social networking and addiction - a review of the psychological literature. Environ. Res. Public Health **8**(9), 3528–3552 (2011)
20. Braun, V., Clarke, V.: Using thematic analysis in psychology. Qual. Res. Psychol. **3**(2), 77–101 (2006)
21. Young, K.S.: Caught in the Net. Wiley, New York (1998)
22. Kim, K.: Internet addiction in Korean adolescents and its relation to depression and suicidal ideation: a questionnaire survey. Int. J. Nurs. Stud. **43**(2), 185–192 (2006)
23. Lino, A., Rocha, A., Sizo, A.: Virtual teaching and learning environments: automatic evaluation with artificial neural networks. Cluster Comput. 1–11 (2017)
24. Roach, Á., Vasconcelos, J.: A framework to analyse the approach adopted in the information systems requirements engineering activity. In: 2004 International Conference on Software Engineering Research and Practice, SERP 2004, Las Vegas, vol. 2, pp. 573–579 (2006)
25. Seijts, G.H., Latham, G.P.: The effect of distal learning, outcome, and proximal goals on a moderately complex task. J. Organ. Behav. **22**(3), 291–307 (2001)
26. Algashami, A., Cham, S., Vuillier, L., Stefanidis, A., Phalp, K., Ali, R.: Conceptualising gamification risks to teamwork within enterprise. In: IFIP Working Conference on the Practice of Enterprise Modelling, pp. 105–120. Springer, Cham (2018)
27. Algashami, A., Shahri, A., McAlaney, J., Taylor, J., Phalp, K., Ali, R.: Strategies and design principles to minimize negative side-effects of digital motivation on teamwork. In: International Conference on Persuasive Technology, pp. 267–278. Springer (2017)
28. Alrobai, A., Dogan, H., Phalp, K., Ali, R.: Building online platforms for peer support groups as a persuasive behavior change technique. In: Ham, J., Karapanos, E., Morita, P., Burns, C. (eds.) Persuasive Technology. PERSUASIVE 2018. LNCS, vol. 10809, pp. 70–83. Springer, Cham (2018)

Improving Cross-Border Educational Services with eIDAS

Tomaž Klobučar[✉]

Jozef Stefan Institute, Jamova 39, 1000 Ljubljana, Slovenia
klobucar@e5.ijs.si

Abstract. The eIDAS regulation and the technical infrastructure for its implementation enable secure cross-border electronic transactions between businesses, organizations, citizens and public authorities. This article shows how eIDAS can be used to improve the reliability of the identification of foreign students, while at the same time reduce the administrative burden of higher education institutions in dealing with those students. The students can use their national identification means for accessing the services, and the required information about them can be transmitted electronically from reliable sources. Several eID4U ("eID for University" project) improvements to existing procedures and services are described in the paper, including an upgrade of the EU reference implementation of the eIDAS node, integration of trusted sources of academic attributes into the eIDAS infrastructure, and three upgraded cross-border e-services based on electronic identities and trusted academic information.

Keywords: Electronic identification · Cross-border service · Infrastructure · Identity provider · Attribute provider · E-education

1 Introduction

The students on exchange, e.g. Erasmus students, are faced with various challenges when accessing foreign educational e-services. Usually, their national identification means (for example, digital certificates) are not recognized as valid abroad, and higher education institutions still require a lot of documents in the paper form. The European Commission aims at improving the quality of student mobility by enabling students to identify themselves in line with the once-only principle and use the services they are entitled to use in the host country, and by allowing secure exchange and verification of student data and academic records [1]. For example, by 2025, all students in Erasmus + mobility should be able to have their national identity and student status recognised automatically across EU Member States, including access to university services (e.g., study materials, subscription services, libraries), arrival abroad [1].

To meet those objectives, the European Commission implements or co-finances various measures, such as Erasmus Without Paper [2, 3], EMREX [4], European Student Card [5], ESMO [6], or StudIES+ [7]. One of them is the eID4U (eID for University) project from the Connecting Europe Facility (CEF) programme [8]. The goal of a 15-month project, which began on February 1, 2018, is to include higher

© Springer Nature Switzerland AG 2019
Á. Rocha et al. (Eds.): WorldCIST'19 2019, AISC 931, pp. 932–938, 2019.
https://doi.org/10.1007/978-3-030-16184-2_88

educational institutions, trusted student data sources, and educational e-service providers in the eIDAS infrastructure. The project involves five EU institutions: Politecnico di Torino (Italy), Jozef Stefan Institute (Slovenia), Graz University of Technology (Austria), Universidad Politecnica de Madrid (Spain), and Universidade de Lisboa (Portugal). The solutions developed will also be suitable for other educational institutions and educational e-services accessed by users from different EU countries.

This paper is organized as follows. The second section briefly presents the legal basis for the identification of foreign users and the technical infrastructure that enables such identification. An example of the use of infrastructure in higher education clearly shows its usefulness for the establishment of secure cross-border services. Section 3 deals with the challenges related to the availability and reliability of student data, for example under the Erasmus+ program, while Sect. 4 describes several eID4U improvements of the infrastructure and services.

2 EU eIDAS Regulation

In 2014, a new legal basis for the provision of secure cross-border electronic transactions was adopted in the EU Member States. It aims to ensure the proper functioning of the EU's internal market and to achieve an adequate level of security for electronic identification and trust services [9]. The EU Regulation on electronic identification and trust services for electronic transactions in the internal market (eIDAS), which entered into force on 1 July 2016, eliminates the existing obstacles to the electronic identification of users from abroad. Defined conditions for mutual recognition of electronic identification means also provide the basis for safer electronic commerce within the EU.

The regulation allows natural and legal persons to use certain national electronic identification means for access to public services, for example, higher education services, in other EU Member States. Examples of the means are ID cards with digital certificates, qualified certificates, mobile identities and other electronic certificates that contain identification data and are intended for validating the e-service user's identity.

2.1 Technical Infrastructure

The eIDAS technical infrastructure provides a technical basis for the implementation of the eIDAS regulation. The infrastructure combines identity, service and attribute providers, and national eIDAS nodes from the EU Member States.

Identity providers (IdP) are organizations that issue electronic identification means within the framework of the notified electronic identification schemes and authenticate users. Their connection to the eIDAS node allows natural or legal persons to use the electronic identification means to access services in other EU countries. Electronic identification means are more or less resistant to misuse and alteration of identities, so the level of trust in the identified e-identity of a service user largely depends on the type of electronic identification used. The assurance level (low, substantial or high) indicates the level of reliability that the electronic identification means determines the person's identity. It depends on the method of proving and verifying the identity of a person at the time of registration, the type of connection between the electronic identification means of natural and legal persons, the procedure of issuing, delivery and activation of electronic identification means, management of the means, resilience to security threats in authentication, management and organization procedures at the identity providers, and technical supervision of the identity providers.

Attribute providers (AP) are entities that manage electronic identity data that go beyond the minimum data set specified in the eIDAS regulation. Additional information (e.g. sector-specific data such as e-education, e-banking, e-health) may be necessary to verify authenticity, in certain circumstances, or grant access to a service for a particular type of user (e.g., students with valid student status).

Service providers (SP), e.g. higher education organizations, provide citizens with online services. It should be noted that the public service providers, e.g. education service providers, requesting a substantial or high assurance level of their users' identification means must from 29 September 2018 recognize the means issued as part of the notified schemes of other EU Member States [9]. For now, this is mandatory only for a German identity card, while in 2019, due to a 1-year transition period, it will also be necessary to recognize the identification means from Belgium, Estonia, Croatia, Italy, Luxembourg, Portugal and Spain.

The fourth infrastructure element is an eIDAS node, which is the central point of trust in a country. On the one hand, it connects national infrastructure with foreign service providers and, on the other hand, national identity, attribute and service providers with the infrastructures of other EU countries. Since all national nodes form a circle of trust, it is sufficient that each service or attribute provider establishes trust only with a node in its own country. Educational service providers offering the eIDAS-enabled services will thus not have to deal with the verification of foreign identification means, but will leave this to the identity providers and national eIDAS nodes.

2.2   Proposed Use of the eIDAS Infrastructure in Higher Education

Figure 1 shows an example of the use of the eIDAS technical infrastructure in higher education. A student from Slovenia wants to apply for a student Erasmus+ exchange in Belgium. In addition, she will use her digital certificate issued in the Slovenian notified scheme for identification, and at the same time provide the necessary evidence of the previous study by electronic means.

The student first tells the service provider from which country she is coming. She is then being redirected by the Belgium national eIDAS node to a similar node in Slovenia, and then to a Slovenian identity provider, that verifies her identity on the basis of a digital certificate. Certified electronic evidence of academic qualifications is obtained from a home higher education institution (attribute provider). The Slovenian national eIDAS node sends the collected data to the Belgium national node, and the data is then forwarded to the service provider. It should be emphasized that for the protection of personal data, the initiator of all actions is the student alone. Likewise, the student selects which personal information (qualifications) should be disclosed to the service provider and explicitly agrees to their disclosure.

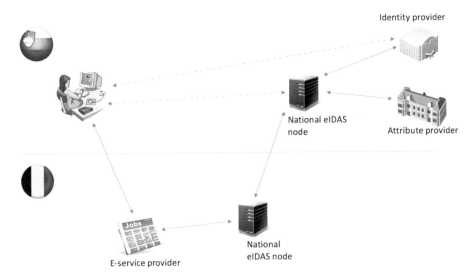

Fig. 1. Example of the use of the eIDAS infrastructure in higher education

3   Student Data

The Erasmus programme allows the student to perform part of the regular study obligations at the partner institution (host) abroad instead of at the home institution. The exchange is a two-step process:

- Student submits the application to the home institution
- Student is registered at the host institution

The purpose of using the eIDAS infrastructure is for students to use their national identification means for registration at the host institution, and enable large amount of the data to come from trusted sources in electronic form, which could reduce the time for entering the required data in the registration forms and for verifying their validity. The data that students must submit at the registration for the Erasmus+ exchange can be divided into four groups: identification data, current study data, past performance data, and information on the proposed exchange.

Some of the required identification data can be provided by the countries from which the user comes from, e.g. through national central population registers. Part of these data are already available in the eIDAS infrastructure itself. The European Union defines a minimum set of identification data that uniformly represent the natural and legal person [10]. Mandatory data for a physical person includes current name, current surname, date of birth, and unique identifier, while optional data are name and surname at birth, place of birth, current address, and gender. EU Member States are obliged to provide mandatory information on users of services, while it is their choice to provide the optional data or not.

The data related to the current and previous study, e.g. current degree, successfully finished studies and acquired competencies, such as foreign language skills, can be obtained from other trustworthy sources, for example, higher education institutions or central student registers, such as the Slovenian Central Evidence System for Higher Education (eVŠ), or from the students themselves. To make this information available, the institutions or central registers must be included in the eIDAS infrastructure. Otherwise, the user must still enter the information in the application form. The last set includes information on the proposed student exchange at the host institution. In this case, data is not yet available in any of the information systems or registers, so they must be provided by each student.

4   Improvements

The proposed eID4U improvements aim at more reliable and simpler user identification through the eIDAS infrastructure and the acquisition of the highest possible volume of data electronically from reliable sources that are part of the infrastructure.

4.1   Enabling the Use of Academic Attributes in the eIDAS Nodes

The current EU DIGIT reference implementations of an eIDAS node (latest versions 1.4.3 and 2.2, both released in September 2018) do not yet allow the identification and treatment of other attributes except those from the eIDAS minimum data set. The first improvement is therefore an upgrade of the eIDAS node (*eIDAS-proxy* and *eIDAS-connector* in Fig. 2). Support (marshaller and changed configuration) for additional, academic attributes related to the current study, the student's home institution, information on current degree, successfully finished studies, and acquired competencies has been included in eID4U in the reference implementation version 1.4.3. Examples of the defined academic attributes are *HomeInstitutionName*, *CurentLevelOfStudy*, *CurrentDegree* or *LanguageProficiency*. The eID4U project partners have already set up modified eIDAS nodes, connected them into a test network, and tested their interoperability, i.e. that the academic attributes are successfully transferred through the network. An example is provided at https://eidas.e5.ijs.si/SP/.

4.2   Integration of Academic Attribute Providers

The second step is integration of the national academic attribute providers into the eIDAS technical infrastructure. AP Connectors are being developed to integrate higher education institutions and central education registries to the national eIDAS nodes. This connection can be achieved either through national eID proxies, such as SPID in Italy [11], or directly with the educational institutions and other trusted sources of academic attributes. In the case of Slovenia, the connector (*SI-CAS proxy* in Fig. 2), which has been integrated into the eIDAS node, connects directly to the central student evidence system eVŠ and uses the eVŠ web services for accessing the academic attributes. Slovenian central authentication system (SI-CAS) plays role of an eID proxy that provides basic identification attributes, such as name, surname and eid.

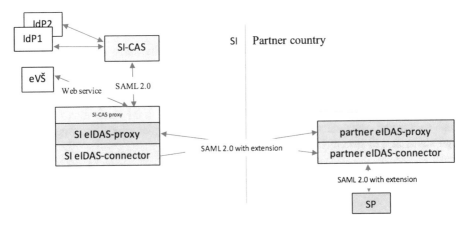

Fig. 2. Schematic representation of the infrastructure

4.3 Upgrade of the Educational E-services

The last step is an upgrade of existing educational e-services and procedures for registering and identifying users from abroad. Again, in Slovenia, for example, educational service providers can connect to the eIDAS node directly (to the *eIDAS-connector* part of the eIDAS node as depicted in Fig. 2) or through SI-CAS.

Student Registration
The website for registering foreign students will provide an additional login method called "eIDAS Login". By clicking on the button, the user will be redirected to the eIDAS infrastructure where she will use her national identity as shown in Fig. 1 and described in Sect. 2.2. In addition, part of the required identification and academic data (as described in Sect. 3) will be automatically obtained from the trustworthy attribute providers. The rest of the data will be, the same as now, entered manually into the web form.

Access to E-services
For a Moodle open source learning management system, a plug-in in PHP has been developed that allows direct connection to the eIDAS node and identity verification using the eIDAS infrastructure. A user is given an appropriate role in the system and access to learning material is provided on the basis of national identification means and verified academic attributes. A Moodle administrator can define which academic attributes are required from a learner for granting access to the learning material.

Wireless Network Access
The last e-service, which is being upgraded, is access to a wireless network at a foreign institution. Similar to the Eduroam network, the service will allow the users to access the wireless network by electronic identification means issued under the notified eIDAS schemes provided that they have relevant academic attributes.

5   Conclusion

This paper presents the eID4U approach on improving the cross-border educational services with eIDAS. The main eID4U contributions are integration of the sector specific attributes (academic attributes) into the EU reference implementation of the eIDAS node, integration of the trusted sources of academic attributes in the eIDAS infrastructure, and upgrade of the educational e-services with the mechanisms for easier registration and authentication of foreign users. Both higher education organizations and students will benefit from the presented solutions.

Acknowledgment. The presented work is a result of the eID4U (eID for University, 2017-EU-IA-0051) project, co-financed by the EU Connecting Europe Facility instrument.

References

1. Commission Staff Working Document Accompanying the document Communication from the Commission to the European Parliament, the Council, the European Economic and Social Committee and the Committee of the Regions on the Digital Education Action Plan (SWD (2018) 12 final) (2018)
2. Erasmus Without Paper project. https://www.erasmuswithoutpaper.eu/ewp-network
3. Mincer-Daszkiewicz, J.: Mobility scenarios supported by the Erasmus Without Paper network. In: EUNIS 2018 (2018)
4. Mincer-Daszkiewicz, J.: EMREX and EWP offering complementary digital services in the higher education area. In: EUNIS 2017 (2017)
5. European Student Card project. http://europeanstudentcard.eu/
6. ESMO project. http://www.esmo-project.eu/
7. StudIES+. https://ec.europa.eu/inea/en/connecting-europe-facility/cef-telecom/2017-de-ia-0022
8. eID4U (eID for University) project. https://ec.europa.eu/inea/en/connecting-europe-facility/cef-telecom/2017-eu-ia-0051
9. Regulation (EU) No 910/2014 of the European Parliament and of the Council of 23 July 2014 on electronic identification and trust services for electronic transactions in the internal market and repealing Directive 1999/93/EC
10. Commission implementing Regulation (EU) 2015/1501 of 8 September 2015 on the interoperability framework pursuant to Article 12(8) of Regulation (EU) No 910/2014 of the European Parliament and of the Council on electronic identification and trust services for electronic transactions in the internal market
11. Berbecaru, D., Lioy, A.: On integration of academic attributes in the eIDAS infrastructure to support cross-border services. In: 22nd International Conference on System Theory, Control and Computing (ICSTCC), pp. 691–696 (2018)

Correction to: Keystroke and Pointing Time Estimation for Touchscreen-Based Mobile Devices: Case Study Children with ASD

Angeles Quezada, Margarita Ramirez Ramírez,
Sergio Octavio Vázquez, Ricardo Rosales, Samantha Jiménez,
Maricela Sevilla, and Roberto Muñoz

Correction to:
Chapter "Keystroke and Pointing Time Estimation
for Touchscreen-Based Mobile Devices: Case Study Children
with ASD" in: Á. Rocha et al. (Eds.): *New Knowledge*
***in Information Systems and Technologies*, AISC 931,**
https://doi.org/10.1007/978-3-030-16184-2_74

In the original version of the book, the chapter "Keystroke and Pointing Time Estimation for Touchscreen-Based Mobile Devices: Case Study Children with ASD" was published with six authors and now an author (Prof. Roberto Muñoz) has been added to this chapter. The correction chapter and the book is now updated with the change.

The updated version of this chapter can be found at
https://doi.org/10.1007/978-3-030-16184-2_74

© Springer Nature Switzerland AG 2019
Á. Rocha et al. (Eds.): WorldCIST'19 2019, AISC 931, p. C1, 2019.
https://doi.org/10.1007/978-3-030-16184-2_89

Author Index

A

Abbas, Ali, 483
Abbas, Muhammad, 472
Abbas, Sohail, 339
Abed, Mourad, 597
Adnan, Awais, 303
Agrebi, Maroi, 597
Aguayo-Canela, Francisco J., 124
Aguiar, Joyce, 587, 607
Aguirre, Jimmy Armas, 24
Ahmed, Naveed, 737
Al Aghbari, Zaher, 900
Alaiz-Moretón, Héctor, 124
Alanis, Arnulfo, 819
Albiol-Pérez, Sergio, 785
Alcarria, Ramón, 175
Aldhayan, Manal, 921
Alfandi, Omar, 483
Algashami, Abdullah, 921
Ali, Raian, 795, 806, 921
Almeida, Ana Leonor, 706
Almeraya, Said, 230
Almeraya, Selin, 230
Almomani, Ameed, 557
Almourad, Mohamed Basel, 921
Altuwairiqi, Majid, 795
Álvarez Sabucedo, Luis, 373
Alves, João, 240
Alves, Rodrigo, 145
Alves, Victor, 577
Amaro, B., 839
Amengual, Esperança, 881
Amin, Adnan, 483
Anfer, Omar, 633
Antipova, Tatiana, 350

Anwar, Sajid, 303, 483
Araújo, João P. B., 492
Arellano, Luis, 219
Artal-Villa, Leyre, 198
Athi, Balaji Ganesh, 438
Au-Yong-Oliveira, Manuel, 186, 726
Azevedo, Fábio, 621

B

Bacca-Maya, Alvaro-Felipe, 785
Barroso, João, 45
Bastos, David, 829
Batista, Josias G., 492
Benavides-Cuéllar, Carmen, 124
Benders, Manon J. N. L., 577
Benítez-Andrades, José Alberto, 124
Berkani, Lamia, 524
Bessa, Maximino, 686, 696
Bibiloni, Antoni, 275, 881
Bispo, Kalil Araujo, 396
Bordel, Borja, 134, 175
Börger, Alexander, 502
Borghei, Amir Hossein, 285
Bosoc, Cristina Sabina, 252
Braga, Arthur P. S., 492
Branco, Frederico, 696, 726
Brandão, Catarina, 764
Buele, Jorge, 208

C

Câmara, Álvaro, 13
Camino Solórzano, Alejandro Miguel, 285
Cañellas, Maria, 275
Cano, Sandra, 785
Caravau, Hilma, 706

© Springer Nature Switzerland AG 2019
Á. Rocha et al. (Eds.): WorldCIST'19 2019, AISC 931, pp. 939–942, 2019.
https://doi.org/10.1007/978-3-030-16184-2

Carchiolo, Vicenza, 361
Cárdenas, Ramón A. Mollineda, 755
Carvalho, Marta, 145
Carvalho, Paulo, 373, 396
Carvalho, Victor, 587, 607
Casero, Jesús Gallardo, 755
Castañón-Puga, Manuel, 547
Castellanos, Esteban X., 208
Castro, Roger, 513
Cavique, Luís, 461
Cham, Sainabou, 921
Chiva, Loredana, 252
Clar, Albert, 275
Close, Emily Arden, 806
Coelho, Florângela Cunha, 716
Coelho, Hugo, 696
Cosmas, John, 303
Costa, Jonatha R., 492
Costa, Nuno, 145
Criollo-C, Santiago, 911
Csabai, István, 663

D
D'Amato, J. P., 839
Dantas, Carina, 706
Del Castillo, Daniel, 219
del Molino, Javier, 275
Delgado, Ramiro, 748
Dias, Gonçalo Paiva, 155
Dobos, László, 663
Domingues, Patrício, 45
Domingues, Ricardo, 145
Dominguez, L., 839
Dutra, João, 706

E
Ehrl, Philipp, 716

F
Faria, Brígida Mónica, 607
Fernandes, João, 577
Fernández-S., Ángel, 208
Ferreira, Leonardo, 13
Ferreira, N. M. Fonseca, 451
Flores-Parra, Josue Miguel, 547
Fonseca, Ema, 186

G
Galațchi, Dan, 103
García-Rodríguez, Isaías, 124
Ginters, Egils, 675
Gomes, Hélder, 155
Gonçalves, Guilherme, 686
Gonçalves, Joaquim, 587, 607

Gonçalves, Ramiro, 420
Graña, Juan, 557
Graña, Manuel, 654
Guerrero, Graciela, 219
Gutierrez-Beltran, Erika-Jissel, 785
Guzmán Rodríguez, Maria Susana, 285

H
Habeeb, Eva, 900
Habib, Sami J., 93, 438
Halász, Gábor, 663
Heppner, Lars, 726

I
Iannello, Stefano, 361
Ibarra-Fiallo, Julio, 75
Intriago-Pazmiño, Monserrate, 75
Ionica, Andreea, 350
Išgum, Ivana, 577
Islam, Naveed, 295
Istrate, Cristiana, 165

J
Jacho-Guanoluisa, Nancy, 785
Jan, Zahoor, 295
Jani, Athraa Juhi, 407
Jariod-Gaudes, Ricardo, 785
Jegundo, Ana Luísa, 706
Jiang, Richard, 472
Jiménez, Samantha, 774
Junior, Antonio B. S., 492
Jurado, Marco, 208

K
Kala Kamdjoug, Jean Robert, 890
Kamdjoug, Jean Robert Kala, 641
Kamel, Ibrahim, 900
Kato, Toshihiko, 385
Khalili, Nadieh, 577
Khan, Muhammad, 303
Khan, Shoab A., 472
Kharoub, Hind, 737
Kiss, Attila, 663
Klobučar, Tomaž, 932
Kokkonen, Tero, 534
Kostoulas, Theodoros, 795

L
Lalia, Saoudi, 867
Lataifeh, Mohammed, 737
Lau, Nuno, 3, 65, 263
Leba, Monica, 350
Leite, Mariana Pinho, 726
Lobo, Joana, 186

Longheu, Alessandro, 361
Lopes, Jorge, 83
Lopez, Roberto, 748
Lopez-Guede, Jose Manuel, 654
Luna, Marco, 513

M
Machado, Ana Marisa, 764
Madani, Sara, 524
Maila-Maila, Fernando, 75
Malgeri, Michele, 361
Mallqui, Jaime Ambrosio, 24
Manuel, Paul, 853
Mareca, Pilar, 134
Marimuthu, Paulvanna N., 93, 438
Marques, Fábio, 155
Marroccia, Mario, 361
Martinez, Leysa Preguntegui, 24
Martins, Ana Isabel, 706
Martins, Ciro, 829
Martins, José, 186, 686
Mascaró, Miquel, 881
Masood, Ghulam, 303
McAlaney, John, 806, 921
Mehboob, Fozia, 472
Mejía, Jezreel, 230
Mekherbeche, Soumeya, 524
Melo, Miguel, 686, 696
Miguez, Alessandro, 110
Mihajlovski, Tamara, 726
Miloslavskaya, Natalia, 317, 328
Mira, E., 839
Moeskops, Pim, 577
Mohammadfarid, Alvansazyazdi, 285
Monreal, Cristina, 557
Moreira, Fernando, 483
Moreira, Rui S., 110, 240
Moreno-P., Rodrigo, 208
Mota, Tânia, 186
Moustafa, Kaddour, 867
Muñoz, Gabriel Fuertes, 755
Muñoz, Roberto, 774

N
Nawaz, Muhammad, 303
Nevavuori, Petteri, 534
Ngassam, Rhode Ghislaine Nguewo, 641
Nida-Ur-Rehman, Qazi, 303
Nikiforov, Andrey, 328
Novais, Paulo, 124, 567

O
Ohzahata, Satoshi, 385
Olaverri-Monreal, Cristina, 198

Oliveira e Sá, Jorge, 587, 607
Oliveira, Alexandra, 607
Oliveira, Inês, 186
Oliveira, Tiago, 567
Oliver, Antoni, 275
Osuna, Nora, 819
Osuna-Millán, Nora, 547

P
Palmer-Rodríguez, Pere, 881
Pelet, Jean-Éric, 431
Pereira, António, 34, 45, 56, 145, 829
Pereira, José Luís, 83
Petre, Ioana, 165, 252
Phalp, Keith, 921
Pinheiro, Paulo, 461
Plaksiy, Kirill, 328
Pluim, Josien, 577
Posso, Richard, 911
Powell, Georgina, 795

Q
Queirós, Alexandra, 706
Quezada, Ángeles, 774, 819
Quezada, Maria, 547
Quiñonez, Yadira, 230
Quintas, João, 706

R
Rabadão, Carlos, 56, 420
Ramírez, Margarita Ramírez, 774
Ramírez-Ramírez, Margarita, 547
Ramos, João, 34, 45, 56
Randazzo, Angelo, 361
Rauf, Abdul, 472
Reis, Arsénio, 34, 56
Reis, José Luís, 621
Reis, Laurinda L. N., 492
Reis, Luís Paulo, 3, 65, 263, 587, 607
Reis, Simão, 65, 263
Renold, Pravin, 438
Reyna, Jorge, 230
Ribeiro, José, 829
Ribeiro, Roberto, 34, 45, 56
Rito Lima, Solange, 373, 396
Riurean, Simona, 350
Rivera Valenzuela, Mario Augusto, 285
Rocha, Alvaro, 350
Rocha, Nelson Pacheco, 706
Rocha, Nelson, 829
Rodrigues, Mário, 829
Rodrigues, Nuno, 34
Rodríguez, Priscila, 748

Rosa, Ana Filipa, 706
Rosales, Ricardo, 547, 774, 819

S

Safadinho, David, 34, 45, 56
Salazar-L., Franklin, 208
Salgado Reyes, Nelson Esteban, 285
Salgado, Consuelo, 819
Sánchez, Eduardo, 557
Sánchez-de-Rivera, Diego, 175
Santos Gago, Juan M., 373
Santos, Leonel, 420
Santos, Rita, 829
Saraiva, A. A., 451
Satoh, Ken, 567
Scheianu, Andrei, 165, 252
Sendi, Mondher, 597
Sevilla, Maricela, 774, 819
Shah, Babar, 295, 483
Sieira, Jorge, 557
Silva, António, 567
Silva, Daniel Castro, 13
Silva, Eliana, 587, 607
Silva, F. V. N., 451
Silva, Fernando, 829
Silva, João Marco C., 373
Silva, Ricardo F., 373
Silva, Thiago Christiano, 716
Simões, David, 3
Soares, Christophe, 110, 240
Soares, Salviano, 451
Sobral, Pedro, 110, 240
Sousa, Jose Vigno M., 451
Souza, Darielson A., 492
Suciu, George, 165, 252
Szalai-Gindl, János Márk, 663

T

Tafokeng Talla, Landry, 890
Tahir, Faryal, 295
Taieb, Basma, 431
Tapia, Freddy, 219
Tasaki, Sota, 385
Tierra, Alfonso, 513
Tolstoy, Alexander, 317, 328
Torres, José M., 110, 240
Torre-Tojal, Leyre, 654

U

Ubilluz, Christian, 748
Ullah, Inayat, 295

V

Valdivia-Salas, Sonsoles, 785
Valente, Antonio, 451
Vasconcelos-Raposo, José, 686, 696
Vázquez, Sergio Octavio, 774
Vega, Pedro, 502

W

Wafo, Sylvain Defo, 641
Wamba, Samuel Fasso, 633
Wamba, Samuel Fosso, 641

Y

Yamamoto, Ryo, 385

Z

Zoican, Roxana, 103
Zoican, Sorin, 103
Zúquete, André, 155

Printed in the United States
By Bookmasters